reproductive biology

in systematics, conservation and economic botany

reproductive biology

in systematics, conservation and economic botany

Edited by

S.J. Owens and P.J. Rudall

The Royal Botanic Gardens, Kew, 1998

Production Editor: S. Dickerson

Cover design by Jeff Eden, page make-up by Media Resources,
Information Services Department, Royal Botanic Gardens, Kew

ISBN 1 900347 62 8

Printed in Great Britain
by Whitstable Litho Printers Ltd., Whitstable, Kent

CONTENTS

Reproductive Biology

PREFACE

This conference was held to celebrate 25 years of research into reproductive biology at the Royal Botanic Gardens, Kew, a research area brought to Kew by Professor Jack and Dr Yolande Heslop-Harrison. The conference was held in their honour. It brought together many of their colleagues and collaborators as well as many others who share in the quest to progress our understanding of plant reproduction.

Jack, a former Director of Kew, set up the Cell Physiology laboratory with his wife Yolande and together with colleagues from all over the world contributed to major advances in our knowledge of the reproductive biology of plants. The work included the discovery of the proteinaceous pellicle covering the cuticle on stigmatic papillae of Caryophyllaceae and Brassicaceae, and a modern classification of stigmatic form supported by a survey of stigmatic surfaces of plants from a wide range of families. When they left Kew, the work was successfully continued in Aberystwyth. Both Jack and Yolande contributed to the Kew conference with penetrating questions and lively and constructive discussion.

Sadly, during the preparation of this volume both Professor Jack Heslop-Harrison and Professor Bruce Knox, a conference contributor, died. Bruce was one of the earliest collaborators with Jack in the Cell Physiology laboratory and his contribution to this volume on pollen allergens built on some of that early work. He, like Jack, was a valued colleague and friend to many staff at Kew.

We would like to thank all those who supported this conference and contributed to its success. We would especially like to thank the other members of the organising committee: Professor Michael D. Bennett, Dr Mike Fay, Dr Eimear NicLughadha and Dr Keith Ferguson, whose advice, experience and encouragement ensured its success.

Simon Owens
Paula Rudall

Barrett, S.C.H., Cole, W.W. and Harder, L.D. (1998). Experimental studies on the effects of floral design and display on plant mating. In: S.J. Owens and P.J. Rudall (Editors). Reproductive Biology, pp. 1–13. Royal Botanic Gardens, Kew.

EXPERIMENTAL STUDIES ON THE EFFECTS OF FLORAL DESIGN AND DISPLAY ON PLANT MATING

SPENCER C.H. BARRETT[1], WILLIAM W. COLE[1] AND LAWRENCE D. HARDER[2]

[1]Department of Botany, University of Toronto, Toronto, Ontario, Canada M5S 3B2
[2]Department of Biological Sciences, University of Calgary, Calgary, Alberta, Canada T2N 1N4

Abstract

Two aspects of floral architecture function primarily to promote mating between plants: floral design largely concerns the structure of flowers whereas floral display involves their arrangement in inflorescences. Here we describe studies using floral manipulations and genetic markers that investigated the influence of floral design and display on pollen dispersal and mating patterns in *Pontederia cordata* and *Eichhornia paniculata*, two closely related, bee-pollinated, tristylous herbs. Two questions were addressed: 1. How does anther position within a flower influence pollen dispersal and male reproductive success? 2. How do different strategies for packaging male and female gametes on an inflorescence affect levels of self-fertilisation? Whether anthers were exserted or inserted within a flower significantly influenced pollen dispersal and male reproductive success. Comparisons of gamete packaging strategies indicated that contrasting distributions of equivalent numbers of pollen and ovules among flowers on an inflorescence resulted in significant differences in selfing rate. Experimental studies of floral design and display offer outstanding opportunitites for investigating the functional basis of variation in reproductive traits.

Introduction

The evolutionary diversification of flowers in many angiosperm families results from natural selection of reproductive characters affecting mating patterns and fertility. Differences in the structure and arrangement of flowers among related species reflects the costs and benefits of alternate means of achieving fitness in a given pollination environment. In animal-pollinated plants, the characteristics of individual flowers including their structure, colour, rewards, and fragrances (floral design) represent the cumulative outcome of pollinator-mediated selection operating during pollinator visits to flowers (reviewed in Waser, 1983; Armbruster, 1996; Galen, 1996; Campbell *et al.*, 1996 Proctor *et al.*, 1996). Of fundamental importance during these brief interactions is the spatial and temporal deployment of male and female sexual organs within flowers. How effectively pollinators contact these structures has an important influence on patterns of pollen dispersal and hence mating opportunities for individual plants (Waser and Price, 1984; Harder, 1990; Kohn and Barrett, 1992a; Campbell *et al.*, 1994). Much work in floral biology has focused on the effects of stigma position relative to anther position, on the amount of seed a plant produces and the incidence of cross- versus self-fertilisation (Barrett, 1979; Thomson and Stratton, 1985; Lloyd and Schoen, 1992; Belaoussoff and Shore,

1995). Far less attention has been paid to how sex-organ position may enhance male fertility through improved pollen dispersal (although see Kohn and Barrett, 1992b; Lloyd and Webb, 1992; Barrett *et al.*, 1996).

Animal-pollinated plants usually expose more than one flower at a time to pollination and thus the number of flowers in bloom and their spatial and temporal arrangement in inflorescences (floral display) will also importantly influence pollen dispersal and mating patterns (Harder and Barrett, 1996a). Inflorescence architecture varies considerably among angiosperm families (reviewed in Weberling, 1989), yet compared with studies of floral structure, the functional consequences of alternative designs have received surprisingly little attention (although see Wyatt, 1982; Schoen and Dubuc, 1990; Fishbein and Venable, 1996). Furthermore, how aspects of floral design and display function in a co-ordinated manner has rarely been investigated, despite the occurrence of repeated evolutionary patterns (e.g. synchronised protandry at the inflorescence level, the association between zygomorphy and vertical inflorescences) among angiosperm families. Because display characteristics of plants can be easily manipulated, the costs and benefits to mating and fertility of alternate designs should in principle be amenable to experimental studies.

Plant attractiveness to pollinators usually increases with display size; however, large displays can promote considerable geitonogamous or intraplant self-pollination (de Jong *et al.*, 1993; Snow *et al.*, 1996). Geitonogamy bears two potential mating costs; one obvious the other less well appreciated. In self-compatible species, geitonogamy can reduce fitness because of inbreeding depression in selfed progeny. Darwin (1876) recognised this possibility and proposed that plants with large floral displays should experience strong selection for self-incompatibility and dioecy, mechanisms that preclude selfing. The less obvious aspect of geitonogamous pollination arises because pollen used in this mode of self-pollination is not available for outcrossing and hence could reduce a plant's reproductive success as a pollen parent (Lloyd 1992; de Jong *et al.*, 1993). This phenomenom, termed geitonogamous pollen discounting, has been demonstrated empirically and has recently been proposed as a potentially important selective force influencing floral design and display (Harder and Barrett, 1995, 1996a).

Here we provide an overview of our recent investigations on the effects of floral design and display on pollen dispersal and mating patterns. We first briefly describe the species used in these studies and their advantages for experimental manipulations. We then address the following two questions: 1. What is the effect of anther position within a flower on pollen dispersal and mating success? 2. How do different strategies for packaging male and female gametes within and between flowers on an inflorescence affect patterns of mating, particularly levels of self-fertilisation? We conclude by suggesting that future experimental work in plant reproductive biology should involve manipulative studies of floral design and display and their contributions to the male and female components of plant fitness.

Experimental Systems

The experiments described below involved *Pontederia cordata* L. and *Eichhornia paniculata* (Spreng.) Solms of the monocotyledonous family Pontederiaceae. *Pontederia cordata* is a clonal perennial native to eastern North America and occurs in a variety of wetland habitats, particularly lake and stream margins. *Eichhornia paniculata* is annual and occurs in ephemeral ponds and ditches in seasonal parts of the Neotropics. Both of these emergent aquatic species produce showy, blue-purple, moderately zygomorphic, tubular, tristylous flowers that are primarily bee-pollinated. Flowers last for 6-8 hours and are produced in spike-like (*P. cordata*) or panicle-like (*E. paniculata*) inflorescences that bloom for 7-10 days depending on size. Although both species are heterostylous, they differ with respect to their compatibility systems and pollen characteristics. *Pontederia cordata* possesses a conventional trimorphic incompatibility system (i.e. self- and intramorph incompatibility) and exhibits conspicuous differences in the size of pollen produced by the three anther levels (Price and Barrett, 1982; Barrett and Anderson, 1985). In contrast, *E. paniculata* is highly self-fertile and pollen from the three anther levels is only weakly differentiated in size (Barrett, 1985; Kohn and Barrett, 1992a).

Several features make these species desirable subjects for investigating the effects of floral design and display on pollen dispersal and mating. First, being tristylous they display more intraspecific variation in the placement of stigmas and anthers than can be found in most other plants. Second, in *P. cordata*, strong pollen-size trimorphism enables the pollen of one morph to be identified on stigmas of another. This allows studies of pollen dispersal without the need for pollen surrogates or emasculation. Finally, in *E. paniculata* rapid seed maturation (10 days), little abortion, and the presence of easily scoreable allozyme polymorphisms in seed make this species particularly useful for mating-system analysis. In the experiments described below the effects of floral design on pollen dispersal were investigated in *P. cordata* under field conditions in Ontario, Canada (further details in Harder and Barrett, 1993, 1996b) and the influence of floral design and display on mating patterns were examined in *E. paniculata* using experimental populations in a garden setting also in Ontario (see Kohn and Barrett, 1992a, b; Barrett *et al.*, 1994; Harder and Barrett, 1995). In both studies bumble bees (primarily *Bombus vagans* Smith) were the main pollinators, although in the field the solitary bee *Melissodes apicata* Lovell and Cockerell also commonly visited *P. cordata*.

Effects of Anther Position on Pollen Dispersal and Mating

The location and structure of anthers in angiosperm flowers varies considerably among different plant species (reviewed in Yeo, 1993; D'Arcy and Keating, 1996). As yet no general theory predicts the optimal anther position for a given floral design. Nevertheless, natural history observations indicate that anther position often varies with pollination syndrome, particularly for tubular flowers (see Grant and Grant, 1965; Harder and Barrett, 1993). Species pollinated by birds frequently possess anthers exserted well outside the floral tube whereas species pollinated by butterflies and moths often have anthers inserted deep within the flower. Among bee-pollinated plants, anthers are usually positioned at the mouth of the perianth tube. In some species anthers are positioned at different locations within a single flower (Murcia, 1990), while in others stamen height varies considerably both within (Price and Waser 1982) and among populations (Belaoussoff and Shore, 1995). Why do these patterns occur and how does anther position influence mating patterns and plant fitness? Below we describe two series of experiments that provide insights into the reproductive consequences of inserted versus exserted anther deployment. For the purpose of this presentation we concentrate on the most distinct comparison (e.g. long- versus short-level anthers) in *P. cordata* and *E. paniculata* although our studies have compared all three stamen levels.

Harder and Barrett (1993) demonstrated differences in the patterns of pollen removal from long- versus short-level anthers of *P. cordata*. By presenting unvisited flowers to freely foraging bees and comparing the number of pollen grains remaining in anthers after 1-4 visits with initial production they estimated the number of pollen grains removed by pollinators. Long-level anthers lost on average 39% of their pollen during the first visit. In contrast, pollen removal from short-level anthers was more stochastic, presumably because of variation in the path traversed by the proboscis while probing for nectar. On average, pollinators removed 28% of the pollen from short-level anthers during a flower's first visit. A second experiment (Harder and Barrett, 1996b) compared the dispersal of pollen from these two anther levels to stigmas in the same position (legitimate pollination) using mid-styled donor plants and a series of either long- or short-styled recipients. The patterns of legitimate pollen dispersal differed significantly between the two anther levels (Fig. 1), with pollen deposition declining much more rapidly from short-level anthers despite much greater (4.5 fold) pollen production at this level (see data in Price and Barrett, 1982; Harder and Barrett, 1993). Differences in susceptibility to pollinator grooming of short versus long-level pollen resulting from their contrasting deposition sites on bees' bodies (proboscis versus thorax and abdomen, see Wolfe and Barrett, 1989) appear to largely account for these different dispersal patterns.

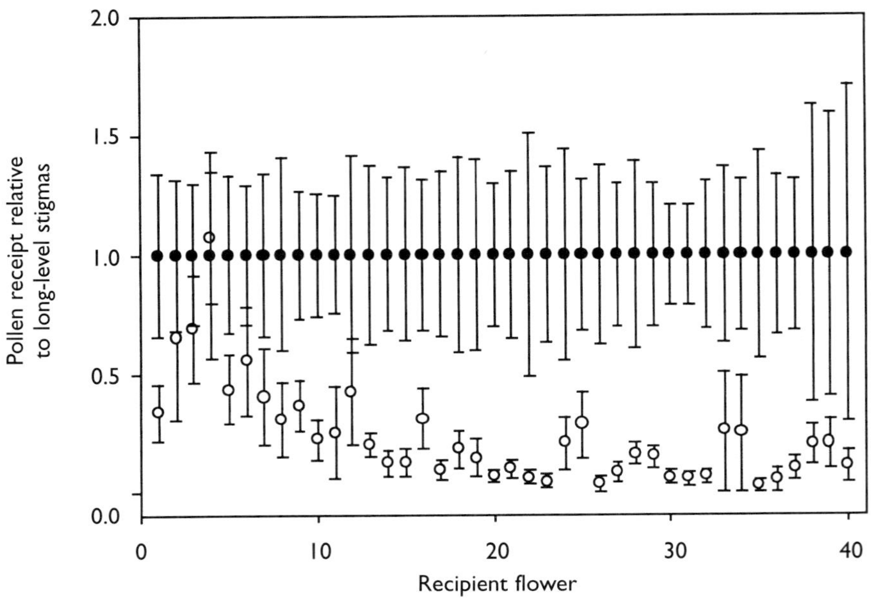

FIG. 1. Contrasting patterns of pollen dispersal from exserted and inserted anther levels in tristylous *Pontederia cordata*. Mean- (±SE) legitimate pollen dispersal of pollen from mid-styled donor plants to long-styled (solid symbols, n = 8 trials) and short-styled (open symbols, n = 8 trials) recipient flowers by bumble bees. Pollen deposition is expressed relative to the mean observed dispersal between long-level anthers and stigmas of long styles.

These studies revealed that the timing of anther dehiscence varied with stamen position. On average, the anthers of the exserted long-level stamens dehisced fully 1.5 h after the flower bud began opening, whereas short-level anthers took 3.2 h before anthers dehisced completely. Harder and Barrett (1993) interpreted this pattern as resulting from an intraflower humidity gradient, with humidity increasing above ambient levels at the base of the floral tube (also see Corbet *et al.*, 1979). Anther dehiscence is thought to result largely from dehydration of anthers (Keijzer, 1987), which likely occurs more rapidly in exposed long-level anthers than in anthers concealed at the base of the floral tube. Because *P. cordata* possesses a trimorphic incompatibility system, so that pollen from one anther level does not compete with pollen from other levels, the staggered dehiscence of different anther levels probably has little relevance to mating. However, in a self-compatible heterostylous plant pollen can participate in matings with any stigma type, so that staggered anther dehiscence could potentially influence mating patterns. We now describe experiments with *E. paniculata* that provides evidence for such an effect.

Marker-gene studies of mating parameters in experimental tristylous arrays of *E. paniculata* by Kohn and Barrett (1992a) revealed unexpected differences in the male fertility of floral morphs. Although maternal outcrossing rates did not differ significantly among morphs (mean outcrossing rate = 0.81), on average short-styled plants sired nearly three times more seeds on mid-styled plants than did long-styled plants, resulting in striking gender differences between the two morphs. In a subsequent experiment using floral manipulations, Kohn and Barrett (1992b) demonstrated that this difference in male fertility between the morphs resulted from the possession of contrasting illegitimate anther types in the two morphs. Removal of long-level anthers from the short-styled morph and short-level anthers from the long-styled morph eliminated the mating asymmetry. With this manipulation, both morphs could only donate pollen to the mid-styled morph from the same anther level (i.e. mid-level anthers) resulting in equivalent male fertility.

Kohn and Barrett (1992b) observed that anther dehiscence in *E. paniculata* differed among stamen levels in a manner similar to *P. cordata* and suggested that pollen precedence could be responsible for the mating asymmetry. Because of early dehiscence, pollinators could transfer pollen from long-level anthers of the short-styled morph to mid-level stigmas before any transfer from short-level anthers had taken place. Pollen competition experiments with *E. paniculata* have shown that early arriving pollen has a siring advantage over later arriving pollen (Graham and Barrett, 1990 and unpubl. data). We tested the hypothesis that earlier dehiscence of long-level versus short-level anthers contributes to the observed mating asymmetry by comparing the male fertility of morphs in arrays exposed to pollinators at 9.00am versus 12.00pm. By 12.00pm *E. paniculata* anthers at all levels have dehisced fully, leaving little opportunity for pollen precedence. When arrays were exposed to pollinators at 9.00am, significant mating asymmetries favouring the short-styled morph were recorded in each of three years (Fig. 2). However, the

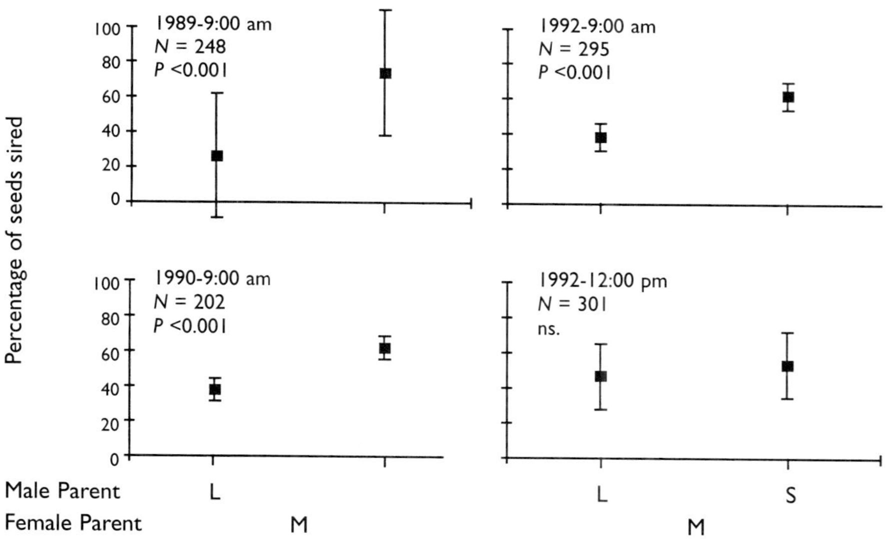

FIG. 2. Mating asymmetries in tristylous arrays of *Eichhornia paniculata*. Bars illustrate the mean (± 95% confidence intervals) percentage of seeds of the mid-styled morph sired by the long- and short-styled morph. Intermorph matings were identified by heterozygosity at *AAT-3* as each morph was homozygous for a different allele at this locus. Four replicate arrays of 36 plants, 12 of each morph, were set up in 1989, 1990 and 1992. Those exposed to pollinators at 9.00AM exhibited significant mating asymmetries (following *G*-tests) favouring the short-styled morph. In the array exposed to pollinators at 12.00PM in 1992 the siring success of the long- and short-styled morphs did not differ significantly.

share of mid-styled ovules fertilised by long versus short-styled plants did not differ significantly when arrays were exposed to pollinators at 12.00pm (1992). This result is consistent with the hypothesis that variation in the timing of anther dehiscence among stamen levels can influence the gender of floral morphs in self-compatible heterostylous species.

Effects of Alternate Gamete Packaging Strategies on Mating Patterns

Flowering plants package male and female gametes in a bewildering variety of spatial and temporal combinations, giving rise to a great diversity of sexual systems (Lloyd and Bawa, 1984). Variation in packaging strategies at the flower, inflorescence, and plant levels all have important consequences for the allocation of resources, mating patterns and plant fitness. Although many flowers are functionally hermaphroditic, pollen presentation and stigma receptivity often occur at differ times (dichogamy) restricting opportunities for intrafloral self-pollination. Because the individual flower is unlikely to be the functional unit of pollination for plants with large displays, differences in the timing of male and female function requires co-ordination at the inflorescence or plant level if self-pollination is to be reduced and cross-pollination enhanced. The modular structure of plants and their formation of multiple reproductive units introduces complex requirements for physiological and developmental regulation of sex function not found in most animal groups.

A particularly striking example of developmental co-ordination involves inflorescence-level dichogamy (Lloyd and Webb, 1986; Bertin and Newman, 1993). Here the sequence of anthesis is such that particular spatial and temporal patterns of sex function reduce the likelihood of self-pollination. For example, in many andromonoecious Umbelliferae and Araliaceae all flowers in an umbel first present pollen so that the entire umbel is in male phase and then some flowers enter female phase (Lovett Doust, 1980; Thomson and Barrett, 1981). At any given time this form of synchronised protandry results in functionally unisexual inflorescences.

A more widespread example of inflorescence-level dichogamy involves cosexual plants with vertical inflorescences (e.g. *Delphinium virescens*, Waddington, 1981; *Digitalis purpurea*, Best and Bierzychudeck, 1982; *Epilobium angustifolium*, Galen and Plowright, 1985), in which flowers are protandrous and flowers open from the bottom to the top of the inflorescence. Because pollinators, such as bees, typically move upwards on inflorescences (Waddington and Heinrich, 1979; Corbet *et al.*, 1981) they usually visit female-phase flowers before male-phase flowers. This flowering pattern has been interpreted as a mechanism to reduce the incidence of self-pollination. To investigate the extent to which such inflorescence-level protandry reduces selfing and to explore the mating consequences of alternative strategies for packaging gametes we used genetic markers to measure rates of self-fertilisation in *E. paniculata* plants in which inflorescence design and display were manipulated experimentally.

We followed protocols for array experiments outlined in Barrett *et al.* (1994). As in that study, arrays were composed of 36 long-styled plants of *E. paniculata* arranged in a six-by-six grid. Use of a single morph removes any complication introduced by heterostyly and resembles a typical self-compatible plant with approach herkogamy, because morphs of *E. paniculata* are fully self- and intramorph compatible. Observations of pollinator foraging indicated that bees most commonly visited flowers near the bottom of an inflorescence and then moved upwards, leaving the plant from the top of the inflorescence (Barrett *et al.*, 1994). This pattern of foraging caused predictable increases in rates of geitonogamous self-fertilisation from bottom to top positions within an inflorescence (see Fig 3. in Barrett *et al.*, 1994). By manipulating the number of flowers on inflorescences of *E. paniculata* we also demonstrated that selfing rates increased with daily display size.

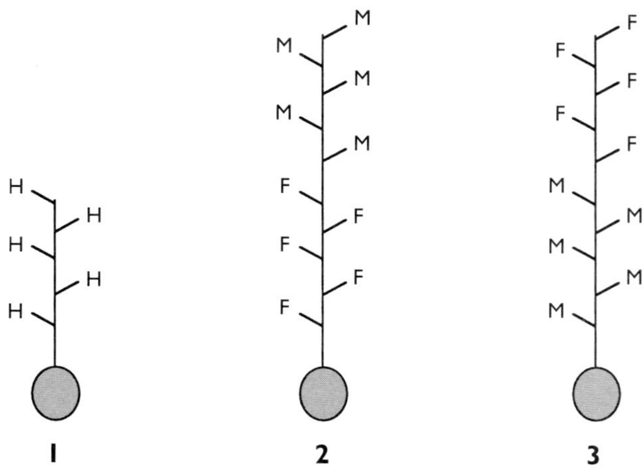

FIG. 3. Schematic diagram of the three inflorescence phenotypes of *Eichhornia paniculata* used in experimental studies described in detail in the text. H, F and M refer to hermaphrodite, female and male flowers, respectively.

We contrasted three floral phenotypes that differed in the presentation of male and female gametes (Fig. 3). The first phenotype involved unmanipulated 5-flowered plants. Because *E. paniculata* has hermaphrodite flowers and no dichogamy, this phenotype simultaneously exposes female and male gametes to mating at the flower level. The second phenotype was 10-flowered with the bottom 5 flowers functionally female (anthers removed) and the top 5 flowers functionally male (pistils removed). Stamens and pistils were removed with fine forceps prior to anthesis. This phenotype simulates plants with the form of protandry commonly found in species with vertical inflorescences, as discussed above. The third phenotype was also 10-flowered, but the five bottom flowers were male and the top five were female. This pattern of sex expression represents a protogynous species, a condition that occurs only rarely in species with vertical inflorescences (Lloyd and Webb, 1986; Bertin 1993), presumably because with bottom to top foraging of pollinators this sequence of sex expression promotes geitonogamy.

Our experiments involved two kinds of arrays with the 5-flowered phenotype contrasted with each of the 10-flowered phenotypes. Arrays contained 50:50 ratios of both types arranged randomly. Note that all three phenotypes produced the same number of gametes, on average, but differed in how gametes were distributed within and among flowers. The phenotypes can be viewed as representing alternative gamete packaging strategies. Within an array all 5-flowered plants were homozygous for an *AAT-3* allele, whereas 10-flowered plants were homozygous for an alternative allele, so that the paternity of each seed could be ascribed to a particular treatment. Electrophoretic assays of seed genotypes for *AAT-3* and randomly occurring *PGI-2* alleles allowed us to assess whether these contrasting sex phenotypes differed in mating patterns: data on the male reproductive success of phenotypes will be presented elsewhere.

The incidence of self-fertilisation differed among the three phenotypes. The 10-flowered dichogamous phenotypes with female flowers at the bottom selfed significantly less than the 5-flowered non-dichogamous plants (paired *t*-test, $t_3 = 4.14$, $P < 0.005$: see Fig. 4). The segregation of sex function in the simulated protandrous phenotype therefore significantly affected mating patterns. This result is undoutedly associated with the foraging behaviour of pollinators, which tended to visit bottom female flowers before top male flowers, hence restricting opportunities for geitonogamous selfing. In contrast, the simulated protogynous phenotype experienced the same selfing frequency as plants with simultaneously hermaphrodite flowers ($t_3 = 1.22$, $P > 0.25$: see Fig. 4) indicating that the two ways of packaging gametes were functionally equivalent at least from the maternal perspective. Bees foraging on the simulated protogynous phenotype visited male flowers at the bottom of inflorescences first, resulting in transfer of self-pollen to female flowers at the top of inflorescences. This phenotype experienced similar levels of selfing to the non-dichogamous phenotype, yet required twice as many flowers for equivalent gamete production. The greater cost incurred in flower production without any apparent mating advantage may partially explain why this floral strategy rarely occurs in nature, although the consequences for male function must also be considered.

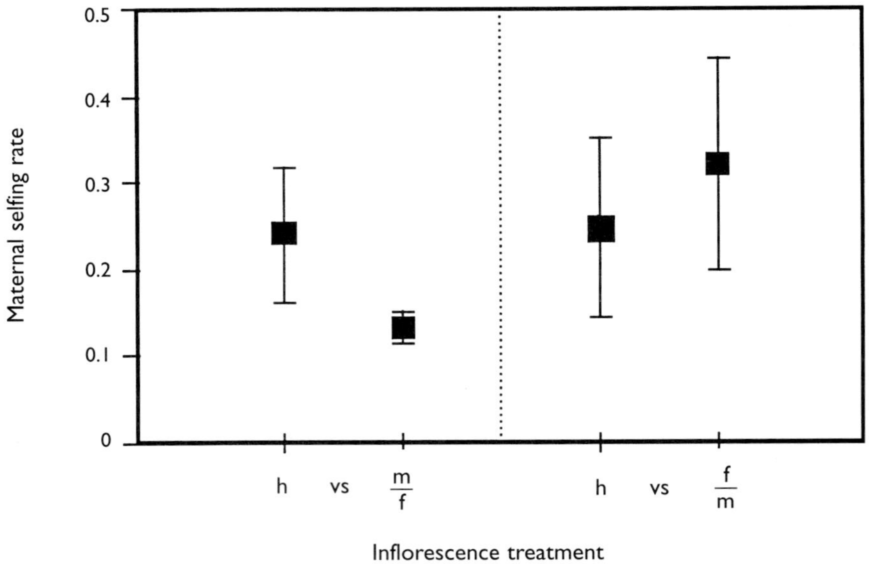

FIG. 4. The influence of different gamete packaging strategies on selfing rates in *Eichhornia paniculata*. Values shown are the mean- (± 95% confidence intervals) frequencies of ovules that were self-fertilised in experimental arrays comparing 5-flowered plants with either of two different 10-flowered phenotypes with contrasting patterns of sex expression. All flowers in 5-flowered plants were hermaphroditic whereas all flowers in 10-flowered plants were unisexual. In the comparison in the left-hand panel the 10-flowered plant had 5 female flowers at the bottom of the inflorescence and 5 male flowers at the top. In the right-hand panel the 10-flowered plants had the reverse order of sex expression. Each comparison involved four replicates.

Discussion

We used floral manipulations and the measurement of mating parameters with genetic markers to examine functional aspects of floral design and display. These investigations differ from many earlier descriptive studies in floral biology which typically used correlative approaches to explore the relation between morphological variation and components of reproductive success. An experimental approach has the advantage that the proximate mechanisms responsible for particular reproductive outcomes can be identified in isolation from unmeasured correlated traits that simultaneously influence reproduction. However, manipulation of floral traits can introduce difficulties if pollinators discriminate against altered phenotypes. Pollinator observations can be used to assess this possibility. In our studies pollinators were apparently indifferent to the floral manipulations performed and hence this problem was not encountered.

Our experiments on the influence of anther position on pollen dispersal and male reproductive success raise several issues concerning the optimal deployment of stamens and floral evolution in general. The frequent observation that in outcrossing species variation in floral traits has a more significant influence on components of male reproductive success than on fruit and seed set, suggests that natural selection on anther position may take evolutionary priority over stigma position (Harder and Barrett, 1993). According to this hypothesis anther position is selected to maximise pollen dispersal and fitness returns through increased male fertility. Once this position is established it determines the location of the stigma to reduce self-pollination and self-interference. Studies on non-heterostylous species into the effects of variation in anther and stigma position on pollen dispersal and pollen receipt would be particularly valuable for assessing this hypothesis. In addition, comparisons of variation in anther and stigma position within and among species would be useful to assess the relative strength of selection acting upon these organs. If anther position takes evolutionary priority over stigma position as argued above we may expect less variation in the former as a result of stronger stabilising selection.

In *P. cordata* the extent of pollen dispersal from exserted anthers was considerably greater than for inserted stamens. This result could have implications for aspects of floral design in non-heterostylous species, because pollen dispersal can play an important role in governing mating patterns (Galen and Rotenberry, 1988; Robertson, 1992; Harder and Barrett, 1996). For example, in mass-flowering species extensive pollen dispersal increases the number and diversity of potential mates by reducing geitonogamous pollen discounting. Our data suggest that the possession of exserted rather than inserted anthers would be selectively advantageous for species that display many flowers simultaneously and are pollinated by long-tongued bees because greater outcrossing opportunities could be realised. Comparative studies could be used to explore whether these evolutionary associations do indeed occur.

Differential anther dehiscence of exserted versus inserted anthers contributes to asymmetries in male fertility between long- and short-styled morphs in tristylous arrays of *E. paniculata* (Fig. 2). Whether the timing of anther dehiscence plays a widespread role in the floral design of animal-pollinated plants does not appear to have been considered. Several patterns that may be relevant involve the associations between pollinator type and anther position discussed above. Large-bodied bees maintain high body temperatures and can forage early in the morning. Such activity patterns may place a premium on early anther dehiscence and would likely place plants with deeply inserted anthers at a competitive disadvantage. Interestingly, most plants pollinated by large-bodied bees present their anthers in the corolla mouth rather than deep within the floral tube. The same argument can also be applied to bird-pollinated species, which commonly possess well exserted stamens. In contrast, butterfly activity is more temperature dependent, so that they usually do not begin foraging until all anthers would likely be dehisced. This could result in little selective disadvantage, at least as far as anther dehiscence, to inserted anthers that typically occur in many butterfly-pollinated plants. Hawkmoths like bees maintain high body temperatures, but they usually forage after dusk, several hours after their flowers have opened and anthers have dehisced. Again there may be little timing disadvantage to their inserted anthers and pollen placement considerations take priority. These speculations are offered to stimulate further consideration of why anthers are often deployed in different locations within a flower depending on pollinator type.

Manipulation of the number and sex expression of flowers on inflorescences of *E. paniculata* enabled us to investigate how different gamete packaging strategies influence mating patterns. Although each of the three floral phenotypes produced equivalent numbers of male and female gametes, these were distributed differently within and among flowers on an inflorescence. The major finding of our experiment was that rates of self-fertilisation were reduced in the phenotype with inflorescence-level protandry compared with the non-dichogamous phenotype. Comparisons of selfing rate between the non-dichogamous phenotype and the phenotype with inflorescence-level protogyny were not significantly different. Preliminary analysis of the male reproductive success of the protandrous phenotype indicates that it had a significantly higher male outcrossing rate than the other two phenotypes (L.D. Harder, S.C.H. Barrett and W.W. Cole, unpubl. data). This suggests that the observed reduction in selfing achieved by inflorescence-level protandry may increase opportunities for outcrossed siring success. If this interpretation is correct it provides further support for the hypothesis that floral mechanisms that decouple the benefits of large floral displays for pollinator attraction from the mating costs associated with geitonogamous pollen discounting are of major importance in animal-pollinated plants (Harder and Barrett, 1995). Inflorescence-level protandry may function in this way.

Since Darwin's pioneering work on the structure and function of orchid flowers (Darwin, 1862) most research in floral biology has focused on the intricate details of floral design in relation to pollination and mating. However, the majority of flowering plants expose many flowers simultaneously to the pollination process and hence functional interpretation of the design of individual flowers cannot be made in isolation from a consideration of the entire floral display. Our experimental studies have exposed several ways in which the interaction between floral design and display influence pollen dispersal and mating patterns and imply that these reproductive traits should evolve in a co-ordinated manner to maximise plant fitness. Yet until recently little attempt has been made to investigate functional associations between alternative design and display characteristics, despite a wealth of variation in most flowering plant groups. We believe that future studies in plant reproductive biology could profitably focus on the joint evolution of floral design and display and their effects on mating patterns. Research that combines both comparative and experimental approaches will likely provide the most useful insights in this regard.

Acknowledgements

We thank Angela Baker and Joshua R. Kohn for assistance with the experimental arrays. Research funded by grants to S.C.H.B. and L.D.H. from the Natural Sciences and Engineering Research Council of Canada. This paper is dedicated to the memory of A. C. Barrett who kept watch over our garden arrays.

References

Armbruster, W.S. (1996). Evolution of floral morphology and function: an integrative approach to adaptation, constraint, and compromise in *Dalechampia* (Euphorbiaceae). In: D.G. Lloyd and S.C.H. Barrett (editors). Floral biology: studies on floral evolution in animal-pollinated plants, pp. 241–272, Chapman and Hall, New York.

Barrett, S.C.H. (1979). The evolutionary breakdown of tristyly in *Eichhornia crassipes* (Mart.) Solms (water hyacinth). *Evolution* **33**: 347–348.

Barrett, S.C.H. (1985). Floral trimorphism and monomorphism in continental and island populations of *Eichhornia paniculata* (Spreng.) Solms (Pontederiaceae). *Biol. J. Linn. Soc.* **25**: 41–60.

Barrett, S.C.H. and Anderson, J.M. (1985). Variation in expression of trimorphic incompatibility in *Pontederia cordata* L. (Pontederiaceae). *Theor. Appl. Genet.* **70**: 355–362.

Barrett, S.C.H., Harder, L.D. and Cole, W.W. (1994). Effects of flower number and position on self-fertilization in experimental populations of *Eichhornia paniculata* (Pontederiaceae). *Funct. Ecol.* **8**: 526–535.

Barrett, S.C.H., Lloyd, D.G., and Arroyo, J. (1996). Stylar polymorphisms and the evolution of heterostyly in *Narcissus* (Amaryllidaceae). In: D.G. Lloyd and S.C.H. Barrett (editors). Floral biology: studies on floral evolution in animal-pollinated plants, pp. 339–376, Chapman and Hall, New York.

Belaoussoff, S. and Shore, J.S. (1995). Floral correlates and fitness consequences of mating-system variation in *Turnera ulmifolia*. *Evolution* **49**: 545–556.

Bertin, R. I. (1993). Incidence of monoecy and dichogamy in relation to self-fertilization. *Amer. J. Bot.* **80**: 557–560.

Bertin, R.I. and Newman, C.M. (1993). Dichogamy in angiosperms. *Bot. Rev.* **59**: 112–152.

Best, L.S. and Bierzychudek, P. (1982). Pollinator foraging on foxglove (*Digitalis purpurea*): a test of a new model. *Evolution* **36**: 70–79.

Campbell, D.R., Waser, N.M., and Price, M.V. (1994). Indirect selection of stigma position in *Ipomopsis aggregata* via a genetically correlated trait. *Evolution* **48**: 55–68.

Campbell, D.R., Waser, N.M., and Price, M.V. (1996). Mechanisms of hummingbird-mediated selection for flower width in *Ipomopsis aggregata*. *Ecology* **77**: 1463–1472.

Corbet, S.A., Wilmer, P.G., Beament, W.L., Unwin, D.M. and Prys-Jones, O.E. (1979). Post-secretory determinants of sugar concentration in nectar. *Pl. Cell Environm.* **2**: 293–308.

Corbet, C., Cuthill, I., Fallows, M., Harrison, T. and Hartley, G. (1981). Why do nectar-foraging bees and wasps work upwards on inflorescences? *Oecologia* **51**: 79–83.

Darwin, C. (1862). On the various contrivances by which British and foreign orchids are fertilised by insects. John Murray, London.

Darwin, C. (1876). The effects of cross and self fertilization in the vegetable kingdom. John Murray, London.

D'Arcy, W.G. and Keating, R.C. (1996). The anther: form, function and phylogeny. Cambridge University Press, Cambridge.

Fishbein, M. and Venable, D.L. (1996). Evolution of inflorescence design: theory and data. *Evolution* **50**: 2165–2177.

Galen, C. (1996). The evolution of floral form: insights from an alpine wildflower, *Polemonium viscosum*. In: D.G. Lloyd and S.C.H. Barrett (editors). Floral biology: studies on floral evolution in animal-pollinated plants, pp. 273–291, Chapman and Hall, New York.

Galen, C. and Plowright, R.C. (1985). The effects of nectar level and flower development on pollen carry-over in inflorescences of fireweed (*Epilobium angustifolium*) (Onagraceae). *Canad. J. Bot.* **63**: 488–491.

Galen, C. and Rotenberry, J.T. (1988). Variance in pollen carryover in animal-pollinated plants: implications for mate choice. *J. Theor. Biol.* **135**: 419–429.

Graham, S.W. and Barrett, S.C.H. (1990). Pollen precendence in *Eichhornia paniculata*: a tristylous species. *Amer. J. Bot.* **77** (6 supplement): 54–55 (Abstract).

Grant, V. and Grant, K. (1965). Flower pollination in the *Phlox* family. Columbia Univ. Press, New York.

Harder, L.D. (1990). Pollen removal by bumble bees and its implications for pollen dispersal. *Ecology* **71**: 1110–1125.

Harder, L.D. and Barrett, S.C.H. (1993). Pollen removal from tristylous *Pontederia cordata*: effects of anther position and pollinator specialization. *Ecology* **74**: 1059–1072.

Harder, L.D. and Barrett, S.C.H. (1995). Mating cost of large floral displays in hermaphrodite plants. *Nature* **373**: 512–515.
Harder, L.D. and Barrett, S.C.H. (1996a). Pollen dispersal and mating patterns in animal-pollinated plants. In: D.G. Lloyd and S.C H. Barrett (editors). Floral biology: studies on floral evolution in animal-pollinated plants, pp. 140–190, Chapman and Hall, New York.
Harder, L.D. and Barrett, S.C.H. (1996b). The role of floral design in pollen dispersal by *Pontederia cordata*. Unpubl. MS.
Jong, T.J. de, Waser, N.M. and Klinkhamer, P.G.L. (1993). Geitonogamy: the neglected side of selfing. *Trends Ecol. Evol.* **8**:321–325.
Keijzer, C.J. (1987). The processes of anther dehiscence and pollen dispersal. I. The opening mechanism of longitudinally dehiscing anthers. *New Phytol.* **105**: 487–498.
Kohn, J.R. and Barrett, S.C.H. (1992a). Experimental studies on the functional significance of heterostyly. *Evolution* **46**: 43–55.
Kohn, J.R. and Barrett, S.C.H. (1992b). Floral manipulations reveal the cause of male fitness variation in experimental populations of *Eichhornia paniculata* (Pontederiaceae). *Funct. Ecol.* **6**: 590–595.
Lovett Doust, J. (1980). Floral sex ratios in andromonoecious Umbelliferae. *New Phytol.* **85**: 265–273.
Lloyd, D.G. (1992). Self- and cross-fertilization in plants. II. Selection of self-fertilization. *Int. J. Pl. Sci.* **153**: 370–380.
Lloyd, D.G. and Bawa, K.S. (1984). Modifications of the gender of seed plants in varying conditions. *Evol. Biol.* **17**: 255–338.
Lloyd, D.G. and Schoen, D.J. (1992). Self- and cross-fertilization in plants. I. Functional dimensions. *Int. J. Pl. Sci.* **153**: 358–369.
Lloyd, D.G. and Webb, C.J. (1986). The avoidance of interference between the presentation of pollen and stigmas in angiosperms. I. Dichogamy. *New Zealand J. Bot.* **24**: 135–162.
Lloyd, D.G. and Webb, C.J. (1992). The selection of heterostyly. In: S.C.H. Barrett (editor). Evolution and function of heterostyly, pp 179–207, Springer-Verlag, Berlin.
Murcia, C. (1990). Effect of floral morphology and temperature on pollen receipt and removal in *Ipomoea trichocarpa*. *Ecology* **71**: 1098–1109.
Price, M.V. and Waser, N.M. (1982). Experimental studies of pollen carryover: Hummingbirds and *Ipomopsis aggregata*. *Oecologia* **54**: 353–358.
Price, S.D. and Barrett, S.C.H. (1982). Tristyly in *Pontederia cordata* (Pontederiaceae). *Canad. J. Bot.* **60**: 897–905.
Proctor, M., Yeo, P. and Lack, A. (1996). The natural history of pollination. Timber Press, Portland, Oregon.
Robertson, A.W. (1992). The relationship between floral display size, pollen carryover and geitonogamy in *Myosotis colensoi* (Kirk) Macbride (Boraginaceae). *Biol. J. Linn. Soc.* **46**: 333–349.
Schoen, D.J. and Dubuc, M. (1990). The evolution of inflorescence size and number: a gamete packaging strategy. *Amer. Naturalist* **135**: 841–857.
Snow, A.A., Spira, T.P., Simpson, R. and Klips, R.A. (1996). The ecology of geitonogamous pollination. In: D.G. Lloyd and S.C.H. Barrett (editors). Floral biology: studies on floral evolution in animal-pollinated plants, pp. 191–216, Chapman and Hall, New York.
Thomson, J.D. and Barrett, S.C.H. (1981). Temporal variation of gender in *Aralia hispida* Vent. (Araliaceae). *Evolution* **35**: 1094–1107.
Thomson, J.D. and Stratton, D.A. (1985). Floral morphology and cross-pollination in *Erythronium grandiflorum* (Liliaceae). *Amer. J. Bot.*, **72**: 433–427.
Waddington, K.D. (1981). Factors influencing pollen flow in bumblebee-pollinated *Delphinium virescens*. *Oikos* 37: 153–159.

Waddington, K.D. and Heinrich, B. (1979). The foraging movements of bumblebees on vertical inflorescences: an experimental analysis. *J. Comp. Physiol.* **134**: 113–117.
Waser, N.M. (1983). Competition for pollination and floral character differences among sympatric plant species: a review of evidence. In: C.E. Jones and R.J. Little, (editors). Handbook of experimental pollination biology, pp. 277–293, van Nostrand Reinhold, New York.
Waser, N.M. and Price, M.V. (1984) Experimental studies of pollen carryover: effects of floral variability in *Ipomopsis aggregata. Oecologia* **62**: 262–268.
Weberling, F. (1989). Morphology of flowers and inflorescences. Cambridge Univ. Press., Cambridge.
Wolfe, L.M., and S.C.H. Barrett. (1989). Patterns of pollen removal and deposition in tristylous *Pontederia cordata* L. (Pontederiaceae). *Biol. J. Linn. Soc.* **36**: 317–329.
Wyatt, R. (1982). Inflorescence architecture: how flower number, arrangement, and phenology affect pollination and fruit-set. *Amer. J. Bot.* **69**: 585–594.
Yeo, P.F. (1993) Secondary pollen presentation: form, function and evolution. Springer, Wien. (Pl. Syst. Evol. Suppl. 6).

Dickinson, H.G., Doughty, J.D., Elleman, C.J. and Hiscock, S.J. (1998). Evolution of angiosperm self-incompatibility mechanisms; data from *Brassica oleracea*. In: S.J. Owens and P.J. Rudall (Editors). Reproductive Biology, pp. 15–29. Royal Botanic Gardens, Kew.

EVOLUTION OF ANGIOSPERM SELF-INCOMPATIBILITY MECHANISMS; DATA FROM *BRASSICA OLERACEA*

HUGH DICKINSON, JAMES DOUGHTY, CAROLE ELLEMAN AND SIMON HISCOCK

Department of Plant Sciences, University of Oxford, South Parks Road, Oxford, OX1 3RB, UK

Abstract

The rapid progress made in the molecular characterisation of the self-incompatibility systems of flowering plants has served to highlight our ignorance of the pathways by which these complex cell-cell signalling systems have evolved. Intense selection pressure appears to have resulted in the recruitment of a number of already-extant systems to the cause of self-sterility, and it is reasonable to assume that SI mechanisms should provide a useful paradigm for establishing rules by which intercellular signalling systems have evolved. Unfortunately, the sheer variety of mechanisms found - ranging from cell surface receptor kinases in *Brassica* to S(SI)-locus specific RNase activity in Solanaceae - has made the identification of common factors difficult. However, the recent discovery of apparently generic pollen-held ligands in *Brassica*, combined with data from interspecific pollinations in Cruciferae and Solanaceae, point to the evolution of interspecific incompatibility systems being a step in the pathway to the development of the sophisticated multiallelic SI mechanisms which characterise modern angiosperms.

Introduction

While the selective pressures acting upon angiosperm pollination mechanisms have long been the subject of argument, there is general agreement that selection will favour those systems promoting out-breeding, and that circumstances exist where rejection of pollen from closely-related individuals - via so called isolating mechanisms - may be advantageous. The relationship between intra- and interspecific incompatibility systems is of obvious interest, both from the viewpoints of commonalties in the molecular mechanisms involved, and similarities in their evolutionary pathways. Intra-specific self-incompatibility (SI) systems are currently best understood (Hiscock *et al.*, 1996); they occur in most angiosperm families and although there has been some debate as to the level of their contribution to the success of the flowering plants, there is little doubt that it has been considerable (Whitehouse, 1950). The molecular data on homomorphic SI systems (i.e. those not involving different forms of flowers, such as heterostyly) which has begun to accumulate in recent years has served to highlight the extreme diversity of the mechanisms involved (Hiscock *et al.*, 1996). Certainly some "fine tuning" of SI mechanisms to evolving pollination strategies would be anticipated, but the discovery (to date) of 4 distinct mechanisms, each differing considerably in their molecular basis (even excluding late-acting systems which seem to operate from yet another physiological standpoint) has been surprising.

Logic demands that any SI system must feature one or more S(incompatibility)-loci, encoding male and female effector molecules. For most species of flowering plants the S-locus is multiallelic, and when pollen and stigma share a particular allele, development of the pollen tube is arrested. Further, all SI systems must be composed of two functional components, a system of recognition, and a mechanism whereby self-pollen is rejected. However, the molecular basis of both recognition and rejection appears to differ very widely between taxa. Thus in the Solanaceae, Rosaceae and the Scrophulariaceae a stylar-secreted S-locus specific RNase (Kao and McCubbin, 1996) interacts with a so-far unidentified pollen factor to arrest self-pollen, in the Papaverales the stigma secretes a 16kDa protein which regulates the development of the pollen tube according to its S-genotype (Franklin *et al.*, 1995), in the Poaceae - where two loci, the S and Z are involved - the male component of the S-locus appears to be a thioredoxin (Li *et al.*, 1994, 1996), while in the Brassicaceae a ligand from incompatible pollen apparently activates a stigmatic receptor kinase which, in a manner so-far undetermined, prevents development of the pollen tube (Nasrallah *et al.*, 1994a, b). Strikingly, there is variation even between plants with seemingly similar pollination systems, for example, plants with dry stigmas may possess any one of three systems described above.

This spectacular diversity in mechanisms is held to result from a range of powerful selective pressures maintaining outbreeding (Charlesworth, 1995), with evolution continually recruiting "new" molecular systems to the cause of identifying and rejecting self-pollen. Many of these mechanisms may, of course, have been fulfilling some other purpose in the plant. Importantly, the evolutionary pathways leading to the SI mechanisms present in extant flowering plant families cannot have been straightforward, both with respect to the development of multiallelic S-loci, and the adaptation and replacement of SI mechanisms during the evolution of new pollination strategies. For example during the transition from wet to dry stigmas, were completely new systems evolved, or were the mechanisms then present "upgraded" to operate in new environments? In an attempt to address some of these questions, we discuss here the types of evolutionary pathway which may have led to functional homomorphic SI mechanisms in the perspective of recent molecular and physiological data from *Brassica oleracea.*

Evolution of self-incompatibility - general considerations

Whilst the evolution of hypothetical SI systems based on the accumulation of homozygosity (if indeed they exist) can easily be understood (Mulcahy, 1979, 1983), this is not the case for the homomorphic multiallelic SI mechanisms which characterise so many flowering plant families. Not only is there spectacular diversity in the processes of recognition and rejection, but the very early stages of their development create considerable problems of logic. The minimum requirements of a functional SI system may be considered as two allelic versions of an "S-locus" carrying male and female determinants, with one determinant linked to some physiological mechanism that results in the arrest of pollen development when alleles are matched (Fig. 1). To expect the evolution of this system in a single step is unreasonable, and leads to the uncomfortable conclusion that the first event must either have been the development of a system that resulted in the general rejection of all pollen, or the appearance of a neutral determinant which, in later steps, could be coupled with an interrelating molecule from the opposite sexual partner. Both possibilities are logically unsound: a universal pollen rejection system would result in no seed set and thus the rapid extinction of the mutation from the population, while a neutral pollen or stigmatic factor would hold no selective advantage, and would also be eliminated. The only reasonable conclusion to be drawn from these observations is that the primary event must have involved the bringing together of two already-present systems at the stigma surface which were themselves being maintained - perhaps in

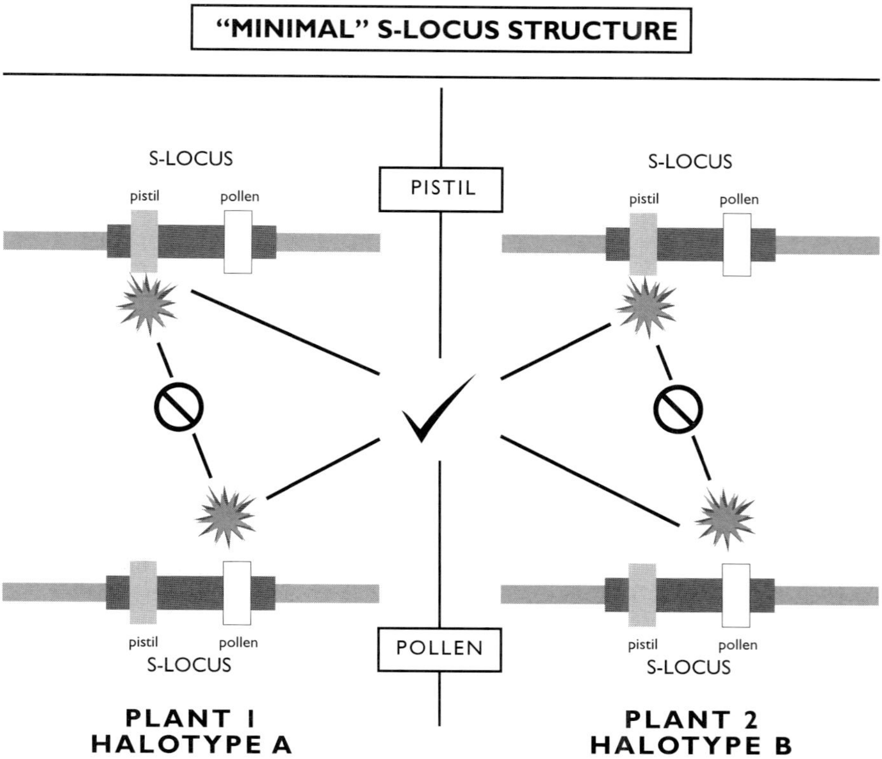

FIG. 1. Minimal requirements for a functional S-locus. Plant 1 of haplotype A possesses a locus containing 2 genes, one expressed in the pistil and the other in the pollen. When the products of these genes interact, pollination is prevented. However, pollination between plant 1 (haplotype A) and plant 2 (haplotype B) is compatible because the pollen-expressed S-locus product from plant 1 fails to interact with the stigma-expressed S-locus product from plant 2 in such a way as to inhibit the pollen tube. The pollen-inhibitory effector system may, of course be linked with either the male- or female-expressed genes of the S-locus.

different tissues - for other selective reasons. One of these systems would thus inhibit pollination, while the other would be capable of suppressing this mechanism and thus allowing pollination to proceed. However, at best, this would generate a system which might contain the components of an outbreeding mechanism but, in the absence of any allelic diversity, it would have no capacity to identify and reject self grains.

Complex models have, however, been advanced to explain the evolution of functional S-loci (see Newbiggin, 1996 for review). One such model (Bateman, 1952; Uyenoyama, 1991) involves an interaction between modifying genes - the existence of which has been demonstrated (Ai *et al.*, 1991) - and so-called proto-S loci. It is proposed that rare alleles of modifying genes so direct expression of the proto-S locus as to identify and reject self pollen, and thus increase their number and become fixed in the population. Certainly this model helps explain the apparent multiple origins of, for example, the RNase-based systems of SI (Newbiggin, 1996), but does little to assist our understanding of the very early events. The model assumes the existence of proto-S loci, the expression of which is modulated by the unlinked modifiers, but fails to explain why they are not eliminated from the population - since in normal circumstances they are insufficient to induce SI and therefore hold no selective advantage.

Interestingly, an increasing body of data has been available for many years which implicates the S-locus in interspecific incompatibility (Hiscock *et al.*, this volume). Certainly as speciation occurs and populations become isolated and "grow apart", structural and molecular differences accumulate with the result that interspecific hybridisation becomes less and less likely. However, above and beyond these "neutral" interspecific barriers - sometimes termed incongruity (Hogenboom, 1975) - there has long been interest in the existence of isolating mechanisms which act to prevent interspecific fertilisations, most notably between closely-related species. Such mechanisms would have selective advantages in certain circumstances, especially where new, useful mutations arise in populations experiencing very different environments from their progenitors (Stebbins, 1950; Grant, 1981). Reciprocal crosses between closely-related species reveals a form of unilateral incompatibility (UI) in many groups of plants (Lewis and Crowe, 1958; Hiscock and Dickinson, 1993; Hiscock *et al.*, this volume). Interestingly, further examination of this effect in a wide range of genera suggests that this unilaterality is in some way connected with the S-locus, for pistils of plants possessing SI systems regularly reject pollen from plants without an SI system, whilst the reciprocal pollination is normally successful. UI of this type has now been demonstrated in plants with gametophytic (pollen compatibility regulated by the S-locus of the vegetative nucleus) and homomorphic and heteromorphic sporophytic (pollen compatibility regulated by the parental S-loci) SI, and the data strongly indicate that pollen not carrying the SI determinant is rejected by a physiological mechanism very closely resembling that employed in the rejection of self-pollen (Hiscock and Dickinson, 1993; Murfett *et al.*, 1996; Hiscock *et al.*, this volume). Significantly, this rejection is not S-allele specific - for the pollen does not express a functional S-locus. The only reasonable conclusion that may be drawn from these observations is that in species with SI, the pollen also carries a factor (which may indeed be part of the S-ligand) which "deactivates" the rejection system (Sampson, 1962; Pandey, 1981; Dickinson *et al.*, 1997). While Sampson (1962) and Pandey (1981) disagree about the precise nature of the male factor which deactivates the SI system in compatible pollinations, the presence of a second male factor brings a new perspective not only to the operation of SI on the stigma surface - but also to its evolution. For example, if this type of "primary" incompatibility (PI) evolved first - before SI itself - would it have any intrinsic advantage? The answer is probably yes, for groups of plants possessing this system would be isolated from their close relatives, and of course from more distinct taxa. Thus in conditions of changing environments a new beneficial mutant will become fixed more effectively in a small gene pool than in the face of gene flow from a large population. Such a view is supported by Pandey (1981) who considers that strong selective pressure for such PI systems would have existed amongst the wind-pollinated gymnosperm-like plants which gave rise to the protoangiosperms.

Importantly, the basic requirements for a PI system are those described above for the primary event in the evolution of SI, namely a simple pollination disabling system carried by one partner, capable of being inhibited by a factor carried by the other (Fig. 2A). If indeed this type of PI evolved first and improved the fitness of those plants possessing it - could it not also have provided the evolutionary springboard for the subsequent development of multiallelic SI systems?

The evolution of multiallelic SI from 'primary' incompatibility (PI) mechanisms

The evolution of a functional multiallelic SI system from a simple PI or isolating mechanism would require a number of major steps. While a PI system assumes the presence

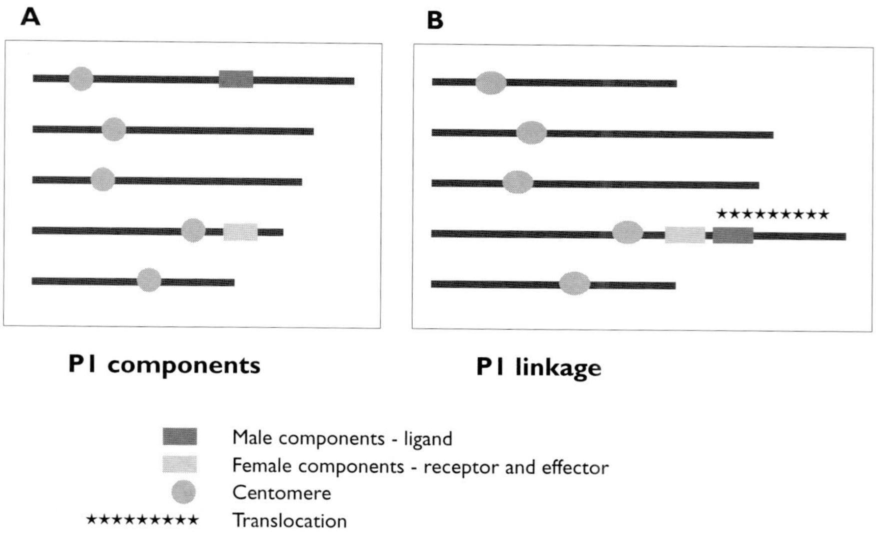

FIG. 2. A. Minimal components of a Primary Incompatibility (PI) system. The male (▬) and female (▬) determinants are encoded by genes carried on different members of the chromosome complement. For the purposes of discussion, the pollination-inhibiting effector system is associated with the female-expressed determinant. B. As 2A, but demonstrating how a translocation can result in the genetic linkage of the male and female PI determinants.

of a system of pollination inhibition in one partner and a "ligand" in the other (Fig. 2A), generation of a functional S-locus demands the generation of polymorphisms in both components and, importantly, polymorphisms which are mirrored in both male and female; clearly, any polymorphism in, say, a male ligand, must be matched in a female receptor system such that self can be recognised (Fig. 3). Most significantly, for the S-locus to function at all the gene(s) encoding the male and female components need to be tightly linked genetically (Figs. 2B and 3).

A possible evolutionary scenario involves at least one component of the PI system - for example the male ligand - belonging to a multigene family which already possesses considerable polymorphisms associated with its original function in the cell. Thus while one member of this male ligand family effectively interacts with and disables a female-expressed pollination-prevention system, other members may not - although they would possess some properties of the active molecule - say binding to a particular site on the female determinant. The effect of this multiplicity of ligands would have an essentially neutral selective effect.

The model, of necessity, also requires polymorphism in a molecule which forms part of the female-expressed rejection system. A combination of the multiplicity and continual polymorphism of the male partner, and an accompanying polymorphism in the female, will generate - at very rare intervals - a member of the male ligand family which modulates the interaction of the functional ligand with the female "receptor" in such a way as to prevent it disabling the rejection system. Where this event occurs on different plants, interfertilisation will of course be prevented, but where it occurs in the same plant - an effective outbreeding system will be generated.

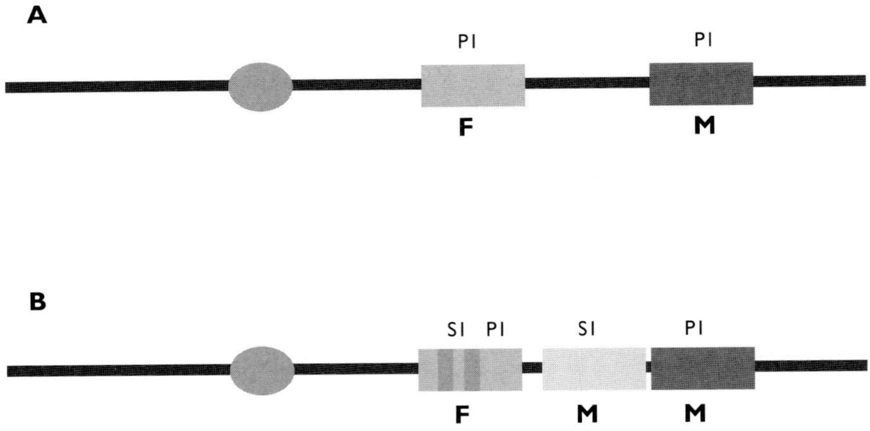

FIG. 3. The development of an SI system from a PI mechanism. A. The gene encoding the male-expressed determinant (M) has become duplicated, and will produce a new protein. B. Variation at domains within the gene encoding the female-expressed determinant (F) result in a polypeptide which will interact with the PI and SI male determinants such that a self SI factor will prevent the PI factor suppressing the pollination-inhibitory system, while a cross SI factor will not (see also Figs. 4 & 5).

In self-pollinations this "new" male element would result in pollen being rejected by the PI system, while cross pollinations (which would involve grains not carrying this element) would be fertile. The fate of this early SI system in subsequent progeny would of course depend on the linkage of the male and female components; only when they are effectively linked could a functional system be inherited. However, assuming that the PI system contributed to the fitness of the individual, there would be selective pressure - at least in the early days of its introduction into the population - for the two components to become linked genetically (Fig. 2B) - otherwise they would be separated and lost through segregation. It is thus implicit in this model that when evolution of the early SI system commenced, the components of the PI mechanism were already linked.

For simplicity, we have in this model proposed that gene duplication only occurs in sequences encoding the male ligand, with the polymorphism in the female receptor/rejection gene product being restricted to domains of a single molecule. However, similar models can easily be advanced involving duplication of the receptor/rejection genes, but they are less well supported by experimental evidence (see below).

There is little doubt than an effective "S"-locus would rapidly become incorporated into a population - until such a stage as it began to have a deleterious effect on cross-pollinations. Such loci would then be maintained at relatively low levels until a sufficient number of different versions (alleles) were generated (Wright, 1939, 1964; Hor Yan Wong, pers. comm.) to enable the elimination of self-compatible plants from the population.

While the model proposed has some merit in logic, the question has to be addressed as to whether there are components of a signalling and linked rejection system which could display sufficient levels of polymorphism. While intercellular signalling systems involved in development may well involve relatively constant molecular species, it is in the interest of mechanisms of defence against pathogens to maintain a considerable degree of molecular diversity - for the pathogens themselves will constantly be modulating their pathogenic strategies. Importantly, while few data presently exist on variability in host receptors, it is nevertheless clear that a number of these receptors, including S-class receptors, are coupled to functional intracellular signalling systems which may stimulate the eventual rejection of the pathogen (Song *et al.,* 1995; Pastuglia *et al.,* 1997).

An important further consideration is that all pollen, and perhaps fungal spores must also emit a generic ligand - or in some other way signal - to activate the rejection-system in the first instance. It is possible that the rejection system is constitutively active, but this is not easily envisaged, and experimental data (see below) suggest that the system is activated by alighting pollen.

Predictions made by the model concerning current incompatibility systems

The model outlined above allows very precise predictions to be made about the types of molecule likely to be involved in systems of inter- and intraspecific incompatibility.

Interspecific incompatibility

Assuming that the PI and SI systems are both present in extant populations, and that the success of the S-loci has eliminated all those SC individuals carrying only PI mechanisms, stigmas from SI species should always reject pollen from close, self-compatible (SC) relatives which would not carry the PI male ligand. Closely related SC plants should cross-pollinate successfully as should matings between SC females and SI males. The results of SI × SI crosses are less easily predicted, for there may be interactions between closely-related alleles from different species, or different specificities of the PI determinant. Although there may be advantages for placing both the PI determinants at the S-locus, technically it would only be necessary for that determinant, linked to the receptor/response system, to be S-linked.

Obviously, these interspecific interrelationships are likely to become progressively more unpredictable the more distantly related the mating partners, and the success of pollinations between distant partners will certainly be determined by structural and physiological incongruities (Hogenboom, 1975).

The model also predicts that the PI/SI rejection system is activated in some way by the contact of the pollen with the stigma; this may involve a specific pollen-borne ligand, or some other consequence of early pollen/stigma interactions, such as the release of oligosaccharides from the stigmatic wall by enzymes released from the pollen. The interactions involved in PI, its activation by a generic pollen ligand and its suppression by the pollen-borne primary compatibility (PI) ligand are shown in Fig. 4.

Intraspecific (self) pollinations

Our present and very incomplete knowledge of SI indicates that the pollination inhibition is effected either through cytotoxic changes induced in the pollen (i.e. in the RNase based system of the Solanaceae; McCubbin *et al.,* 1997) or by means of events commencing in the stigmatic papillae (i.e. in *Brassica,* where a receptor kinase-stimulated response rejects self pollen, Nasrallah *et al.,* 1994a, b). The model predicts that in both pollen and stigma there should be polymorphic S-locus specific determinants, one of which should be linked to (or form part of) the response mechanism; the model makes no assumptions about whether one determinant is likely to belong to a more polymorphic gene family than the other, but unless a multiplicity of response systems is invoked, a single response-linked factor is assumed to carry both PI and SI specific sites, whereas the other "determinant" may take the form of a number of products of a multigene family.

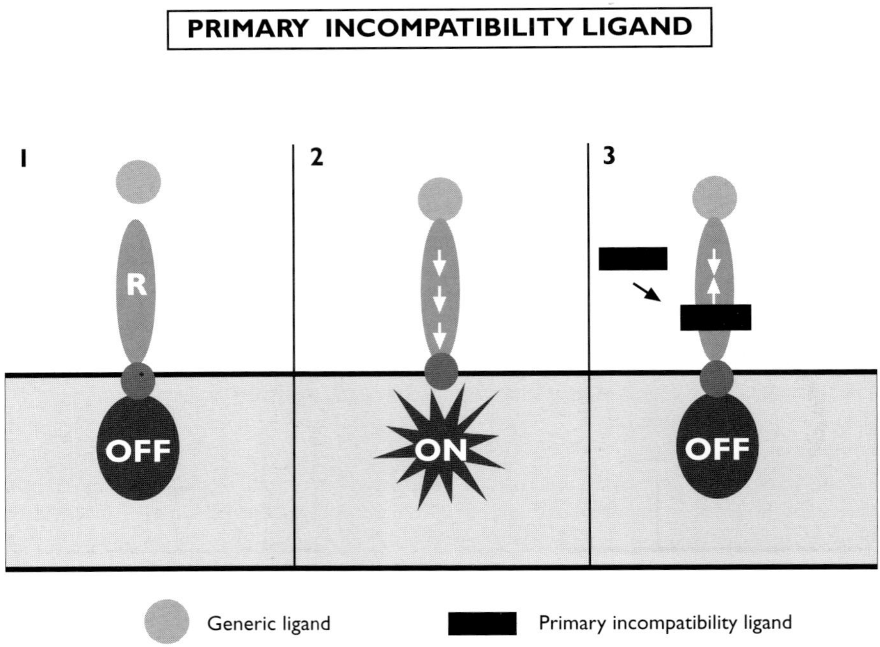

FIG. 4. A molecular basis for a PI system. 1. & 2. The arrival of pollen stimulates the release of a generic pollen ligand factor which activates a pollen rejection system (R) in the stigma (perhaps a transmembrane receptor kinase, such as the SRK in *Brassica*). 3. The primary incompatibility ligand from the pollen interacts with the rejection system, such that it fails to be activated by the generic pollen ligand.

The model predicts that, following a self-pollination, the S-locus determinant from one partner should in some way modulate the interaction of the PI determinant (from the same partner) with the factor (in the other partner) linked to the response system, which would carry both PI and SI sites (Fig. 5). It is possible that those PI and SI determinants not linked to the response system may also constitute domains of the same molecule and, under these circumstances, both PI and SI determinants would map to the S-locus. Where the PI and SI determinants are separate molecules and unlinked, the genes encoding them are nevertheless predicted to belong to the same polymorphic multigene family.

While the model makes relatively simple predictions about gametophytically controlled SI systems (where the pollen compatibility is controlled by its own genome, and thus one set of determinants is expressed), the situation is considerably complicated in sporophytic systems where the pollen carries two sets of determinants. Thus, in gametophytic SI two sets of stigmatic determinants interact with one set of pollen factors, in sporophytic systems there are two sets of each. Here, for the model to hold, it must be assumed that the stoichiometric relationship between PI and SI factors - rather than their presence or absence - forms the basis of the regulation of the response/rejection system.

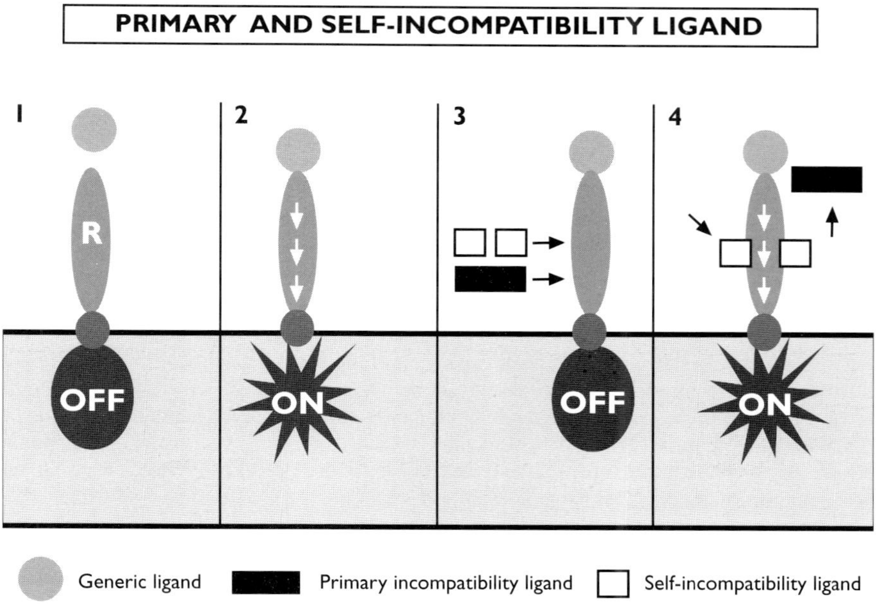

FIG. 5. Competition between pollen-borne SI and PI determinants (see also Fig. 4). 1. & 2. A generic signal released by the pollen activates a stigmatic pollen rejection system (R). 3. & 4. Normally this event is suppressed by a PI ligand from the pollen (see Fig. 4), but a self SI ligand interacts with the pollen rejection system is such a way (shown here as displacing the PI factor) as to prevent the inhibitory action of the PI determinant. A cross SI determinant fails to affect the PI interaction in this way (not shown).

Inter- and intraspecific incompatibility in *Brassica oleracea*

Brassica oleracea represents perhaps the best current paradigm for incompatibility systems. With regard to SI, much is known of the molecular mechanisms involved in the female-expressed recognition and rejection systems (Nasrallah *et al.,* 1991, 1993, 1994a) and new data are available on pollen-borne determinants (Dickinson *et al.,* 1997). While interspecific incompatibility has not been extensively studied, there are data from a number of investigations relating to the interrelationships between *B. oleracea* and its close relatives (Sampson, 1962; Hiscock and Dickinson, 1993; Hiscock *et al.*, this volume).

Interspecific incompatibility

Investigations into interspecific interrelations between *Brassica oleracea* and its close relatives (Sampson, 1962; Hiscock and Dickinson. 1993; Hiscock *et al.,* this volume) have revealed unilaterality in pollinations between species with and without S-loci. As is the case with other genera, stigmas from species possessing SI regularly reject pollen from SC species. However, while in plants with gametophytically-controlled SI, pollinations between SI relatives are frequently compatible (Lewis and Crowe, 1958), this is not the case in *Brassica* ssp. where these pollinations give highly variable results (Sampson, 1962; Hiscock and

Dickinson, 1993). Importantly, the structure and physiology of the pollen rejection mechanism activated in these unilateral pollinations are indistinguishable from those operating in the SI system (Hiscock and Dickinson, 1993; Hiscock *et al.*, this volume).

Self-incompatibility

SI in Brassicaceae (Cruciferae) is sporophytically determined, with self pollen being inhibited on the stigmatic surface (Dickinson and Lewis, 1973). Some 80 S-alleles are known in *Brassica oleracea* (Ockendon, 1974) which have been divided into two classes (I and II) on the basis of their ability to reject self pollen (Nasrallah *et al.*, 1991). Since a number of genes are now known to be located at the S-locus, S-alleles are conventionally termed S-haplotypes.

On pollination compatible grains adhere to the dry stigmatic surface and access stigmatic water through pores in the cuticle; self pollen fails to adhere (Stead *et al.*, 1980), and hydrates only slightly (Zuberi and Dickinson, 1985; Dickinson, 1995), the degree of hydration dependent on the strength of the S-haplotypes involved. Both cross and self pollinations induce a stigmatic response which takes the form of a local expansion of the stigmatic cell wall beneath the grain (Elleman and Dickinson, 1996). In an elegant series of molecular studies, Nasrallah *et al.*, (1991, 1993, 1994a) have identified a number of genes expressed at the papillar surface, two of which are located on the S-locus. *SLG* encodes a *S-L*ocus *G*lycoprotein which is S-haplotype specific and secreted into the papillar cell wall; other versions of *SLG* also exist which feature transmembrane domains (Tantikanjana *et al.*, 1993; Delorme *et al.*, 1995) and it is assumed that these proteins are anchored in the plasma membrane. *SRK* encodes a *S*-locus specific *R*eceptor *K*inase; this transmembrane kinase is also inserted into the stigmatic plasma membrane and its extracellular domain exhibits very high homology to the *SLG* of the same haplotype (Stein *et al.*, 1991). Persuasive evidence has been reported (Nasrallah *et al.*, 1994b) for the activation of *SRK* following a self pollination to result in the rejection of the pollen. While *SLG* and *SRK* are assumed to encode the female determinants of SI, far less is known of S-locus specific factors carried by the pollen. A recent series of bioassays have confirmed the long-held view (Dickinson and Lewis, 1973; Heslop-Harrison *et al.*, 1974) that the male determinant of SI is carried in the pollen grain coating (Dickinson and Elleman, 1994; Dickinson, 1995; Stephenson *et al.*, 1997), but the molecule itself has yet to be positively identified. Boyes and Nasrallah (1995) have reported the presence of *SLA*, an anther-expressed gene at the S-locus of one S-haplotype which exhibits a complex transcription pattern suggesting that *SLA* may either encode one or two small proteins. Work on self-sterile lines of the tetraploid *B. napus* (Yu *et al.*, 1996) has identified a further anther-expressed gene at the S-locus, which also potentially encodes a small protein, but the sequence of which does not appear to vary from S-haplotype to S-haplotype. Fractionation of the pollen coating of *B. oleracea* has revealed the presence of a family of small, highly-charged proteins, one of which interacts with SLG in an *in vitro* gel-shift assay (Doughty *et al.*, 1993). This interaction is, however, not generally S-locus specific.

The relationship between inter- and intraspecific incompatibility in Brassica oleracea

While commonalities between the pollen rejection mechanism in UI and SI are of interest, they do not themselves constitute unequivocal evidence of any evolutionary link between the two systems. However, more recent data concerning the family of small, highly-charged pollen coat proteins suggest that these proteins (termed **P**ollen **C**oating **P**roteins, class **A**; **PCP-A**s) may play a role in both processes. The amino acid sequence of PCP-A1, the member of this family which interacts *in vitro* with the *SLG* gene product in a gel-shift assay (Doughty *et al.*, 1993), has now been obtained for a number of S-haplotypes (J. Doughty, pers. comm.). Apart from a few single amino acid changes, this sequence appears to be identical for all haplotypes - indicating that it cannot fulfil the role of male determinant (Dickinson *et al.*, 1997). Since PCP-A1 is clearly present in the pollen coating, and is released onto the stigmatic surface, there is a strong probability that it penetrates the fissures in the papillar cuticle and enters the stigmatic apoplast.

While little is known of conditions in the stigmatic apoplast, examples from other cell types suggest that the environment will be such that PCP-A1 will interact with the *SLG* gene product - as it does *in vitro*. Preliminary data also indicate that PCP-A1 also interacts with the *SRK* gene product, which would be predicted because of the close similarity - within individual S-haplotypes - between the gene products of *SLG,* and the extracellular domain of *SRK* (Stein *et al.,* 1991). PCP-A1 may thus play a wider role in pollen/stigma interactions in *Brassica oleracea* - possibly acting as the pollen held PI determinant discussed above. The model predicts that this factor should be a member of a polymorphic multigene family, and indeed there are at least 12 members of the PCP-A family present in the pollen coatings (Doughty *et al.,* submitted). Further, Southern hybridisations using a probe based on another member of the family (Stanchev *et al.,* 1996) reveal the presence of a number of related sequences in the genome. In contrast with some other members of the PCP-A family, *PCP-A1* appears to be present as a single copy in the haploid genome (Doughty *et al.,* submitted). The model also requires that members of this family should at some stage have played some role unrelated to incompatibility, and it is interesting that there is very close homology between members of the PCP-A family and the so-called defensin group of proteins held to be important in resistance to pathogen attack (Terras *et al.,* 1993). It is thus feasible that the expression of PCP-A type proteins in the anther originally protected the pollen against infection, and these small highly-charged proteins were later recruited into a role in interspecific incompatibility. Present evidence suggests that PCP-A1 is not S-linked (Doughty *et al.,* submitted), but, as is discussed above, once components of the PI and SI systems have become fixed in the population, there is no *prima facie* reason why the PI male factor should be present at the S-locus.

PCP-A1 may thus act as a PI male factor to disable a "defence" mechanism - first expressed, through a mutation, at the stigma surface some 40 million years ago (Uyenoyama, 1991). It follows that its transfer to the surface of pollen of an SC plant, should result in acceptance by the *Brassica oleracea* stigma - as far as other "incongruities" permit. The pollen of *Arabidopsis thaliana* fails to develop on the stigma of *Brassica oleracea*, unless the papillae are treated with cycloheximide (or other protein synthesis inhibitor) a suppressor of the rejection mechanism in both PI and SI in *Brassica oleracea* (Hiscock and Dickinson, 1993). Pollen grain coating from *Brassica oleracea*, which is rich in PCP-A1, has been experimentally applied to the surface of the *Arabidopsis* pollen, and these grains used to pollinate *Brassica oleracea* stigmas (Dickinson *et al.,* 1997). Encouragingly, transfer of the coating to the SC *Arabidopsis* pollen confers the ability to develop on the surface of the SI stigmatic papilla of *Brassica oleracea.* Certainly, the *Arabidopsis* grains do not develop fully and penetrate the stigmatic cuticle, but do hydrate to a level, and with a frequency expected of compatible *Brassica* pollen. The model also predicts that the determinants involved in PI and SI should be related; however, as has already been stated, PCP-A1 is not S-linked, and its interactions with *SLG* gene products from a range of S-haplotypes do not, in general differ (Doughty *et al.,* 1993). Nevertheless data from an *in vivo* bioassay convincingly demonstrate that a protein fraction principally containing PCP-A1 and other members of the PCP-A family is capable of determining S-specificity (Stephenson *et al.,* 1997; Dickinson *et al.,* 1997) and, significantly, highly-purified fractions of PCP-A1 itself are not active in this bioassay (Dickinson *et al.,* 1997). The possibility thus remains that it is not PCP-A1, but another member of the PCP-A family which confers S-specificity on the pollen coat fraction described above. Current research is focused on the molecular analysis of components of the bioassay-active fraction, other than PCP-A1. As is the case for PI (see above), the model predicts that possession of the SI-specific members of the PCP-A family should confer S-specificity on pollen coatings. To test this, the heterologous experiments described above, in which *Arabidopsis* pollen was treated with *Brassica* pollen coating, were expanded to include coatings from different S-haplotypes (Dickinson *et al.,* 1997). In these circumstances, the *Arabidopsis* pollen hydrated on the *Brassica* stigmatic surface in accordance with the S-haplotype carried by the interacting partners. If, of course, the male determinant of SI in *Brassica oleracea* emerges as *SLA* (Boyes and Nasrallah 1995), or the gene in *Brassica napus* described by Yu *et al.,* (1996), the model will, of necessity, require radical reinterpretation, although those parts concerning PCP-A1 and the establishment of PI will still have relevance.

The model predicts that the response-linked determinant, which is the *SLG/SRK* gene product complex in *Brassica oleracea*, should be active in both PI and SI, and available data tends to confirm such an inference (Hiscock and Dickinson, 1993). However, new results from work on members of the Solanaceae, where self-pollen is rejected by an "S"-RNase based system (Lee *et al.,* 1994; Murfett *et al.,* 1994) demonstrate conclusively that in some interspecific pollinations, pollen is rejected by the S-RNase that is also involved in the rejection of self pollen (Murfett *et al.,* 1996), although, in other interspecific pollinations, different mechanisms may be involved.

Events in *Brassica oleracea* and a general model for incompatibility

The data from inter- and intraspecific pollinations in *Brassica oleracea* thus provide general support for the model described above. In no way do they supply conclusive evidence, and further results are needed in particular areas - specifically confirmation that PCP-A1 interacts with the *SRK* gene product, and molecular analysis of the PCP-A proteins in the bioassay-active fraction and the genes encoding them. At this early stage in the development of the model, the complexities resulting from the sporophytic inheritance of pollen compatibility in *Brassica oleracea* SI have not been addressed, nor any effect on the model caused by the apparent difference in allelic strength between S-haplotypes (Nasrallah *et al.,* 1991).

Nevertheless, the model, uniquely, provides an evolutionary basis for the *de novo* development of S-loci, and explains the apparent interrelationship between SI and UI. It predicts the prior evolution of PI, which is held to have operated as an isolating mechanism in early populations. Data from present day *Brassica oleracea* identify a class of proteins carried by the pollen coating, perhaps originally possessing a role in defending the anther against pathogens, different members of which may act as the male determinants in PI and SI. The utility of the model will clearly depend on the support (or otherwise) it receives from new data from inter- and intra-specific pollination in *Brassica* and other genera.

Acknowledgements

Financial support from the UK BBSRC, NATO and the Leverhulme Foundation is acknowledged. The authors also wish to thank Suzanne Dixon for invaluable technical assistance, Sarah Compton for work with *Arabidopsis* as part of her student project, and Ann Rogers for help with the manuscript.

References

Ai, Y.J., Kron, E. and Kao, T-H. (1991). *S*-alleles are retained and expressed in a self-compatible cultivar of *Petunia inflata*. *Molec. Gen. Genet.* **230**: 353–358.

Bateman, A.J. (1952). Self-incompatibility systems in angiosperms. I. Theory. *Heredity* **6**: 285–310.

Boyes, C.D. and Nasrallah, J.B. (1995). An anther-specific gene encoded by an S-locus haplotype of *Brassica* produces complementary and differentially regulated transcripts. *Pl. Cell* **7**: 1283–1294.

Charlesworth, D. (1995). Multi-allelic self-incompatibility polymorphisms in plants. *Bioessays* **17**: 31–38.

Delorme, V., Giranton, J., Hatzfeld, Y., Friry, A., Heizman, P., Ariza, M.J., Dumas, C., Gaude, T. and Cock, M. (1995). Characterization of the *S* locus genes, *SLG* and *SRK*, of *Brassica S3* haplotype: identification of a membrane-localized protein encoded by the *S* locus receptor kinase gene. *Plant J.* **7**: 429–440.

Dickinson, H.G. (1995). Dry stigmas, water and self-incompatibility in *Brassica*. *Sexual Pl. Reprod.* **8**: 1–10.

Dickinson, H.G. and Lewis, D. (1973). The formation of the tryphine coating the pollen grains of *Raphanus* and its properties relating to the self-incompatibility system. *Proc. Roy. Soc. London, Ser. B, Biol. Sci.* **184**: 149–165.

Dickinson, H.G. and Elleman, C.J. (1994). Pollen hydrodynamics and self incompatibility in *Brassica oleracea*. In: A.G. Stephenson and T.-H. Kao (editors). Pollen pistil interactions and pollen tube growth. Penn State Symposium, May 1994. pp. 45–61. Rockeville, Maryland. (American Society of Plant Physiologists Series, 12).

Dickinson, H.G., Doughty, J., Hiscock, S.J., Elleman, C.J. and Stephenson, A.G. (1997). Pollen-stigma interactions in *Brassica*. In: A. Greenland, E. Meyerowitz and M. Steer (editors). Control of plant development: genes and signals. Oxford Univ. Press (Society for Experimental Biology symposium series).

Doughty, J.D., Hedderson, F., McCubbin, A. and Dickinson, H.G. (1993). Interaction between a coating-borne peptide of the *Brassica* pollen grain and S(incompatibility)-locus linked stigmatic glycoproteins. *Proc. Natl. Acad. Sci. USA* **90**: 467–471.

Doughty, J., Dixon, S., Hiscock, S.J. and Dickinson, H.G. (submitted). Cloning and characterisation of the gene for PCPA1, a pollen-coating borne peptide which interacts with S-locus glycoproteins from *Brassica oleracea*..

Elleman, C.J. and Dickinson, H.G. (1996). Identification of pollen components regulating key pollination-specific responses in the stigmatic papillae of *Brassica oleracea*. *New Phytol.* **133**: 197–205.

Elleman, C.J., Franklin-Tong, V. and Dickinson, H.G. (1992). Pollination in species with dry stigmas: the nature of the early stigmatic response and the pathway taken by pollen tubes. *New Phytol.* **121**: 413–424.

Franklin, F.C.H., Lawrence, M.J. and Franklin-Tong, V.E. (1995). Cell and molecular biology of self-incompatibility in flowering plants. *Int. Rev. Cytol.* **158**: 1–64.

Grant, V. (1981). Plant speciation. Columbia University Press, New York.

Heslop-Harrison, J., Knox, R.B. and Heslop-Harrison, Y. (1974). Pollen-wall proteins: exine-held fractions associated with the incompatibility response in Cruciferae. *Theor. Appl. Genet.* **44**: 133–177.

Hiscock, S.J. and Dickinson, H.G. (1993). Unilateral incompatibility within the Brassicaceae; further evidence for the involvement of the self-incompatibility (S)-locus. *Theor. Appl. Genet.* **86**, 744–753.

Hiscock, S.J., Kües, U. and Dickinson, H.G. (1996). Molecular mechanisms of self-incompatibility in flowering plants and fungi - different means to the same end. *Trends Cell Biol.* **6**: 421–428.

Hiscock, S.J., Doughty J. and Dickinson, H.G. (1998). Unilateral incompatibility and the *s* (self-incompatibility)-locus. In: S.J. Owens and P.J. Rudall (editors). Reproductive biology, pp. 31–46, Royal Botanic Gardens, Kew.

Hogenboom, N.G. (1975). Incompatibility and incongruity: two different mechanisms for the non-functioning of intimate partner relationships. *Proc. Roy. Soc. Lond. Ser. B. Biol. Sci.* **188**: 361–375.

Kao, T-H, and McCubbin, A. G. (1996). How flowering plants discriminate between self and non-self pollen to prevent inbreeding. *Proc. Natl. Acad. Sci. USA* **93**: 12059–12065.

Lee, H-S., Huang, S. and Kao, T-H. (1994). Direct evidence that S-proteins are responsible for rejection of incompatible pollen in *Petunia inflata*. *Nature* **367**: 560–563.

Lewis, D. and Crowe, L.K. (1958). Unilateral interspecific incompatibility in flowering plants. *Heredity* **12**: 233–256.

Li, X., Nield, J., Hayman, D. and Langridge, P. (1994). Cloning a putative self-incompatibility gene from the pollen of the grass *Phalaris coerulescens*. *Pl. Cell* **6**: 1923–1932.

Li, X., Nield, J., Hayman, D. and Langridge, P. (1996). A self-fertile mutant of *Phalaris* produces an S protein with reduced thioredoxin activity. *Plant J.* **10**: 505–513.

McCubbin, A.G., Chung, Y-Y. and Kao, T-H. (1997). A mutant S_3 RNase of *Petunia inflata* lacking RNase activity has an allele-specific dominant negative effect on self-incompatibility interactions. *Pl. Cell* **9**: 85–95.

Mulcahy, D.L. (1979). The rise of angiosperms: a genecological factor. *Science* **206**, 20–23.

Mulcahy, D.L. and Mulcahy, G.B. (1983). Gametophytic self-incompatibility re-examined. *Science* **220**: 1247–1251.

Murfett, J. *et al.* (1994). S-RNase expressed in transgenic *Nicotiana* causes S-allele specific pollen rejection. *Nature* **367**: 563–566.

Murfett, J., Strabala, T.J., Zurek, D.M., Mou, B., Beecher, B. and McClure, B.A. (1996). S-RNase and interspecific pollen rejection in the genus *Nicotiana*: multiple pollen-rejection pathways contribute to unilateral incompatibility between self-incompatible and self-compatible species. *Pl. Cell* **8**: 943–958.

Nasrallah, J.B. and Nasrallah, M.E. (1993). Pollen-stigma signalling in the sporophytic self-incompatibility response. *Pl. Cell* **5**: 1325–1335.

Nasrallah, J.B., Nishio, T. and Nasrallah, M.E. (1991). The self-incompatibility genes of *Brassica*: expression and use in genetic ablation of floral tissues. *Annual Rev. Pl. Physiol. Pl. Molec. Biol.* **42**: 393–422.

Nasrallah, J.B., Stein, J.C., Kandasamy, M.K. and Nasrallah, M.E. (1994a). Signalling the arrest of pollen tube development in self-incompatible plants. *Science* **266**: 1505–1508.

Nasrallah, J.B., Rundle, S.J. and Nasrallah, M.E. (1994b). Genetic evidence for the requirement of the *Brassica* S-locus receptor kinase gene in the self-incompatibility response. *Plant J.* **5**: 373–384.

Newbiggin, E. (1996). The evolution of self-incompatibility: a molecular voyeur's perspective. *Sexual Pl. Reprod.* **9**: 357–361.

Ockendon, D.J. (1974). Distribution of self-incompatibility alleles and breeding structure of open-pollinated cultivars of Brussels sprouts. *Heredity* **33**: 159–171.

Pandey, K.K. (1981). Evolution of unilateral incompatibility in flowering plants: further evidence in favour of twin specificities controlling intra- and interspecific incompatibility. *New Phytol.* **89**: 705–728.

Pastuglia, M., Roby, D., Dumas, C. and Cock, M.J. (1997). Rapid induction by wounding and bacterial infection of an S-gene family receptor-like kinase gene in *Brassica oleracea*. *Pl. Cell* **9**: 49–60.

Sampson, D.R. (1962). Intergeneric pollen-stigma incompatibility in the Cruciferae. *Canad. J. Genet. Cytol.* **4**: 38–49.

Song, W-Y., Wang, G-L., Chen, L-L., Kim, H-S., Pi, L-Y., Holsten, T., Gardner, J., Wang, B., Zhai W-X., Zhu, L-H., Farquet, C. and Ronald, P. (1995). A receptor kinase-like protein encoded by the rice disease resistance gene Xa21. *Science* **270**: 1804–1806.

Stanchev, B.S., Doughty, J., Scutt, C.P., Dickinson, H.G. and Croy, R.R.D. (1996). Cloning of *PCP1*, a member of a family of pollen coat protein (*PCP*) genes from *Brassica oleracea* encoding novel cysteine-rich proteins involved in pollen-stigma interactions. *Plant J.* **10**: 303–313.

Stead, A.D., Roberts, I.N. and Dickinson, H.G. (1980). Pollen-stigma interaction in *Brassica oleracea*; the role of stigmatic proteins in pollen adhesion. *J. Cell Sci.* **42**: 417–423.

Stebbins, G.L. (1950). Variation and evolution in plants. Columbia Univ. Press, New York.

Stein, J.C, Howlett, B., Boyes, D.C., Nasrallah, M.E. and Nasrallah, J.B. (1991). Molecular cloning of a putative receptor protein kinase gene encoded at the self-incompatibility locus of *Brassica oleracea*. *Proc. Natl. Acad. Sci. USA* **88**: 8816–8820.

Stephenson, A., Doughty, J., Dixon, S., Elleman, C.J., Hiscock, S.J. and Dickinson, H.G. (1997). The male determinant of sporophytic self-incompatibility in *Brassica* is located in the pollen coating. *Plant J.* **12**: 1351–1359.

Tantikanjana, T. *et al.* (1993). An alternative transcript of the S-locus glycoprotein gene in a class II pollen-recessive self-incompatibility haplotype of *Brassica oleracea* encodes a membrane-anchored protein. *Pl. Cell* **5**: 657–666.

Terras, F.R.G., Torrekens, S., Van Leuven, F., Osborn, R.W., Vanderleyden, J., Cammue, B.P.A. and Broekaert, W.F. (1993). A new family of basic cysteine-rich plant antifungal proteins from Brassicaceae species. *FEBS Lett.* **316**: 233–240.

Uyenoyama, M.K. (1991). On the evolution of genetic incompatibility systems, VI: a 3-locus modifier model for the origin of gametophytic self-incompatibility. *Genetics* **128**: 453–469.

Whitehouse, H.L.K. (1950). Multiple allelomorph incompatibility of pollen and style in the evolution of angiosperms. *Ann. Bot. n.s.* **14**, 198–216.

Wright, S. (1939). The distribution of self-sterility alleles in populations. *Genetics* **24**: 538–552.

Wright, S. (1964). The distribution of self-incompatibility alleles in populations. *Evolution* **18**: 609–619.

Yu, K., Schafer, U., Glavin, T.L., Goring, D.R., and Rothstein, S.J. (1996). Molecular characterization of the *S* locus in two self-incompatible *Brassica napus* lines. *Pl. Cell* **8**: 2369–2380.

Zuberi, M.I. and Dickinson, H.G. (1985). Pollen stigma interaction in *Brassica*. III. Hydration of the pollen grains. *J. Cell Sci.* **76**: 321–336.

Hiscock, S.J., Doughty J. and Dickinson, H.G. (1998). Unilateral incompatibility and the *s* (self-incompatibility)-locus. In: S.J. Owens and P.J. Rudall (Editors). Reproductive Biology, pp. 31–46. Royal Botanic Gardens, Kew.

UNILATERAL INCOMPATIBILITY AND THE *S* (SELF-INCOMPATIBILITY)-LOCUS

SIMON J. HISCOCK, JAMES DOUGHTY AND HUGH G. DICKINSON

Dept. Plant Sciences, University of Oxford, South Parks Rd., Oxford OX1 3RB, UK

Abstract

In this paper we will review evidence from the literature and present new data that indicate that genes which control self-incompatibility in flowering plants also exert a role in the control and regulation of certain types of pre-zygotic interspecific incompatibility, most notably the phenomenon of unilateral incompatibility. Before presenting this evidence however, a brief review of current knowledge of self-incompatibility (SI) is necessary.

Self-incompatibility systems in flowering plants

Despite the apparent diversity of SI systems in flowering plants, in terms of the number of loci involved and whether the SI phenotype is associated with distinct floral morphologies (heteromorphic SI) or not (homomorphic SI), the majority of SI systems fall into one of two genetically defined categories (de Nettancourt, 1977). In gametophytic SI (GSI) systems the incompatibility phenotype of the pollen is determined by its own haploid genome, whereas in sporophytic SI (SSI) the incompatibility phenotype of the pollen is determined by the diploid genome of the parent plant, through the transfer of sporophytically synthesised S products from the anther tapetum to the coating of the pollen grain during pollen development in the anther (Dickinson and Lewis, 1973, Heslop-Harrison *et al.*, 1974, Stephenson *et al.*, 1997). GSI is typical of the Solanaceae, Rosaceae, Scrophulariaceae, Papaveraceae and Poaceae while SSI, in its homomorphic form, is found in the Brassicaceae, Asteraceae and Convolvulaceae, and in its heteromorphic form in a great many families, including the Primulaceae, Linaceae and Lythraceae. In the majority of cases both GSI and SSI are each controlled by a single locus - the *S* (self-incompatibility)-locus, which is multi-allelic (de Nettancourt, 1977). Notable exceptions are the grasses (Poaceae), where two multiallelic loci, *S* and *Z* control GSI, and species with heteromorphic SSI (*Primula*, for instance) where a single diallelic *S* locus (*S* and *s*) controls the SI reaction. Classical genetic studies of SI predict that each *S* allele should be bipartite, having both a male and female specificity component. Such predictions lead logically to the 'dimer' hypothesis of *S* gene action (Lewis, 1965), which proposed that identical *S* gene products from pollen and pistil dimerise to form an active 'inhibitor complex' that causes the cessation of pollen growth.

Recent molecular genetic analyses of SI have identified molecules, predominantly from female tissues, that are involved in GSI and SSI (see Franklin *et al.*, 1995; Dodds *et al.*, 1996; Hiscock *et al.*, 1995a, 1996 for recent reviews), and the absence of such molecules from pollen tends to refute the dimer hypothesis in its purest form. Also emergent from these studies is the intriguing realisation that GSI operates through at least three different molecular pathways. In the Solanaceae, Rosaceae and Scrophulariaceae S-RNases mediate the female part of the SI reaction (Dodds *et al.*, 1996), presumably by degrading pollen RNA molecules, while in the Papaveraceae a Ca^{2+}-based signal transduction system involving differential protein phosphorylation and altered gene expression in the pollen, is central to the SI reaction (Franklin-Tong *et al.*, 1993, Rudd *et al.*, 1996), and in the grasses a pollen-expressed thioredoxin has been implicated in the SI response (Li *et al.*, 1994, 1996). Less is known about the molecular diversity (if any) of SSI because molecular studies have focussed exclusively on *Brassica.* In *Brassica* it has been demonstrated that two stigmatically-expressed genes *SLG* (*S*-Locus Glycoprotein) and *SRK* (*S*-Receptor Kinase) are required for fully functional SSI (Nasrallah and Nasrallah, 1993, Nasrallah *et al.*, 1994a, b) and it is presumed that a pollen-coating-held *S*-ligand binds to the receptor domain of SRK to initiate a phosphorylation cascade that culminates in the localised inhibition of pollen tube development (Nasrallah *et al.*, 1994; Dickinson, 1994). With the possible exception of the grasses, the molecular identity of the male component(s) of the various SI systems remains to be determined.

Unilateral incompatibility

Unilateral interspecific incompatibility (UI) was first described by Harrison and Darby (1955) in the genus *Antirrhinum.* They extended the earlier crossing studies of Mather (1947) to include six self-incompatible (SI) species of *Antirrhinum* and two self-compatible (SC) species. Different combinations of crosses between the eight species revealed that when self-compatible species were used as the female parent, cross pollinations were always compatible, ie with pollen from both SC species and SI species, whereas when SI species were used as the female parent, only crosses with other SI species were compatible; crosses with SC plants as the pollen donor were always incompatible. Thus, crosses between SI species and SC species were described as 'unilaterally' incompatible.

The concept of interspecific UI was greatly expanded by Lewis and Crowe (1958) when they reported UI to be extremely widespread among angiosperms and able to manifest itself in intergeneric crosses, as well as interspecific crosses, and even (exceptionally) to transgress family boundaries, for instance in crosses between members of the Scrophulariaceae and the Solanaceae. In their extensive paper Lewis and Crowe (1958) reported the results of a large series of crosses amongst SI and SC species and genera from the Onagraceae, Solanaceae, and Brassicaceae (Cruciferae) which confirmed the finding of Harrison and Darby (1955) that UI almost always occurs in SI × SC crosses. They also reviewed literature from plant breeding studies and highlighted records of UI in crosses within the Linaceae and Primulaceae (which have heteromorphic SSI) and within the grasses (Poaceae). On the basis of this extensive survey they proposed a series of 'rules' for interspecific crosses (summarised in Fig. 1), namely that SI × SC crosses always exhibit unilateral incompatibility, and that SC × SC crosses and SI × SI crosses are always compatible. Clearly, the common denominator in all cases of UI, described by the authors, is the possession of an SI system by the plant rejecting the pollen. This strong correlation between SI and UI lead them to propose the hypothesis that UI is controlled by the *S*-locus.

Notable exceptions to the UI 'rules' were highlighted from the Onagraceae and Brassicaceae, where certain self-compatible species were found to be incompatible with pollen from other SC species but compatible (as pollen) with SI pistils. These exceptions to the 'rules' were explained in terms of the anomalous self-compatible species having recently evolved from species with fully functional SI, such that the recognition and rejection functions of the *S*-locus were not yet fully eroded in terms of their function in UI, despite loss

Pistil \ Pollen	SI	SC	Sc*
SI	+	-	+
SC	+	+	+
Sc*	+	-	+

* Category designated on basis of results obtained for Onagraceae and Brassicaceae.
+ = compatible; - = incompatible

FIG. 1. Summary of the interspecific cross-compatibility 'rules' proposed by Lewis and Crowe (1958) on the basis of results obtained for the Solanaceae, Scrophulariaceae, Onagraceae and Brassicaceae. SI × SC crosses and Sc × SC crosses are unilaterally incompatible, while all other cross combinations are compatible. The Sc category represents a self-compatible species of recent origin having evolved from a self-incompatible species. Lewis and Crowe (1958) proposed that SC species evolve from SI species by a stepwise series of mutations at the *S*-locus (SI - Sc - Sc' - SC) that eventually leads to the SC state which lacks the power of pollen discrimination.

of SI function. These recently evolved SC species, designated 'Sc', were considered to be an early stage in a series of step-wise mutations at the *S*-locus which eventually lead to the full SC condition, as typified by indiscriminatory acceptance of pollen from other SC species as well as from SI species.

The Lewis and Crowe hypothesis for *S*-locus control of UI was further developed and greatly expanded by Pandey, working with *Nicotiana*, who in a series of papers during the 1960s and 1970s promoted the concept of the *S*-gene complex (see Pandey, 1968, 1969, 1973, 1979, 1981). Pandey proposed that this complex encodes two specificities - a 'primary' specificity element which is responsible for controlling interspecific recognition/ incompatibility and a 'secondary' specificity element which controls intraspecific SI (Fig. 2A). He further proposed that the primary specificity had evolved first in gymnosperms as a mechanism to limit the number of futile inter-species pollinations incurred by plants during the promiscuous process of wind pollination. Secondary specificity (SI) then evolved later, probably in the proto-angiosperms, by duplication and modificational 'fine tuning' of this male/female recognition system such that the early angiosperms were endowed with the ability to recognise and reject self-pollen, as well as pollen from related species (Fig. 2B). Such a significant pre-adaptation to outcrossing would have been a distinct evolutionary advantage to the predominantly insect pollinated angiosperms and could well have played a major role in the explosive adaptive radiation of flowering plants during the Cretaceous (Whitehouse, 1950). Using the *S*-gene complex as a model, and assuming various levels of erosion of the complex during the evolution of SC from SI, Pandey (1969, 1981) was able to explain all 'exceptions' to the original 'rules' described by Lewis and Crowe (1958). Such 'exceptions' to the 'rules' were (and still are) frequently used by other authors as evidence disclaiming a mechanistic model for UI based on control by the *S*-locus, and to promote an alternative model, that of 'incongruity' (Hogenboom, 1975; Mutschler and Leidl, 1994). According to the incongruity model interspecific incompatibilities (including UI) are simply the manifestation of the degree

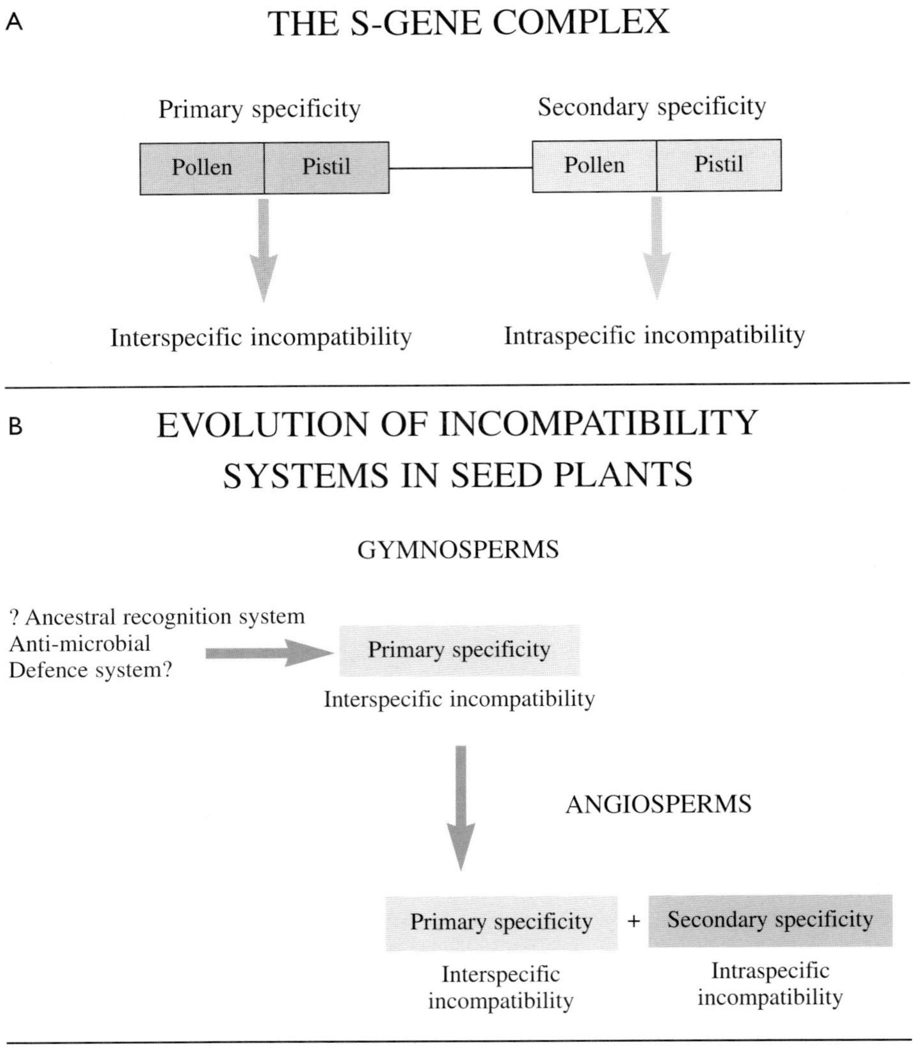

FIG. 2. A. The *S*-gene complex as proposed by Pandey (1968, 1969, 1979, 1981) to account for intraspecific and interspecific incompatibility in the Solanaceae particularly, and flowering plants in general. The primary specificity region of the *S*-locus controls interspecific incompatibility, including unilateral incompatibility and the secondary specificity region controls self-incompatibility. B. Hypothetical scheme proposed by Pandey (1981) for the evolution of incompatibility systems in seed plants. Primary specificity, regulating interspecific matings was proposed to have evolved early in the history of seed plants (in the gymnosperms) as a mechanism to prevent the futile matings that would arise as a consequence of the non-specific nature of wind pollination. Secondary specificity (SI) then evolved in the ancestors of flowering plants from this primary male-female recognition system by a process of gene duplication and modification to form an allelic series of '*S*'-genes. Natural selection would presumably have favoured any physiological mechanism which could promote outcrossing by restricting self-pollination during the opportunistic visits of insect pollinators to the first hermaphrodite flowers. The primary recognition system may have evolved through the modification of an existing recognition system such as a defence system against microbial pathogens.

of evolutionary divergence of the two species involved and, as such, are under the control of many genes rather than being under the specific control of a particular locus or loci. Such a model, while undoubtedly holding good for wide intergeneric and interfamilial crosses, fails to account satisfactorily for the high level of unilaterality observed in crosses between closely related species and genera. Indeed within the genus *Lycopersicon*, which Hogenboom (1975) used as the primary source for his data supporting the incongruity model, earlier classical genetic studies on the inheritance of UI clearly supported a relationship between UI and SI, even though they also implicated the involvement of other genes, not linked to the *S*-locus (McGuire and Rick, 1954; Martin, 1964, 1967, 1968).

The genetic and circumstantial plant breeding evidence for a more than coincidental relationship between SI and UI was clear to Pandey, as it had been to Lewis and Crowe, but many were and still are sceptical about accepting the possibility of a role for *S*-genes in UI (Hogenboom, 1975; Mutschesler and Leidl, 1994; Kandasamy *et al.*, 1994, Leidl *et al.*, 1996). Towards the end of his 20 year quest to establish incontrovertible evidence for the SI/UI model, Pandey (1979) provided the most direct evidence yet put forward for such a model. This genetic evidence came from a series of crosses carried out between SC *Nicotiana glauca* and a rare SI race of the same species. Crosses between the two races were unilaterally incompatible but 'hybrids' from the cross SC race × SI race are fertile, which made it possible to follow the segregation of SI and UI in subsequent generations and in backcrosses. Even though the crosses were intraspecific rather than interspecific, the cytology and timing of the UI reaction appeared identical to that of normal interspecific UI, and Pandey (1979) was able to show conclusively for *N. glauca* that the ability of a plant to resist pollination by a SC plant segregated with the SI phenotype. These findings have since received support from genetic mapping studies, which indicate that the UI function of the pollen in *Lycopersicon pennellii* is determined by three major loci, one of which maps to, or very near, the *S*-locus (Chetelat and DeVerna, 1991), and from genetic transformation studies involving cloned S-RNases, which have shown that in *Nicotiana* certain UI pathways are dependent on the presence of an active S-RNase (Murfett *et al.*, 1996, see final section). These latter studies do not support an omnificent role for the *S*-locus in UI, as postulated by Lewis and Crowe (1958) and Pandey (1981), but they do imply a significant role for the *S*-locus in UI.

Unilateral incompatibility in species with sporophytic self-incompatibility

The majority of studies carried out on UI have focussed on species from the Solanaceae, which possess GSI, while similar studies on species with SSI are far fewer. Lewis and Crowe (1958) reported that their 'rules' of interspecific incompatibility were applicable to species from the Brassicaceae, but their crossing study used a limited number of species and only two were SC, the majority being SI or Sc. In a much more extensive investigation into intergeneric crossing within the Brassicaceae (particularly involving species from the tribe Brassiceae) Sampson (1962) confirmed that SI × SC UI was the normal situation in this family, and that SC × SC crosses were usually compatible. However, in contrast to the findings of Lewis and Crowe (1958), Sampson (1962) showed that crosses involving only SI species were frequently incompatible, and often unilaterally so. Sampson proposed that the stigmas of SI plants presented a 'fundamental barrier' to pollen from other species, and that certain closely related species possessing the same type of SI system could, on occasion, overcome this barrier by virtue of their pollen containing a molecular 'key' similar enough to that of the SI plant upon whose stigma their pollen had landed. Sampson explained that a 'species area' and an '*S*-allele area' governed this complementary pollen-stigma interaction. Such areas may be envisaged to be different molecules or different domains within the same molecule. Sampson explained further using a 'lock' and 'key' analogy that it may be possible for certain pollen molecules, representing small 'keys' (for small 'locks'), to also work on larger 'locks' (meant for large 'keys'), whereas the opposite scenario, involving large 'keys' and small 'locks' could not work - hence unilateral incompatibility. In essence this model with its inherent flexibility is the same as that proposed by Pandey (1981) for the Solanaceae, although in Pandey's model SI × SI crosses are usually compatible.

Clearly the findings of Sampson (1962) for the Brassicaceae are at some variance with those presented earlier by Lewis and Crowe (1958) for the same family and with those presented by Pandey (1981) and others for the Solanaceae. Sampson's results clearly suggest a fundamental difference between UI in a family with SSI and families with GSI, in that SI × SI incompatibility and UI is the rule rather than the exception for the Brassicaceae.

To address this apparent dichotomy in the 'rules' of UI for species with GSI and SSI we decided to carry out an extensive intergeneric crossing study within the Brassicaceae paying particular attention to SI × SI crosses (Hiscock and Dickinson, 1993). Crosses were carried out involving 25 species from 10 of the 19 tribes of the Brassicaceae (Schulz, 1936) and compatibility was assessed in terms of pollen tube growth in the stigma. Our results (summarised in Fig. 3) were clearly in accordance with those of Sampson (1962) and revealed a higher than 'exceptional' incidence of incompatibility (which was often unilateral) in crosses between SI species. Our results also disclosed a hierachy among SI species, with respect to the ability of their stigmas to resist pollination by other SI species and the ability of their pollen to develop on the stigmas of other SI species (Fig. 4). A similar relationship between the relative success of a species in crossing when present as male partner or female partner had been previously described in the Brassicaceae by Harberd (1976) for 29 species with unspecified incompatibility phenotype. The different behaviour of SI × SI crosses in species with SSI and GSI does suggest a fundamental difference in the way that UI operates within the Brassicaceae and Solanaceae, but our data, and those of Sampson, clearly do not refute the model of UI proposed by Lewis and Crowe (1958) and Pandey (1969, 1981) because UI is still correlated with the presence of an SI system in the pistil with which pollen is incompatible. However the data from the Brassicaceae predict that interspecific incompatibility and UI will operate through pathways with differing dependencies on *S*-gene products.

		Pollen	
		SI	SC
Pistil	SI	-/+	-
	SC	+	+

+ = compatible; - = incompatible

FIG. 3. Table summarising data for interspecific and intergeneric crosses in the Brassicaceae (Hiscock and Dickinson, 1993). 11 SI species and 14 SC species were crossed in different combinations and compatibility in the pollen-pistil interaction assessed in terms of pollen tube growth on and in the stigma using fluorescence microscopy after staining pollen tubes with aniline blue. A minimum of 5 replicates was used in each pollination. The outcome of SI × SI crosses was somewhat unpredictable - approximately 60% of crosses were incompatible and 63% of reciprocal SI × SI crosses were unilaterally incompatible.

Crossability hierarchy among SI species of the Brassicaceae

SI species	Stigmatic inhibitions (incompatible/ total pollinations)	Pollen compatibilities (compatible/ total pollinations)	UI ratio*
Brassica oleracea	9/10	10/11	0.82
Aubretia deltoidea	5/8	5/7	0.45
Raphanus sativus	5/8	7/10	0.44
Cakile maritima	4/7	3/7	0.25
Matthiola parviflora	7/9	2/6	0.25
Rapistrum rugosum	5/6	1/4	0.21
Iberis sempervirens	3/5	2/6	0.20
Capsella grandiflora	0/6	1/6	0.17
Cardamine pratensis	0/8	1/6	0.17
Cardamine bulbifera	3/5	2/9	0.13
Moricandia arvensis	3/6	1/6	0.09

FIG. 4. Crossability hierarchy among SI species from the Brassicaceae expressed as a UI ratio. The UI ratio, obtained by multiplying stigmatic incompatibilities (expressed as a fraction) by pollen compatibilities (expressed as a fraction), reflects the correlation between the ability of a species with SI to resist pollination by other species with SI and the ability of the pollen of the SI species to develop on the stigmas of other species with SI. For example, *Brassica oleracea*, at the top of the hierarchy is unreceptive to most pollens but its own pollen is able to grow and develop on the stigmas of almost all other SI species tested.

Physiology of unilateral incompatibility in the Brassicaceae

To obtain information on the possible mechanism(s) of UI in the Brassicaceae we carried out a series of simple physiological experiments to determine whether UI is sensitive to treatments that overcome SSI in *Brassica* species.

SSI, like GSI, is developmentally regulated and only manifests itself late in floral development, becoming strongest towards and just after anthesis (de Nettancourt, 1977). The onset of SI has been correlated with the appearance of proteins encoded by the *S*-locus in the tissues of the pistil (Nasrallah *et al.*, 1985, Cornish *et al.*, 1987) - in *Brassica* expression of *SLG* and *SRK* are optimal at this time (Nasrallah *et al.*, 1994a). If pistils from small buds of SI plants are self-pollinated with mature pollen, the pollen grows freely into the pistil and fertilisation is followed by full seed-set; such a technique is commonly used by plant breeders to obtain inbred homozygous *S*-lines. We took buds from SI species and pollinated them with pollen from SC species, with which they are normally unilaterally incompatible, and assessed the ability of the SC pollen to penetrate the stigma and grow within the style (Hiscock and Dickinson, 1993 and Fig. 5). Our data clearly show that UI is almost totally absent in small buds which lack a functional SI system. Thus UI like SI appears to be developmentally regulated, suggesting that the onset of UI coincides with the onset of SI and the appearance of *S*-proteins in the pistil.

We also investigated the effect of treating mature SI stigmas with cycloheximide prior to pollination with pollen from SC species (Hiscock and Dickinson, 1993 and Fig. 6). This potent inhibitor of protein synthesis overcomes SI in *Brassica* in a way which is not yet understood (Ferrari and Wallace, 1976, Sarker *et al.*, 1988). Our data showed that the inhibitor is also able to overcome the UI stigmatic barrier, because pollen tubes penetrated the stigma and grew freely in the style, albeit at a lower frequency than after self-pollinations following

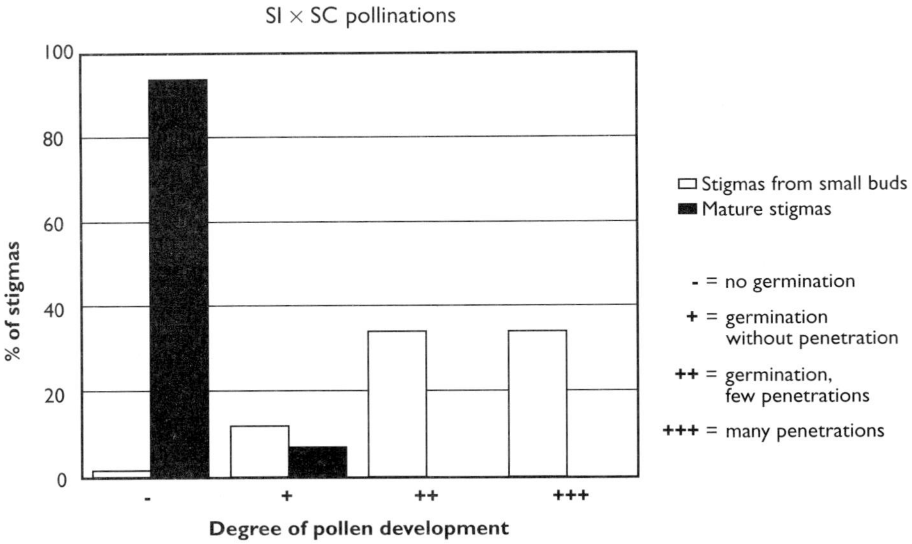

FIG. 5. Effect of stigmatic maturity on normally incompatible interspecific pollinations in the Brassicaceae. Pollinations were carried out on stigmas taken from small buds and on mature stigmas from open flowers (Hiscock and Dickinson 1993). The results were obtained from crosses using stigmas from the SI species *Brassica oleracea*, *Iberis sempervirens* and *Cardamine pratensis* and pollen from the SC species *Arabidopsis thaliana*, *Alliaria petiolata* and *Capsella bursa-pastoris*. A minimum of 5 pollinations was carried out for each cross combination and pollen growth was assessed as described in Fig. 3. The histogram is based on 65 bud pollinations and 50 control pollinations on mature stigmas.

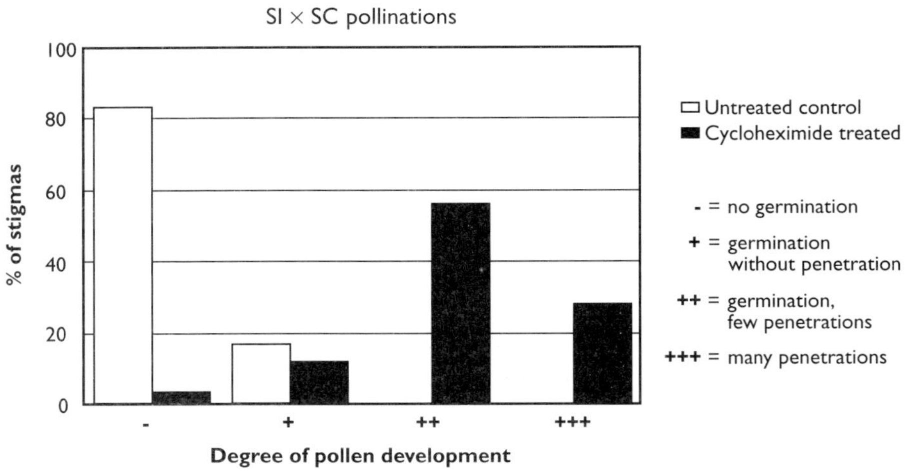

FIG. 6. Effect of cycloheximide treatment of pistils on normally incompatible interspecific pollinations in the Brassicaceae. Prior to pollination flowers were treated with 0.2 mM cycloheximide as described in Hiscock and Dickinson (1993). SI and SC species used were the same as for Fig. 5 and the histogram summarises the results of 70 SI × SC pollinations after cycloheximide treatment and 50 control (untreated) pollinations.

cycloheximide treatment (Hiscock and Dickinson, 1993). Interestingly the pollen tubes appeared to cease growth in the style earlier than in self-pollinations following cycloheximide treatment - this effect is assumed to be due to the inhibitor entering the pollen tube once it is within the style. Thus UI like SI is sensitive to and overcome by cycloheximide indicating further commonalities between the metabolic events of SI and UI.

The proteinaceous stigmatic pellicle in interspecific pollen-stigma interactions

Species from the Brassicaceae possess 'dry' stigmas (Heslop-Harrison and Shivanna, 1977) which lack a secreted germination fluid for pollen to develop in. The epidermal cells of such 'dry' stigmas have been shown to be covered by a continuous cuticle which is overlaid by a very thin proteinaceous layer, termed the 'pellicle' (Mattson *et al.*, 1974, Fig. 7). A characteristic of the pellicle is its strong non-specific esterase activity and the presence of glycoproteins within it (Mattson *et al.*, 1974; Heslop-Harrison and Heslop-Harrison, 1975; Heslop-Harrison, 1977). Removal of the pellicle using dilute solutions of proteases, such as pronase, has been shown to prevent compatible intraspecific pollinations in the Brassicaceae, Iridaceae and Caryophyllaceae (Shivanna *et al.*, 1978; Heslop-Harrison, 1977; Heslop-Harrison and Heslop-Harrison, 1975). Such findings lead to the proposal that the pellicle plays a major role in pollen-stigma recognition and might even provide an inactive cutinase precursor to aid pollen tube penetration, or alternatively might activate a cutinase precursor secreted by the pollen tube (Knox *et al.*, 1976; Heslop-Harrison and Heslop-Harrison, 1975). Our recent finding (Hiscock *et al.*, 1994) that the pollen of *Brassica* contains an active cutinase in the intine of the pollen wall questions the suggestion of cutinase activation by the pellicle and focuses attention on a role for the pellicle in some other complementary interaction with pollen-held molecules required for a compatible pollen-stigma interaction. Such a complementation may be part of the 'species area' envisaged by Sampson (1962).

To address this question we treated stigmas of various species from the Brassicaceae with dilute pronase and pollinated them with pollen from other species with which they were previously found to be compatible (Hiscock and Dickinson, 1993). Our results then and subsequently (Fig. 8) clearly show that an intact pellicle is required for compatible interspecific pollinations, because pollen tubes failed to penetrate the stigma in the absence of a pellicle. Interestingly we found this also to be true for stigmas from small buds where removal of the pellicle from, for instance a *Brassica* bud stigma prevented penetration of pollen from *Arabidopsis* (Hiscock and Dickinson 1993). Such findings immediately suggest that the 'species area' and 'SI area' proposed by Sampson (1962) are most likely to be situated on different molecules because the bud stigmas do not possess a functional SI system. In the light of these findings and previous studies a more thorough characterisation of the pellicle and its protein components is clearly required.

Conspectus

The phenomenon of unilateral interspecific incompatibility has puzzled plant breeders, plant geneticists and plant reproductive biologists alike for many years. Two schools of thought have emerged on the possible control and regulation of this very widespread and important reproductive barrier in flowering plants. The first school (Lewis and Crowe, 1958; Pandey, 1969, 1981) believes that genes which reside at the *S*-locus, or genes which are closely linked to the *S*-locus control UI, whereas the second school (Hogenboom, 1975; Mutschler and Leidl, 1994) believes that UI is instead under polygenic control and is independent of genes at the *S*-locus; genetic 'mismatch' during the pollen-stigma interaction being due entirely to evolutionary divergence of the two species concerned. It is clear from our survey of the

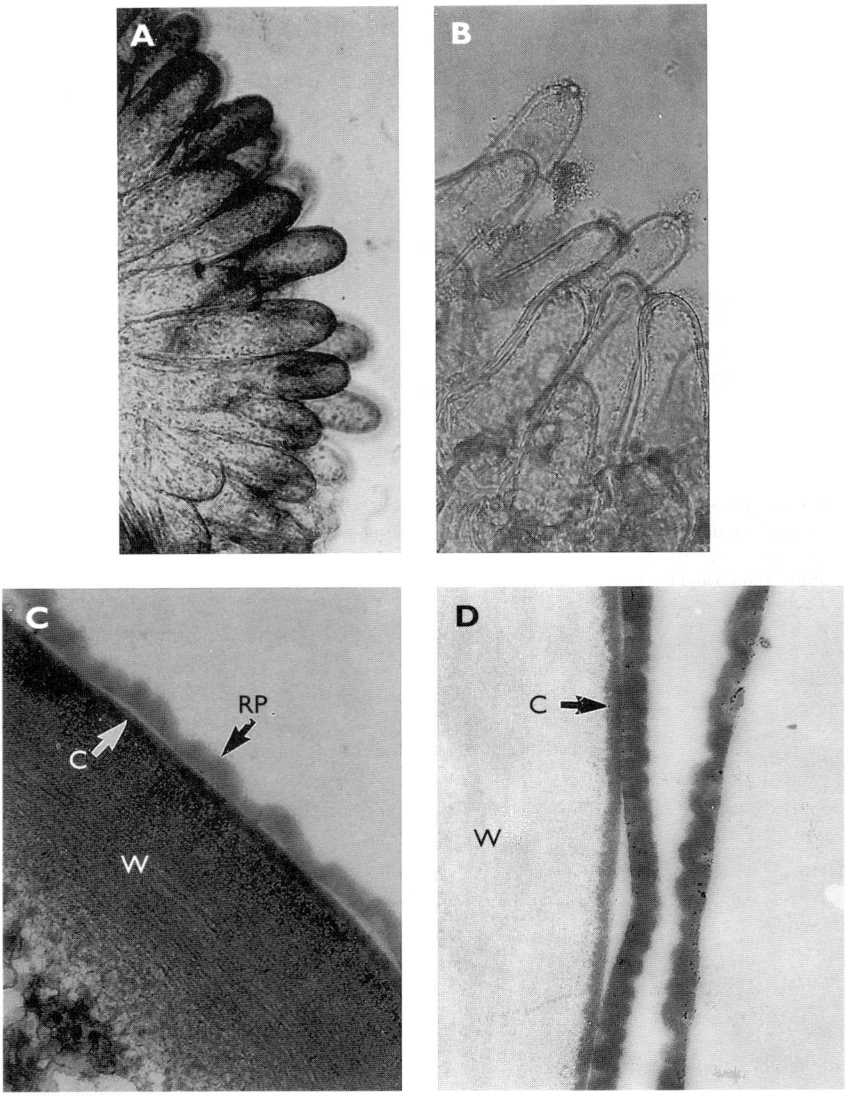

FIG. 7. Visualisation of the proteinaceous pellicle covering the stigmatic papillar cells of *Brassica oleracea*. a) and b): light micrographs of squashed stigma; c) and d): electron micrographs of transverse sections through papillar cells. A. shows stigmatic papillae after staining for non-specific esterase activity using α-naphthyl acetate and hexazatised pararosaniline (Mattson *et al.* 1974). Dark reaction deposits can be seen on the papillar cells which reflects the strong non-specific esterase activity of the pellicle (×400). B. As a) but with α-naphthyl acetate omitted from the reaction mix as a control (×400). C. Transverse section through papillar cell wall. Material was fixed in 4% paraformaldehyde and stained as in a) to visualise non-specific esterase activity. The section was also stained with 0.5 % uranyl acetate and 2.7% lead citrate. RP = esterase reaction product, C = cuticle, W = cell wall (×30,000). D. As c) but section not stained with uranyl acetate and lead citrate to allow clearer visualisation of the pellicle/esterase reaction product and the cuticle (C). The esterase reaction product on the right has dissociated from its papilla (×35,000).

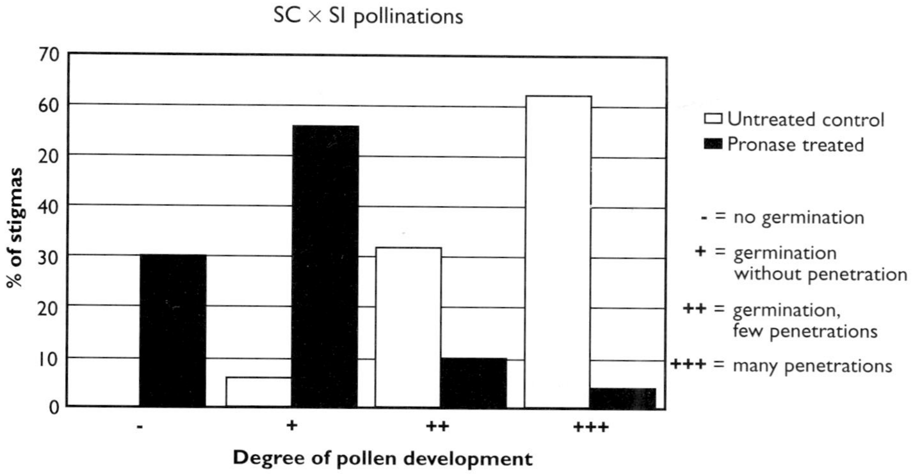

FIG. 8. Effect of removing the proteinaceous pellicle on normally compatible interspecific pollinations in the Brassicaceae. Prior to pollination pistils from SC species were treated with 0.1mg ml^{-1} pronase as described in Hiscock and Dickinson (1993). SI and SC species were the same as for Figs. 5 and 6 and the histogram summarises the results of 58 interspecific pollinations on pronase-treated stigmas and 50 control pollinations on untreated stigmas.

literature that there is a large consensus of opinion in favour of a role, at least in part, for the *S*-locus in the control of many forms of UI. Indeed it is impossible to ignore the striking correlation between the presence of an SI system in a species and its ability to be unilaterally incompatible with closely related SC species. This observed correlation is supported by Pandey's (1979) genetic correlation between SI and UI in *Nicotiana glauca*, where he showed that the UI phenotype segregates with the *S*-locus. These data from intraspecific crosses strongly suggest that a similar genetic correlation must exist with respect to interspecific UI.

Support for this hypothesis has recently been forthcoming from molecular genetic studies carried out on species of *Nicotiana* in the laboratory of Bruce McClure. Murfett *et al.* (1996) focussed on a SC race of *N. alata* (a species which is normally strongly SI) that appeared to differ from the SI race only by its inability to synthesise S-RNase. SI × SC inter-race crosses were unilaterally incompatible and the SI race was able to resist pollination by pollen from SC species such as *N. plumbaginifolia*, while the SC race accepted such pollen. To test whether S-RNase synthesis was the absolute requirement for inhibition of pollen from *N. plumbaginifolia* it would be necessary to transform SC *N. alata* with an S-RNase construct to determine whether this conferred the ability to reject the *N. plumbaginifolia* pollen. Unfortunately the SC *N. alata* was not amenable to transformation so the S-RNase construct was instead transformed into *N. plumbaginifolia* and the primary transformant crossed with the SC *N. alata* to generate a hybrid with an active S-RNase gene. This hybrid was capable of rejecting pollen from SC *N. plumbaginifolia* and also pollen from SC *N. tabacum* indicating a direct relationship between the presence of an active S-RNase in a style and the ability of that style to reject pollen from SC species. Surprisingly the *N. plumbaginifolia* primary transformants, while not being capable of rejecting pollen from SC *N. plumbaginifolia*, were capable of rejecting pollen from SC *N. tabacum*, demonstrating that UI in *Nicotiana* must operate through more than one pathway, and that the dependence of the different pathways on the S-RNase does vary. Indeed, in their study Murfett *et al.*, (1996) identified three different pathways to UI in *Nicotiana*: Inhibition of SC *N. tabacum* pollen by the *N. plumbaginifolia*+S-RNase primary transformant was S-RNase-dependent, but inhibition of the same pollen by the SC race of *N. alata* (which lacks the S-RNase) had to be S-RNase-independent, while inhibition of SC *N. plumbaginifolia* pollen by SI *N. alata* requires the S-RNase but also requires additional factors present in the pistil of *N. alata*, which are not present in the pistil of *N. plumbaginifolia* because the primary transformant *N. plumbaginifolia*+S-RNase was incapable of inhibiting pollen of the SC race of *N. alata*. Interestingly the same research group have demonstrated that S-RNases specifically are required for inhibition of SC *N. plumbaginifolia*, pollen by the transgenic hybrid because stylar expression of an *E. coli* RNase gene in the *Nicotiana* hybrid was not sufficient to cause rejection of the *N. plumbaginifolia* pollen (Beecher *et al.* 1998).

These molecular data from *Nicotiana* have provided unparalleled support for the hypothesis of a role for the *S*-locus in the regulation of UI. The finding of an S-RNase-independent pathway to UI in the cross *N. alata* (SC) × *N. tabacum* (SC) indicates that UI can operate through multiple pathways, but does not preclude an involvement of other *S*-locus-linked genes (which have yet to be identified) in UI, because the SC race of *N. alata* is most probably recently derived from an SI race of *N. alata* so it falls into the 'Sc' category designated by Lewis and Crowe (1958) which is expected to contain only a partially disfunctional *S*-locus (Lewis and Crowe, 1958; Pandey, 1981), and might therefore lack only a functional S-RNase gene, any other *S*-genes remaining fully functional.

Returning to our findings in the Brassicaceae, the strong similarities between the physiology of SI and UI provide very convincing, albeit circumstantial, evidence for a common route to pollen tube inhibition in both these forms of pollen-stigma incompatibility. The absence of SI and UI in stigmas from small buds of species which normally have strong SI and UI, suggests that genes expressed at the *S*-locus may regulate UI as well as SI. The finding that a SC mutant of *Brassica oleracea*, which lacks a functional *SRK*, appears to retain UI with SC *Arabidopsis thaliana* (Kandasamy *et al.*, 1994) needs careful consideration. According to the Lewis and Crowe (1958) model this mutant, like the S-RNase-less mutant of *N. alata* (Murfett *et al.*, 1996) should be placed in the 'Sc' category of self-compatible plants which have evolved recently from SI plants and therefore retain other *S*-genes which are still fully functional; indeed *SLG* expression levels are unaffected in this *SRK* mutant (Nasrallah *et al.*, 1994b). However, given that *SRK* is expected to play a central role in SI (Nasrallah *et al.*, 1994a; Dickinson, 1994) and also possibly in UI (Dickinson *et al.*, 1998a, b), the *SRK*-less mutant needs more extensive investigation through crossing with a wider selection of SC crucifers.

It is worth considering at this point our results from treating stigmas with pronase to remove the pellicle: this treatment prevents stigmas of *Arabidopsis* accepting pollen of *Brassica* and similarly prevents bud stigmas of *Brassica* accepting pollen of *Arabidopsis*. Without such treatment these pollen-stigma interactions are usually compatible. While such data is open to a number of explanations (Hiscock and Dickinson, 1993), a possible interpretation of these findings is that pellicle-held proteins are required for compatibility of pollen and stigma and in the absence of these proteins compatibility breaks down, and this breakdown occurs in the presence and absence of *SRK*. The implications of this hypothesis for interspecific incompatibility and UI make characterisation of and cloning the genes encoding pellicle proteins a high priority.

Failure so far to identify the pollen-held S molecule, which is expected to interact with *SRK* to determine *S*-specificity (Nasrallah *et al.*, 1994a; Dickinson, 1994, 1998a) makes it difficult to formulate a convincing molecular model for the role of *S*-locus products in UI. However, we have identified a family of small pollen-coating-borne proteins, designated PCPs (pollen coat polypeptides) which may play a fundamental role in pollen-stigma recognition reactions in *Brassica* (Doughty *et al.*, 1993, 1998; Hiscock *et al.*, 1995b; Stephenson *et al.*, 1997). One PCP species, PCP–A1, is known to interact specifically with SLGs in vitro (Doughty *et al.*, 1993, 1998) but the significance of this interaction, in the context of SI, is unclear because PCP–A1 is not linked to the *S*-locus and varies very little between *S*-genotypes (Dickinson *et al.*, 1998a; Doughty *et al.*, 1998). However, a heterogeneous mixture of approximately 6 PCPs (which includes PCP–A1) has been shown recently to be responsible for conferring *S*-specificity to the pollen, indicating that this mixture contains the pollen *S*-determinant (Stephenson *et al.*, 1997; Dickinson *et al.*, 1998a). Interestingly this mixture of PCPs is also able to overcome the barrier to hydration of pollen from SC *Arabidopsis thaliana* on stigmas of *B. oleracea* (Dickinson *et al.*, 1998a, b), suggesting that PCPs are able to overcome at least part of the UI system which rejects *Arabidopsis* pollen. However, the failure of hydrated *Arabidopsis* grains to penetrate the *Brassica* stigma indicates a requirement for the presence or even the absence of other pollen or stigma (perhaps pellicle-held) identity molecules in order for full pollen-stigma compatibility to be achieved.

We have attempted here to promote the hypothesis of a role for *S*-locus-specific genes in the control and regulation of interspecific incompatibilities such as UI. From the data presented it is clear that genes linked to the *S*-locus (or 'elements of the *S*-gene complex') do play a role in these interspecific events, but the situation appears more complex than was envisaged by Lewis and Crowe (1958) and Pandey (1981), in that a model for UI and interspecific incompatibility based entirely on control by the *S*-locus now seems very unlikely in most systems studied in any detail. The, so far, unique molecular studies of UI in *Nicotiana* (Murfett *et al.*, 1996, Beecher *et al.*, 1998) indicate that UI operates through a number of different pathways which vary in their dependency on *S*-gene products. Further molecular-based studies of UI in other genera, especially those with sporophytic SI systems, are clearly required before we can begin to understand the complexities of the involvement of *S*-gene products in interspecific incompatibility and UI.

Acknowledgements

The authors thank the UK BBSRC for financial support, Suzanne Dickson for technical assistance, Liz Paul for help in preparing the figures and Carole Elleman and James Cunningham for helpful discussions.

References

Beecher, B. Murfett, J. and McClure, B.A. (1998). RNaseI from *Escherichia coli* cannot substitute fro S-RNase in rejection of *Nicotiana plumvaginifolia* pollen. *Plant Mol. Biol.* **36**: 553–563.

Chetelat, R.T. and De Verna, J.W. (1991). Expression of unilateral incompatibility in pollen of *Lycopersicon pennellii*. *Theor. Appl. Genet.* **82**: 704–712.

Cornish, E.C., Pettit, J.M., Bonig, I. and Clarke, A.E. (1987). Developmentally controlled expression of a gene associated with self-incompatibility in *Nicotiana alata*. *Nature* **326**: 99–102.

de Nettancourt, D. (1977). Incompatibility in angiosperms. Springer-Verlag, Berlin. (Monographs on theoretical and applied genetics, no. 3).

Dickinson, H.G. (1994). Simply a social disease? *Nature* **367**: 517–518.

Dickinson, H.G. and Lewis, D. (1973). The formation of the tryphine coating the pollen grains of *Raphanus* and its properties relating to the self–incompatibility system. *Proc. Roy. Soc. London Ser. B, Biol. Sci.* **184**: 149–165.

Dickinson, H.G., Doughty, J., Hiscock, S.J., Elleman, C.J. and Stephenson, A.G. (1998a). Pollen-stigma interactions in *Brassica*. In: A. Greenland, E. Meyerowitz and M. Steer (editors). Control of plant development: genes and signals. Society for Experimental Biology symposium series. Oxford University Press.

Dickinson, H.G., Doughty, J., Elleman, C.J. and Hiscock, S.J. (1998b). Evolution of angiosperm self-incompatibility mechanisms; data from *Brassica oleracea.* In: S.J. Owens and P.J. Rudall (editors). Reproductive biology, pp. 15–29, Royal Botanic Gardens, Kew.

Dodds, P.N., Clarke, A.E. and Newbigin, E. (1996). A molecular perspective on pollination in flowering plants. *Cell* **85**: 141–144.

Doughty, J., Hedderson, F., McCubbin, A. and Dickinson, H.G. (1993). Interaction between a coating-borne peptide of the *Brassica* pollen grain and stigmatic S (incompatibility)-locus-specific glycoproteins. *Proc. Natl. Acad. Sci. USA* **90**: 467–471.

Doughty, J., Dixon, S., Hiscock, S.J., Willis, A.C. and Dickinson, H.G. (1998). PCP–A1, a *Brassica* pollen coat protein which binds the *S*-(Incompatibility) locus glycoprotein is related to plant defensins and is the product of gametophytic gene expression. *Pl. Cell* (in press).

Ferrari, T.E. and Wallace, D.H. (1976). Pollen protein synthesis and control of self-incompatibility in *Brassica*. *Theor. Appl. Genet.* **48**: 243–249.

Franklin, F.C.H., Lawrence, MJ. and Franklin-Tong, V.E. (1995). Cell and molecular biology of self-incompatibility in flowering plants. *Int. Rev. Cytol.* **158**: 1–64.

Franklin-Tong, V.E., Ride, J.P., Read, D.R., Trewavas, A.J. and Franklin, F.C.H. (1993). The self-incompatibility response in *Papaver rhoeas* is mediated by cytosolic free calcium. *Pl. J. Cell Molec. Biol.* **4**, 163–177.

Harberd, D.J. (1976). Cytotaxonomic studies of *Brassica* and related genera. In: J.G. Vaughan, A.J. MacLeod and B.M.G. Jones (editors). The biology and chemistry of the Cruciferae, pp. 47–68, Academic Press, London.

Harrison, B.J. and Darby, L.A. (1955). Unilateral hybridisation. *Nature* **176**: 952.

Heslop-Harrison, J. and Heslop-Harrison, Y. (1975). Enzymic removal of the proteinaceous pellicle of the stigma papilla prevents pollen tube entry in the Caryophyllaceae. *Ann. Bot.* **39**: 163–165.

Heslop-Harrison, J., Knox, R.B. and Heslop-Harrison, Y. (1974). Pollen-wall proteins: exine-held fractions associated with the incompatibility response in Cruciferae. *Theor. Appl. Genet.* **44**: 133–177.

Heslop-Harrison, Y. (1977). The pollen-stigma interaction: pollen tube penetration in *Crocus*. *Ann. Bot.* **41**: 913–922.

Heslop-Harrison, Y. and Shivanna, K.R. (1977). The receptive surface of the angiosperm stigma. *Ann. Bot.* **41**: 1233–1258.

Hiscock, S.J. and Dickinson, H.G. (1993). Unilateral incompatibility within the Brassicaceae - further evidence for an involvement of the self-incompatibility (*S*)-locus. *Theor. Appl. Genet.* **86**: 744–753.

Hiscock, S.J., Dewey, F.M., Coleman, J.O.D. and Dickinson, H.G. (1994). Identification and localization of an active cutinase in the pollen of *Brassica napus* L. *Planta* **193**: 377–384.

Hiscock, S.J., Kües, U. and Stahl, U. (1995a). Sexual reproduction in plants: self-incompatibility as a mechanism promoting outbreeding and gene flow. *Progr. Bot.* **56**: 276–300.

Hiscock, S.J., Doughty, J., Willis, A.C. and Dickinson, H.G. (1995b). A 7-kDa pollen coating-borne peptide from *Brassica napus* interacts with *S*-locus glycoprotein and *S*-locus-related glycoprotein. *Planta* **196**: 367–374.

Hiscock, S.J., Kües, U. and Dickinson, H.G. (1996). Molecular mechanisms of self-incompatibility in flowering plants and fungi - different means to the same end. *Trends Cell Biol.* **6**: 421–428.

Hogenboom, N.G. (1975). Incompatibility and incongruity: two different mechanisms for the non-functioning of intimate partner relationships. *Proc. Roy. Soc. London, Ser. B., Biol. Sci.* **188**: 361–375.

Kandasamy, M.K., Nasrallah, J.B. and Nasrallah, M.E. (1994). Pollen-pistil interactions and developmental regulation of pollen tube growth in *Arabidopsis*. *Development* (Cambridge). **120**: 3405–3418.

Knox, R.B., Clarke, A.E., Harrison, S., Smith, P. and Marchalonis, J.J. (1976). Cell recognition in plants: determinants of the stigma surface and their pollen interactions. *Proc. Natl. Acad. Sci. USA* **73**: 2788–2792.

Lewis, D. (1965). A protein dimer hypothesis on incompatibility. In: S.J. Geerts (editor) Genetics today: proceedings of the XI International Congress of Genetics, The Hague, Netherlands, Sept. 1963, pp. 657–663, Pergamon, Oxford.

Lewis, D. and Crowe, L.K. (1958). Unilateral interspecific incompatibility in flowering plants. *Heredity* **12**: 233–256.

Li, X., Nield, J., Hayman, D. and Langridge, P. (1994). Cloning a putative self-incompatibility gene from the pollen of the grass *Phalaris coerulescens*. *Pl. Cell* **6**: 1923–1932.

Li, X., Nield, J., Hayman, D. and Langridge, P. (1996). A self-fertile mutant of *Phalaris* produces an S protein with reduced thioredoxin activity. *Plant J.* **10**: 505–513.

Liedl, B.E., McCormick, S. and Mutschler, M.A. (1996). Unilateral incongruity in crosses involving *Lycopersicon pennellii* and *L. esculentum* is distinct from self-incompatibility in expression, timing and location. *Sexual Pl. Reprod.* **9**: 299–308.

Martin, F.W. (1964). The inheritance of unilateral incompatibility in *Lycopersicon hirsutum*. *Genetics* **50**: 459–469.

Martin, F.W. (1967). The genetic control of unilateral incompatibility between two tomato species. *Genetics* **56**: 391–398.

Martin, F.W. (1968). The behaviour of *Lycopersicon* incompatibility alleles in an alien genetic milieu. *Genetics* **60**: 101–109.

Mather, K. (1947). Species crosses in *Antirrhinum* I. Genetic isolation of the species *majus*, *glutinosum* and *orontium*. *Heredity* **1**: 175–186.

Mattsson, O., Knox, R.B., Heslop-Harrison, J. and Heslop-Harrison, Y. (1974). Protein pellicle of stigmatic papillae as a probable recognition site in incompatibility reactions. *Nature* **247**: 298–300.

McGuire, D.C. and Rick, C.M. (1954). Self-incompatibility in species of *Lycopersicon* Sect. *Eriopersicon* and hybrids with L. esculentum. *Hilgardia* **23**: 101–124.

Murfett, J., Strabala, T.J., Zurek, D.M., Mou, B., Beecher, B. and McClure, B.A. (1996). S-RNase and interspecific pollen rejection in the genus *Nicotiana*: multiple pollen-rejection pathways contribute to unilateral incompatibility between self-incompatible and self-compatible species. *Pl. Cell* **8**: 943–958.

Mutschler, M.A. and Leidl, B.E. (1994). Interspecific crossing barriers in *Lycopersicon* and their relationship to self-incompatibility. In: E.G. Williams, A.E. Clarke and R.B. Knox (editors). Genetic control of self-incompatibility and reproductive development in flowering plants pp. 164–188, Kluwer Academic Publishers, London.

Nasrallah, J.B. and Nasrallah, M. E. (1993). Pollen-stigma signalling in the sporophytic self-incompatibility response. *Pl. Cell* **5**: 1325–1335.

Nasrallah, J.B., Doney, R.C. and Nasrallah, M.E. (1985). Biosynthesis of glycoproteins involved in the pollen-stigma interaction of incompatibility in developing flowers of *Brassica oleracea* L. *Planta* **165**: 100–107.

Nasrallah, J.B., Stein, J.C., Kandasamy, M.K. and Nasrallah, M.E. (1994a). Signalling the arrest of pollen tube development in self-incompatible plants. *Science* **266**: 1505–1508.

Nasrallah, J.B., Rundle, S.J. and Nasrallah, M.E. (1994b). Genetic evidence for the requirement of the *Brassica* *S*-locus receptor kinase gene in the self-incompatibility response. *Plant J.* **5**: 375–384.

Pandey, K.K. (1968). Compatibility relationships in flowering plants: role of the *S*-gene complex. *Amer. Naturalist* **102**: 475–489.

Pandey, K.K. (1969). Elements of the *S*-gene complex. V. Interspecific cross-compatibility relationships and theory of the evolution of the *S*-complex. *Genetica* **40**: 447–474.

Pandey, K.K. (1973). Phases in *S*-gene expression, and *S*-allele interaction in the control of interspecific incompatibility. *Heredity* **31**: 381–400.

Pandey, K.K. (1979). The genus *Nicotiana*: evolution of incompatibility in flowering plants. In: J.G. Hawkes, R.N. Lester and A.D. Skelding (editors). Biology and taxonomy of the Solanaceae, pp 421–434, Academic Press, London. (Linnean Society symposium series, no. 7.)

Pandey, K.K. (1981). Evolution of unilateral incompatibility in flowering plants: further evidence in favour of twin specificities controlling intra- and interspecific incompatibility. *New Phytol.* **89**: 705–728.

Rudd, J.J., Franklin, F.C.H., Lord, J.M. and Franklin-Tong, V.E. (1996). Increased phosphorylation of a 26 kDa pollen protein is induced by the self-incompatibility response in *Papaver rhoeas*. *Pl. Cell* **8**: 713–724.

Sampson, D.R. (1962). Intergeneric pollen-stigma incompatibility in the Cruciferae. *Canad. J. Genet. Cytol.* **4**: 38–49.

Sarker, R.H., Elleman, C.J. and Dickinson, H.G. (1988). Control of pollen hydration in *Brassica* requires continued protein synthesis, and glycosylation is necessary for intraspecific incompatibility. *Proc. Natl. Acad. Sci. USA* **85**: 4340–4344.

Schulz, O.E. (1936). Cruciferae In: A. Engler and H. Harms (editors). Die natürlichen Pflanzenfamilien 2, vol. 17b, pp 227–658, Engelmann, Leipzig.

Shivanna, K.R., Heslop-Harrison, J. and Heslop-Harrison, Y. (1978). The pollen-stigma interaction: bud pollination in the Cruciferae. *Acta Bot. Neerl.* **27**: 107–119.

Stephenson, A., Doughty, J., Dixon, S., Elleman, C.J., Hiscock, S.J. and Dickinson, H.G. (1997). The male determinant of self-incompatibility in *Brassica oleracea* is located in the pollen coating. *Pl. Journal.* 12, 1351–1359.

Whitehouse, H.L.K. (1950). Multiple allelomorph incompatibility of pollen and style in the evolution of angiosperms. *Ann. Bot.* n.s. **14**, 198–216.

Keijzer, C.J. (1998). Tracing the role of the cytoskeleton during sperm cell transport and fertilisation in lily using high resolution SEM: techniques and first result. In: S.J. Owens and P.J. Rudall (Editors). Reproductive Biology, pp. 47–56. Royal Botanic Gardens, Kew.

TRACING THE ROLE OF THE CYTOSKELETON DURING SPERM CELL TRANSPORT AND FERTILISATION IN LILY USING HIGH RESOLUTION SEM: TECHNIQUES AND FIRST RESULTS

C.J. KEIJZER

Dept. Plant Cytology and Morphology, Agricultural University
Arboretumlaan 4, 6703 BD Wageningen, The Netherlands

Abstract

High resolution scanning electron microscopy was used to study the role of cytoskeletal elements in generative and sperm cell transport, both in pollen tubes and embryo sacs of the lily. Therefore, we developed techniques firstly to isolate and dry-cleave pollen tubes and secondly to precisely freeze-fracture through the egg apparatus and the central cell of embryo sacs. As for the pollen tubes, the first observations show cortical microfilaments connected to organelles and bundles of microtubules in the more central part of the tubes. Clear interconnections between any of these skeletal elements and the generative cell have not been found so far. In the embryo sacs, microtubules were found radiating from the central cell nuclei, possibly also interconnecting them. A fusion found between a sperm nucleus and a central cell nucleus was not accompanied by the presence of cytoskeletal elements.

Introduction

In the last two decades the number of studies focused on the cytoskeleton have increased, thanks to the development of specific immunolabelling techniques both for fluorescence, confocal laser scanning and transmission electron microscopy (Lloyd, 1991; De Ruijter *et al.*, 1997). With the use of relatively thick sections compared with TEM sections, the two microscope types give an insight into the spatial arrangement of these structures, thus facilitating explanations as to their functions in the cell. Thanks to the development of high resolution scanning electron microscopes, cytoskeleton studies using this instrument became possible, which have given rise to a small number of reports (Blackmore *et al.*, 1984; Toyoshima *et al.*, 1986; Fukudome and Tanaka, 1992; Vesk *et al.*, 1994; Reichelt *et al.*, 1995; Chen *et al.*, 1995).

Earlier studies were mainly focused on optimal preservation, where the unpredictability of the plane of freeze-fracturing (which is necessary to open the cells) was no problem. In contrast, our studies on the roles of the cyto- and nuclear skeletons in sperm cell transport and fertilisation, respectively, required pollen tubes and particular cells in the embryo sac to be opened. In the present study we present such techniques and preliminary observations on the fertilisation process.

The role of the cytoskeleton in sperm cell transport, especially in pollen tubes, has been the subject of earlier studies (reviewed by Mascarenhas, 1993; Derksen *et al.*, 1995 and Cai *et al.*, 1996). However, direct associations have not been found so far. The present study focuses on finding such contacts.

Materials and Methods

Preparing the pollen tubes

Flowers of the lily (*Lilium longiflorum* Thunb.) cv. Gelria were pollinated with pollen of cv. White American. Two days after pollination the styles were fixed by injecting 1% osmium tetroxide in 0.07M phosphate buffer (pH 7.2) into the stylar canal for 10 minutes, followed by the buffer for 30 minutes, both at room temperature. Subsequently the styles were dehydrated through a graded ethanol series and critical point dried over carbon dioxide. They were cut longitudinally into two halves from which bundles of pollen tubes were lifted carefully using forceps. These bundles were stuck onto stubs using conductive tape and the pollen tubes were dry-cleaved by gently touching them with a piece of sellotape. If desired, the cytoplasm was subsequently macerated by placing the stub in aqueous 0.1% osmium tetroxide in 0.07M phosphate buffer (pH 7.2) for 1-3 days at 4°C (according to Tanaka and Mitsushima, 1984), followed by ethanol dehydration and critical point drying. Finally the specimens were sputter-coated with 2 nm platinum and observed in a JEOL 6300F field emission SEM at 5 kV.

Preparing the embryo sacs

For the embryo sacs we developed a two-step freeze-fracturing technique that can exactly open the central cell and the egg apparatus of the lily, preserving both cyto- and nuclear skeletal elements and membranous organelles.

Longitudinal ovary sectors with rows of ovules attached to the placenta were fixed in paraformaldehyde-ethanol-MSB (5 : 10 : 85 vol%) for 3h at room temperature. The microtubule stabilising buffer (MSB) was prepared according to Sanz *et al.* (1995). Subsequently the specimens were rinsed 3 times for 3h in the buffer and stepwise brought into 70% ethanol in a petri dish. Visually controlled by a preparation microscope, 3-5 ovules on a piece of placenta were isolated and (in a sawing movement) carefully pushed on the edge of a piece of razor blade, in such a way that this knife longitudinally sectioned the micropyles up to the nucellus. The razor blade with the attached specimen was then taken out of the ethanol, which was sucked off using filter paper, and subsequently plunged into nitrogen-cooled liquid propane with the specimen downward to ensure a rapid freezing process. Subsequently the razor blade was placed on a nitrogen-cooled brass block with the frozen specimen downward. The latter was freeze-fractured by slamming the razor blade with a small hammer. The fractured ovules were allowed to thaw in 70% ethanol, stepwise brought into 100% ethanol and critical point dried via liquid carbon dioxide. After being mounted on stubs they were sputter-coated with 2 nm platinum and observed in the SEM mentioned above.

Results

Figs 1-9 are high-resolution SEM images of the dry-cleaved pollen tubes. Fig. 1 shows a number of pollen tubes at low magnification, showing the different levels of cleavage that can be found: apart from uncleaved sectors, sectors might either have lost their (upper) cell wall, their (upper) plasma membrane or have been cleaved deeper inside the cytoplasm. In Fig. 2 the wall has been removed shortly behind the (swollen) tip, thus showing the plasma membrane, which can be recognised as an irregular plane with little holes due to the penetration of ice crystals during freezing. Figs 3A–C show the gradual maceration of the cytoplasm using a diluted osmium solution. In Fig. 3A the plasma membrane has been removed, showing organelles embedded in a rather compact cytosol. Figs 3B and 3C show the same situation after maceration with diluted osmium tetroxide during 1 and 3 days respectively, by which

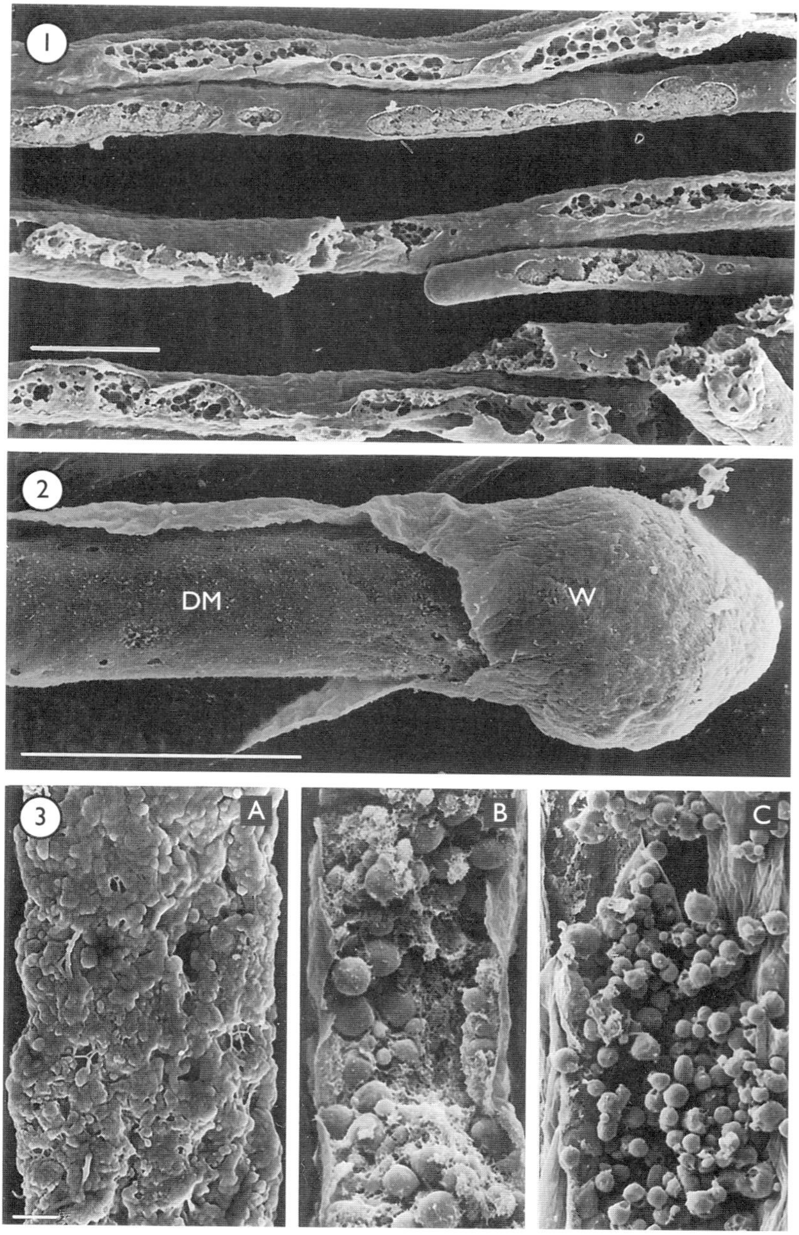

FIGS. 1–3. 1: Pollen tubes on an SEM stub have been dry-cleaved at different depths. Scale bar = 10 μm; 2: A pollen tube tip still carries its wall (w), from the subapical part the wall has been removed, by which the plasma membrane (pm) has become visible. Scale bar = 10 μm; 3: Subsequent stages of cytoplasmic maceration with diluted osmium tetroxide. 3A: dry-cleaving has removed the wall and the plasma membrane and the cytoplasm shows a mixture of cytosol, cytoskeletal elements, membranous organelles and vacuoles. 3B: after 1 day of maceration part of the cytosol and cytoskeleton have disappeared, The membranous organelles are still present, some of them have floated away. 3C: after 3 days of maceration only membranous structures have remained. Scale bar = 1 μm.

time the cytosol gradually disappears. In Fig. 3C merely membranous organelles have remained. Fig. 4 shows a detail of a pollen tube that has been fractured just below the plasma membrane in the cortical cytoplasm. Many longitudinal microfilaments can be seen, identified by their width of about 10 nm in the higher magnification in Fig. 5. They appear to be connected with vesicle-sized organelles (Figs 4 and 5). Deeper inside the cytoplasm bundles of microtubules can be found (Fig. 6), their association with other cell components being unclear. Figs 7 and 8 show the vegetative nucleus and the generative cell, respectively, identified by their overall morphology, known from fluorescence microscopy. A detail of a generative cell shows microtubules lining its surface (Fig. 9).

Figs 10-15 are images of the two-step freeze-fractured ovules. Fig. 10 shows two fractured ovules, in which the result of the sectioning step can be distinguished from the fracturing step. The fracture has opened the embryo sacs. Fig. 11 shows part of a central cell in which both cyto- and nuclear skeletal elements are visible. In Fig. 12 a bundle of thin cytoplasmic tubules can be seen in the central cell, which, due to their diameter can be identified as microtubules. Fig. 13 shows a fracturing plane through the two nuclei of a central cell. Microtubules can be seen running between the nuclei, although it is not clear from this image whether they also interconnect the nuclei (that will fuse during fertilisation). Fig. 14 shows (early) contact of a sperm nucleus with one of the two central cell nuclei. Cytoskeletal elements cannot be found associated with any of these nuclei.

Discussion

1. Techniques

The techniques presented here, although requiring improvement, offer good possibilities to trace and investigate tiny structures, important during fertilisation in lily. Compared with the TEM, the third dimension is indispensable to study different cytoskeletal elements and their configurations, and there is a greater chance of finding tiny sperm cells inside the female target cells.

Dry cleaved pollen tubes. The plane of cleavage is made randomly, but the large quantity of pollen tubes allows for sufficient images of the different depths inside these cells. Cleavage is generally close to the periphery of the tubes: either between the wall and the plasma membrane, between the plasma membrane and the cytoplasm, or in the cortical cytoplasm close to the plasma membrane. Cleaving the cytoplasm near the longitudinal axis of the tubes is rare, but rehydration and maceration of specimens in osmium tetroxide removes the cytosol and accordingly allows for a view deeper into the cells (Barnes and Blackmore, 1984). However, this maceration also removes the cytoskeleton and therefore cannot be used in cytoskeletal studies, as Tanaka and Mitsushima (1984) and Barnes and Blackmore (1984) also found. Apart from this disadvantage, the possibility to rehydrate the specimens probably opens ways for subsequent heavy metal staining and labelling procedures, to be observed using backscatter SEM (Toyoshima *et al.*, 1986; Reichelt *et al.*, 1995; Chen *et al.*, 1995). A disadvantage of the dry cleavage method is the absence of cleaved organelles. Mitochondria, for example, cannot be distinguished from plastids or larger vesicles.

Ovules. The two-step (i.e. sectioning followed by freeze-fracturing) technique presented here limits the problem of the randomly chosen pathway of the knife during freeze-fracturing, as we can direct the knife between the borders of the lily embryo sac. Since our studies focus on the process of fertilisation (including fusion of nuclei and the role of the different skeletons

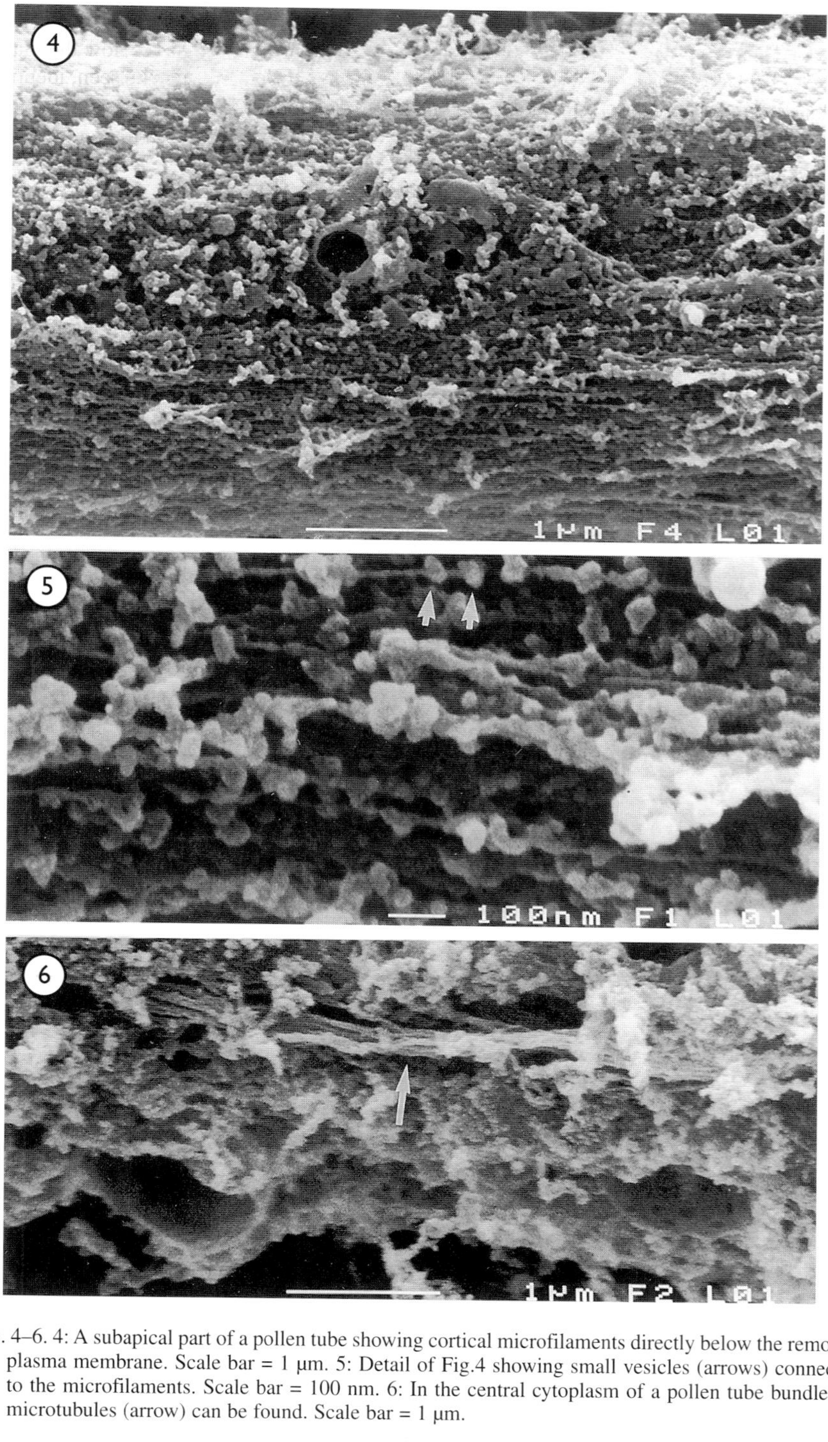

FIGS. 4–6. 4: A subapical part of a pollen tube showing cortical microfilaments directly below the removed plasma membrane. Scale bar = 1 μm. 5: Detail of Fig.4 showing small vesicles (arrows) connected to the microfilaments. Scale bar = 100 nm. 6: In the central cytoplasm of a pollen tube bundles of microtubules (arrow) can be found. Scale bar = 1 μm.

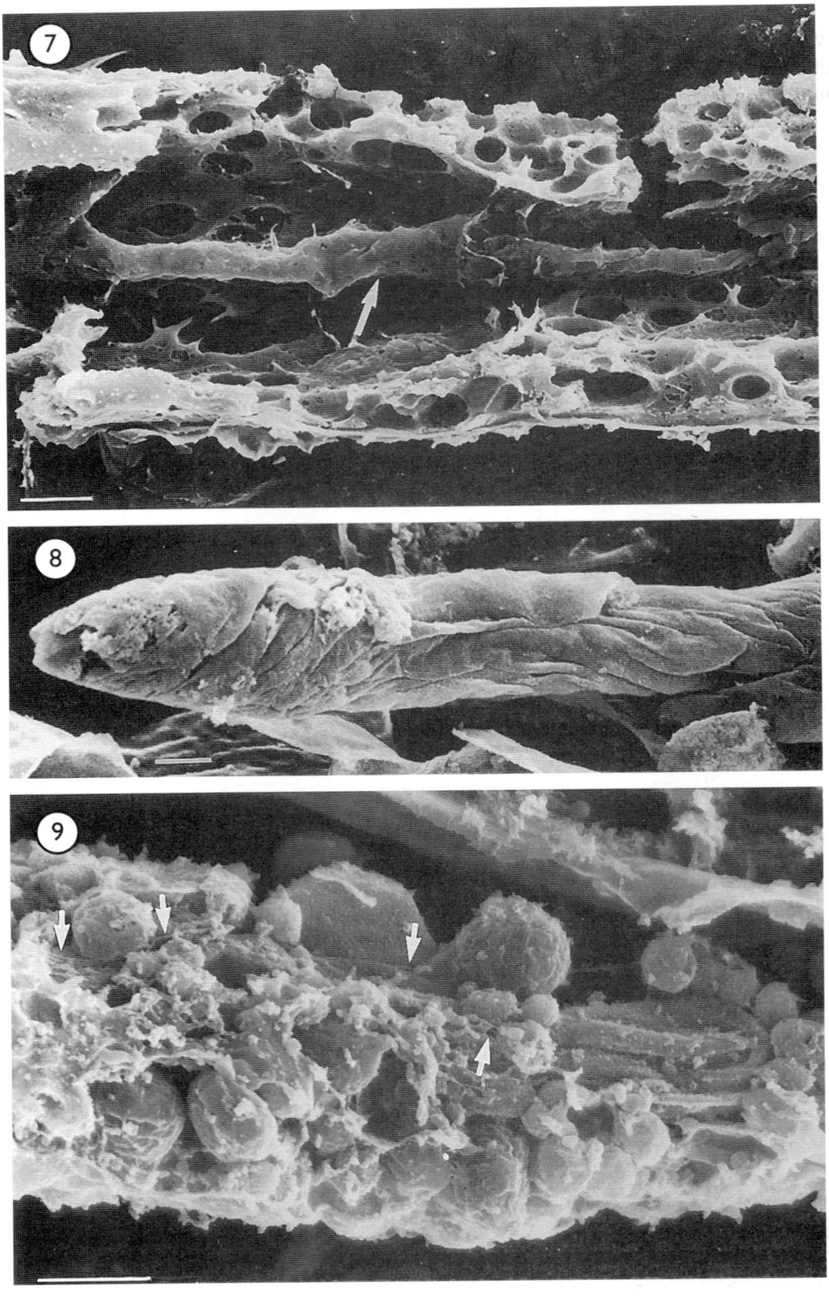

FIGS. 7–9. 7: In the axial cytoplasm of a pollen tube the vegetative nucleus (arrow) is visible. In the surrounding cytoplasm many vacuoles can be seen. Scale bar = 1 μm. 8: In the axial cytoplasm of a pollen tube the generative cell is visible. Scale bar = 1 μm. 9: Detail of a generative cell, embedded in an organelle-saturated cytoplasm and lined with microtubules (arrows). Scale bar = 1 μm.

therein), this limitation has to be narrowed down to predictably fracturing a nucleus. Therefore it would be advantageous to visualise the sperm, egg and central cell nuclei during the sectioning step, which can be facilitated by pollination with DAPI-stained pollen and UV-illumination (Willemse and Keijzer, 1990).

Transport and attraction mechanisms for sperm cells and nuclei before and during fertilisation. These techniques reveal both nuclei and cytoskeletal elements. Although we could fracture membranous structures in the ovules (nuclei, vacuoles and organelles), the cross-fractured membranes themselves frequently showed contamination at higher magnifications, seldomly showing the unit membrane, as in the techniques of Tanaka and Mitsushima (1984) and Barnes and Blackmore (1984). To improve this, experiments are underway to combine the fracturing technique presented here with an existing fixation procedure for both fractured membranous organelles and skeletal elements for SEM (Fukudome and Tanaka, 1992).

Instead of the frequently-used 50% DMSO (Tanaka and Mitsushima, 1984; Barnes and Blackmore, 1984; Fukudome and Tanaka, 1992), we used 70% ethanol as a cryo-protectant (Sasaki, 1988) due to its low surface tension, necessary for easily handling the specimens during the microscope-controlled sectioning step.

2. Role of cytoskeleton in transport and attraction of cells and nuclei: first impressions

In the pollen tubes both microfilaments and microtubules could be identified, given their width (known from TEM images). Also, (dividing) generative cells and vegetative nuclei could be recognised, from comparisons with images from fluorescence microscopy and TEM. An association between the cytoskeletal elements on the one hand and generative or sperm cells on the other could not so far be clearly demonstrated, although these structures were seen close to each other. Type and location of the different cytoskeletal elements found here confirm results from differential interference contrast, fluorescence and confocal laser scanning microscopical studies on pollen tubes (reviewed by Mascarenhas, 1993; Derksen *et al.*, 1995 and Cai *et al.*, 1996). Since cleavage planes close to the longitudinal axis of the pollen tubes are scarce, the amounts of observed generative and sperm cells (and thus their associations with the cytoskeleton) are relatively small if compared with the large numbers of pollen tubes per stub. Experiments are in progress to solve this problem.

The fractured ovules show associations between the different embryo sac nuclei and microtubules. Remarkably, microfilaments were seldom observed. Further study in larger amounts of ovules will focus on their presence and possible role in nuclear migration. In contrast, in their thorough TEM study on fertilisation in lily, Janson and Willemse (1995) found sperm and central cell nuclei associated with strands of endoplasmic reticulum and not with cytoskeletal elements.

Up to now, sperm cells were traced inside embryo sacs in only three cases, none of them associated with cytoskeletal elements, thus confirming the findings of Janson and Willemse (1995). In the degenerated synergid (i.e. the sperm cell receptor), the compact cytoplasmic structure is a complicating factor to find structures like sperm cells and cytoskeletal elements.

To our knowledge, *in situ* images of the nuclear skeleton in the SEM have not previously been produced. This skeleton belongs to the nuclear matrix, a complex of structures that remains after extracting the membranes and the nucleic acids (Verheijen *et al.*, 1988). Our technique demonstrates this, partly due to the huge nuclei of the lily embryo sac with their widely spaced chromatin distribution. This skeleton might play a role in chromosome migrations and therefore become interesting for intranuclear studies during karyogamy.

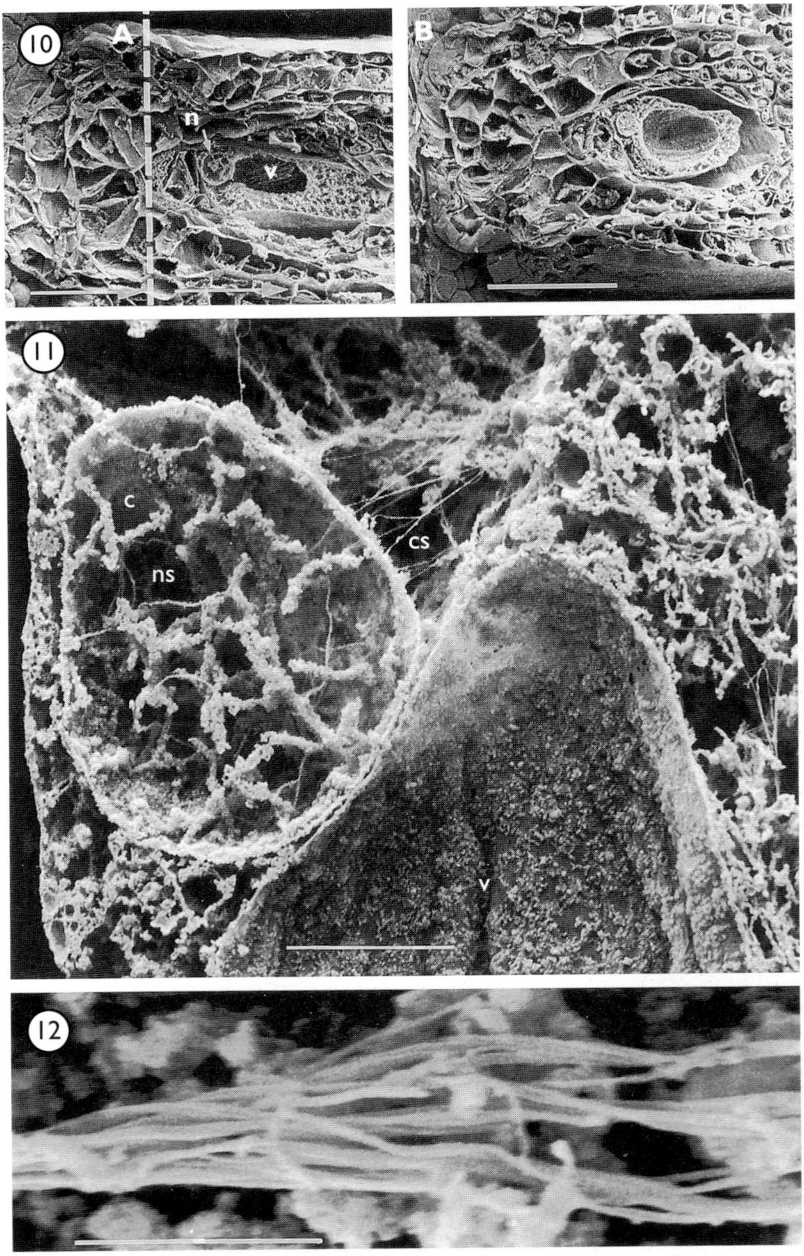

FIGS.10–12. 10: Two ovules, showing the result of the two-step freeze-fracturing method. From the left margin of Fig.10A up to the dotted line the sectioning has taken place (identified by the flattened integument cells), from the dotted line to the right side of the photograph the fracturing has occurred. A similar pattern can be seen in Fig.10B. Both figures 10A and 10B show a prominent central cell, with a central vacuole (v) and one of its nuclei (n) at the micropylar side. Scale bar = 100 µm 11: Detail of a central cell, showing a nucleus with chromosomes (c) and elements of the nuclear skeleton (ns), cytoskeletal elements (cs) and part of the vacuole (v). Scale bar = 10 µm. 12: A bundle of microtubules can be seen crossing the cytoplasm of a central cell. Scale bar = 1 µm.

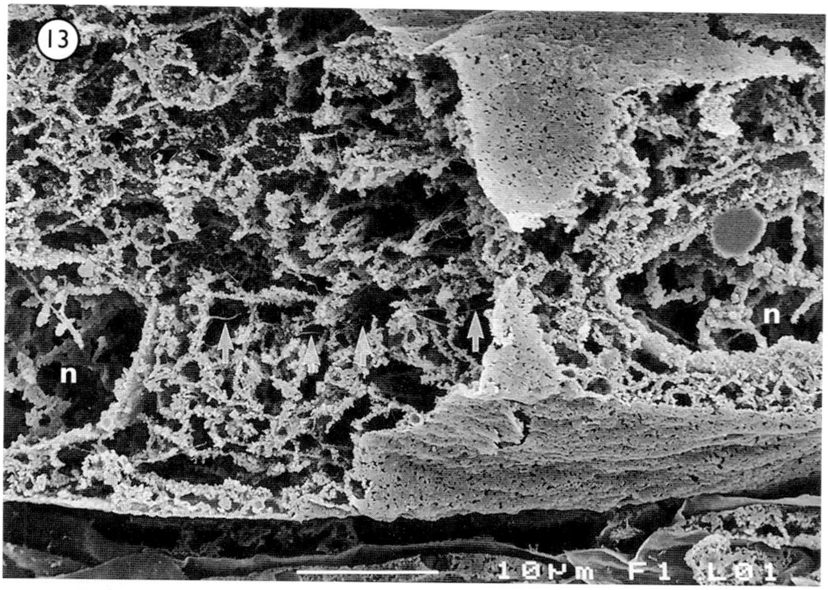

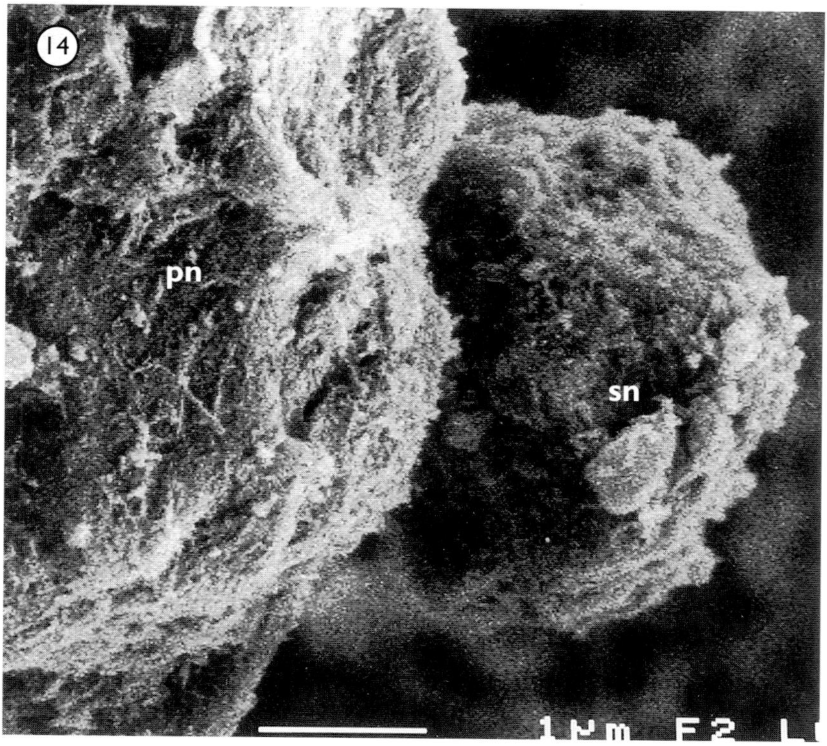

FIGS. 13–14. 13: Detail of a central cell, showing microtubules (arrows) in between its two nuclei (n). Scale bar = 10 µm. 14: Inside a central cell one of the polar nuclei (pn) contacts a sperm nucleus (sn). Scale bar = 1 µm.

Acknowledgements

The author thanks professor M.T.M. Willemse for critically reading the text and S. Massalt for printing the photographs.

References

Barnes, S. and Blackmore, S.H. (1984). Scanning electron microscopy of chloroplast ultrastructure. *Micron Microscop. Acta* **15**: 187–194.

Blackmore, S., Barnes, S.H. and Claughter, D. (1984). Scanning electron microscopy of cytoskeletal components in *Aucuba japonica* leaves. *J. Ultrastruct. Res.* **86**: 215–219.

Cai G., Moscatelli A., Del Casino C. and Cresti M. (1996). Cytoplasmic motors and pollen tube growth. *Sexual Pl. Reprod.* **9**: 59–64.

Chen, Y., Centonze, V.E., Verkhovsky, A. and Borisky, G.G. (1995). Imaging of cytoskeletal elements by low temperature high resolution scanning electron microscopy. *J. Microscopy* **78**: 67–76.

Derksen, J., Rutten, T., Van Amstel, T., De Win, A., Doris, F. and Steer, M. (1995). Regulation of pollen tube growth. *Acta Bot. Neerl.* **44**: 93–119.

De Ruijter, N.C.A., Pietrusiewicz, J., Montijn, B., Oud, J.L., Schel, J.H.N. and Van Lammeren, A.A.M. (1997). Spatial limitations induce spindle tilting and result in oblique phragmoplasts in *Vicia faba* L. root tip cells, but do not result in oplique cell walls. *Acta Bot. Neerl.*, **46**(3): 279–290.

Fukudome, H. and Tanaka, K. (1992). A method for simultaneously revealing both the cytoskeleton and membranous cell organelles for scanning electron microscopy and its application to rat tissues *J. Electron Microscopy* **41**: 357–363.

Janson, J. and Willemse, M.T.M. (1995). Pollen tube penetration and fertilisation in *Lilium longiflorum* (Liliaceae). *Amer. J. Bot.* **82**: 186–196.

Lloyd, C.W. (1991). The cytoskeletal basis of plant growth and form. Academic Press, London, UK.

Mascarenhas, J. (1993). Molecular mechanisms of pollen tube growth and differentiation. *Pl. Cell* **5**: 1303–1314.

Reichelt, S., Ensikat, H.J., Barthlott, W. and Volkmann, D. (1995). Visualization of immunogold-labeled cytoskeletal proteins by scanning electron microscopy. *Eur. J. Cell Biol.* **67**: 89–93.

Sasaki, K. (1988). A simple method to observe intracellular organelles with the scanning electron microscope. *J. Electron Microscopy* **37**: 171–173.

Tanaka, K. and Mitsushima, A. (1984). A preparation method for observing intracellular structures by scanning electron microscopy. *J. Microscopy* **133**: 213–222.

Toyoshima, S., Saito, T., Yamaguchi, J. (1986). Immuno-scanning electron microscopy of macrophage cytoskeleton by using colloidal gold and backscattered electron imaging mode. *J. Electron Microscopy* **35**: 247–258.

Verheijen, R., Van Venrooy, W. and Ramaekers, F. (1988). The nuclear matrix: structure and composition. *J. Cell Sci.* **90**: 11–36.

Vesk, P.A., Rayns, D.G. and Vesk, M. (1994). Imaging of plant microtubules with high resolution scanning electron microscopy. *Protoplasma* **182**: 71–74.

Willemse, M.T.M. and Keijzer, C.J. (1990). Tracing pollen nuclei in the ovary and ovule of *Gasteria verrucosa* (Mill.) H. Duval after pollination with DAPI-stained pollen. *Sexual Pl. Reprod.* **3**: 219–224.

Willemse, M.T.M. and Wittich, P.E. (1998). Pollination and fertilisation in *Gasteria verrucosa*: interaction between pollen tube and ovary. In: S.J. Owens and P.J. Rudall (Editors). Reproductive Biology, pp. 57–69. Royal Botanic Gardens, Kew.

POLLINATION AND FERTILISATION IN *GASTERIA VERRUCOSA*: INTERACTION BETWEEN POLLEN TUBE AND OVARY

M.T.M. WILLEMSE AND P.E. WITTICH

Dept. Plant Cytology and Morphology, Agricultural University, Arboretumlaan 4, NL-6703 BD Wageningen, The Netherlands

Abstract

The interactions in the interfaces between pollen and pistil of *Gasteria* are studied with special attention to the interaction between pollen tubes and placental fluid, and between pollen tube and ovule.

Placental cells in the ovary produce the placental fluid. This fluid contains carbohydrates and proteins which are used during pollen tube growth. The enzyme invertase, present in the placental cells and placental fluid, is probably involved in the carbohydrate composition of the fluid. An indication for the arrest of the production of placental fluid is the deposition of callose in the transfer walls of ageing placental cells, the passing of pollen tubes is not an indication. A component in the stylar and placental fluid that appears as a result of the presence of pollen tubes is probably trehalose. Pollen tubes are attracted to the ovule by a proteinaceous substance in the micropylar exudate. This exudate originates from the synergids and contains also some free calcium. However, this calcium does not seem to be involved in attraction. In the micropylar canal mono- and disaccharides are found which might be produced by invertase activity in the micropylar nucellar cells and synergids. Mono- and disaccharides, and consequently invertase activity, could be involved in the growth of pollen tubes into the micropyle and through the nucellus into a synergid.

The rate of swelling of *in vitro* cultured ovules depends on the presence of pollen tubes near the ovules. The pollen tubes function as a nurse-factor for the ovules, but an extra stimulus for the integumental growth seems to be pollen tube penetration. In the ovary interactions exist between the pollen tubes and the placental fluid, and between the pollen tube and the micropylar exudate. The pollen tube growth, the way of attraction, and the micropylar ingrowth are discussed.

Introduction

Pollination signals are necessary for the interaction between the germinating pollen and pistil, resulting in fertilisation of the ovule. In *Gasteria* this interaction takes place in the interfaces of the pistil, i.e. the stigmatic exudate, the first solid wall part of the style, the liquid in the hollow style, the placental fluid, and finally the micropylar exudate. Signals from the pollen coat stimulate the water supply to the pistil. This completes the pollen tube pathway and results in the penetration of the ovule by the pollen tube in the case of cross pollination (Willemse, 1996). Pollination signals prepare the interface and the ovule for a successful fertilisation.

This study is focused on the interactions in the ovary between pollen tubes and placental fluid and between pollen tubes and micropylar exudate.

The placenta of the ovary develops elongated papillar glandular cells below the ovules. These placental cells produce a high amount of the placental fluid which fills the placental locule around the ovules. These placental cells show before and during the receptive period of the pistil osmiophilic granules and have on their tip a transfer wall. The placental fluid contains sucrose, fructose and glucose as main carbohydrates as well as proteins and glycoproteins (Willemse and Franssen-Verheijen, 1986).

Before arrival of the pollen tube in the placental locule, there is an increase of the locular fluid volume together with a decrease of the osmolarity of this fluid. This change in volume and osmolarity points to an increase of water in the placental locule due to a pollination signal. This results in a connection between the stylar fluid and the placental fluid, which completes the pollen tube pathway (Willemse and Franssen-Verheijen, 1992; Willemse, 1996).

The pollen tubes grow at about 2 mm per hour through the placental fluid. In the placental locule the callosic secondary pollen tube wall is formed again after its absence in the stylar canal. During their growth the pollen tubes affect and break down the glycoproteins of the placental fluid (Willemse and Franssen-Verheijen, 1988, 1992; Willemse and Vletter, 1995). With respect to these data, the presence of callose in the wall of the placental cells is studied here, as well as the changes in carbohydrate composition of the placental fluid during pollen tube growth.

In the micropyle a droplet of a micropylar exudate is present which originates from the large filiform apparatus of the synergids. This filiform apparatus shows remarkable changes during synergid development and the exudate seems to be pressed out into the micropylar canal. This exudate contains an attractant for pollen tubes (Franssen-Verheijen and Willemse, 1990, 1993).

The stigmatic tissue and the ovular micropyle are sites for a chemotropic effect on growing pollen tubes within the pistil (Heslop-Harrison and Heslop-Harrison, 1986). Pollen tube attraction towards the micropyle and micropylar penetration implicates an activity of the synergids (Van Went and Willemse, 1984; Russell, 1992). Other ovular tissues, such as the inner integument and nucellar cap (Tilton, 1977) or the nucellus (Bruun and Oleson, 1989) can also function in pollen tube attraction. Proteins, carbohydrates like mucopolysaccharides (Chao, 1970), and free calcium are mentioned as possible chemotrophic attractants for the pollen tube into the micropyle (Mascarenhas, 1993). Calcium shows an effect on pollen tube attraction (Malho *et al.*, 1995) as does a glycoprotein in the style of *Nicotina* (Cheung *et al.*, 1995).

In *in vitro* studies using pearl millet pistil diffusates, germinated pollen shows chemotropy on glucose, calcium and an acidic ovarian protein of low molecular weight. In that study a defined stigmatic pellet washing also points to the action of a protein (Reger *et al.*, 1992a). Reger *et al.* (1992b) demonstrated the chemotrophic response of pollen tubes to glucose, which was hydrolysed from sucrose by tissue-bound invertase. In *Gasteria*, during *in vitro* pollen tube penetration an attractant of proteinaceous nature is present in the micropylar exudate (Willemse *et al.*, 1995).

Based on these data, the influence of invertase and calcium on a possible attraction mechanism is studied. Pollen tube behaviour in the placental fluid, the attraction to the micropyle and ingrowth of the pollen tube into the micropyle are discussed.

Material and Methods

Plants of *Gasteria verrucosa* (Mill.) Duval were cultured in a greenhouse at 20–23°C and 16h/day artificial light.

Several experiments were preformed to elucidate the interactions between the pollen tube and its pathway. To investigate the interaction between pollen tube and placental cells, the cells were tested for callose changes. To study the interaction between pollen tubes and carbohydrates in the pathway carbohydrates were analysed *in vivo* and *in vitro*. The micropylar exudate was tested for calcium as a possible attractant for pollen tubes. To investigate the possible role of different carbohydrates in the micropyle several carbohydrates were tested for their influence on pollen growth, and the presence and activity of invertase was localised along the pollen tube pathway. Protein analysis was preformed 20 hours after pollination to test the influence of pollen signals in the ovule, and to detect an influence of germinating pollen or penetrated pollen into the micropyle, the swelling of ovules was measured *in vitro*.

Histochemistry on fresh sections

Callose was stained by basic aniline blue stain (Eschrich and Currier, 1964). Chlortetracycline, which reacts with (free) *calcium* that is present in the vicinity of hydrophobic sites as membranes (CTCf), fluphenazine which reacts with the calcium/calmoduline complex (FPZf; Timmers *et al.* 1989), fura-2 and fluo-3 for free *calcium* were used in a 10^{-4} M demi water solution with 6% sucrose to adjust the osmolarity.

Histochemistry on fixed sections

For staining *mono-* and *disaccharides*, ovules were dissected from the placental tissue, fixed, and stained according to the method of Okamoto and co-workers (Gabe, 1976). Invertase activity could be shown by the precipitation of tetrazolium salt, using the method described by Doehlert and Felker (1987). The presence of invertase could be demonstrated on sections of methacrylate embedded ovaries, using antibodies raised against grape soluble invertase. The procedure followed was according to the protocol of Baskin *et al.* (1992). The antibody was a gift from H.P. Ruffner (Ruffner *et al.*, 1995).

Collection of fluid interfaces

The stylar and placental fluid, with and without geminating pollen, and the stigmatic fluid were used to detect the presence of the different carbohydrates in a Dionex Spectra Physics HPLC system, using a ED40:IAMP detector. To collect the stylar fluid, the pollen coating fluid, and the placental fluid, the same methods were used as described by Willemse and Vletter (1995). The stigmatic exudate was collected with a small pipette.

Pollen germination and pollen tube penetration *in vitro*

To examine the effect of trehalose on pollen germination, a 5% sucrose with 300 ppm boric acid, 300 ppm calcium nitrate, 200 ppm magnesium sulphate and 100 ppm potassium nitrate medium was enriched with 1% trehalose. For comparison, media of 5% sucrose were enriched with 1% sucrose, glucose, fructose, ribose, or raffinose. To test the behaviour of the pollen tubes with respect to the penetration in the micropyle as well as the growth of ovules after pollen tube penetration, the *in vitro* micropylar penetration system was used (Willemse *et al.* 1995). To test the possibility of a function in the attraction of pollen tubes by calcium in the micropylar exudate, calcium was not added to the medium. To culture the ovules a 6% sucrose medium with 0.5% agar and enriched with Murashige Skoog Medium (Imperial) was used.

Protein electrophoresis

Proteins were extracted from ovules before pollen tube penetration and 20 hours after penetration *in vivo*. For electrophoresis 20 ovules were macerated in water with protein inhibitors (Boehringer Manheim) at 40°C. The supernatant was separated as described by Willemse and Vletter (1995).

Results

Interaction between the placental cells or the placental fluid and the pollen tubes

The placental cells below the ovules are elongated cells with a relative large vacuole (Fig. 1). In a flower with stigmatic exudate, which is a sign for receptivity, about 4% of the placental cells show a positive reaction for callose in the tip of their wall (Fig. 2). The number of cells reacting for callose staining increases during the wilting of the flower to about 55%. No influence of the arrival of pollen tubes from self or cross pollination on the callose deposition could be determined.

The placental cells produce the placental fluid in which the pollen tubes grow. The fluid interfaces in the pistil, the stigmatic droplet, the stylar canal fluid, and the placental fluid, were analysed for the presence of carbohydrates. The stylar as well as the placental fluid were analysed with and without passing pollen tubes. The results are presented in Graph 1.

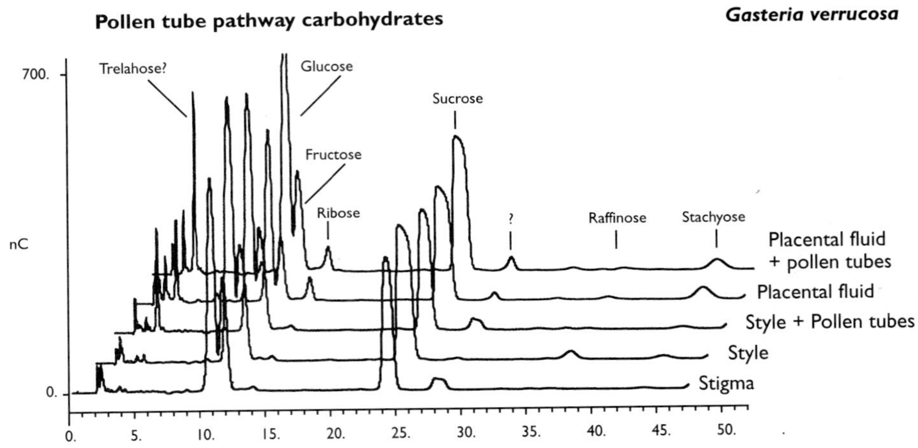

GRAPH 1. HPLC analysis of stigmatic, stylar, and placental fluid before and after pollination.

The stigmatic droplet, stylar canal fluid, and placental fluid all show a comparable composition of sucrose, fructose, and glucose as main components. Less abundant are ribose, raffinose and stachyose. A carbohydrate, probably trehalose, increases in the presence of the pollen tubes in the style and ovarian locule.

In vitro pollen germination experiments with *Gasteria* on a sucrose medium enriched with trehalose, showed 80% germination. The addition of sucrose instead of trehalose resulted in 74% germination, ribose 79%, glucose 28%, fructose 16%, and raffinose in 8%.

Interaction between the micropylar exudate and pollen tubes

The interaction with the micropylar exudate consists of a chemotrophic directed growth of the pollen tubes into the micropyle. Next to proteins as possible attractants, calcium, carbohydrates and an invertase activity of the pollen tubes are postulated in the literature.

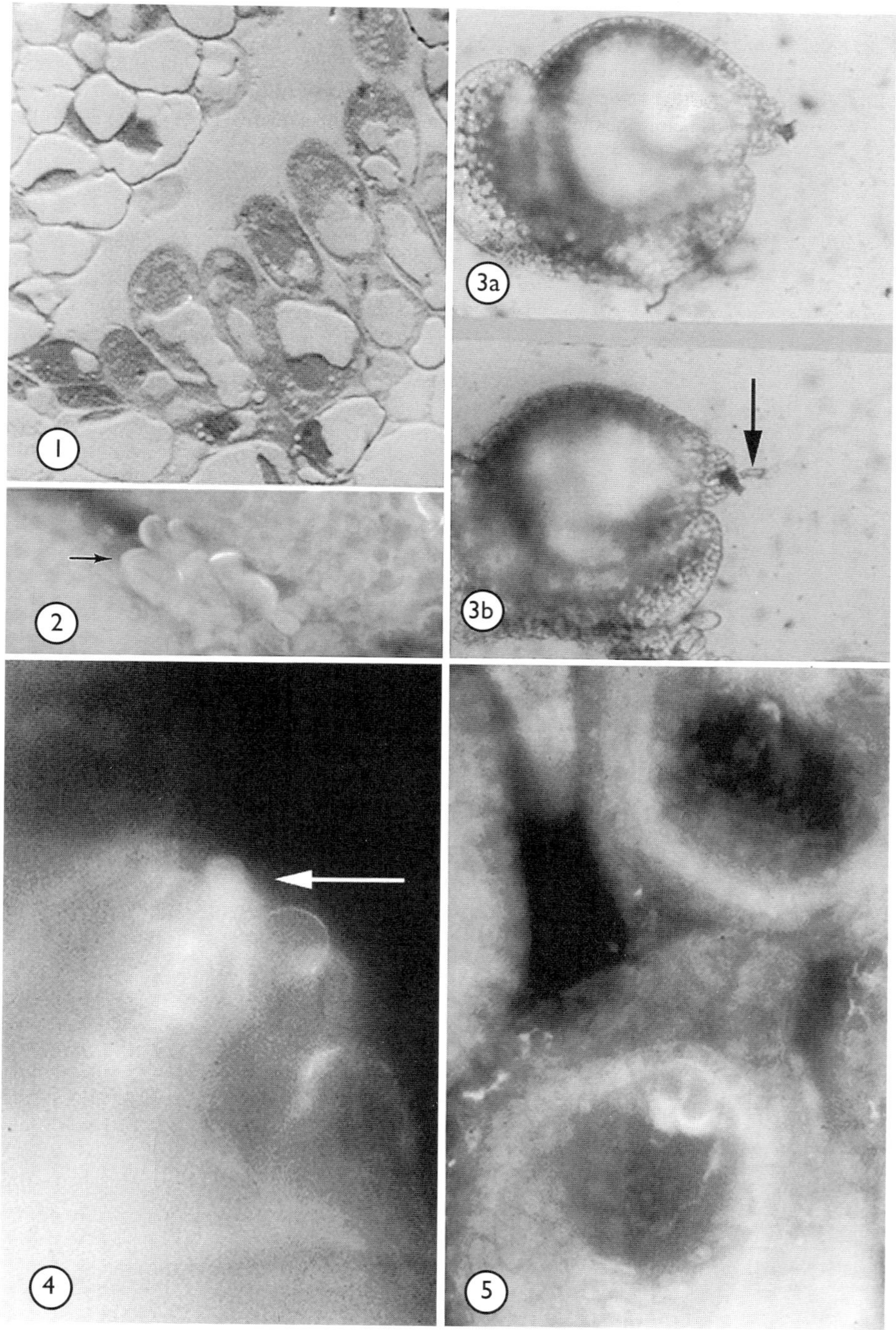

FIGS. 1–5. 1. Section of placental cells, ×400. 2. Callose in the tip part of the placental cells, see arrow. × 400. 3. Coomassie blue stained micropylar exudate, without (A) and with pollen tube (B), × 40. 4. Fluo-3 stained micropylar exudate indicating the presence of free calcium, see arrow, × 500. 5. CTCf stained section of ovule, with intensive reaction in the egg apparatus, × 100.

In the micropylar exudate of *Gasteria* proteins are present (Fig. 3). The micropylar exudate also showed a positive reaction with fura-3 and fluo-3 as fluorochromes, indicating the presence of free calcium (Fig. 4). Free calcium present in the vicinity of hydrophobic sites was detected by CTCf in the egg apparatus and central cell (Fig. 5). In the egg cell the cytoplasm positioned near the synergids was clearly positive (Fig. 6). Both synergids show also a positive reaction (Fig. 7). Calmodulin was detected with FPZf at the same place where CTCf reacted. So the micropylar exudate as well as the egg apparatus can be considered as sites with calcium.

Using the *in vitro* pollen tube penetration system on a medium without calcium, the pollen tubes grew slowly and made a relative short pollen tube. The micropylar exudate is the main place of free calcium and a stimulated attraction can be expected. Table 1 presents the results of three different experiments on pollen tube penetration in the micropyle on a calcium free medium compared with a calcium containing medium. In both conditions there is a mean penetration of 17%.

TABLE 1. The influence of calcium in a medium on the micropylar penetration of pollen tubes. The percentage of micropylar penetration and in brackets the total number of tested ovules are given.

Medium with calcium	Medium without calcium
20% (170)	20% (120)
15% (155)	14% (127)
16% (184)	19% (120)
Mean values:	
= 17% ± 3% (509)	= 17% ± 4% (367)

The carbohydrates present in the micropylar exudate as demonstrated by a PAS reaction (Verheijen and Willemse, 1993) indicate the presence of polysaccharides on this place of attraction. Staining of mono- and disaccharides showed a ring- like reaction in the chalazal part of the nucellus and in the micropylar nucellus. In the narrow part of the funnel-shaped micropyle mono- and disaccharides were found (Fig. 8).

Invertase activity could be detected in the whole ovule (Fig. 9) and a high activity was found in the placental cells (Figs. 9, 10). A remarkable feature is the positive immunocytological reaction with the invertase antibody outside the young placental cells during the tetrad stage of ovular development (Fig. 11), which point to an extracellular invertase. The synergids (Fig. 12) but not the egg cell (Fig. 13) also showed invertase activity, together with the nucellar cells near the micropyle. This position of invertase activity is near the micropylar pollen tube pathway.

Interaction with ovules and penetrated pollen tubes

The pollen coating components stimulate pollen tube penetration (Willemse, 1996). After approximately 15 hours, the pollen tubes arrived in the micropyle and within the next 20 hours pollen tubes transferred both sperm cells into a synergid. The two dimensional protein patterns extracted from ovules before pollen tube penetration and 20 hours after penetration were similar.

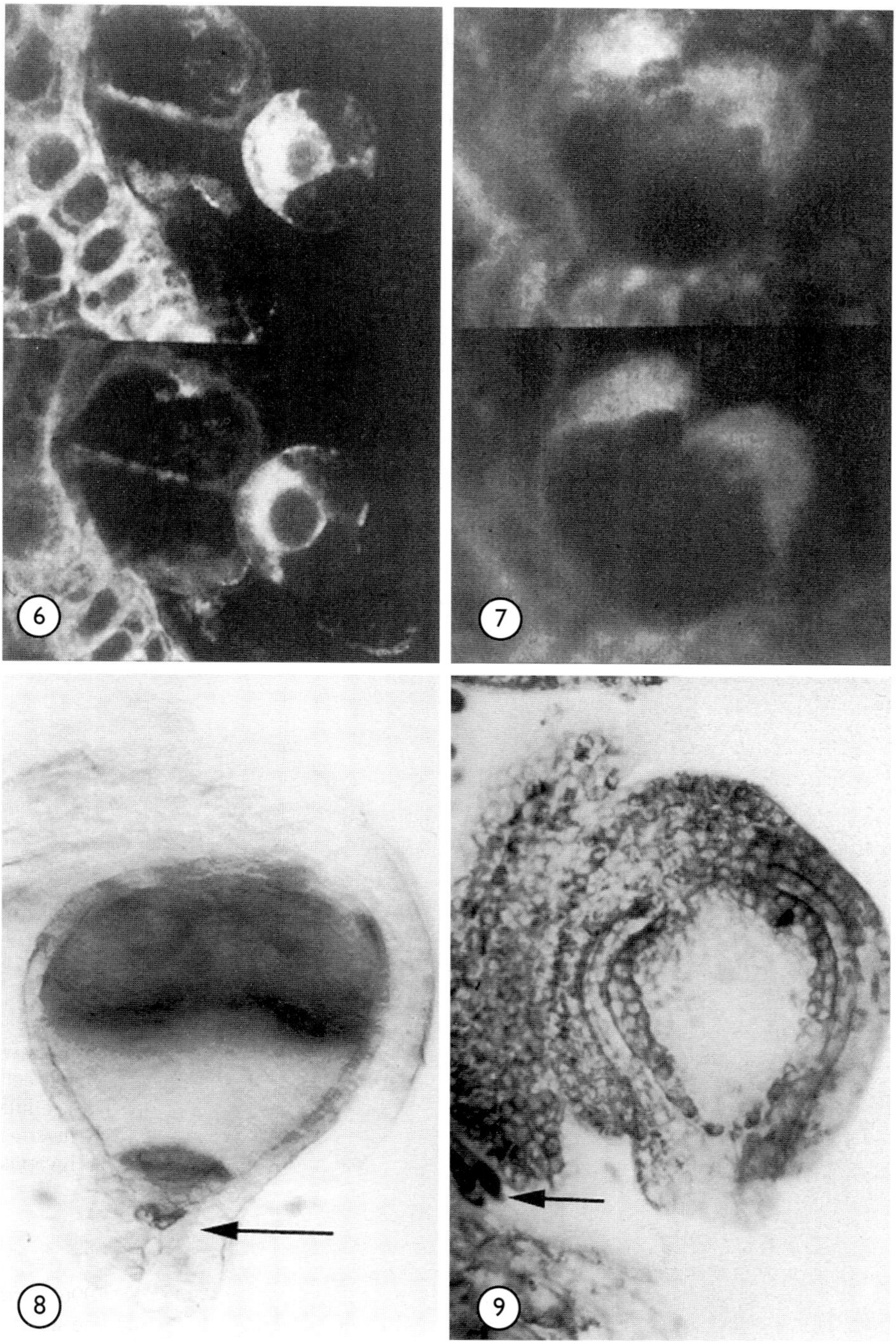

FIGS. 6–9. 6. Confocal image of a CTCf stained egg cell, × 850. 7. Confocal image of CTCf stain synergids, × 850. 8. Ovule stained on the presence of monosaccharide, note the reaction in the narrow canal of the micropyle (arrow), × 40. 9. Survey of the invertase activity on a section of an ovule, note the placental cells (arrow), × 40.

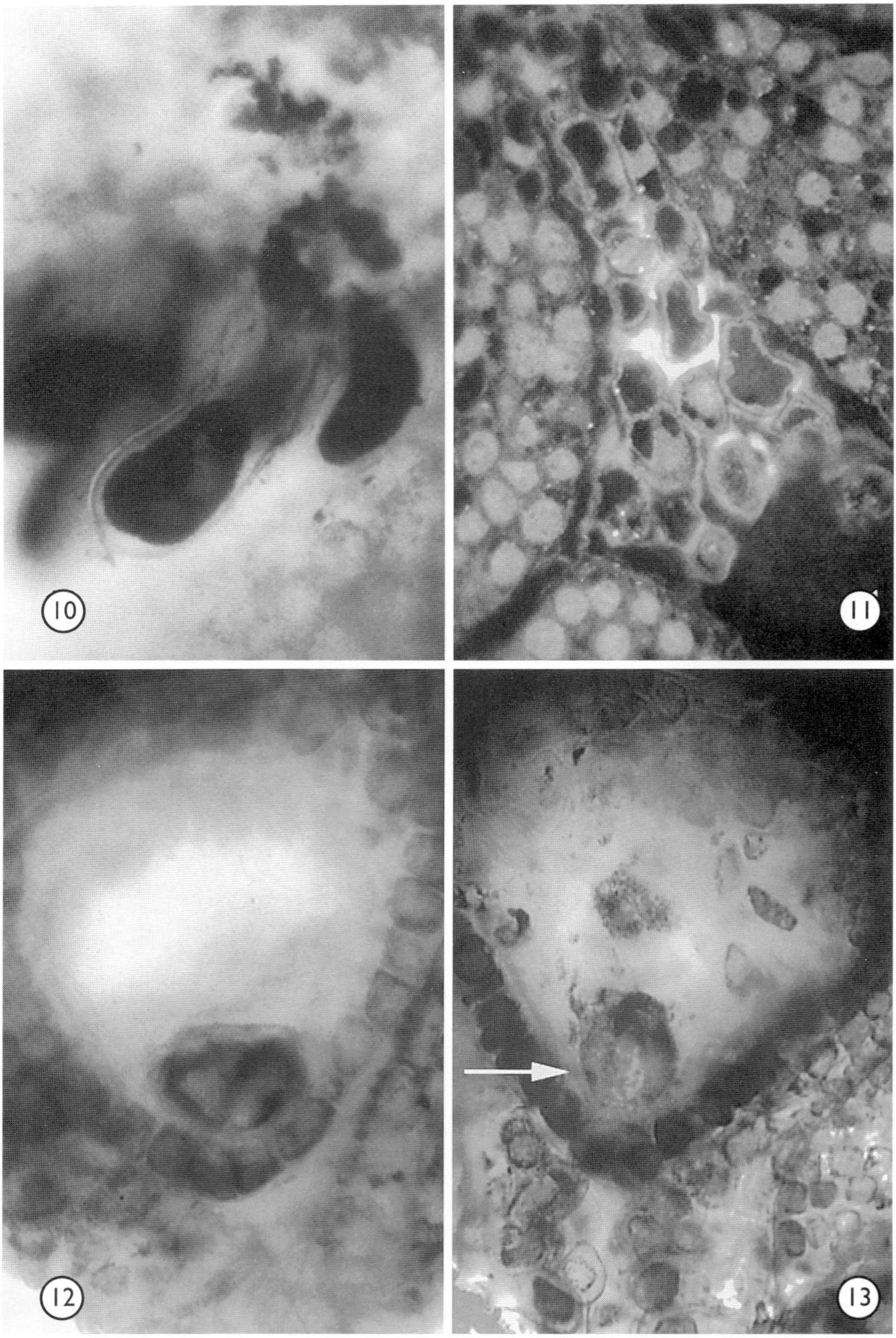

FIG. 10–13. 10. Invertase activity in the placental cells, × 400. 11. The presence of invertase, shown by immunohistochemistry, outside young placental cells, × 400. 12. Invertase activity in the synergids, note the reaction in the micropylar nucellus cells, ×400. 13. Invertase activity in synergids but not in the egg cell (arrow), note the reaction in the micropylar nucellus cells, × 400.

Earlier data shown by Plyushch *et al.* (1995) point to a swelling of lily ovules after pollen tube penetration *in vitro* as a sign of pollen tube arrival, due to an enlargement of mainly the integumental cells.

To investigate the rate of ovule swelling in *Gasteria* a population of ovules was put on the *in vitro* pollination culture system. Two parts of the ovules collected were separated and pollinated *in vitro* with self and cross pollen respectively. After one day the penetrated self and cross pollinated ovules were selected and cultured on culture medium for 60 days. The unpenetrated ovules of the self and cross pollinated ovules were also cultured. Some ovules were not pollinated and used as control for the rate of swelling in the culture. After 60 days the ovules were separated into four classes: a diameter of 0.3–0.5 mm, 0.5–0.8 mm, 0.8–1.2 mm, and 1.2 mm or more. Graph 2 represents the percentage of the classes of swollen ovules in culture. Compared with the control, the unpenetrated ovules have more swollen ovules in class 0.8–1.2 mm and 1.2 mm or more. The penetrated ovules after cross pollination show a comparable pattern in the distribution over the classes, with a small increase in the classes 0.8–1.2 mm and 1.2 mm or more. The penetrated ovules with self pollen show a higher percentage of ovules in class 1.2 or more, compared with the cross pollinated and unpenetrated ovules.

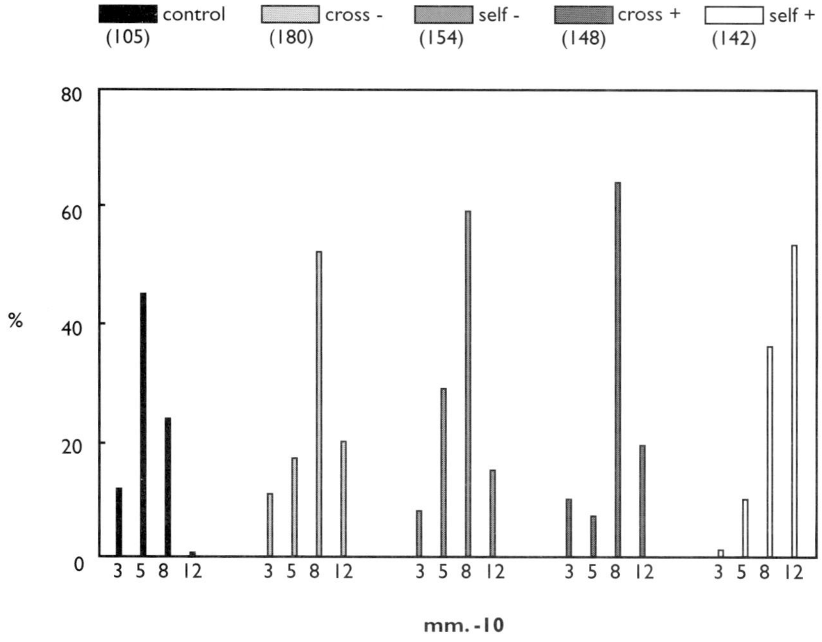

GRAPH 2. Percentage of several classes of swollen ovules of *Gasteria verrucosa* in culture: Numbers in brackets represent the total number of used ovules. - = no pollen tube penetration, + = with pollen tube penetration. Four classes of diameter: (3) = 0.3-0.5, (5) = 0.5-0.8, (8) = 0.8-1.2, (12) = 1.2 and more mm.

Discussion and conclusion

Placental cells, placental fluid and pollen tubes

In the ovary, pollen tubes pass the placental fluid. The placental cells which produce the placental fluid, deposit callose in the tip part of the cell at the site of the transfer wall. The presence of callose may stop the transfer of placental fluid. The increasing number of placental cells with callose might be a sign of the cessation of production of placental fluid. Until now no response of callose synthesis by the placental cell could be related to the presence of pollen tubes in the placental fluid.

In the placental cell the conversion of sucrose into glucose and fructose can take place, marked by the high intensity of the invertase reaction. Maybe a part of the sucrose in the placental cell is converted and excreted, with the consequence that the placental fluid contains sucrose, glucose and fructose. However, it is possible that invertase is also active outside the placental cell. Immunolocalisation showed invertase in the placental fluid at an early stage of development. In that case the placental cell could be considered as a place for production and transfer of invertase. In plants, invertases can be released from the wall matrix and set free (Faye and Ghorbel, 1983). As in the flower of *Lilium* invertases are always glycoproteins (Miller and Ranwala, 1994). Maybe the earlier demonstrated glycoproteins in the placental fluid (Willemse and Vletter, 1995), whose presence could be affected by pollen tubes, also represent invertases.

Passing pollen tubes probably stimulate the presence of trehalose in the stylar and placental fluid. The origin of the accumulation of this product is unknown. It could be produced by the stylar or placental cells or it could be split from glycoproteins present in the fluid. Pollen of *Lilium* shows trehalase activity (Gussin and McCormack, 1970) which points to the possibility of breakdown of the trehalose. *In vitro*, ribose and trehalose may function as nutrients, as osmolic regulators, or both. On the other hand, glucose, fructose and raffinose in the placental fluid may suppress *in vitro* pollen germination. Commonly the use of fructose and also glucose results in a lower germination of pollen, while raffinose usually gives good results (Vasil, 1987). The comparable complex of carbohydrates in the stylar fluid as well as the placental fluid show *in vivo* a difference in the velocity of pollen tube growth. Pollen tubes grow faster in the style than in the placental fluid (Willemse and Franssen-Verheijen, 1988).

There clearly exists an interaction between pollen tubes and placental fluid, but not between pollen tubes and the placental cells. The placental fluid contains sugars as nutrients for pollen tubes and in this fluid the pollen tube regains a callose wall. The pollen tubes are also able to affect the placental fluid proteins (Willemse and Franssen-Verheijen, 1992b). These effects could be related to differences between stylar and placental proteins. Also the presence of signal molecules related to the recognition of self or cross pollen cannot be excluded.

Pollen tube attraction and growth into the micropyle

Although proteinaceous substances may play a role in the pollen tube attraction to the micropylar exudate (Willemse, 1995), other products such as calcium, invertase or carbohydrates should be included.

In *Gasteria* the effect of calcium seems neither to stimulate nor inhibit micropylar penetration of the pollen tube, even though calcium is found to be present in the egg apparatus and micropylar exudate. Also the function of calcium in the election of one of the synergids for pollen tube penetration could not be detected since there was an equal localisation of calcium in both synergids.

Carbohydrates are present in a high amount and diversity in the placental fluid as well as in the micropylar exudate. The release of glucose as a result of pollen tube invertase activity may occur (Reger *et al.*, 1992b), therefore it seems unlikely that this micropylar invertase activity functions in pollen tube attraction. The presence of mono- and disaccharides in the micropyle is remarkable. They may function as a source of nutrients for the further pollen tube growth through the micropyle and nucellus towards one of the synergids. Although the micropylar exudate, which originates from the filiform apparatus of the synergids, guarantees an attractive and continuous pathway to one of the synergids, pollen tube growth slows strongly in the micropyle and still needs nutrients as carbohydrates.

There is also invertase activity in the nucellar cells and synergids, which are positioned in the final part of the pollen tube pathway. This may indicate that there is a conversion of sucrose into glucose and fructose for pollen tube nutrition in this area of the nucellus. Pollen tubes are also able to grow over the nucellus as is found during multiple penetration of pollen tubes. This points to a sufficient external energy source besides the nutrients provided by the egg apparatus.

Influence of pollen tube penetration on ovular protein synthesis and ovular swelling

In contrast to the data of Deurenberg (1977), showing RNA and protein synthesis as a response of pollination before fertilisation in *Petunia*, *Gasteria* seems not to react within 20 hours of the pollination signal with newly synthesised proteins.

There is an influence of germinating pollen and pollen tubes that have penetrated the micropyle pollen on the rate of swelling of the ovule. The presence of pollen tubes may be considered as a nurse culture for unpenetrated ovules. But the difference evoked by penetrating pollen tubes may point to a signal from the pollen tube on the growth of the integuments of the ovule.

In summary, there exists a complex interaction between the pollen tubes and the ovary via the different interfaces. The difference in composition of interfaces such as placental fluid and micropylar exudate reveals the heterogeneity of the pollen tube pathway in the pistil. A successful fertilisation seems to depend on pollination signals and a balanced interplay between cross pollen and pistil via the different interfaces.

Acknowledgements

The authors thank H.P. Ruffner for the gift of the invertase antibodies; S. de Bruin for the HPLC analysis; C. Arts-Reinders for the ovule culture; S. Massalt for making the photographs and G. van Geerenstein for culturing the plants.

References

Baskin, T.I., Busby, C.H., Fowke, L.C., Sammut, M. and Gubler, F (1992). Improvements in immunostaining samples embedded in methacrylate: localization of microtubules and other antigens throughout developing organs in plants of diverse taxa. *Planta* **187**: 405–413.

Bruun, L. and Oleson, P. (1989). A structural investigation of the ovule in sugar beet, *Beta vulgaris*: the micropylar nucellus. *Nordic J. Bot.* **9**: 81–87.

Chao, C. (1970). Further cytological studies of a periodic acid- Schiff's substance in the ovules of *Paspalum orbiculare* and *Paspalum congiflorum*. *Amer. J. Bot.* **64**: 921–930.

Cheung, A.Y., Hong, W. and Hen-Ming, W. (1995). A floral transmitting tissue-specific glycoprotein attracts pollen tubes and stimulates their growth. *Cell* **82**: 383–393.

Deurenberg, J.J.M. (1977). Differentiated protein synthesis with polysomes from *Petunia* ovaries before fertilization. *Planta* **133**: 201–206.

Doehlert, D.C. and Felker, F.C. (1987). Characterization and distribution of invertase activity in developing maize (*Zea mays*) kernels. *Physiol. Pl.* **70**: 51–57.

Eschrich, W. and Currier, H.B. (1964). Identification of callose by its diachrome and fluorochrome reactions. *Stain Technol.* **39**: 303–307.

Faye, L. and Ghorber, A. (1983). Studies on beta-fructosidase from radish seedlings. III. Comparative studies on soluble and wall bound forms. *Pl. Sci. Lett.* **29**: 49–60.

Franssen-Verheijen, M.A.W. and Willemse, M.T.M. (1990). The ovule of *Gasteria verrucosa* at receptivity of the stigma: an ultrastructural study. *Acta. Bot. Neerl.* **39**: 53–63.

Franssen-Verheijen, M.A.W. and Willemse, M.T.M. (1993). Micropylar exudate in *Gasteria* (Aloaceae) and its possible function in pollen tube growth. *Amer. J. Bot.* **80**: 253–262.

Gabe, M. (1976). Histological techniques. Masson, Paris.

Gussin, A.E.S. and McCormack, J.H. (1970). Trehalase and the enzymes of trehalose biosynthesis in *Lilium longiflorum* pollen. *Phytochemisty* **9**: 1915–1920.

Heslop-Harrison, J. and Heslop-Harrison Y. (1986). Pollen tube chemotropism: fact or delusion? In: M. Cresti and R. Dallai (editors). Biology of reproduction and cell motility in plants and animals, pp 169–174, Universitá di Siena, Siena.

Malho, R., Read N.D., Trewavas A.J. and Pais M.S. (1995). Calcium channel activity during pollen tube growth and reorientation. *Pl. Cell* **7**: 1173–1184.

Mascarenhas, J.P. (1993). Molecular mechanisms of pollen tube growth and differentiation. *Pl. Cell* **5**: 1303–1314.

Miller, W.B. and Ranwala, A.P. (1994). Characterization and localization of three soluble invertase forms from *Lilium longiflorum* flower buds. *Physiol. Pl.* **92**: 247–253

Plyushch, T.A., Willemse, M.T.M., Franssen-Verheijen, M.A.W. and Reinders, M.C. (1995). Structural aspects of *in vitro* pollen tube growth and micropylar penetration in *Gasteria verrucosa* (Mill.) H. Duval and *Lilium longiflorum* Thunb. *Protoplasma* **187**: 13–21

Reger, B.J., Chaubal, R. and Pressey, R. (1992a). Chemotropic responses by pearl millet pollen tubes. *Sexual Pl. Reprod.* **5**: 47–56.

Reger, B.J., Pressey, R. and Chaubal, R. (1992b). *In vitro* chemotropism of pearl millet pollen tubes to stigmatic tissue: a response to glucose produced in the medium by tissue-bound invertase. *Sexual Pl. Reprod.* **5**: 201–205.

Ruffner, H.P., Hürlimann, M. and Skrivan, R. (1995) Soluble invertase from grape berries: purification, deglycosylation and antibody specificity. *Pl. Physiol. Biochem.* (Montrouge) **33**: 25–31

Russell, S.D. (1992). Double fertilization. *Int. Rev. Cytol.* **140**: 357–388.

Tilton, V.R. (1977). The nucellar epidermis and micropyle of *Ornithogalum caudatum* (Liliaceae) with a review of these structures in other taxa. *Canad. J. Bot.* **58**: 1872–1884

Timmers, A.C.J., de Vries, S.C. and Schel, J.H.N. (1989). Distribution of membrane-bound calcium and activated calmodulin during somatic embryogenesis of carrot (*Daucus carota* L.). *Protoplasma* **153**: 24–29

Van Went, J.L. and Willemse, M.T.M. (1984). Fertilization, In: B.M. Johri (editor), Embryology of angiosperms, pp. 274–317, Springer Verlag, Berlin.

Vasil, I.K. (1987). Physiology and culture of pollen. *Int. Rev. Cytol.* **107**: 127–174.

Willemse, M.T.M. (1996). Progamic phase and fertilization in *Gasteria verrucosa* (Mill.) H. Duval: pollination signals. *Sexual Pl. Reprod.* **9**: 348–352.

Willemse, M.T.M. and Franssen-Verheijen, M.A.W. (1986). Stylar development in the open flower of *Gasteria verrucosa* (Mill.) H. Duval. *Acta Bot. Neerl.* **33**: 297–309.

Willemse, M.T.M. and Franssen-Verheijen, M.A.W. (1988). Pollen tube growth and its pathway in *Gasteria verrucosa* (Mill.) H. Duval. *Phytomorphology* **38**: 127–132.

Willemse, M.T.M. and Franssen-Verheijen, M.A.W. (1992) Pollen tube growth and ovule penetration in *Gasteria verrucosa* (Mill) H. Duval. In: E. Otaviani, D.L. Mulcahy, M. Sari Gorla and G. Bergamini Mulcahy (editors). Angiosperm pollen and ovules, pp. 168–173, Springer, New York.

Willemse, M.T.M. and Vletter, A. (1995). Appearance and interaction of pollen and pistil pathway proteins in *Gasteria verrucosa* (Mill.) H. Duval. *Sexual Pl. Reprod.* **8**: 161–167

Willemse, M.T.M., Plyushch, T.A. and Reinders, M.C. (1995). *In vitro* micropylar penetration of the pollen tube in the ovule of *Gasteria verrucosa* (Mill.) H. Duval and *Lilium longiflorum* Thunb.: conditions, attraction and application. *Pl. Sci.* **108**: 201–208.

Richards A.J. (1998). Lethal linkage and its role in the evolution of plant breeding systems. In: S.J. Owens and P.J. Rudall (Editors). Reproductive Biology, pp. 71–83. Royal Botanic Gardens, Kew.

LETHAL LINKAGE AND ITS ROLE IN THE EVOLUTION OF PLANT BREEDING SYSTEMS

A.J. Richards

Department of Agricultural and Environmental Science, Ridley Building, University of Newcastle upon Tyne NE1 7RU, UK.

Abstract

Chromosomes or parts of chromosomes protected from recombination will accumulate disadvantageous recessive mutants in association with other genes linked to these chromosomes. This may explain why apomicts which are non-recombinational on the female side and accumulate a large genetic load, but are pseudogamous rarely survive as diploids. When a non-recombinational chromosome segment normally occurs heterozygously, as for *Y* chromosomes (dioecy) and *S* chromosomes (heteromorphy), recessive mutants will be protected from selection and might increase in number and frequency. Evidence is presented that such *S*-linked mutants commonly occur in heteromorphic *Primula*, so that *SS* homozygotes occur very rarely. This finding may explain the association between recombinational homostyly and polyploidy, and the scarcity of recombinational short homostyles. It also has implications for the origin of heterostyly. If a newly arisen short-style mutant arose in a selfing homostyle in close linkage to a recessive lethal mutant, it would reach frequency equilibrium, and might satisfy conditions which would favour the later evolution of other heteromorphic characteristics.

Introduction

Outbreeding plants accumulate a genetic load of recessive mutants, many of which are harmful. Recessive mutants are not expressed phenotypically when they occur as heterozygotes, so that they are protected from the effects of natural selection in the sporophyte generation except when they occur homozygously (Charlesworth & Charlesworth, 1987). Consequently, each may persist at a low frequency in the population, depending on chance effects and mutation rates. However, a large proportion of this mutational load may not survive a sexual reproductive cycle in diploid angiosperms. Sex in diploids involves a haploid generation during which recessive genes are not protected from expression by dominants. As up to 70% of flowering plant genes may be phenotypically expressed during the haploid gametophyte generation (pollen-tube or embryo-sac; Ottaviano & Mulcahy, 1989; Mascarenhas, 1990), at least some mutants will be 'screened out' at this stage.

In contrast, polyploid plants carry many gene loci which are polysomic (present more than twice) in the sporophyte generation, so that they occur at least twice (disomically) in the gametophyte. For these loci, recessive mutants are much less likely to be expressed phenotypically in the gametophytes of polyploids, and so tend to be protected from selection during sexual reproduction. Also, where gene loci are present more than twice in polyploids, recessive mutants appear in the homozygous condition less frequently than they do in diploids. Consequently, each mutant is likely to persist at a higher frequency than in diploids, so that the genetic load of polyploids tends to be higher than that of diploids.

In the next section I shall consider briefly the consequences of this genetic load in apomicts, and in later sections I shall explore the possibility that a genetic load accumulates on the *S* chromosome of dimorphic species, and that *SS* inviability may have played a role in the evolution of dimorphic mating systems.

Polyploidy and apomixis

In most plants, recessive mutants do not remain linked to a particular chromosome or gene for many generations, as they are recombined at meiosis onto homologous chromosomes derived from other parental lines to form linkage associations with other genes. However, in the absence of meiotic recombination, for instance as a result of gametophytic apomixis, recessive mutants accumulate on the chromosomes on which they arise (Maynard Smith, 1978; Mogie, 1992). Apomicts also lack a meiotic segregation of the chromosomes, so that a mutant gene is inherited by all the offspring of a mother in which the mutation arises. As long as no haploid gametophytic stage occurs during reproduction (also true for polyploids) this recessive genetic load will remain protected from selection. However, the majority of aposporous apomicts rely on a pseudogamous fertilisation of the primary endosperm nucleus to occur before the endosperm can develop to nurture the seed (Asker, 1980; Richards, 1997). If pseudogamous apomicts are diploid, the recessive genetic load expressed in the haploid pollen tube should cause it to fail, so preventing the pseudogamous development of the endosperm and the subsequent development of the seed. Most gametophytic apomicts probably evolved from ancestors which required pseudogamous fertilisation of the primary endospermous nucleus, even if the endosperm is now able to develop autonomously (Mogie, 1992). Perhaps as a consequence, gametophytic apomicts are almost always polyploid (Asker & Jerling, 1992; Richards, 1997).

Obligate apomixis is an unusual condition (Gustafsson, 1946-47; Richards, 1997; Asker & Jerling, 1992). Many apomicts coexist in a genetic continuum with sexuals and are facultative, able to undertake limited sexual reproduction although sometimes only as male parents (Mogie, 1992). In facultatively apomictic *Taraxacum* (Richards, 1970; Muller, 1972; den Nijs et al., 1987), *Ranunculus auricomus* (Nogler, 1984), and *Panicum* (Savidan & Pernes, 1982), sexuals are commonly diploid, and apomicts almost invariably polyploid.

In contrast, sporophytic apomicts are mostly diploid (Richards, 1997). In these, sexual embryos commonly coexist with, or even stimulate the production of, asexual embryos. Consequently, sporophytic apomictic lines frequently encounter gametophytic selection so that harmful recessive mutants are unlikely to accumulate as a result of apomixis, and such lines can thrive as diploids.

Heterogametic chromosomes

In apomixis, where a reductional meiosis is absent, the whole genome generally avoids the effects of genetic recombination, so that many recessive mutants accumulate in association with particular genes. In sexual plants, chromosomes or parts of chromosomes which are protected from recombination, for instance as a result of inversion heterozygosity, are also likely to accumulate recessive mutants. The likelihood of a recessive mutant becoming associated with such a non-recombinant zone will depend on the size of the zone; the genetic load will be much higher if the whole chromosome is non-recombinant than if only a small part of the chromosome is so protected. A mutant within a non-recombinant zone will occur in association with a particular gene which also occurs in the protected area at a much higher frequency than if it had been freely recombined (i.e. it will 'hitchhike' with it; Maynard Smith, 1978).

Coadapted genes such as those which control primary and secondary gender linked together on sex chromosomes *(X* and *Y)* should only function correctly when protected from recombination (Darlington, 1932). Consequently, such linkage groups might form targets for non-recombinable recessive mutants (Charlesworth, 1978). In heteromorphic species too, coadapted genes occur in linkage groups (*S* and *s*; Dowrick, 1956; Lewis & Jones, 1993) and if these are also protected from recombination, they might become associated with mutants.

Recessive mutants which arise on a chromosome which normally occurs only in a heterogametic phase (the *Y* chromosome for gender control, and the *S* chromosome for floral dimorphy) will be protected from selection if a wild-type dominant occurs in the heterologous (*X* or *s*) chromosome. Consequently, mutants which arise on *Y* or *S* chromosomes might attain much higher frequencies than those arising on autosomes. As the homozygous genotypes *YY* or *SS* rarely if ever occur in most dioecious or floral dimorphy systems, recessive mutants linked to the *Y* or *S* chromosomes never normally occur homozygously, so they are protected from selection. Conversely, mutants linked to the *X* or *s* chromosome will be expressed in the *XX* or *ss* phase at the square of their frequency, and so that they can be selected against if disadvantageous.

There is abundant evidence that sex chromosomes, or linkage groups which control gender, are normally protected from recombination, so that intersexes rarely if ever occur (Westergaard, 1958). In contrast, there is no evidence that the *S/s* linkage group which controls floral dimorphy is protected from recombination. Indeed, recombinants within the dimorphy linkage group have been reported for several genera with floral dimorphy (*Limonium, Pemphis, Primula, Turnera* and *Armeria,* summarised in Lewis & Jones, 1993). However, *S/s* linkage group recombination seems to be an unusual occurrence, at least in *Primula* in which the dimorphy linkage group seems to include at least eight recombinable loci (Kurian & Richards, 1997). These loci are very tightly linked into a zone of the chromosome which Lewis & Jones (1993) estimate may be only 0.57 map units long. As these authors estimate that this linkage group may lie within 3 map units of the centromere, this zone may also suffer restricted rates of chiasma formation. As yet there is little information on linkage groups which control floral dimorphy in other genera. Nevertheless, if recombination within the *S/s* linkage group occurred commonly, floral dimorphy would be unlikely to survive unless non-recombinant phenotypes were much more successful that the recombinants, so it is reasonable to assume that *S/s* linkage group recombination is typically a rare event, perhaps due to the loci being closely linked.

As yet, only one example is known in which the dimorphy linkage group may be fully protected from recombination by inversion heterozygosity. In several species classified within *Primula* section *Muscarioides,* Bruun (1932) noted that the thrum (*Ss*) morph is heterozygous for an inversion which is homozygous in the (*ss*) pin morph.

Segments of *Y* and *S* chromosomes on which recessive mutants are normally protected from selection and recombination are expected to accumulate such mutants and each may attain a high frequency on these chromosomes. Thus, when exceptions to normal mating patterns occur amongst species which are dioecious or have a floral dimorphy of a kind which should produce some homozygous (*YY* or *SS)* offspring, it is expected that such offspring would be inviable. One consequence of this may be the rarity of *YY* homozygotes amongst subdioecious species with *X/Y* genetic control, but in which self-fertile *XY* hermaphrodites occur (Westergaard, 1958).

Homozygotic SS inviability in distylous Primulas.

Floral dimorphy is widespread, being found in 18 orders, 25 families and at least 160 genera of flowering plants. Typically these are insect-pollinated perennials with narrow or tube-shaped flowers and fused petals. Floral dimorphy usually describes conditions where genes controlling two mating-types are linked to several genes which control the morphology of floral organs, as a so-called coadaptive 'supergene' (e.g. Lewis & Jones, 1993). The majority of dimorphic species are distylous, as in *Primula,* where one mating-type has long styles ('pin') and the other short styles ('thrum') with, frequently, reciprocally placed anthers (Fig. 1; Ganders, 1979; Richards, 1997; Barrett *et al.,* 1993). In many instances, long-style plants have smaller, more numerous pollen and longer stigmatic papillae than those with short styles, and the long-style morph is usually under recessive (*ss*) control (Richards, 1997).

Pin and thrum *Primulas* invariably set less seed after illegitimate selfs or crosses, compared with that set after legitimate crosses made between the floral morphs (Darwin, 1877; Wedderburn & Richards, 1990). Nevertheless, the incompatibility imposed by the mating type is not absolute. Most *Primula* species can set some seed after illegitimate crosses or selfs amongst pins and amongst thrums. Pin selfs or illegitimate crosses frequently lead to true-breeding pin (*ss*) lines, especially in cultivation (Richards, 1993). After thrum (*Ss)* selfs or illegitimate crosses, 25% of the offspring might be expected to be thrum homozygotes *SS* which on further selfing would form true breeding thrum lines. However, true-breeding thrums seem to be extremely rare in *Primula*, having only been recorded once for a highly inbred horticultural line of *P. sinensis* Sabine (Mather & DeWinton, 1941). In this case, thrum homozygotes (*SS*) were estimated to be about 70% as viable as thrum heterozygotes (*Ss*), providing the first reported case of a lowered viability amongst *SS* homozygotes. This low viability may be caused by the *S*-linked genetic load.

Because illegitimate thrum × thrum crosses and thrum selfs tend to be relatively infertile, it is not surprising that reports of other conditions which may result from homogametic inviability come from self-fertile recombinant strains. Self-fertile homostyle primroses (*P. vulgaris* Huds.) apparently arose from distylous plants after recombination between the loci *G/g, A/a* and *P/p* in the *S/s* distyly linkage group (Dowrick, 1956). Heterozygous long homostyle primroses have the putative recombinant genotype *gAP/gap* (Richards, 1997; Kurian & Richards, 1997) having inherited the major part of the *S/s*-linkage group from the thrum (*S* = *GAP*) chromosome. Crosby (1949) quoted the unpublished results of Catcheside, showing that although homozygous long homostyles (*SS*) occur, these are only about 65% as viable as heterozygotes (*Ss*). It seems likely that part of the genetic load of recessive mutants on the *S* = *GAP* chromosome may have entered the recombinant long homostyle *gAP* chromosome, resulting in low viability amongst *gAP/gAP* homozygotes. The long homostyle chromosome has not become fixed in any primrose population, and Curtis & Curtis (1985) showed that it suffers from frequency dependent selection, populations never exceeding the ratio 80/20 long homostyle/pin. According to Curtis & Curtis' calculations (1985), low viability amongst *SS* homostyles can explain some, but not all of this selection, however.

The first proof of *SS* homozygote inviability in *Primula* comes from another self-fertile recombinant strain, termed P*, in the hybrid *P.* × *tommasinii* Godr. (Kurian & Richards, 1997). P* gives rise to plants which are superficially similar to pins (*ss*), but which carry at least three dominant alleles, including that for large thrum-type pollen, which seem to have been recombined from the *S* chromosome. Thus, they have the putative genotype *gaP /gap* where the *P/p* locus controls pollen size alone. Although normal thrums are *Ss*, normal thrum pollen is of a uniform, large size, so that the genetic control of pollen size is under sporophytic control with *S* dominant to *s*. However, the pollen grain size phenotype of P* plants is expressed gametophytically, half the grains of P* being the size typical of thrums, and half of the much smaller size typical of pins, the size distributions being non-overlapping. Both sizes of pollen grain grew equally well on P* and pin stigmas, i.e. the male mating type of P* was thrum. Consequently, the genetic control of male mating type together with that of pollen size and a female attribute have apparently been recombined from the thrum chromosome.

As P* plants all had pollen of two sizes, all were apparently heterozygous at a locus controlling pollen size so that 25% of the 189 offspring raised from P* selfs (*gaP/gap* × *gaP* × *gap*) should have been P* homozygotes *gaP/gaP* with uniformly large pollen. However, none were of this type. Also, none of the P* plants were true-breeding. When P* pollen was crossed onto pin stigmas, 50% of the offspring were invariably of the P* phenotype, showing that all P* were heterozygous (Table 1). When P* were selfed, offspring segregated at 2 P* : 1 pin, showing clearly that P* homozygotes were inviable.

TABLE 1. Inheritance of the P* phenotype in *Primula* × *tommasinii* after selfs and crosses onto pin mothers; bulked results for seedlings raised from ten different fathers and mothers in each case (after Kurian & Richards, 1997).

[1]Tested against a 2:1 ratio; [2] tested against a 3:1 ratio; [3] tested against a 1:1 ratio.

	Type of offspring			
Type of cross	P*	Pin	Total	X^2
P* self	50	31	81	0.88[1] NS 7.61[2] P<0.006 4.46[2] P<0.02
P × P*	139	133	272	0.13[3] NS

These results strongly suggest that one or more lethal recessive genes must have been recombined, together with several others, from the *S* chromosome onto the P* chromosome. Any alternative explanation is much more unlikely, requiring that that several new dominant mutants (or, just conceivably, one new mutant with manifold pleiotropic effects), and a lethal recessive mutant invariably occur in association, and never alone.

There are other, indirect, indications that a genetic load of disadvantageous recessive mutants linked to the *S* chromosome may be a typical feature of distylous *Primula.* For instance, 6% of the 426 *Primula* species are recombinant long homostyles (*gAP/gap*) (Wedderburn & Richards, 1992), but the reciprocal recombinant short homostyle morph (genotype *Gap/gap*) which should arise at the same frequency is much rarer, only one case being known (*P. septemloba* Franchet var. *minor* Ward; Richards, 1993). A comparable disparity is also found in distylous *Limonium* (one species is known to be long homostyle in comparison to 15 short homostyle species). *Limonium* differs from *Primula* and most other distylous genera in that the short-style morph is dominant to the long-style type. Lewis & Jones (1993) concluded that in *Limonium* (and by implication, *Primula*) a recessive deleterious gene is linked to the dominant female-determining allele *G* so that selfing homostyle recombinant chromosomes containing this allele (long homostyle *Gap* in *Limonium* and short homostyle *Gap* in *Primula*) are selected against in the homozygous phase. If this explanation is true, it suggests that deleterious mutants are generally linked to *G* rather than to *AP*, which might suggest that they arose on the *GAP* chromosome at an early stage in its evolution, as models suggest that short-style mutants *G* may have arisen before *A* and *P* (see below).

However, two other arguments can be adduced to explain the rarity of short homostyles in *Primula*. For instance, short homostyles with a stigma inserted in the thrum position within the floral tube, are more likely to suffer high levels of autogamous selfing than are long homostyles, which have been estimated to receive about 10% outcrossed pollen (Crosby, 1949; Piper & Charlesworth, 1986). This might result in inbreeding depression being particularly associated with the short homostyle chromosome. Also, short homostyles have an automatic genetic disadvantage compared to long homostyles, as long homostyles can act as male parents to coexisting pins, but short homostyles are incompatible as male parents to pins.

It is also possible that the overwhelming association between recombinational long homostyle selfers in *Primula* and allopolyploidy (80% of homostyle species are polyploid, but only 8% of heterostyle outcrossers are) is associated with *S* = *GAP*-linked mutants (Richards, 1997). Allopolyploid *Primulas* may have a high proportion of 'fixed heterozygosity', as in the hexaploid *P. scotica* (Glover & Abbott, 1995), which is presumably caused by cosegregation of homoeologous chromosomes derived from parental diploids which fail to form multivalents at meiosis. If mutant alleles coexisted in an allopolyploid with non-mutant homologous alleles residing on homoeologous chromosomes, this so-called 'fixed heterozygosity' would prevent deleterious genes linked to the recombined *S* = *gAP* chromosome from becoming effectively homozygous in polyploid selfers, although they would prevent the formation of diploid selfers.

SS Inviability and the Evolution of Distyly

Charlesworth & Charlesworth's (1979) model

Two models have been proposed which invoke quite different selective principles. The first (Charlesworth & Charlesworth, 1979) considered that distyly evolved in response to offspring fitness, by providing an escape from the inbreeding depression typical of plants which undergo bouts of self- fertilisation. They assumed that the founder would have been a selfing long homostyle whose offspring suffered inbreeding depression (Fig. 1). Only a short-style (thrum) mutant would have succeeded, starting from this condition; a short-anther (pin) mutant which arose first would still be highly selfed, giving rise to offspring of low fitness, and would in addition suffer reduced male fitness in comparison with its homostyle founder. Essentially, the system would be unstable, so that there would be no conditions in which a polymorphism of thrums and homostyles could equilibrate.

There are two main disadvantages to this model. Firstly, no examples of homostylous species with diallelic self-incompatibility are known, but there are a number of distylous species which lack diallelic self-incompatibility (Richards, 1997; Lloyd & Webb, 1993a). This suggests that morphological distyly may have usually evolved before the incompatibility, so that these examples may represent intermediate evolutionary conditions in which diallelic incompatibility has not evolved. Secondly, the thrum condition is in most cases dominant to the pin, suggesting that it evolved after, not before, the pin (Richards, 1997).

Lloyd & Webb model

Lloyd & Webb (1993a, b) suggested instead that distyly evolved from mutations which responded to selection for efficiency in male function (Fig. 2), rather than for offspring fitness. They suggested that the founder was potentially self-fertile, but bore flowers showing approach herkogamy (Fig. 2), i.e. of the 'pin' configuration, so that many of the offspring resulted from outcrosses. Their model is based on Darwin's (1877) assumption that pollen travels disassortatively to stigmas positioned in the flower at the same level as the anthers from which it originates. Thus, a short-styled ('thrum') mutant in this 'pin' founder would have been

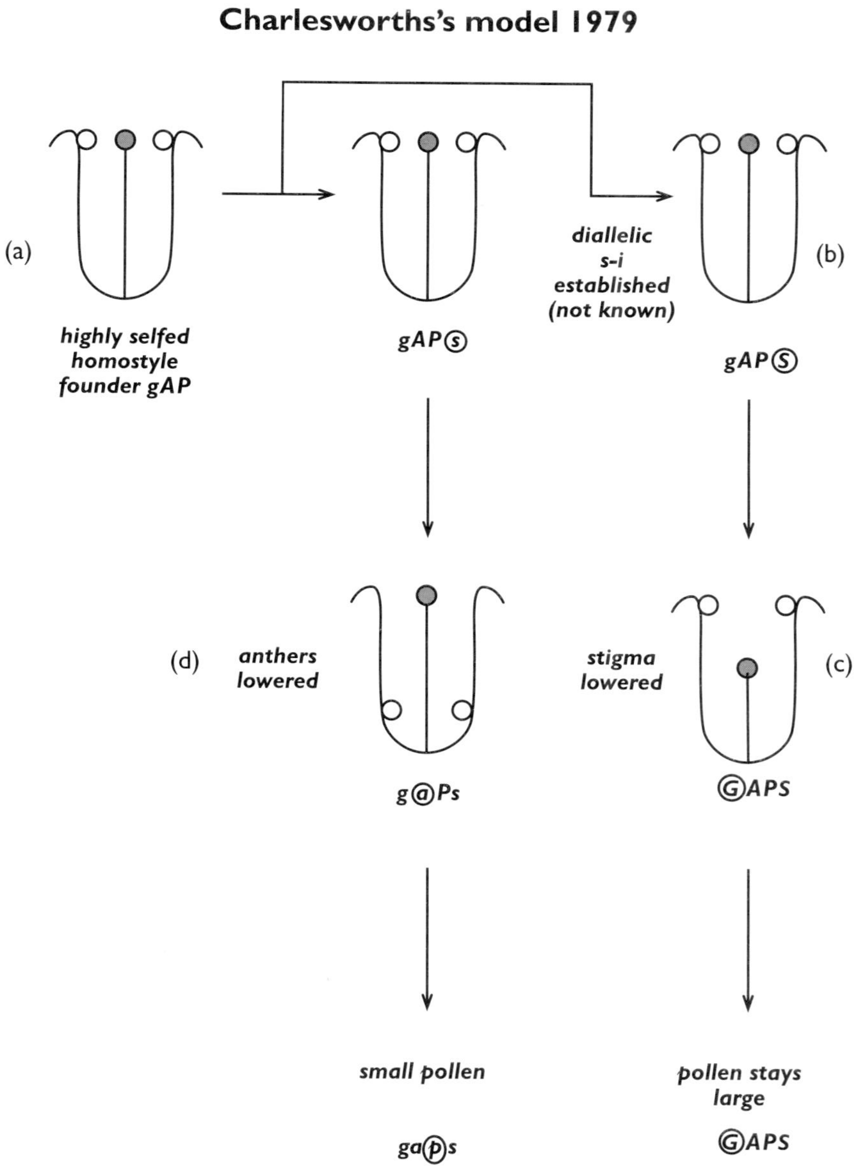

FIG. 1. The Charlesworth & Charlesworth (1979) model for the evolution of distyly. (a) a long homostylous self-fertile ancestor *gAP/gAP* develops a diallelic incompatibility (b) to give two mating types *gAPS* and *gAPs*. A mutation which lowers the stigma to the thrum position *GAPS* (c) would subsequently be favoured as more of its pollen would be transferred to cross-compatible long stigmas and less to incompatible short stigmas. A further mutation linked to *s* which lowered anthers of long styled plants into a pin position *gaPs* (d) would then succeed as a more efficient male parent to short styled thrums. A further mutation linked to the pin chromosome which reduced pollen volume and consequently increased pollen number *gaps* would then be favoured by male sexual selection.

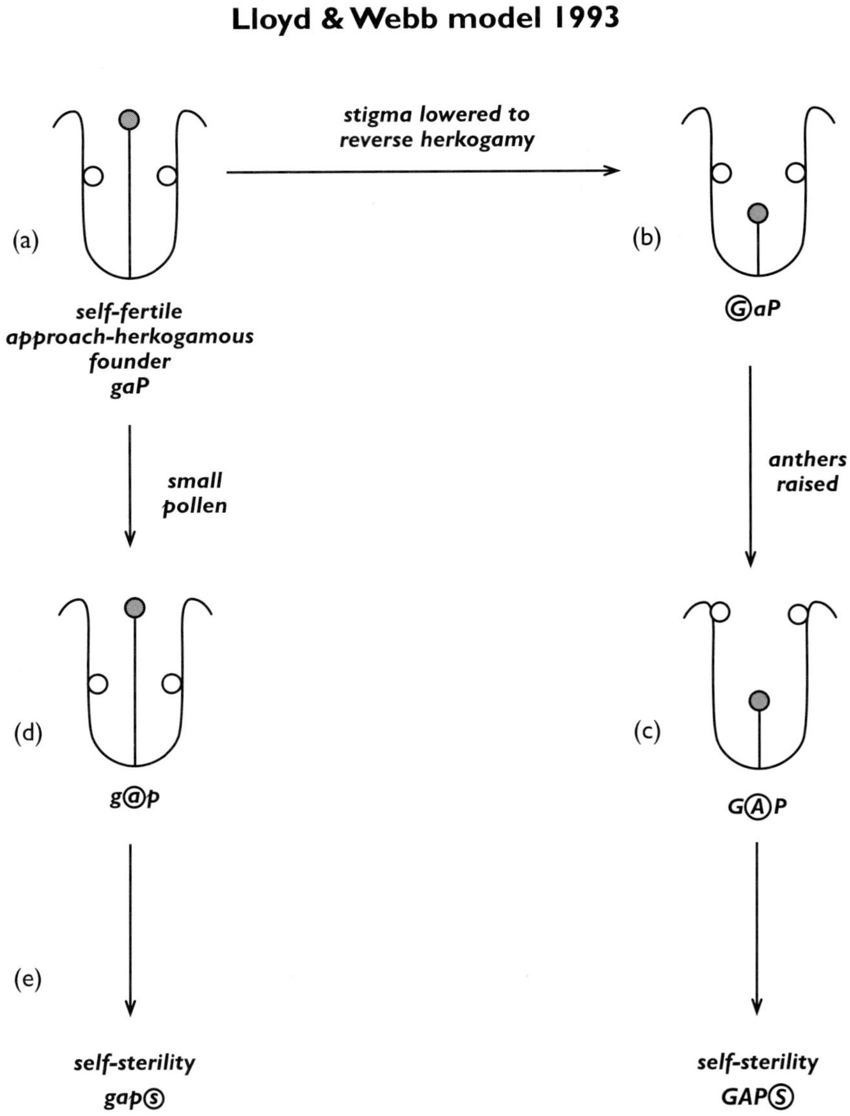

FIG. 2. The Lloyd & Webb (1993 a,b) model for the evolution of distyly. (a) the ancestral condition, approach herkogamy (*gaP/gaP*) in self-fertile but substantially outcrossing plants. (b) A stigma-height polymorphism (*GaP/gaP; gaP/gaP*) is selected for by selection for male function efficiency. (c) An anther height polymorphism (*GAP/gaP; gaP/gaP*) is subsequently favoured by increasing the efficiency of pollen flow to other-type stigmas in a frequency-dependent equilibrium.

A pollen size polymorphism (*GAP/gap; gap/gap*) (d) and a diallelic incompatibility (*GAPS/gaps; gaps/gaps*) (e) linked to these chromosomes would be favoured by male function selection, and selection for high levels of outcrossing respectively.

favoured as a more efficient male parent to the original pin morph (Fig.2). Unlike the Charlesworths' model, where thrum frequency is unstable, when the founder is 'pin' the male advantage imposed by the thrum condition would be frequency dependent; as thrums become more frequent, the pin stigmas on which thrum pollen can outcompete pin pollen become scarcer, reducing the male fitness gain of thrums. Consequently, the thrum condition would be stable and would equilibrate in the population. A further mutation linked to the pin which further lowered anthers to a position in the flower equivalent to that of the thrum stigma would also be favoured by male sexual selection, and would also reach equilibrium through frequency dependent selection. In their model, diallelic incompatibility would have evolved later than the distyly, responding to selection which favoured further reductions in selfing above those resulting from the original herkogamy.

An alternative model for the evolution of distyly

The Lloyd and Webb model has many attractive features, and may explain the evolution of distyly in most cases. However, after comparing species classified in *Primula* section *Sphondylia*, Al Wadi & Richards (1993) suggested that distylous members of this genus may at times have evolved from selfing homostyles (*gAP/gAP*) rather than from outcrossing herkogamous (*gaP/gaP*) ancestors, in the manner originally suggested by Charlesworth & Charlesworth (1979). *P. boveana* Decesn. may represent an intermediate evolutionary condition, for in this species thrum and long homostyle morphs coexist in a self-fertile species, and the thrum chromosome does not predominate (Fig. 3). In the very limited wild population sample available for this species (n = 20), 55% of individuals were interpreted as thrum and 45% as long homostyle.

In order to explain the evolution of distyly from a homostylous ancestor in the absence of incompatibility, the key question is whether the frequency of a newly arising successful outcrossing thrum chromosome *GAP* should necessarily be unstable in the absence of frequency dependent selection.

To invoke inbreeding depression amongst the offspring of a selfing homostyle, it is necessary to suggest that these homostyles carried a genetic load of deleterious recessive genes (Charlesworth & Charlesworth, 1987). If the new thrum mutant arose in tight linkage or in conditions of restricted recombination with one or more of these deleterious recessives, the chromosome on which it arose, rendered successful by the high fitness of its outcrossing offspring, would nevertheless rarely if ever occur in the homozygous condition (Fig. 4). The extent to which the new thrum chromosome *lGAP* (where *l* denotes one or more recessive genes of deleterious effect) could recombine its deleterious genetic load with the original long homostyle chromosome *gAP* in heterozygotes would depend at least in part on the linkage of *l* and *G*. If this was tight, thrums would usually occur only as heterozygotes, so that the thrum chromosome *lGAP* would occur in equilibrium with the homostyle chromosome *LgAP*, as in *P. boveana*. This condition should favour a mutant arising in the long homostyle chromosome *LgAP* which lowered the anther position in the flower to that equivalent to the thrum stigma (*LgaP*) as envisaged by Charlesworth & Charlesworth (1979) and Lloyd & Webb (1993a). Plants homozygous for this 'pin' chromosome (*LgaP/LgaP*) would be fitter than long homostyles *LgAP/LgAP* or *LgAP/LgaP* as they would be more efficient pollen donors to thrums than their homostyle ancestors and, as a result of their approach herkogamy, a higher proportion of their offspring than those of their homostyle ancestors would result from outcrosses (Fig. 4). Consequently, the 'pin' chromosome *LgaP* should invade at the expense of the homostyle chromosome *LgAP*, to establish a frequency-dependent equilibrium with the thrum chromosome *lGAP* as detailed by Lloyd & Webb (1993a).

Once this distyly had evolved, a further mutation linked to the pin chromosome *LgaP* which reduced pollen size and consequently increased pollen number in pin anthers *Lgap* would also

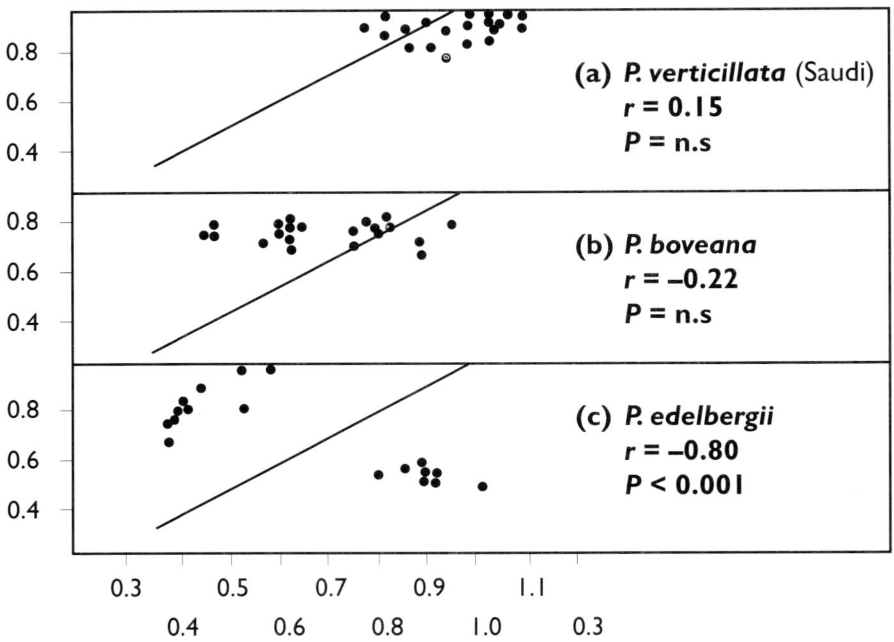

FIG. 3. The development of distyly, as apparently illustrated by three self-fertile species classified within *Primula* subgenus *Sphondylia*. Mean style length relative to corolla tube length (x axis); mean anther position relative to corolla tube length (y axis). (a) The long homostylous *P. verticillata* from southern Arabia. (b) *P. boveana* endemic to the Sinai, which is dimorphic for style length, but monomorphic for anther position, so that 'thrums' and 'long homostyles' coexist. (c) The fully distylous *P. edelbergii* from Afghanistan. (After Al Wadi & Richards, 1993).

have been favoured by sexual selection for male function in pins. A smaller proportion of pollen grains are removed from the 'hidden' pin anthers in comparison with thrum anthers, and the thrum stigma, in the corresponding position in the flower to the pin anther receives a smaller proportion of the available pollen (Piper & Charlesworth, 1985). In fact there are examples of distylous plants which lack an anther position dimorphy (so that pins, putatively, are *LgAp/LgAp)*, or which lack a pollen size dimorphy (pins, putatively are *LgaP/LgaP)*, so that neither of these modifications to male function in pins have proved essential to the evolution of distyly in every case (Richards, 1993, 1997).

Conclusion

The accumulation of mutants can influence the evolution of breeding systems whose function depends on the limitation of recombination. It has long been recognised that harmful recessive mutants tend to accumulate on the heterogametic *Y* chromosome on which recombination is necessarily restricted, but it has not been acknowledged that an analogous accumulation might be expected to occur on the heterogametic *S* chromosome of dimorphic breeding systems. As yet there is little evidence that recombination is in fact restricted within the *S* supergene, and

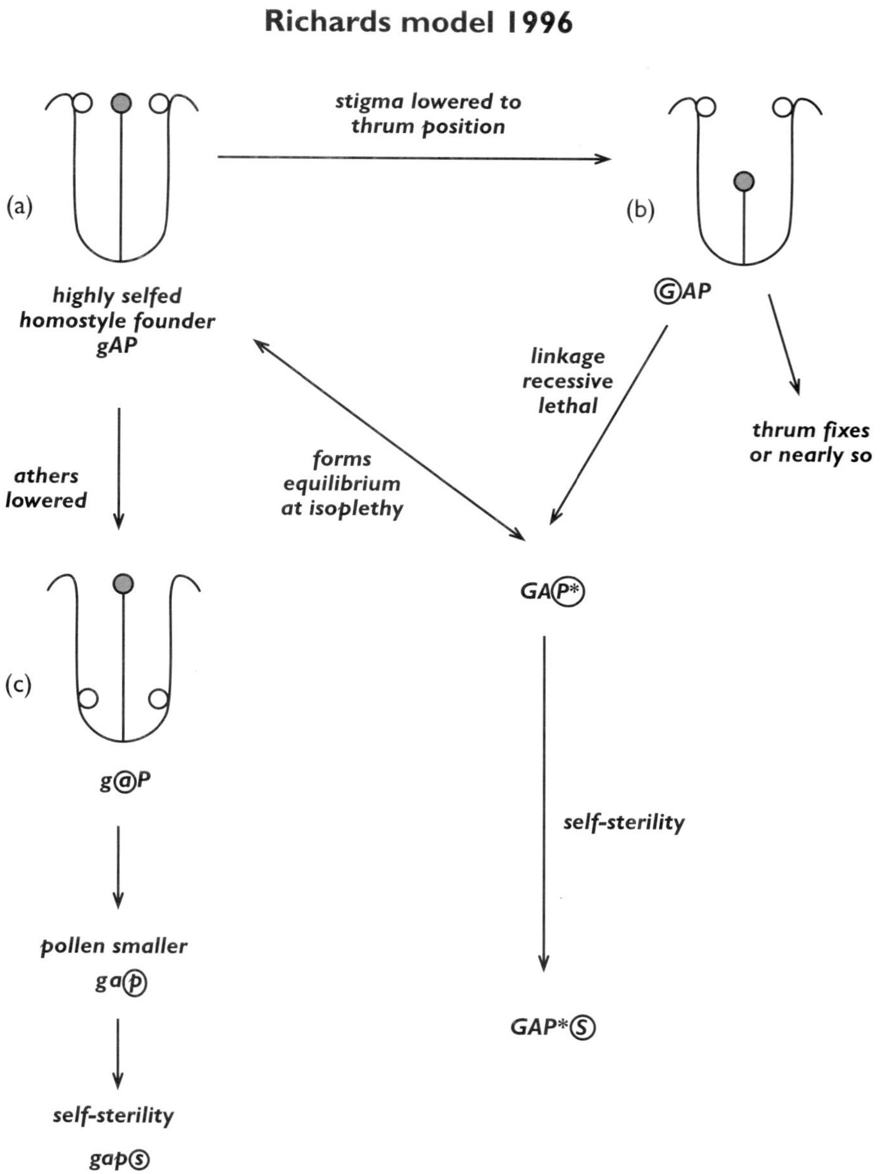

FIG. 4. The Richards scheme for the evolution of distyly. (a) a long homostylous self-fertile ancestor *gAP/gAP* carries a genetic load of one or more deleterious recessive mutants *lgAP/LgAP* at low frequency. (b) a short-style mutant *G* arises in close linkage to, or in restricted recombination with *l*, so that an equilibrium of thrums *lGAP/LgAP* and long homostyles *LgAP/LgAP* is established in the absence of inviable *lGAP/lGAP* homogametes. (c) Selection for male function efficiency amongst long homostyles *LgAP/LgAP* favours the establishment of a short anther mutation *a* in linkage to give rise to pins *LgaP/LgaP*. A pollen size polymorphism (*lGAP/Lgap;Lgap/Lgap)* and a diallelic incompatibility (*lGAPS/Lgaps; Lgaps/Lgaps*) linked to these chromosomes would be favoured by male function selection, and selection for high levels of outcrossing respectively.

recombinants within this linkage group have been reported. Nevertheless, such recombinants are apparently rare, and it seems likely that dimorphic breeding systems would have been unable to persist if challenged by free recombination.

Deleterious recessives indeed appear to be linked to the *S* chromosome in some species of *Primula*. In some cases these lead to a complete inviability of the *SS* homozygote; in others *SS* homozygotes have lowered viability in comparison with heterozygotes *Ss*. Indirect evidence suggests that deleterious gene *S*-linkage may be common to most primulas, and may occur in other genera with floral dimorphy.

Because a restriction of recombination within the *S* supergene will inevitably lead to an accumulation of harmful recessives within the region of restriction, it is impossible to be sure whether deleterious gene *S*-linkage has been involved in the evolution of distyly, or whether it has merely accumulated as a consequence of the restriction of recombination within the *S* linkage group implicit in the successful function of distyly. This is a classic 'chicken or egg' situation. However, deleterious gene *S*-linkage may satisfy conditions favouring S chromosome equilibrium. Such a population in equilibrium between self-fertile thrum and long homostyle individuals is apparently found in *Primula boveana* (Fig. 3). The establishment of such an equilibrium would encourage the later evolution of such features as low anther position and small numerous pollen typical of the 'pin' morphs of many distylous species.

Acknowledgements

I should like to thank my former research associates Valsa Kurian and Hussen Al-Wadi for producing much of the data which led to the development of ideas reported in this paper. I should also like to thank an anonymous reviewer.

References

Al Wadi, H. and Richards, A.J. (1993). Primary homostyly in *Primula* L. subgenus *Sphondylia* (Duby) Rupr. and the evolution of distyly in *Primula*. *New Phytol.* **124**: 329–338.

Asker, S. (1980). Gametophytic apomixis: elements and genetic regulation. *Hereditas* (Lund) **93**: 277–293.

Asker, S. and Jerling, L. (1992). Apomixis in plants. CRC Press, Boca Raton, Florida.

Barrett, S.C.H. (1993). Heterostylous genetic polymorphisms: model systems for evolutionary analysis. In: S.C.H. Barrett (editor). Evolution and function of heterostyly, pp. 1–29, Springer-Verlag, Berlin.

Bruun, H.G. (1932). Cytological studies in *Primula*. *Symb. Bot. Upsal.* **1**: 1–239.

Charlesworth, B. (1978). Model for evolution of Y chromosomes and dosage compensation. *Proc. Natl. Acad. Sci. USA* **75**: 5618–5622.

Charlesworth, D. and Charlesworth, B. (1979). A model for the evolution of distyly. *Amer. Naturalist* **114**: 467–498.

Charlesworth, D. and Charlesworth, B. (1987). Inbreeding depression and its evolutionary consequences. *Annual Rev. Ecol. Syst.* **18**: 237–268.

Crosby, J.L. (1949). Selection of an unfavourable gene complex. *Evolution* **3**: 212–230.

Curtis, J. and Curtis, C.F. (1985). Homostyle primroses revisited. 1. Variation in time and space. *Heredity* **54**: 227–234.

Darlington, C.D. (1932). Recent advances in cytology. Churchill, London.

Darwin, C. (1877). The different forms of flowers on plants of the same species. John Murray, London.

Dowrick, V.P. (1956). Heterostyly and homostyly in *Primula obconica*. *Heredity* **10**: 219–236.
Ganders, F.R. (1979). The biology of heterostyly. *New Zealand J. Bot.* **17**: 607–635.
Glover, B.J. and Abbott, R.J. (1995). Low genetic diversity in the Scottish endemic *Primula scotica* Hook. *New Phytol.* **129**: 147–153.
Gustafsson, A. (1946–47). Apomixis in higher plants. I-III. *Acta Univ. Lund.* **42**: 1–67; **43**: 69–179, 183–370.
Kurian, V. and Richards, A.J. (1997). A new recombinant in the 'S' supergene in *Primula*. *Heredity* **78**: 383–390.
Lewis, D. and Jones, D.A. (1993). The genetics of heterostyly. In: S.C.H. Barrett (editor). Evolution and function of heterostyly, pp. 129–150, Springer-Verlag, Berlin.
Lewis, K.R. and John, B. (1963). Chromosome marker. Churchill, London.
Lloyd, D.G. and Webb, C.J. (1993a). The evolution of heterostyly. In: S.C.H. Barrett (editor). Evolution and function of heterostyly, pp. 151–178, Springer-Verlag, Berlin.
Lloyd, D.G. and Webb, C.J. (1993b). The selection of heterostyly. In: S.C.H. Barrett (editor). Evolution and function of heterostyly, pp. 179–209, Springer-Verlag, Berlin.
Mascarenhas, J.P. (1990). Gene activity during pollen development. *Annual Rev. Pl. Physiol. Pl. Molec. Biol.* **41:** 317–338.
Mather, K. and de Winton, D. (1941). Adaptation and counter-adaptation of the breeding system in *Primula*. *Ann. Bot.* n.s. **5**: 297–311.
Maynard Smith, J. (1978). The evolution of sex. Cambridge University Press, Cambridge.
Mogie, M. (1992). The evolution of asexual reproduction in plants. Chapman Hall, London.
Muller, U. (1972). Zytologisch-embryologische Beobachtungen an *Taraxacum* Arten aus der Sektion *Vulgaria* Dahlst. in der Schweiz. *Ber. Geobot. Inst. ETH. Stiftung Rubel* **41**: 48–55.
Nijs, J.C.M. den, Menken, S.B.M. and Vlot, L. (1987). Gene flow in a di-triploid mixed stand of *Taraxacum* (Asteraceae) in the Odenwald, BRD, as measured by isozyme analysis. In: W. Greuter, B. Zimmer and H.D. Behnke (editors). Abstracts 14th International Botanical Congress, Berlin, July 24 to August 1 1987, p. 307, [The Congress], Berlin.
Nogler, G. A. (1984). Genetics of apospory in apomictic *Ranunculus auricomus*. V. Conclusion. *Bot. Helv.* **94**: 411–422.
Ottaviano, E. and Mulcahy, D.L. (1989). Genetics of angiosperm pollen. *Advances Genet.* **26:** 1–64.
Piper, J.G. and Charlesworth, B. (1986). The evolution of distyly in *Primula vulgaris*. *Bot. J. Linn. Soc.* **29**: 123–137.
Richards, A.J. (1970). Eutriploid facultative agamospermy in *Taraxacum*. *New Phytol.* **69:** 761–774.
Richards, A.J. (1993). *Primula*. Batsford, London.
Richards, A.J. (1997). Plant breeding systems (ed. 2). Chapman Hall, London.
Savidan, Y. and Pernes, J. (1982). Diploid-tetraploid-dihaploid cycles and the evolution of *Panicum maximum* Jacq. *Evolution* **36**: 596–600.
Wedderburn, F.M. and Richards, A.J. (1990). Variation in within-morph incompatibility inhibition sites in heteromorphic *Primula* L. *New Phytol.* **116**: 149–162.
Wedderburn, F.M. and Richards, A.J. (1992). Secondary homostyly in L.; evidence for the model of the 'S' supergene. *New Phytol.* **121**: 649–655.
Westergaard, M. (1958). The mechanism of sex determination in dioecious flowering plants. *Advances Genet.* **9**: 217–281.

Ronse Decraene, L.P. and Smets, E.F. (1998). Meristic changes in gynoecium morphology, exemplified by floral ontogeny and anatomy. In: S.J. Owens and P.J. Rudall (Editors). Reproductive Biology, pp. 85–112. Royal Botanic Gardens, Kew.

MERISTIC CHANGES IN GYNOECIUM MORPHOLOGY, EXEMPLIFIED BY FLORAL ONTOGENY AND ANATOMY

LOUIS P. RONSE DECRAENE AND ERIK F. SMETS

Katholieke Universiteit Leuven, Laboratory of Plant Systematics, Botanical Institute, Kardinaal Mercierlaan 92, B-3001 Heverlee, Belgium

Abstract

A survey of parameters related to carpel number and insertion is given for angiosperms. Using examples from floral ontogeny and anatomy, and a literature survey it is demonstrated that number, position and increases versus reductions of carpels follow settled trends with systematic significance. Different pathways can be postulated starting from isomerous, syncarpous gynoecia. Reductions of carpels are analysed, with special emphasis on pseudomonomerous gynoecia. A secondary increase in carpel number may occur by an addition of extra whorls, by the protrusion of false septa, or by polymerisation within an existing whorl. The basic carpel number in several families (monocots, Magnoliidae and part of Hamamelidae excluded) is five, and reductions can be visualised leading to 4, 3, 2, and 1 carpels. Loss of carpels is not random, but is correlated with changes affecting the whole flower, especially the androecium. Secondary increases are related to a mainly lateral multiplication of stamens, or to changes in merosity of the whole flower. In Magnoliidae, part of Hamamelidae, and monocots, variation in carpel number is high and evolved mainly along a trimerous or dimerous pattern, but trends run along lines similar to other angiosperms.

Introduction

Contrary to morphological interpretations of the carpel, carpel merosity has not been approached in a consistent way. Information is most often limited to an enumeration of the number of carpels in the flower (e.g. Cronquist, 1981; Endress, 1990; Taylor, 1991). Such summing-up bears little information for comparative morphological studies and can sometimes be highly misleading. Since Eichler's Blüthendiagramme (1875-1878), a masterwork in floral morphology, few recent studies have evaluated the exact position of carpels in the flower, and certainly not on a large scale. One reason for this may be that carpel position can vary, accommodating to the available space in the flower (e.g. Leins, 1964; Eckert, 1966; van Heel, 1966; Gelius, 1967). For example, within a single family an isomerous gynoecium may have carpels in an antepetalous or antesepalous position (e.g. Sterculiaceae, Caryophyllaceae). Eckert (1966) argued that carpel primordia arise as far as possible from the floral centre, viz. where there is maximum space for their initiation. Therefore, it is the level of development of the preceding androecial whorl that determines the position of the carpels. A second reason for the neglect in studying carpel merosity is that carpels are more subject to reduction than other organs. Indeed, while other organ whorls may be isomerous, carpels are

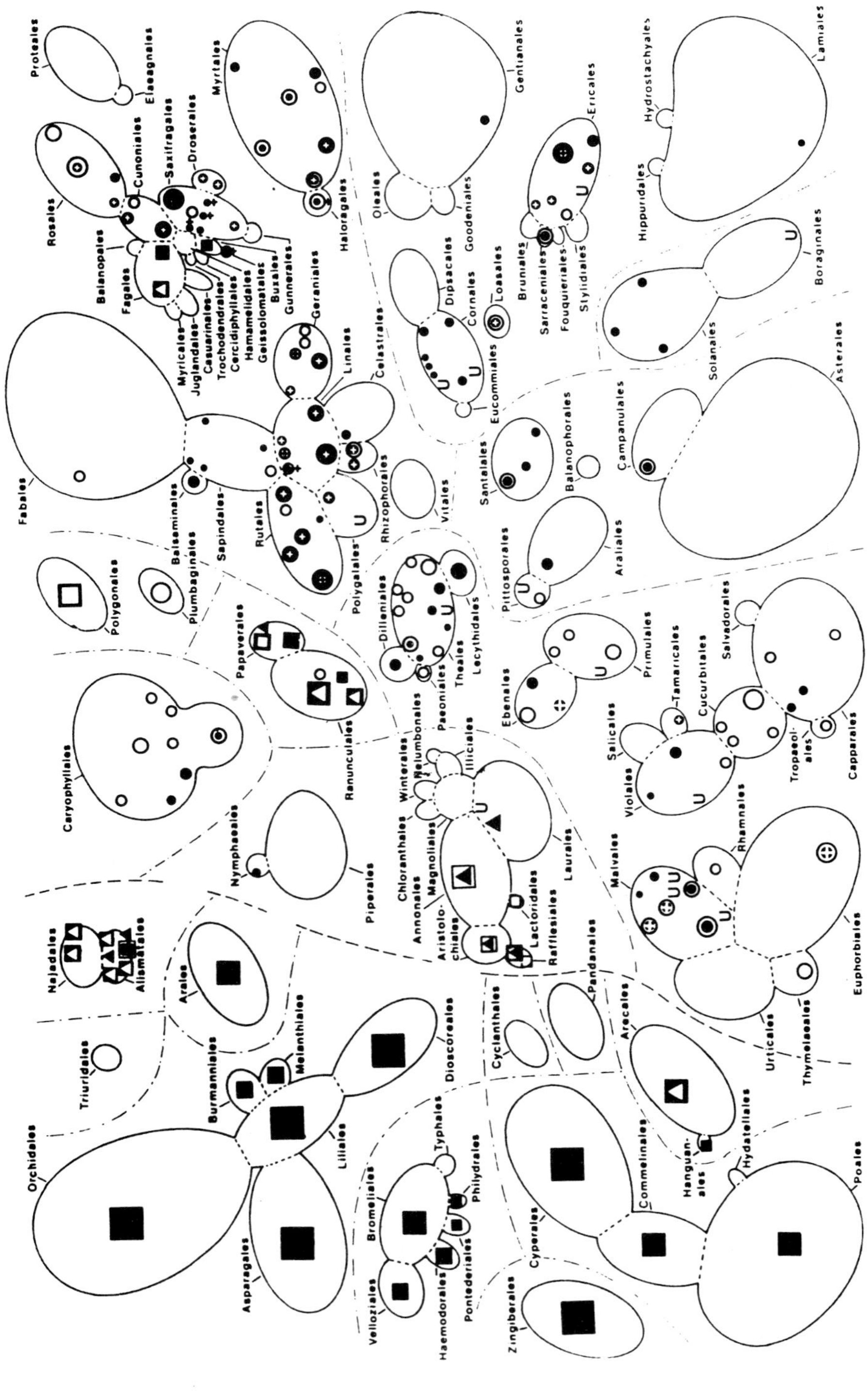

Proteales
Elaeagnales
Myrtales
Gentianales
Hydrostachyales
Lamiales
Rosales
Cunoniales
Saxifragales
Droserales
Haloragales
Oleales
Goodeniales
Ericales
Hippuridales
Balanopales
Fagales
Myricales
Juglandales
Casuarinales
Trochodendrales
Cercidiphyllales
Hamamelidales
Geissolomatales
Buxales
Gunnerales
Geraniales
Dipsacales
Cornales
Loasales
Bruniales
Sarraceniales
Fouquieriales
Stylidiales
Boraginales
Linales
Celastrales
Eucommiales
Solanales
Asterales
Fabales
Balsaminales
Plumbaginales
Sapindales
Rutales
Rhizophorales
Vitales
Santalales
Balanophorales
Campanulales
Polygonales
Polygalales
Dilleniales
Paeoniales
Theales
Lecythidales
Pittosporales
Araliales
Primulales
Salvadorales
Tamaricales
Cucurbitales
Papaverales
Ranunculales
Ebenales
Salicales
Violales
Tropaeolales
Capparales
Caryophyllales
Winterales
Nelumbonales
Illiciales
Nymphaeales
Chloranthales
Annonales
Magnoliales
Laurales
Malvales
Rhamnales
Piperales
Aristolochiales
Lactoridales
Rafflesiales
Euphorbiales
Najadales
Alismatales
Arales
Pandanales
Arecales
Urticales
Thymelaeales
Triuridales
Burmanniales
Melanthiales
Dioscoreales
Cyclanthales
Orchidales
Liliales
Typhales
Philydrales
Commelinales
Hanguanales
Hydatellales
Poales
Asparagales
Bromeliales
Velloziales
Haemodorales
Pontederiales
Cyperales
Zingiberales

often reduced to three, two or a single unit. This reduction is also related to a stronger tendency to fusion, possibly because space is more confined in the upper regions of the flower, where carpels occur.

A major difference in carpel merosity occurs between Magnoliids-monocots on the one hand, and Caryophyllids, Dilleniids, Asterids and Rosiids on the other. The former have spiral flowers with fluctuating numbers, or an essentially trimerous or dimerous *Bauplan*, and petals that are mostly of tepaline origin (Fig. 2). Gynoecia are often apocarpous and the number of carpels ranges from one to almost 2000 (Endress, 1990). The latter are pentamerous or tetramerous and usually have petals with a different morphological nature than the sepals (Fig. 3). Gynoecia are mostly syncarpous and have between two and five carpels. This number is reduced compared to apocarpous gynoecia, because of the positional constraints of whorls and the building of a functional compitum (Endress, 1990, 1994). These differences must be taken into account, as the position of the carpels depends on these parameters. In other words, a pentamerous flower with an isomerous gynoecium in antesepalous position will not be derived from the same ancestry as a trimerous flower with an isomerous gynoecium opposite the outer tepals (see also Ronse Decraene & Smets, 1994).

It is important to evaluate the process of carpel initiation, its development and its mature stages to provide basic material for comparative studies. Therefore, in a review of carpel merosity, based on ontogenetic and anatomical evidence, two questions take priority: (1) Can the number of carpels and their position in the flower have potential systematic value? In other words, can we find guidelines for evaluating phylogenetic trends in changes in carpel merosity? (2) Which carpel number and position is plesiomorphic for a taxon, and which is derived?

Methods

The use of Dahlgrenograms (Dahlgren G., 1989a,b) to plot the distribution of characters in the Magnoliatae has not been practised widely, although it has the potential to test the value of characters in a convincing way. We have used Dahlgrenograms extensively for plotting androecial characters (see e.g. Ronse Decraene & Smets, 1993, 1995). For the gynoecium a similar methodology can be applied. The number and position of carpels depend upon the merosity of the flower and upon the position of other organs, especially the androecium (Ronse Decraene & Smets, 1994). Based on our own observations on available material and from a study of the literature, the following parameters were analysed and eventually plotted on Dahlgrenograms:

1. carpels isomerous with the rest of the flower (i.e. having the same number of parts as other floral whorls and corresponding to the merosity of the flower; Fig. 1)
antepetalous, antesepalous, opposite outer tepals, opposite inner tepals or alternitepalous

FIG. 1. Dahlgrenogram (Dahlgren, 1989b) depicting the distribution of isomerous gynoecia in the angiosperms. Hollow circles: carpels antesepalous. Full circles: carpels antepetalous. Superposed circles: both antesepalous and antepetalous carpels occurring in the same family. Full squares: carpels opposite outer tepals. Hollow squares: carpels opposite inner tepals or alternitepalous. Triangles: carpels opposite both outer and inner tepals. U: position of carpels unknown. Crosses: presence of obdiplostemony.

TABLE 1. Examples of known taxa with a probable duplication of number of carpels compared to the basic plan of previous whorls.

Family	Genera or species mentioned	Number of carpels in the family	Position of carpels	Reference
Neuradaceae	*Neurada*	10	one whorl, alternating with sepals and petals (false septa)	Murbeck (1916), Ronse Decraene & Smets (1996)
Leeaceae (Vitaceae)	*Leea*	2-3(4)	4-6(8) locules (false septa)	Gerrath *et al.* (1990)
Ericaceae	*Gaylussacia*	(2)-4-5(-10)	10, in one whorl, alternating with sepals and petals (false septa)	Palser (1961)
Ebenaceae	–	2-10	6-8(-10)-locular with 3-4(-5) styles (false septa)	Baillon (1892), Saunders (1939), Ng (1991)
Zygophyllaceae	*Kallstroemia*	5-10(-12)	10(-12) in one whorl	Engler (1931)
Tiliaceae	*Clappertonia ficifolia*	(2-)5-8-(20)	8 carpels alternating with sepals and petals	van Heel (1966)
Sterculiaceae	*Ruizia*	(1-)5(-60)	apparently 10 carpels in one whorl	van Heel (1966)
Ochnaceae	*Ochna*	2-5-10-(15)	10, arranged in pairs	Saunders (1939)
Empetraceae	*Empetrum*	2-9	in one whorl, double (triple) the number of other whorls	Eichler (1878), Saunders (1939), Cronquist (1981)
Dilleniaceae	*Dillenia, Hibbertia crenata*	(1-)5(-20)	two whorls of five (H); all carpels in one whorl (D)	Payer (1857), Wilson (1965, 1973)
Linaceae	–	(2)3-5	partially 10-locular (false septa)	Narayana & Rao (1976a, b, 1977a, b)
Theaceae	*Laplacea*	(2)3-5(-10)	?	Melchior (1925)
Araliaceae	*Schefflera, Reynoldsia*	(1-)2,3,5 (-100)	double the number of other organs	Eyde & Tseng (1971)
Solanaceae	*Nicotiana quadrivalvis, Lycopersicon esculentum*	2(3-5),>5	8-10 in two whorls or 4 diagonally placed (N); more than 10 in one whorl (L)	Eichler (1878)
Coriariaceae	*Coriaria*	5(-10)	unknown, apparently in a single whorl	Cronquist (1981)
Nolanaceae	*Nolana*	(3)5(10)	2 (more) concentric series of 5, "originating phyletically by tangential division" (Cronquist 1981: 894)	Saunders (1939), Cronquist (1981)

TABLE 1 continued

Celastraceae	*Siphonodon*	2-5(-10)	10, in two whorls	Croizat (1947, in van Heel, 1978)
Malvaceae	*Malvaviscus, tribe Ureneae*	(1-)5(-many)	10, in two whorls	Eichler (1878), van Heel (1966, 1978), Crozier & Tucker (1993)
Punicaceae (Lythraceae)	*Punica*	7-9(-15)	1-3 whorls (5+5, 5+3, 5+5+3)	Payer (1857), Eichler (1878), Figs 12-13
Myoporaceae	?	2	divided by supernumerary partitions in (3) 4-10 uniovulate compartments	Cronquist (1981)
Plantaginaceae	*Plantago*	2	appearing 4-locular (false septa)	Cronquist (1981)
Verbenaceae	*Duranta, Geunsia*	2,(3,4,5)	8 (D)-10 (G) locular (4-5 false septa)	Bocquillon (1861), Cronquist (1981)
Convolvulaceae	*Mina, Calonyction*	2(3-5)	four-locular (false septa)	Weberling (1989)
Brassicaceae	*Nasturtium hispidum, N. globosum*	2	Four-carpellate (division of placentas?)	Eichler (1878), Schulz (1936)
Callitrichaceae	*Callitriche*	2	false septa	Leins & Erbar (1988)
Loganiaceae	*Anthocleista*	2-3	false septa	Cronquist (1981)
Bignoniaceae	*Tourrettia*	2	false septa	Cronquist (1981)
Lennoaceae	?	6-14	false septa (flowers 5-10-merous)	Cronquist (1981)
Goodeniaceae	*Scaevola porocarya*	2-4	2 lateral locules slightly lower than 2 median ones.	Cronquist (1981)
Pedaliaceae	*Sesamen, Josephinia*	2 (4?)	2 false septa, appearing late in ontogeny, in *Josephinia* with 8 uniovulate locelli	Singh (1960), Cronquist (1981)
Juglandaceae	*Juglans*	2(-3)	seemingly 4-8 locular (false septa)	Weberling (1989)
Batidaceae	*Batis*	2	4-locular (false septa)	van Heel (1958), Eckardt (1959)
Cistaceae	*Cistus ladaniferus*	3(5-10)	increase to ten carpels	Janchen (1925)
Rhizophoraceae	*Crossostylis*	2-5(-6),(-20)	stigmatic lobes fused per two, or four groups of stigmatic lobes	Setoguchi *et al.* (1996)

2. carpel number lower than the number of parts of other whorls in the flower (Fig. 20)
four carpels diagonally or orthogonally inserted
three carpels inserted with three orientations
two carpels, medianly or transversally placed
a single carpel, directed adaxially, abaxially, or obliquely (pseudomonomerous, viz. with the presence of sterile carpels or truly monomerous, viz. without traces of other carpels; Fig. 21, Table 3)

3. carpel number exceeding the number of parts of other whorls
carpel number double the number of parts of other whorls (inserted in one whorl or in two alternating whorls; Table 1)
carpel number high (with acyclic insertion, inserted in a single whorl, or in multiple whorls; Table 2)

It may sometimes be difficult to recognise the position of the carpels for a number of families, especially among Magnoliidae, which have a low synorganisation with a high and variable number of parts (Endress, 1990). In taxa with a whorled arrangement of the inner organs and with fewer parts, carpel position and number tends to be more or less stable, or changes occur along well-settled patterns.

Isomerous gynoecia

When the number of carpels is identical to the number of parts of other whorls of organs in the flower, one can describe the gynoecium as isomerous. In pentamerous or tetramerous flowers carpels are inserted opposite the sepals or the petals (Figs 1, 3). Similarly in trimerous flowers with two tepaline whorls, carpels may be situated opposite the inner tepals (often distinguished as petals), opposite the outer tepals, or in alternitepalous position (Figs 1, 2). We prefer to retain these configurations as separate expressions of isomery, although they may be closely similar in appearance. The presence of five or four isomerous carpels is scattered over the four subclasses Dilleniidae, Rosidae, Asteridae and Caryophyllidae of Cronquist (1981; Fig. 1). Some families have taxa with either antesepalous or antepetalous insertion, sometimes in equal proportions (e.g. Malvales, Myrtales), or with one type of insertion dominating and the other occurring exceptionally (e.g. Crassulaceae: Eichler, 1878; Ericaceae: Leins, 1964). Carpels are mainly in antepetalous position in certain orders, while others are mainly antesepalous (Fig. 1). Obdiplostemony (taken from Ronse Decraene & Smets, 1995) is almost exclusively associated with an antepetalous carpel position (except for Humiriaceae and Ledocarpaceae, Fig. 1). On a general basis the isomerous carpel position can give an indication of affinity, especially at the family level. It demonstrates the basic nature of this configuration, with easy switches between an antesepalous and antepetalous position.

The configuration of trimerous flowers with isomerous carpels (as in monocots, some hamamelids and Magnoliidae) demonstrates another trend. Isomerous carpels are inserted mainly opposite the outer tepals, rarely opposite the inner (*Bocagea* of Annonaceae) or in alternation with tepals (e.g. Lactoridaceae, Aristolochiaceae). Both configurations rarely exist together, with one or the other as an exception (e.g. *Meconella* in Papaveraceae). The majority of monocots have their carpels opposite the outer tepals in a very stable configuration (Fig. 1; see below). This induced Dahlgren *et al.* (1985) to interpret this condition as ancestral for the monocots.

TABLE 2. Taxa with a high number of carpels exceeding duplication, due to a secondary increase

Family	Genus	Carpel number	Position of carpels	Number of stamens	Authority
Araliaceae	*Tupidanthus, Reynoldsia, Plerandra, Peekeliopanax*	20-100	single whorl	numerous, in a single or more whorls	Philipson (1970), Eyde & Tseng (1971)
Eucryphiaceae	*Eucryphia*	4-up to 18	single whorl	numerous	Dickison (1972a)
Rhizophoraceae	*Crossostylis*	up to 20	single whorl, alternating with stamens	few to numerous, in a single whorl	Juncosa (1988), Setoguchi *et al.* (1996)
Sonneratiaceae (Lythraceae)	*Duabanga, Sonneratia*	4-up to 20	single whorl, higher number correlated with more stamens	numerous, uni-, biseriate	Jayaweera (1967)
Aquifoliaceae	*Ilex*	up to 24	single whorl	low, haplostemonous	Loesener (1942)
Thymelaeaceae	*Lethedon*	up to 12	single whorl?	numerous, more or less in a single whorl	Domke (1934)
Bombacaceae	*Adansonia*	more than 10	single whorl, not related to other whorls	very high number on circular mound	van Heel (1966)
Gyrostemonaceae	*Gyrostemon, Tersonia, Codonocarpus*	5-21	single-two whorls, not related to the perianth	no stamens; numerous in male flowers	Eckardt (1971), Hufford (1996)
Paracryphiaceae	*Paracryphia*	8-15	single whorl	8 stamens	Dickison & Baas (1977)
Caryocaraceae	*Anthodiscus*	8-up to 20	single whorl	numerous	Dickison (1990a)
Rutaceae	*Citrus, Aegle*	13-20	single whorl; in "navel oranges" second whorl, not related to other whorls	numerous, in a single whorl	Leins (1967), Lord & Eckard (1985), Fig. 10

TABLE 2 continued

Quiinaceae	*Touroulia*	7-11	?	no stamens, numerous in male flowers	Engler (1925)
Medusagynaceae	*Medusagyne*	17-25	single whorl, not related to the perianth	very numerous	Dickison (1990b)
Malvaceae	tribes Malopeae, Malveae	many	single whorl no relation with other whorls	numerous	van Heel (1966, 1995), Sattler (1973)
Scyphostegiaceae	*Scyphostegia*	8-13(-16)	single whorl	no stamens, few in male flowers	van Heel (1967)
Sapotaceae	*Chrysophyllum, Achras*	14 (up to 30)	single whorl, in intervals of stamens and staminodes (A) or paired in between (C)	matching number of carpels	Eichler (1878), Hartog (1878), Saunders (1939)
Cactaceae	most genera	3-many	single whorl	numerous, upper alternating with carpels	Payer (1857), Ross (1982), Hofmann (1993)
Anacardiaceae	*Pleiogynium*	ca.12	two (?) whorls in pistillate flower; five carpellodes in staminate flower	same number as carpels and alternating	Wannan & Quinn (1991)
Rosaceae	subfamily Rosoideae	numerous	several ± regular whorls on carpel bearer	numerous 5(-10) merous whorls	Kania (1973), Sattler (1973), Fig. 11
Actinidiaceae	*Actinidia*	15->30	single whorl, distorted	numerous in single whorl, distorted	Dickison (1972b), van Heel (1987), Figs. 8-9

TABLE 3. Distribution of pseudomonomerous gynoecia in angiosperms

Doubtful presence of pseudomonomerous gynoecia in *Didymeles* (Didymelaceae; no indication in Cronquist, 1981), Centrolepidaceae (no indication in Dahlgren *et al*., 1985), *Aucuba* (Aucubaceae: Eckardt, 1937) and *Hippuris* (Hippuridaceae: Leins & Erbar, 1988).

Family	**Genus (if applicable)**	**Original no. carpels**	**No. ovules**	**Presence of zygomorphy**	**Reference**
Pontederiaceae	*Pontederia, Reussia*	3, median abaxial fertile	1	–	Cronquist (1981), Dahlgren *et al.* (1985)
Stemonaceae	*Stemona, Pentastemona*	3-1, arising as one structure (controversial)	>1-numerous	–	van Heel (1992)
Restionaceae	–	3-2, sterile locules or vascular traces	1	+	Dahlgren *et al.* (1985)
Poaceae	–	2-3, stylodial stubs, gynoecial primordia, vascular traces, presence of 3 stigmas	1	–	Sattler (1973), Dahlgren *et al.* (1985)
Ruscaceae	*Ruscus*	1-3 lobed primordial structure	2	–	Sattler (1973)
Sparganiaceae	*Sparganium*	2-3, 2 stylodial branches, 1 empty locule abaxially	1	–	Eckardt (1937), Cronquist (1981), Dahlgren *et al.* (1985)
Araceae	*Spathicarpa, Anchomanes*	2	1	–	Eckardt (1937), Barabé & Chrétien (1985), Barabé *et al.* (1986)
Marantaceae	*Thalia, Marantochloa*	1-2 empty locules, median abaxial displaced and fertile	1	+	Baillon (1860), Dahlgren *et al.* (1985)
Hydatellaceae	*Trithuria*	3-2, 3 slits and rudimentary pistil bundles	1	–	Dahlgren *et al.* (1985)
Arecaceae	*Ptychosperma*	3, median abaxial fertile	1	–	Uhl (1976)
Gyrostemonaceae	*Cypselocarpus, Gyrostemon tepperi*	2-7 (more)	1	–	Eckardt (1937), Hufford (1996)
Rhamnaceae	?	2	1	–	Cronquist (1981)
Urticaceae	–	2, median abaxial fertile, single (2) style	1	–	Eckardt (1937), Cronquist (1981)
Cecropiaceae	–	2, median abaxial fertile,	1	–	Cronquist (1981)
Moraceae	–	2(3), median abaxial fertile, 2 styles	1	–	Eckardt (1937), Cronquist (1981), Weberling (1989)
Ulmaceae	–	2(3), median abaxial fertile, 2 styles	1	–	Eckardt (1937), Okamoto *et al.* (1992), Omori & Terabayashi (1993)

TABLE 3 continued.

Cannabaceae	–	2, median abaxial fertile, 2 styles	1	–	Payer (1857), Eckardt (1937), Cronquist (1981)
Leitneriaceae	*Leitneria*	2, median adaxial fertile, 1 style	1	–	Cronquist (1981)
Thymelaeaceae	*Pimelea, Daphne, Phaleria*	2	1	–	Eckardt (1937), Heinig (1951), Cronquist (1981)
Lythraceae	*Pleurophora*	2, median abaxial reduced	>1	+	Eichler (1878)
Eucommiaceae	*Eucommia*	2, 2 stigmas present	2, 1 sterile	–	Eckardt (1957), Cronquist (1981)
Araliaceae	*Arthrophyllum*	2 (only one locule present but 2 ventral bundles)	2, 1 sterile	–	Philipson (1970), Eyde & Tseng (1971)
Chrysobalanaceae	*Stylobasium*, most species	3, fertile carpel opposite sepal 3, style gynobasic	2	+/–	Eichler (1878), Cronquist (1981)
Polygalaceae	*Securidacea*	2, median-posterior reduced	1	+	Krüger & Robbertse (1988)
Vochysiaceae	*Erisma*	3, latero-abaxial fertile	1–2	+	Cronquist (1981), Kopka & Weberling (1983)
Krameriaceae	*Krameria*	2, median adaxial reduced	2	+	Milby (1971), Cronquist (1981)
Corynocarpaceae	*Corynocarpus*	2, Fertile carpel opposite sepal 1; evidence of 2nd carpel (controversial)	1	–	Baillon (1872), Narayana *et al.* (1986), Philipson (1987)
Anacardiaceae	*Harpephyllum*, tribe Rhoeae	5,4,3, median abaxial fertile, sterile locules	single ovule	+	Payer (1857), Sattler (1973), Wannan & Quinn (1991), Figs. 18-19
Alangiaceae	*Alangium*	2, abaxial-median fertile	1	–	Eyde (1968)
Casuarinaceae	*Casuarina*	2, median abaxial fertile, 2 styles	2 (+2 abortive)	–	Cronquist (1981)
Nyctaginaceae	*Pisonia*	2?, presence of sterile stylar appendage	1	–	Vanvinckenroye *et al.* (1993)
Loasaceae	*Gronovioideae, Petalonyx*	3, 2 stylar lobes	1	–	Hufford (1989)
Icacinaceae	*Gomphandra, Gonocaryum, Phytocrene, Citronella*	2-3, pseudoseptum, trilobed stigma, or protuberances	1(2)	–	Sleumer (1960)
Nyssaceae	*Nyssa, Camptotheca*	2-3, unilocular with single or 2-3 branched style	1	–	Moser (1968), Cronquist (1981)
Cornaceae (including Aralidiaceae, Griseliniaceae)	*Aralidium, Griselinia*	3(4) or 2(3) small, sterile locules without ovules (A); 1 locule with three styles (G).	1	–	Philipson (1967), Philipson & Stone (1970)

TABLE 3 continued.

Valerianaceae	–	3, posterior-lateral carpel fertile	1	+/–	Eichler (1878), Sattler (1973), Hofmann & Göttmann (1990)
Dipsacaceae	–	2, or uncertain (2,3,4?) number of original carpels	1	+/–	Cronquist (1981), Hofmann & Göttmann (1990), Roels & Smets (1996)
Caprifoliaceae (including Viburnaceae)	*Viburnum, Linnaea, Symphoricarpos*	3, posterior lateral carpel fertile (V); 4, 2 median sterile (S)	1; 1-many abortive in sterile loci	+/–	Eichler (1878), Cronquist (1981), Roels & Smets (1996)
Morinaceae	*Morina*	2, one obsolete	1	+	Eckardt (1937), Hofmann & Göttmann (1990)
Stylidiaceae	*Stylidium*	2, median posterior occ. sterile or suppressed	numerous	+	Eichler (1878), Cronquist (1981), Erbar (1992)
Calyceraceae	*Acicarpha*	2?, 5 antepetalous ledges, one ovule fertile	–	(+)/–	Cronquist (1981), Erbar (1993)
Balanophoraceae	*Balanophora*	2-3, single style	1-2	–	Cronquist (1981)
Rubiaceae	*Theligonum*	2, median abaxial fertile, vestigial septum	single ovule	(+)	Rutishauser *et al.* (1997)
Goodeniaceae	*Velleia, Scaevola* sp.	1 locule (V), or 2 (lateral) vestigial and 2 (median) fertile locules (S)	1-many	+	Cronquist (1981)
Solanaceae	*Henoonia*	2, unilocular	single ovule	+	Cronquist (1981)
Boraginaceae	*Rochelia, Harpagonella*	2, median-adaxial reduced	two ovules	+	Junell (1938, sensu Weberling 1989), Hilger (1984)
Gesneriaceae	*Hemiboea*	2, median-adaxial sterile	numerous	+	Weber (1971, sensu Weberling, 1989)
Globulariaceae	–	2, median-adaxial sterile or suppressed	1(2)	+	Eckardt (1937), Cronquist (1981)
Verbenaceae	*Phryma, Lantana*	2, abaxial sterile (P), anterior suppressed (L)	1-2	+	Bocquillon (1861), Cronquist (1981)

Primitively polymerous gynoecia

In taxa belonging to Magnoliidae and monocots (Alismatales) the number of carpels is often high, arranged along a spiral, or in multiple whorls (Fig. 4). When low, the carpels can be unordered, as in Winteraceae (e.g. *Drimys*). There are cases of taxa with a high number of carpels in multiple trimerous whorls (e.g. *Sagittaria* in Alismataceae, *Brasenia* in Cabombaceae, *Castanea* in Fagaceae, Lardizabalaceae). Cases with two trimerous carpel whorls are even more frequent (Fig. 1) and are accompanied by taxa with a single carpel whorl opposite inner or outer tepals. They are especially frequent in the Najadales and Alismatales and are often found in families with high fluctuations in number and position of floral parts. With the broadening of the floral apex, carpels may be inserted in a single whorl (e.g. *Nymphaea* in Nymphaeaceae, *Platystemon* in Papaveraceae). The high number of carpels in these taxa may be plesiomorphic, but may also be the result of a secondary increase (see below). The tendency for carpel fluctuation is related to apocarpy and low synorganisation of the flower (Endress, 1990).

Secondary increase of carpels

False septa

False septa are invaginations of tissue within the locule which do not correspond to the lateral parts of a carpel; in other words these are outgrowths of the median carpel surface (Weberling, 1989). However, in Brassicaceae, walls formed between the opposite placentae are also described as "false septa" (Fig. 5). This difference is explained as a case of septation in paracarpous gynoecia, contrary to the formation of false septa in ovaries which are already completely or partially septate (Weberling, 1989). Several taxa have some sort of invagination dividing the ovary in numerous sections (Table 1). The occasional presence of false partitions can become a means to increase the carpel number. This may start ontogenetically as the building of short invaginations in the locule (e.g. in Aizoaceae: Fig. 6, Rosaceae: Murbeck, 1916, Hydrangeaceae: pers. obs., *Lagerstroemia*: Tobe & Raven, 1990, Linaceae: Narayana & Rao, 1976a,b, 1977; Juglandaceae: Weberling, 1989). These invaginations may then become aligned with the septa and divide each locule in two equal chambers (Table 1). As these outgrowths may arise late in ontogeny, they can be readily identified as "false septa" (e.g. Gerrath *et al.*, 1990). In other cases no difference can be made between false and true septa; one can only assume their origin in an ancestral condition. In support of this assumption is the fact that the number of carpels is "doubled" and that they occupy an intermediate position in relation to other organ whorls of the

FIGS. 2–7. Figs. 2–3: isomerous gynoecia. Fig. 2. *Cocculus laurifolius* DC. (Menispermaceae, 276 Lm); trimerous flower with carpels opposite outer tepals. ao: outer stamen. ai: inner stamen. Fig. 3. *Tribulus terrestris* L. (Zygophyllaceae, 502 Lm), obdiplostemonous flower with carpels opposite petals (p). Fig. 4. *Clematis recta* L. (Ranunculaceae, 438lb): flower with many spirally arranged carpels continuing sequence of stamens (a); one parastichy shown (black dots). Figs. 5–7: false septa. Fig. 5. *Barbarea vulgaris* R.Br. (Brassicaceae, Smets 178). Transverse section through middle of ovary showing false septum (fs) arising from placentas (pl). Fig. 6. *Mesembryanthemum nodiflorum* L. (Aizoaceae, 317 Lt). Transverse section in upper part of ovary showing septa (s) alternating with false septa (arrow). Fig. 7. *Neurada procumbens* L. (Neuradaceae, 310 Lt). Part of flower showing petals (p), antepetalous (ap), antesepalous stamens (as), and five of ten carpels alternating with stamens. Scale bars = 100 μm, except Fig. 4 = 200 μm, Fig. 5 = 50 μm.

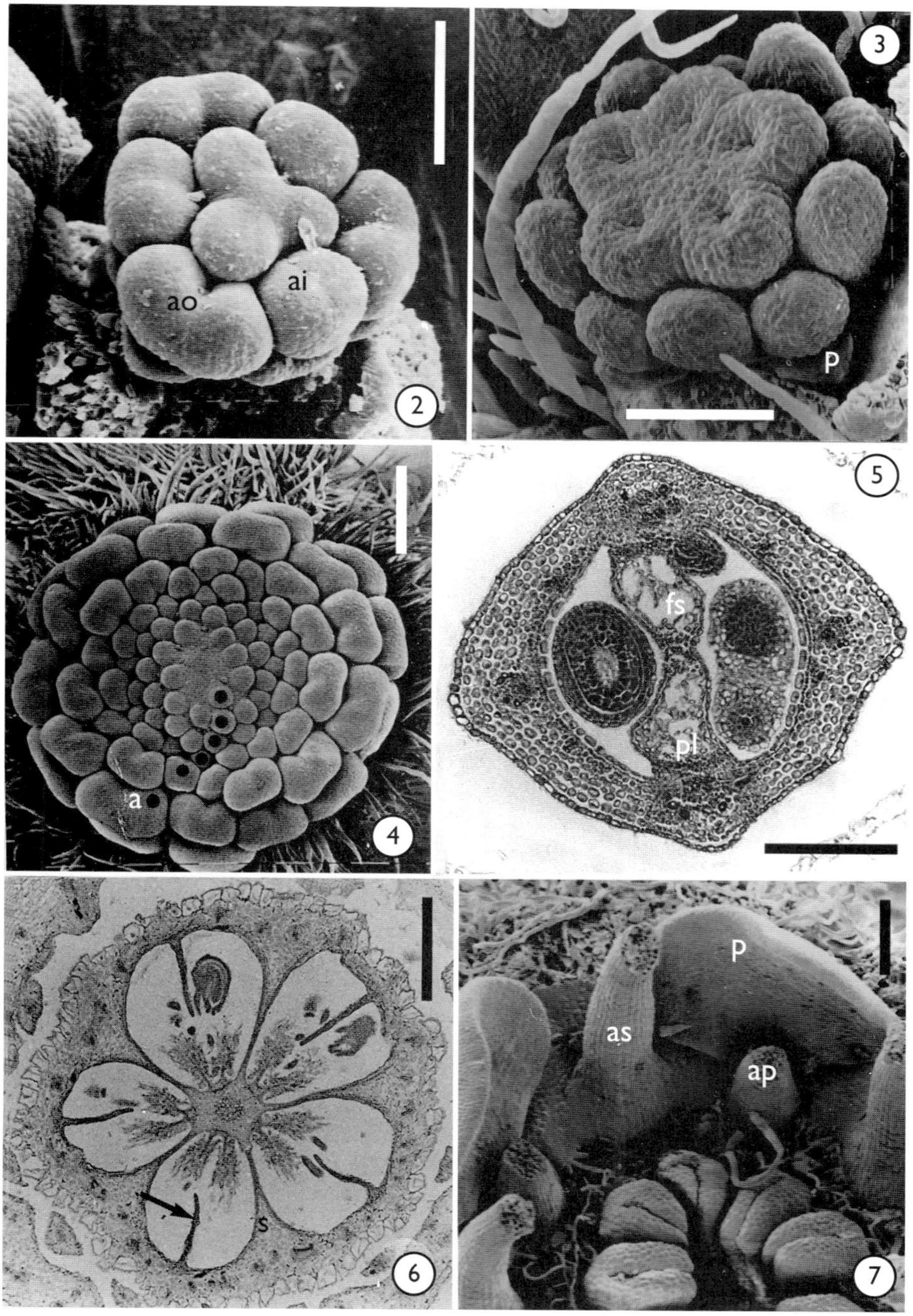
ao
ai
2
3
P
4
a
5
fs
pl
6
s
7
P
as
ap

flower, compared with taxa having no doubling. An example of this is Neuradaceae with ten carpels (Fig. 7); a phylogenetic link may be possible with Rosaceae (Maloideae) having partial invaginations (Murbeck, 1916; Ronse Decraene & Smets, 1996). Another example is *Gaylussacia* (Ericaceae: Palser, 1961). The gynoecia of the Boraginaceae and Lamiaceae are classic examples of the formation of false septa. Two median carpels are initiated as in most Asteridae (Payer, 1857). Each carpel contains two ovules, which become separated by the enfolding of the carpel wall leading to the formation of four locules. Such false septa become indistinguishable from real septa in some Convolvulaceae (e.g. *Calonyction*, *Mina*: Weberling, 1989). The development of false septa, unidentifiable from real septa, will double the number of carpels in one whorl. If occurring repeatedly, a high number of carpels can arise.

Secondary increase of carpels within a whorl

In certain taxa, especially with a secondary stamen increase, the number of carpels tends to be unstable, often fluctuating between five and more. A number of families scattered over Rosidae and Dilleniidae, but also some Alismatales have taxa with a high number of carpels inserted in a single whorl (Figs 8-10; Table 2). Endress (1981), Sattler (1973), and Sattler & Singh (1978) interpreted these as the result of a subdivision of primary primordia in the flower. However, van Heel (1995) did not observe a division of primary carpel primordia in pluricarpellate Malopeae but instead a continuous gynoecium meristem: a high number of carpels arise on a pentagone which is antepetalous in *Kitaibelia* and antesepalous in *Malope*. The difference is influenced by the different degree of contortion of the flower and androecium of both species, and carpels arise where there is sufficient space left alternating to the main portions of the androecium. As shown by van Heel (1966) there is a relationship between floral contortion and the position of the carpels in Malvales. We also studied a number of taxa (*Citrus*, *Actinidia*, *Alcea*: Figs 8-10) and could not observe traces of primary primordia. Carpels arise very rapidly (perhaps more rapidly in certain sectors of the flower) or even simultaneously in a single whorl; they are pushed into a wave-line (some Malvaceae) or become displaced by lateral pressures (*Actinidia*). Very often other members of the family have isomerous gynoecia (e.g. *Saurauia* in Actinidiaceae). Another possibility for carpel increase within a whorl results from a lateral division of the placentas; this process may be at the base of carpel multiplication in the Capparaceae (pers. obs.).

An increase of carpels within a whorl tends to be correlated with a secondary (mostly lateral) increase of stamens (Table 2). In many cases the numerous stamens arise in lateral direction from antesepalous forerunners (*Actinidia*: Figs 8-9, van Heel, 1987; *Citrus*: Fig. 10; Lord & Eckard, 1985). This is also related with the development of a broadly flattened floral apex, occupying a considerable space in the flower (Fig. 9).

FIGS. 8–13. Examples of secondary increase of carpels. Figs. 8–9. *Actinidia melanandra* Franch. (Actinidiaceae, 123 Lb). Fig. 8. Early stage of initiation of carpels in one whorl. Stamens have developed laterally from primordia alternating with the petals (p). Fig. 9. Older bud, with crowded displaced stamens and irregularly whorled carpels. Fig. 10. *Citrus aurantifolia* (Christm.) Swingle (Rutaceae, P153). Initiation of seven carpels in one whorl, with laterally increased stamens from primordia alternating with petals (p). Fig. 11. *Rubus ulmifolius* Schott. (Rosaceae, 291 Lt). Development of numerous carpel whorls on receptacular dome; outer whorls alternate more or less regularly. Figs. 12–13. *Punica granatum* L. (Punicaceae, 311 Lt). Fig. 12. Early initiation of two carpel whorls. Fig. 13. Longitudinal section of flower at ovule initiation. Outer style (so) with outer placenta (po) inserted above inner style (si) and inner placenta (pi). Scale bars = 100 μm.

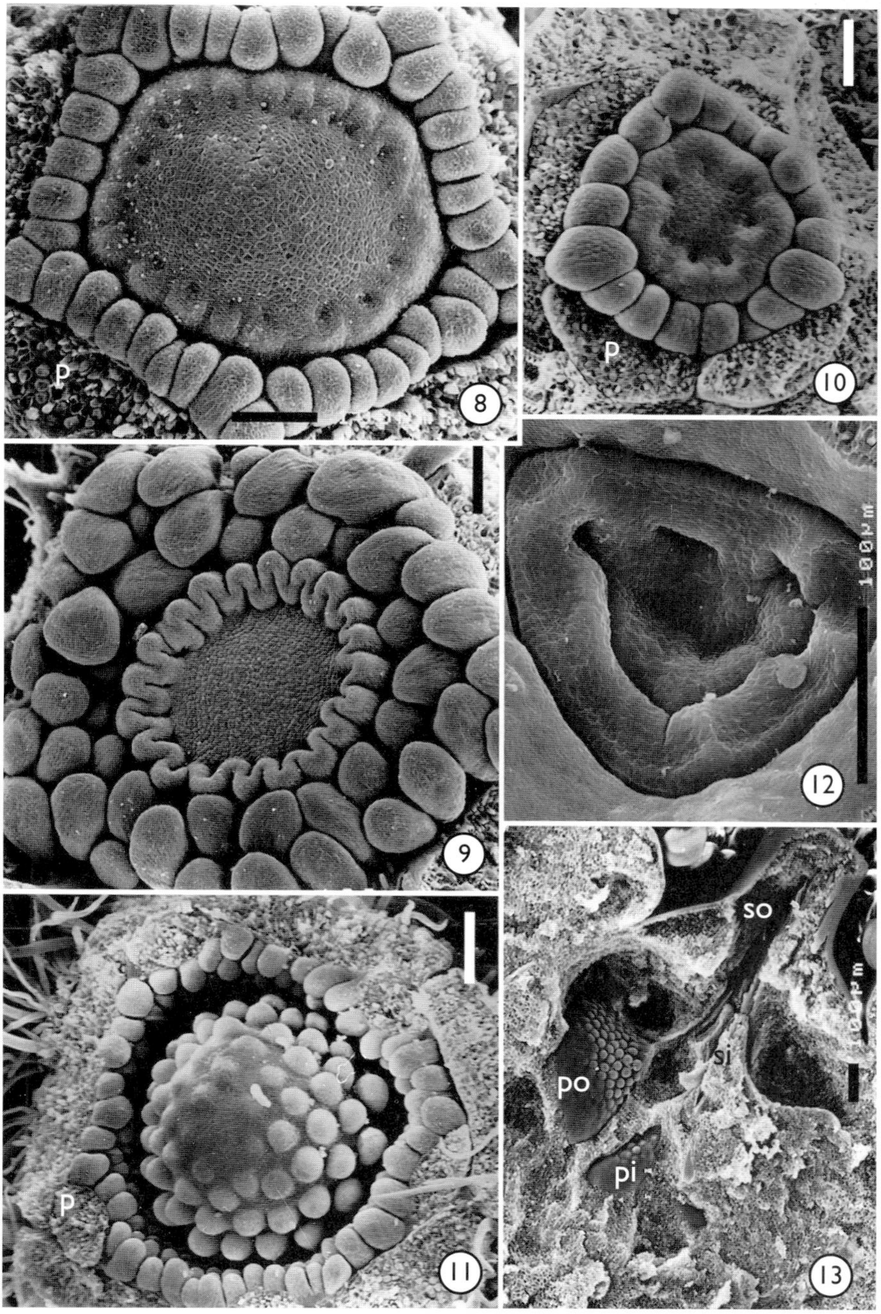
P
8
10
9
100µm
12
so
si
po
pi
11
13

Increase in number of carpel whorls

Rosaceae-Rosoideae often have a very high number of carpels, far exceeding the five or fewer carpels of other subfamilies. Kania (1973) related this development to a swelling of the base of the receptacle becoming a carpel bearer with a very high number of carpels. In *Rubus* (Fig. 11) carpels arise as regular whorls on the dome, becoming irregular higher up. The structure strongly resembles an apocarpous gynoecium. In *Rubus* the first carpels arise pairwise next to the petals without evidence of complex primordia. Sattler (1973) interpreted the development of the gynoecium of *Fragaria* as five primary gynoecial primordia with numerous secondary carpel primordia.

Contrary to the doubling of carpels induced by false septa, the presence of two carpel whorls is relatively rare (Table 1). A double carpel number, placed in antesepalous and antepetalous position has been observed in *Hibbertia* (Dilleniaceae: Wilson, 1965), *Malvaviscus* and other Ureneae (Malvaceae: van Heel, 1978; Crozier & Tucker, 1993), *Siphonodon* (Celastraceae: Croizat, 1947, in van Heel, 1978), but also in *Punica* (Punicaceae: Payer, 1857; Figs 12-13). In *Punica*, the outer whorl may be pentamerous or hexamerous, while the inner is trimerous. In some cases a third whorl has been observed (Payer, 1857). This phenomenon has been explained by an extended longitudinal growth of the periphery of the ovary (Eichler, 1878). In Malvaceae ten carpels may be initiated as two whorls of five with the inner one sterile (tribe Ureneae: van Heel, 1978).

Carpel reductions

Two possibilities for carpel reduction must be considered: (1) A carpel fails to develop beyond a primordial stage (Fig. 14), and (2) Carpel primordia may become laterally concrescent and grow as a single structure. This may be a teratology, matching other cases reported for the perianth and androecium (see e.g. Ronse Decraene & Smets, 1994). We observed floral buds of *Koelreuteria paniculata* (Sapindaceae) with either three (most common), or two carpels. In a few cases buds had two laterally concrescent carpel primordia (Fig. 15).

In most cases, however, flowers with reduced carpel numbers occur together with isomerous gynoecia and there is no ontogenetic evidence of a progressive reduction. It is not known whether the initiation versus non-initiation of a carpel is influenced by developmental processes, size and position of floral buds on the inflorescence, or genetic changes. These topics should be studied more extensively. Reductions of carpels tend to be generally correlated with meristic changes of the whole flower. There is a strong correlation between pentamerous flowers and the presence of three carpels with two positions: two carpels are placed abaxially, more or less opposite sepals 1 and 3 and the third is pointing in an adaxial direction, more or less opposite sepal two; the opposite with two adaxial carpels and a single

FIGS. 14–19. Examples of flowers with reduced carpel numbers. Fig. 14. *Gypsophila acutifolia* L. (Caryophyllaceae, 192 Lb). Gynoecium with three carpels, one primordium strongly reduced (arrow). Fig. 15. *Koelreuteria paniculata* Laxm. (Sapindaceae, 363 Lm). Abnormally developed gynoecium with three carpels, two of which are laterally fused without trace of placentation; arrows show placentas. Fig. 16. *Acer pseudoplatanus* L. (Aceraceae, 86 Ll). Gynoecium of two transversally arranged carpels. Numbers give sequence of sepal initiation. Fig. 17. *Actaea rubra* (Ait.) Willd. (Ranunculaceae, 80 Lb) with truly monomerous gynoecium. Figs 18–19. *Cotinus coggygria* Scop. (Anacardiaceae, 152 Lb). Fig. 18. Unilateral initation of four carpels with progressive increase in sterility towards adaxial side. Fig. 19. Mature flower with three carpels. Note three equally developed styles, but only one fertile carpel (asterisk). s: sepal. p: petal. Scale bars = 100 μm, except Fig. 19 = 1 mm.

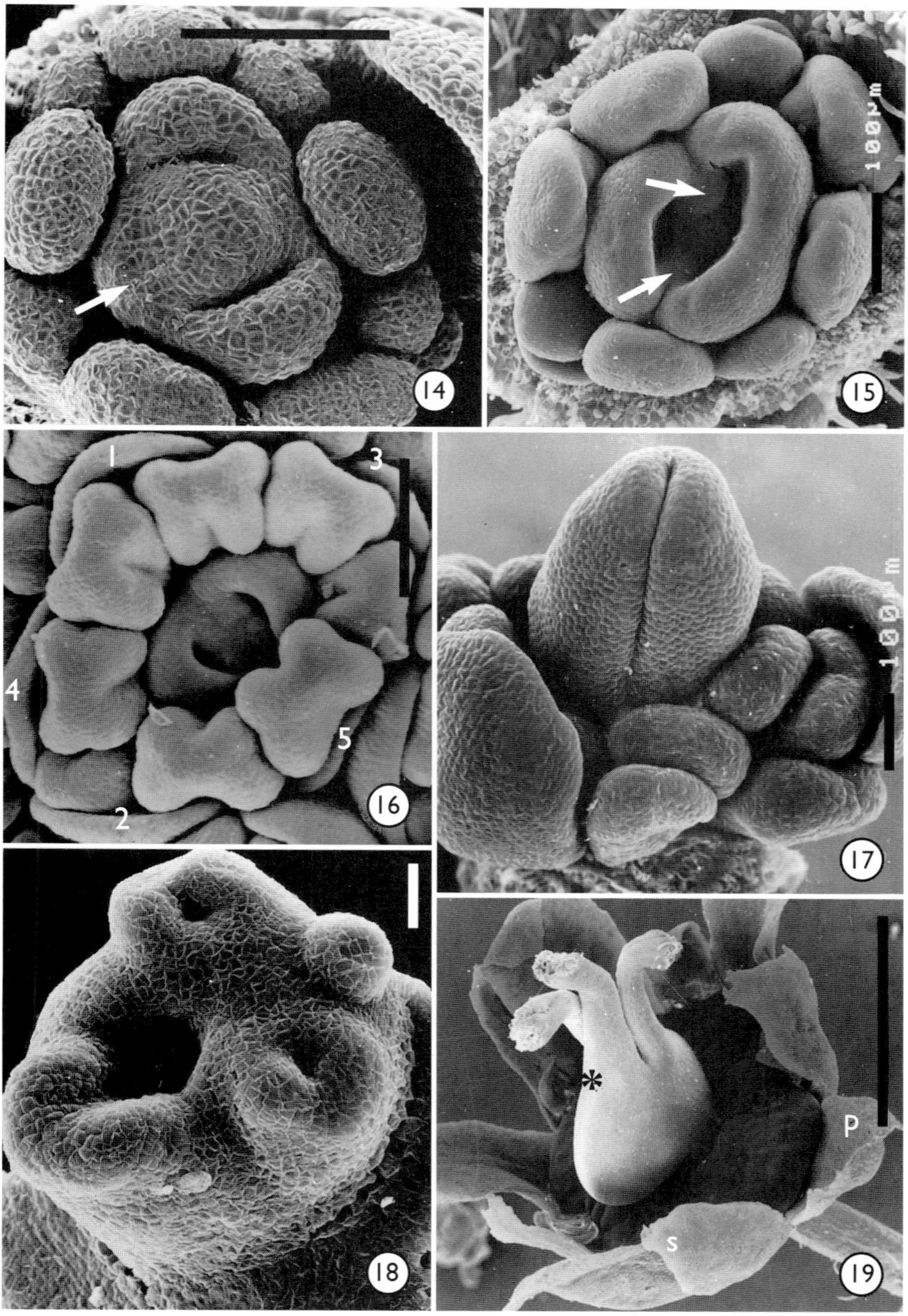
100 µm
1
3
4
5
2
16
14
15
100 µm
17
18
P
S
19

abaxial is less frequent and may be linked with changes in the position of the sepals related to the inflorescence axis. This is also the common position of the three carpels of most trimerous flowers. Three transversally placed carpels (two on one side of the sagittal line and the third on the other side) are restricted to Moringaceae and Sapindaceae, in relation to a peculiar form of zygomorphy (pers. obs., Fig. 20).

When flowers are tetramerous with a reduced number of carpels, these will always be two in number, most often arranged in a median position. Tetramerous flowers with three carpels were not observed. Pentamerous flowers with two (most often median) carpels are common, especially in Asteridae (Fig. 20). The presence of two carpels is often correlated with the reduction of one posterior stamen and the fusion of two posterior petals (e.g. Scrophulariaceae). This can ultimately lead to tetramerous flowers (e.g. Plantaginaceae: see Ronse Decraene & Smets, 1994). Transversally placed carpels occur less frequently among the dicots (Figs 16, 20); they tend to be associated mainly with strongly reduced flowers (e.g. Betulaceae, Callitrichaceae, Salicaceae, Oleaceae, Gunneraceae) or with dimerous flowers (e.g. Fumariaceae, Dialypetalanthaceae). It may be suggested that the position of the carpels, transversally or medianly, depends on the available space in the flower and on the position of the whorl preceding the carpels. In Brassicaceae, most space remains in the transversal plane, in alternation with the four upper stamens. A few families have both positions (e.g. Grossulariaceae, Achatocarpaceae, Podostemaceae), while it is variable in others (e.g. Apocynaceae, Campanulaceae). The occurrence of four carpels, not isomerous with the perianth tends to be related with two positions in the flower: either diagonally or orthogonally. In some cases four carpels may develop as the result of loss of one carpel (as in *Parnassia*), or be the result of a secondary increase (as in *Capparis micracantha*).

For certain families a reduction series can be constructed by comparing genera with a different arrangement of the carpels. Examples are Anacardiaceae (Wannan & Quinn, 1991), Rosaceae (Eichler, 1878), Verbenaceae (Bocquillon, 1861), and Araliaceae (Philipson, 1970; Eyde & Tseng, 1971). The reduction of carpels follows a common trend in Dipsacales, especially for Caprifoliaceae and Valerianaceae. Some genera of Caprifoliaceae are characterised by an isomerous gynoecium in antepetalous position or they have all their locules fertile (e.g. *Weigelia, Diervilla, Lonicera, Leycesteria*). In others one (*Triosteum*) to two locules (*Dipelta, Symphoricarpos, Linnaea, Kolkwitzia, Abelia*) are sterile. Some Valerianaceae and all Dipsacaceae have pseudomonomerous gynoecia and no traces of reduced carpels (see Sattler, 1973; Hofmann & Göttmann, 1990; Roels & Smets, 1996). An interesting fact is the presence of small ovules within sterile locules, outnumbering the single fertile ovule. This may be suggestive of a late event in the sterilisation process. More ontogenetic research should clarify this.

Monocarpellate gynoecia (Figs 17, 21)

Monocarpellate ovaries can be the result of a decrease in the number of carpels in an apocarpous gynoecium (many Magnoliidae), or the result of a reduction of all but one carpel in a syncarpous ovary (Weberling, 1989). The presence of one carpel may also be related to a strong reduction of the reproductive structures which are aggregated into pseudanthia (most

FIG. 20. Dahlgrenogram (Dahlgren, 1989b): distribution of reduced carpel numbers. Four carpels, through loss of one carpel (with orthogonal insertion: squares; with diagonal insertion: diamonds). Three carpels, through loss of two carpels (in pentamerous flowers only) triangles, with three orientations. Two carpels, through loss of three (in pentamerous flowers) or two carpels (in tetramerous flowers) bars, with median, transverse or oblique orientations. U: position unknown.

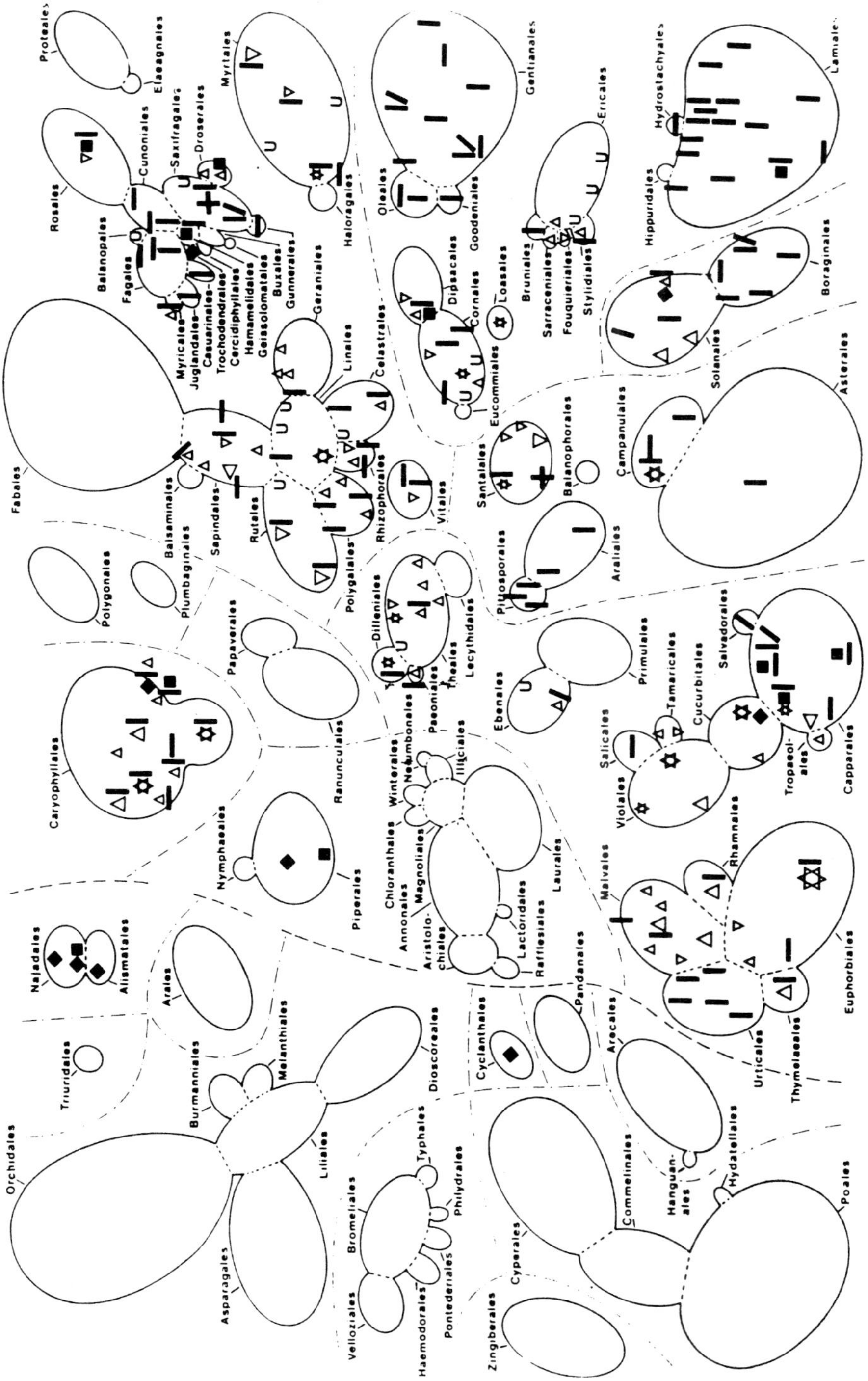
Proteales
Elaeagnales
Myrtales
Gentianales
Ericales
Hydrostachyales
Lamiales
Rosales
Cunoniales
Saxifragales
Droserales
Haloragales
Oleales
Goodeniales
Hippuridales
Balanopales
Fagales
Myricales
Juglandales
Casuarinales
Trochodendrales
Cercidiphyllales
Hamamelidales
Geissolomatales
Buxales
Gunnerales
Geraniales
Dipsacales
Cornales
Loasales
Bruniales
Sarraceniales
Fouquieriales
Stylidiales
Boraginales
Linales
Celastrales
Eucommiales
Solanales
Asterales
Fabales
Balsaminales
Sapindales
Rutales
Rhizophorales
Vitales
Santalales
Balanophorales
Campanulales
Polygalales
Pittosporales
Araliales
Polygonales
Plumbaginales
Papaverales
Dilleniales
Theales
Lecythidales
Paeoniales
Ebenales
Primulales
Tamaricales
Cucurbitales
Salvadorales
Caryophyllales
Ranunculales
Winterales
Nelumbonales
Illiciales
Salicales
Violales
Tropaeolales
Capparales
Nymphaeales
Piperales
Chloranthales
Annonales
Magnoliales
Aristolochiales
Lactoridales
Rafflesiales
Laurales
Malvales
Rhamnales
Euphorbiales
Najadales
Alismatales
Arales
Pandanales
Urticales
Thymelaeales
Triuridales
Burmanniales
Melanthiales
Dioscoreales
Cyclanthales
Arecales
Orchidales
Liliales
Typhales
Philydrales
Bromeliales
Commelinales
Hanguanales
Hydatellales
Poales
Asparagales
Velloziales
Haemodorales
Pontederiales
Cyperales
Zingiberales

Najadales, some Cyperaceae). When only one carpel occurs in the flower it is usually oriented towards the inflorescence axis (in 95% of the cases), rarely away from it (e.g. Cercidiphyllaceae, *Actaea* in Ranunculaceae: Fig. 17, *Cissampelos* in Menispermaceae: Eichler, 1878; Fig. 21). In a few taxa the single carpel can be oriented variably (median, oblique or transversal; e.g. Lauraceae, Berberidaceae).

A common cause for the occurrence of unicarpellate flowers is the development of monosymmetry. The Fabales are the clearest example of this with a widespread occurrence of a single carpel oriented in median adaxial direction. There are only few cases with more than one carpel, such as *Amherstia* with two carpels as a teratology (see Tucker, 1987; van Heel, 1993). *Archidendron* (Mimosaceae) probably represents the plesiomorphic condition for the order with five antesepalous carpels (van Heel, 1993).

In Magnoliidae the gynoecia are prevailingly apocarpous and unicarpellate (see Endress, 1990, 1994; e.g. Lauraceae, Myristicaceae, Berberidaceae). However, the number of carpels (with ovules) can apparently be increased ontogenetically, as in *Monodora* (Annonaceae), where a single carpel with a ventral slit produces c. 6 double rows of ovules on its flanks (Leins & Erbar, 1979). Endress (1981, 1990, 1994) suggests that groups with paracarpous gynoecia and parietal placentation have been derived in this way. Family pairs of closely related paracarpous and unicarpellate forms include Canellaceae / Myristicaceae, Saururaceae / Piperaceae, and perhaps Papaveraceae / Berberidaceae (Endress, 1990, 1994).

Pseudomonomerous gynoecia

There are several known cases where a number of carpels is initiated in the flower with a subsequent abortion of all carpels, except one (Table 3). A prerequisite for this trend is the presence of a low number of carpels (two or three, rarely up to five: Fig. 21; e.g. Anacardiaceae, Calyceraceae). Pseudomonomerous gynoecia are frequently described as monomerous because the participation of reduced carpels can often only be detected by means of ontogenetic, anatomical and comparative research (Weberling 1989). An excellent review of pseudomonomerous gynoecia is given by Eckardt (1937), but more studies are needed. Ontogenetic studies can demonstrate whether pseudomonomerous gynoecia are derived from gynoecia with axial or parietal placentation (Okamoto *et al.* 1992). Another reason for the importance of detecting pseudomonomerous gynoecia is the link they make between pluricarpellate gynoecia and strictly monomerous ones. Without evidence of one or more additional carpels one cannot decide whether a gynoecium has become monomerous through reduction of a pluricarpellate, coenocarpous gynoecium, or through loss of other carpels, which were apocarpous (see also Eyde, 1975, regarding the affinities of Proteaceae and Elaeagnaceae).

A number of taxa produce equally developing carpel primordia. However, only a single locule has fertile ovules and the other remain sterile or empty (e.g. in Erythroxylaceae, Aextoxicaceae, Quiinaceae, Celastraceae, Brunoniaceae, *Balanites* and *Nitraria* of Zygophyllaceae). This condition may be seen as one possible prerequisite for pseudomonomery. Pseudomonomery may also be the consequence of the development of monosymmetric flowers (e.g. in Zingiberales, Anacardiaceae, Chrysobalanaceae, Krameriaceae, Vochysiaceae, some Boraginaceae). In other cases (e.g. Urticales, Araceae), it may be caused by an extreme reduction of the whole flower. Interestingly, pseudomonomerous

FIG. 21. Dahlgrenogram (Dahlgren, 1989b): distribution of truly monomerous (hollow circles) and pseudomonomerous gynoecia (full circles) in angiosperms. Numbers give original number of carpels. Arrows indicate orientation of carpel suture relative to inflorescence / plant axis.

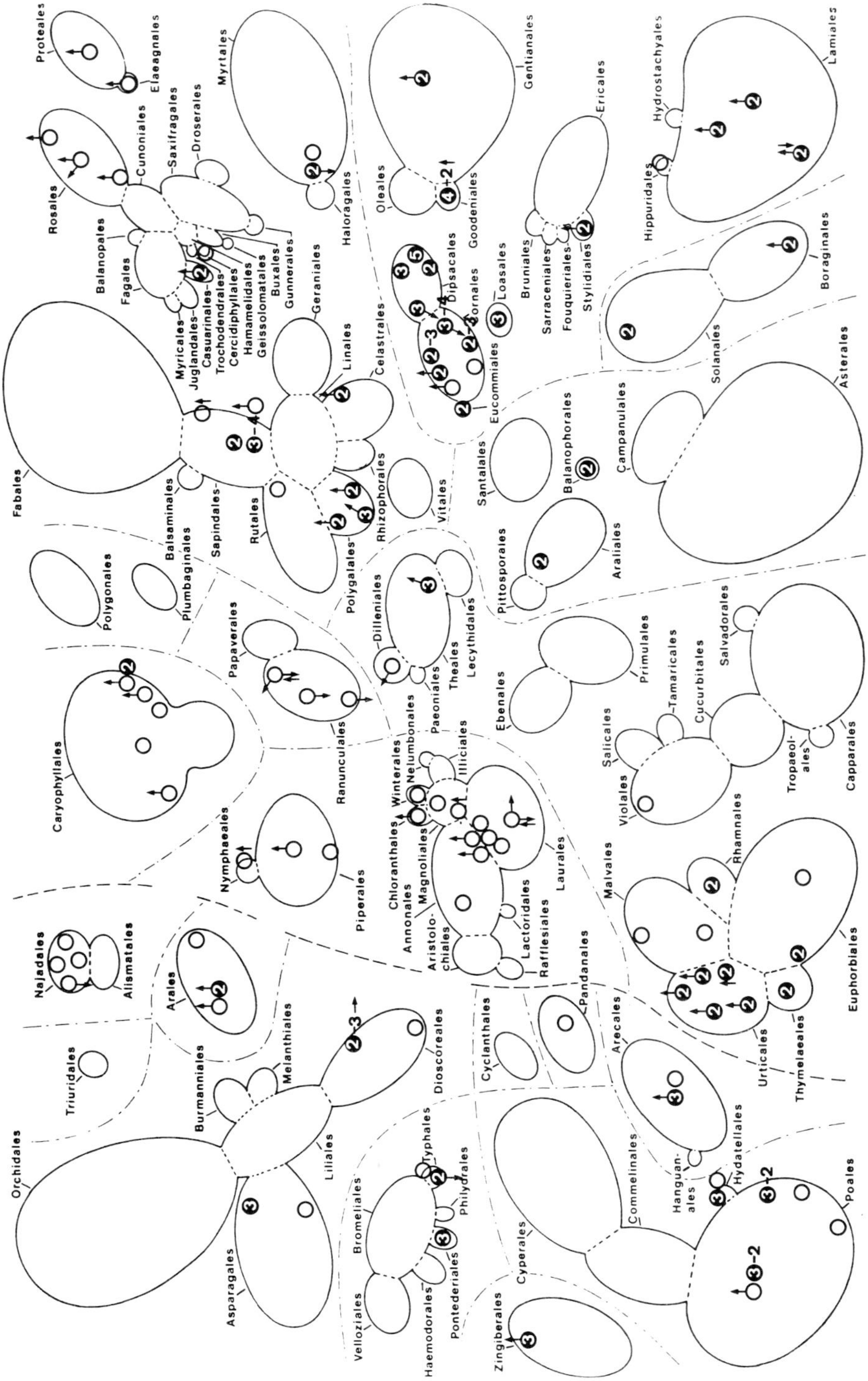
Proteales
Elaeagnales
Myrtales
Gentianales
Ericales
Hydrostachyales
Lamiales
Rosales
Cunoniales
Saxifragales
Droserales
Haloragales
Oleales
Goodeniales
Hippuridales
Balanopales
Fagales
Myricales
Juglandales
Casuarinales
Trochodendrales
Cercidiphyllales
Hamamelidales
Geissolomatales
Buxales
Gunnerales
Geraniales
Dipsacales
Cornales
Loasales
Bruniales
Sarraceniales
Fouquieriales
Stylidiales
Boraginales
Linales
Celastrales
Eucommiales
Solanales
Asterales
Fabales
Balsaminales
Sapindales
Rutales
Rhizophorales
Polygalales
Vitales
Santalales
Balanophorales
Campanulales
Polygonales
Plumbaginales
Pittosporales
Araliales
Papaverales
Dilleniales
Paeoniales
Theales
Lecythidales
Ebenales
Primulales
Tamaricales
Cucurbitales
Salvadorales
Caryophyllales
Ranunculales
Winterales
Nelumbonales
Illiciales
Salicales
Violales
Tropaeolales
Capparales
Nymphaeales
Piperales
Chloranthales
Magnoliales
Annonales
Aristolochiales
Lactoridales
Rafflesiales
Laurales
Malvales
Rhamnales
Euphorbiales
Najadales
Alismatales
Arales
Pandanales
Urticales
Thymelaeales
Triuridales
Melanthiales
Burmanniales
Dioscoreales
Cyclanthales
Arecales
Orchidales
Liliales
Asparagales
Typhales
Philydrales
Bromeliales
Haemodorales
Pontederiales
Velloziales
Commelinales
Hanguanales
Hydatellales
Poales
Cyperales
Zingiberales

gynoecia are absent from Magnoliidae. However this has been debated for the unicarpellate Lauraceae and Berberidaceae, which some authors consider to be derived from trimerous gynoecia (e.g. Eichler, 1878; Saunders, 1939; Singh & Singh, 1985). Ontogenetically there seems to be no evidence for this (e.g. Endress, 1972; Payer, 1857). Singh & Singh (1985) observed a triangular primordial dome before onset of initiation of the single carpel in *Cinnamomum camphora* (Lauraceae) and interpreted this as evidence for a pseudomonomerous gynoecium. More investigations should clarify this point of view. Another controversial case is *Corynocarpus* (Corynocarpaceae). Philipson (1987) questioned the existence of pseudomonomery because he could find no evidence for it, except for some knoblike primordia in early ontogeny. However, Narayana *et al.* (1986) found vascular evidence for a second carpel.

Urticales are a classic example of pseudomonomery. There is a progressive reduction series of one carpel and its accompanying vascular system running from Ulmaceae to Urticaceae (Eckardt 1937; Omori & Terabayashi, 1993; Okamoto *et al.*, 1992). Two (rarely three) carpels become initiated in Ulmaceae, Moraceae, and Cannabaceae, but only one develops further while the other disappears in the surrounding tissue (e.g. *Zelkova*: Okamoto *et al.*, 1992; *Cannabis, Ficus* : Payer, 1857). In Urticaceae there is no ontogenetic trace of the other carpel (Payer, 1857; Sattler, 1973). In Anacardiaceae pseudomonomerous gynoecia are correlated with the development of monosymmetric flowers, as taxa with isomerous gynoecia are regular (Wannan & Quinn, 1991). This process can be followed ontogenetically by the unilateral initiation of three (-four) primordia (e.g. *Cotinus*: Fig. 18). Only a single ovule is initiated on the cross-zone between the carpels. All carpels produce styles of different length (Fig. 19; also in *Rhus*: Payer, 1857; Sattler, 1973). In *Pistacia vera* the initiation of the two lateral carpels is much retarded (Hormaza and Polito, 1996) and in *Mangifera indica* there is only a single carpel without traces of sterile carpels (Payer, 1857). The gynoecium has become truly monomerous. Another striking example is *Theligonum cynocrambe* L. (Rubiaceae). This species has strongly reduced pistillate flowers with a single ovule and gynobasic style. It has been described by most authors as unicarpellate and is aberrant for Rubiaceae. However, Rutishauser *et al.* (1997) demonstrated the presence of a residual septum dividing the ovary into two incomplete locules. The unilocular ovaries of other Rubiaceae (e.g. *Otiophora, Calanda*) still have traces of a second carpel. The presence of a reduced septum has also been observed in the pseudomonomerous gynoecium of *Spathicarpa* (Araceae: Barabé & Chrétien, 1986). The pseudomonomerous gynoecium of *Rochelia* (Boraginaceae) can only be detected by comparison with related taxa (Hilger, 1984).

Correlation of carpel merosity with other organs

The merosity of the whole flower has a decisive influence on the number of carpels. Pentamerous flowers, if isomerous, will always have five carpels and if they have a reduced carpel number, these will be most often three in number, arranged as 1/2, or two with median insertion; tetramerous flowers will have four carpels, if isomerous, or two medianly inserted carpels. There is thus a strong arithmetic correlation between the merosity of the other floral whorls and the number of carpels. The main reason for this may be the synorganisation of the flower, regulating the number of parts and arrangement of each whorl, as space is limited (Endress, 1990, 1994).

This study and earlier reports demonstrate that the androecium can strongly influence the number and position of the carpels. Charlton & Posluszny (1991) gave an excellent survey of carpel position and its variation in *Potamogeton*. The most common number of carpels is four and they are positioned diagonally in the flower in alternation with the stamens. This implies that the stamens have a positional effect on the carpel initiation. Reductions to three, two or

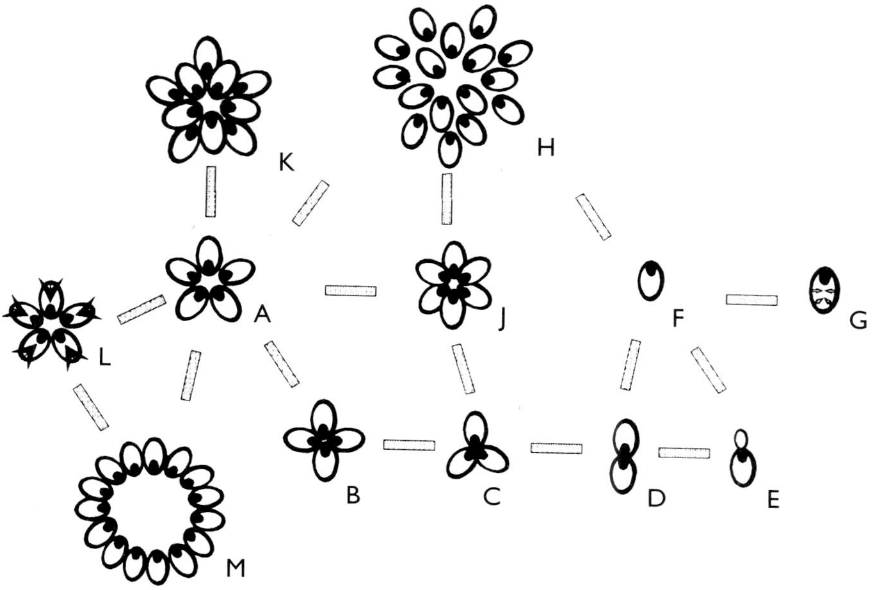

FIG. 22. Diagrammatic presentation of possible trends in the evolution of the number of carpels; E, pseudomonomery; H, acyclic arrangement of numerous carpels; J: two trimerous carpel whorls; G: secondary increase by the proliferation of placentae; L: formation of false septa; M: increase within a whorl; K, increase by the addition of a whorl; A,B,C,D,F are self-explanatory.

one depend on the area of the gynoecium available for carpel initiation; increases to five or more carpels depend on the continuous growth of the apex. Other examples are given in Ronse Decraene & Smets (1994). Setoguchi *et al.* (1996) observed that in *Crossostylis* (Rhizophoraceae) the number of stigmatic lobes corresponds with the number of stamens which have become laterally multiplied. However, lobes are fused groupwise at the base, corresponding with the number of perianth parts. A developmentally coordinated increase also occurs in Sapotaceae (Eichler, 1878; Hartog, 1878) where the calyx is tetramerous, but the rest of the flower is in octomerous or hexamerous whorls (e.g. *Lucuma*). An increase in merosity can also affect the whole flower, except for the carpels, as e.g. the octomerous flower of *Bruguiera* (Rhizophoraceae) with four carpels (Baillon, 1875), and the 16-merous flower of *Lafoensia* (Lythraceae) with only two carpels (Eichler, 1878).

When isomerous with the other floral whorls, the carpels can stand in antesepalous or antepetalous position. In pentamerous flowers of Dilleniidae, Caryophyllidae and Rosidae there is a strong correlation between obdiplostemony and antepetalous carpel position (Fig. 1; see also Ronse Decraene & Smets, 1995). This corresponds to a maximum use of space, as the antepetalous stamens are displaced to the periphery. In monocotyledons, carpels are most commonly three in number (or two in dimerous flowers) and they are invariably positioned opposite the outer tepals. This position is very stable and mostly does not change by reductions in the androecium (e.g. loss of inner stamens in Iridaceae, Mayacaceae; outer stamens in Burmanniales). In Alismatales and Najadales the position and number of carpels is unfixed, and follows rules of alternation more strictly than in other monocotyledons.

Conclusions

Fig. 22 illustrates possible relations between carpel numbers. Several factors influence changes in position or number. There are two main trends: one leading from five carpels to four, three, two and finally one carpel; and the other starting from numerous spirally arranged carpels and leading to six, three, or a single carpel. No postulates are made about the direction of trends. Secondary increases are possible at all levels. It is important to know the directions of trends by studying individual cases, because there are many reversals and variations possible. An interesting example of diverging carpel numbers within a species is *Pleiogynium* (Anacardiaceae: Wannan & Quinn, 1991): staminate flowers have five carpellodes, while this number is increased to 12 in pistillate flowers. It remains important to know the merosity of the gynoecium, including its position in the flower, directions of evolutionary trends, and possible reversals for purposes of phylogenetic studies. Indeed, the difference between few and many carpels can be more complex than it looks.

Acknowledgements

We thank Peter Vanvinckenroye for providing photographs of *Citrus* and the director of the National Botanic Garden of Belgium in Meise for permission to collect floral material and for use of the facilities. Our thanks go to Prof. S. Owens, Dr. P. Rudall, and other co-organisers of the Conference in Reproductive Biology at Kew for their invitation to present a lecture and submit this manuscript. We acknowledge travel grants from the National Fund for Scientific Research of Belgium (NFWO), the Vlaamse Leergangen Leuven, and the Vlaamse Wetenschappelijke Stichting. This study was supported by research grants (project no. 2.0038.91: SEM, and project no. G.0143.95: general research project) from the NFWO. The first author is a postdoctoral fellow of the NFWO.

References

Baillon, H. (1860). Mémoire sur la symétrie et l'organogénie florale des Marantées. *Adansonia* **1**: 306–327.

Baillon, H. (1872). Deuxième étude sur les Mappiées. *Adansonia* **10**: 261–282.

Baillon, H. (1875). Sur l'androcée des Rhizophoracées. *Bull. Mens. Soc. Linn. Paris* **1(8)**: 58–59.

Baillon, H. (1892). Ebénacées. In: Histoire des Plantes, 11, pp. 221–229. Hachette, Paris.

Barabé, D. and Chrétien, L. (1985). Anatomie florale de *Spathicarpa sagittifolia* (Araceae). *Beitr. Biol. Pflanzen* **60**: 403–412.

Barabé, D., Chrétien, L. and Forget, S. (1986). Vascularisation de la fleur pistillée d'*Anchomanes difformis* Engl. *Feddes Repert.* **97**: 831–835.

Bocquillon, H. (1861). Revue du groupe des Verbénacées. *Adansonia* **2**: 81–165.

Charlton, W.A. and Posluszny, U. (1991). Meristic variation in *Potamogeton* flowers. *Bot. J. Linn. Soc.* **106**: 265–293.

Cronquist, A. (1981). An integrated system of classification of flowering plants. Columbia University Press, New York.

Crozier, T.S. and Tucker, S.C. (1993). Floral organogenesis in *Malvaviscus arboreus* var. *drummondii* (Malvaceae). *Amer. J. Bot.* **80(6 suppl.)**: 21. (abstract)

Dahlgren, G. (1989a). The last Dahlgrenogram. System of classification of the dicotyledons. In: K. Tan (editor). The Davis and Hedge Festschrift, pp. 249–261. Edinburgh University Press, Edinburgh.

Dahlgren, G. (1989b). An updated angiosperm classification. *Bot. J. Linn. Soc.* **100**: 197–203.
Dahlgren, R.M.T., Clifford, H.T. and Yeo, P.F. (1985). The families of the monocotyledons. Structure, evolution and taxonomy. Springer Verlag, Berlin.
Dickison, W.C. (1972a). Comparative anatomy of Eucryphiaceae. *Amer. J. Bot.* **65**: 722–735.
Dickison, W.C. (1972b). Observations on the floral morphology of some species of *Saurauia*, *Actinidia*, and *Clematoclethra*. *J. Elisha Mitchell Sci. Soc.* **88**: 43–54.
Dickison, W.C. (1990a). A study of the floral morphology and anatomy of the Caryocaraceae. *Bull. Torrey Bot. Club* **117**: 123–137.
Dickison, W.C. (1990b). The morphology and relationships of *Medusagyne* (Medusagynaceae). *Pl. Syst. Evol.* **171**: 27–55.
Dickison, W.C. and Baas, P. (1977). The morphology and relationships of *Paracryphia*. *Blumea* **23**: 417–438.
Domke, W. (1934). Untersuchungen über die systematische und geographische Gliederung der Thymelaeaceen nebst eines Neubeschreibung ihrer Gattungen. *Biblioth. Bot.* **27, 3**: 1–151.
Eckardt, T. (1937) Untersuchungen über Morphologie, Entwicklungsgeschichte und systematische Bedeutung des pseudomonomeren Gynoeceums. *Nova Acta Leop.* n.f. **5**: 1–112.
Eckardt, T. (1957). Zur systematischen Stellung von *Eucommia ulmoides*. *Ber. Deutsch. Bot. Ges.* **69**: 487–498.
Eckardt, T. (1959). Das Blütendiagramm von *Batis* P.Br. *Ber. Deutsch. Bot. Ges.* **72**: 411–418.
Eckardt, T. (1971). Anlegung und Entwicklung der Blüten von *Gyrostemon ramulosus* Desf. *Bot. Jahrb. Syst.* **90**: 434–446.
Eckert, G. (1966). Entwicklungsgeschichtliche und Blütenanatomische Untersuchungen zum Problem der Obdiplostemonie. *Bot. Jahrb. Syst.* **85**: 523–604.
Eichler, A.W. (1875–1878). Blüthendiagramme I and II. W. Engelmann, Leipzig.
Endress, P.K. (1972). Zur vergleichenden Entwicklungsmorphologie, Embryologie und Systematik bei Laurales. *Bot. Jahrb. Syst.* **92**: 331–428.
Endress, P.K. (1981). Gynoecia with ontogenetical carpel multiplication and their evolutionary significance. Abstracts International Botanical Congress, 13th Sidney 1981. Abstr. 137.
Endress, P.K. (1990). Evolution of reproductive structures and functions in primitive angiosperms (Magnoliidae). *Mem. New York Bot. Gard.* **55**: 5–34.
Endress, P.K. (1994). Diversity and evolutionary biology of tropical flowers. Cambridge University Press, Cambridge.
Engler, A. (1925). Quiinaceae. In: A. Engler and K. Prantl (editors). Die natürlichen Pflanzenfamilien, 2 Aufl., Band 21, pp. 106–108, W. Engelmann, Leipzig.
Engler, A. (1931). Zygophyllaceae. In: A. Engler and K. Prantl (editors). Die natürlichen Pflanzenfamilien, 2 Aufl., Band 19a, pp. 144–184, W. Engelmann, Leipzig.
Erbar, C. (1992). Floral development of two species of *Stylidium* (Stylidiaceae) and some remarks on the systematic position of the family Stylidiaceae. *Canad. J. Bot.* **70**: 258–271.
Erbar, C. (1993). Studies on the floral development and pollen presentation in *Acicarpha tribuloides* with a discussion of the systematic position of the family Calyceraceae. *Bot. Jahrb. Syst.* **115**: 325–350.
Eyde, R.H. (1968). Flowers, fruits, and phylogeny of Alangiaceae. *J. Arnold Arbor.* **49**: 167–192.
Eyde, R.H. (1975). The bases of angiosperm phylogeny: floral anatomy. *Ann. Missouri Bot. Gard.* **62**: 521–537.
Eyde, R.H. and Tseng, C.C. (1971). What is the primitive floral structure of Araliaceae? *J. Arnold Arbor.* **52**: 205–239.
Gelius, L. (1967). Studien zur Entwicklungsgeschichte an Blüten der Saxifragales sensu lato mit besonderer Berücksichtigung des Androeceum. *Bot. Jahrb. Syst.* **87**: 253–303.
Gerrath, J.M., Lacroix, C.R. and Posluszny, U. (1990). The developmental morphology of *Leea guineensis* II. Floral development. *Bot. Gaz.* **151**: 210–220.

Hartog, M.A. (1878). On the floral structure and affinities of Sapotaceae. *J. of Bot.* **16**: 65–72.

Heinig, K.H. (1951). Studies in the floral morphology of the Thymelaeaceae. *Amer. J. Bot.* **38**: 113–132.

Hilger, H.H. (1984). Wachstum und Ausbildungsformen des Gynoeceums von *Rochelia* (Boraginaceae). *Pl. Syst. Evol.* **146**: 123–139.

Hofmann, U. (1993). Flower morphology and ontogeny. In: H.–D. Behnke and T.J. Mabry (editors). Caryophyllales. Evolution and systematics, pp. 123–166. Springer, Berlin.

Hofmann, U. and Göttmann, J. (1990). *Morina* L. und *Triplostegia* Wall. ex DC. im Vergleich mit Valerianaceae und Dipsacaceae. *Bot. Jahrb. Syst.* **111**: 499–553.

Hormaza, J.I. and Polito, V.S. (1996). Pistillate and staminate flower development in dioecious *Pistacia vera* (Anacardiaceae). *Amer. J. Bot.* **83**: 759–766.

Hufford, L.D. (1989). Structure of the inflorescence and flower of *Petalonyx linearis* (Loasaceae). *Pl. Syst. Evol.* **163**: 211–226.

Hufford, L.D. (1996). Developmental morphology of female flowers of *Gyrostemon* and *Tersonia* and floral evolution among Gyrostemonaceae. *Amer. J. Bot.* **83**: 1471–1487.

Janchen, E. (1925). Cistaceae. In: A. Engler and K. Prantl (editors). Die natürlichen Pflanzenfamilien, Aufl. 2, Band 21, pp. 289–313, W. Engelmann, Leipzig.

Jayaweera, D.M.A. (1967). The genus *Duabanga*. *J. Arnold Arbor.* **48**: 89–100.

Juncosa, A.M. (1988). Floral development and character evolution in Rhizophoraceae. In: P. Leins, S.C. Tucker and P.K. Endress (editors). Aspects of floral development, pp. 83–101. Cramer, Berlin.

Kania, W. (1973). Entwicklungsgeschichtliche Untersuchungen an Rosaceenblüten. *Bot. Jahrb. Syst.* **93**: 175–246.

Kopka, S. and Weberling, F. (1983). Zur Morphologie und Morphogenese der Blüte von *Vochysia acuminata* Bong. ssp. *laurifolia* (Warm.) Stafleu (Vochysiaceae). *Beitr. Biol. Pflanzen* **59**: 273–302.

Krüger, H. and Robbertse, P.J. (1988). Floral ontogeny of *Securidaca longepedunculata* Fresen (Polygalaceae) including inflorescence morphology. In: P. Leins, S.C. Tucker and P.K. Endress (editors). Aspects of floral development, pp. 159–167, Cramer, Berlin.

Leins, P. (1964). Entwicklungsgeschichtliche Studien an Ericales-Blüten. *Bot. Jahrb. Syst.* **83**: 57–88.

Leins, P. (1967). Die frühe Blütenentwicklung von *Aegle marmelos* (Rutaceae). *Ber. Deutsch. Bot. Ges.* **80**: 320–325.

Leins, P. and Erbar, C. (1979). Zur Entwicklung der Blüten von *Monodora crispata* (Annonaceae). *Beitr. Biol. Pflanzen* **55**: 11–22.

Leins, P. and Erbar, C. (1988). Einige Bemerkungen zur Blütenentwicklung und systematischen Stellung der Wasserpflanzen *Callitriche*, *Hippuris* und *Hydrostachys*. *Beitr. Biol. Pflanzen* **63**: 157–178.

Loesener, T. (1942). Aquifoliaceae. In: A. Engler. and K. Prantl (editors). Die natürlichen Pflanzenfamilien Aufl. 2, Band 20b, pp. 36–86. W. Engelmann, Leipzig.

Lord, E.M. and Eckard, K.J. (1985). Shoot development in *Citrus sinensis* L. (Washington navel orange). I. Floral and inflorescence ontogeny. *Bot. Gaz.* **146**: 320–326.

Melchior, H. (1925). Theaceae. In: A. Engler and K. Prantl (editors). Die natürlichen Pflanzenfamilien Aufl. 2, Band 21, pp. 109–154. W. Engelmann, Leipzig.

Milby, T.H. (1971). Floral anatomy of *Krameria lanceolata*. *Amer. J. Bot.* **58**: 569–576.

Moser, W. (1968). Der Blütenbau der *Davidia* und *Camptotheca*. *Vierteljahrsschr. Naturf. Ges. Zürich* **113**: 158–186.

Murbeck, S.V. (1916). Über die Organisation, Biologie und Verwandtschaftlichen Beziehungen der Neuradoideen. *Acta Univ. Lund.* n.s. Section 2 **12(6)**: 1–29.

Narayana, L.L. and Rao, D. (1976a). Contributions to the floral anatomy of Linaceae 6. *J. Jap. Bot.* **51**: 92–96.

Narayana, L.L. and Rao, D. (1976b). Contributions to the floral anatomy of Linaceae 7. *J. Jap. Bot.* **51**: 349–352.
Narayana, L.L. and Rao, D. (1977a). Contributions to the floral anatomy of Linaceae 8. *J. Jap. Bot.* **52**: 56–59.
Narayana, L.L. and Rao, D. (1977b). Contributions to the floral anatomy of Linaceae 9. *J. Jap. Bot.* **52**: 231–234.
Narayana, L.L., Satyanarayana, P. and Prakasa Rao, P.S. (1986). The floral anatomy of *Corynocarpus laevigatus* (Corynocarpaceae). *Phytomorphology* **36**: 325–329.
Ng, F.S.P. (1991). The relationships of the Sapotaceae within the Ebenales. In: T.D. Pennington (editor). The genera of Sapotaceae, pp. 1–13, Royal Botanic Gardens, Kew.
Okamoto, M., Kosuge, K. and Fukuoka, N. (1992). Pistil development and parietal placentation in the pseudomonomerous ovary of *Zelkova serrata* (Ulmaceae). *Amer. J. Bot.* **79**: 921–927.
Omori, Y. and Terabayashi, S. (1993). Gynoecial vascular anatomy and its systematic implications in Celtidaceae and Ulmaceae (Urticales). *J. Pl. Res.* **106**: 249–258.
Palser, B.F. (1961). Studies of floral morphology in the Ericales V. Organography and vascular anatomy in several United States species of the Vacciniaceae. *Bot. Gaz.* **123**: 79–111.
Payer, J.B. (1857). Traité d'organogénie comparée de la fleur. V. Masson, Paris.
Philipson, W.R. (1967). *Griselinia* Forst. fil. - anomaly or link. *New Zealand J. Bot.* **5**: 134–165.
Philipson, W.R. (1970). Constant and variable features of the Araliaceae. In: N.K.B. Robson, D.F. Cutler and M. Gregory (editors). New research in plant anatomy, pp. 87–100, Academic Press, London. (*Bot. J. Linn. Soc.* **63** suppl. 1).
Philipson, W.R. (1987). *Corynocarpus* J.R. & G. Forst. - an isolated genus. *Bot. J. Linn. Soc.* **95**: 9–18.
Philipson, W.R. and Stone, B.C. (1980). The systematic position of *Aralidium* Miq. - a multidisciplinary study. *Taxon* **29**: 391–416.
Roels, P. and Smets, E.F. (1996). A floral ontogenetic study in the Dipsacales. *Int. J. Plant Sci.* **157**: 203–218.
Ronse Decraene, L.P. and Smets, E.F. (1993). The distribution and systematic relevance of the androecial character polymery. *Bot. J. Linn. Soc.* **113**: 285–350.
Ronse Decraene, L.P. and Smets, E.F. (1994). Merosity in flowers: origin, definition and taxonomic significance. *Pl. Syst. Evol.* **191**: 83–104.
Ronse Decraene, L.P. and Smets, E.F. (1995). The distribution and systematic relevance of the androecial character oligomery. *Bot. J. Linn. Soc.* **118**: 193–247.
Ronse Decraene, L.P. and Smets, E.F. (1996). The floral development of *Neurada procumbens* L. (Neuradaceae). *Acta Bot. Neerl.* **45**: 229–241.
Ross, R. (1982). Initiation of stamens, carpels, and receptacle in the Cactaceae. *Amer. J. Bot.* **69**: 369–379.
Rutishauser, R., Ronse Decraene, L.P., Smets, E. and Mendoza-Heuer, I. (1997). *Theligonum cynocrambe* - the developmental morphology of a peculiar rubiaceous herb. *Pl. Syst. Evol.* (in press).
Sattler, R. (1973). Organogenesis of flowers, a photographic text-atlas. University of Toronto Press, Toronto.
Sattler, R. and Singh, V. (1978). Floral organogenesis of *Echinodorus amazonicus* Rataj and floral construction of the Alismatales. *Bot. J. Linn. Soc.* **77**: 141–156.
Saunders, E.R. (1939). Floral morphology, a new outlook, with special reference to the interpretation of the gynoecium. vol. II. Heffer and Sons, Cambridge.
Schulz, O.E. (1936). Cruciferae. In: A. Engler and K. Prantl (editors). Die natürlichen Pflanzenfamilien, 2 Aufl., Band 17b, pp.227–658. W. Engelmann, Leipzig.
Setoguchi, H., Ohba, H., and Tobe, H. (1996). Floral morphology and phylogenetic analysis in *Crossostylis* (Rhizophoraceae). *J. Pl. Res.* **109**: 7–19.

Singh, S.P. (1960). Morphological studies in some members of the family Pedaliaceae - I. *Sesamen indicum* DC. *Phytomorphology* **10**: 65–82.
Singh, V. and Singh, A. (1985). Floral organogenesis in *Cinnamomum camphora*. *Phytomorphology* **35**: 61–67.
Sleumer, H. (1942). Icacinaceae. In: A. Engler and K. Prantl (editors). Die natürlichen Pflanzenfamilien, 2 Aufl., Band 20b, pp. 322–396. W. Engelmann, Leipzig.
Taylor, D.W. (1991). Angiosperm ovules and carpels: their characters and polarities, distribution in basal clades, and structural evolution. *Postilla* **208**: 1–40
Tobe, H. and Raven, P.H. (1990). Comparative reproductive morphology of "*Orias*" and *Lagerstroemia* (Lythraceae). *Flora, Morphol. Geobot. Oekophysiol.* **184**: 177–185.
Tucker, S.C. (1987). Floral initiation and development in Legumes. In: C.H. Stirton (editor). Advances in Legume Systematics 3, pp. 184–239, Royal Botanic Gardens, Kew.
Uhl, N.W. (1976). Developmental studies in *Ptychosperma* (Palmae). II. The staminate and pistillate flowers. *Amer. J. Bot.* **63**: 97–109.
van Heel, W.A. (1958). Additional investigations on *Batis argillicola* van Royen. *Nova Guinea n.s.* **9**: 1–7.
van Heel, W.A. (1966). Morphology of the androecium in Malvales. *Blumea* **13**: 177–394.
van Heel, W.A. (1967). Anatomical and ontogenetic investigation on the morphology of the flowers and the fruit of *Scyphostegia borneensis* Stapf (Scyphostegiaceae). *Blumea* **15**: 107–125.
van Heel, W.A. (1978). Morphology of the pistil in Malvaceae-Ureneae. *Blumea* **24**: 123–137.
van Heel, W.A. (1987). Androecium development in *Actinidia chinensis* and *A. melanandra* (Actinidiaceae). *Bot. Jahrb. Syst.* **109**: 17–23.
van Heel, W.A. (1992). Floral morphology of Stemonaceae and Pentastemonaceae. *Blumea* **36**: 481–499.
van Heel, W.A. (1993). Floral ontogeny of *Archidendron lucyi* (Mimosaceae), with remarks on *Amherstia nobilis* (Caesalpiniaceae). *Bot. Jahrb. Syst.* **114**: 551–560.
van Heel, W.A. (1995). Morphology of the gynoecium of *Kitaibelia vitifolia* Willd. and *Malope trifida* L. (Malvaceae-Malopeae). *Bot. Jahrb. Syst.* **117**: 485–493.
Vanvinckenroye, P., Cresens, E., Ronse Decraene, L.P. and Smets, E. (1993). A comparative floral developmental study in *Pisonia, Bougainvillea* and *Mirabilis* (Nyctaginaceae) with special emphasis on the gynoecium and floral nectaries. *Bull. Jard. Bot. Nat. Belg.* **62**: 69–96.
Wannan, B.C. and Quinn, C.J. (1991). Floral structure and evolution in the Anacardiaceae. *Bot. J. Linn. Soc.* **107**: 349–385.
Weberling, F. (1989). Morphology of flowers and inflorescences. Cambridge University Press, Cambridge.
Wilson, C.L. (1965). The floral anatomy of the Dilleniaceae. I. *Hibbertia* Andr. *Phytomorphology* **15**: 248–274.
Wilson, C.L. (1973). The floral anatomy of the Dilleniaceae. II. Genera other than *Hibbertia*. *Phytomorphology* **23**: 25–42.

Cocucci, A. and Sérsic, A. (1998). Evidence of rodent pollination in subtropical South America. In: S.J. Owens and P.J. Rudall (Editors). Reproductive Biology, pp. 113–121. Royal Botanic Gardens, Kew.

EVIDENCE OF RODENT POLLINATION IN SUBTROPICAL SOUTH AMERICA

ANDREA COCUCCI AND ALICIA SÉRSIC

Instituto Multidisciplinario de Biología Vegetal, Casilla de Correo 495, 5000 Córdoba, Argentina

Abstract

Several vegetative and floral features of *Cajophora coronata* (Loasaceae) suggest adaptation to pollination by rodents: (1) mammal guard represented by an indument of strongly pungent hairs; (2) geoflory; (3) white corolla; (4) open flowers with copious low concentration nectar; (5) abundant pollen; (6) maximum pollen and nectar presentation in the afternoon hours and in the night. Palynological analysis revealed pollen loads of *C. coronata* on the nostrils and whiskers of captured rodents (*Graomys griseoflavus*, fam. Muridae). Pollen and anther remains were also found in faeces sampled in the surroundings. Additional evidence includes rodent footprints obtained by placing smoked plates beneath the flowers, which revealed flower visitation during the night. Floral traits strongly suggest pollination by terrestrial mammals in other subtropical South American species, for example in *Bromelia urbaniana* (Bromeliaceae).

Introduction

The possible existence of flowers pollinated by mammals other than bats was first suggested by Kerner von Marilaun (1895). Later, Porsch (1934, 1935, 1936a, 1936b) reported that a number of arboreal or climbing mammals visit flowers to take nectar, pollen, or flower parts. These reports included monkeys, squirrels, bears, several kinds of possums and even marsupial gliders. Flower visitors do not necessarily act as pollinators and many of them are known to be more destructive than beneficial to the plant, but in Australia, arboreal marsupials and notably the honey possum (*Tarsipes*, a flower specialist), are reported to feed on and regularly pollinate *Melaleuca* and *Banksia*, amongst other plants. Despite this, pollination by non-flying mammals (NFM) was considered a curiosity of subordinate role in the biology of flowers. There were also no clear signs of adaptations of flowers to NFMs, so a syndrome could not be clearly defined. The subject was not given much further attention until forty years later, but has since become an important issue in the ecology and evolutionary biology of flowers.

Reports now exist on pollination by lemurs in Madagascar (Kress *et al.*, 1992, Nilsson *et al.* 1993, Sussmann & Raven 1978), prosimians in tropical Africa (Grünmeier, 1990), rodents in South Africa (Rourke & Wiens, 1977, Wiens *et al.* 1983) and Costa Rica (Lumer, 1980, Lumer & Schoer, 1986), marsupials in Australia and New Guinea (Collins & Rebelo, 1987; Goldingay *et al.*, 1991; Hopkins, 1992; Rourke & Wiens, 1977) and marsupials, procyonids, and primates in a number of neotropical plants (Gribel, 1988; Janson *et al.*, 1981; Prance, 1980; Steiner, 1981; Vieira *et al.* 1991). The plants involved in these cases do not always have obvious flower adaptations to NFM-pollination. For example, in tropical environments, arboreal non-flying mammals may act as alternative pollinators to bats and birds of obvious chiropterophilous (*Parkia, Mabea, Pseudobombax, Adansonia, Ceiba, Quararibea*) or ornithophilous flowers (*Combretum, Banksia, Strongylodon*) respectively. Thus flowers visited and even pollinated by non-flying mammals may exhibit flower syndromes typical of other pollination types, and flowers may even function on open pollination systems operated also by other animals such as bats and birds. However, for reasons outlined below, we have defined an idealised flower type for non-flying mammals.

TABLE 1: Traits for NFM-pollinated plants

Pollination units	single flowers in many-flowered inflorescence
	whole inflorescence
	strongly attached to stems
	bowl-shaped
	with a large stigma-nectar distance
	with stiff flower parts (wiry styles)
	with mammal guards (structures discouraging mammals from chewing blossoms)
Mechanisms of pollen placement	explosive pollen release
	pump or bellows-like mechanism
Exposition of pollination units	geoflorous and cryptic
Anthesis	short-lived flowers
	during night or day
Phenology	flowering in winter or when resources are low
Colour	dark
	dull
	light
	pale yellow
Odour	nutty
	strong musky
	yeasty
	like sour milk
Nectar	copious
	replenishable
	sucrose-rich
	concentration between 14% and 20%
Pollen	abundant

Various authors have suggested a combination of traits for NFM-pollinated plants (Table 1). Obviously no blossom can exhibit all of these traits together. Adaptive traits to pollination by non-flying mammals depend on evolutionary constraints within each lineage. Plants provide a stable organisation pattern, the ground plan, inherent to their systematic provenance. Expression of adaptations to NFM-pollination would act at a more superficial level (flower form, nectar quality, colour, odour, etc) without blurring the ground plan.

Adaptations to terrestrial (as opposed to arboreal) mammals can be readily distinguished by geoflory, as in some South African and Australian Proteaceae demonstrated to be pollinated by non-flying mammals (Rourke & Wiens, 1977). We therefore searched for the possible existence of pollination by terrestrial mammals in subtropical South America in environments where bat flowers and flower-visiting bats are absent. Exploratory studies, mainly in subtropical zones of Argentina, suggested two species that exhibited a combination of many of the syndrome characters (Table 1). These species belong to unrelated families and very different plant communities. This paper presents preliminary evidence of rodent pollination of *Cajophora coronata* (Arn.) H. et A. (Loasaceae) and suggests pollination by non-flying mammals based on the floral features of *Bromelia urbaniana* (Mez) L. B. Smith (Bromeliaceae). Detailed accounts on the floral biology of these species have been or will be published elsewhere (Cocucci & Sérsic, 1998 and in prep.).

Cajophora coronata (Loasaceae)

This species grows in the Andes of Peru, Bolivia, Northern Chile and Argentina between 2,000 and 5,000 m (Sleumer, 1955). It is part of a subnival alpine plant community constituted predominantly by tussock grasses, low-growing often cryptically-coloured rosette herbs and flat cushions. Upright leaves arise from a horizontal, partly underground herbaceous stem and form more or less globular clusters. Flowers arise at ground level and lie nearly horizontal around the leaf clusters. Leaves and outside flower surfaces are densely covered by strongly pungent hairs.

The pentamerous flowers show an open bowl-shaped white corolla. Each hooded petal encloses a bundle of approximately 33 indehiscent stamens, alternating with five staminodial complexes, each of which consists of one external saccate staminodium and two internal subulate elements. These complexes are situated in coincidence with five nectary pillows. These pillows are interpreted by Brown & Kaul (1981) as members of a split disc on the upper part of the ovary. Nectar accumulates under the outer saccate staminodia and may come out as five drops from their entrances. The massive inferior ovary ends in a straight style with three triangular and connivent stigmatic branches. The flowers are odourless.

Anthesis

The open flower stage lasts about ten days, during which the corolla opens gradually and fertile parts show a markedly protandrous pattern. When flowers are just open, all 160 stamens are placed inside the hooded petals but they start lifting during the first day of anthesis. The staminate (male) phase lasts about 8 days. Anthers become dehiscent and the stamens gradually take an upright position in the centre of the flower. Stamens lift at an average rate of 20 per flower each day.

Stamen movement was not triggered by touching the staminode scales as shown by Schlindwein (1995) for bee pollinated *Cajophora* species and other Loasaceae. Instead, in *C. coronata* they move spontaneously following a daily rhythm. Each day stamens lift up mainly in the afternoon, pollen presentation reaching a maximum between 20:00 and 24:00. At the end of the male phase all fertile stamens are elevated and placed around the pistil forming a conical bundle. The male phase gives place to the female phase when stamens fold back simultaneously, to the inside of the hooded petals. At this point the stigma is exposed for the first time and shows an increased secretory activity, which lasts at least 2 days. The three stigmatic branches remain folded, their receptive surfaces being located at the margins of each branch.

Nectar secretion and visitors

Small quantities of nectar are detectable during the day, but most production starts after 20:00, and reaches a maximum of about 150 microlitres per flower between 21:00 and 06:00. Nectar concentration ranges from 7.5% to 49% (average 16.03%) in standing crop flowers.

Diurnal visitors include pollen-collecting bees and a variety of nectar-feeding birds. Both females and males of high altitude bees of the genera *Megachile* and *Anthidium* were constant morning visitors. The females visited the flowers only in the staminate phase and actively collected pollen. Neither females nor males were seen sucking nectar, but even if they had, the long nectar-anther or nectar-stigma distance precludes pollination by these small animals. Therefore, contrary to Arroyo *et al.* (1982), *Cajophora coronata* is not bee-pollinated.

Among nectar feeding-birds five different species were seen visiting the flowers: two hummingbirds (*Oreotrochilus leucopleurus* Gould and *Sappho sparganura sapho* [Lesson]) and three passerines (*Poospiza hypochondria affinis* Berlepsch, *Phrygilus gayi caniceps* Burmeister, and *Sicalis olivascens olivascens* [D'Orbigny & Lafresnaye]). To reach the nectar under the staminode scales the birds twist their heads to the centre of the flower, thus striking the stamens and style. Abundant pollen could be seen at the base of the beaks of these birds.

Although rodent activity was observed in the evening and during the night, no nocturnal visitations to the flowers were observed directly. However indirect evidence shows that rodents do visit *C. coronata*. To trap them, 100 snap-traps baited with toasted bread and bacon were set on a 10 × 10 m square grid. Three individuals of leaf eared mice (*Graomys griseoflavus* (Waterhouse, 1837; fam. Muridae, subfamily Sygmodontinae) were captured during the night. Palynological analysis of the samples obtained from their fur revealed pollen of *C. coronata* on the nostrils and whiskers of all captured individuals. Pollen and anther remains were found also in faeces collected near the plants and mouse dwellings.

To further demonstrate visits by terrestrial mammals, their footprints were obtained by placing smoked glass plates just in front of the flowers. The footprints obtained corresponded to the foot sole pattern of the captured species. The position of the footprints allows reconstruction of the pose adopted by the rodents in front of the flowers (Fig. 1). Sitting on their hind legs and tail, they must have had the rear part of their bodies more or less perpendicular to the flower axis. The footprints of the front legs reveal that the rodent had the front body partly twisted towards the flower entrance. Occasionally, flower damage that can be attributed to rodents was also observed.

Pollination system of *Cajophora coronata*

Loasaceae show an evolutionary radiation of floral types including species pollinated by bees, hawkmoths and birds (Braun & Kaul, 1981; Thomson & Ernst, 1967; Schlindwein, 1995). Now mammal pollination can be added (Fig. 1). The pollination system of *C. coronata* is clearly not a closed one (Cocucci & Sérsic, in press); birds apparently play an important role. Because of the timing of pollen presentation, birds may not be optimally used as pollen vectors. Hummingbirds play a key role in the pollination of *C. macrocarpa* Urb. et Gilg and *C. lateritia* (Hook.) Klotzsch. *C. macrocarpa* is probably a close relative of *C. coronata*, from vegetative and floral features, with intermediate hybrids between them (Sleumer, 1955).

In the plant community to which *C. coronata* belongs, no other plant with a similar flower syndrome was found. Ornithophilous plants are also absent. Thus *C. coronata* is probably the main nectar source for the local flower-visiting birds and rodents (Cocucci & Sérsic, in press). *Oreotrochilus* also takes nectar from *Adesmia* (Fabaceae) a plant clearly adapted to and visited by bees such as *Megachile* (pers. obs.).

Schlindwein (1995) presented evidence suggesting that bee-pollinated *Cajophora* species and other related Loasaceae (*Blumenbachia* and *Loasa*) are narrowly adapted to oligolectic Colletidae (species of *Bicolletes* and *Diphaglossa*): (1) The distribution centre of Colletidae is the same as that of the Loasaceae. (2) The Colletidae mouth parts fit into the staminode scales, thus activating tygmonastic movement of the stamens. (3) The time lag between stimulus and complete stamen uplift corresponds to the visitation intervals of the bees. Bees visiting within shorter or longer intervals cannot optimally profit from the pollen resources.

The retarded stamen uplift would also be in agreement with trap-lining behaviour by hummingbirds (Janzen, 1968), such as may be the case for *C. lateritia* and *C. macrocarpa* visitors. In *C. coronata*, (sensitive?) staminode appendages are hidden inside the staminode scales and stamen uplift is autonomous. In agreement with Schlindwein (pers. com.) a scenario can be proposed in which the initially melittophilous Loasaceae is adapted to pollination by vertebrates. In such an evolutionary shift the tygmonastic stamen movement would still be suitable for trap-lining hummingbirds. Under certain circumstances (scarcity of bees) evolution could have selected for forms pollinated by vertebrates, but it is not clear which circumstance could have involved rodents in pollination. Rodents seem less suited pollen vectors than birds, since they cover shorter distances in their foraging tours and more frequently clean their fur by grooming and licking and eat pollen. Pollen loss could be counterbalanced by the plant, with the already existing strategy of surplus pollen production that would be initially needed to compensate pollen taken by oligolectic bees. Considerable pollen depletion could be tolerated. A similar situation is known for many bat-pollinated plants (V. Helversen, 1993). It is also not clear why *C. coronata* exhibits more rodent-pollination features than ornithophilous features when in the study site birds seem to be more frequent flower visitors. This observed imbalance could possibly be an artefact of poor conditions for night observation and because rodents may be more easily affected than birds by the interference of observers. It is also feasible that a higher frequency of bird visits is a local or temporal event (Cocucci & Sérsic, 1998).

FIG. 1. Reconstruction based on footprints of how the pollinator rodent *Graomys griseoflavus* visits *Cajophora coronata*. Flower diameter c. 45 mm.

Bromelia urbaniana (Bromeliaceae)

This plant is locally dominant in the lowland subtropical Chaco forest. The rhizomatous rosette plants form dense understorey populations and grow often near to rodent nesting sites. We do not have any direct evidence that mammals visit the flowers of this species, but it fits well with the idealised NFM-pollination syndrome (Table 1), taking into account the ground plan of a bromeliad flower.

Plants flower at the beginning of spring when in the Chaco forest the growing season has not started and other floral resources are scarce. Inflorescences have few flowers that open at ground level in the centre of the rosette. Flowers are broadly bowl-shaped and white, stamens are exserted and purplish in colour. This purplish colour is probably cryptic in darkness. Colour and form contrast with the typical hummingbird-pollinated bromeliads. A copious jelly-like nectar is produced mainly in the evening. The septal nectary is large and resembles that of chiropterophilous bromeliads. An odour reminiscent of garlic and sour milk is produced in the evening. Flower life span is 5-7 days. Anthers start opening during the 1st and 2nd day. Stamens progressively move out of the flower's center. Nectar accumulates until days 3 to 4. The jelly-like nectar can only be taken by licking. Few insects, mainly flies are able to take viscous secretions but nectar quantities are more substantial than needed for such small insects. There is a possible mammal guard represented by very stiff and sharp hooked prickles at the leaf margins that are turned to the inside of the rosette. When trying to smell the flowers we could witness how effective is this terrible defence, to which the former genus name *Deinacanthon* is applied.

Pollination system of *Bromelia urbaniana*

Vertebrate pollination is dominant in tropical Bromeliaceae. Most species are hummingbird-pollinated (Sazima *et al.*, 1995a) and several species of *Aechmea* and *Encholirium* are pollinated by phyllostomatid bats (Helversen, 1993; Sazima *et al.*, 1989, 1995, 1996; Vogel, 1969). In subtropical zones where atmospheric epiphytes (subfamily Tillandsioideae) become more diverse, entomophily seems to prevail in the family. Flower-visiting bats are only sporadic south of the tropic of Capricorn and apparently absent in the Chaco forest. Passerines and hummingbirds may still be important for some terrestrial bromeliads at these latitudes (notably the Pitcairnioideae genera *Deuterocohnia*, *Dyckia*, *Puya*). Without the possibility of bats as pollen vectors, adaptation to non-flying mammals was probably the closest alternative for the pollination of large nectar-rich flowers.

Several other species of *Bromelia* and *Gregia* have inflorescences at ground level and cryptic flowers; it will be interesting to know if they share with *B. urbaniana* other features that we relate to NFM-pollination.

If our suggestion that *Bromelia urbaniana* is NFM-pollinated is corroborated, it may be rewarding to search for convergent examples among other monocotyledon families with a ground plan similar to that of Bromeliaceae. Such an example may be present in the Cape flora where rodents are already known to act as flower pollinators. The flowers of *Massonia latifolia* L. (Hyacinthaceae) have several traits strongly reminiscent of *Bromelia urbaniana*. The plant is geoflorous with broadly bowl-shaped greenish-yellow flowers. The fleshy, fused perianth, the strong, stiff inwardly curved stamens and the wiry straight style make the flowers extraordinarily robust. Nectar is so copiously produced that it almost overflows the flower cup. A syrup-like scent is also produced. Although flower visitors have not been seen, this species has been tentatively classified as ornithophilous because of flower sturdiness and nectar richness (Vogel, 1954). *Androcymbium longipes* (Colchicaceae) provides a further potential case from the Cape Flora.

Rodents as pollinators

To our knowledge, this paper represents the first evidence of flower visitation by rodents in South America. In the New World, pollination by rodents has previously been reported only for the tropical epiphyte *Blakea* (Melastomataceae) from Costa Rica (Lumer, 1980; Lumer & Schoer, 1986). In the south-western Cape (South Africa), rodents are involved in the pollination of a specialised group of geoflorous *Protea* species. As in other cases of NFM-pollination, the mutualistic relationship between rodents and the flowers is highly asymmetrical, with no clear evidence for plant-pollinator co-evolution (Wiens *et al.*, 1983; Nilsson *et al.*, 1993), so that there are no obvious anthophilous traits of the rodents. Six species of rodent visitors to *Blakea* belong to genera of the subfamily Sygmodontinae *Oryzomys*, *Peromyscus*, *Reithrodontomys* and *Scotinomys* (Lumer, 1986). Five species captured near and carrying pollen of *Protea* belong to four rodent genera (Fam. Muridae, subfamily Murinae): *Acomys*, *Aethomys*, *Praomys* and *Rhabdomys*. A fifth genus belongs to the order Macroscelidea: *Elephatulus* (Wiens *et al.* 1983).

Our findings suggest that pollination by non-flying mammals may be more widespread than hitherto believed, with respect to both geographical and taxonomic distribution. Observations on pollination by lemurs in Madagascar indicate that NFM-pollination is an archaic system from the early Cretaceous (Sussmann & Raven, 1978), consistent with findings in *Ravenala* (*Strelitziaceae*) by Kress *et al.* (1992). The same conclusion is tempting with evidence provided for *Strongylodon* (Nilsson *et al.* 1993) and *Adansonia* (Baum, 1995). Pollination in *Parkia madagascariensis* R. Viguier is also an important issue in this context (Grünmeier, 1990). Such interactions, which can be regarded as living fossils (Sussman & Raven, 1978), are apparently relicts of systems more widespread during the Cretaceous. They probably represent the forerunners required by the hypothesis explaining the parallel association of certain amphitropical taxa (*Mucuna, Parkia*) to bats and flying foxes (Vogel, 1969). Other flowers regularly pollinated by arboreal vertebrates in the South American tropics do not fall into a class of flowers specially adapted to pollination by non-flying mammals. In an environment where flower-visiting bats are dominant pollinators in the adaptive zone of large nectar-rich canopy flowers, this situation seems the logical consequence.

Flowers associated with non-flying mammals in the Neotropics mostly do not have special adaptations. Only *Blakea* exhibits traits apparently aimed at the pollinating mice. On the other hand, pollination by non-flying mammals in South Africa (Wiens *et al.* 1983), subtropical South America and to some extent Australia seem to be more recent events. Wiens *et al.* (1983) suggested that in *Protea* it evolved from bird-pollinated prototypes. Interestingly the same is likely for Loasaceae and Bromeliaceae. Nectarivory in birds is a trait that was not established until the Miocene (Sussmann & Raven, 1978). The large, sturdy and nectar rich bird-pollinated flowers would have favored a shift to NFM-pollination. In *Cajophora* this dichotomy even seems to be *in statu nascendi* as is suggested by the existence of a red-flowered sister species of *C. coronata*, the possible existence of hybrids between both, and the obvious importance birds still have in the pollination of *C. coronata*.

References

Arroyo, M.T.K., Primack, R. and Armesto, J. (1982). Community studies in pollination ecology in the high temperate Andes of Central Chile. I. Pollination mechanisms and attitudinal variation. *Amer. J. Bot.* **69**: 82–97.

Baum, D. (1995). The comparative pollination and floral biology of baobabs (*Adansonia*-Bombacaceae). *Ann. Missouri Bot. Gard.* **82**: 322–348.

Brown, D. and Kaul, R. (1981). Floral structure and mechanism in Loasaceae. *Amer. J. Bot.* **68**: 361–372.

Collins, B.G. and Rebelo, A.G. (1987). Pollination biology of the Proteaceae in Australia and southern Africa. *Austral. J. Ecol.* **12**: 387–421.

Cocucci, A.A. and Sérsic, A.N. (1998). Evidence of rodent pollination in *Cajophora coronata* (Loasaceae). *Pl. Syst. Evol.* **211**: 113–128

Goldingay, R.L., Carthew, S.M. and Whelan, R.J. (1991). The importance of non-flying mammals in pollination. *Oikos* **61**: 79–87.

Gribel, R. (1988). Visits of *Caluromys lanatus* (Didelphidae) to flowers of *Pseudobombax tomentosum* (Bombacaceae): a probable case of pollination by marsupials in central Brazil. *Biotropica* **20**: 344–347.

Grünmeier, R. (1990). Pollination by bats and non-flying mammals of the African tree *Parkia bicolor* (Mimosaceae). *Mem. New York Bot. Gard.* **55**: 83–104.

V. Helversen, O. (1993). Adaptations of flowers to pollination by Glossophagine bats. In: W. Barthlott *and others* (editors). Animal-plant interactions in tropical environments, pp. 41–59, Museum Koenig, Bonn.

Janson, C., Terborgh, J. and Emmons, L. (1981). Non-flying mammals as pollinating agents in the Amazonian forest. *Biotropica* **13**: 1–6.

Janzen, D. H. (1968). Reproductive behavior in the Passifloraceae and some of its pollinators in Central America. *Behaviour* **32**: 33–48.

Hopkins, H.C. (1992). The radiation of *Mucuna* in New Guinea and the role of birds, bats and possums as floral visitors. In: B. Schrire (editor) International Legume Conference [3rd: 1992: Kew], Poster Sessions, Abstr. 14, Royal Botanic Gardens, Kew.

Kerner von Marilaun, A. (1891). Pflanzenleben. Bd. 2. Leipzig.

Kress, W., Schatz, G., Andrianifahanana, M. and Morland, H. (1992). Pollination of *Ravenala madagascariensis* (Strelitziaceae) by lemurs in Madagascar: evidence for an archaic coevolutionary system? *Amer. J. Bot.* **81**: 542–551.

Lumer, C. (1980). Rodent pollination of *Blakea* (Melastomataceae) in a Costa Rican cloud forest. *Brittonia* **32**: 512–517.

Lumer, C. and Schoer, R. (1986). Pollination of *Blakea austin-smithii* and *B. penduliflora* (Melastomataceae) by small rodents in Costa Rica. *Biotropica* **18**: 363–364.

Nilsson, A., Rabakonandrianina, E, Pettersson, B. and Grünmeier, R. (1993). Lemur pollination in the malagasy rainforest liana *Strongylodon craveniae* (Leguminosae). *Evol. Trends Pl.* **7**: 49–56.

Prance, G.T. (1980). A note on the probable pollination of *Combretum* by *Cebus* monkeys. *Biotropica* **12**: 239.

Porsch, O. (1934). Säugetiere als Blumenausbeuter und die Frage der Säugetiereblume I. *Biol. Gen.* **10**: 657–685.

Porsch, O. (1935). Säugetiere als Blumenausbeuter und die Frage der Säugetiereblume II. *Biol. Gen.* **11**: 171–188.

Porsch, O. (1936a). Säugetiere als Blumenausbeuter und die Frage der Säugetiereblume III. *Biol. Gen.* **12**: 1–21

Porsch, O. (1936b). Säugetiereblumen. *Forsch. & Fortschr.* **12**: 207.

Rourke, J. and Wiens, D. (1977). Convergent floral evolution in South African and Australian Proteaceae and its possible bearing on pollination by nonflying mammals. *Ann. Missouri Bot. Gard.* **64**: 1–17.

Sazima, I., Vogel, S. and Sazima, M. (1989). Bat pollination of *Encholirium glaziovii*, a terrestrial bromeliad. *Pl. Syst. Evol.* **168**: 167–179.

Sazima, I., Buzato, S. and Sazima, M. (1996). An assemblage of hummingbird-pollinated flowers in a montane forest in southeastern Brazil. *Bot. Acta* **109**: 149–160.

Sazima, M., Buzato, S. and Sazima, I. (1995). Bat pollination of *Vriesia* in southeastern Brazil. *Bromelias, Revista Soc. Brasil. Bromélias* **2(4)**: 29–37.
Schlindwein, C. (1995). Wildbienen und ihre Trachtpflanzen in einer Südbrasilianischen Buschlandschaft: Fallstudie Guaritas, Bestäubung bei Kakteen und Loasaceen. Verlag Ulrich E. Grauer, Stuttgart.
Sleumer, H. (1955). Die Loasaceae Argentiniens. *Bot. Jahrb. Syst.* **76**: 411–472.
Steiner, K. (1981). Nectarivory and potential pollination by a neotropical marsupial. *Ann. Missouri Bot. Gard.* **68**: 505–513.
Sussmann, R. and Raven, P. (1978). Pollination by lemurs and marsupials: an archaic coevolutionary system. *Science* **200**: 731–736.
Vieira, M. F., de Carvalho-Okado, R. M., and Sazima, M. (1991). The common opossum, *Didelphis marsupialis* as a pollinator of *Mabea fistulifera* (Euphorbiaceae). *Ci. & Cult.* **43**: 390–393.
Wiens, D., Rourke, J., Casper, B., Rickart, E., LaPine, T., Peterson, C. and Channing, A. (1983). Nonflying mammal pollination of southern african Proteas: a non-coevolved system. *Ann. Missouri Bot. Gard.* **70**: 1–31.
Vogel, S. (1954). Blütenbiologische Typen als Elemente der Sippengliederung. *Bot. Stud.* **1**: 1–338.
Vogel, S. (1969). Chiropterophilie in der neotropischen Flora. Neue Mitteilungen III. *Flora, B* **158**: 289–323.

Linder, H.P. (1998). Morphology and the evolution of wind pollination. In: S.J. Owens and P.J. Rudall (Editors). Reproductive Biology, pp. 123–135. Royal Botanic Gardens, Kew.

MORPHOLOGY AND THE EVOLUTION OF WIND POLLINATION

H.P. LINDER

Bolus Herbarium, Botany Department, University of Cape Town, Rondebosch 7700, South Africa

Abstract

The wind pollination syndrome is clearly characterised by a number of morphological features, involving inflorescence structure, perianth structure, ovaries and the pollen. An analysis of some 65 transitions from biotic to wind pollination shows that some of these characteristics (reduction in the number of ovules, reduction in the perianth, loss of perianth colour, smooth pollen, evolution of dioecy, loss of nectaries) often happen after the transition from biotic to wind pollination. These are interpreted as adaptations to wind pollination (i.e. features that evolve in response to wind pollination, and that apparently make wind pollination more efficient). However, there is a second set of features, namely small, simple flowers and more or less dry pollen, that appear to be always present before wind pollination evolves, or at least arise at the time of the evolution of wind pollination. The presence of sticky pollen and large or complex flowers are interpreted as constraints, that would prohibit the evolution of wind pollination. On this basis a model for the evolution of wind pollination can be constructed, which suggests that the intermediate, ambophilous situation, is an unspecialised, basic condition, from which either wind or insect pollination can evolve. Either leads to specialisations which makes a transition to the other unlikely.

Introduction

A recent survey indicates that wind pollination is found in some 10% of the Angiosperms (Linder, in prep.), derived from at least 65 transitions from biotic to wind pollination. Wind pollination is usually explained in terms of environmental factors (Whitehead, 1969; Faegri & van der Pijl, 1979; Regal, 1982); however, it is evident that wind pollinated taxa are found in habitats regarded as unsuitable for anemophily (Bawa & Crisp, 1980), while insect pollinated plants are found in habitats more suitable for wind pollination (e.g. Kevan, 1972). It is therefore clear that there is some 'phylogenetic conservatism' in the evolution of pollination syndromes.

The morphological characteristics of wind pollinated plants have been well described by Faegri & van der Pijl (1979). The flowers are small, and clustered in dense inflorescences often on the outside of plants; the perianth is reduced or absent and when present it is often dully coloured. Nectaries are usually absent, but when they are present, they tend to be non-functional. The anthers are often borne on long filaments outside the flowers, produce abundant pollen and have explosive dehiscence mechanisms. The stigma surface is usually expanded, often variously elaborated as brushes or feathers, and is sometimes enlarged and sticky and usually projecting beyond the perianth. The flowers tend to be dichogamous, and the sexes are often separated, either monoecious or dioecious. The pollen grains are small and smooth, with a dry surface (Wodehouse, 1935; Hesse, 1979a-c). There is frequently a reduction in the number of ovules, often to a single ovule per ovary.

However, it is not immediately evident which characters evolve as a consequence of anemophily, and which floral structures are prerequisite for the evolution of wind pollination (Kaplan & Mulcahy, 1971). Characters which evolved as a consequence of anemophily can be described as adaptations to wind pollination, in that they evolve in response to wind pollination, and function to make wind pollination more efficient (Gould & Vrba, 1982; Baum & Larson, 1991). Characters which are prerequisites for the evolution of wind pollination need to be in place before wind pollination can evolve. They are not adaptations to wind pollination, as they evolve before wind pollination; in the terminology of Gould & Vrba (1982) they would be exaptations. They open up the possibility of wind pollination, and the absence of the these characters would prohibit the evolution of wind pollination. These characters could be regarded as 'key innovations' (Wake & Larson, 1987; Larson *et al.*, 1981; Baum & Larson, 1991) or 'key mutations' (Simpson, 1953), as they open up a new adaptive zone. They differ, at least in some cases, in that they cannot be demonstrated to be 'innovations' or 'mutations', but may be relictual characters. Key innovations are frequently interpreted as innovations which result in increases in the speciation rate (Hodges & Arnold, 1995). Here I did not investigate the speciation effects of wind pollination, but rather the innovations that appear necessary for wind pollination to evolve.

The morphological attributes of wind pollination are partitioned into adaptations, exaptations, and key innovations, using the distinctions developed above. Although the original data set includes all the angiosperms, in this paper I deal only with the higher hamamelid clade, and the poalean clade of the monocotyledons.

Methods and caveats

Each transition from biotic to wind pollination was determined by an iterative procedure. First the largest entirely anemophilous taxa were located (either orders, or families, or genera). Then the sister taxon to this group was inspected to determine whether it is anemophilous or biotically pollinated. If biotically pollinated, then the transition to anemophily is postulated to occur at the node connecting the two taxa. If anemophilous, then the next sister taxon is inspected, until the transition is found. For this procedure phylogenetic information on the relationship between wind pollinated and biotically pollinated taxa is needed; this information was gleaned from the angiosperm phylogenies of Chase *et al.* (1993) and Hufford (1992), as well as the more restricted phylogenies of Les & Haines (1995), Hufford & Crane (1989), Olmstead *et al.* (1992), Duval *et al.* (1993), Chase *et al.* (1995), Linder & Kellogg (1995) and Stevenson & Loconte (1995). The approach used does not require the reconstruction of a complete phylogenetic hypothesis for the angiosperms: of interest are only the phylogenetic relationships of anemophilous taxa.

The major morphological features of the wind pollination syndrome were related to these transitions, to determine whether the feature appeared before, at or after the transition. As the majority of the transitions to wind pollination are found within the rosids (including the Hamamelidae), particular attention was paid to the partial cladogram of Chase *et al.* (1993), which illustrates the nesting of the hamamelids (*sensu* Cronquist, 1981) within the rosids (Figs. 1–3) and the commelinid clade of the monocots, which contains the vast majority of wind pollinated species (Fig. 4). Families were used as the terminal units, and where the families are polymorphic, an attempt was made to estimate the ancestral condition for the family. This is less than ideal, as the analysis should investigate each transition, irrespective of the position in the phylogeny. However, this is obviously not feasible, as for most families no phylogenetic hypotheses exist.

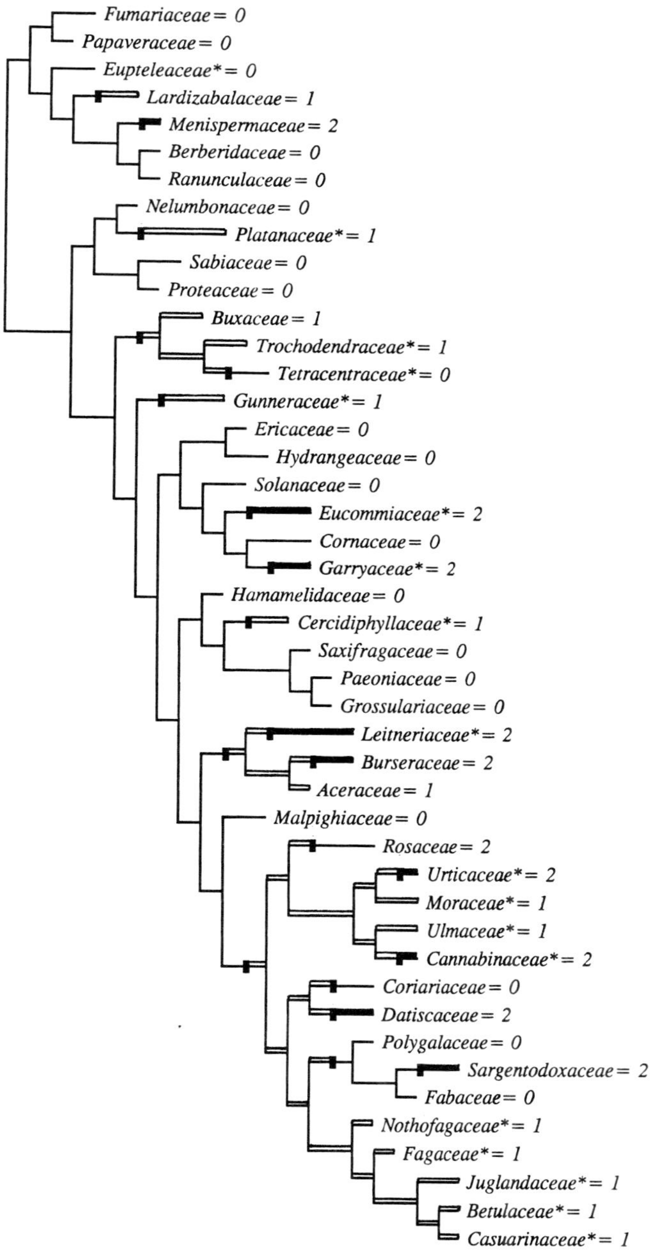

FIG. 1. Reproductive organisation of wind pollinated taxa and allies, optimised onto a cladogram of the lower rosids based on Chase *et al.* (1993). Wind pollinated taxa are indicated by asterisks, bisexual flowered taxa by 0, monoecious taxa by 1 and dioecious taxa by 2. Characters are optimised using DELTRAN, with thin lines indicating bisexual flowers, diclinous lineages by hollow lines and dioecious lineages by thick, solid lines. The correlation between flower sexuality and pollination mode is poor (Table 1), but wind pollinated taxa are rarely bisexual.

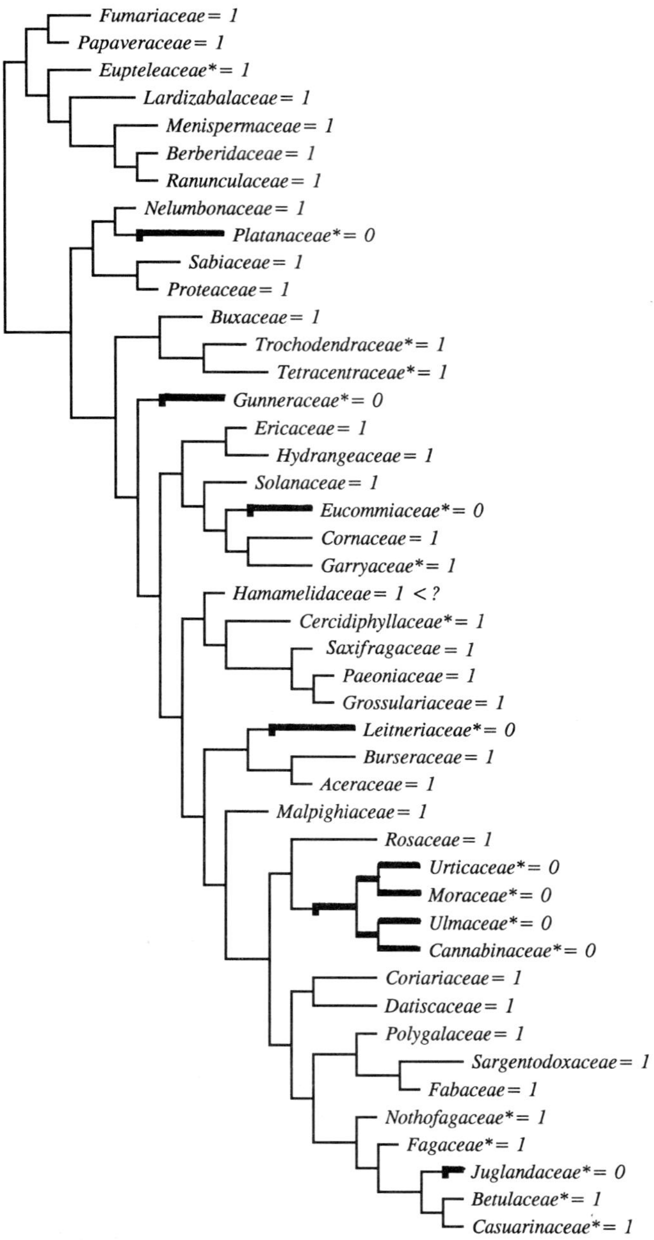

FIG. 2. Ovule number optimised onto a cladogram of the lower rosids based on Chase *et al.* (1993). Wind pollinated taxa are indicated by asterisks, uniovulate ovaries are indicated by 0 and lineages by thick lines; multiovulate ovaries by 1 and lineages by thin lines. Transitions from multi- to uniovulate ovaries indicated by bars. Note that all uniovulate taxa are wind pollinated, while not all wind pollinated taxa are uniovulate. The uniovulate condition evolved after wind pollination in a number of lineages.

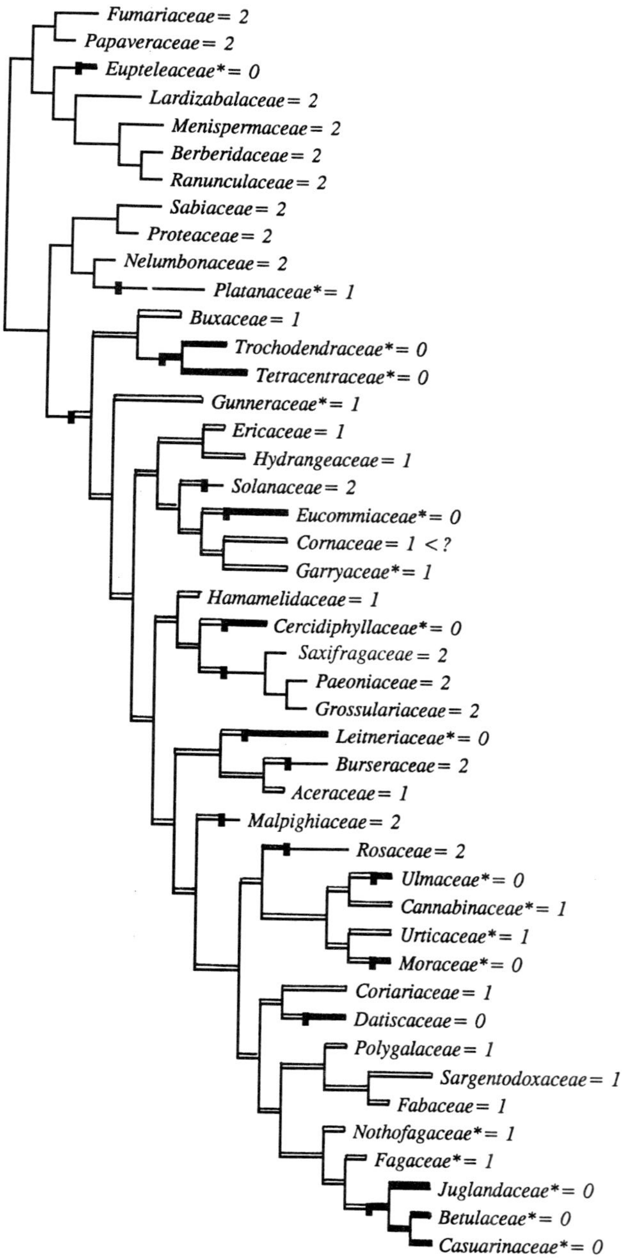

FIG. 3. Perianth development optimised onto a cladogram of the lower rosids based on Chase *et al.* (1993). Wind pollinated taxa are indicated by asterisks; families without perianths indicated by 0 and lineages by thick, solid lines; families with small perianths by 1 and lineages by hollow lines; large perianths by 2 and lineages by thin lines. There are no wind pollinated taxa with large perianths, and in many cases the reduction in the perianth appears to be coincident with the evolution of wind pollination. Complete loss of perianth occurs in several wind pollinated lineages.

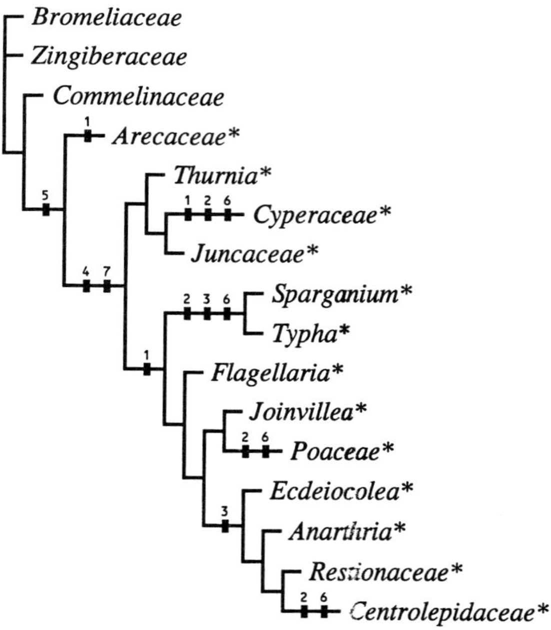

FIG. 4. Phylogeny of the poalean monocots (based on Linder & Kellogg, 1995), with wind pollinated families are indicated by asterisks. Character transitions shown on the tree are (1) ovules from several per carpel to solitary per carpel; (2) carpels from two or three to solitary one; (3) flowers from bisexual to unisexual; (4) pollen from sulcate to porate; (5) perianth from coloured to colourless; (6) perianth from complete, with six parts in two whorls, to various reduced and (7) loss of nectaries. Loss of perianth colour occurs at the base of the wind pollinated lineage (if palms are interpreted as wind pollinated), while porate pollen and loss of nectaries is at the base of the obligately wind pollinated lineage. The other reductions (ovule number, carpels and number of perianth parts) occur within the wind pollinated lineage.

The correlations between the morphological changes and the transitions to wind pollination were investigated using the concentrated-changes test of Maddison (1990), as implemented in MacClade vers. 3 (Maddison & Maddison, 1992). The test evaluates the association of changes in one character with another variable. The test is sensitive to the inclusion or exclusion of taxa (Maddison, 1990), and this is potentially problematic in the higher Hamamelid tree, which contains only a sample of the families. It should be less problematic in the commelinid monocots, which include all the families of the clade. Ideally this should be calculated for a cladogram of all the angiosperm families: this is currently not feasible.

Results

Not surprisingly, the morphological characters are, with the exception of dioecious flowers, significantly correlated with the evolution of anemophily (Table 1). There are, however, large differences in whether anemophily in the two analysed test cases evolves in the absence of the character, and only pollen stickiness and perianth reduction appear to have to be in place before wind pollination can evolve, although at the level of this analysis the two features often evolve together. The other features (reduction to single ovules, loss of nectaries, dioecy) all frequently evolve after wind pollination, although in some cases they may evolve without wind pollination, or even before wind pollination.

TABLE 1. Evaluation of characters associated with anemophily in the hamamelid and commelinid clades. The second column indicates the probability that the distribution of the character is not related to the distribution of wind pollination on the cladogram, as calculated by the concentrated-changes test of Maddison (1990). This is calculated only on the hamamelid clade.

Character	Significance	Changes to anemophily after or at morphological changes/total changes	Number of changes in characters relative to change to anemophily. The first column refers to morphological changes not with associated pollination mode changes.			
			without	before	at	after
Ovules solitary	0.0	4/11	0	0	5	3
Pollen dry	0.0	11/11	0	0	11	0
Nectaries lost	0.027		3	1	7	1
Perianth medium or small sized	0.225	11/11	0	3	5	0
Perianth small	0.0	6/11	1	0	6	3
Flowers imperfect	0.021	10/11	3	2	7	0
Plants dioecious	0.446	3/11	3	0	3	4

Discussion

Pollen characteristics

Pollen characteristics are often correlated with anemophily. Anemophilous plants invariably have dry pollen, while biotically pollinated plants have sticky or dry pollen (Pohl, 1929b; Wodehouse, 1935; Faegri & van der Pijl, 1979; Hesse, 1980; Vroege & Stelleman, 1990; Pacini & Franchi, 1993). Stickiness, which is easily measured by the scatter of pollen on a microscope slide, is a complex character, affected by the quantity of pollenkitt, its electron density, and the position of its deposition. In Poaceae pollen is dry due to a very limited pollenkitt production, much of which is 'wasted' on the anther walls (Hesse, 1980). In other wind pollinated taxa, the pollenkitt is made 'inactive' by secreting it into enlarged infratectal cavities (Hesse, 1978, 1979a-c, 1980). The combination of these ways of varying the effects of the pollenkitt accounts for the subtle changes in the stickiness of pollen, which was first reported by Pohl (1929b), and documented in detail by Vroege & Stelleman (1990).

Dry pollen is essential for effective wind pollination, and since pollenkitt is universal in angiosperms (Hesse, 1980), the sticky effect of the pollenkitt first has to be reduced before wind pollination can evolve. Hesse (1979c) suggested that dry pollen is found in flowers in which the pollen is dusted over the pollinators. Buzz pollination would be a subset of this type, which is common in Ericaceae and other porate-anther taxa, as is found in *Calluna vulgaris* L. (Beijerrinck, 1940). The pollen has to be dry to scatter over the pollinator, and is retained in the indumentum of the pollinator. This could contribute to the frequent evolution of wind pollination in the Ericaceae s.l. A second mechanism could operate where the pollen gradually dries out in the anthers. In the initial stages of anthesis, the pollen is sticky and suitable for entomophily, while in the later stages wind dispersal may become effective. The pollen of ambophilous taxa is often of intermediate stickiness, suggesting that this mechanism may well facilitate the transition to wind pollination. This hypothesis is in need of corroboration, and there is scope for extending the studies of Hesse in documenting the transition to dry pollen. The distribution of dry pollen in non-anemophilous taxa needs to be documented, and the relationship between the evolution of anemophily and the presence of dry pollen in the sister taxa to anemophilous species needs to be explored.

Most wind pollinated taxa have smooth pollen, but anemophily evolved in Asteraceae, a family in which echinate pollen is common. Wodehouse (1935) elegantly demonstrated the reduction in the pollen spines with wind pollination. The reduction is evident from morphological observations, as even the apparently smooth pollen of *Artemisia* still appear to have the remnants of spines. A curious exception is found in *Arceuthobium*, where the pollen spines are well developed: apparently to retain the pollen in the anthers (Player, 1979).

Flower sexuality

Wind pollinated taxa have flowers in which the sexes do not co-occur: they are either dioecious, or at least dichogamous if bisexual. Faegri & van der Pijl (1979) suggested that this serves to eliminate selfing, as wind dispersed pollen is always most dense around the pollen-producing plant, rapidly becoming less dense with greater distance from the pollen source. Hermaphroditic flowers will therefore have a very high rate of self-pollination, which, even if there is an incompatibility mechanism, could lead to the clogging of the stigmas. In *Leucadendron* (Proteaceae) dioecy evolved before anemophily, as the whole genus is dioecious, but only a few species are anemophilous (Williams, 1972), suggesting that dioecy here may have permitted the evolution of anemophily. However, in many other lineages dioecy follows the evolution of anemophily (e.g. Fig. 1). In the poalean group, where the whole group is anemophilous, dioecy evolved independently and later in Restionaceae, many species of Poaceae, and some genera of Cyperaceae (Fig. 4). In these cases, dichogamy may have been the precondition for the evolution of anemophily, but it is not clear whether dicliny generally precedes anemophily, or *vice versa* (Kaplan & Mulcahy, 1971; Faegri & van der Pijl, 1979). Kaplan & Mulcahy (1971) suggested that dicliny evolved first, and that with pollen-feeding pollinators this leads to inefficient pollination, shifting the selective balance in favour of anemophily. This is contrary to the general pattern which argues for environmental factors selecting for anemophily, rather than the morphology of the plants.

Ovule number

The relationship between ovule numbers and pollination mode was reviewed in detail by Pohl (1929a). The ovaries of anemophilous plants are frequently uniovulate, but there are numerous exceptions. In many taxa several carpels are present, but only one is fertile, thus indicating that the reduction is often from a multiovulate to a uniovulate condition (Fig. 2, 4), and there appears to be little doubt that the uniovulate condition is an adaptation to wind pollination. Pohl (1929a) suggested that this can be explained in terms of the pollen dispersal characteristics, with wind pollen usually dispersed as monads, while biotically dispersed pollen arrives at the stigmas in clumps. Although some lineages are entirely uniovulate, in others the reduction series from many to a single ovule occurs within the anemophilous lineage. This is exemplified by the poalean lineage (Fig 4). Throughout the lineage the ovaries are fundamentally tricarpellary. The basal Juncaceae have numerous ovules in each locule, while in all other families each carpel is uniovulate. In several families (Cyperaceae, Poaceae,

Typhaceae, many Restionaceae) a single functional carpel is found (pseudomonomery), and this reduction has happened several times (Linder & Kellogg, 1995; Linder 1992a, b). The reduced ovule number is therefore an adaptation to wind pollination, rather than a necessary prerequisite for the evolution of wind pollination. The functional advantage of uni-ovulate carpels in anemophilous plants is not clear. Pohl's hypothesis would suggest that wind-pollinated carpels are more likely to capture single pollen grains, than biotic carpels. This has not been tested, but Honig *et al.* (1992) and Linder & Midgley (1996) reported large pollen loads on the stigmas of anemophilous plants.

Nectaries

The absence of nectaries in anemophilous plants is shown in many cases as evolving later than anemophily. Faegri & van der Pijl (1979) indicated that *Cannabis* and *Urtica* have relictual nectaries; Player (1979) added *Arceuthobium* to this list; Grasshof & Beaman (1970) described functional nectaries in presumably anemophilous species of *Eupatorium*. This feature therefore appears to be a result of anemophily, rather than facilitating its evolution. The presence of nectaries is sometimes used to argue for biotic pollination, but this assumes that structures are lost the moment they lose their functions.

Perianth

Plants with showy flowers are usually assumed to be biotically pollinated, although there is no obvious reason why anemophilous plants should have dull flowers. Wind pollination does seem to be associated with several perianth features. The first is reduction in the perianth (Figs. 3, 4). This is taken to an extreme in Poaceae, where the perianth is reduced to three or two fleshy lodicules, completely enclosed by the palea and lemma, and in Cyperaceae, where it is often reduced to a set of bristles. Among many of the wind-pollinated rosids the perianth may be completely absent. Anemophily evolved particularly commonly in the magnoliids, caryophyllids, and rosids - these are all groups in which the perianth is either generally poorly developed, or not differentiated into two highly specialised whorls, as it is in asterids and many of the lilioid monocots. In lineages like *Erica*, which are sympetalous and often possess showy perianths, wind pollination evolved only in groups with small flowers and tiny perianths. There are no cases where anemophily co-occurs with large showy perianths. Anemophily clearly evolves most commonly in taxa with small or no perianths, and with open flowers. This is strongly corroborated by data from ambophilous taxa. These are all taxa with very similar, open flowers, characterised by Arecaceae in monocots and *Salix* in dicots. This observation is consistent with the high frequency of anemophily in groups with simple flowers, and its rarity in groups with more complex flowers.

There appear to be no cases of the evolution of anemophily in groups with zygomorphic flowers. Zygomorphy is common in the asterids, and anemophily evolved at least 11 times in this group, but always from families with actinomorphic flowers, such as Oleaceae, Asteraceae, Rubiaceae, but not from the Lamiales or Scrophulariales, despite their frequent appearance in high latitude habitats suitable for the evolution of anemophily. Similarly, large, zygomorphic flowered families like Fabaceae have produced almost no anemophilous species. Apparently a striking exception would be Proteaceae, which have zygomorphic flowers: but in these taxa the perianth is simple and small, and the reproductive structures exceed the perianth even in biotically pollinated species.

The argument is somewhat weaker in monocotyledons, where zygomorphic flowers are relatively rare. However, anemophily never evolved in the 'petaloid' monocotyledons, which often have specialised pollination systems: these are usually associated with showy or zygomorphic flowers.

Model of the evolution of anemophily

The model

Plants with numerous small, open flowers, without a perianth, or with a small perianth, and with dry pollen, may often be both wind and insect pollinated. This is exemplified by *Hormathophylla spinosa* (Cruciferae) which has small, open flowers, densely clustered at the ends of the branches (Gómez & Zamora, 1996) and by *Tilia* (Tiliaceae; Hesse, 1979). Although these flowers are visited and pollinated by insects, a substantial amount of pollen is also distributed by wind. If, for some reason, the insect pollination fails, features which improve wind pollination, even at the cost of effective biotic pollination, are at a selective advantage. This then results in adaptations to wind pollination, which involves the evolution of the other characters of the syndrome: smooth pollen, a reduction in the number of ovules, dicliny or dioecy. These characteristics may make anemophily more efficient, but they are not essential for wind pollination. These are 'optional adaptations' to wind pollination. In various lineages, the reduction series after the establishment of anemophily differ. In some cases these characteristics may have evolved before anemophily, then they would be exaptations. This is exemplified by dioecy in *Leucadendron*. Clearly, the features of anemophily are not restricted to anemophilous taxa, and may have evolved earlier. Thus the classification of characteristics as 'adaptations' and 'exaptations' is lineage dependent, and features which may be exaptations in one lineage (i.e. dioecy in *Leucadendron*) may be adaptations in others (i.e. in the poalean lineage).

This model therefore starts with an ambophilous situation, where a small perianth and numerous small flowers, in conjunction with dry pollen, act as key innovations to facilitate the evolution of wind pollination. If wind pollination evolves, adaptations to wind pollination make the return to insect pollination more and more unlikely. Similarly, specialisation of the flowers for biotic pollination make the evolution of wind pollination more unlikely.

The frequency of wind pollination in groups with unspecialised perianths is accounted for in this model. The model can be tested by investigating the characteristics of ambophilous taxa, and the morphology of taxa which show reversals from wind to insect pollination.

Reversals

Anemophily represents a reduction series involving the loss of nectaries, showy flowers and scents capable of attracting pollinators, and culminating in unisexual flowers. With a reduction series like this, it is tempting to assume that once the structures required for attracting and rewarding pollinators are lost, they cannot readily be acquired again, and that, therefore, reversals from anemophily to biotic pollination do not occur. However, phylogenetic analysis suggests at least some reversions from anemophily to entomophily:

(1) *Ficus* (Urticales). Here the highly reduced, unisexual, anemophilous flowers are aggregated into synconia, and the synconia offer ovules, rather than nectar, as reward (Stebbins, 1970). Figs have evolved a highly specific wasp pollination system.

(2) Cyperaceae. *Dichromena*, widespread in the neotropics, attracts stingless bees with its white involucral bracts, and apparently the bees are rewarded with pollen (Leppik, 1955).

(3) Poaceae. *Paspalum dilatatum* in Oklahoma is at least partially pollinated by insects (Adams *et al.*, 1981). The remarkably large pollen grains do not travel far on the wind, are actively collected by bees, and stigmas with bees excluded contained less pollen grains than stigmas exposed to both wind and bees. Soderstrom & Calderón (1971) suggested that several rainforest understorey grasses may be entomophilous, as they are extensively visited by insects, and there is no evidence of wind to disperse the pollen.

It is evident that there are many different routes from anemophily back to biotic pollination. An analysis of the different structures that could be used to attract and reward pollinators would be interesting, and is indicative of the morphological and evolutionary plasticity of angiosperms.

Conclusion

Wind pollination has evolved numerous times, and from the current information it appears that dry pollen and simple, open perianths are structures or innovations that have to be in place before wind pollination can evolve. The other characteristics of the wind pollination syndrome evolve as later adaptations to the system. These adaptations result in the loss of the means to attract and reward pollinators, so that reversal to wind pollination often involves complex transfers of functions. Wind pollination is therefore seen as a specialisation from a basic ambophilous system, characterised by simple, open, flowers with somewhat sticky pollen.

Acknowledgments

Wendy Hitchcock for putting the data together; FRD for funding; Jeremy Midgley, Mike Crisp and Tony Verboom for discussion, and two anonymous referees for several insightful comments.

References

Adams, D.E., Perkins, W.E. and Estes, J.R. (1981). Pollination systems in *Paspalum dilatatum* Poir. (Poaceae): an example of insect pollination in a temperate grass. *Amer. J. Bot.* **68**: 389–394.

Baum, D.A. and Larson, A. (1991). Adaptation reviewed: a phylogenetic methodology for studying character macroevolution. *Syst. Zool.* **40**: 1–18.

Bawa, K.S. and Crisp, J.E. (1980). Wind-pollination in the understorey of a rain forest in Costa Rica. *J. Ecol.* **68**: 871–876.

Beijerinck, W. (1940). *Calluna*. A monograph on the Scotch heather. *Verh. Kon. Ned. Akad. Wetensch., Afd. Natuurk., Tweede Sect.* **38(4)**: 1–180.

Chase, M.W., *and others* (1993). Phylogenetics of seed plants: an analysis of nucleotide sequences from the plastid gene *rbc*L. *Ann. Missouri Bot. Gard.* **80**: 528–580.

Chase, M.W., Stevenson, D.W., Wilkin, P. and Rudall, P.J. (1995). Monocot systematics: a combined analysis. In: P.J. Rudall, P.J. Cribb, D.F. Cutler and C.J. Humphries (editors). Monocotyledons: systematics and evolution, pp.685–730, Royal Botanic Gardens, Kew.

Cronquist, A. (1981). A integrated system of classification of flowering plants. Columbia University Press, New York.

Duvall, M.R., *and others* (1993). Phylogenetic hypotheses for the monocotyledons constructed from *rbc*L sequence data. *Ann. Missouri Bot. Gard.* **80**: 607–619.

Faegri, K. and Van der Pijl, L. (1979). The principles of pollination ecology. Pergamon Press, Oxford.

Gómez, J.G. and Zamora, R. (1996). Wind pollination in high-mountain populations of *Hormathophylla spinosa* (Cruciferae). *Amer. J. Bot.* **83**: 580–585.

Gould, J.S. and Vrba, E.S. (1982). Exaptation - a missing term in the science of form. *Paleobiology* **8**: 4–15.

Grashoff, J.L. and Beaman, J.H. (1970). Studies in *Eupatorium* (Compositae). III. Apparent wind pollination. *Brittonia* **22**: 77–84.

Hesse, M. (1978). Entwicklungsgeschichte und Ultrastructur von Pollenkitt and Exine bei nahe verwandten entomophilen und anemophilen Angiospermen Sippen: Ranunculaceae, Hamamelidaceae, Platanaceae und Fagaceae. *Pl. Syst. Evol.* **130**: 13–42.

Hesse, M. (1979a). Ultrastruktur und Verteilung des Pollenkitts in der insekten- und windblütigen Gattung *Acer* (Aceraceae). *Pl. Syst. Evol.* **131**: 277–289.

Hesse, M. (1979b). Entwicklungsgeschichte und Ultrastructur von Pollenkitt und Exine bei nahe verwandten entomo- und anemophilen Angiospermen: Polygonaceae. *Flora, Morphol. Geobot. Oekophysiol.* **168**: 558–577.

Hesse, M. (1979c). Entwicklungsgeschichte und Ultrastructur von Pollenkitt und Exine bei nahe verwandten entomo- und anemophilen Angiospermen: Salicaceae, Tiliaceae und Ericaceae. *Flora, Morphol. Geobot. Oekophysiol.* **168**: 540–557.

Hesse, M. (1980). Entwicklungsgeschichte und Ultrastructur von Pollenkitt und Exine bei nahe verwandten entomophilen und anemophilen Angiospermensippen der Alismataceae, Liliaceae, Juncaceae, Cyperaceae, Poaceae und Araceae. *Pl. Syst. Evol.* **134**: 229–267.

Hodges, S.A. and Arnold, M.L. (1995). Spurring plant diversification: are floral nectar spurs a key innovation? *Proc. Roy. Soc. London, Ser. B. Biol. Sci.* **262**: 343–348.

Honig, M.A., Linder, H.P. and Bond, W.J. (1992). Efficacy of wind pollination: pollen load size and natural microgametophyte populations in wind-pollinated *Staberoha banksii* (Restionaceae). *Amer. J. Bot.* **79**: 443–448.

Hufford, L.D. (1992). Rosidae and their relationship to other nonmagnoliid dicotyledons: a phylogenetic analysis using morphological and chemical data. *Ann. Missouri Bot. Gard.* **79**: 218–248.

Hufford, L.D. and Crane, P.R. (1989). A preliminary phylogenetic analysis of the 'lower' Hamamelidae. In: P.R. Crane and S. Blackmore (editors). Evolution, systematics, and fossil history of the Hamamelidae, volume 1: introduction and 'Lower' Hamamelidae, pp. 175–192, Clarendon Press, Oxford.

Kaplan, S.M. and Mulcahy, D.L. (1971). Mode of pollination and floral sexuality in *Thalictrum*. *Evolution* **25**: 659–668.

Kevan, P.G. (1972). Insect pollination of high Arctic flowers. *J. Ecol.* **60**: 831–847.

Larson, A., Wake, D.B., Maxson, L.R. and Highton, R. (1981). A molecular phylogenetic perspective on the origins of morphological novelties in the salamanders of the tribe Plethodontini (Amphibia, Plethodontidae). *Evolution* **35**: 405–422.

Leppik, E. (1955). *Dichromena ciliata*, a noteworthy entomophilous plant among Cyperaceae. *Amer. J. Bot.* **42**: 455–458.

Les, D.H. and Haines, R.R. (1995). Systematics of subclass Alismatidae: a synthesis of approaches. In: P.J. Rudall, P.J. Cribb, D.F. Cutler and C.J. Humphries (editors). Monocotyledons: systematics and evolution, pp. 353–377. Royal Botanic Gardens, Kew.

Linder, H.P (1992a). The structure and evolution of the female flower of the African Restionaceae. *Bot. J. Linn. Soc.* **109**: 401–425

Linder, H.P. (1992b). The gynoecia of Australian Restionaceae: morphology, anatomy and systematic implications. *Austral. Syst. Bot.* **5**: 227–245

Linder, H.P. and Kellogg, E.A. (1995). Phylogenetic patterns in the commelinid clade. In: P.J. Rudall, P.J. Cribb, D.F. Cutler and C.J. Humphries (editors). Monocotyledons: systematics and evolution, pp. 473–496, Royal Botanic Gardens, Kew.

Linder, H.P. and Midgley, J.J. (1996). Anemophilous plants select pollen from their own species from the air. *Oecologia* **108**: 85–87.

Maddison, W.P. (1990). A method for testing the correlated evolution of two binary characters: are gains or losses concentrated on certain branches of a phylogenetic tree? *Evolution* **44**: 539–557.

Maddison, W.P. and Maddison, D.R. (1992). MacClade. Analysis of phylogeny and character evolution, version 3. Sinauer Associates, Inc., Sunderland.

Olmstead, R.G., Michaels, H.J., Scott, K.M. and Palmer, J.D. (1992). Monophyly of the Asteridae and identification of their major lineages inferred from DNA sequences of *rbc*L. *Ann. Missouri Bot. Gard.* **79**: 249–265.

Pacini, E. and Franchi, G.G. (1993). Role of the tapetum in pollen and spore dispersal. *Pl. Syst. Evol., Suppl.* **7**: 1–11.

Player, G. (1979). Pollination and wind dispersal of pollen in *Arceuthobium*. *Ecol. Monogr.* **49**: 73–87.
Pohl, F. (1929a). Beziehungen zwischen Pollenbeschaffenheit, Bestäubungsart und Fruchtknotenbau. *Beih. Bot. Centralbl.* **46**: 247–285.
Pohl, F. (1929b). Kittstoffrest auf der Pollenoberfläche windblütiger Pflanzen. *Beih. Bot. Centralbl.* **46**: 286–305.
Regal, P.J. (1982). Pollination by wind and animals: ecology of geographic patterns. *Annual Rev. Ecol. Syst.* **13**: 497–524.
Simpson, G.G. (1953). The major features of evolution, Columbia University Press, New York.
Soderstrom, T.R. and Calderón, C.E. (1971). Insect pollination in tropical rain forest grasses. *Biotropica* **3**: 1–16.
Stebbins, G.L. (1970). Adaptive radiation of reproductive characteristics in Angiosperms, I: pollination mechanism. *Annual Rev. Ecol. Syst.* **1**: 307–326.
Stevenson, D.W. and Loconte, H. (1995). Cladistic analysis of monocot families. In: P.J. Rudall, P.J. Cribb, D.F. Cutler and C.J. Humphries (editors). Monocotyledons: systematics and evolution, pp.543–578, Royal Botanic Gardens, Kew.
Vroege, P.W. and Stelleman, P. (1990). Insect and wind pollination in *Salix repens* L. and *Salix caprea* L. *Israel J. Bot.* **39**: 125–132.
Wake, D.B. and Larson, A. (1987). Multidimensional analysis of an evolving lineage. *Science* **238**: 42–48.
Whitehead, D.R. (1969). Wind pollination in the angiosperms: evolutionary and environmental considerations. *Evolution* **23**: 28–35.
Williams, I.J.M. (1972). A revision of the genus *Leucadendron* (Proteaceae). *Contr. Bolus Herb.* **3**:1–425.
Wodehouse, R.P. (1935). Pollen grains. McGraw-Hill, New York.

Owens, S.J. and Bennett, M.D. (1998). The significance of wide crossing in systematics. In: S.J. Owens and P.J. Rudall (Editors). Reproductive Biology, pp. 137–146. Royal Botanic Gardens, Kew.

THE SIGNIFICANCE OF WIDE CROSSING IN SYSTEMATICS

SIMON J. OWENS AND MICHAEL D. BENNETT

Royal Botanic Gardens, Kew, Richmond, Surrey, TW9 3AB, UK

Abstract

Systematists have used the presence and effectiveness of barriers to fertilisation and interbreeding to investigate generic and species relatedness and to aid the definition of a genus and a species. Knowledge of genotype, especially ploidy level and breeding system, is essential in the interpretation of wide crossing results with respect to investigating systematic relationships. Self incompatibility is particularly significant since it reduces both fertilisation and seed set. Studies of fertilisation and early embryo development in wide crosses are also highly desirable. For systematic use and for evolutionary study there may also be family-specific considerations. Taxonomic judgements can mean that some species in some families may be as far apart genetically as some genera in other families. This may lead to surprising success when wide crosses are made in some families (e.g. Solanaceae). In other families (e.g. Poaceae, Orchidaceae) their success cannot be explained simply on the basis of taxonomic judgement.

Introduction

This paper discusses the results of wide crossing in several interspecific and intergeneric combinations carried out under controlled pollination conditions, and examines the application of the data to understanding phylogenetic relationships. A wide cross is a cross at or above the species level which may occur naturally or may have been created by manual pollination under controlled experimental conditions.

Modern biotechnology allows the production of organisms combining genomes from surprisingly distantly related species using protoplast fusion technologies. Indeed, some such organisms are sufficiently stable to develop to maturity (Tibok *et al.*, 1995) and hence may offer new opportunities for gene transfer in plant breeding, provided the usual major problems of pollen and egg sterility can be overcome. However, strictly speaking, such novel genome combinations produced by techniques unknown in nature are not wide crosses and hence are outside our present remit. Moreover, as yet they have no known use for systematics. Discussions in this paper are therefore restricted to where pollen has been deposited on the stigma and where gametes are delivered to the embryo sac by siphonogamy.

There is a consensus of opinion that most higher plant genera and species exist because species reproduce sexually and that genera and species remain distinct because they do not interbreed (Stebbins, 1959; Maynard Smith, 1989). If barriers to fertilisation and interbreeding establish genera and species, can taxonomists rely on the presence and effectiveness of such barriers to investigate generic and species relatedness and perhaps to aid the definition of a genus and a species? Put simply, is the extent of the barrier to successful wide crossing a useful and reliable indication of the extent of the taxonomic distance between two genera or two species?

There have been many attempts to make wide crosses experimentally and to interpret the results in relation to systematics and evolution (e.g. Grant, 1958; Frederiksen and von Bothmer, 1989; Gradziel *et al.*, 1993; van Creij *et al.*, 1997). In plants, wide crossing occurs naturally and numerous spontaneous hybrids are well documented. Indeed, the occurrence of natural interspecific hybrids is known in all major groups of plants and in all well studied floras (Grant, 1971) and is therefore thought to contribute significantly to the evolution of new species. Many naturally occurring hybrids are polyploids and the majority of these are allopolyploids which is taken to confirm a hybrid origin (Clausen *et al.*, 1945). Natural hybrids ought to suggest close relationships since they imply that there are few barriers to seed set and plant establishment. Careful and detailed studies of wide crossing have clearly contributed to taxonomic understanding in many plant groups and the need for such work will continue. For example, the genus *Hordeum* has been divided into two genera by some authors (Löve, 1982, 1984; Dewey, 1984) but on the basis of F1 hybrids this generic delimitation was not supported by von Bothmer and Jacobsen (1986) and the genus remains as a single entity in most classifications. In other wide crosses, characteristics such as low crossability and lack of meiotic pairing were used by Frederiksen and von Bothmer (1989) to indicate that *Taeniatherum* is a distinct genus, only distantly related to other genera in the tribe Triticeae.

Interspecific hybrids

(i) Wide crosses in *Hordeum* (Poaceae)

The effect of genotype is clearly identified by crosses in the genus *Hordeum* (Poaceae: Triticeae) and particularly in crosses between self compatible *H. vulgare* and self incompatible *H. bulbosum* at the diploid level (2x=14). Fertilisation occurs following the cross and the parental genomes combine normally in the zygote. However, the resulting pattern at post-zygotic early embryo development depends very much upon which clone of *H. bulbosum* is used (Subramanyam and Kasha, 1973; Bennett *et al.*, 1976). Some clones regularly display elimination of the entire haploid *H. bulbosum* genome during early embryo development and monoploid plantlets are produced after embryo rescue. However, when other diploid clones of *H. bulbosum* are used, hybrids with 2x=14 combining both parental genomes are formed, though some show infrequent loss of a single *H. bulbosum* chromosome in occasional cells (Simpson *et al.*, 1980). Indeed, *H. bulbosum* clones can be ranked from strongly eliminated to weakly eliminated to non-eliminated clones in their combinations with other species. The implication of this result for systematic studies is clear: unless reasonable steps are taken to test the fidelity of differing genotypes in differing combinations, then the interpretation would be of a significantly distant relationship with the former result and of a closer relationship with the latter result.

Hordeum bulbosum is regarded as the species most closely related to *H. vulgare* within the genus *Hordeum*, yet it is now known to possess three loci affecting uniparental chromosome loss in their wide hybrids, whose action in early embryo development can wholly prevent seed production in many interspecific combinations. Two important lessons can be learnt. Firstly, it is unwise to draw any conclusion regarding the taxonomic distance between two species based on wide crossing results for a single combination of parental species genotypes. Secondly, investigations of ovules from pollinated florets to examine embryo sac development are always highly desirable when studying wide hybrids.

(ii) Wide crosses in *Gibasis* (Commelinaceae)

The effect of the self incompatibility (SI) system is exemplified by a study of interspecific crosses carried out on *Gibasis*, a genus of Commelinaceae from Mexico. The species are almost exclusively self incompatible, both as diploids and as tetraploids (Table 1), with incompatibility controlled by a gametophytic, single gene system (Owens, 1977, 1980). All three species used are from section *Gibasis*. The results show that a majority of the crosses are incompatible. This predominantly negative result presents a problem when assessing species relationships because it is essential to determine whether the SI gene is operating alone or whether other genes are actively involved. If self incompatibility is involved, then the same alleles must be present in the different individuals of the species even though individuals were chosen at random and from different geographical sites. This may indicate recent differentiation of the species. There is some evidence that the crosses using the different species show different pollen tube responses on the stigma or in the style and that the self incompatibility gene is not operative. Such a result suggests that interspecific incompatibility controls the success of the cross. However, the result needs further investigation using controlled environments (particularly temperature) to determine whether pollen tube responses are truly different from those in the self incompatible combinations.

TABLE 1. Results of crosses between three species (five cytotypes) of *Gibasis* (Commelinaceae) (data from Owens, 1977, 1980)

	G. pulchella (2×)	*G. karwinskyana* (2×)	*G. karwinskyana* (4×)	*G. pellucida* (2×)	*G. pellucida* (4×)
G. pulchella (2×)	+/-	-	-	nt	-
G. karwinskyana (2×)	-	-	nt	+	-
G. karwinskyana (4×)	-	-	-	-	-
G. pellucida (2×)	-	+	+	-	-
G. pellucida (4×)	-	-	+	-	+/-

- pollen tube arrest
+ pollen tubes penetrate ovule
nt cross not tested

Only two species had self incompatible and self compatible genotypes. Diploid *G. pulchella* has both normal cytotypes and complex heterozygotes (Kenton *et al.*, 1988), the latter cytotypes being almost cleistogamous. Complex heterozygotes were not used in the crosses. Some tetraploid cytotypes of *G. pellucida* were self compatible but they did not invoke unilateral incompatibility responses (Harrison and Darby, 1955; Lewis and Crowe, 1958; Hiscock and Dickinson, 1998). Unilateral incompatibility is present in one interspecific cross combination, between tetraploid *G. karwinskyana* and diploid *G. pellucida*.

The results show some compatibility between genotypes of *G. pellucida* and *G. karwinskyana*, which suggests a closer relationship between these two species than between either species and *G. pulchella*. Such a relationship is further supported by similar karyotype morphology (Jones *et al.*, 1975).

(iii) Wide crossing in *Solanum* (Solanaceae)

A more recent and potentially very interesting observation was made in the genus *Solanum* (Solanaceae) concerning the interspecific cross between cultivated *Solanum tuberosum* and *S. phureja* (Wilkinson *et al.*, 1994). Both are self compatible species that have been used for many years in plant breeding programmes, particularly for producing dihaploids. Almost all dihaploid plants of *S. tuberosum* (with a chromosome number of 2n=24) are made by pollinating tetraploid *S. tuberosum* (4x=48) with clones of diploid *S. phureja* (2x=24) known as 'dihaploid inducers'. Double fertilisation can occur to produce triploid or tetraploid hybrids, but occasionally both sperm nuclei fertilise the secondary endosperm nucleus and dihaploids are produced. Although the resultant progeny are not all dihaploid, dihaploids can readily be artificially selected, even though the zygote normally fails. It was believed until recently that dihaploids are formed by parthenogenesis and hence are genetically pure *S. tuberosum*. Recent work using RAPD primers, microsatellites and leaf protein analysis has suggested that *S. phureja* DNA is present and is expressed in the dihaploid *S. tuberosum*. Moreover, genomic *in situ* hybridisation confirmed the presence of *S. phureja* DNA as small segments on three chromosomes of *S. tuberosum*. These translocations appeared to be stable (Wilkinson *et al.*, 1994), must have occurred somatically and hence may represent a natural transformation system. The presence of six translocations in one somatically doubled regenerant suggested (1) that alien segments can be stable during both chromosome duplication and subsequent DNA replication and (2) that direct introgression from the inducer species into tetraploid *S. tuberosum* occurred without the formation of intermediate triploid or tetraploid hybrids. Such interspecific hybrid material was produced under controlled conditions, but it is tantalising to speculate, particularly with a polyploid, whether there is a potential evolutionary significance to such events in nature. If wide crossing results in the delivery of gametes and fertilisation in a polyploid, even if the zygote fails there may still be a possibility of genomic translocation and the development of a dihaploid propagule. Thus there may be a cryptic and largely unsuspected role for wide crosses in angiosperm evolution. If so, such events will have implications for systematics and further work is needed to explore this exciting possibility.

Intergeneric crosses

Some families allow wide crosses successfully among genera (Whitson *et al.*, 1914; Laurie *et al.*, 1990); such data should contribute to any systematic analysis. However, data on the frequency of intergeneric crosses in any one family are sporadic and dispersed in the literature. To estimate frequencies, the numbers of specimens of intergeneric crosses that are held in the Herbarium of the Royal Botanic Gardens, Kew have been assessed. This is a biased sample because the Kew Herbarium supports much active work on Orchidaceae (502 representatives) but little on Rosaceae (seven representatives) or Brassicaceae (no representatives), families recorded as having frequent natural and synthetically produced intergeneric crosses. Intergeneric hybrids in members of Orchidaceae are numerous but many have been produced by controlled breeding, while the tedious preparatory requirements for herbarium specimens in Aloaceae (two representatives) may have significantly depressed their representation in Kew's Herbarium. The holdings of specimens of intergeneric hybrids in Poaceae are fewer but are often of wild origin and field botanists do find that grass identification is sometimes made difficult by apparent hybridisation between species or genera. There is little doubt that the number and scope of hybrids reported is a considerable under-estimation for Poaceae.

The genera of Aloaceae (Asphodelaceae-Alooideae) show considerable morphological similarity. Bayer (1982) stated that existing genera portray natural groups and the genera reflect the diversity of the group, despite weak rationalisation in terms of morphological and other criteria. Leaf surface characters and floral morphology are commonly used in distinguishing the genera. Among the other criteria are stability of chromosome number (x=7) and similarity in gross chromosome structure (Brandham,1971; Takahashi *et al.*, 1997), characters that may support the sinking of some of the genera, particularly where hybrids can be made. Natural interspecific hybrids are common where species grow together (Reynolds, 1950) but crosses between the genera *Aloe*, *Gasteria* and *Haworthia* have been made artificially, albeit with some difficulty (Brandham, 1983, 1990). Many of the intergeneric hybrids are remarkably stable and although some may be pollen sterile, others are partially pollen fertile. One possible reason for the high numbers of intergeneric hybrids and for their partial fertility may be associated with the stability of chromosome number and structure (Brandham, 1971; Takahashi *et al.*, 1997) even though it is clear that DNA amounts differ among many species and genera. In addition, some of the observed chromosome rearrangements influence the fertility of the hybrids (Brandham, 1983, 1990). It may also simply suggest that the generic limits in this group need revision like those in orchids (see below).

The characteristics of the breeding system may, in addition, enable the development of such hybrid combinations. In self incompatible species (controlled by at least two multi-allelic genes; Brandham and Owens, 1978) pollen tubes are observed growing into the ovary and penetrating the micropyle prior to fertilisation failure. Such a mechanism is certain to be less efficient in preventing crossing when compared with systems where pollen tube inhibition occurs at a much earlier stage, on the stigma or in the upper part of the style. In this case wide crossing confirms what has already been deduced from morphological and chromosomal data, namely a closer relationship between *Gasteria* and *Aloe* in comparison with *Haworthia*.

There has long been debate among systematists working on Orchidaceae that distinctive morphology may mask close genetic relatedness and in turn raise the question of generic delimitation. Such debate has been fuelled where fertile multi-intergeneric hybrids exist (Grant, 1958). Orchid systematists would argue that structural change in orchid flowers, a character suite commonly used in their classification, is often driven by the association of orchid species with specific pollinators and that the morphological variation which exists or which has evolved genuinely merits the placing of some groups of species into different genera. Recently, Pridgeon *et al.* (1997) and Bateman *et al.* (1997) performed significant generic recircumscription of the subtribe Orchidinae, based on a detailed phylogeny derived from nuclear internal transcribed spacer (ITS) DNA sequences. This is now also supported by *trn*L data (Bateman pers comm. 1998). Genera in this subtribe have frequently been reported to be involved in successful intergeneric crosses, but if the taxonomic revisions of Bateman *et al.* (1997) are accepted then the number of natural intergeneric hybrids reduces considerably. Many of these genera are represented by self compatible species, which may go some way towards explaining the propensity for successful hybridisation. The reproductive mechanism of the orchids, though not unique, is unusual. Pollen germination and pollen tube growth to the ovary often takes more than one week and fertilisation occurs only after lengthy development of the ovules triggered by the compatible pollination. Floral abscission appears to start at anthesis and requires a signal after a compatible pollination to stop its progress (Owens *et al.*, 1994; Y. Guo and S.J. Owens unpublished data). All these factors seem to mitigate against successful wide hybridisation but in reality they do not. Interestingly, where self incompatibility is found in orchid genera, the frequency of interspecific and intergeneric hybrids is far fewer (Y. Guo and S.J. Owens unpublished data).

Attempts to make artificial wide hybrids between cereal species in different genera have a very long history. For example, Wilson reported success in the production of hybrids between *Triticum aestivum* (breadwheat) and *Secale cereale* (rye) as early as 1876. These particular hybrids were sterile and it was another 15 years before Rimpau (1891) reported obtaining fertile hybrids. It is clear that genotype again exerts a strong influence on the relative success with which intergeneric *Triticale* hybrids can be made. In the cross between *Festuca arundinacea* and *Dactylis glomerata*, genotype considerably influences the success of the cross; gametic incompatibility ensures that the majority of cross combinations are unsuccessful. Five hybrid plants have been obtained after utilising embryo rescue techniques (Matzk, 1981). Notwithstanding these hybrids, however, the two genera are classified in the same tribe of the family; their wide cross success supports their relatedness and hence their classification in the same tribe.

Following earlier reports by Zenkteller and Nitzsche (1984) in a remarkable range of intergeneric plant crosses, Laurie and Bennett (1989) confirmed their observation of embryo development in *Triticum aestivum* (breadwheat) pollinated by *Zea mays.* Laurie and Bennett (1989) also showed that intergeneric zygote formation occurs in a remarkably high proportion of florets for this particular species combination. Formation of hybrid zygotes is rapidly followed by elimination of the genome of *Zea mays* during early embryo development in this species combination, so that haploid wheat plants can be obtained by embryo rescue techniques. Indeed, wide crossing with *Zea mays* has become the method of choice for producing haploids of breadwheat for plant breeding purposes. Laurie *et al.* (1990) attempted wide crosses, predominantly between genera in the tribes Triticeae and Panicoideae, over several years (Table 2). This is evidenced by the number of intergeneric crosses they have succeeded in obtaining to at least the post-zygotic stage, primarily though not exclusively using *Zea mays* as the male parent (*Z. mays* is now known to be a tetraploid). Data from these combinations are among the most detailed that exist for any wide crosses.

TABLE 2. Intergeneric crosses in Poaceae (Pooideae-Triticeae × Panicoideae-Andropogoneae, exception asterisked) listed by Laurie *et al.* (1990)

Diploids:			
	Aegilops comosa	×	*Zea mays*
	A. mutica	×	*Z. mays*
	A. umbellulata	×	*Z. mays*
	A. speltoides	×	*Z. mays*
	A. squarrosa	×	*Z. mays*
	Triticum thaoudar	×	*Z. mays*
	T. urartu	×	*Z. mays*
	Hordeum vulgare	×	*Z. mays*
	S. cereale	×	*Z. mays*
Tetraploids:			
	A. crassa	×	*Z. mays*
	A. triaristata	×	*Z. mays*
	T. dicoccoides	×	*Z. mays*
	T. dicoccum	×	*Z. mays*
	T. timopheevii	×	*Z. mays*
	T. durum	×	*Z. mays*
Hexaploids:			
	A. crassa	×	*Z. mays*
	A. triaristata	×	*Z. mays*
	T. aestivum	×	*Z. mays*
	**T. aestivum*	×	*Pennisetum glaucum*
	T. aestivum	×	*Sorghum bicolor*

Several aspects of these results bear on their value to systematics. Fertilisation frequencies were variable, with diploids giving the lowest values (0-28%), tetraploids intermediate (2-48%) and hexaploids the highest, between (14-69 %). The ease with which such fertilisation takes place begs several questions, not least those related to the genotypes chosen for the crosses, the significance of self compatibility in most but not all combinations and (allo)polyploidy in such combinations.

The characteristics of pollen tube growth in the grasses may also play a role in the success of pollen tube growth in intergeneric crosses, since most of the reserves required for grass pollen tube growth may already be present in the vegetative cell at the time of pollen dispersal (Heslop-Harrison, 1979). Since the life of the zygotes and embryos in such crosses is short, often being associated with chromosome elimination, the effect of such striking intergeneric crosses on the evolution of the grasses appeared at first sight to be insignificant. However, more recent results for wide crosses between hexaploid *Avena sativa* (2n=42) and *Zea mays* (Riera-Lizarazu *et al.*, 1996) reporting the production of both haploid oats and of spontaneously doubled lines of oat with stable disomic *Z. mays* addition chromosomes (2n=44), poses important new questions about the potential role of at least some wide crosses in grass evolution.

The results of wide hybridisation in these grasses show that although barriers to pollen tube growth and fertilisation are lacking, separation of genomes on the bases of chromosome number and DNA amount supports the traditional taxonomic groupings.

Conclusions

Although the results of wide crossing may influence taxonomic judgement, the absence of crossing barriers may have greater significance. For example, *Hordeum spontaneum* was treated as a separate species from *H. vulgare* but due to the complete lack of sterility barriers between the two, *H. spontaneum* is now considered a subspecies of *H. vulgare* (Bowden, 1959; Giles and von Bothmer, 1985; von Bothmer and Jacobsen, 1985; Scholz, 1986). There appear to be several significant factors which promote the success of wide crossing, notably the genotype, the lack of self incompatibility barriers in at least one parent and the occurrence of polyploidy in at least one parent. Genotype, however, appears to be crucial in any cross combination as the example of chromosome elimination in *Hordeum* demonstrates. Since the choice of genotype within the seed-parent species significantly affects success in interspecific crosses, it may reinforce the proposal of a more genotype specific control than taxonomic divergence (Gradziel *et al.*, 1993). This issue is difficult to resolve.

What is important in evaluating the taxonomic distance between two genera or species based on wide crossing results is that study of phenomena beyond simply examination of pollen tube growth or seed set is essential. Clearly studies of, for example, ovules from pollinated flowers to examine embryo sac development are highly desirable; they expose chromosome size (DNA amount) differences and phenomena such as chromosome elimination, which are now known to be determined genotypically. These characteristics are significant items to be taken into account when inferring generic or species relationships.

In many cases barriers to hybridisation become greater as relationships between species or genera become increasingly distant (Hogenboom, 1975). However, the above-mentioned crosses, particularly between genera of Poaceae, are exceptions to this. Wheat and maize and oats and maize are poles apart taxonomically, yet hybridise readily.

Knowledge of genotype (especially ploidy level) and incompatibility system are significant when drawing conclusions about systematic relationships. Self incompatibility is particularly relevant in relation to a reduction in both the occurrence of fertilisation and of seed set in any wide cross.

Acknowledgements

We are grateful to Richard Bateman (Royal Botanic Garden, Edinburgh) for discussion of the manuscript.

References

Bateman, R.M., Pridgeon, A.M. and Chase, M. (1997). Phylogenetics of subtribe Orchidinae (Orchidoideae, Orchidaceae) based on nuclear ITS sequences. 2. Infrageneric relationships and reclassification to achieve monophyly of *Orchis sensu stricto. Lindleyana* **12**: 113–141.

Bayer, M.B. (1982). The new *Haworthia* handbook. National Botanic Gardens of South Africa, Kirstenbosch.

Bennett, M.D., Finch, R.A. and Barclay, I.R. (1976). The time rate and mechanism of chromosome elimination in *Hordeum* hybrids. *Chromosoma* **54**: 175–200.

Bothmer, R. von and Jacobsen, N. (1985). Origin, taxonomy and related species. In: D.C. Rasmusson (editor). Barley, pp. 19–56, American Society of Agronomy, Washington. (Monograph no. 26.)

Bothmer, R. von and Jacobsen, N. (1986). Interspecific crosses in *Hordeum* (Poaceae). *Pl. Syst. Evol.* **153**: 49–64.

Bowden, W.M. (1959). The taxonomy and nomenclature of wheats, barleys and ryes and their wild relatives. *Canad. J. Bot.* **37**: 657–684.

Brandham, P.E. (1971). The chromosomes of the Liliaceae. II. Polyploidy and karyotype variation in the Aloineae. *Kew Bull.* **25**: 381–399.

Brandham, P.E. (1983). Evolution in a stable chromosome system. In: P.E. Brandham and M.D. Bennett (editors). Kew Chromosome conference II, pp. 251–260, George Allen and Unwin, London.

Brandham, P.E. (1990). Meiotic crossing-over between sites of opposite sides of the centromere of homeologues is frequent in hybrid Aloaceae. *Genome* **33**: 170–176.

Brandham, P.E. and Owens, S.J. (1978). The genetic control of self-incompatibility in the genus *Gasteria* (Liliaceae). *Heredity* **40**: 165–169.

Clausen, J., Keck, D.D. and Heisey, W.M. (1945). Experimental studies on the nature of species. II Plant evolution through amphiploidy and autopolyploidy, with examples from the Madiinae. Washington. (Carnegie Institute Washington Publications. 564.)

Dewey, D.R. (1984). The genomic system of classification as a guide to intergeneric hybridization with the perennial Triticeae. In: J.P. Gustafson (editor). Gene manipulation in plant improvement, pp. 209–279, Plenum Press, New York.

Frederiksen, S. and von Bothmer, R. (1989). Intergeneric hybridisation between *Taeniatherum* and different genera of Triticeae, Poaceae. *Nordic. J. Bot.* **9**: 229–240.

Giles, B.E. and von Bothmer, R. (1985). The progenitor of barley (*Hordeum vulgare* ssp. *spontaneum*)-its importance as a gene resource. *Sveriges Utsädesforën. Tidskr.* **95**: 53–61.

Gradziel, T.M., Okada, M. and St. Clair, D.A. (1993). Pollen-tube hypertrophy in ovules of *Solanum* and *Lycopersicon* following interspecific crosses. *Sexual Pl. Reprod.* **6**: 147–152.

Grant, V. (1958). The regulation of recombination in plants. Exchange of genetic material: mechanisms and consequences. *Cold Spring Harbor Symp. on Quant. Biol.* **23**: 337–363.

Grant, V. (1971). Plant speciation. Columbia University Press, New York.

Harrison, B.J. and Darby, L.A. (1955). Unilateral hybridisation. *Nature* **176**: 952.

Heslop-Harrison, J. (1979). Pollen-stigma interaction in grasses: a brief review. *New Zealand J. Bot.* **17**: 537–546.

Hiscock, S.J. and Dickinson, H.G. (1998). Unilateral incompatibility and the S (self-incompatibility) locus. In: S.J. Owens and P.J. Rudall (editors). Reproductive biology, pp. 31–46, Royal Botanic Gardens, Kew.

Hogenboom, N.G. (1975). Incompatibility and incongruity: two mechanisms for the non-functioning of intimate partner relationships. *Proc. Roy. Soc. London, Ser. B., Biol. Sci.* **188**: 361–375.

Jones, K., Papes, D. and Hunt, D.R. (1975). Contributions to the cytotaxonomy of the Commelinaceae, II. Further observations on *Gibasis geniculata* and its allies. *Bot. J. Linn. Soc.* **71**: 145–166.

Kenton, A.Y., Owens, S.J. and Langton, D. (1988). The origin of ring-formation and self-compatibility in *Gibasis pulchella* (Commelinaceae). In: P.E. Brandham (editor). Kew Chromosome Conference III, pp. 75–84, HMSO, London.

Laurie, D.A. and Bennett, M.D. (1989). The timing of elimination in hexaploid wheat × maize crosses. *Genome* **32**: 953–961.

Laurie, D.A., O'Donoughue, L.S. and Bennett, M.D. (1990). Wheat × maize and other wide sexual hybrids: their potential for genetic manipulation and crop improvement. In: J.P. Gustafson (editor). Gene manipulation in plant improvement II, pp. 95–126. Plenum Press, New York.

Lewis D. and Crowe, L.K. (1958). Unilateral interspecific incompatibility in flowering plants. *Heredity* **12**: 233–256.

Löve, A. (1982). Generic evolution of the wheatgrasses. *Biol. Zentralbl.* **101**: 199–212.

Löve, A. (1984). Conspectus of the Triticeae. *Feddes Repert.* **95**: 452–521.

Matzk, F. (1981). Successful crosses between *Festuca arundinacea* Schreb. and *Dactylis glomerata* L. *Theor. Appl. Genet.* **60**: 119–122.

Maynard Smith, J. (1989). Macroevolution. Evolutionary genetics. Oxford University Press.

Owens, S.J. (1977). Self-incompatibility in the genus *Gibasis* section *Heterobasis* D.R. Hunt (Commelinaceae). *Heredity* **39**: 365–371.

Owens, S.J. (1980). The genetics of self-incompatibility in *Gibasis karwinskyana* (Commelinaceae). *Incompat. Newslett.* **12**: 36–40.

Owens, S.J., Guo, Y. and Clifford, S.C. (1994). Pollination, fertilization and the major barriers to hybridization. In: Proceedings of the 14th World Orchid Congress, Glasgow, 1993, pp. 148–152, HMSO, Edinburgh.

Pridgeon, A.M., Bateman, R.M., Cox, A.V., Hapeman, J.R. and Chase, M. (1997). Phylogenetics of subtribe Orchidinae (Orchidoideae, Orchidaceae) based on nuclear ITS sequences. 1. Intergeneric relationships and polyphyly of *Orchis sensu lato*. *Lindleyana* **12**: 89–109.

Reynolds, G.W. (1950). The Aloes of South Africa. The Aloes book fund, Mbabane, Swaziland.

Riera-Lizarazu, O., Rines, H.W. and Phillips, R.L. (1996). Cytological and molecular characterization of oat × maize partial hybrids. *Theor. Appl. Genet.* **93**: 123–135.

Rimpau, W. (1891). Kreuzungsprodukte landwirtschaftlicher Kulturplanzen. *Landw. Jahrb.* **20**: 335–371.

Scholz, H. (1986). Die Entstehung der Unkraut-Gerste *Hordeum vulgare* subsp. *agriocrithon* emend. *Bot. Jahrb. Syst.* **106**: 419–426.

Simpson, E., Snape, J.W. and Finch, R.A. (1980). Variation between *Hordeum bulbosum* genotypes in their ability to produce haploids of barley, *Hordeum vulgare*. *Z. Pflanzenzücht.* **85**: 205–211.

Stebbins, G.L. (1959). The synthetic approach to problems of organic evolution. Genetics and twentieth century Darwinism. *Cold Spring Harbor Symp. Quant. Biol.* **24**: 305–311.

Subramanyam, N.C. and Kasha, K.J. (1973). Selective chromosome elimination during haploid formation in barley following interspecific hybridisation. *Chromosoma* **42**: 111–125.

Takahashi, C., Leitch, I.J., Ryan, A., Bennett, M.D. and Brandham, P.E. (1997). The use of *in situ* hybridisation (GISH) to show the transmission of recombinant chromosomes in a partially fertile bigeneric hybrid, *Gasteria lutzii* × *Aloe aristata* (Aloaceae), to its progeny. *Chromosoma* **105**: 342–348.

Tibok, A., Power, J.B. and Davey, M.R. (1995). Progress in protoplast technology for woody angiosperms. In: S. Jain, P. Gupta and R. Newton (editors). Somatic embryogenesis in woody plants. Vol. 1, pp. 143–166, Kluwer Academic, Dordrecht.

Van Creij, M.G.M., Kerchoffs, D.M.F.J. and Van Tuyl, J.M. (1997). Interspecific crosses in the genus *Tulipa* L.: identification of pre-fertilization barriers. *Sexual Pl. Reprod.* **10**: 116–123.

Whitson, J., Robert, J. and Williams, H.S. (editors) (1914). Luther Burbank. His methods and discoveries and the practical implications. Volume 12, pp. 107–154, Luther Burbank Press, London.

Wilkinson, M.J., Bennett, S.T., Clulow, S.A., Guillaume, A., Harding, K. and Bennett, M.D. (1994). Evidence for somatic translocation during potato dihaploid induction. *Heredity* **74**: 146–151.

Wilson, A.S. (1876). On wheat and rye hybrids. *Trans & Proc. Bot. Soc. Edinburgh* **12**: 826–828.

Zenkteler, M. and Nitzsche, W. (1984). Wide hybridization experiments in cereals. *Theor. Appl. Genet.* **68**: 311–315.

Southworth, D. and Cresti, M. (1998). Evolutionary relationships among sperm cells of land plants. In: S.J. Owens and P.J. Rudall (Editors). Reproductive Biology, pp. 147–157. Royal Botanic Gardens, Kew.

EVOLUTIONARY RELATIONSHIPS AMONG SPERM CELLS OF LAND PLANTS

D. SOUTHWORTH[1] AND M. CRESTI[2]

[1]Department of Biology, Southern Oregon State College, Ashland, OR 97520, USA
[2]Dipartimento di Biologia Ambientale, Università di Siena, Via P. A. Mattioli 4, I-53100, Siena, Italy

Abstract

Beyond obvious differences in presence or absence of flagella among flagellated and non-flagellated sperm cells in land plants, cytoskeletal patterns and nuclear structure have similarities that distinguish sperm from somatic cells. The cytoskeleton of flowering plant sperm consists of microtubule bundles arranged obliquely around the nucleus. Microtubule bundles branch and rejoin along the axis of the sperm. This pattern is initiated in the generative cell, one division before sperm formation, a situation parallel to spermatogenous cell development in plants with flagellated sperm. We hypothesise that the flowering plant sperm cytoskeleton is homologous to the cytoskeleton of flagellated sperm in more primitive vascular plants.

Introduction

Sperm cells in flowering plants are highly specialised with microtubule bundles, condensed chromatin, small cytoplasmic volume and absence of cell walls (Russell, 1984; McConchie *et al.*, 1985, 1987; Mogensen and Rusche, 1985; Palevitz and Cresti, 1988; Southworth and Knox, 1989; Cresti *et al.*, 1992; Palevitz and Tiezzi, 1992; Knox *et al.*, 1993). Flowering plant sperm differ from flagellated sperm of non-seed plants in their lack of flagella however, close examination reveals similarities in pattern of cytoskeleton and nuclear structure.

In this paper, we hypothesise that the cytoskeleton of flagellated sperm is homologous to that of non-flagellated sperm. Similarities in structure of mature sperm and in pattern of development would support this hypothesis. Another possibility is that non-flagellated sperm are homologous to precursor cells at an earlier developmental stage such as spermatogenous cells that precede formation of flagellated sperm in ferns or gymnosperms (Friedman and Gifford, 1988; Gifford and Foster, 1989). This would be a form of developmental heterochrony. Alternatively, flagellated and non-flagellated sperm may be homoplasic, arising by diverse patterns of modification of somatic cells. Similarity among non-flagellated sperm and somatic cells would support this hypothesis.

Descriptions of Flagellated and Non-flagellated Sperm

Flagellated Sperm in Vascular Plants

The only extant seed plants that have flagellated sperm are *Ginkgo* and Cycadales (Figs. 1–14; Gifford and Foster, 1989; Paolillo, 1981; Norstog, 1987; Li *et al.*, 1989). Sperm of

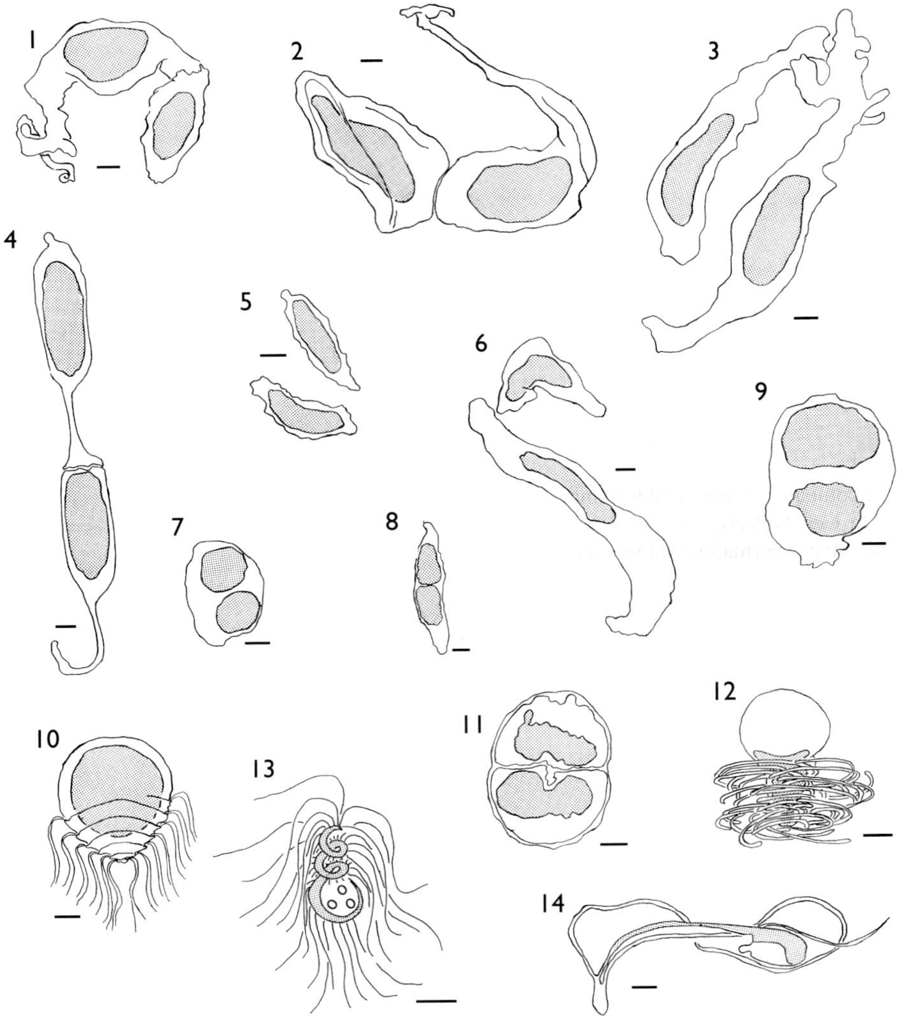

FIGS. 1–14. Comparison of morphology of sperm of land plants. Nuclei are stippled. 1–6. Bar=2 μm. 1: *Plumbago zeylanica*. From serial reconstruction in pollen (Russell, 1984). 2: *Brassica campestris*. From serial reconstructin in pollen (McConchie *et al.*, 1985). 3: *Gerbera jamesonii*. From serial reconstruction in pollen (Provoost *et al.*, 1988). 4: *Nicotiana tabacum*. From serial reconstruction in pollen tube (Yu *et al.*, 1992). 5: *Hordeum vulgare*. From serial reconstruction in pollen (Mogensen and Rusche, 1985). 6: *Zea mays*. From serial reconstruction in pollen (McConchie *et al.*, 1987). 7: *Gnetum gnemon*. From light-microscope section in pollen (Carmichael and Friedman, 1996). Bar=2 μm. 8: *Agathis australis*. From thin section in ovule (Owens *et al.*, 1995). Bar=40 μm. 9: *Picea glauca*. From thin section in ovule (Dawkins and Owens, 1993). Bar=10 μm. 10: *Zamia pumila*. From whole cell (Norstog, 1986). Bar=40 μm. 11: *Ginkgo biloba*. From thin section in ovule (Li *et al.*, 1989). Bar=10 μm. 12: *Marsilea vestita*. From drawing of free sperm cell (Myles and Hepler, 1977). Bar=5 μm. 13: *Pteris vittata*. From drawing and light micrograph of free sperm cell (Kuroiwa *et al.*, 1988). Bar=10 μm. 14: *Blasia pusilla* (Bryophyta) from whole-mount preparation (Carothers, 1973). Bar=2 μm.

Pterophyta and related groups are flagellated and resemble sperm of *Ginkgo* and cycads, but with fewer flagella (Garbary *et al.*, 1993). Numbers of flagella vary, from two in bryophytes and lycopods to 1,000 in *Ginkgo* and up to 40,000 in cycads (Fig. 10) (Duckett and Bell, 1971; Norstog, 1975; Paolillo, 1981; Renzaglia and Duckett, 1988, 1991; Maden *et al.*, 1996).

In ferns, the sperm cell and its nucleus are spiral. Chromatin is condensed, filling the nucleus with tightly packed DNA. Highly basic protamines replace histones in sperm of *Marsilea*, *Pteridium*, and *Marchantia* (Reynolds and Wolfe, 1978, 1984).

Cytoplasm is reduced to a thin layer around the nucleus. Reduced cytoplasmic volume results from pinching off an enucleate cytoplasmic body containing numerous organelles in ferns (Myles and Hepler, 1982) or from progressive elimination of cytoplasm in bryophytes (Renzaglia and Duckett, 1988). In cycads and *Ginkgo*, the cytoplasmic body remains attached posteriorly to the coiled flagellar band (Norstog, 1986, 1990).

The microtubular cytoskeleton consists of flagella with basal bodies, a multilayered structure consisting of a spline of parallel microtubules subtended for the anterior portion of its length by a lamellar strip, and cytoplasmic microtubles, including posterior extensions of spline microtubules, without apparent order. The microtubule spline, the outermost layer of the multilayered structure, extends along the coil of the nucleus posteriorly. In the spline, microtubules in a single layer are connected side to side. At the posterior end of the spline, microtubules diverge and terminate.

In *Ceratopteris*, acetylated α-tubulin, associated with stable tubulin arrays, was found in mid-stage spermatids and in microtubule splines and flagella of mature sperm, but not in other cytoplasmic microtubules nor in earlier stages of spline or flagellar development (Hoffman and Vaughn, 1995a, b). Tyrosinated α-tubulin, associated with dynamic microtubule arrays, occurred in all types of microtubules throughout spermiogenesis (Hoffman and Vaughn, 1995a, b). The lamellar strip was labelled by antibodies to centrin, a calcium-binding protein found in microtubule-organising centres of centrosomes, but was not labelled by antibodies to α- or β-tubulin (Vaughn *et al.*, 1993; Hoffman and Vaughn, 1995a, b). Centrin also occurred in amorphous material between basal bodies, but not in axonemes.

Mutations, developmental abnormalities and chemical treatments resulted in variation of cytoskeletal structures. Non-motile sperm mutants in male-sterile cultures of *Ceratopteris* lacked emergent flagella (Duckett *et al.*, 1979). Sperm did not retain a spiral form. Flagella emerged irregularly in disorganised arrays of singlet, doublet and triplets. In the cytoplasm, microtubules formed an unorganised mesh of mostly single microtubles. The microtubule spline was relatively normal. This single mutation did not change fern sperm into flowering plant sperm, yet it showed that mutations in cytoskeletal genes allowed cells to divide but prevented formation of organised flagella. In some flagella of *Ginkgo*, two to four axonomes occurred within one flagellar membrane (Li *et al.*, 1989). *Ceratopteris* sperm treated with taxol developed numerous cross-linked microtubule bundles in the cytoplasm not associated with nucleus or mitochondria as would be found with the microtubule spline (Hoffman and Vaughn, 1995a).

Land plants with flagellated sperm do not maintain a centriolar lineage as animals do. In spermatogenous cells prior to the last division leading to spermatid formation, a pair of multiple centrioles, the blepharoplasts, form *de novo* (Gifford and Lin, 1975; Marc and Gunning, 1986; Norstog, 1986: Hoffman and Vaughn, 1995b). Following final cell division, the blepharoplast gives rise to multilayered structure, microtubule spline and basal bodies of developing spermatids.

Non-flagellated sperm in vascular plants

Shapes of sperm in flowering plants vary from spheroidal to highly elongate with one or more thin extensions (Figs. 1–6., 15–16; Mogensen and Rusche, 1985; Rusche, 1988; Yu *et al.*, 1992). Cytoplasmic volume decreases by a process of cytoplasmic diminution by pinching off vesicles (Mogensen and Rusche, 1985; Yu *et al.*, 1992; Yu and Russell, 1992). Sperm nuclei were often elongate, but not spiralled (Figs. 1–6, 16). Chromatin is condensed and fluoresces brightly with DNA-fluorochromes (Figs. 16,17). In *Lilium*, sperm and generative cells form male-specific histones that are variant of histones H2B and H3 (Ueda and Tanaka, 1995a, b).

The microtubular cytoskeleton of sperm and generative cells is distinctive and differs from microtubule arrangements found in somatic cells. Microtubules do not radiate from the nuclear envelope as is common in somatic cells and in reproductive cells up to the single-celled microspore stage of pollen development (Sheldon and Dickinson, 1984; Southworth and Jernstedt, 1995). In sperm and generative cells, bundles of microtubules, with a predominantly axial orientation, branch and rejoin to form a slightly twisted basket or cage of microtubule bundles in the thin layer of cytoplasm surrounding the nucleus (Figs. 21, 22; Palevitz and Cresti, 1988; Cresti *et al.*, 1992, Palevitz and Tiezzi, 1992; Pierson and Cresti, 1992; Knox *et al.*, 1993).

The cytoskeleton of sperm develops from the cytoskeleton of generative cells (Figs. 17–19) (Palevitz and Cresti, 1989; Palevitz, 1990; Liu and Palevitz, 1991, 1992; Palevitz and Tiezzi, 1992; Palevitz, 1993; Yu and Russell, 1993; Zhang *et al.*, 1995). Sperm have fewer microtubules and a less well organised microtubular cytoskeleton than generative cells. In generative cell division, acetylated α-tubulin was localised in kinetochores (Åström, 1992). Localisation of γ-tubulin in *Nicotiana* differed from that of α- and β-tubulin with a punctate pattern along microtubule bundles, a layer around the nuclei at telophase and a patch on sperm nuclei following telophase, possibly marking assembly sites (Palevitz *et al.*, 1994).

Microtubules in sperm and generative cells often have arms resembling curved dynein arms of flagella. The arms exhibit various arrangements including bridges between two microtubules, between a microtubule and a vesicle or plasma membrane, and blind extensions from one microtubule without evident connections to another structure (Figs. 20, 23; Cresti *et al.*, 1990; Del Casino *et al.*, 1992).

Microtubules influence cell shape in sperm and generative cells. Elongation of generative cells is correlated with a change in arrangement of microtubules from a meshwork to parallel bundles (Palevitz and Cresti, 1989; Zhou and Yang, 1991). Change in shape of isolated sperm from elongate to spheroidal is correlated with loss of microtubule bundles and of polymerised microtubules (Russell, 1991; Theunis *et al.*, 1991; Palevitz and Tiezzi, 1992). Apparently, microtubules in flowering plant sperm are particularly labile. Only in high osmotica was the spindle shape maintained (Southworth and Knox, 1989).

Evidence for actin filaments in generative cells and sperm is variable with conflicting reports (Taylor *et al.*, 1989; Palevitz and Liu, 1992; Knox *et al.*, 1993). In generative cells preserved by high-pressure freezing and freeze-substitution for transmission electron microscopy, actin appeared as single filaments within microtubule bundles, as fusiform bodies with short filaments, and as monomers throughout the cytoplasm, but did not occur as filament bundles (Hess *et al.*, 1995). Profilin, which binds actin monomers and prevents polymerisation, was present throughout the generative cell.

Non-flagellated sperm in Coniferales and Gnetales

Sperm of Coniferales vary in number of nuclei. In *Agathis* and *Pseudotsuga*, sperm were binucleate (Owens and Morris, 1991; Owens *et al.*, 1995). In *Picea*, a partial cell plate formed during mitosis of the spermatogenous cell (Dawkins and Owens, 1993). In *Taxus*, sperm were reported as uninucleate and released as naked nuclei (Pennell and Bell 1986). In *Ephedra* and in *Gnetum*, sperm cells were binucleate with chromatin not strongly condensed (Friedman, 1990, 1991; Carmichael and Friedman, 1995, 1996). The cytoskeleton is unknown in non-flagellated sperm of gymnosperms. Sperm cells in Coniferales and Ephedrales have not been studied *in vitro*. Fixation of sufficient quality to fix microtubules in sperm is difficult because of the thickness of female tissues.

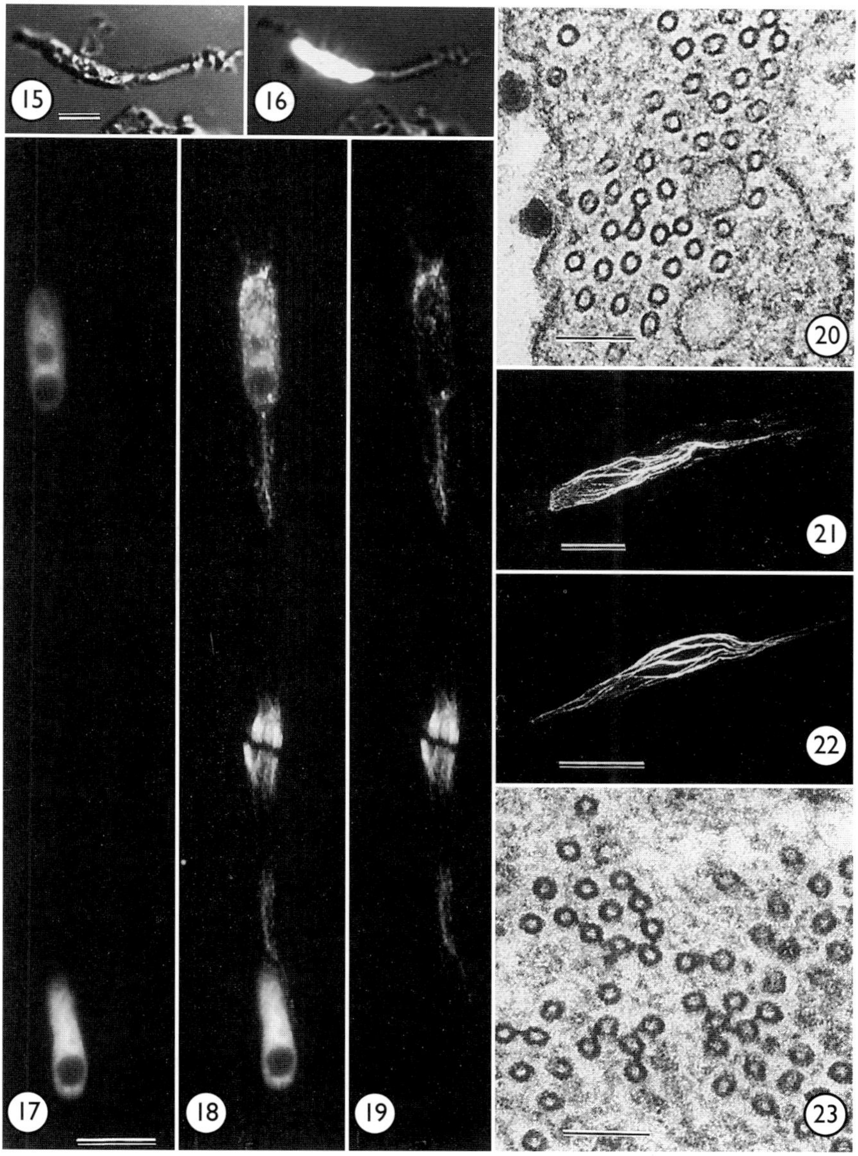

FIGS. 15–16. *Gerbera jamesonii* sperm cell with extensions. From Southworth and Knox (1989). Bar=2 µm. 15: Differential interference contrast microscopy. 16: DAPI florescence of sperm nucleus. Figs. 17–19. *Gagea lutea* sperm cell pair following mitosis in the pollen tube (Zhang *et al*., 1995). 17: DAPI fluorescence of sperm nuclei. 18: Double staining with DAPI and fluorescent antibody to α-tubulin. 19: Immunofluorescence with antibodies to α-tubulin by confocal scanning laser microscopy. Bar=25 µm. 20: Microtubule bundles in cross-section of sperm in the pollen tube of *Brassica oleracea* (glutaraldehyde-osmium fixation, M. Cresti, unpublished). Bar=0.1 µm. FIGS. 21–22. Microtubule bundles in generative cells in pollen tubes of *Crinum norei* stained with antibodies to α-tubulin, fluorescence by confocal scanning laser microscopy, (Y-Q. Li, unpublished). Bar=2 µm. 21: Generative cell with one extension. 22: Generative cell with two extensions. 23: Microtubule bundles in cross-section of generative cell in the pollen tube of *Iris pseudacorus* (glutaraldehyde-tannic acid fixation, M. Cresti, unpublished). Bar=0.1 µm.

Discussion

Evolution of sperm structure has been considered among land plants with flagellated sperm (Paolillo, 1981). Comparative ultrastructural studies of cytoskeletal development in male gametes have been useful in elucidating bryophyte phylogeny (Duckett and Renzaglia, 1988).

Among flowering plants, less attention has been paid to sperm evolution. Palevitz and Tiezzi (1992) suggested that the cytoskeleton of generative cells and sperm might be derived from the multilayered structure or from axonemes of ancestors with flagellated sperm similar to that of *Ginkgo* and cycads. Tiezzi *et al.*, (1992) noted that interconnected microtubules in bundles were a common feature of male gametes in a wide range of organisms including flowering plants, gymnosperms, lower plants and animals. However, microtubule bundles in flowering plant sperm differ significantly from flagellar axonomes. Flowering plant sperm lack flagella and distinct microtubule-organising centres such as blepharoplasts, centrioles or basal bodies. No organised multilayered structure, microtubule spline or blepharoplast-like osmiophilic material has been observed. Microtubules do not occur regularly as doublets nor in a 9 + 2 arrangement.

The hypothesis that the cytoskeleton in flowering plant sperm is a remnant of and homologous to the cytoskeleton of flagellated plant sperm is supported by the following evidence. An unusually large number of microtubules in non-flagellated sperm, as compared to somatic plant cells, is organised in bundles. These bundles resemble neither flagella nor common somatic arrangements of microtubules. Irregularity of bundle size, not some precise number of microtubules per bundle, suggests a remnant function, perhaps due to a dispersed or disorganised initiation centre. Presence of a large number of microtubules indicates the capacity to form microtubules, but not to organise them in a precise manner. Membranous extensions containing cross-linked microtubules superficially resemble a flagellum.

Presence of a distinctive cytoskeleton in both generative cells and sperm parallels the situation in flagellated plants whose spermatogenous cells form blepharoplasts which rise to the flagellated sperm cytoskeleton. The developmental origin of the cytoskeleton of sperm begins in the cell one division before spermiogenesis.

No active function for the sperm cytoskeleton in flowering plants is known. It is not attached in a regular way to the plasma membrane; sperm do not divide; no cell wall is deposited; it is not required for fusion; however, its presence confers an elongated shape on the cell. It seems unlikely that a complex structure, the result of male gamete-specific gene expression, would disappear completely, to be replaced by another system of microtubule bundles organised in a slightly different manner. Similar structural patterns are likely to be homologous and derived by relatively simple modifications. It is probable that during evolution of higher plants, the potential to organise a motile system, represented in its earliest stages by the blepharoplast, was largely lost. The aspect remaining is the arrangement of microtubules in bundles.

Morphological evidence supports the hypothesis that the cytoskeleton of sperm in flowering plants is derived from the internal cytoskeleton and axonemes of swimming sperm of primitive land plants. Microtubular cytoskeletons of non-flagellated and flagellated sperm have unusual arrangements of microtubules including large numbers of parallel microtubules aligned in groups oriented along the axis of the cell. Cytoskeletons of generative cells in flowering plants and spermatogenous cells of gymnosperms and ferns also have structural similarities that one would expect in homologous cells in that both differentiate from somatic cells and initiate a microtubule pattern that will result in the specialised sperm cytoskeleton.

Future Investigations

Examination of sperm cytoskeletons in a wider range of taxa, including primitive angiosperms and male-sterile mutants might uncover unusual or variant patterns. Descriptions of the cytoskeleton of sperm in Coniferales and Ephedrales would show whether these taxa possess the pattern of microtubule bundles as in flowering plants.

Quantitative descriptions of the cytoskeleton in flowering plant generative cells and sperm would be valuable. Counts of microtubules per sperm and per bundle would show the extent of consistency within a species and variability among species. Quantitative comparisons of microtubules in generative cells and sperm in the same species would show whether numbers of microtubules are halved during mitosis or are controlled by other factors.

More complete descriptions of the cytoskeleton, including location of modified tubulins and microtubule-associated proteins, would facilitate comparison. For purposes of evaluating homology, it would help to know if microtubule bundles in mature sperm include acetylated α-tubulin and therefore resemble microtubule splines and flagella. Antibodies to centrin and other components of centrioles, basal bodies and blepharoplasts could identify dispersed microtubule-organising centres (Doonan *et al.*, 1986; Marc and Gunning, 1986, 1988; Marc *et al.*, 1988; Vaughn *et al.*, 1993).

Improvement of *in vitro* techniques would be valuable. In most cases, isolation of sperm results in disassembly of microtubule bundles. Although cells remain alive and may divide or fuse with an egg, they lose their distinctive cytoskeleton. Techniques that maintain the cytoskeleton would allow isolation of microtubule bundles which would permit determination of microtubule polarity. *In vitro* methods for gymnosperm pollen development would facilitate an examination of the cytoskeletal properties of these sperm.

A molecular phylogeny comparing cellular components such as proteins, germ line promoters, and gene sequences of cytoskeletal components among motile sperm, generative cells and non-motile sperm would provide more conclusive evidence for evolutionary homologies. An awareness of possible homology, particularly of the cytoskeleton and nucleus, among flagellated and non-flagellated sperm types should lead to research of a more broadly comparative nature.

Acknowledgements

We thank Peter Bell (London), Darleen DeMason (Riverside, California) and Karen Renzaglia (Carbondale, Illionois) for comments and Barbara Boone for diagrams. This work was supported in part by NSF grant IBN-9418178.

References

Åström, H. (1992). Acetylated α-tubulin in the pollen tube microtubules. *Cell Biol. Int. Rep.* **16**: 871–881.

Carmichael, J.S. and Friedman, W.E. (1995). Double fertilization in *Gnetum gnemon*: the relationship between the cell cycle and sexual reproduction. *Pl. Cell* **7**: 1975–1988.

Carmichael, J.S. and Friedman, W.E. (1996). Double fertilization in *Gnetum gnemon*: (Gnetaceae): its bearing on the evolution of sexual reproduction within the Gnetales and the anthophyte clade. *Amer. J. Bot.* **83**: 767–780.

Carothers, Z.B. (1973). Studies of spermatogenesis in the Hepaticae. IV. On the blepharoplast of *Blasia. Amer. J. Bot.* **60**: 819–828.

Cresti, M., Blackmore, S. and Van Went, J.L. (1992). Atlas of sexual reproduction in flowering plants. pp. 62–83. Springer, New York.

Cresti, M., Mugia, M. and Theunis, C.H. (1990). Microtubule organization in sperm cells in the pollen tubes of *Brassica olearacea* L. *Protoplasma* **154**: 151–156.

Dawkins, M.D. and Owens, J.N. (1993). In vitro and in vivo pollen hydration, germination and pollen-tube growth in white spruce, *Picea glauca* (Moench) Voss. *Int. J. Pl. Sci.* **154**: 506–521.

Del Casino, C., Tiezzi, A., Wagner, V.T. and Cresti, M. (1992). The organization of the cytoskeleton in the generative cell and sperm of *Hyacinthus orientalis*. *Protoplasma* **168**: 41–50.

Doonan, J.H., Lloyd, C.W. and Duckett, J.G. (1986). Anti-tubulin antibodies locate the blepharoplast during spermatogenesis in the fern *Platyzoma microphyllum* R.Br.: a correlated immunofluorescence and electron-microscopic study. *J. Cell Sci.* **81**: 243–265.

Duckett, J.G. and Bell, P.R. (1971). Studies on fertilization in archegoniate plants. I. Changes in the structure of the spermatozoids of *Pteridium aquilinum* (L.) Kuhn during entry into the archegonium. *Cytobiologie* **4**: 421–426.

Duckett, J.G., Klekowski, E.J., Jr. and Hickok, L.G. (1979). Ultrastructural studies of mutant spermatozoids in ferns. I. The mature nonmotile spermatozoid of mutation 230X in *Ceratopteris thalictroides* (L.) Brongn. *Gamete Res.* **2**: 317–343.

Friedman, W.E. (1990). Sexual reproduction in *Ephedra nevadensis* (Ephedraceae): further evidence of double fertilization in a nonflowering seed plant. *Amer. J. Bot.* **77**: 1582–1598.

Friedman, W.E. (1991). Double fertilization in *Ephedra trifurca*, a non-flowering seed plant: the relationship between fertilization events and the cell cycle. *Protoplasma* **165**: 106–120.

Friedman, W.E. and Gifford, E.M. (1988). Division of the generative cell and late development in the male gametophyte of *Ginkgo biloba*. *Amer. J. Bot.* **75**: 1434–1442.

Garbary, D.J., Renzaglia, K.S. and Duckett, J.G. (1993). The phylogeny of land plants: A cladistic analysis based on male gametogenesis. *Pl. Syst. Evol.* **188**: 237–269.

Gifford, E.M. and Foster, A.S. (1989). Morphology and evolution of vascular plants. W.H. Freeman, New York.

Gifford, E.M. and Lin, J. (1975). Light microscope and ultrastructural studies of the male gametophyte in *Ginkgo biloba*: the spermatogenous cell. *Amer. J. Bot.* **62**: 974–981.

Hess, M.W., Mitterman, K., Luschnig, C. and Valenta, R. (1995). Immunocytochemical localisation of actin and profilin in the generative cell of angiosperm pollen: TEM studies on high-pressure frozen and freeze-substituted *Ledebouria socialis* Roth (Hyacinthaceae). *Histochem. Cell Biol.* **104**: 443–451.

Hoffman, J.C. and Vaughn, K.C. (1995a). Post-translational tubulin modifications in spermatogeneous cells of the pteridophyte *Ceratopteris richardii*. *Protoplasma* **186**: 169–182.

Hoffman, J.C. and Vaughn, K.C. (1995b). Using the developing spermatogenous cells of *Ceratopteris* to unlock the mysteries of the plant cytoskeleton. *Int. J. Pl. Sci.* **156:** 346–358.

Knox, R.B., Zee, S.Y., Blomstedt, C. and Singh, M.B. (1993). Male gametes and fertilization in angiosperms. *New Phytol.* **125**: 679–694.

Kuroiwa, H., Sugai, M. and Kuroiwa, T. (1988). Behaviour of chloroplasts and chloroplast nuclei during spermatogenesis in the fern, *Pteris vittata* L. *Protoplasma* **146**: 89–100.

Li, Y., Wang, F.H. and Knox, R.B. (1989). Ultrastructural analysis of the flagellar apparatus in sperm cells of *Ginkgo biloba*. *Protoplasma* **149**: 57–63.

Liu, B. and Palevitz, B.A. (1991). Kinetochore fiber formation in dividing generative cells of *Tradescantia*. Kinetochore reorientation associated with the transition between lateral microtubule interactions and end-on kinetochore fibers. *J. Cell Sci.* **98**: 475–482.

Liu, B. and Palevitz, B.A. (1992). Anaphase chromosome separation in dividing generative cells of *Tradescantia*. Changes in microtubule organization and kinetochore distribution visualized by antitubulin and CREST immunocytochemistry. *Protoplasma* **166**: 122–133.

Maden, A.R., Renzaglia, K.S. and Whittier, D.P. (1996). Ultrastructure of the spermatozoid of *Lycopodium obscurum* (Lycopodiaceae). *Amer. J. Bot.* **83**: 419–429.

Marc, J. and Gunning, B.E.S. (1986). Immunoflurescent localization of cytoskeletal tubulin and actin during spermatogenesis in *Pteridum aquilinum* (L.) Kuhn. *Protoplasma* **134:** 163–177.

Marc, J. and Guning, B.E.S. (1988). Monoclonal antibodies to a fern spermatozoid detect novel components of the mitotic and cytokinetic apparatus in higher plant cells. *Protoplasma* **142**: 15–42.

Marc, J., Gunning, B.E.S., Hardman, A.R., Perkin, J.L. and Wick, S.M. (1988). Monoclonal antibodies to surface and cytoskeletal components of the spermatozoid of *Pteridium aquilinum*. *Protoplasma* **142**: 5–14.

McConchie, C.A., Jobson, S. and Knox, R.B. (1985). Computer-assisted reconstruction of the male germ unit in pollen of *Brassica campestris*. *Protoplasma* **127**: 57–63.

McConchie, C.A., Hough, T. and Knox, R.B. (1987). Ultrastructural analysis of the sperm cells of mature pollen of maize, *Zea mays*. *Protoplasma* **139**: 9–19.

Mogensen, H.L. and Rusche, M.L. (1985). Quantitative analysis of barley sperm: occurrence and mechanism of cytoplasm and organelle reduction and the question of sperm dimorphism. *Protoplasma* **128**: 1–13.

Myles, D.G. and Hepler, P.K. (1977). Spermiogenesis in the fern *Marsilea*: microtubules, nuclear shaping and cytomorphogenesis. *J. Cell Sci.* **23**: 57–83.

Myles, D.G. and Hepler, P.K (1982). Shaping of the sperm nucleus in *Marsilea*: a distinction between factors responsible for shape generation and shape determination. *Developm. Biol.* **90**: 238–252.

Norstog, K. (1975). The motility of cycad spermatozoids in relation to structure and function. In: J.G. Duckett and P.A. Racey (editors). The biology of the male gamete, pp. 135–142, Academic Press, London.

Norstog, K. (1986). The blepharoplast of *Zamia pumila* L. *Bot. Gaz.* **147**: 40–46.

Norstog, K. (1987). Cycads and the origin of insect pollination. *Amer. Sci.* **75**: 270–279.

Norstog, K. (1990). Spermatozoids of *Microcycas calocoma*: ultrastructure. *Bot. Gaz.* **151**: 275–284.

Owens, J.N. and Morris, S.J. (1991). Cytological basis for cytoplasmic inheritance in *Pseudotsuga menziesii*. II. Fertilization and proembryo development. *Amer. J. Bot.* **78**: 1515–1527.

Owens, J.N., Catalano, G.L., Morris, S.J. and Aitken-Christie, J. (1995). The reproductive biology of Kauri (*Agathis australis*). II. Male gametes, fertilization and cytoplasmic inheritance. *Int. J. Pl. Sci.* **156**: 404–416.

Palevitz, B.A. (1990). Kinetochore behaviour during generative cell division in *Tradescantia virginiana*. *Protoplasma* **157**: 120–127.

Palevitz, B.A. (1993). Organization of the mitotic apparatus during generative cell division in *Nicotiana tabacum*. *Protoplasma* **174**: 25–35.

Palevitz, B.A. and Cresti, M. (1988). Microtubule organization in the sperm of *Tradescantia virginiana*. *Protoplasma* **146**: 28–34.

Palevitz, B.A. and Cresti, M. (1989). Cytoskeletal changes during generative cell division and sperm formation in *Tradescantia virginiana*. *Protoplasma* **150**: 54–71.

Palevitz, B.A. and Liu, B. (1992). Microfilaments (F-actin) in generative cells and sperm: an evaluation. *Sexual Pl. Reprod.* **5**: 89–100.

Palevitz, B.A. and Tiezzi, A. (1992). The organization, composition and function of the generative cell and sperm cytoskeleton. *Int. Rev. Cytol.* **140**: 149–185.

Palevitz, B.A., Liu, B. and Joshi, H.C. (1994). γ-tubulin in tobacco pollen tubes: association with generative cell and vegetative microtubules. *Sexual Pl. Reprod.* **7**: 209–214.

Paolillo, D.J. (1981). The swimming sperms of land plants. *BioScience* **31**: 367–373.

Pennell, R.I. and Bell, P.R. (1986) The development of the male gametophyte and spermatogenesis in *Taxus baccata* L. *Proc. Roy. Soc. Lond, Ser. B. Biol. Sci.* **228**: 85–96.

Pierson, E.S. and Cresti, M. (1992). Cytoskeleton and cytoplasmic organization of pollen and pollen tubes. *Int. Rev. Cytol.* **140**: 73–125.

Provoost, E., Southworth, D. and Knox, R.B. (1988). Three-dimensional reconstruction of sperm cells and vegetative nucleus in pollen of *Gerbera jamesonii*. In: H.J. Wilms and C.J. Keijzer (editors). Plant sperm cells as tools for biotechnology, pp. 69–73, Pudoc, Wageningen.

Renzaglia, K.S. and Duckett, J.G. (1988). Different developmental processes underlie similar spermatozoid architecture in mosses, liverworts and hornworts. *J. Hattori Bot. Lab.* **64**: 219–236.

Renzaglia, K.S. and Duckett, J.G. (1991). Towards an understanding of the differences between the blepharoplasts of mosses and liverworts, and comparisons with hornworts, biflagellate lycopods and charophytes: a numerical analysis. *New Phytol.* **117**: 187–208.

Reynolds, W.F. and Wolfe, S.L. (1978). Changes in basic protein during sperm maturation in a plant, *Marchantia polymorpha. Exp. Cell Res.* **116**: 269–274.

Reynolds, W.F. and Wolfe, S.L. (1984). Protamines in plant sperm. *Exp. Cell Res.* **152**: 443–448.

Rusche, M.L. (1988). The male germ unit of *Zea mays* in the mature pollen grain. In: H.J. Wilms and C.J. Keijzer (editors). Plant sperm cells as tools for biotechnology, pp. 61–67, Pudoc, Wageningen.

Russell, S.D. (1984). Ultrastructure of the sperm of *Plumbago zeylanica* II. Quantitative cytology and three-dimensional organization. *Planta* **162**: 385–391.

Russell, S.D. (1991). Isolation and characterization of sperm cells in flowering plants. *Annual Rev. Pl. Physiol. Pl. Molec. Biol.* **42**: 189–204.

Sheldon, J.M. and Dickinson, H.G. (1984). Determination of patterning in the pollen wall of *Lilium henryi. J. Cell. Sci.* **63**: 191–208.

Southworth, D. and Jernstedt, J. (1995). Pollen exine development and microtubule rearrangement in *Vigna unguiculata* (Fabaceae): A model for pollen wall patterning. *Protoplasma* **187**: 79–87.

Southworth, D. and Knox, R.B. (1989). Isolation of sperm cells from *Gerbera jamesonii* pollen. *Pl. Sci.* **60**: 273–277.

Taylor, P., Kenrick, J., Li, Y., Kaul, V., Gunning, B.E.S. and Knox, R.B. (1989). The male germ unit of *Rhododendron*: quantitative cytology, three-dimensional reconstruction, isolation and detection using fluorescent probes. *Sexual Pl. Reprod.* **2**: 254–264.

Theunis, C.H., Pierson, E.S. and Cresti, M. (1991). Isolation of male and female gametes in higher plants. *Sexual Pl. Reprod.* **4**: 145–154.

Tiezzi, A., Pierson, E.S., Theunis, C.H., Ciampolini, F., Cai, G.P., Bartalesi, A. and Cresti, M. (1992b). The motile apparatus of sperm cells in angiosperms: correlation with lower plants, gymnosperms and animals. In: B. Baccetti (editor). Comparative spermatology after 20 years, pp. 1017–1020, Raven Press, New York (Serono symposium, vol. 75).

Ueda, K. and Tanaka, I. (1995a). The appearance of male gamete-specific histones gH2B and gH3 during pollen development in *Lilium longiflorum. Developm. Biol.* **169**: 210–217.

Ueda, K. and Tanaka, I. (1995b). Male-specific H2B and H3 histones, designated gH2B and gH3 in *Lilium longiflorum. Planta* **197**: 289–295.

Vaughn, K.C., Sherman, T.D. and Renzaglia, K.S. (1993). A centrin homologue is a component of the multilayered structure in bryophytes and pteridophytes. *Protoplasma* **175**: 58–66.

Yu, H.S. and Russell, S.D. (1992). Male cytoplasmic diminution and male germ unit in young and mature pollen of *Cymbidium goeringii*: a 3-dimensional and quantitative study. *Sexual Pl. Reprod.* **5**: 169–181.

Yu, H.S. and Russell, S.D. (1993). Three-dimensional ultrastructure of generative cell mitosis in the pollen tube of *Nicotiana tabacum. Eur. J. Cell Biol.* **61**: 338–348.

Yu, H.S., Hu, S.Y. and Russell, S.D. (1992). Sperm cells in pollen tubes of *Nictiana tabacum* L.: three-dimensional reconstruction, cytoplasmic diminution, and quantitative cytology. *Protoplasma* **168**: 172–183.

Zhang, H.Q., Bohdanowicz, J., Pierson, E.S., Li, L.Q., Tiezzi, A. and Cresti, M. (1995). Microtubular organization during asymmetrical division of the generative cell in *Gagea lutea*. *J. Pl. Res.* **108**: 269–276.

Zhou, C. and Yang, H.Y. (1991). Microtubule changes during the development of generative cells in *Hippeastrum vittatum* pollen. *Sexual Pl. Reprod.* **4**: 293–297.

Blackmore, S. and Crane, P.R., (1998). The evolution of apertures in the spores and pollen grains of embryophytes. In: S.J. Owens and P.J. Rudall (Editors). Reproductive Biology, pp. 159–182. Royal Botanic Gardens, Kew.

THE EVOLUTION OF APERTURES IN THE SPORES AND POLLEN GRAINS OF EMBRYOPHYTES

STEPHEN BLACKMORE[1] AND PETER R. CRANE[2]

[1]Dept. Botany, The Natural History Museum, Cromwell Road, London, SW7 5BD, UK
[2]Dept. Geology, The Field Museum, Roosevelt Road at Lake Shore Drive, Chicago, IL 60605, USA

Abstract

The origin and evolution of apertures in the haploid microspores and pollen grains of land plants are reviewed from a systematic and ontogenetic perspective. Five key innovations are discussed: the origins of proximal, distal, equatorial, compound and multiple apertures. Proximal apertures, which are not invariably associated with proximal germination, are present in the early evolution of land plants and occur in some 'bryophytes' (although hepatics may be primitively inaperturate based on ultrastructural criteria), in 'pteridophytes' and in the prepollen of some seed plants. Based on current ideas of seed plant relationships, distal apertures have arisen more than once, perhaps in three or four lineages of seed plants, and are also present in certain microspores that possess proximal triradiate marks. Distal apertures are characterised by localised thinnings of the ectexine, in contrast to proximal apertures which involve modifications of layers homologous to the endexine of most seed plants. The transformation, within the angiosperms, from distal to equatorial apertures appears to be correlated with evolutionary changes in the processes of meiosis. Compound apertures, in multilayered pollen walls, form where the presence of onci (lens shaped, pectocellulosic bodies) disrupt endexine deposition. Within angiosperms numerous lineages have independently evolved higher numbers of apertures from the basic monosulcate and tricolpate forms. The morphological diversity of apertures appears to be correlated with evolutionary modifications to the processes of meiosis and pollen development. Functional correlations with reproductive biology can be proposed for some, but not all, of these changes.

Introduction

Apertures are one of the most important features of pollen grains and spores of land plants, both in terms of their biological function and their systematic significance. Apertures are specialised areas, formed during development, where the pollen or spore wall is often thinner and where germination of the gametophyte generally takes place. In addition, apertures may function in the uptake of water, in the harmomegathic accommodation of hydrodynamic volume changes (Wodehouse, 1935) and as sites for the transfer of water and recognition substances (Heslop-Harrison, 1976; Muller, 1979; Chaloner, 1976). Besides their manifestation in the exine, apertures are often represented internally as localised modifications of the cellulosic intine that may include channels, thickenings or multiple layers.

This paper presents an interpretation of the patterns of occurrence of different forms of aperture in the systematic context of land plants. Our purpose is to explore the homologies between apertural forms and provide insights into the evolutionary acquisition of reproductive processes in the life cycle. Important questions that we wish to address include: how have the different kinds of apertures originated among land plants and have they arisen more than once? We attempt to derive generative models by comparing development, from the earliest ontogenetic stages through to maturity, in land plants exhibiting a wide variety of apertures. To be of value, such models must be sufficiently robust to explain the diversity of form encountered in nature and should also be open to testing through further observation and experimentation. For example, based on developmental evidence Blackmore and Crane (1988) proposed a model for pollen dispersal in tetrads, a feature known from more than fifty families (Walker and Doyle, 1975). Subsequent observations (Preuss *et al.*, 1994) showed that a developmental mutant of *Arabidopsis thaliana* conforms to the predictions of this model in having tetrads with greatly reduced special cell walls.

An important concept in describing the diversity of aperture configurations is that of polarity (Erdtman, 1952, 1969), defined with respect to position and orientation in the meiotic tetrad (Wodehouse, 1935; Huynh, 1976). Proximal apertures are situated on the surface facing towards the centre of the tetrad and distal apertures on the surface facing outwards from the centre of the tetrad (Fig. 1). Within this basic framework, the background to our study of apertures is provided by an extensive literature that includes many valuable comparative reviews, some of which incorporate hypotheses of aperture evolution (see for example, Wodehouse, 1936; Walker and Doyle, 1975; Van Campo, 1976; Thanikaimoni, 1986; Doyle and Hotton, 1991). Studies of the evolution of life histories have also clarified the nature of correlations between phylogenetic groups, apertural features and aspects of reproductive biology such as homospory, heterospory, the seed habit, siphonogamy and double fertilisation (Chaloner, 1970; Chaloner and Hemsley, 1991; Friedman, 1993; Bateman and DiMichele, 1994). Overviews of pollen and spore ontogeny (Dickinson, 1976; Blackmore and Crane, 1988; Blackmore and Knox, 1990) and accounts of aperture development are provided for many species of 'bryophytes' (Brown and Lemmon, 1981, 1983, 1986; Neidhart, 1979),

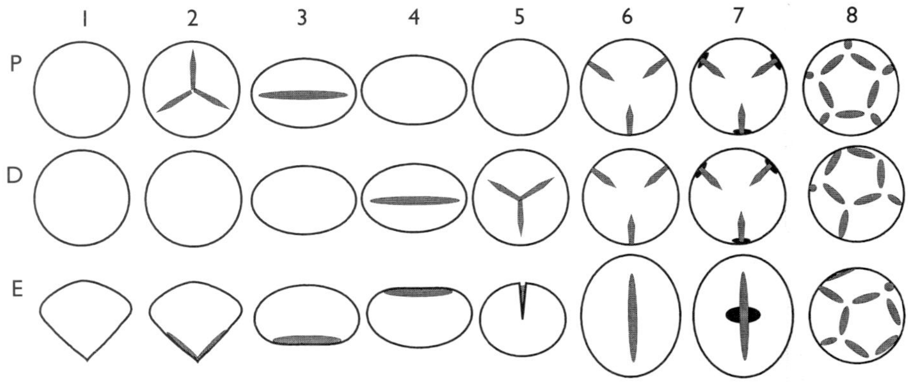

FIG. 1. Diagram summarising the diversity of aperture configurations in land plants. 1 = inaperturate spore; 2 = trilete spore; 3 = monolete spore; 4 = monosulcate pollen; 5 = trichotomosulcate pollen; 6 = tricolpate pollen; 7 = tricolporate pollen; 8 = pantocolpate pollen. The five key innovations discussed in the text relate to proximal apertures (2-3), distal apertures (4-5), equatorial apertures (6-7), compound apertures (7) and multiple apertures (8). P = proximal view, D = distal view, E = equatorial view. Ectoapertures are lightly shaded, endoapertures have darker shading.

'pteridophytes' (Lugardon, 1976; Buchen and Sievers, 1981; Mogensen, 1983; Uehara and Kurito, 1989a,b,1991; Uehara *et al.*, 1991), 'gymnosperms' (Singh, 1978) and angiosperms (Huynh, 1976; Shivanna and Johri, 1985; Knox, 1984). In this paper we discuss five major morphological innovations: proximal apertures, distal apertures, equatorial apertures, compound apertures and multiple apertures (Fig. 1) focusing primarily on extant taxa. The distribution of the five innovations among extant land plants and extinct 'seed ferns' is summarised in the cladogram shown in Fig. 2.

Proximal apertures

Proximal apertures exhibit a variety of forms but all are positionally identical in being situated on the surface of the microspore which is directed towards the centre of the meiotic tetrad during development. Among extant plants proximal apertures are found in many 'bryophytes', most 'pteridophytes' and, in a very rudimentary form, in a few seed plants. Charophytes, the closest relatives of land plants, do not have apertures in their meiotic products, although they may possess sporopollenin (Blackmore and Barnes, 1987a). A sporopollenin wall is not present around the meiotic products of charophytes.

'Bryophytes'

A variety of apertures occur in 'bryophytes'. In the putatively basal group, the liverworts (Mishler and Churchill, 1984, 1985; Kenrick and Crane, 1997a,b) the spores are globose and clearly defined apertures are generally absent (Inoue, 1960; Mogensen, 1983; Brown and Lemmon, 1986, 1990; Brown *et al.* 1986). Localised modifications of the intine that would be indicative of apertures are apparently unknown in hepatics (Brown *et al.*, 1982; Mogensen, 1983). The spores of hepatics may thus normally be considered inaperturate even though they may show three weak contact faces on the proximal spore surface. A few taxa, such as *Riccia sorocarpa,* have equatorial wall thinnings, or 'pseudo-apertures', that apparently do not serve as sites for germination (Denizot 1976). Sorsa (1976), Brown and Lemmon (1990) and others have pointed out fundamental differences in exine ultrastructure between hepatics, which have a relatively simple, lamellated exine, and hornworts and mosses which have more complex, multilayered walls. In hornworts the apertures are trilete and are presumed to develop by the folding of lamellated exine layers, although detailed ultrastructural studies have been hindered by problems of fixation and resin infiltration (Brown and Lemmon, 1990). In mosses a proximal triradiate aperture is clearly present in a few groups, some of which (for example, *Sphagnum*, Brown *et al.*, 1982) are among the earliest diverging lineages in the group. However, in other near-basal groups a distinct trilete aperture is absent (for example in *Andreaea*, Brown and Lemmon, 1984) as it is in most 'higher', peristomate mosses. Most peristomate mosses have a simpler aperture, a leptoma consisting of thinnings in both the exine and intine of the proximal surface (Brown and Lemmon, 1990). The intine is usually modified at the apertures, but may be either thicker or thinner (Olesen and Mogensen, 1978). However, in others such as *Ditrichum* and *Trematodon* they consist of specially formed areas where the sporoderm becomes folded during development (Brown and Lemmon, 1980, 1981).

'Pteridophytes'

Among 'pteridophytes', the distinct trilete or monolete apertures seen in most lycopods and ferns are generally formed by localised folding of the developing exospore in a manner comparable to that of mosses such as *Sphagnum* (Lugardon, 1976; Uehara and Kurito, 1989a, 1991).

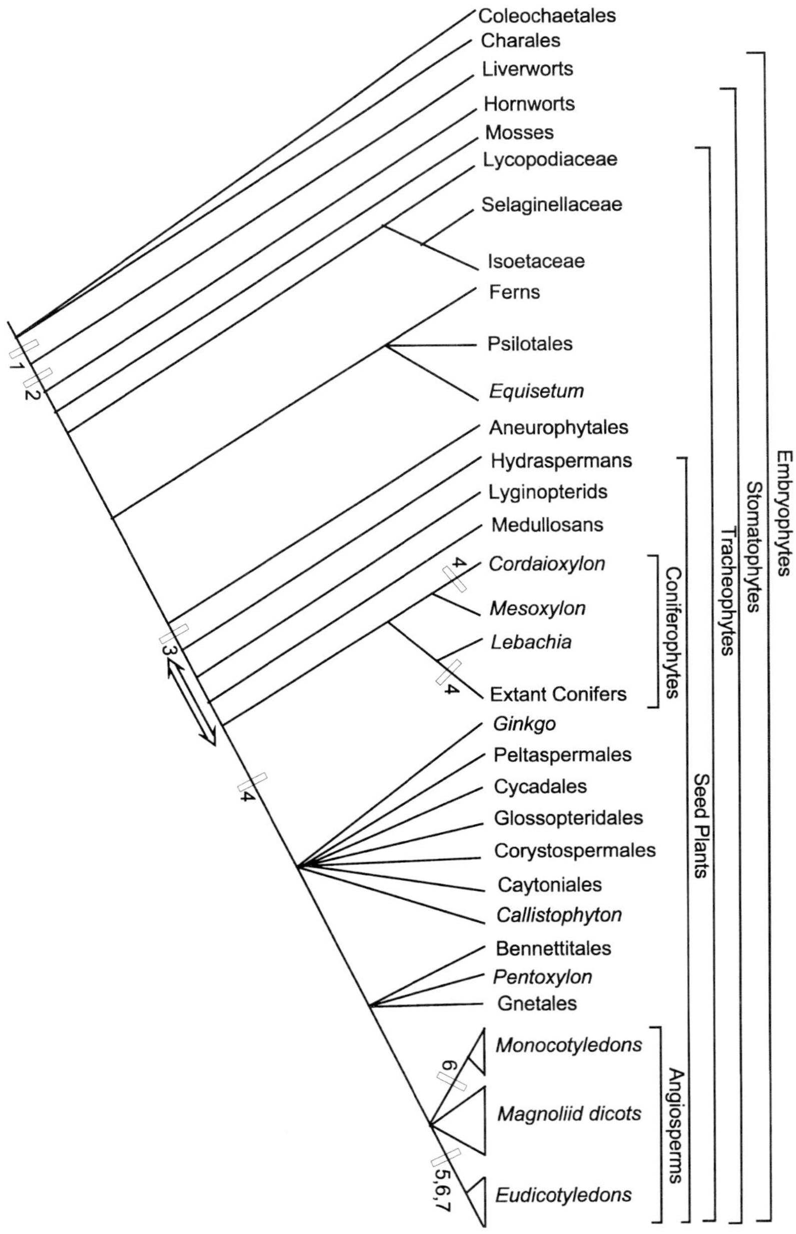

FIG. 2. Summary cladogram of relations among the major groups of land plants discussed in this paper and selected extant and fossil taxa. Apertural and other innovations are indicated on the cladogram as follows: 1 = possession of sporopollenin walled spores; 2 = presence of proximal apertures; 3 = seeds; 4 = presence of distal apertures; 5 = presence of equatorial or equatorial-derived apertures; 6 = presence of multiaperturate pollen grains in some taxa; 7 = presence of endoapertures in some taxa. The occurrence of prepollen is indicated by the large arrow. Cladogram derived from Kenrick and Crane (1997b).

A variety of developmental mechanisms are involved in the formation of proximal apertures. Superficial effects of contact between the developing microspores, such as angular contact faces, are not themselves considered to be apertures. Proximal triradiate apertures *sensu stricto* in both 'bryophytes' and 'pteridophytes' arise through the folding of lamellations of the developing exospore. The position and orientation of these folds is correlated with the symmetry of the tetrad which is established by the orientation of the spindles during meiosis (Sladkov, 1964; Verma and Khullar, 1976; Brown and Lemmon, 1990). The transformation from trilete to monolete proximal apertures, which has occurred within lycopods and also several times among ferns, appears to be directly correlated with the nature of meiosis. Simultaneous cytokinesis, without a dyad stage, generally results in a tetrahedral tetrad in which all four meiotic products are in direct contact and appears to be reflected in a distinct trilete aperture. In contrast, successive cytokinesis with a distinct dyad stage involves contact between pairs of daughter cells and produces monolete spores. The polarity of the individual spores seems to be imparted by the spindles of the second meiotic division. Occasionally, monolete spores are apparently associated with simultaneous partitioning and mixtures of monolete and trilete spores have been observed in the sporangia of some ferns including *Pityrogramma hybrida* (Erdtman and Praglowski, 1959), *Loxogramme lanceolata* (Nayar, 1963), *Pteris longifolia* (Devi, 1974) and *Pteris vittata* L. (Devi and Khare, 1985). In these species with both monolete and trilete spores relative timing of spindle formation and cytokinesis may be less tightly constrained than in other taxa. In *Isoetes*, the microsporocytes divide successively to produce microspores whereas megasporocytes undergo simultaneous cytokinesis and are trilete (Brown and Lemmon, 1990; Uehara *et al.*, 1991).

Developmental studies of *Equisetum* (Lugardon, 1969; Uehara and Kurito, 1989b) have shown a process of aperture definition that appears to be unique among pteridophytes. The apertures are small, circular and ulcerate or pore-like thinnings of the exine and phylogenetic analyses that include fossils suggest that the absence of a clear trilete aperture in *Equisetum* is the result of secondary loss (Kenrick and Crane, 1997b).The deposition of exospore lamellae is locally interrupted near to the proximal pole by an oncus-like plug of material which Lugardon referred to as the 'sub-apertural obturator'. An analagous mechanism appears to contribute to the definition of endoapertures in flowering plants.

Germination polarity

Based on current knowledge, proximal apertures, in the form of folded areas of exine usually associated with modified intine, appear to be a synapomorphy of the stomatophytes *sensu* Kenrick and Crane (1997a,b). They are absent from hepatics, but present in some mosses, all hornworts and many tracheophytes. Germination polarity is difficult to determine in inaperturate spores or pollen grains, except where the occurrence of tetrad dispersal enables it to be observed at maturity but interestingly distal germination appears to be widespread in hepatics. It is ubiquitous in extant hepatic genera that have inaperturate spores united in permanent tetrads (Inoue, 1960) and occurs in *Riccia* (Duthie and Garside, 1936, 1939; Udar, 1958) and *Sphaerocarpos* (Doyle, 1975; Mogensen, 1983). In *Sphaerocarpos donnellii* distinct races possessing either monads or cohesive tetrads exist (Doyle, 1975) that differ only in the extent of the special cell wall deposited between the developing microspores (Siler, 1934). The spores of both monad and tetrad races germinate distally (Doyle, 1975). Similarly, in *Riccia rautanenii* the monads have weakly defined contact marks on the proximal faces but germination is distal, as it is in species of the genus with spores in tetrads (Duthie and Garside, 1939; Erdtman, 1965). Distal germination has also been described in monads of *Monocarpus* (*Carrpos*, Proskauer, 1961), *Riella* (Proskauer, 1955; Mogensen, 1983; Nehira, 1983) and *Stephensoniella* (Mehra and Kachroo, 1952; Nehira, 1983). There appear to be no reports of distal germination among basal stomatophytes and vascular plants that have spores with true proximal apertures formed by the folding of exospore lamellae. This is consistent with the traditional view that structurally complex apertures evolved as predetermined sites for germination.

Distal apertures

Among extant plants, distal apertures or forms derived from them, are characteristic of most non-angiosperm seed plants ('gymnosperms'), monocotyledons and magnoliid dicotyledons (Fig. 2). In contrast, among other extant groups the presence of a proximal aperture is always associated with the free-sporing condition ('pteridophyte' reproduction). However, in several groups of extinct seed plants this correlation breaks down. In the earliest seed plants, such as hydraspermans, lyginopterids and medullosan 'seed ferns', as well as certain conifers and cordaites, the pollen clearly retains a proximal suture and, although direct evidence is lacking, may also have retained the ability to germinate through the proximal surface. Pollen of this kind, with a 'pteridophytic' level of organisation, has been termed prepollen (Schopf, 1938; Chaloner, 1970). Evidence from phylogenetic analyses that include fossil plants (Crane, 1985, 1990; Doyle and Donoghue, 1986; Nixon *et al.* 1994) indicates that the distal apertures may have evolved within as many as three or four lineages of seed plants (Fig. 2): within cordaites, within conifers, perhaps in the lineage leading to extant *Ginkgo,* and also in the lineage that includes *Callistophyton* and a variety of other late Palaeozoic and Mesozoic seed plants that ultimately gave rise to the Gnetales and angiosperms (Crane, 1990).

The occurrence of several parallel patterns of transition from proximal apertures and germination to distal suggests that the morphogenetic changes involved may not be complex. The occurrence of palynomorphs with both proximal and distal apertures indicates that the two developmental mechanisms are not mutually exclusive. The implication is that proximal and distal apertures may have performed different functions, at least initially, and that the transition from proximal to distal germination may have occurred gradually in several different lineages (Blackmore and Crane, 1988).

Among extant plants, spores with distal thinnings and proximal apertures occur in *Osmunda* (Wodehouse, 1935; Bobrov, 1966; Kuprianova, 1974). Interestingly, germination in the Osmundaceae is bipolar; a rhizoidal initial emerging from the proximal surface and a prothallial initial from the distal surface (Nayar and Kaur, 1968; Stokey and Atkinson, 1956). Among extant conifers proximal triradiate marks have been reported in *Abies* (Wodehouse, 1935; Erdtman, 1965; Kuprianova, 1974; Bagnall, 1975), *Dacrycarpus* (Ueno, 1957; Cranwell, 1961; Huynh and Sampson, 1983; Pocknall, 1981), *Pseudotsuga* (Erdtman, 1965) and *Tsuga* (Ueno, 1957). In each of these cases, the triradiate proximal mark is most conspicuous in immature pollen grains (Ueno, 1957; Cranwell, 1961; Huynh and Sampson, 1983). Among angiosperms, reports of proximal triradiate marks are rare, although pollen of *Trapa natans* was shown by Erdtman (1943) to have a putative indication of such a feature during development. Some monocotyledons and magnoliid dicotyledons also have proximal apertures but in these groups this aperture configuration is most parsimoniously interpreted as derived. For example, *Commelinantha* has two proximal furrows in addition to a distal sulcus (Huynh, 1976; Rowley and Dahl, 1962) and annonaceous monads with proximal apertures that function in germination (Jonker, 1974) are probably derived from permanent tetrads which have retained reduced proximal walls (Le Thomas *et al.*, 1986).

Several previous discussions of the origin of distal apertures have suggested that they represent a fundamentally new kind of feature with a different functional and ontogenetic basis from proximal apertures. Kuprianova (1967, 1974), for example, suggested that distal apertures originally evolved as an adaptation for the harmomegathic accommodation of volume changes associated with the degree of cytoplasmic hydration and were preadapted as sites for germination. She interpreted the bipolar germination of *Osmunda* spores (Bobrov, 1966; Nayar and Kaur, 1968) as evidence of a tendency towards distal apertures even among ferns. In this connection it is interesting that according to current analyses of seed plant relationships the several transitions from proximal to distal apertures all seem to have occurred in the seed plant clade characterised by platyspermic seeds and saccate pollen (Crane, 1985, 1990; Doyle and Donoghue, 1986). In the saccate pollen of extant conifers such as that of *Picea* and *Pinus*, the thin distal area buckles on dehydration forcing the two lateral sacci to bend inward covering the distal pole. It is conceivable that a thin distal area in Palaeozoic saccate pollen, even that with a ring-like single saccus, may have facilitated a similar harmomegathic change, and would have been preadapted for co-option for distal germination (an example of exaptation, *sensu* Gould and Vrba, 1982). Based on the above considerations we agree with Kuprianova's suggestion that the tendency of some pteridophytes to produce distal thinnings in the spore exine may provide an analogy for the early phases in the evolution of distal apertures although ultrastructural studies might be expected to reveal differences in both the wall layers affected and the mechanism of aperture definition. Distal thinnings, such as occur in *Osmunda* are the result of locally reduced exine deposition and, unlike the proximal apertures of most 'pteridophyte' spores, they are not formed by folding of the developing wall, and they thus reflect the appearance of novel morphogenetic control mechanisms.

The principal ontogenetic mechanism that results in aperture definition, which has been described from angiosperms and at least some conifers (for example, *Pinus*, Dickinson, 1976), is the formation of apertural shields of endoplasmic reticulum. These are thought to locally inhibit the deposition of primexine in the early tetrad stage resulting in areas where ectexine is lacking or greatly reduced (Dickinson, 1976; Heslop-Harrison, 1968; Schmid *et al.* 1996). Ectexine is a wall layer that is thought to have no strict homologue in 'pteridophyte' spore walls and it is probable that the exospore of 'pteridophytes' is homologous with the endexine of seed plants (Blackmore and Barnes, 1987a; Blackmore, 1990). Thus distal apertures should also be recognised as features associated with the evolution of a new exine layer (Fig. 3).

Equatorial apertures

In contrast to monocotyledons, magnoliid dicotyledons and non-angiospermous seed plants (Fig. 2), eudicotyledons (eudicot subclasses Ranunculidae, Hamamelidae, Dilleniidae, Rosidae, Asteridae, *sensu* Takhtajan, 1969; Walker and Walker, 1984) are characterised by pollen with three colpi spaced equidistantly around the equator of the grain, or some modification of this basic triaperturate type (Van Campo, 1976; Walker and Doyle, 1975). Tricolpate or tricolpate-derived pollen is a synapomorphy of the 'higher', non-magnoliid dicotyledons (Crane, 1990) or eudicots (Doyle and Hotton, 1991). The only families of Magnoliidae *sensu stricto* in which tricolpate pollen grains have been reported are Illiciaceae and Schisandraceae (Walker, 1974, Walker and Doyle, 1975;Walker and Walker, 1984) although the homologies of apertures in these taxa are uncertain (see below). In *Illicium* there are three colpi that join at each pole, but in *Schisandra grandiflora* fusion occurs only distally, and the aperture is a distal trichotomosulcus. In *Kadsura* and other species of *Schisandra* three additional short, equatorial colpi are intercalated between the three longer, distally fused apertures.

There is general agreement that similar developmental processes are responsible for both distal and equatorial apertures (Blackmore and Crane, 1988) and that the formation of triaperturate grains reflects duplication and repositioning of a developmental process that was formerly restricted to the distal surface. Evidence for this comes from the ontogenetic similarity of aperture formation in monosulcate and triaperturate taxa. In addition, although both forms may be present in a single flower of *Nelumbo* (Kuprianova, 1979) they never occur in the same grain. Recent observations of fresh material of *Nelumbo nucifera* from Takamatsu,

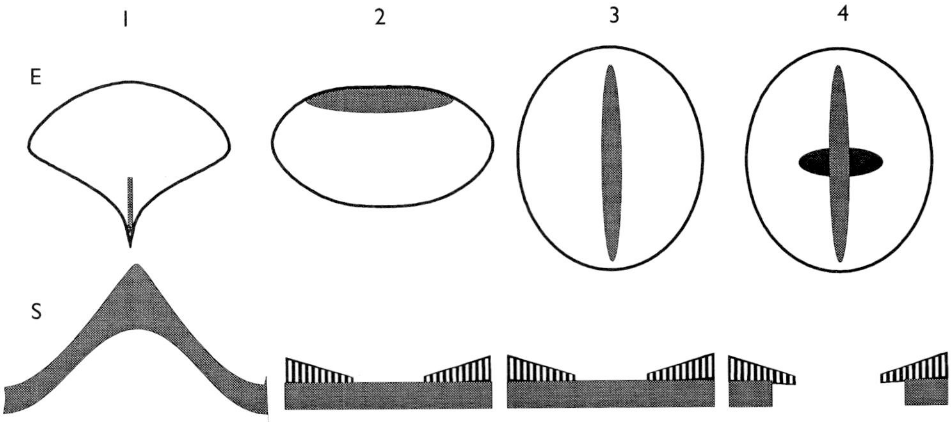

FIG. 3. Diagram summarising the exine layers affected by different aperture types on the assumption that the exospore of pteridophytes is homologous with endexine of seed plants (see text for details). 1 = trilete spore; 2 = monosulcate pollen; 3 = tricolpate pollen, 4 = tricolporate pollen. Proximal trilete apertures involve folded exospore layers (1) whereas distal and equatorial ectoapertures (2-4) affect the ectexine (a layer with no homologue in spores) and endoapertures (4) affect the endexine (which is homologous to the exospore). E = equatorial view, S = sectional view through exine. Ectoapertures are lightly shaded; endoapertures have darker shading. In sectional views, columellate ectexine is in black and white; endexine and exospore are shaded light grey.

Japan (Blackmore, unpublished observations of ca. 5000 grains from a single flower) confirmed that monosulcate grains are present among typically tricolpate grains, but represent only 3% of the total, with spiraperturate grains accounting for a further 1-2%. Flynn and Rowley (1971) failed to detect the usual thinnings of the primexine in *N. nucifera* that could be interpreted as developing apertures. It may be that this species has a more specialised mode of aperture definition but this requires further investigation.

Four principal pathways have been proposed for the origin of triaperturate pollen grains (Fig. 4). Wodehouse (1935, 1936) suggested an origin from the three-armed apertures of *Schisandra,* which he interpreted as a proximal trilete aperture with three shorter colpi between the arms (Fig. 4A). Subsequently it was shown that the trichotomosulcus of *Schisandra* is distal (Erdtman, 1952; Walker, 1974), but several authors have proposed that loss of continuity between the three arms of the trichotomosulcus could provide a form intermediate to tricolpate pollen (Straka, 1963; Kuprianova, 1967; Muller, 1970; Jonker, 1974, 1977). However, there is a significant limitation to this hypothesis because, as Walker (1974) and Walker and Doyle (1975) pointed out, the arms of the trichotomosulcus in *Schisandra* are not oriented in the positions which correspond to typical tricolpate apertures (see Fig. 5) because they follow 'Garside's Rule' (twelve apertures per tetrad, in six pairs, Erdtman, 1969; Garside, 1946) and not 'Fischer's Rule' (twelve apertures per tetrad in four sets of three, Wodehouse, 1935; Erdtman, 1969). This route to typically arranged tricolpate apertures (Fischer's Rule) would therefore not only involve the extension and isolation of the arms of the trichotomosulcus but also a significant change in microspore symmetry (Fig. 4A). The issue of aperture orientation does not preclude the possibility that equatorial apertures are derived from branching sulci in certain taxa. For example, the derivation of tri-ulcerate from trichotomosulcate pollen is well documented in the Palmae (Erdtman and Singh, 1957; Thanikaimoni, 1970, 1986) and indicates the way in which such transitions might occur. Nevertheless, the evidence currently available does not support this hypothesis at the level of the magnoliid-eudicot transition.

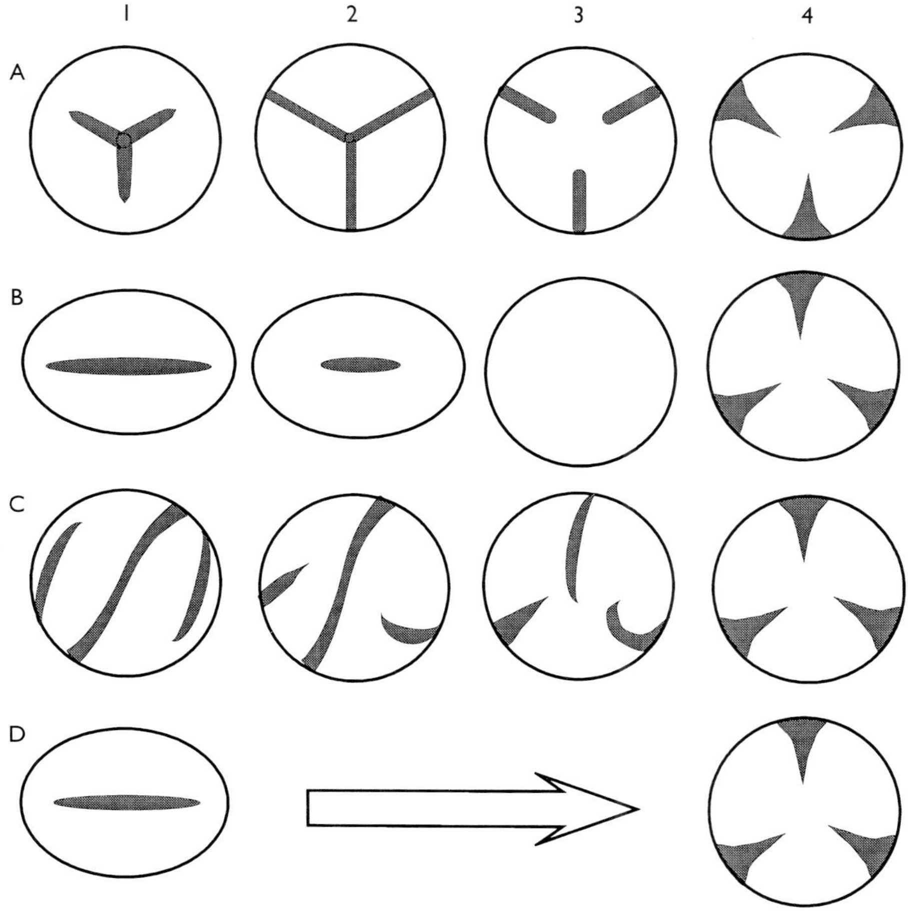

FIG. 4. Four theories concerning the origin of tricolpate apertures. A. Gradual evolution from a distal trichotomosulcate aperture (A1) by extension (A2) and subsequent separation (A3) of the arms of the trichotomosulcus to form equatorial colpi oriented according to Garside's Rule (A4). B. Gradual evolution from a distal monosulcate aperture (B1) by reduction (B2) and loss (B3) of the sulcus and subsequent *de novo* origin of tricolpate apertures (B4). C. Gradual evolution from irregularly spiraperturate grains (C1) by separation into several apertures (C2), increasing restriction to an equatorial position (C3) and stabilisation as tricolpate apertures (C4). D. Rearrangement and multiplication of developmental processes resulting in change from distal sulcus (D1) or trichotomosulcus to equatorial tricolpate pollen (D4) without intermediate forms. Apertures are shaded, all grains are shown in distal polar view.

A second hypothesis (Fig. 4B), that equatorially triaperturate pollen arose via inaperturate intermediates, has been advanced by Meeuse (1965), Muller (1970), Walker (1974) and Walker and Doyle (1975). Muller (1970) suggested that the loss of apertures could be the result of a neotenic change causing the microspores to be released prior to the establishment of apertures. This would imply that the triaperturate condition represented the origin of an entirely new kind of aperture, and as pointed out by Muller would be consistent with the important role envisaged for neoteny in other aspects of angiosperm evolution. Muller also noted that the intermediate inaperturate type pollen may also be difficult to detect in the fossil record because of the reduced exine frequently associated with inaperturate pollen grains (see for example, Stone *et al.*, 1981, Kress, 1986).

A third hypothesis (Fig. 4C), that equatorially triaperturate pollen arose from spiraperturate forms, is suggested by the observation of anomalies in symmetry and aperture arrangement in Lower Cretaceous angiosperm pollen from Gabon (Doyle,1980; see also, Doyle and Hotton, 1991). The extension of Sulci, accompanied by the breakdown of symmetry and the development of spiralisation, may have permitted the possibility of subsequent modification to a number of separate apertures, no longer located at the distal pole. Doyle and Hotton (1991) thus hypothesised that the ancestors of tricolpate eudicots may have had spiraperturate pollen although they accepted that there was no clear vestige of this process among extant eudicots. Spiralisation of apertures appears to have arisen independently in about five monocotyledonous and four dicotyledonous families (Furness, 1985) where it is associated with a variety of ontogenetic modifications including irregular meiotic division. Furness pointed out that, as in inaperturate pollen, the intine of spiraperturate pollen is often very thick and she concluded that spiraperturate grains are always relatively derived within the families that exhibit them.

Although gradualistic hypotheses of the transformation from monosulcate to triaperturate pollen grains have predominated, there are also several explanations involving abrupt developmental changes (for example by Takhtajan, 1948, cited in Sladkov, 1962). These developmental theories tend to emphasise physical properties, such as surface tension between microspores in the tetrad stage, as determinants of tetrad and aperture symmetry (Kuprianova, 1979; Melville, 1981).

One feature of microspore meiosis that shows considerable variation is the extent of development of the special cell wall. Deposition begins prior to meiosis and is extended by the addition of new layers during the process of division, particularly during cytokinesis (Longly and Waterkeyn, 1979). The special cell wall can also play a role in contributing to the pattern of ornamentation (Waterkeyn and Bienfait, 1970; Barnes and Blackmore, 1986) and in the formation of permanent tetrads (Blackmore and Crane, 1988). Although it contributes to the maintenance of the specific tetrad configuration after meiosis, the special cell wall itself plays only a minor part in its determination since otherwise normal tetrads, and pollen grains with normal apertures, form in taxa that lack such a wall (Knox and McConchie, 1986; Blackmore and Crane, 1988). The special cell wall also seems insufficiently fluid for surface tension to cause the rearrangement of the microspores within a tetrad (as had been suggested by Wodehouse, 1929, 1936; Melville, 1981) and thereby control the number and arrangement of apertures.

While the special cell wall may play a role in maintaining rather than establishing tetrad geometry, much stronger evidence suggests that the position and relative orientation of the meiotic spindles is the most important determinant and early indicator of division plane and tetrad configuration. Although the loose correlation of distal apertures with successive cytokinesis and equatorial or globally dispersed apertures with simultaneous cytokinesis had often been noted (Thompson, 1917; Tammes, 1930; Wodehouse, 1928, 1935, 1936; Drahowzal, 1936; Garside, 1946; Melville, 1981) it was not generally considered that such differences in meiosis were important developmental determinants. Wodehouse (1936) expressed this very strongly, concluding that 'the simultaneous division of the tetrad with its resultant tetrahedral arrangement is prevalent among higher dicotyledons, and the successive division is prevalent among primitive dicotyledons, monocotyledons and gymnosperms. But there are so many exceptions that this can scarcely be called a rule'. In contrast to this view, we consider that variations in meiosis play a major role in determining aperture position and symmetry and that the number of apparent exceptions decreases as the full diversity of meiotic variation is documented. It is therefore important to emphasise that although the fundamental processes of meiosis are the same in all groups (Darlington, 1977) genetically controlled variations do exist that have important developmental consequences and that may also be of great systematic value.

Polarity is initially imparted to microsporocyte cells by the formation of the first (heterotypic) meiotic spindle. Geitler (1935) considered the relative orientations of the second (homotypic) meiotic spindles to be characteristic of species, although he recognised that each species may exhibit a restricted range of relative spindle orientations. That spindle orientation and other aspects of meiosis, such as cytoplasmic cleavage, are under direct genetic control is confirmed by the occurrence of mutants where meiosis is interrupted at various stages (Frankel and Galun, 1977; Rees, 1961; Sjödin, 1970; Welzel, 1954). Most significantly in relation to microspore polarity, Peloquin (1983) has shown that meiotic mutants involving single genes in *Solanum,* which modify the orientation of the second meiotic spindles, produce dyads or triads rather than the typical tetrahedral tetrads of the wild type. Similar triads have been observed in meiotic mutants of the fern *Ceratopteris* (Hickok, 1977) and a growing number of meiotic and microspore mutants have been documented in *Arabidopsis thaliana* (see for example, Preuss *et al.*, 1993, 1994; Chaudhury *et al.*, 1994; Twell, this volume).

In 'bryophytes' and 'pteridophytes' with monoplastidic meiosis (that is in mosses, hornworts, *Selaginella* and in megaspores of *Isoetes*), sporocyte polarity is established in meiotic prophase when the single plastid divides into four plastids which play a role in the establishment of the spindle, determine its tetrahedral arrangement and thus control the subsequent planes of meiocyte partitioning (Brown and Lemmon, 1980, 1982a,b, 1990). Interestingly, the spindles of certain higher plants also form tetrahedrons with the four spindle poles connected by six groups of microtubules although this is not associated directly with plastid behaviour. These elaborate cytoskeletal structures are formed by the production of supplementary spindles that have been observed in higher dicotyledons by workers using optical microscopy (for example, Gates, 1920; Gates and Rees, 1921), electron microscopy (Heslop-Harrison, 1971a) and immunofluorescence (Hogan, 1987). Other similarities between the division processes in higher and lower plants include the polarised movements of organelles, including plastids, observed during meiosis in a number of taxa including *Pteridium* (Sheffield and Bell, 1979), *Ginkgo* (Wolniak, 1976) and *Larix*, *Tradescantia* and *Impatiens* (Rodkiewicz *et al.*, 1984, 1986).

In addition to the details of spindle formation, further features of meiosis that exhibit considerable variation are the mode and tempo of cytoplasmic partitioning (see Fig. 5 for a summary). Variation in the timing and mechanism of cytokinesis has been reviewed by Farr (1916), Maheshwari (1949, 1950), Murty (1964), Longly and Waterkeyn (1979), Sampson (1970), Rudall *et al.* (1997) and others but there is relatively little comparative work using ultrastructural and immunofluorescence techniques. Three principal methods of cytoplasmic cleavage may be distinguished in the meiotic products: centrifugally extending cell plates, centripetal constriction furrows or a combination of both cell plates and furrows (Fig. 5). The timing of cytokinesis may be successive, involving a dyad stage after telophase I, or may occur simultaneously after telophase II (Guignard, 1915). In flowering plants with successive cytokinesis, the deposition of the new wall generally involves centrifugal cell plates but sometimes involves constriction furrows (for example, in *Pseudowintera*; Sampson, 1963). From the standpoint of microspore ontogeny, a dyad stage precludes the possibility of further interactions between the two daughter nuclei, or the two spindles of the second meiotic division. If the heterotypic spindle axes are parallel then linear, rhomboidal or tetragonal tetrads result whereas if they are perpendicular T-shaped or decussate tetrads are produced. Whatever the arrangement of the tetrad, seed plants with successive meiosis, and therefore with a distinct dyad stage, invariably have distal apertures (Fig. 5A). Although, exceptionally, additional apertures may also be present in a few taxa these never form a regular triaperturate arrangement around the equator. This situation is comparable to that in monolete spores, where successive meiosis constrains subsequent aspects of spore development.

Simultaneous cytokinesis is often treated as a single uniform process in the literature, but at least two different categories of simultaneous cytokinesis have been distinguished and more detailed investigations may document a nearly continuous spectrum of developmental modes that result from relatively minor differences in the relative timing of aspects of meiosis. The most complete analysis of this diversity has been provided by Sampson (1963,1969, 1970; and see Bhandari, 1984 for a review). At one end of the spectrum, sometimes referred to as 'modified simultaneous' (Farr, 1916; Murty, 1964), ephemeral cell plates form after the first meiotic division but then disperse before partitioning into dyads occurs (Fig. 5B-C). Cytoplasmic cleavage subsequently involves the simultaneous formation of either constriction furrows (Fig. 5B) or cell plates (Fig. 5C). In some cases, such as *Pseudowintera traversii* (Sampson, 1970) ephemeral cell plates are also present for a short period after the second nuclear division, although they are not involved in the final cytoplasmic partitioning. The presence of ephemeral cell plates, especially after the first meiotic division, appears to be effective in isolating the two spindles of the second meiotic division in a manner analagous to a dyad wall. Depending on the relative orientation of the spindles in the second division linear, T-shaped, tetragonal, rhomboidal or decussate developmental tetrads may result, as in a plant with a strictly successive type of division. The pollen grains produced have distal apertures, or are inaperturate.

At the other end of the spectrum, in simultaneous cytokinesis *sensu stricto* (Fig. 5D-E), the second meiotic division follows the first without an ephemeral wall or cell plate being established. In this situation the spindles of the first and second meiotic division appear to interact forming 'supplementary spindles' such that the complete spindle has a tetrahedral form. In most cases the cytoplasm is partitioned by centripetal constriction furrows (Fig. 5E) and typical tricolpate pollen grains with apertures positioned according to Fischer's Rule are produced. Rather rarely, simultaneous partitioning involves centrifugal cell plates (Fig. 5D) as in some Proteaceae (Blackmore and Barnes, 1995). Cytoplasmic cleavage by centrifugal cell plates is correlated with the three arms of the trichotomosulcus or the three pores corresponding to Garside's Rule (Fig. 5D). However, it is important to recognise that there are many minor variations in the details of meiosis (see for example, Rudall *et al.* 1997). For example, some members of the Proteaceae have successive cytokinesis with a distinct dyad stage and the pollen grains are biporate (Blackmore and Thiele, 1988). Given the systematic position of the Proteaceae, within the eudicots, we conclude that such biporate forms are derived, as is the unusual Garside's Rule arrangement of triaperturate pollen. Simultaneous partitioning by cell plates had also been reported in *Helleborus* (Waterkeyn, 1962) but the accompanying drawings appear to show the typical features of meiosis in eudicots with a tetrahedral spindle and cytoplasmic furrowing. A further report, by Hayashi (1960), from the Illiciaceae could more accurately be described as involving ephemeral cell plates.

FIG. 5. Diagram summarising five developmental trajectories (A-E) through meiosis in angiosperms from microsporocytes (column 1), first nuclear division (column 2), initiation of cytokinesis (column 3), formation of dyads (column 4), second nuclear division (column 5), cytokinesis (column 6) to formation of tetrads (column 7). A. Successive meiosis by cell plates (e.g. *Lilium*) to form a distinct dyad stage (A4) and distal apertures (A7). B. Modified simultaneous meiosis (e.g. *Magnolia*) with furrows after second division (B6) to give distal apertures. C. Simultaneous meiosis by cell plates with incomplete cell plates after first (C3) and ephemeral cell plates after second (C5-6) nuclear divisions (e.g. *Pseudowintera*). The final cytokinesis is by cell plates and the apertures are distal. D. Simultaneous meiosis with cytokinesis by cell plates (e.g. *Grevillea*) involves supplementary spindles (D5) which are subsequently bisected by centrifugal cell plates; the apertures are equatorial and follow Garside's Rule. E. Simultaneous meiosis by centripetal cytoplasmic furrows involves supplementary spindles (E5) and equatorial apertures following Fischer's Rule. Shaded circles = nuclei; narrow lines (columns 2 and 5) ending in X = longitudinal axis of meiotic spindle; small white arrows = centrifugal cell plates; black arrows = centripetal constriction furrows; large white arrow = absence of a developmental stage (heterochrony); unshaded ovals (column 6) = apertures.

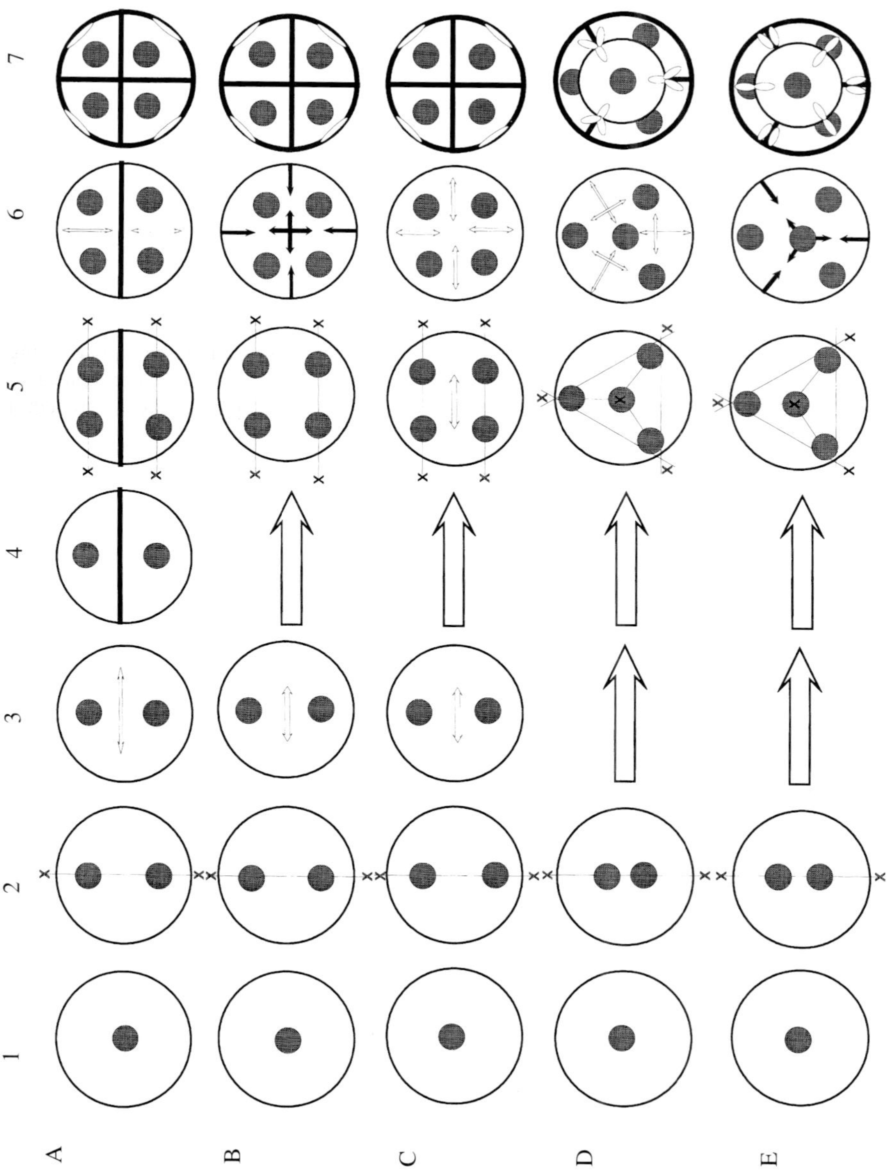
A
B
C
D
E
1
2
3
4
5
6
7

In our view it is highly likely that the relationship between microspore polarity and the symmetrical disposition of apertures is, like tetrad configuration itself, closely related to the symmetry established by the meiotic spindle. However, the range of taxa that have been examined in detail is limited and the precise linkage to aperture initiation has yet to be firmly demonstrated in each case. Aperture determination by apertural shields coincides with the transition from a meiotic to an interphase cytoskeleton as evidence from chemical or physical disruption have shown (Heslop-Harrison, 1971b; Sheldon and Dickinson, 1983). In *Triticum*, Dover (1972, 1973) demonstrated that apertures are sited at the poles of the spindles of the second meiotic division (as in Fig. 5A). The same also appears to apply wherever an ephemeral cell plate is present (Fig. 5B-C) regardless of the final mode of cytoplasmic cleavage. Equatorial apertures, rather than forming at the poles of the spindles, are initiated in the division plane perpendicular to the spindle axis. In both Fischer's Rule and Garside's Rule they form at the last points of cytoplasmic contact between daughter cells. These points occur in pairs at six points on the tetrad surface in Fischer's Rule and in sets of three at four points on the tetrad surface in Garside's Rule.

The distribution of different types of meiosis in angiosperms has been the subject of brief reviews (Maheshwari, 1949, 1950; Bhandari, 1984) and a few more detailed surveys (Murty, 1964; Davis, 1966) but interpretation of the literature is complicated or confused by the tendency to classify cytokinesis as either successive or simultaneous without recognising other variants, and especially the modified simultaneous category (as for example in Davis, 1966). Nevertheless, we agree with Sampson (1969, 1970) in interpreting meiosis of microsporocytes by successive cytokinesis, with one centrifugal cell plate forming after the first division and two more after the second division as plesiomorphic in angiosperms. Acceleration in the rate of division may have led to the second nuclear division commencing prior to the completion of the first cell plate (corresponding to the modified simultaneous type) and ultimately to the loss of cell plates. Alternatively, dissolution of the first meiotic spindle and cytoplasmic partitioning may have been progressively delayed.

Concomitant with the loss of centrifugal cell plates has been the origin of centripetal constriction furrows as an alternative mode of cell cleavage. There are cytoskeletal transformations associated with these changes in tempo and mode of cytoplasmic partitioning. In successive cytokinesis the microtubular spindle of the first division is disassembled and subsequently two new spindles form during the second division. In simultaneous cytokinesis the first meiotic spindle may not fully disassociate, and where cell plates are absent or highly reduced after the first nuclear division, microtubules may be directly incorporated into the second spindle resulting in a tetrahedral configuration.

Compound apertures

Compound apertures (Fig. 3), comprising a separate ectoaperture affecting the ectexine and endoaperture in the endexine (Van Campo, 1958) are restricted to non-magnoliid dicotyledons. However, their absence in monocotyledons may be simply a reflection of the reduction or absence of endexine in many taxa.

The development of endoapertures has been documented in many studies and invariably involves the development of onci, lens shaped bodies of pectocellulosic material, that interrupt the deposition of white line centred lamellae (see for example, Blackmore and Barnes, 1987b). Typically, a few of these lamellae extend into the material of the oncus.

Once symmetry and polarity have been established during the tetrad stage they generally remain constant and the development of endoapertures normally correlates with the position of ectoapertures. However, this is not always the case and examples of heterocolpate pollen grains with more ectocolpi than endoapertures are relatively widespread (Erdtman, 1952; Faegri and Iversen. 1975). Compound apertures, especially in the family Compositae, sometimes have a third aperture component, the mesoaperture, lying between the ectoaperture and endoaperture (Dimon, 1971; Blackmore, 1982).

Multiple apertures

As Wodehouse (1929, 1935), Huynh (1968, 1976) and others have shown, in terms of their symmetry, small increases in aperture number generally involve intercalation of apertures between those reflecting the primary symmetry of the pollen grain. Higher numbers tend towards a more even global distribution (pantoporate or pantocolpate).

The distribution of multiaperturate pollen grains in monocotyledonous (for example in many Alismatales) and dicotyledonous taxa (including Amaranthaceae, Chenopodiaceae, Caryophyllales and other 'centrosperms', Convolvulaceae and Malvaceae) clearly shows that they have numerous independent origins, from pollen with either distal or equatorial apertures. Among monocotyledons multiaperturate pollen may be derived from inaperturate, trichotomosulcate or monosulcate grains (Muller, 1970; Thanikaimoni, 1986). The transition from a single distal sulcus to polyporate or polyforate pollen grains with additional apertures, has occurred in numerous groups including the Alismatales except for *Butomus* which is monosulcate (Chanda *et al.* 1989), *Sarcandra* (Chloranthaceae), Trimeniaceae and Aristolochiaceae. Rudall *et al.* (1997) provided one of the few detailed systematic reviews of meiosis and microsporogenesis, in Asparagales. They suggested that reversals between simultaneous and successive cytokinesis may have occurred within Asparagales and concluded that, within this group, trichotomosulcate pollen is associated with simultaneous cytokinesis.

In the non-magnoliid dicotyledons (eudicots) increases in aperture number by the intercalation of additional equatorial apertures have been recognised as the start of the 'successiform' series of Van Campo (1967, 1976). In Malvaceae, Christensen (1986) concluded that increasing the number of equatorial apertures leads to instability in aperture distribution and increases are accommodated by a switch in to a regular pattern of hexagons or polygons. Van Campo (1976) listed the principal families in which such transitions occur and they are mainly present in the orders Caryophyllales, Geraniales, Papaverales, Podostemales, Polygonales and Ranunculales. Several explanations have been proposed for the equidistant hexagonal spacing of pantoaperturate multiple apertures. Melville (1982) suggested that they might reflect the physical effect of chemical diffusion, although no such chemical pattern was demonstrated. Rodgers and Harris (1969) suggested that aperture positioning was related to cytoplasmic volume in *Linum*.

There have been relatively few developmental studies of multiaperturate pollen and none designed to facilitate direct comparisons with closely related monosulcate or tricolpate taxa. Additional interpolated equatorial apertures frequently occur in pollen grains of many plants, such as *Alnus* (Faegri and Iversen, 1975) or *Sonchus* (Blackmore, 1984). This suggests that the changes in development required to produce them are relatively minor. Discussing further increases in aperture number in the successiform series, Van Campo (1976) cited Wodehouse's explanation that the highly symmetrical polyporate and polycolpate grains originate in response to surface tension and contact between microspores during development. Heslop-Harrison (1972) invoked similar factors in his discussion of space-filling pattern forming processes involved in the establishment of the surface ornamentation of developing microspores (for example in the positioning of spines in *Tagetes* pollen, Heslop-Harrison, 1969). Whatever the factors involved it is clear that the symmetry of many multiaperturate grains is independent of that initially established in microspores by the meiotic spindle and requires an alternative explanation. One possibility is that aperture definition does not occur until after the complete dissolution of spindle microtubules. Freed from the constraint imposed by the polarity and symmetry of the meiotic cytoskeleton, the interphase cytoskeleton may influence the positioning of apertural shields. Spirally arranged multiple apertures and spiraperturate pollen grains are interesting cases where aperture definition clearly does not follow the symmetry of the meiotic cytoskeleton and the tetrad. Spiralisation might perhaps be imparted by spirally arranged cortical microtubules, which are frequently present early in plant cell interphase, but further studies would be necessary to confirm this.

Discussion and conclusions

From this review, it is apparent that apertures form through a variety of developmental processes that do not all share a single origin. The basal group of land plants, hepatics, appear to lack true apertures although their proximal faces often bear the impression of contact during development. Where the polarity of germination has been determined in hepatics, it appears to be distal. Proximal apertures appear to have arisen in stomatophytes as novel features associated with proximal germination. Their development involves the modification of exine layers formed from white line centred lamellae. Structurally complex, pre-formed, proximal apertures are present in mosses, hornworts and basal groups of tracheophytes. Derived forms of proximal apertures, in *Equisetum*, involve novel developmental processes.

Distal apertures appear to have arisen independently in several groups of tracheophytes that already possessed proximal apertures. They appear to be correlated with the origin of more complex exine stratification and specifically with exine layers derived from primexine (ectexine, see Blackmore, 1990). Apertural shields that locally disrupt the deposition of primexine are the basic mechanism by which distal apertures are defined. Equatorial apertures and forms derived from them occur in eudicotyledons and also involve the action of apertural shields. However, the developmental factors influencing the number and spatial arrangement of these apertures are not fully elucidated. Detailed comparative studies of meiosis and the early stages of pollen development are needed. Based on present data it appears that the organisation of the meiotic spindle, the relative timing of cytokinesis and the mode of cytoplasmic partitioning are the principal determinants of distal and equatorial apertures.

Among land plants there has been considerable evolutionary diversification in the processes of meiosis which has not yet been fully charted. The present terminology used to categorise meiosis inadequately encompasses its full diversity across the broad spectrum of embryophytes. In most ‘bryophytes’ cytokinesis is simultaneous (Neidhart, 1979) and quadripolar spindles are frequently formed (Brown and Lemmon, 1984, 1990). In ‘pteridophytes’ cytokinesis may be simultaneous or successive, with the latter clearly being derived (Verma and Khullar, 1976). Among seed plants the replacement of cell plates by constriction furrows has been considered derived, with successive division generally regarded as primitive in angiosperms (Davis, 1966). In contrast, Maheshwari (1949, 1950) concluded that successive meiosis was derived, because of the widespread occurrence of simultaneous cytokinesis in lower plants. The presence of simultaneous cytokinesis in many flowering plants apparently represents a reversion to a condition resembling that in ‘bryophytes’ that is, nevertheless, relatively derived at the more inclusive level of seed plants. Functional arguments have been presented for the adaptive significance of simultaneous cytokinesis in angiosperms. For example, Heslop-Harrison (1966) suggested that, by prolonging the continuity of cytoplasm between daughter cells, simultaneous cytokinesis enables more precise coordination of the early stages of pollen development. This interpretation is consistent with the widely recognised trend towards the acceleration of development and morphological reduction of the male gametophyte in land plants.

Endoapertures appear to have arisen within the eudicots but it is evident from their absence among the basal members of many families (see for example, Wodehouse, 1935, Erdtman, 1952) that they have arisen more than once. Similarly, multiple apertures have arisen independently at least in monocotyledons and eudicotyledons.

Pollen aperture characters are among the most important characters for systematic analysis at a variety of levels in the taxonomic hierarchy (Erdtman, 1952; Van Campo, 1976). The major transitions discussed in this paper constitute a relatively small number of characters, but include several that are critical to the resolution of relationships between major groups of land plants. While morphological assessments provide useful preliminary indicators of the homology of microspore features, these hypotheses need to be tested with comparative studies of pollen and spore development in key extant taxa. An understanding of the developmental basis of microspore apertures is also critical to the use of these characters in phylogenetic analyses. The recognition, for example, that proximal apertures are fundamentally different from distal apertures, and that equatorial apertures represent a multiplication and repositioning of ontogenetic processes responsible for distal apertures, has important implications for how these features should be coded.

The overarching need that is highlighted by this review is for detailed comparative studies of meiosis and pollen development in a selection of extant taxa carefully chosen to sample and adequately represent the diversity of living land plants. Among the specific issues that need to be addressed are: (i) more detailed studies of spore apertures and development in 'bryophytes', which may improve interpretations of the spore fossil record, and (ii) clarification of the pattern of aperture evolution in extant and fossil coniferophytes (especially the transition from proximal to distal apertures). Clarifying the developmental basis of the shift among angiosperms from a single distal aperture to multiple apertures dispersed in various patterns (triaperturate, the successiform trend and other configurations) also promises to be a productive area for future research.

Acknowledgements

We are grateful to Simon Owens for the invitation to take part in this conference. This paper is dedicated to Professors J. and Y. Heslop-Harrison for their outstanding contributions to the field of plant reproductive biology. During the early stages of this paper, S.B. benefited from a Bass Visiting Scientist grant from the Field Museum.

References

Bagnall, C.R. (1975). Species distinction among pollen grains of *Abies*, *Picea* and *Pinus* in the Rocky Mountain area (a scanning electron microscope study). *Rev. Palaeobot. Palynol.* **19**: 203–220.

Barnes, S.H. and Blackmore, S. (1986). Some functional features during pollen development. In: S. Blackmore and I.K. Ferguson (editors). Pollen and spores: form and function, pp. 71–80, Academic Press, London.

Bateman, R. M. and DiMichele, W. A. (1994). Heterospory: the most iterative key innovation in the evolutionary history of the plant kingdom. *Biol. Rev. Cambridge Philos. Soc.* **69**: 345–417.

Bhandari, N.N. (1984). The microsporangium. In: B.M. Johri (editor). The embryology of angiosperms, pp. 53–157, Springer-Verlag, Berlin.

Blackmore, S. (1982). The apertures of Lactuceae pollen. *Pollen & Spores* **24**: 453–462.

Blackmore, S. (1984). The northwest European pollen flora. 32. Compositae: Lactuceae. *Rev. Palaeobot. Palynol.* **42**: 45–85.

Blackmore, S. (1990). Sporoderm homologies and morphogenesis in land plants, with a discussion of *Echinops sphaerocephala* (Compositae). *Pl. Syst. Evol., Suppl.* **5**, 1–12.

Blackmore, S. and Barnes, S.H. (1987a). Embryophyte spore walls: origin, development and homologies. *Cladistics* **3**: 185–195.

Blackmore, S. and Barnes, S.H. (1987b). Pollen wall morphogenesis in *Tragopogon porrifolius* (Compositae: Lactuceae) and its taxonomic significance. *Rev. Palaeobot. Palynol.* **52**: 233–246.

Blackmore, S. and Barnes, S.H. (1995). Garside's Rule and the microspore tetrads of *Grevillea rosmarinifolia* A. Cunn. and *Dryandra polycephala* Benth. (Proteaceae). *Rev. Palaeobot. Palynol.* **85**: 111–121.

Blackmore, S., and Crane, P.R. (1988). Systematic implications of pollen and spore ontogeny. In: C.J. Humphries, (editor). Ontogeny and systematics. pp. 83–115, Columbia University Press. New York.

Blackmore, S. and Knox, R.B. (1990). Microspores: ontogeny and evolution. Academic Press, London.

Blackmore, S. and Thiele, K. (1988). Successive cytokinesis during microsporogenesis in the Proteaceae. In: R.B. Knox, M.B. Singh and L.F. Troiani (editors) Pollination '88, pp. 47–49, University of Melbourne, Melbourne.

Bobrov, A.E. (1966). A contribution to the spore morphology of Osmundaceae. *Bot. Zhurn.* **51**: 1452–1460.

Brown, R.C. and Lemmon, B.E. (1980). Ultrastructure of sporogenesis in a moss, *Ditrichum pallidum*. II. Metaphase I to the tetrad. *Bryologist* **67**: 918–934.

Brown, R.C. and Lemmon, B.E. (1981). Aperture development in spores of the moss, *Trematodon longicollis* Mx. *Protoplasma* **106**: 273–287.

Brown, R.C. and Lemmon, B.E. (1982a). Ultrastructure of meiosis in the moss *Rhyncostegium serrulatum*. I. Prophasic microtubules and spindle dynamics. *Protoplasma* **110**: 23–33.

Brown, R.C. and Lemmon, B.E. (1982b). Ultrastructural aspects of moss meiosis: review of nuclear and cytoplasmic events during prophase. *J. Hattori Bot. Lab.* **53**: 29–39.

Brown, R.C. and Lemmon, B.E. (1983). Microtubule organisation and morphogenesis in young spores of the moss *Tetraphis pellucida* Hedw. *Protoplasma* **116**: 115–124.

Brown, R.C. and Lemmon, B.E. (1984). Spore wall development in *Andreaea* (Musci: Andreaeaopsida). *Amer. J. Bot.* **71**: 412–420.

Brown, R.C. and Lemmon, B.E. (1986). Spore wall development in the liverwort, *Haplomitrium hookeri*. *Canad. J. Bot.* **64**: 1174–1182.

Brown, R.C. and Lemmon, B.E. (1990). Sporogenesis in bryophytes. In: S. Blackmore and I.K. Ferguson (editors). Pollen and spores: form and function, pp. 1–10, Academic Press, London.

Brown, R.C., Lemmon, B.E. and Carothers, Z.B. (1982). Spore wall development in *Sphagnum lescurii*. *Canad. J. Bot.* **60**: 2394–2409.

Brown, R.C., Lemmon, B.E. and Renzaglia, K.S. (1986). Sporocytic control of spore wall pattern in liverworts. *Amer. J. Bot.* **73**: 593–596.

Buchen, B. and Sievers, A. (1981). Sporogenesis and pollen grain formation. In: O. Kiermayer (editor). Cytomorphogenesis in plants, pp. 349–376, Springer-Verlag, Wien.

Chaloner, W.G. (1970). The evolution of miospore polarity. *Geosci & Man* **1**: 47–56.

Chaloner, W.G. (1976). The evolution of adaptive features in fossil exines. In: I.K. Ferguson and J. Muller (editors). The evolutionary significance of the exine, pp. 1–14, Academic Press, London.

Chaloner, W.G. and Hemsley, A.R. (1991). Heterospory: cul-de-sac or pathway to the seed? In: S. Blackmore and S.H. Barnes (editors). Pollen and spores, patterns of diversification. 151–167, Clarendon Press, Oxford. (Systematics Association Special Volume; no. 44).

Chanda, S., Nilsson, S. and Blackmore, S. (1989). Phylogenetic trends in the families of the Alismatales with reference to pollen grains. *Grana* **27**: 257–272.

Chaudhury, A.M., Lavithis, M., Taylor, P.E., Craig, S., Singh, M.B., Signer, E.R., Knox, R.B. and Dennis, E.S. (1994). Genetic control of male fertility in *Arabidopsis thaliana*: structural analysis of premeiotic developmental mutants. *Sexual Pl. Reprod.* **7**: 17–28.

Christensen, P.B. (1986). Pollen morphological studies in the Malvaceae. *Grana* **25**: 95–117.

Crane, P.R. (1985). Phylogenetic analysis of seed plants and the origin of angiosperms. *Ann. Missouri Bot. Gard.* **72**: 716–793.

Crane, P.R. (1990). The phylogenetic context of microsporogenesis. In: S. Blackmore and R.B. Knox (editors). Microspores: evolution and ontogeny, pp. 11–41, Academic Press, London.

Cranwell, L.M. (1961). Coniferous pollen types of the southern hemisphere. I. Aberration in *Acmopyle* and *Podocarpus dacrydioides*. *J. Arnold Arbor.* **42**: 416–423.

Darlington, C.D. (1977). Meiosis in perspective. *Phil. Trans. Ser. B.* **277**: 185–189.

Davis, G.L. (1966). Systematic embryology of the angiosperms. John Wiley and Sons, New York.

Denizot, J. 1976. Remarques sur l'édification des différentes couches de la paroi sporale à exine lamellaire de quelques Marchantiales et Sphaerocarpales. In: I.K. Ferguson and J. Muller (editors). The evolutionary significance of the exine, pp. 185–210, Academic Press, London.

Devi, S. (1974). On the occurrence of bilateral spores in cultivated *Pteris longifolia* L. *Grana* **14**: 1–13.

Devi, S. and Khare, P.B. (1985). Circular laesurae in *Pteris vittata* L. *Rev. Palaeobot. Palynol.* **44**: 147–150.

Dickinson, H.G. (1976). Common factors in exine deposition. In: I.K. Ferguson and J. Muller (editors). The evolutionary significance of the exine, pp. 67–89, Academic Press, London.

Dimon, M.Th. (1971). Problèmes généraux souleves par l'étude pollinique de Composées mediterranéenes. *Naturalia Monspel., Ser. Bot.* **22**: 129–144.

Dover, G.A. (1972). The organisation and polarity of pollen mother cells of *Triticum aestivum*. *J. Cell Sci.* **11**: 699–711.

Dover, G.A. (1973). On pores and pairing. In: J. Wahrman and K.R. Lewis (editors). Chromosomes today 4, pp. 197–204, Wiley, New York.

Doyle, W.T. (1975). Spores of *Sphaerocarpos donnellii*. *Bryologist* **78**: 80–84.

Doyle, J.A. (1980). Evidence on the origin of tricolpate pollen, lower Cretaceous, Gabon. In: International Organization of Palaeobotany and Linnean Society of London. Abstracts, International Palaeobotanical Conference, Reading, England 1980, 14.

Doyle, J.A. and Donoghue, M.J. (1986). Seed plant phylogeny and the origin of angiosperms: an experimental cladistic approach. *Bot. Rev.* **52**: 321–431.

Doyle, J.A. and Hotton, C.L. (1991). Diversification of early angiosperm pollen in a cladistic context. In: S. Blackmore and S.H. Barnes (editors). Pollen and spores: patterns of diversification, pp. 169–195, Clarendon Press, Oxford.

Drahowzal, G. (1936). Beiträge zur Morphologie und Entwicklungsgeschichte der Pollenkörner. *Österr. Bot. Z.* **85**: 241–269.

Duthie, A.V. and Garside. S. (1936). Studies in South African Ricciaceae. I. Three annual species: *R. plana* Taylor, *R. cupulifera* sp. nov., and *R. curtisii* T.P. James. *Trans. Roy. Soc. South Africa* **24**: 93–133.

Duthie, A.V. and Garside, S. (1939). Studies in South African Ricciaceae. II. *Trans. Roy. Soc. South Africa* **27**: 17–28.

Erdtman, G. (1943). An introduction to pollen analysis. Ronald Press, New York.
Erdtman, G. (1952). Pollen morphology and pant taxonomy. Angiosperms. Hafner, New York.
Erdtman, G. (1965). Pollen and spore morphology, plant taxonomy. Gymnospermae, Bryophyta (An introduction to palynology. III). Almqvist and Wiksell, Stockholm.
Erdtman, G. (1969). Handbook of palynology. Hafner Publishing Co., New York.
Erdtman, G, and Praglowski, J. (1959). Six notes on pollen morphology and pollen morphological techniques. *Bot. Not.* **112**: 175–181.
Erdtman, G. and Singh, G. (1957). On the pollen morphology in *Sclerosperma manii. Bull. Jard. Bot. Etat.* **27**: 217–220.
Faegri, K. and Iversen, J. (1975). Textbook of modern pollen analysis. Munksgaard, Copenhagen.
Farr, C.H. (1916). Cytokinesis of the pollen-mother-cells of certain dicotyledons. *Mem. New York Bot. Gard.* **6**: 253–317.
Flynn, J.J. and Rowley, J.R. (1971). The primexine of *Nelumbo nucifera. Experientia* **27**: 227–228.
Frankel, R. and Galun, E. (1977). Pollination mechanisms, production and plant breeding. *Monogr. Theor. Appl. Genet.* **50**: 211–225.
Friedman, W.E. (1993). Evolutionary history of the seed plant male gametophyte. *Trends Ecol. Evol.* **8**: 15–20.
Furness, C.A. (1985). A review of spiraperturate pollen. *Pollen & Spores* **27**: 307–319.
Garside, S. (1946). The developmental morphology of the pollen of Proteaceae. *J. S. African Bot.* **12**: 27–34.
Gates, R.R. (1920). A preliminary account of the meiotic phenomena in the mother cells and tapetum of lettuce (*Lactuca sativa*). *Proc. Roy. Soc. London, Ser. Biol. Sci.* **91**: 216–223.
Gates, R.R. and Rees, E.M. (1921). A cytological study of pollen development in *Lactuca. Ann. Bot.* **35**: 365–398.
Geitler, L. (1935). Beobachtungen über die erste Teilung im Pollenkorn der Angiospermen. *Planta* **24**: 361–386.
Gould, S.J. and Vrba, E.S. (1982). Exaptation - a missing term in the science of form. *Paleobiology* **8**: 4–15.
Guignard, L. (1915). Sur la formation du pollen. *Compt. Rend. Hebd. Séances Acad. Sci.***160**: 428–433.
Hayashi, Y. (1960). On the microsporogenesis and pollen morphology in the family Magnoliaceae. *Sci. Rep. Tohoku Univ., Ser. 4 Biol.* **26**: 45–52.
Heslop-Harrison, J. (1966). Cytoplasmic continuities during spore formation in flowering plants. *Endeavour* **25**: 65–72.
Heslop-Harrison, J. (1968). The pollen grain wall. *Science* **161**: 230–237.
Heslop-Harrison, J. (1969). The origin of surface features of the pollen wall of *Tagetes patula* as observed by scanning electron microscopy. *Cytobios* **2**: 177–186.
Heslop-Harrison, J. (1971a). Wall pattern formation in angiosperm microsporogenesis. *Symp. Soc. Exp. Biol.* **25**: 277–300.
Heslop-Harrison, J. (1971b). Sporopollenin in the biological context. In: J. Brooks, P.R. Grant, M. Muir, P. van Gijzel and G. Shaw (editors). Sporopollenin, pp. 1–30, Academic Press.
Heslop-Harrison, J. (1972). Pattern in plant cell walls: morphogenesis in miniature. *Proc. Roy. Inst. Gt. Brit.* **45**: 335–351
Heslop-Harrison, J. (1976). The adaptive significance of the exine. In: I.K. Ferguson and J. Muller (editors). The evolutionary significance of the exine, pp. 27–37, Academic Press, London.
Hickok, L.G. (1977). The cytology and derivation of a temperature sensitive meiotic mutant in the fern *Ceratopteris. Amer. J. Bot.* **64**: 552–563.
Hogan, C.J. (1987). Microtubule patterns during meiosis in two higher plant species.

Protoplasma 138: 126–136.
Huynh, K.-L. (1968). Etude de l'arrangement du pollen dans la tétrade chez les angiosperms sur la base de données cytologiques. *Ber. Schweiz. Bot. Ges.* **78**: 151–191.
Huynh, K.-L. (1976). Arrangement of some monosulcate, disulcate, trisulcate, and tricolpate pollen types in the tetrads, and some aspects of evolution in the angiosperms. In: I.K. Ferguson and J. Muller (editors). The evolutionary significance of the exine, pp. 101–124, Academic Press, London.
Huynh, K.-L. and Sampson, F.B. (1983). Tetrad arrangement of the trisaccate pollen of *Dacrycarpus dacrydioides* (Podocarpaceae) and developmental morphology of the triradiate ridge. *Grana* **22**: 1–9.
Inoue, H. (1960). Studies in spore germination and the earlier stages of gametophyte development in the Marchantiales. *J. Hattori Bot. Lab.* **23**: 148–191.
Jonker, F.P. (1974). Reflections on pollen evolution. *Advances Pollen-Spore Res.* **1**: 50–61.
Jonker, F.P. (1977). Spermatophytes and pre-spermatophytes, ovules and pre-ovules, pollen and pre-pollen - a comparison of pollen and seed evolution. *Advances in Pollen-Spore Res.* **2**: 6–13.
Kenrick, P. and Crane, P. R. (1997a). The origin and early evolution of plants on land. *Nature.* **389**: 33–39.
Kenrick, P. and Crane, P. R. (1997b). The origin and early diversification of land plants: a cladistic study. Smithsonian Institution Press, Washington. (Smithsonian series in comparative evolutionary biology).
Knox, R.B. (1984). The pollen grain. In: B.M. Johri (editor). Embryology of angiosperms, pp. 191–297, Springer-Verlag, Berlin.
Knox, R.B. and McConchie, C.A. (1986). Structure and function in compound pollen. In: S. Blackmore and I.K. Ferguson (editors). Pollen and spores: form and function, pp. 265–285, Academic Press, London.
Kress, W.J. (1986). Exineless pollen structure and pollination systems of tropical *Heliconia* (Heliconiaceae). In: S. Blackmore and I.K. Ferguson (editors). Pollen and spores: form and function, pp. 329–345, Academic Press, London.
Kuprianova, L.A. (1967). Apertures of pollen grains and their evolution in angiosperms. *Rev. Palaeobot. Palynol.* **3**: 73–80.
Kuprianova, L.A. (1974). On the evolutionary levels in the morphology of pollen and spores. *Advances Pollen-Spore Res.* **1**: 31–49.
Kuprianova, L.A. (1979). On the possibility of the development of tricolpate pollen from monosulcate. *Grana* **18**: 1–4.
Le Thomas, A., Morawetz, W. and Waha, M. (1986). Pollen of palaeo- and neotropical Annonaceae: definition of the aperture by morphological and functional characters. In: S. Blackmore and I.K. Ferguson (editors). Pollen and spores: form and function, pp. 375–388, Academic Press, London.
Longly, B. and Waterkeyn, L. (1979). Étude de la cytocinese. III. Les cloissonnements simultanées et successifs des microsporocytes. *Cellule* **73**: 65–80.
Lugardon, B. (1969). Sur la structure fine de parois sporales d'*Equisetum maximum* Lamk. *Pollen & Spores* **11**: 449–474.
Lugardon, B. (1976). Sur la structure fine de l'exospore dans les divers groupes de ptéridophytes actuelles (microspores et isospores). In: I.K. Ferguson and J. Muller (editors). The evolutionary significance of the exine, pp. 231–250, Academic Press, London.
Maheshwari, P. (1949). The male gametophyte of angiosperms. *Bot. Rev.* **15**: 1–75.
Maheshwari, P. (1950). An introduction to the embryology of angiosperms. McGraw-Hill, New York.
Meeuse, A.D.J. (1965). The message of the pollen grains. *Advancing Frontiers Pl. Sci.* **11**: 112–124.

Mehra, P.N. and Kachroo, P. (1952). Sporeling germination studies in Marchantiales. II. *Stephensoniella brevipedunculata* Kash. *Bryologist* **55**: 59–64.

Melville, R. (1981). Surface tension, diffusion and the evolution and morphogenesis of pollen aperture patterns. *Pollen & Spores* **23**: 179–203.

Mishler, B.D. and Churchill, S.P. (1984). A cladistic approach to the phylogeny of the 'bryophytes'. *Brittonia* **36**: 406–424.

Mishler, B.D. and Churchill, S.P. (1985). Transition to a land flora: phylogenetic relationships of the green algae and bryophytes. *Cladistics* **1**: 305–328.

Mogensen, G.S. (1983). The spore. In: R.M. Schuster (editor). New manual of bryology. pp. 325–342, Hattori Botanical Laboratory, Japan.

Muller, J. (1970). Palynological evidence on the early differentiation of the angiosperms. *Biol. Rev. Cambridge Philos. Soc.* **45**: 417–450.

Muller, J. (1979). Form and function in angiosperm pollen grains. *Ann. Missouri Bot. Gard.* **66**: 596–632.

Murty, Y.S. (1964). Cytokinesis and microspore tetrad. In: P.K.K. Nair (editor). Advances in Palynology, pp. 176–202, National Botanic Gardens, Lucknow.

Nayar, B.K. (1963). Spore morphology of *Loxogramme*. *Grana Palynol.* **4**: 388–392.

Nayar, B.K and Kaur, S. (1968). Spore germination in homosporous ferns. *J. Palynol.* **4**: 1–14.

Neidhart, H.V. (1979). Comparative studies of sporogenesis in bryophytes. In: G.C.S. Clarke, and J.G. Duckett (editors). Bryophyte systematics, pp. 251–280, Academic Press, London.

Nehira, K. (1983). Spore germination, protonema development and sporeling development. In: R.M. Schuster (editor). New manual of bryology, pp. 347–385, Hattori Botanical Laboratory, Japan.

Nixon, K.C., Crepet, W.L., Stevenson, D. and Friis, E.M. (1994). A re-evaluation of seed plant phylogeny. *Ann. Missouri Bot. Gard.* **81**: 484–533.

Olesen, P. and Mogensen, G.S. (1978). Ultrastructure, histochemistry and notes on germination stages of spores of selected mosses. *Bryologist* **81**: 393–416.

Peloquin, S.J. (1983). Genetic engineering with meiotic mutants. In: D.L. Mulcahy and E. Ottaviano (editors). Pollen: biology and implications for plant breeding, pp. 311–316, Elsevier, New York.

Pocknall, D.T. (1981). Pollen morphology of the New Zealand species of *Dacrydium* Solander, *Podocarpus* L'Heritier and *Dacrycarpus*. *New Zealand J. Bot.* **19**: 67–95.

Preuss, D., Lemieux, B., Yen, G. and Davis, R.W. (1993). A conditional sterile mutation eliminates surface components from *Arabidopsis* pollen and disrupts cell signalling during fertilization. *Genes & Developm.* **7**: 974–985.

Preuss, D., Seung, Y.R. and Davis, R.W. (1994). Tetrad analysis possible in *Arabidopsis* with mutation of the QUARTET (QRT) genes. *Science* **264**: 1458–1460.

Proskauer, J. (1955). The Sphaerocarpales of South Africa. *J. S. African Bot.* **21**: 63–75.

Proskauer, J. (1961). On *Carrpos* I. *Phytomorphology* **11**: 359–378.

Rees, H. (1961). Genotypic control of chromosome form and behaviour. *Bot. Rev.* **27**: 288–314.

Rodgers, C.M. and Harris, B.D. (1969). Pollen exine deposition: a clue to its control. *Amer. J. Bot.* **56**: 1209–1211.

Rodkiewicz, B., Kudlicka, K.D. and Stobiecka, H. (1984). Patterns of amyloplast distribution during microsporogenesis in *Tradescantia*, *Impatiens* and *Larix*. *Acta Soc. Bot. Poloniae* **53:** 437–441.

Rodkiewicz, B.K., Bednara, J., Mostowska, A., Duda, E. and Stobiecka, H. (1986). The change in disposition of plastids and mitochondria during microsporogenesis and sporogenesis in some higher plants. *Acta Bot. Neerl.* **35**: 209–215.

Rowley, J.R. and Dahl, A.O, (1962). The aperture of the pollen grains in *Commelinantha*. *Pollen & Spores* **4**: 221–232.

Rudall, P.J., Furness, C.A., Chase, M.W. and Fay, M.F. (1997). Microsporogenesis and pollen sulcus type in Asparagales (Lilianae). *Canad. J. Bot.* **75**: 408–430.
Sampson, F.B. (1963). The floral morphology of *Pseudowintera*, the New Zealand member of the vesselless Winteraceae. *Phytomorphology* **13**: 403–423.
Sampson, F.B. (1969). Cytokinesis in pollen mother cells of angiosperms, with emphasis on *Laurelia novae-zelandiae* (Monimiaceae). *Cytologia* **34**: 627–634.
Sampson, F.B. (1970) Unusual features of meiosis in pollen mother cells of *Pseudowintera traversii* (Buchan.) Dandy (Winteraceae). *Beitr. Biol. Pflanzen*, **47**: 71–77.
Schmid, A.M., Eberwein, R.K. and Hesse, M. (1996). Pattern morphogenesis in cell walls of diatoms and pollen grains: a comparison. *Protoplasma* **193**: 144–173.
Schopf, J.M. (1938). Spores from the Herrin (no. 6) coal bed in Illinois. *Rep. Invest. Illinois State Geol. Surv.* **50**: 1–55.
Sheffield, E. and Bell, P.R. (1979). Ultrastructural aspects of sporogenesis in a fern, *Pteridium aquilinum* (L.) Kuhn. *Ann. Bot.* **44**: 393–405.
Sheldon, J.M. and Dickinson, H.G. (1983). Determination of patterning in the pollen wall of *Lilium henreyi*. *J. Cell. Sci.* **63**: 191–208.
Shivanna, K.R. and Johri, B.M. (1985). The angiosperm pollen. structure and function. Wiley Eastern Limited, New Delhi.
Siler, M.B. (1934). Development of spore walls in *Sphaerocarpos donnellii*. *Bot. Gaz.* **95**: 563–591.
Singh, H. (1978). Embryology of gymnosperms. Gebrüder Bornträger, Berlin.
Sjödin, J. (1970). Induced asynaptic mutants in *Vicia faba* L. *Hereditas* **66**: 215–232.
Sladkov, A.N. (1962). Morphology of the pollen and spores of recent plants of the USSR. In: A.N. Sladkov (editor). K pervoi Mezhdunarodnoi Palinogicheskoi Konferentsii, Takson, ShA. [International Conference on Palynology, (1st: 1962: Tucson, Arizona)] Doklady sovetskikh palinologov. pp. 5–16, Izd-vo Akademii Nauk SSSR, Moscow. [Translated 1969 by U.S. Dept. Commerce, IPST cat. No. 5109.]
Sladkov, A.N. (1964.) Succession of spore tetrads and hexads. *Nauchnye Dokl. Vysshei Shkoly Biol. Nauki* **3**: 91–98. [Translated 1969 by U.S. Dept. Commerce, IPST cat. No. 5109.]
Sorsa, P. (1976). Spore wall structure in Mniaceae and some adjacent bryophytes. In: I.K. Ferguson and J. Muller (editors). The evolutionary significance of the exine. pp. 211–229. Academic Press, London.
Stokey, A.G. and Atkinson, L.R. (1956). The gametophyte of the Osmundaceae. *Phytomorphology* **6**: 19–40.
Stone, D.E., Sellers, S.C. and Kress, W.J. (1981). Ontogenetic and evolutionary implications of a neotonous exine in *Tapeinochilos* (Zingiberales: Costaceae). *Amer. J. Bot.* **68**: 49–63.
Straka, H. (1963). Über die mögliche phylogenetische Bedeutung der Pollenmorphologie der Madagassischen *Bubbia perrieri* R.Cap. (Winteraceae). *Grana Palynol.* **4**: 355–360.
Takhtajan, A. (1948). Morfologicheskaya evolyutsia pokrytosemennykh. Morphological evolution of the Angiospermae. Izd-vo Moskovskogo Obschestva Ispytatelei Prirody, Moscow.
Takhtajan, A. (1969). Flowering plants, origin and dispersal. Oliver and Boyd, Edinburgh. [Translated from the Russian by C. Jeffrey].
Tammes, P.M.L. (1930). On the origin of number and arrangement of the places of exit on the surface of pollen grains. *Recueil Trav. Bot. Néerl.* **27**: 1–84.
Thanikaimoni, G. (1970). Les palmiers: palynologie et systématique. *Trav. Sect. Sci. Techn. Inst. Franç. Pondichéry* **11**: 1–286.
Thanikaimoni, G. (1986). Pollen apertures: form and function. In: S. Blackmore and I.K. Ferguson (editors). Pollen and spores: form and function, pp. 119–136, Academic Press, London.
Thompson, D'A. W. (1917). Growth and Form. Cambridge University Press, Cambridge.

Twell, D. Cell polarity, asymmetric division and cell fate determination in developing pollen. In: S.J. Owens and P.J. Rudall (Editors). Reproductive Biology. pp. 197–218. Royal Botanic Gardens, Kew.
Udar, R. (1958). Culture studies in the genus *Riccia* (Mich.) L. III. Sporeling germination in *R. trichocarpa* Howe - a reinvestigation. *J. Indian Bot. Soc.* **37**: 70–74.
Uehara, K. and Kurito, S. (1989a). Ultrastructural study of spore wall morphogenesis in *Ophioglossum thermale* Kom. var. *nipponicum* (Miaybe et Kudo) Nishida. *Bot. Mag. (Tokyo)* **102**: 413–427.
Uehara, K. and Kurito, S. (1989b). An ultrastructural study of spore wall morphogenesis in *Equisetum arvense*. *Amer. J. Bot.* **76**: 939–951
Uehara, K. and Kurito, S. (1991). Ultrastructural study on spore wall morphogenesis in *Lycopodium clavatum* (Lycopodiaceae). *Amer. J. Bot.* **78**: 24–36.
Uehara, K., Kurito, S., Sahashi, N. and Ohmoto, T. (1991). Ultrastructural study on microspore wall morphogenesis in *Isoetes japonica* (Isoetaceae). *Amer. J. Bot.* **78**: 1182–1190.
Ueno, J. (1957). Relationships of the genus *Tsuga* from pollen morphology. *J. Inst. Polytechn. Osaka City Univ, Ser. D, Biol.* **8**: 191–196.
Van Campo, M. (1958). Palynologie Africaine. *Bull. Inst. Franç. Afrique Noire Ser A. Sci. Nat.* **20**: 753–759.
Van Campo, M. (1967). Pollen et classification. *Rev. Palaeobot. Palynol.* **3**: 65–71.
Van Campo, M. (1976). Patterns of pollen morphological variation within taxa. In: I.K. Ferguson, and J. Muller (editors) The evolutionary significance of the exine. pp. 125–135. Academic Press, London.
Verma, S.C. and Khullar, S.P. (1976). Adaptive significance of simultaneous cytokinesis in pteridophytes. *Phytomorphology* **26**: 96–102.
Walker, J.W. (1974). Aperture evolution in the pollen of primitive angiosperms. *Amer. J. Bot.* **61**: 1112–1137.
Walker, J.W. and Doyle, J.A. (1975). The bases of angiosperm phylogeny: palynology. *Ann. Missouri Bot. Gard.* **62**: 644–723.
Walker, J.W. and Walker, A.G. (1984). Ultrastructure of lower Cretaceous angiosperm pollen and the origin and early evolution of flowering plants. *Ann. Missouri Bot. Gard.* **71**: 464–521.
Waterkeyn, L. (1962). Les parois microsporocytaires de nature callosique chez *Helleborus* et *Tradescantia*. *Cellule* **62**: 224–255.
Waterkeyn, L. and A. Bienfait. (1970). On a possible function of the callosic special cell wall in *Ipomoea purpurea* (L.) Roth. *Grana* **10:** 13–20.
Welzel, G. (1954). Entwicklungsgeschichte-genetische Untersuchungen an pollensterilen Mutanten von *Petunia*. *Z. Indukt. Abstammungs-Vererbungsl.* **86**: 35–53.
Wodehouse, R.P. (1928). The phylogenetic value of pollen grain characters. *Ann. Bot.* **42**: 891–934.
Wodehouse, R.P. (1929). The origin of symmetry patterns of pollen grains. *Bull. Torrey Bot. Club* **56**: 339–350.
Wodehouse. R.P. (1935). Pollen grains, their structure, identification and significance in science and medicine. McGraw-Hill, New York.
Wodehouse, R.P. (1936). Evolution of pollen grains. *Bot. Rev.* **2**: 67–84.
Wolniak, S.M. (1976). Organelle distribution and apportionment in the microsporocyte of *Ginkgo biloba* L. *Amer. J. Bot.* **63**: 251–258.

Pacini, E. and Franchi, G.G. (1998). Pollen dispersal units, gynoecium and pollination. In: S.J. Owens and P.J. Rudall (Editors). Reproductive Biology, pp. 183–195. Royal Botanic Gardens, Kew.

POLLEN DISPERSAL UNITS, GYNOECIUM AND POLLINATION

ETTORE PACINI[1] AND GIAN GABRIELE FRANCHI[2]

[1]Dept. of Environmental Biology, University of Siena, Via P. A. Mattioli 4, 53100 Siena, Italy
[2]Dept. of Pharmaceutical Chemistry and Technologies, University of Siena, Via P. A. Mattioli 4, 53100 Siena, Italy

Abstract

Pollen grains may be dispersed singly (in monads) or grouped in four ways to form pollen dispersal units: 1. filiform pollen grains that tangle together; 2. grains adhering together by virtue of viscous substances; 3. grains connected by filaments; 4. grains joined by common walls. Types of pollen dispersal units are compared with mode of pollination. The types of pollen dispersal unit of plants with anemophilous pollination are fewer in number and composed of fewer grains than the pollen dispersal units of plants with zoophilous pollination, in which the mechanisms holding the grains together also attach the units to pollinator bodies. When dispersal units consist of thousands of pollen grains (compound pollen), special structures have developed for attaching them to insect bodies. Families such as Labiatae, Compositae and Poaceae are extremely uniform in terms of pollen dispersal units, whereas others such as the Orchidaceae have a variety of types. Evolutionary changes in the angiosperm gynoecium reflecting the evolution of pollen dispersal units are also described.

Introduction

The aim of this study is to review the various types and characteristics of pollen dispersal units and to compare them among themselves and with the mode of dispersal and pistil structure with which they are associated. Observations and examples are original unless followed by references.

In gymnosperms, the pollen landing site is close to the micropyle and the pollen grains which land there compete for the single ovule. The ovule may contain more than one female gamete, but only one embryo is produced per seed (Singh, 1978). In primitive angiosperms (Magnoliidae) the landing site is the stigma, but as in gymnosperms, each carpel contains only one ovule (apocarpic ovary). The typical distance that the pollen tube travels, however, is greater than in gymnosperms.

In gymnosperms, pollination is almost exclusively anemophilous and pollen is dispersed in monads. It is generally agreed that the earliest angiosperms were entomophilous (Friis & Crepet, 1987). Pollen can be attached to the body of insects in two ways: (a) during flight the insect becomes charged electrostatically so that on arrival at the flower it collects pollen without even touching the anthers (Corbet et al., 1982; Erickson & Buchman, 1983; Chaloner, 1986); (b) by means of a substance which coats the pollen grains and holds them together

(Pacini & Franchi, 1993). This adhesive agent is also responsible for uniting pollen grains in massulae that are transported as a unit. If the unit does not break up, it reaches the stigma intact (Pacini & Franchi, 1993; Lisci et al., 1996). If the pollen is held together in another way, there may be a special appendage as in Orchidaceae and Asclepiadaceae (Faegri & van der Pijl, 1979). In case b, the vector must touch the anthers.

In certain more evolved angiosperms, we find syncarpic ovaries with many ovules (up to several thousand in a few cases eg. Cucurbitaceae). In such cases, pollen transport in large groups becomes a necessity for the fertilization of a sufficient number of ovules.

Types of pollen dispersal units

When pollen grains are not dispersed singly, they may be held together in one or more of the following ways (Fig 1):

1. Filiform pollen grains form a tangle, as in monocots with submarine pollination (McConchie & Knox, 1989).
2. Grains adhere together by virtue of a pollen coat of viscous material.
3. Grains are connected by filaments of sporopollenin, viscin threads or other material.
4. Grains are joined by common walls.

Adhesion mechanisms for monad pollen

Three types of viscous materials, pollenkitt, tryphine and elastoviscin may be produced in the anther from the transient layer of tissue known as the tapetum (Pacini & Franchi, 1993). Pollenkitt is derived from the degeneration of the tapetum and from elaiosomes and spherosomes (Pacini & Franchi, 1993) and is hydrophobic in character (Dobson, 1988). Tryphine seems to occur only in the Cruciferae. It is derived from the early degeneration of tapetal cells so that organelles which are still recognizable are released into the loculus (Pacini & Franchi, 1993). Pollenkitt and tryphine are both derived from degeneration of the tapetum, but in the first case degeneration is slow and total, whereas in the second case it is sudden and partial. This process has been called *in situ* degeneration in the case of pollenkitt and *extra situm* degeneration in the case of tryphine (Pacini & Franchi, 1991). The functions of pollenkitt and tryphine seem to be substantially the same, however tryphine never coexists with orbicules (Pacini & Franchi, 1993). In *Cucurbita pepo*, pollenkitt has the principal function of sticking the pollen to the insect body, since the pollen grains are large and are transported singly (Nepi & Pacini, 1993). For smaller pollen grains, pollenkitt also has the function of forming clumps with a variable number of grains. In *Mercurialis annua*, an anemophilous plant in which pollen dispersal has been studied under natural and experimental conditions, clumps of up to 44 grains have been observed (Lisci *et al.*, 1996). On impact at the stigma, the clumps break up into monads because the degree of adherence is insufficient to withstand the impact. Pollenkitt and tryphine are also necessary to stick the pollen together to form a compact load in the pollen basket (corbicula) of bees. Another possible function of pollenkitt and tryphine is to hold sporophytic proteins inside the exine ornamentation as part of the stigma/pollen recognition system (Heslop-Harrison & Shivanna 1977). Apart from its perfume, if any, the chemicophysical properties of pollenkitt do not change during exposure to air. The tetrads of *Calluna vulgaris*, on the other hand, dry as the flower ages and are carried by wind (Percival, 1965). A similar phenomenon was observed by us in *Erica arborea*, in which the anthers emerge from the corolla during the anemophilous phase (Franchi & Pacini, 1996).

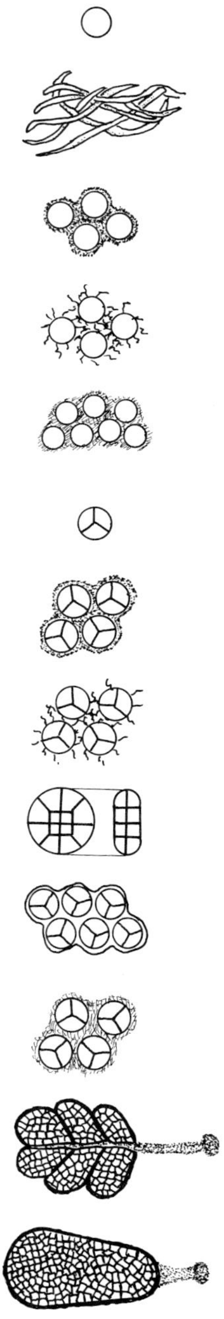

A - Monad pollen as in gymnosperms, Betulaceae, Urticaceae and Poaceae.

B - A tangle of elongated monad pollen as in Zosteraceae (McConchie & Knox 1989).

C - Monad pollen grains grouped by pollenkitt, Euphorbiaceae, Labiatae, Compositae and *Pterostylis concinna* (Orchidaceae; Fitzgerald et al. 1994); or tryphine as in Cruciferae.

D - Monad pollen grouped by viscin threads as in *Oenothera, Fuchsia* (Oenotheraceae) and *Jacqueshuberia* (Caesalpiniaceae; Patel et al. 1985).

E - Monads grouped by elastoviscin as in *Apostasia wallachii* (Orchidaceae; Schill & Wolter 1986).

F - Tetrad of pollen as in Typhaceae and Juncaceae.

G - Tetrads grouped by pollenkitt as in Ericaceae and some Araceae.

H - Tetrads grouped by viscin threads as in *Rhododendron hemitrichotum* (Ericaceae; Waha 1984).

I - Polyads as in *Acacia* (Leguminosae: Mimosaceae; Kenrick and Knox 1989) or *Chimaphila* sp. (Pyrolaceae; Takahashi 1986).

J - Groups of tetrads united together by a thin external layer of callose (soft pollinium A) as in *Polystachia pubescens* (Schlag & Hesse 1993).

K - Tetrads grouped by elastoviscin (soft pollinium B) as in some members of tribes Orchideae and Neottieae (Yeung 1987b).

L -Tetrads grouped by common walls forming a massula; exine is present only on exterior of each massula; massulae are kept together by the stipe or other devices consisting of elastoviscin (dotted) which also sticks the pollinarium to pollinator body, by means of the viscidium, (soft pollinium C) as in *Loroglossum hircinum* (Pandolfi & Pacini 1995).

M - Tetrads grouped in a compact pollinium with exine on the outside; the pollinarium adheres to the body of the pollinator by means of a viscidium in orchids (Yeung & Law 1987) and a corpusculum in the Asclepiadaceae (Dunnenbaum & Schill 1991).

FIG. 1. Different types of pollen dispersal units with examples. Pollen grains are not drawn to scale. The family Orchidaceae has many types of pollen dispersal units.

Elastoviscin is formed as unsaturated lipid droplets by the tapetal cytoplasm from the microspore stage onwards. Before anthesis, the droplets fuse and flow into the loculus, glueing the pollen together. Elastoviscin is much more viscous than pollenkitt and tryphine and the number of pollen grains held together is greater (Schill & Wolter, 1985; Yeung & Law, 1987a; Wolter *et al.*, 1988). It is only found in orchids, where it is produced by a localized group of tapetal cells (Schill & Wolter, 1985). Another viscous material composed of polysaccharides and lipids, produced by the connective tissue of the anther, was described by De Frey *et al.* (1992) in *Tylosema esculentum* (Caesalpiniaceae).

Viscin threads, instead, are thin ropes of sporopollenin, continuous with the outer exine layer (Hesse, 1984) of monad and tetrad pollen. They form on the microspore when it is released from the callosic wall (Takahashi & Skvarla, 1990). Pollen-connecting threads derived from various parts of the anther, but not composed of sporopollenin, are found in *Heliconia* (Heliconiaceae) and *Strelitzia* (Strelitziaceae) and like viscin threads, are responsible for clump formation (Hesse, 1986; Rose & Barthlott, 1995).

Compound pollen

In many species pollen is presented to pollinating agents as compound pollen. The most common and simplest pollen dispersal unit of this type is the tetrad. Some or almost all members of the following families have pollen dispersed in tetrads: Agavaceae, Annonaceae, Apocynaceae, Araceae, Asclepiadaceae, Begoniaceae, Berberidaceae, Bignoniaceae, Bromeliaceae, Caesalpiniaceae, Clethraceae, Cornaceae, Cucurbitaceae, Cynomoriaceae, Datiscaceae, Droseraceae, Empetraceae, Ericaceae, Gentianaceae, Goodeniaceae, Guttiferae, Hydrocharitaceae, Juncaceae, Lactoridaceae, Marcgraviaceae, Mimosaceae, Monimiaceae, Myrtothamnaceae, Nepenthaceae, Nympheaceae, Oenotheraceae, Orchidaceae, Pedaliaceae, Periplocaceae, Philydraceae, Proteaceae, Pyrolaceae, Rubiaceae, Sapindaceae, Saxifragaceae, Solanaceae, Tiliaceae, Typhaceae, Velloziaceae, Winteraceae (Pacini & Franchi, 1996). The four pollen grains are derived from the same microspore mother cell; the tetrads can be uniform, as in most cases, or of different types (Walker & Doyle, 1975): for instance rhomboidal, decussate or linear in some species of the genus *Periploca* (Verhoeven & Venter, 1994) or in pollinia as in orchids (Yeung, 1987b). In tetrad pollen, the degree to which the four grains are fused may vary: they may be barely joined in the central part by exine bridges, or have much exine in common. The tetrad is the basic unit for more complex pollen dispersal units such as polyads and pollinia. When pollen is dispersed in tetrads, the latter may in turn be held together in clumps by pollenkitt, elastoviscin or viscin threads (Fig. 1 G, H, K).

Polyads (Fig. 1I) are formed by the union of 2-8 tetrads in which each grain is surrounded by its own exine and intine (Fitzgerald *et al.*, 1993). They are typical of Mimosaceae. Species with 16 grains are the most common (Kenrick & Knox, 1989). Pollen of *Chimaphila* (Pyrolaceae) species is characterized by polyads composed of a variable number of tetrads (Takahashi, 1986).

When tetrads are joined together and the overall number of pollen grains is greater than 32, the term massula is used. A pollinium is composed of many massulae. The tetrads composing a massula may be held together in different ways: by elastoviscin (Fig. 1K), a thin external callosic wall (Fig. 1J), or by different types of wall fusion (exine and/or intine).

The term "pollinium" is used to indicate a pollen dispersal unit typical of certain groups of orchids and Asclepiadaceae, in which pollen grains produced by the anther are normally carried away together, often by a single insect (Yeung, 1987b). An orchid flower has two pollinia, sometimes independent, sometimes joined at their base to form a pollinarium (Endress, 1996). The pollinium may consist of many massulae held together by a stalk or similar structure (Fig. 1L), as in *Loroglossum hircinum* (Pandolfi & Pacini, 1995), or may be a unit consisting of many tetrads packaged and held together by a thick external layer of exine

(Fig. 1M), as in members of the Orchidaceae, *Cleisostoma racemifera* and *Laelia autumnalis* (Zavada, 1990) or in Asclepiadaceae such as the genera *Asclepias* and *Coralluma* (Dannenbaum & Schill, 1991). In the Orchidaceae, the number of grains per pollinium may range from 40,000 to 4 million (Schill *et al.*, 1992); in Asclepiadaceae the number is much lower (Dannenbaum & Schill, 1991). Pollinia may be soft (or mealy) as distinct from compact (or hard). The pollinator takes up the whole pollinium but at subsequent flowers, massulae may be left on different stigmas (soft pollinia) or the whole (compact) pollinium on a single stigma. The Asclepiadaceae only have compact pollinia.

Massula structure may vary in that the tetrads composing it may have thin exine, or the exine may only extend around each massula. In compact pollinia, the exine is prevalently external and coats the externally orientated surfaces of the tetrads of the outer layer, which is why it is not continuous (Dannenbaum & Schill, 1991).

When the tetrads join to form larger compound pollen, modifications occur due to compaction: (a) mature pollen grains are small in volume and grow little during development (Pandolfi & Pacini, 1995; Pandolfi *et al.*, 1993); (b) the tetrads are of different types and are disposed in different ways so that the space occupied is a minimum; in massulae and compact pollinia, the tetrads are packed together with little space between them (Pandolfi & Pacini, 1995); (c) the pollinia of orchids have a form which facilitates pollen tube emission - the massulae of soft pollinia crumble after sticking to the stigma, whereas the compact pollinia forming the pollinarium of *Calypso bulbosa* are flat and composed of few layers of tetrads. This allows the contemporaneous emission of many pollen tubes (Proctor & Harder, 1994); (d) when the dispersal unit consists of many grains with common walls, only some of which are external, as in pollinia, there are no apertures. In the Asclepiadaceae, the tubes emerge from a predetermined area of the pollinium (Sage & Williams, 1995), in orchids with pollen in massulae, they are emitted from all sides but only after the tetrads have separated, which in *Loroglossum hircinum* is about 24 h after pollination (Pandolfi & Pacini, 1995); (e) pollen grains are exposed to pollinators for a long time, in some cases more than a month (Eisikowitch *et al.*, 1987; Endress, 1995), and hence there must be physiological adaptations to maintain pollen viability and stigma receptivity (Pacini & Franchi, 1996); (f) the generative cell of mature pollen dispersal units of orchids is always spherical, unlike the classical spindle shape of all other pollen dispersal units. It becomes fusiform only when the tube is emitted (Pandolfi *et al.*, 1993; Pandolfi & Pacini, 1995). Since much time elapses between when orchid pollen lands on the stigma and when the pollen tube is emitted, a spherical shape may require less energy for long conservation. Moreover, the second haploid mitosis occurs as soon as the generative cell enters the pollen tube (Pandolfi & Pacini, 1995).

Pollination type

The types of pollen dispersal unit are compared with the main types of pollination in Table 1. In strictly anemophilous plants, pollen leaves the anther as soon as it opens; the pollen is in monads or tetrads, orbicules are present and pollenkitt absent. In some anemophilous plants, such as *Betula* (Betulaceae), the pollen is sticky (Percival, 1965) because it is coated by pollenkitt. This occurs in anemophilous genera belonging to prevalently entomophilous families, such as *Thalictrum* (Ranunculaceae), *Mercurialis* and *Ricinus* (Euphorbiaceae), *Lycopersicon* (Solanaceae), and *Artemisia* and *Ambrosia* (Compositae).

Hydrophilous pollination may occur at the water surface, especially in freshwater (e.g. *Vallisneria* and *Elodea*) or underwater (e.g. *Posidonia* and *Amphibolis*). In the first case, the pollen grains are isodiametric and dispersed in monads; in the second they are elongated and when the anther opens, form large masses that break up in the current (McConchie & Knox, 1989; Ackerman, 1995).

TABLE 1. Types of pollen dispersal units in relation to types of pollination. Many types of pollen dispersal units are dispersed by insects. Insects vary in their specificity and are attracted by different mechanisms. Pollen held together by pollenkitt is the most versatile pollen dispersal unit.

	types of pollination				
Types of pollen dispersal unit	anemophilous	hydrophilous	entomophilous	ornithophilous	chiropterophilous
A - monad pollen	x	x[1]			
B - extremely elongated monad pollen forming a tangle		x[2]			
C - monad pollen grains grouped by pollenkitt or tryphine	x		x	x	x
D - monad pollen grains grouped by viscin threads			x		
E - monads pollen grains grouped by elastoviscin			x		
F - tetrad pollen	x				
G - tetrad pollen grouped by pollenkitt	x		x		
H - tetrads grouped by viscin threads			x	x	
I - polyads			x	x	
J - groups of tetrads united by a thin external layer of callose (soft pollinium A)			x		
K - tetrads grouped by elastoviscin (soft pollinium B)			x		
L - tetrads grouped by common walls to form a massula (soft pollinium C)			x		
M - tetrads grouped in a compact pollinium with external exine			x		

[1] *surface pollination*
[2] *submarine pollination*

The pollen of entomophilous plants is always dispersed in aggregates, albeit in different ways. In anemophilous species with pollenkitt-coated pollen, it leaves the anther when there is sufficient breeze to overcome the forces that hold the grains together and to the anther (Lisci *et al.*, 1996). The pollen adheres to the body of animals that touch the anther or by electrostatic attraction (Corbet *et al.*, 1982). Pollen adheres to the stigma in random positions; only in the case of the pollinia of the Asclepiadaceae does it have to land in a certain manner in order to germinate (Kevan *et al.*, 1989). Irrespective of the way in which the pollen attaches to the animal, not all parts of the body become covered in pollen; this depends on the type of animal, the type of flower and the animal's approach to the flower.

Flowers that are pollinated by animals such as birds and larger insects usually have pollen with viscin threads (Endress, 1996). These threads cause the pollen to join in festoons that stick or hang from the bodies of the pollinators (Table 1). When honeybees collect the pollen of *Oenothera*, which has viscin threads, they are unable to pack the pollen because of these threads, and festoons of pollen hang from the bee in flight. On the other hand, other bees of the genus *Onograndrena* succeed in packing this pollen by virtue of specialized tibial hairs. Such bees are specific pollinators of the genus *Oenothera* (Percival, 1965).

Polyads (as in the genus *Acacia*) adhere to the hairs of pollinators by means of electrostatic charge: pollenkitt or other adhesion mechanisms are absent. This pollen however has orbicules that favour exit of pollen from the anther (Pacini & Franchi, 1993).

This type of pollen dispersal unit is very versatile because it can be transported by insects and birds (Ford & Forde, 1976).

When pollen grains constitute a pollinium, the insect is attracted to the flower by mechanisms of seduction or deception (Dafni, 1987). The pollen does not adhere directly to the insect body, but hangs off it, being attached by a viscidium in the case of orchids and by the corpusculum in the case of the Asclepiadaceae. These structures may adhere to the upper part of the insect body in various positions, even to the eyes (Nilsson *et al.*, 1992). The insect can load a number of pollinia and when it visits a flower, it may lose the pollinia from a previous flower and gain others. If the pollinium is of the soft type, several massulae or the whole pollinium may be lost. In tropical habitats, owing to the richness of living forms, pollinia may be transported by vectors belonging to several animal groups (Johnson 1996).

The gynoecium

Stigma

Stigmas are described as wet or dry on the basis of the presence or absence of exudate (Heslop-Harrison & Shivanna, 1977). Pollen can also be described as wet or dry in relation to the presence or absence of pollen coating. Table 2 gives some examples of types of stigma and pollen. When other means of adhesion are provided, different situations can occur: for example, stigmas are wet when the pollen has viscin threads, as in *Oenothera*, *Fuchsia* and *Rhododendron* (Heslop-Harrison & Shivanna, 1977) or when the pollen forms soft or hard pollinia, as in orchids.

Stigma structure is often feathery when pollen is in monads and pollination is anemophilous. The stigma surface is smaller when pollination is entomophilous and the number of ovules per ovary is small. The Euphorbiaceae, which are prevalently entomophilous and have 3-6 ovules and nectaries, are an interesting example. However, there are anemophilous species such as *Ricinus communis* (monoecious) and *Mercurialis annua* (dioecious), both of which have larger stigma surfaces than in entomophilous species of the same family. The number of ovules per ovary is three in *Ricinus communis* and two in *Mercurialis annua*. Wind pollination and reduced ovule number have evolved together in other families: Ranunculaceae, Rhizophoraceae, Hamamelidaceae and Commelinaceae (Kress, 1981).

TABLE 2. Types of stigma and pollen coating

	dry stigma	**wet stigma**
uncoated pollen	Moraceae Poaceae	*Lycopersicon esculentum*
coated pollen	Euphorbiaceae Malvaceae Compositae Cruciferae *	*Olea europaea* *Prunus avium* *Citrus limon*

* In the Cruciferae, the pollen coat is tryphine

Number of ovules per ovary

The more ovules per ovary, the more pollen grains must fall on the stigma, even if the pollen-ovule ratio varies widely in relation to different factors (Cruden, 1977). With anemophilous pollination, much airborne material and pollen from plants of the same species fall on the stigma. This genomic diversity means that gametophytic competition is strong. In entomophilous species with only one or few ovules (e.g. Compositae, Boraginaceae, Labiatae, Araceae), the visit of even one insect may be sufficient to bring enough pollen to ensure fertilization. If the flower has many ovules and many visits are required, the insect must be rewarded (Nepi & Pacini, 1993).

In plants with pollen dispersal units consisting of a high number of pollen grains, there is a decrease in the pollen-ovule ratio (Cruden, 1977; Cruden & Jensen, 1979). In Australian species of the genus *Acacia*, this ratio ranges from 0.8 to 4 (Kenrick & Knox, 1989). All this obviously leads to a decrease in pollen competition, which is high in species like maize that have only one ovule per carpel and a stigma with a large surface area (Ottaviano & Mulcahy, 1989; Table 3).

Proctor & Harder (1994) investigated whether pollen load influenced seed production in three orchid species with different types of pollen dispersal units: *Cypripedium calceolus* (monads grouped by elastoviscin), *Amerorchis rotundifolia* (soft pollinium) and *Calypso bulbosa* (compact pollinium). Only in the latter species was the mean number of pollinia deposited on the stigma, 2.53, proportional to the number of seeds produced.

TABLE 3. Some steps in the evolution of the gynoecium related to pollen dispersal units and the degree of stigmatic and gametophytic competition. Families regarded as evolved (*) solve reproductive problems by different strategies.

gynoecium structure		**stigma and gametophytic competition**		**examples**
no. of carpels	**no. of ovules per carpel**	**Stigma size**	**Gametophytic competition**	
many carpels	one ovule	small stigma	high competition	*Ranunculus* sp (Ranunculaceae)
few carpels	few ovules	small stigma	high competition	*Eranthis hiemalis* (Ranunculaceae)
few carpels widely fused, stigmas are free	few ovules	small stigma	medium competition	*Nigella damascena* (Ranunculaceae)
few fused carpels	few ovules	big stigma	medium competition	*Malus domestica* (Rosaceae) *Pancratium maritimum* (Iridaceae)
few fused carpels	many ovules	big stigma	low competition	Papaveraceae, Cucurbitaceae, Orchidaceae*
one carpel	one ovule	small or big stigma	high competition	Poaceae*, Compositae*

In gymnosperms the mature male and female gametophytes vary in cell number from species to species (Singh, 1978). In angiosperms, on the other hand, there is a transition from plants with apocarpic ovaries with several monospermic carpels, to plants in which the carpels fuse and there are many ovules. After this there are two evolutionary lines, either the number of ovules increases to very high values, or decreases to one (Endress, 1994). Correlations have been sought between the number of ovules per ovary and the number of pollen grains in the pollen dispersal unit (Kress, 1981). In a study of 49 families, 25 with monads and 24 with other types of pollen dispersal unit, the mean number of ovules per ovary was 13.1 and 16.7 respectively.

In all these cases, male and female competition varies. When the pollen-ovule ratio is very high, the pollen grains compete among themselves to fertilize, as for example in uniovulate anemophilous species. Another aspect of male gametophytic competition is that the seeds produced may have the same or different paternity. In plants with anemophilous pollination, multiple paternity tends to be more frequent than in entomophilous species (Kress, 1981). When the pollen dispersal units contains a number of pollen grains similar or not much greater than the number of ovules, gametophytic competition hardly exists (Table 3) and the fruit has seeds of more uniform sizes, but also sometimes, more aborted seeds (Kress, 1981). When fewer pollen grains than the number of ovules fall on the stigma, there is strong female gametophytic competition, which is particularly evident when the ovules are arranged in one or more rows (as in legumes). Further competition can also occur after fertilization, in the form of seeds competing for nutrients from the mother plant; such competition often leads to aborted seeds in certain positions.

Conclusions

Strictly anemophilous families such as Poaceae are uniform in that they always have pollen in monads. Among entomophilous families Ranunculaceae and Compositae always have clumped pollen with pollenkitt; Cruciferae with tryphine. Leguminosae s.l. are all zoophilous and have various types of pollen dispersal units: monads with pollenkitt, monads with viscin threads and polyads (Mimosaceae). The orchids, which have nearly all types of pollen dispersal units, are a special case.

Many evolutionary convergences are found among pollen dispersal units, even in quite distant taxa. The mucilage produced by anther connective tissue in *Tylosema esculentum* (Caesalpiniaceae) is similar to pollenkitt and tryphine and has the same function (De Frey *et al.*, 1992). The free filaments produced by the anther of certain Heliconiaceae and Strelitziaceae are similar to viscin threads (Rose & Barthlott, 1995). Orchidaceae and Asclepiadaceae have a series of evolutionary convergences: both have many different types of pollen dispersal units, the pollinium of Asclepiadaceae is similar to that of Orchidaceae, even in external morphology, and both have pollen that remains viable for a long time (Eisikowitch *et al.*, 1987; Endress, 1996). However they differ in that: (a) the pollinium of Asclepiadaceae contains fewer pollen grains; (b) orchids do not have apertures, whereas Asclepiadaceae pollinia have areas where exine is absent, from which all the pollen tubes are emitted; (c) the orchid generative cell is spherical (Pandolfi & Pacini, 1995) and that of Asclepiadaceae is spindle-shaped (Dannenbaum & Schill, 1991).

Modes of pollen adesion to pollinator bodies may be summarized as follows, acting singly or in combination:

1. Viscous substances: pollenkitt, tryphine, elastoviscin;
2. Thin ropes interconnecting grains: they may be continuous with exine (viscin threads) or derived from various parts of the anther;

3. Grains are trapped by the rough or hairy surface of the pollinators;
4. Pollen grains form a pollinium which adheres to the pollinator body by means of a viscidium (Orchidaceae) or a corpusculum (Asclepiadaceae);
5. Electrostatic charges on the pollinator body surface attract pollen grains;
6. Pollinator body is sticky.

Thirteen types of pollen dispersal unit are identified in the present study (Fig. 1): some are very common, occurring in many families; others are less common or restricted to a few families. The most common types are monads and monads grouped with pollenkitt, followed by tetrads. The other types can be regarded as specialities of a few families, or as in the case of polyads, of a few genera of a single family. Orchidaceae has by far the greatest number of types of pollen dispersal units (Pacini & Franchi, 1996); unfortunately there is no data on the number of ovules per ovary, which means that we do not know if the number of ovules per ovary increases with increasing number of pollen grains per pollen dispersal unit.

As the number of pollen grains composing a pollen dispersal unit increases, the paternal multiplicity of the seeds decreases. In Asclepiadaceae, quintuple paternity is possible because the gynoecium is designed to receive up to five pollinia. However in *Asclepias syriaca*, one pollinium is sufficient to fertilize all the ovules of an ovary (Gold & Share, 1995) because the number of grains in a pollinium is 40% more than the number of ovules in an ovary (Wyatt & Broyles, 1992). The compact pollinium is unique among the angiosperms: in all the other plants which have more than one ovule per ovary, especially anemophilous species, the seeds all have the same mother but the paternity is different. The more ovules per ovary, the greater this diversity. On the other hand, in the case of the compact pollinium, all seeds may also have the same parents.

Acknowledgements

This research was supported by a grant from CNR, program "Reproductive Ecology" and MURST. We are grateful to Dr. B.M. Parkinson for helpful comments and linguistic improvement of the manuscript.

References

Ackerman, J.D. (1995). Convergence of filiform pollen morphogenesis in seagrasses: functional mechanisms. *Evol. Ecol.* **9**: 139–153.

Bransett, D.L., Bridgewater, F.E., Jett, J.B. and Matthews, F.R. (1985). Theoretical impact of pollen viability and distribution on the number of strobili for controlled pollinations in loblolly pine. In: Proceedings of the 18th Southern Forest Tree Improvement Conference, pp. 194–203, Southern Forest Experimental Station and University of Missouri, Gulf Park, Mo.

Chaloner, W.G. (1986). Electrostatic forces in insect pollination and their significance in exine ornament. In: S. Blackmore and I.K. Ferguson (editors). Pollen and spores: form and function, pp. 103–108, Academic Press, London.

Corbet, S.A., Beament, J. and Eisikowitch, D. (1982). Are electrostatic forces involved in pollen transfer? *Pl. Cell Environm.* **5**: 125–129.

Cruden, R.W. (1977) Pollen-ovule ratios: a conservative indication of breeding systems in flowering plants. *Evolution* **31**: 32–46.

Cruden, R.W. and Jensen, K.G. (1979). Viscin threads, pollination efficiency and low pollen-ovule ratio. *Amer. J. Bot.* **66**: 875–879.

Dafni, A. (1987). Pollination in *Orchis* and related genera: evolution from reward to deception. In: J. Arditti (editor). Orchid biology: reviews and perspectives, vol. IV, pp. 80–103, Comstock Publishing Associates, Ithaca, N.Y.

Dannenbaum, C. and Schill, R. (1991). Die Entwicklung der Pollentetraden und Pollinien bei den Asclepiadaceae. *Biblioth. Bot.* **34 (141):** 1–138.

De Frey, H.M., Coetzer, L.A. and Robbertse, P.J. (1992). A unique anther-mucilage in the pollination biology of *Tylosema esculentum*. *Sexual Pl. Reprod.* **5**: 298–303.

Dobson, H.E.M. (1988). Survey of pollen and pollenkitt lipids chemical cues for flower visitors. *Amer. J. Bot.* **75**: 170–182.

Dobson, H.E.M. (1991). Pollen and flower fragrances in pollination. *Acta Hort.* **288**: 313–323.

Eisikowitch, D., Kevan, P.G., Fowle, S. and Thomas, K. (1987). The significance of pollen longevity in *Asclepias syriaca* L. under natural conditions. *Pollen & Spores* **29**: 121–128.

Endress, P.K. (1994). Diversity and evolutionary biology of tropical flowers. Cambridge University Press, Cambridge.

Endress, P.K. (1996). Diversity and evolutionary trends in angiosperm anthers. In: W.G. D'Arcy and R.C. Keating (editors). The anther: form, function and phylogeny, pp. 92–110, Cambridge University Press, Cambridge.

Erickson, E.H. and Buchman, S.L. (1983). Electrostatics and pollination. In: C.E. Jones and R.J. Little (editors). Handbook of experimental pollination biology, pp. 173–184, Scientific and Academic Editions, New York.

Faegri, K. and van der Pijl, L. (1979) The principles of pollination ecology. Pergamon Press, Oxford.

Fitzgerald, M.A., Barnes, S.H., Blackmore, S., Calder, D.M. and Knox, R.B. (1994). Pollen development and cohesion in a mealy and hard type of orchid pollinium. *Int. J. Pl. Sci.* **155**: 481–491.

Fitzgerald, M.A., Calder, D.M. and Knox, R.B. (1993). Character states of development and inhibition of cohesion between compound pollen grains of *Acacia paradoxa*. *Ann. Bot.* **71**: 51–59.

Ford, H.A. and Forde, N. (1976). Birds as possible pollinators of *Acacia pycnantha*. *Austral. J. Bot.* **24**: 793–795.

Franchi, G.G. and Pacini, E. (1996). Types of pollination and seed dispersal in Mediterranean plants. *Giorn. Bot. Ital.* **130**: 579–585.

Friis, E.M. and Crepet, W.E. (1987). Time of appearance of floral features. In: E.M. Friis, W.G. Chaloner and P.R. Crane (editors). The origin of angiosperms and their biological consequences, pp. 145–179, Cambridge University Press, Cambridge.

Gold, J.J. and Shore, J.S. (1995). Multiple paternity in *Asclepias syriaca* using a paired-fruit analysis. *Canad. J. Bot.* **73**: 1212–1216.

Guinet, Ph. (1986). Geographic patterns of the main pollen characters in genus *Acacia* (Leguminosae), with particular reference to subgenus Phyllodineae. In: S. Blackmore and I.K. Ferguson (editors). Pollen and spores: form and function, pp. 297–311, Academic Press.

Heslop-Harrison, Y. and Shivanna, K.R. (1977). The receptive surface of the angiosperm stigma. *Ann. Bot.* **41**: 1233–1258.

Hesse, M. (1984). An exine architecture model for the viscin threads. *Grana* **23**: 69–75.

Hesse, M. (1986). Nature, form and function of pollen connecting threads in angiosperms. In: S. Blackmore and I.K. Ferguson (editors). Pollen and spores: form and function, pp. 109–119, Academic Press, London.

Johnson, S.D. (1996). Bird pollination in South African species of *Satyrium* (Orchidaceae). *Pl. Syst. Evol.* **203**: 91–98.

Kenrick, J. and Knox, R.B. (1989). Pollen-pistil interactions in Leguminosae (Mimosoideae). In: C.H. Stirton and J.L. Zarucchi (editors). Advances in legume biology. *Monogr. Syst. Bot. Missouri Bot. Gard.* **29**: 127–156.

Kevan, P.G., Eisikowitch, D. and Rathwell, B. (1989). The role of nectar in the germination of pollen in *Asclepias syriaca* L. *Bot. Gaz.* **150**: 266–270.
Kress, W.J. (1991). Sibling competition and evolution of pollen unit, ovule number and pollen vector in angiosperms. *Syst. Bot.* **6**: 101–112.
Lisci, M., Cardinali, G. and Pacini, E. (1996). Pollen dispersal and role of pollenkitt in *Mercurialis annua* L. (Euphorbiaceae). *Flora, Morphol. Geobot. Oekophysiol.* **191** 131–141.
McConchie, C.A. and Knox, R.B. (1989). Pollination and reproductive biology of seagrasses. In: A.W.D. Larkum, A.J. McComb and S.A. Shepherd (editors). Biology of seagrasses, pp. 74–111, Elsevier, Amsterdam.
Nepi, M. and Pacini. E. (1993). Pollination, pollen viability and pistil receptivity in *Cucurbita pepo*. *Ann. Bot.* **72**: 527–536.
Nilsson, L.A., Rabakonandrianina, E., Razananaivo R. and Randriamanindry J.-J. (1992). Long pollinia on eyes: hawk-moth pollination of *Cynorkis uniflora* Lindley (Orchidaceae) in Madagascar. *Bot. J. Linn. Soc.* **109**: 145–160.
Ottaviano, E. and Mulcahy, D.L. (1989). Genetics of angiosperm pollen. *Advances Genet.* **26**: 1–64.
Pacini, E. and Franchi, G.G. (1991). Diversification and evolution of tapetum. In: S. Blackmore and S.H. Barnes (editors). Pollen and spores - patterns of diversification, pp. 301–316, Clarendon Press, Oxford. (Systematics Association special volume, 44).
Pacini, E. and Franchi, G.G. (1993). Role of tapetum in pollen and spore dispersal. *Pl. Syst. Evol., Suppl.* **7**: 1–11.
Pacini, E. and Franchi, G.G. (1996). Some cytological, ecological and evolutionary aspects of pollination. *Acta Soc. Bot. Poloniae* **65**: 11–16.
Pandolfi, T. and Pacini, E. (1995). The pollinium of *Loroglossum hircinum* (Orchidaceae) between pollination and pollen tube emission. *Pl. Syst. Evol.* **196**: 141–151.
Pandolfi, T., Pacini, E. and Calder, D.M. (1993). Ontogenesis of monad pollen in *Pterostylis plumosa* (Orchidaceae, Neottioideae). *Pl. Syst. Evol.* **186**: 175–185.
Patel, V., Skvarla, J., Ferguson, I.K., Graham, A. and Raven, P.H. (1985). The nature of threadlike structures and other morphological characters in *Jacqueshuberia* pollen (Leguminosae: Caesalpinioideae). *Amer. J. Bot.* **72**: 407–413.
Percival, M.S. (1965). Floral biology. Pergamon Press, Oxford.
Proctor, H.C. and Harder, L.D. (1994). Pollen load and seed production in three orchid species. *Canad. J. Bot.* **72**: 249–255.
Rose, M.-J. and Barthlott, H.W. (1995). Pollen connecting threads in *Heliconia* (Heliconiaceae). *Pl. Syst. Evol.* **195**: 61–65.
Sage, T.L. and Williams, E.G. (1995). Structure, ultrastructure, and histochemistry of the pollen tube pathway in the milkweed *Asclepias exaltata* L. *Sexual Plant Reprod.* **8**: 257–265.
Schill, R. and Wolter, M. (1985). Ontogeny of elastoviscin in the Orchidaceae. *Nordic J. Bot.* **5**: 575–580.
Schill, R. and Wolter, M. (1986). On the presence of elastoviscin in all subfamilies of the Orchidaceae and the homology to pollenkitt. *Nordic J. Bot.* **6**: 321–324.
Schill, R., Dannenbaum, C. and Neyer, P. (1992). Quantitative Untersuchungen an Orchideenpollinien. *Bot. Jahrb. Syst.* **114**: 153–171.
Schlag, M. and Hesse, M. (1993). Morphogenesis of the sporoderm in *Polystachia pubescens* (Orchidaceae). *Grana* **32**: 22–28.
Singh, H. (1978). Embryology of gymnosperms. Gebrüder Borntraeger, Berlin. (Encyclopedia of plant anatomy, spezieller Teil, Bd 10, T.2)
Takahashi, H. (1986). Pollen polyads and their variation in *Chimaphila* (Pyrolaceae). *Grana* **25**: 161–169.

Takahashi, H. and Skvarla, J. (1990). Pollen development in *Oenothera biennis* (Onagraceae). *Amer. J. Bot.* **77**: 1142–1148.
Verhoeven, R.L. and Venter, H.J.T. (1994). Pollen morphology of *Periploca* (Periplocaceae). *S. African J. Bot.* **60**: 198–202.
Waha, M. (1984). Zur Ultrastruktur und Function Pollenverbindender Faden bei Ericaceae und anderen Angiospermenfamilien. *Pl. Syst. Evol.* **147**: 189–203.
Walker, J.W. and Doyle, J.A. (1975). The bases of angiosperm phylogeny: palynology. *Ann. Missouri Bot. Gard.* **62**: 664–723.
Wolter, M., Seuffert, C. and Schill, R. (1988). The ontogeny of pollinia and elastoviscin in the anthers of *Doritis pulcherrima* (Orchidaceae). *Nordic J. Bot.* **8**: 77–88.
Wyatt, R. and Broyles, S.B. (1992). Hybridization in North American *Asclepias*. III Isozyme evidence. *Syst. Bot.* **17**: 640–648.
Yeung, E.C. (1987a). Mechanisms of pollen aggregation into pollinia in *Epidendrum ibaguense* (Orchidaceae). *Grana* **26**: 47–52.
Yeung, E.C. (1987b). Development of pollen and accessory structures in orchids. In: J. Arditti (editor). Orchid biology: reviews and perspectives, vol. IV, pp. 194–225, Comstock Publishing Associates, Ithaca, N.Y.
Yeung, E.C. and Law, S.K. (1987). The formation of hyaline caudicle in the vandoid orchids. *Canad. J. Bot.* **65**: 1459–1464.
Zavada, M. (1990). A contribution to the study of pollen wall ultrastructure of orchid pollinia. *Ann. Missouri Bot. Gard.* **77**: 785–801.

Twell, D. and Howden, R. (1998). Cell polarity, asymmetric division and cell fate determination in developing pollen. In: S.J. Owens and P.J. Rudall (Editors). Reproductive Biology, pp. 197–218. Royal Botanic Gardens, Kew.

CELL POLARITY, ASYMMETRIC DIVISION AND CELL FATE DETERMINATION IN DEVELOPING POLLEN

DAVID TWELL AND ROSS HOWDEN

Department of Biology, University of Leicester, UK

Abstract

Double fertilisation in angiosperms is critically dependent upon the functional specialisation of the large vegetative cell and small generative cell formed as a result of the highly asymmetric division known as pollen mitosis I. The generative cell divides to form the two sperm cells, while the vegetative cell remains undivided and forms the pollen tube, through which the sperms are transported to the ovule. The emphasis of this chapter is to discuss evidence for differential cell fate determination at pollen mitosis I and the expression of cell polarity required to achieve asymmetric division. There is now overwhelming evidence that the symmetry of division at pollen mitosis I plays a critical role in determining the differential fate of the two daughter cells. Current models which relate division symmetry to differential gene expression and chromatin dispersion are discussed. Components of the polarity apparatus required to execute correct asymmetric division are examined, including the role of the cytoskeleton in controlling nuclear migration, spindle axis determination and the synthesis of the specialised generative cell plate. Molecular-genetic strategies are described which have lead to the isolation of gametophytic mutants affecting microspore division and polarity, thereby identifying genetic components of the polarity system. Cell fate analysis in such mutants is consistent with the proposed models of differential cell fate determination in response to division asymmetry. These results highlight the value of genetic approaches in dissecting microspore polarity and differential cell fate determination during pollen development.

Introduction

Asymmetry in the fate of daughter cells is commonly established by the unequal division of intrinsically polar mother cells (Horvitz and Herskowitz, 1992). In plants, intrinsically asymmetric divisions occur during zygotic division and embryogenesis (Mayer *et al.*, 1991), stomatal complex development (Palevitz, 1982) and during pollen development (Eady *et al.*, 1995). Our interest is in elucidating the mechanisms establishing cell polarity, asymmetric cell division and cell fate during pollen development. The aims of this chapter are to discuss the significance of asymmetric cell division for correct pollen differentiation, how asymmetric division leads to differential cell fate, how microspores develop the necessary polarity required for asymmetric division, and to summarise the results of new approaches that are being adopted to identify key genes involved in these processes.

Pollen cell lineage and development

The completion of meiosis of the diploid microsporocytes within the anther loculus marks the initiation of a unique process of cellular differentiation which ultimately leads through a simple cell lineage to the formation of a large number of male gametophytes shed from the plant as pollen grains (Fig. 1A). In the majority of angiosperms this pathway involves only a single asymmetric mitotic cell division, pollen mitosis I (PMI), which serves as a developmental marker for the termination of microspore development and the initiation of pollen development. After pollen germination, the generative cell (GC) completes a further mitotic division in the pollen tube to form the two sperm cells. In most graminaceous species and certain dicots including *Arabidopsis thaliana* GC division occurs precociously prior to pollen being shed (Brewbaker, 1967). Since major cytological processes, such as vacuolation, cell wall synthesis, microspore nuclear migration and the accumulation of metabolic reserves are often conserved between species which shed pollen in a bicellular or tricellular condition (Schrauwen *et al.*, 1990; Regan and Moffatt, 1990; Eady *et al.*, 1995; Owen and Majaroff, 1995), it is likely that many gametophytically expressed gene regulatory systems are also conserved. Detailed molecular-genetic analysis of pollen development in the model species *Arabidopsis thaliana* is therefore expected to provide the most efficient route to isolate gametophytic genes involved in asymmetric division at PMI and the control of differential cell fate. Details of pollen cell lineage relationships and key differentiation steps in the pathway of pollen development in *Arabidopsis* are illustrated in Fig. 1.

The tapetum and anther loculus provide essential metabolites and a suitable environment in which differentiation of the microspores can occur (reviewed in Pacini, 1990). Indeed, many sporophytically determined mutations act in the tapetum and anther wall and lead to male sterility (Chaudhury, 1993). However, at later stages of both microspore and pollen development apparently normal differentiation can be achieved in several species in *in vitro* culture if simple essential nutrients are provided (Tanaka and Ito, 1980, 1981; Kyo and Harada, 1986; Menito Moreno *et al.*, 1988; Tupy *et al.*, 1991; Dupuis and Pace, 1993). Thus, the male gametophyte from its inception at PMI, and perhaps to a lesser extent the haploid microspore, is capable of directing its own genetic programme to achieve maturity. This degree of autonomy is reflected in the highly active haploid gene expression programme required for the specialised structure and function of the pollen grain. The patterns and diversity of haploid gene expression during male gametophyte development have been comprehensively reviewed (Mascarenhas, 1992; McCormick, 1993; Twell, 1994).

The asymmetric cell division in pollen

A key event in the differentiation of the angiosperm male gametophyte is the unique asymmetric mitotic division of the haploid microspore termed pollen mitosis I (PMI). This highly asymmetric division produces two unequal daughter cells, the vegetative and generative cells, that have dramatically different structures and developmental fates. The large vegetative cell (VC) constitutes the bulk of the pollen cytoplasm and contains an abundance of stored metabolites required for rapid pollen tube extension. In contrast, the diminutive generative cell (GC) is completely enclosed by the VC cytoplasm and contains relatively few organelles and stored metabolites (Fig. 1). The extreme dimorphism in cytoplasmic volume, organelle, protein and RNA contents clearly suggests that these cells have very different synthetic and genetic activities (reviewed in Mascarenhas, 1975; Sunderland and Huang, 1987).

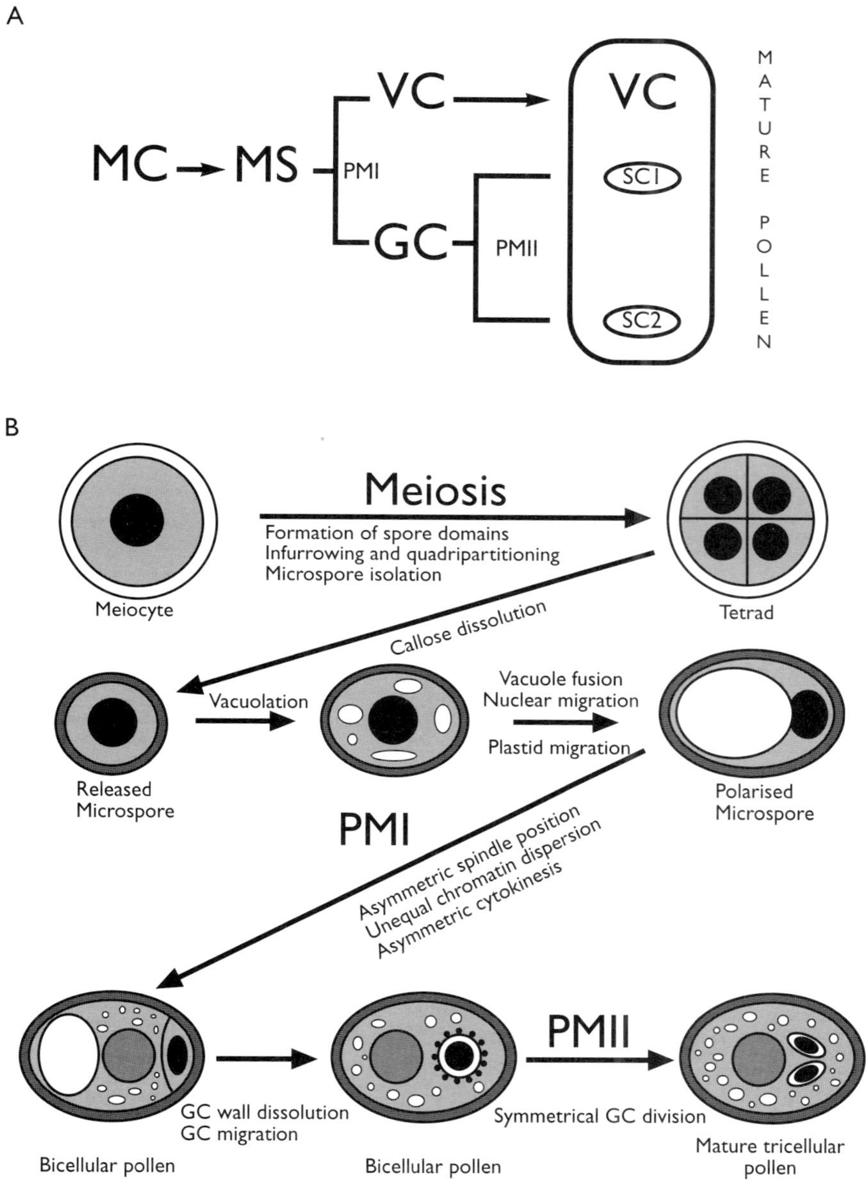

FIG. 1. Pollen cell lineage and differentiation. A. Simple cell lineage relationships in the pathway of pollen development. B. Major cytological events which occur during microspore and pollen development in *Arabidopsis* (Owen and Makaroff, 1995). MC = meiocyte; MS = microspore; VC = vegetative cell; GC = generative cell; SC = sperm cell; PMI = pollen mitosis 1; PMII = pollen mitosis 2.

Differential gene expression in vegetative and generative cells

Direct evidence for the presence of specific transcripts within the vegetative cytoplasm of a number of species has been provided by the *in situ* hybridisation of cDNA clones to developing, mature or germinated pollen (Hanson *et al.*, 1989; Raghavan, 1989; Ursin *et al.*, 1989; Brown and Crouch, 1990; Olmedilla *et al.*, 1991; Reijnen *et al.*, 1991; Theerakulpisut *et al.*, 1991; Shen and Hsu, 1992). However, in these studies it was not possible to determine whether transcripts were also present in the generative or sperm cell cytoplasm. Direct evidence for differential gene expression in the vegetative and GCs was obtained by linking the promoter of the tomato *lat52* gene to a GUS fusion protein containing the tobacco etch virus Nia nuclear-targeting signal. The *lat52* promoter was shown to be activated specifically from the vegetative nucleus of tobacco, and not in the GC, neither during pollen development or tube growth (Twell, 1992). The VC-specific activity of the *lat52* promoter was also shown to be maintained in *Arabidopsis* (Eady *et al.*, 1994), and VC-specific activity was similarly demonstrated for the promoter of the tobacco pollen-specific Ntp303 gene (Weterings, 1994).

Several studies have shown that generative and sperm cells actively synthesise RNA and protein at the onset of germination (Mascarenhas, 1966; Reynolds and Raghavan 1982; Haskell and Rogers, 1985; Zhang *et al.*, 1993). More detailed studies have shown that isolated GCs express unique basic histone-like proteins (Ueda and Tanaka, 1995a, b) and possess their own pool of translatable mRNAs some of which encode proteins unique to the GC (Blomstedt *et al.*, 1996). Recently the first definitive identification of a gene expressed specifically in the GC has been obtained by *in situ* hybridisation of a cDNA designated clone #231 isolated from a cDNA library prepared from embryogenic microspores of *Brassica napus* (Havlicky *et al.*, 1996). Taken together these data now provide compelling evidence for the unique gene expression programmes of the vegetative and generative cells.

The importance of asymmetric division for differential cell fate

Experimental evidence which supports the significant role of division asymmetry at PMI has come from studies in which symmetrical divisions have been induced by the application of microtubule inhibitors such as colchicine (Tanaka and Ito, 1981; Zaki and Dickinson, 1991), or by physical means such as centrifugation (Terasaka and Niitsu, 1987). The two apparently identical cells resulting from symmetrical division do not posses the nuclear characteristics of the GC but more closely resemble VCs. Further insight into the differentiated state of the daughter cells resulting from unequal and equal division at PMI has been gained through the application of transgenic plants expressing *gus* activity under the control of the VC-specific *lat52* promoter to monitor cell fate (Twell, 1992; Eady *et al.*, 1994, 1995).

When division at PMI was blocked in tobacco microspores matured in the presence of colchicine uninucleate ‘pollen grains’ were produced which were capable of *in vitro* germination (Eady *et al.*, 1995). These results demonstrated first, that a functional GC is not required for pollen maturation, germination or pollen tube growth; and second since the *lat52-gus* marker was activated in uninucleate ‘pollen grains’, that VC-specific gene expression can be activated independently of division at PMI. This uncoupling of late pollen gene activation from the division process demonstrates that these processes, although normally activated in concert, are subject to independent control. Late gene activation is therefore likely controlled by gametophytically-expressed transcription factors which accumulate in the developing microspore and which normally reach a threshold of activity at PMI by a pathway independent of cytokinesis (see Fig. 2).

When symmetrical divisions at PMI are induced with limiting concentrations of colchicine, *lat52-gus* is activated in both daughter cells, and both daughter nuclei show a dispersed ‘vegetative-like’ chromatin appearance (Eady *et al.*, 1995). Therefore symmetrical

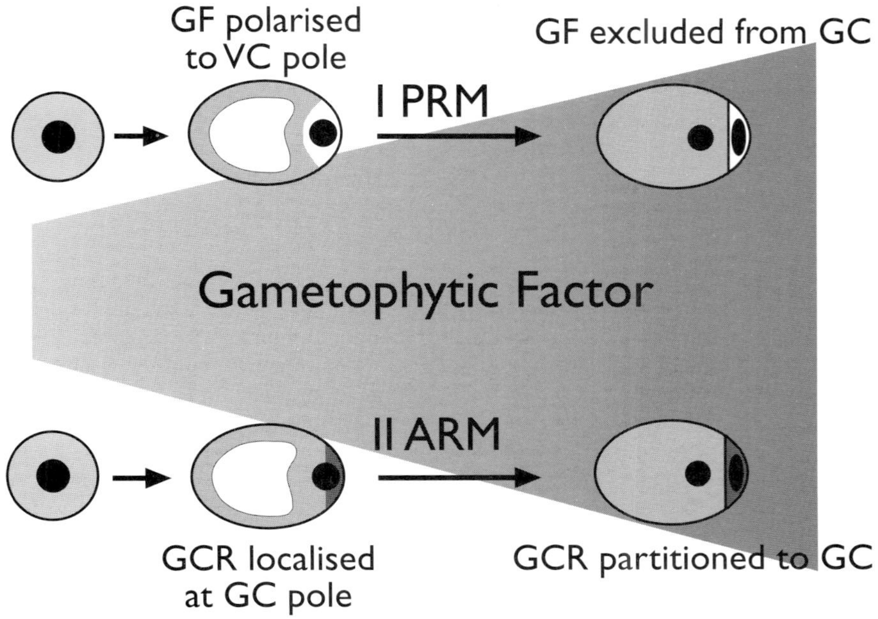

FIG. 2. Mechanisms involved in cell fate determination in pollen. In both models, VC-specific gene expression is activated through the accumulation of a gametophytic transcription factor (GF). The repression of VC-specific genes and chromatin condensation in the GC then results either from exclusion of GF from the GC pole (I. Passive Repression Model (PRM)), or from the asymmetric localisation of a generative cell repressor (GCR) protein at the GC pole (II. Active Repression Model (ARM)). VC = vegetative cell; GC = generative cell.

division produces apparently identical daughter cells which express vegetative cell fate. These data firmly establish that the symmetry of division at PMI plays a vital role in determining GC differentiation, including the repression of VC-specific genes and chromatin dispersion.

Mechanisms of differential cell fate determination

How then does asymmetric division result in the repression of VC gene expression specifically in the GC and the differential dispersion of chromatin within the vegetative and generative nuclei?

Repression of VC gene expression in the GC

Two general models have been proposed to account for the differential activation of VC-specific genes as a result of asymmetric cell division (Eady *et al.*, 1995). Both models are based upon the underlying proposition that the activation of VC genes is the default pathway resulting from the accumulation of a gametophytic transcription factor (GF) to threshold levels at PMI. These models provide alternative mechanisms, both involving the polarised distribution of gametophytic regulatory factor(s) to explain how VC-specific genes are normally repressed in the GC as a result of asymmetric division.

In the passive repression model (PRM) a gametophytic factor (GF) is excluded from a cytoplasmic domain at the GC-pole (Fig. 2). Repression of VC-specific gene activity then results from the absence of GF in the nascent GC. In the active repression model (ARM) a generative cell repressor (GCR) which is concentrated at the GC-pole and acts to block VC-specific gene activation by GF in the GC (Fig. 2). These models provide a useful framework for designing new experimental approaches and evaluating new evidence in this field.

Differential chromatin dispersion

The ARM and PRM must also take into account the highly condensed nature of the GC chromatin. The compact GC chromatin structure is likely to limit the access of transcription factors to many genes and may therefore be directly responsible for the maintenance of the transcriptionally repressed condition of the GC. One clue as to the origin of this differential chromatin structure comes from the observation that differential dispersion of the VC and GC chromatin is observed from late anaphase to late telophase during mitosis (Terasaka and Tanaka, 1974; Terasaka, 1982). Chromatin dispersion therefore appears to be differentially controlled by processes occurring within the same cytoplasm *before* the chromosomes are enclosed within the daughter nuclei. This suggests polarity in the microspore cytoplasm immediately before division and that during microspore development a ‘GC domain’ is established at the future GC pole.

Within this ‘GC domain’ specific factors must exist which act to limit chromatin dispersion. Unequal chromatin behaviour could therefore involve a cytoplasmic gradient of a GC pole localised ‘condensation factor’ which may be equivalent to the GCR proposed in the ARM. Histone or histone-associated factors are putative candidates for such a GCR, in particular because of their known ability to regulate chromatin packaging. Evidence for differences in histone content between the VC and GC nuclei was first reported in studies of the differential cytochemical staining of histones in pollen nuclei of *Lilium* (Jalouzot, 1969) and *Paeonia* (Sauter, 1969). Recent studies have confirmed and extended these findings in that two histone variants gH2B (lysine rich) and gH3 (arginine rich) purified from GC nuclei of lily have been shown to be present specifically in GC and sperm cell nuclei (Ueda and Tanaka, 1995a, b).

Evidence for microspore polarity

Microspore polarity is ultimately realised at the completion of asymmetric division. The approach adopted in the following discussion is to trace back from this point to the origins of polarity by systematically identifying cellular components and processes which must be co-ordinated to achieve asymmetric division.

Spindle asymmetry and position

The most advanced expression of polarity comes from the observation that the spindle is always asymmetric at PMI being blunt at the GC pole and sharp at the VC pole (Geitler, 1935; Brumfield, 1941; Heslop-Harrison, 1968). Since the cell plate is formed in an asymmetric position closer to the GC pole, spindle asymmetry is therefore likely to be involved in positioning the cell plate through the directed transport and fusion of phragmoplast vesicles. Spindle asymmetry itself may be caused by the close proximity of the microspore nucleus to the cell wall limiting spindle extension at the GC pole (Brumfield, 1941; Brown and Lemmon, 1991d).

Treatment of microspores with microtubule (MT) inhibitors or cold temperature leads to the formation of a central spindle and symmetrical divisions (Terasaka and Niitsu, 1987). The processes which initiate microtubule organising centre (MTOC) activity and the origin of the spindle poles are poorly understood in plants. Given the absence of structurally distinct centrosomes in plants and the observation that isolated nuclei can act as MTOC *in vitro* (Stoppin *et al.*, 1994), the nuclear membrane may contain a diffuse site with MTOC activity. It follows then that the position of the microspore nucleus would dictate the site of spindle assembly.

Spindle pole axis

The orientation of the spindle axis has a key role in determining division asymmetry. Centrifugation, cold and caffeine treatments were used to reposition the nucleus in *Pinus* microspores (Terasaka and Niitsu, 1987). Repositioning led to symmetric divisions resulting from a centrally located spindle, and symmetrical divisions in which both daughter nuclei were positioned near the wall. The second type of symmetrical divisions are consistent with an asymmetric spindle position with its axis parallel to the microspore wall rather than perpendicular to the wall which normally occurs. Clearly then, both spindle position and orientation needs to be tightly controlled to generate asymmetric division.

What then triggers the assembly of the spindle and defines the axis of the spindle poles? Brumfield (1941) observed that the appearance of the vegetative spindle pole preceded that of the generative pole, and postulated that this was the result of the reduced amount of cytoplasm at the generative pole, caused by the flattening of the prophase nucleus against the wall. Extending this postulate, the greater concentration of cytoplasmic factors, including tubulin, required for spindle assembly at the inner face of the nucleus would lead first to the inward and central positioning of the VC spindle pole. Fixation of the spindle axis perpendicular to the microspore wall could then be determined by factors which complete spindle bipolarity. Recent studies of the localisation of spindle assembly in *Tradescantia* have confirmed that the spindle begins to construct itself in the vegetative pole and that all of the available tubulin in the form of MTs are localised in the cytoplasm at the vegetative face of the nucleus immediately before spindle assembly (Terasaka and Niitsu, 1990; 1995).

Nuclear migration

Given the role of the nucleus in dictating the site of spindle assembly, a key process in generating polarity is the control of nuclear migration to the wall. Two factors appear to be important in controlling nuclear migration, vacuolation and the appearance of stage-specific MT arrays. In the majority of species examined a single large vacuole is formed as a result of the fusion of small vacuoles, during which time the microspore nucleus migrates to a site adjacent to the wall. In species such as *Tradescantia* (Terasaka and Niitsu, 1990) and tobacco (Twell *et al.*, 1993; Eady *et al.*, 1995) a second nuclear migration occurs to the future GC pole. This second nuclear migration (NM-2) has been shown to be sensitive to colchicine whereas the first (NM-1) is insensitive (Terasaka and Niitsu, 1990). In these species simple physical growth of the vacuole may account for NM-1 while NM-2 is dependent on a microtubule system which develops after formation of the single large vacuole.

In *Tradescantia* MT arrays appear to shift dynamically over the nucleus during NM-2 as a perinuclear cap at the future vegetative pole. In species such as *Brassica napus* (Hause *et al.*, 1991) and *Arabidopsis* (Owen and Makeroff, 1995) which possess rather spherical microspores it has not been possible to determine whether two nuclear migrations occur. Here the nucleus could arrive at the microspore wall simply as a result of physical displacement by the growing vacuole, or by a more directed MT-dependent transport system. MT localisation studies in *Brassica napus* showed that perinuclear arrays and cortical MT bundling were observed to develop in association with nuclear migration (Hause *et al.*, 1991).

Direct evidence for a specialised MT system that may be involved in nuclear migration comes from studies of orchid species. In *Phalaenopsis* a specialised MT array termed the generative pole microtubule system (GPMS) appears at the future GC pole prior to nuclear migration (Brown and Lemmon, 1991c, 1992). Here vacuolation does not occur so nuclear migration may be solely controlled by interactions of nuclear associated proteins with the GPMS.

Further evidence for the role of MTs in the positioning of the premitotic nucleus comes from studies of premitotic cultured tobacco BT2 cells in which the nucleus migrates from the periphery to the central region of the cell. Treatment with cytoskeleton destablising drugs showed that premitotic nuclear migration was dependent on MTs (Katsuta *et al.*, 1990). Although nuclear migration in premitotic cells is commonly dependent on MTs, maintenance of nuclear and spindle position in BY2 cells also involves actin microfilaments (MFs). Earlier studies of MFs in microspores clearly showed that MFs were associated with the margins of the growing cell plate at PMI but only diffusely staining patterns were observed at earlier stages (Hause *et al.*, 1991). Recently, by applying improved immunolocalisation procedures and the anti-MF drug cytochalasin D MFs were shown to be necessary for maintenance of nuclear position in cultured *Brassica* microspores (Gervai *et al.*, 1994). MFs initially radiated out from the centrally located microspore nucleus but became asymmetrically localised to the vegetative face of the acentric nucleus. However, MTs have also been strongly implicated in the maintenance of the acentric microspore nuclear position (Tanaka and Ito, 1981; Terasaka and Niitsu, 1990) and have been observed tethering the nucleus to the plasma membrane (Hause *et al.*, 1991; Brown and Lemmon, 1991c, 1992). Therefore both components may contribute to the maintenance of asymmetric nuclear position.

Polarity determinants

Although further detailed structural and functional studies of MT and MF organisation need to be carried out it is clear the MT-dependent systems play a key role in establishing the ability of the nucleus to migrate. The organisation of such MT-systems must respond to a polarity signal or determinant within the cytoplasm. This polarity signal could be a cytoplasmic gradient of a factor such as a microtubule motor or a regulator of MT assembly localised at the GC pole. Such a gradient could in turn by established by the localisation of its mRNA at the GC pole. There is considerable evidence for the determination of protein gradients by the MT directed transport or translation of mRNA species in different systems (Evans *et al.*, 1994; Pokrywka and Stephenson, 1995; Kloc and Etkin, 1995). Evidence for directed mRNA transport in plants has been obtained in developing *Fucus* embryos which show MF dependent-actin mRNA localisation at the thallus pole of the polarised zygote followed by its redistribution to the first cell plate by a mechanism insensitive to inhibitors (Bouget *et al.*, 1996).

The origin of polarity determinants

This leaves the important question of the origin and nature of the signals which execute early events in generating polarity. First, is it possible that polarity may be generated during microspore development by external signals? If external signals operate then such signals could not be polarised with regard to the anther locule since in most species microspores are often randomly orientated with regard to the anther locule and tapetum. However a localised determinant at the future GC pole could act as a polarising factor if it were localised in the microspore wall, plasma membrane or in a cortical cytoplasmic domain. The activation of such a localised determination by external conditions in the anther locule could then initiate a signal transduction cascade leading to polarisation of the MTs and nuclear migration.

Studies of cell fate determination in *Fucus* zygotes have recently shown that cell fate determinants are localised in the cell wall and act to maintain the differentiated state of thallus and rhizoid cells following asymmetric division (Berger *et al.*, 1994). Cell polarisation in animal cells and yeast commonly involves reorganisation of both MFs and MTs which are regulated by a hierarchy of GTPases and protein kinases (Glotzer and Hyman, 1995). Similar signalling cascades can be envisioned to operate to signal polarity and establish MT reorganisation in the microspore.

The localisation of a microspore polarity determinant could be an early event generated during meiosis or tetrad formation or it could be generated during microspore development as a result of intrinsically determined events associated with early gametophytic gene expression. With regard to the first possibility, cytokinesis following meiosis in most dicot species is simultaneous with all intersporal walls developing together. This results in the quadripartitioning of the postmeiotic cytoplasm into four equal 'spore domains' (Brown and Lemmon, 1991a, b and references therein). Intrinsic polarity within the spore domains, which could ultimately direct nuclear migration, could be generated as part of the process of generating the spore domains.

Detailed studies of the cytoskeleton and phragmoplast development have revealed radial arrays of MTs emanating from the postmeiotic nuclei which interact and define the boundaries of the spore domains and the sites of phragmoplast development (Brown and Lemmon, 1991a, b). The localisation of actin microfilaments in the multipolar spindle and MFs radiating from the four nuclei also indicates their involvement in phragmoplast development (Van Lammeren *et al.*, 1989; Schopfer and Hepler, 1991). Together, such systems could establish polarity by generating a cortical cytoplasmic domain or by delivering polarity determinants to membrane or well locations as discussed above. Direct evidence for early cytoplasmic polarity is provided by observations which show that future aperatural sites are defined by the presence of endosplasmic reticulum (ER) closely associated with the plasma membrane in the tetrad. These polar ER configurations reflect the tetrad cleavage planes and appear to be dependent upon spindle orientation at meiosis I and II, since disruption of meiotic spindles with colchicine and/or centrifugation prevents aperture formation or leads to apertures in incorrect positions often with unusual patterning features (reviewed in Heslop-Harrison, 1971).

In species which shed pollen in permanent tetrads it has been possible to determine the polarity of nuclear migration. Nuclear migration can either be towards the inner (radial) or to the outer (distal) walls of tetrad, but is usually constant within a species (Geitler, 1935). So although the axis of polarity may differ between species nuclear migration appears to be predetermined towards a fixed site. This raises the question of whether the ability of the microspore to achieve asymmetric division is dependent upon the nucleus arriving at this site? Asymmetric division in *Pinus* microspores was shown to occur at any site along the radial or distal walls following nuclear displacement by centrifugation or cold treatments (Terasaka and Niitsu, 1987). Therefore, the formation of an asymmetric spindle and asymmetric division was not dependent on arrival of the nucleus at a particular location. The proposed role of the system generating regular polarity then is first, to ensure nuclear migration to the wall and second, to ensure that the correct spindle axis is established to achieve asymmetric division.

Sporophytic mutants as tools to analyse microspore polarity and the action of gametophytic mutations

Two recently isolated sporophytic mutants in *Arabidopsis* affecting male gametophyte structure and development are described which have particular value as tools for the analysis of microspore polarity and male gametophytic mutants.

quartet

Two mutant loci have previously been reported, *quartet 1(qrt1)* and *qrt2* which shed pollen in a tetrad configuration, wherein all four meiotic products remain joined in a characteristic tetrahedral array (Preuss *et al.*, 1994). Despite this early defect the cytoplasmic contents of each microspore remain discrete and pollen matures correctly. Furthermore, in plants hemizygous for introduced *lat52-gus* gene fusion in the *qrt1/qrt1* background, only two of the four members of the tetrad stain positively for GUS activity. The *quartet* phenotype is particularly valuable in characterising male gametophytic mutants since in a *qrt* background gametophytic mutants will show the phenotype in two of the four pollen grains within a tetrad (Chen and McCormick, 1996). The *qrt* mutants may be of further value in determining whether microspore nuclear migration occurs towards a predetermined site in *Arabidopsis* since the four meiotic products remain attached apparently in a fixed tetrahedral arrangement. In the screen described below (Section 8) we have isolated 7 independent mutants which show the *quartet* phenotype out of a population of 10,000 M2 individuals. At this stage it is unknown whether these newly identified mutants are allelic to *qrt1* or *qrt2* or represent new *qrt* loci. *qrt1* maps to chromosome 5 (position 84 $\pm$ 1.5) and *qrt2* to chromosome 3 (position 3 $\pm$ 2).

stud/tetraspore

A class of reduced fertility mutants of *Arabidopsis* which result in the formation of large multicellular pollen grains have recently been reported. These mutants have been independently classified as two allelic series in genes termed *STUD* (Hulskamp *et al.*, 1997) and *TETRASPORE (TES*; Spielman *et al.*, 1997) respectively. *stud/tes* mutations are all sporophytic monogenic recessive and prevent the formation of separating callose walls after meiosis II. This initially creates coenocytic microspores (CM) composed of four nuclei within a common cytoplasm. Similar mutants have been described in soybean (*male sterile-1*(*ms1*); Albertsen and Palmer, 1979) and alfalfa (*jumbo pollen* (jp); McCoy and Smith, 1983). In *stud/tes* mutants up to four sperm cell pairs have been observed in mature pollen grains suggesting that the four nuclei can each undergo nuclear migration and asymmetric division within the common cytoplasm. The *jp* and *stud/tes* mutants demonstrate that cytokinesis and microspore isolation after meiosis II are not prerequisites for microspores to achieve polarity (nuclear migration and asymmetric division), and suggests that polarity determinants can operate locally and independently within each spore domain. Four large multi-nucleate pollen mutants with similar phenotypes to *stud/tes* mutants were identified in the screen described below (Section 8) which were shown to be allelic to *tes* (Howden and Twell, unpublished). Both *STUD* and *TES* map to the lower arm of chromosome 3 close to *gl-1* and therefore are likely to represent the same gene.

Mutants at the *STUD/TES* locus may find application in the analysis of the action of gametophytic mutations. If male gametophytic lethal (*mgl*) mutants are completely penetrant it will not be possible to isolate homozygous individuals and therefore test whether gametophytic mutants also have a sporophytic phenotype. In the *stud/tes* mutant background however the lethal mutation may be complemented in the common cytoplasm by the product of the wild type allele (assuming a diffusible product) and it would be possible to isolate a homozygous (*mgl/mgl*) individual. If however the presence of the wild type product did not complement the mutant phenotype this would suggest that these gametophytic factors act locally in spore domains within the common cytoplasm.

Gametophytic mutations in *Arabidopsis*

Evidence for gametophyte-specific mutations in *Arabidopsis* is very limited. Redei (1965) described a mutation, *gametophyte factor 1*, which mapped to the upper arm of chromosome

2, whose transmission was completely blocked through the female but showed reduced (32%) transmission through pollen. More recently Springer *et al.* (1995) identified a transposonmediated 'gene-trap' mutant termed *prolifera* which showed reduced (50%) transmission through the female with only slightly reduced transmission through pollen. The *PROLIFERA* gene was shown to be related to the MCM2-3-5 family of proteins which are required for the initiation of DNA replication in yeast. This suggests that DNA replication during male and female gametophyte development may be controlled by different genes.

The first gametophytic mutant of *Arabidopsis* shown to specifically affect male gametophyte development has been termed *sidecar pollen* (*scp*; Chen and McCormick, 1996). Plants heterozygous for the *scp* mutation show 50% aborted pollen in the Landsberg *erecta* background but the Nossen background 48% of pollen are aborted and 7% are shed in a divided condition with an extra uninucleate vegetative-like cell. This extra 'vegetative' cell is attached to a cell containing a vegetative-like nucleus and two sperm-like nuclei. Other than being joined the extra vegetative cell appears to mature relatively normally in that the *lat52-gus/nia* (Twell, 1992) and *lat59-gus* (Twell *et al.*, 1990) markers are activated and germination occurs *in vitro*. This mutant appears to undergo an extra symmetrical cell division at the microspore stage and it is proposed that this may result either from the mis-orientation of the spindle apparatus during PMI or the premature division of an unpolarised microspore. Therefore *SIDECAR POLLEN* may encode a component required for nuclear migration and/or asymmetric spindle orientation, or a repressor of cell division normally active until the microspore achieves polarity. Further detailed analysis of the precise time of production of the extra cell in *scp* mutants should shed light on these hypotheses and why *scp* carrying pollen frequently aborts. *SIDECAR POLLEN* has been mapped to chromosome 3 near the markers ngao and BGLI.

Screening for male gametophytic mutants

Given the success of the genetic approach in identifying molecular components of the polarity generating mechanisms and cell fate determinants in other systems (Horvitz and Herskowitz, 1992; Rhyu and Knoblich, 1995; Kraut *et al.*, 1996) this should be a powerful approach for the dissection of microspore polarity and determination. We are using several molecular-genetic screens described below to isolate gametophytic mutants using *Arabidopsis thaliana* (Fig. 3).

Direct visible screening

Approximately 10,000 M2 plants from an EMS mutagenised population (Nossen (No) background) were screened for visible male gametophytic mutants by examining mature pollen grains after DAPI staining using fluorescence microscopy. At maturity *Arabidopsis* pollen is tricellular containing a large vegetative nucleus and two smaller sperm cells with highly condensed intensely staining nuclei (Fig. 4A). M2 seedlings were planted in a 96 'microtitre plate' arrangement and at anthesis mature pollen was collected from each plant in a microtitre plate and stained with DAPI. Pollen was initially examined at low magnification to detect obvious morphological aberrations including *gemini* (divided or twinned) pollen. Examination at higher magnification under UV illumination was then used to identify any mutants affecting nuclear composition and cell division.

Male gametophytic lethal (*mgl*) mutants

As expected in such a screen, a high frequency (~1%) of putative male gametophytic lethal mutations were detected in which approximately 50% of the pollen grains were aborted. This phenotypic class is likely to include mutations in a diverse array of genes requires for basic

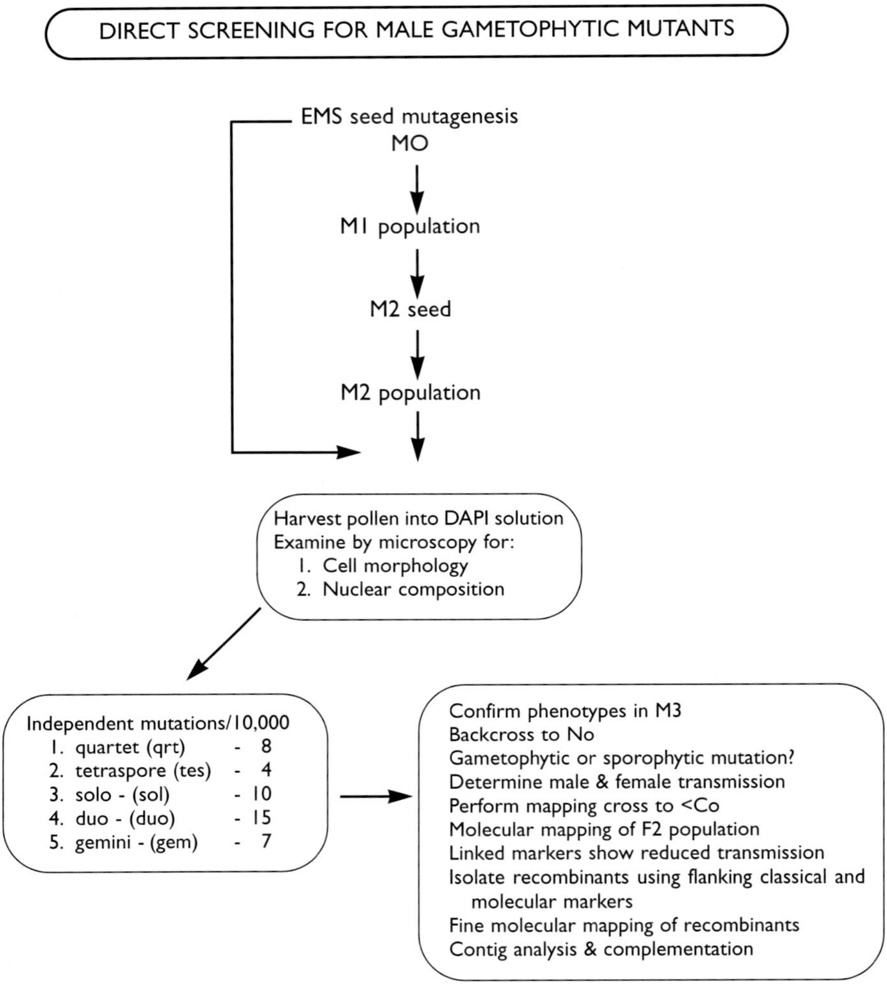

FIG. 3. Flow diagram showing steps involved in direct visible screening for male gametophytic mutants and their further characterisation aimed towards map-based cloning. (No = Nossen; M1 = ...)

microspore cell function and metabolism. The size and apparent homogeneity of this major class make it difficult to gain insight into the role of such genes without further detailed phenotypic analysis. Perhaps the best way to make progress in understanding the role of such genes in microspore development would be to use the gene-tagging stategies described below to directly clone such genes though the identification of 'tagged' *mgl* mutants.

Gametophytic cell division mutants

As predicted a range of mutant phenotypes were observed that affected the stereotypical pollen cell divisions. These division mutants were classified into three groups based upon the proportion of pollen grains showing particular phenotypes. All three groups contain mutants resulting from gametophytically acting mutations since: (1) plants are fertile with normal seed set; (2) phenotypes are observed in 50% or less of the pollen population; (3) heterozygous individuals show the phenotype; (4) transmission of the mutant allele is blocked or reduced through the pollen parent.

***solo* mutants (unicellular pollen, failing to divide at PMI)**

In *solo* pollen mutants approximately 50% of mature pollen grains contain a single nucleus, indicating failure of cytokinesis at PMI (Figs. 4 and 5). These mutations which affect the ability of the microspore to divide may therefore affect a structural or regulatory component in the cellular division process. As such these mutants may be thought of as cell cycle mutants in that they affect the ability of the microspore to complete its cell cycle. Components of the spindle apparatus or including factors regulating the assembly and/or function of the microtubule system are putative candidates. Further studies of the nuclear migration and progression of the cell cycle using microdensitometry (Binarova *et al.*, 1993; Carmichael and Friedman, 1995) in wild type and *solo* microspores should lead to further insight into the effect of these mutations. If *solo* mutants also fail to express polarity *solo* genes may provide a link between the expression of cell polarity and cell cycle progression.

Despite the possession of a single nucleus *solo* microspores undergo maturation including cell expansion and the synthesis of a dense cytoplasm (Fig. 4C). The GUS positive staining of mutant *solo* pollen grains carrying the *lat52-gus/nia* gene shows that *solo* grains correctly activate the VC-specific promoter *lat52* (Fig. 4D). Therefore in *solo* pollen the single nucleus adopts the fate of the vegetative nucleus of wild type pollen grains. These data confirm that VC fate is the default condition of the microspore nucleus in the absence of division at PMI (Eady *et al.*, 1995).

***duo* mutants (bicellular pollen, failing to divide at PMII)**

The *duo* pollen phenotype (Fig. 4) is indicative of pollen which has failed to enter or complete division at PMII. In some cells the chromosomes appear to have initiated condensation in prophase but division has arrested prior to alignment on the metaphase plate. In *duo116* approximately 50% of the pollen population show the *duo* phenotype consistent with a gametophytically acting mutation (Fig. 5). These phenotypes suggest that the mutation may be a structural or regulatory component required for cell division, but that such components would specifically function in the symmetrical second division of the generative cell. *DUO* genes may therefore be specifically expressed in the GC. The low frequency of symmetrically divided pollen grains observed in *duo116* (Fig. 5) suggests that *DUO116* may also participate in regulating division asymmetry at PMI. Incorporation of the *lat52-gus/nia* marker into *duo116* showed that the VC correctly activates *lat52* in the absence of a functional GC and that GC fate is unaltered in that the arrested GC correctly fails to activate *lat52* (Fig. 4).

***gemini* mutants (twin-celled pollen, more symmetrical division at PMI)**

The *gemini* class of mutants display the widest range of pollen phenotypes which are illustrated in Fig. 5. In *gemini* mutants the symmetry of division at PMI is affected such that symmetrical and partial divisions frequently occur (Figs. 4 and 5). In symmetrically divided cells both daughter cells correctly activate the *lat52-gus/nia* marker and show dispersed nuclear chromatin (Fig. 4G and H) in accord with previous observations which demonstrate that induced symmetrical divisions result in the default VC fate in both daughter cells (Eady *et al.*, 1995). Initial characterisation of the *gemini* mutants by aniline blue staining reveals that the persistent cell walls present at pollen maturity are composed of callose and are often seen to connect with the pollen grain's wall even if these internal walls are incomplete. Such phenotypes may be expected if the spindle position or axis is altered.

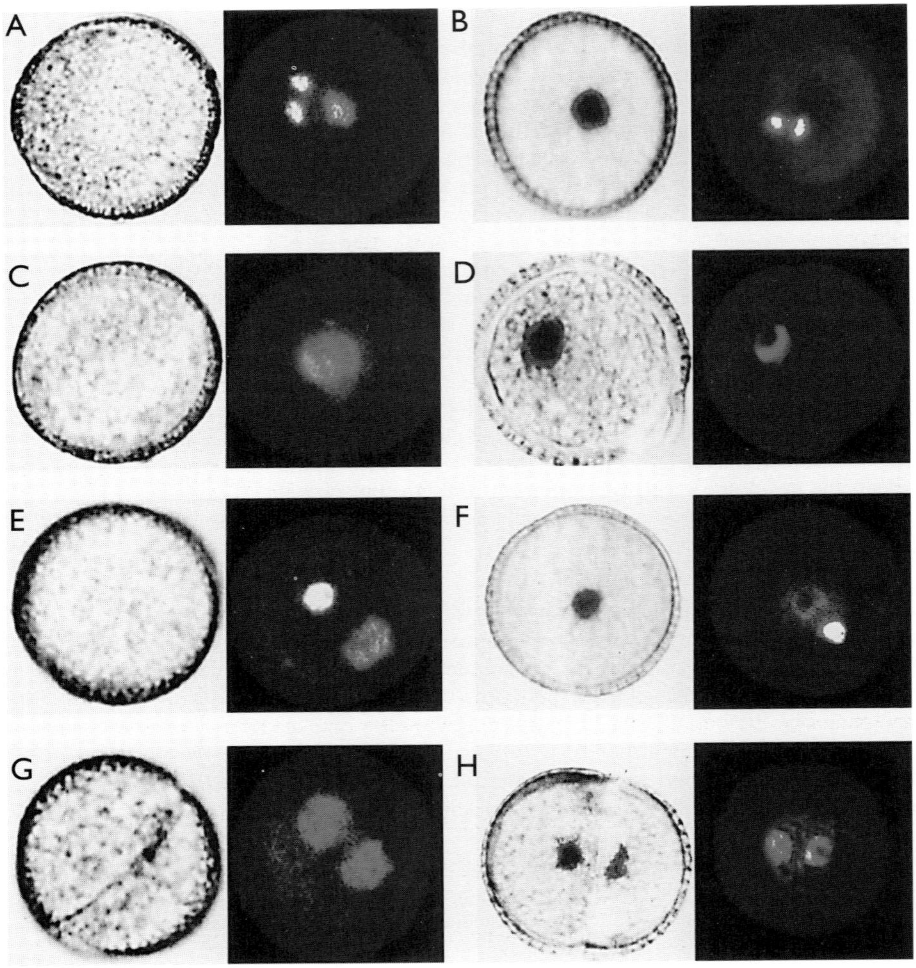

FIG. 4. Photographic images of mature pollen isolated from wild type and gametophytic mutants of *Arabidopsis*. Bright field and corresponding DAPI stained epifluorescence images are shown in the left and right panels of A.-H. respectively. A, B. wild type pollen; C, D. *solo* mutant; E, F. *duo* mutant; G, H. *gemini* mutant. B, D, F. and G. show nuclear-localised GUS staining in wild type, *solo*, *duo* and *gemini* mutants respectively containing the *lat52-gus/nia* VC fate marker.

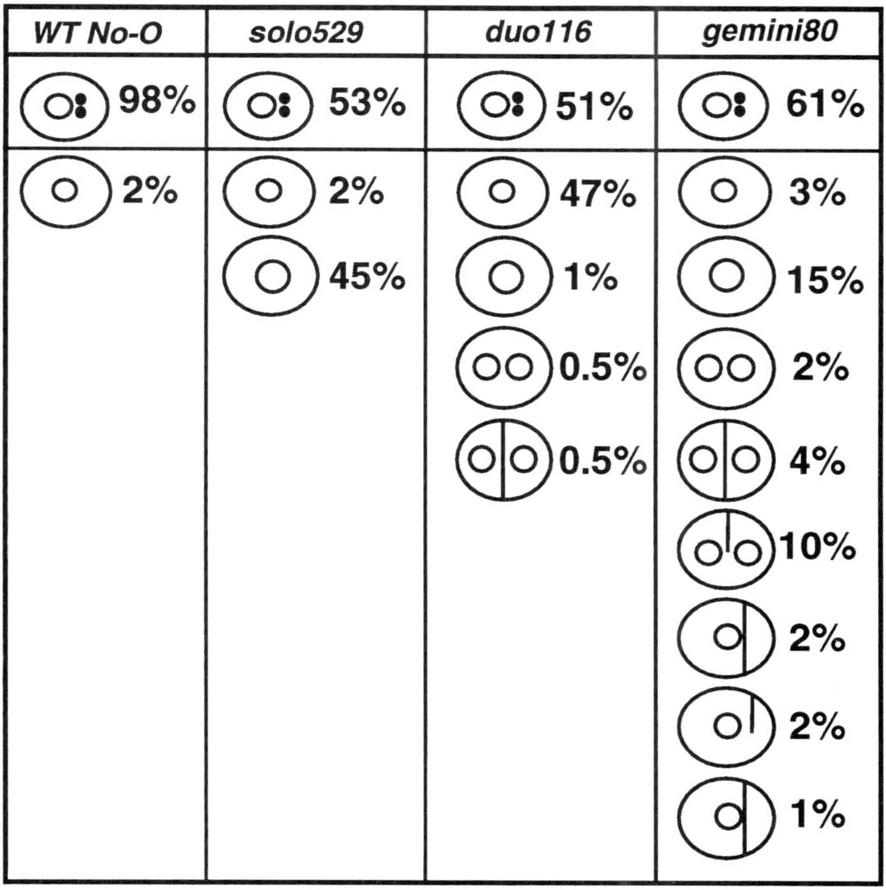

FIG. 5. Diagram showing the frequencies of aberrant pollen phenotypes observed in wild type (Nossen) and a representative of each of the phenotypic mutants classes *solo*, *duo*, and *gemini*.

Normally the microspore nucleus at incipient PMI is positioned against the microspore wall. The spindle axis is set up perpendicular to the wall such that the cell plate is initiated parallel to the outer wall and is subsequently connected to the outer wall by centrifugal growth of the cell plate at its margins (Hause *et al.*, 1991; Owen and Makeroff, 1995). However, if the spindle axis was rotated 90° in *gemini* mutants becoming parallel to the microspore wall, delivery of phragmoplast vesicles and the centrifugal growth of the phragmoplast would connect the cell plate to the outer wall. The inward growth of the cell plate would most likely be inefficient and fail. As discussed above if MTOCs are associated with the nucleus and serve to anchor the spindle poles, spindle pole position may be influenced by interactions between the nucleus and cortical domains or factors linked to the plasma membrane.

A further interesting phenotype in the *gemini* mutants is that nuclear division (karyokinesis) is often uncoupled from cellular division (cytokinesis). This commonly results in a smaller enucleate portion of the microspore cytoplasm being walled off from a larger cell containing one or two nuclei (Fig. 5). These 'uncoupling' phenotypes demonstrate that although cytokinesis and nuclear division are normally closely coupled these processes can operate independently. This concept has been elegantly demonstrated in experiments in which chromosomes were physically removed from the mitotic spindle of grasshopper spermatocytes. Chromosomeless cells still executed normal cytokinesis to produce enucleate daughters demonstrating that cleavage can operate independently of the presence of chromosomes (Zhang and Nicklas, 1996). *GEMINI* genes may therefore function as 'checkpoint' genes ensuring close coupling of karyokinesis and cytokinesis. It is also possible that *gemini* phenotypes result directly from defects in components of the intracellular targeting machinery required for the asymmetric positioning of the cell plate.

Sporophytic mutants which affect cell plate synthesis but act in the early embryo and developing seedlings have been identified. These include the 'pattern formation' mutants *gnom/emb30* (Mayer *et al.*, 1993), *fass* (Torres Ruiz and Jurgens, 1994) and *knolle* (Lukowitz *et al.*, 1996). Both the *gnom* and *knolle* mutants result from defects in putative components of the secretory apparatus. The GNOM protein shows sequence similarity to yeast Sec7 which is involved in protein transport through the golgi (Shevell *et al.*, 1994), while KNOLLE is similar to T-SNARE protein which are involved in vesicle docking (Luckowitz *et al.*, 1996). Both mutants disrupt the normally regular planes and symmetry of division in the developing embryo and *knolle* in particular shows cells with partial walls and multi-nucleate cells with partial walls and multi-nucleate cells which appear to result from incomplete cytokinesis. If similar proteins function in the positioning and synthesis of the cell plate at PMI it is possible that *GEMINI* genes encode gametophyte-specific versions of these proteins.

It is apparent that *GEMINI* genes could encode proteins which function at several points in the pathway which signals and executes microspore polarity. Further detailed phenotypic analysis and the molecular cloning of *GEMINI* genes is now required to begin to define the role and identity of molecular components comprising this pathway.

Screening for T-DNA and transposon-tagged gametophytic mutants

A potentially rapid and systematic, although labour intensive, approach for the isolation of important genes involved in gametophytic development is provided by random gene-tagging by T-DNA or transposon-mediated insertional inactivation (reviewed in Roberts, 1996). Since the inactivation of many important genes controlling gametophytic development is likely to lead to reduced transmission through the male or female gametes, it is possible to screen for the reduced transmission of a resistance marker carried by a T-DNA or a transposon. For example, if a single locus insertion carrying a seedling expressed resistance marker inactivates a postmeiotically-expressed gene essential for male gametophytic development, then the ratio of resistant:sensitive seedlings in selfed progeny would be 1:1 rather than the expected 3:1 (Fig. 6). This approach initially does not allow one to distinguish between mutations affecting transmission through male or female gametes, but is attractive in that: (1) it will identify important gametophytically expressed genes required for the complete developmental pathway of male and female gametophyte development, including gamete recognition and fusion; (2) genes identified in this way are very likely to be tagged or very closely linked to a molecular marker; and (3) it is simple to carry out given the availability of large populations of transgenic lines containing a high proportion of single locus T-DNA or transposon insertions.

We have screened approximately 2,000 T-DNA lines and 1,500 transposon containing lines. In both approaches approximately one percent of the lines show segregation distortion (~1:1) for antibiotic resistance. This suggests that both T-DNA and transposon tagging of important gametophytic genes through segregation distortion is similarly efficient depending on the availability or ease with which seed from large numbers of individuals hemizygous for single

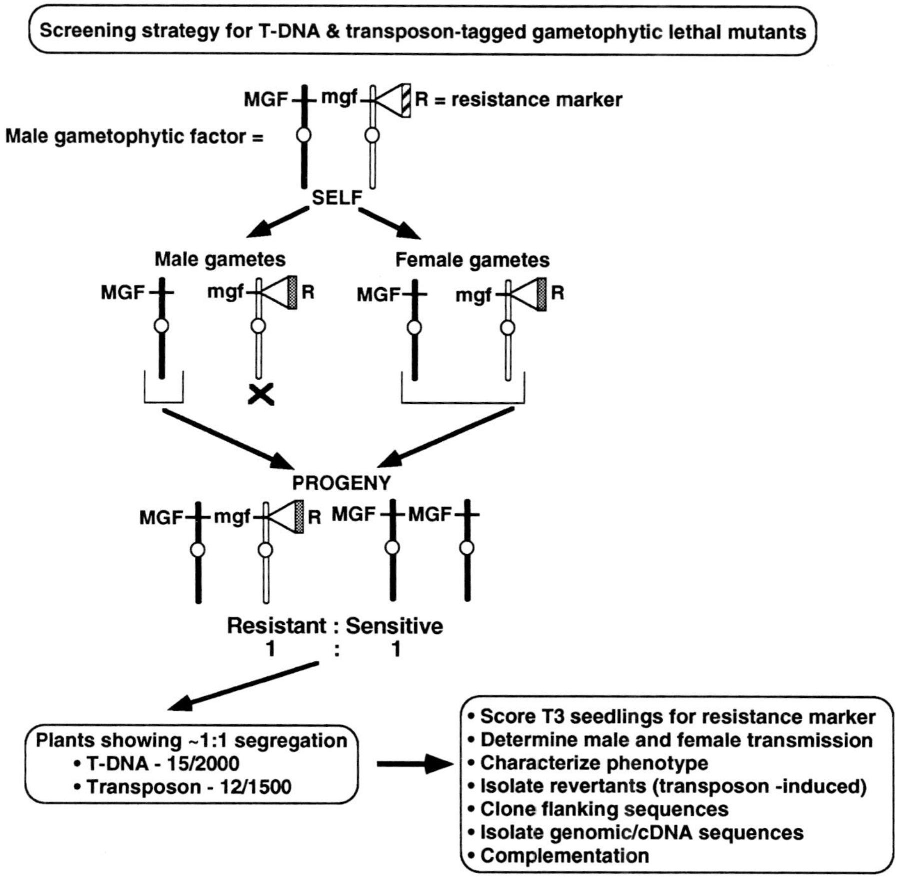

FIG. 6. Flow diagram showing a screening strategy for the isolation of T-DNA and transposon-tagged gametophytic lethal mutations and steps in their subsequent analyses.

locus insertions can be generated. In a proportion (~20%) of the T-DNA lines which have shown reliable 1:1 segregation ratios, further analysis has shown that occasionally some siblings are generated which segregate normally (3:1) for the resistance marker. This suggests that the T-DNA may not be the direct cause of the reduced transmission by insertional inactivation, and that T-DNA may have inserted at a site closely linked to a gametophytic mutation. For both tagged and T-DNA linked mutations direct access to the gene is most efficiently achieved by the cloning of plant DNA flanking the insertion using Inverse-PCR (Thomas, 1996) or TAIL-PCR (Liu *et al.*, 1995) procedures. If a concerted effort could be made to identify large numbers of random T-DNA or transposon induced gametophytic mutations many genes with important cellular functions could be isolated in a similar strategy to that being developed for assigning the function of embryo lethal mutations through random T-DNA tagging (Errampalli *et al.*, 1991).

Perspective

The identification and characterisation of genes which play essential roles in the control of asymmetric cell division and cell fate in developing pollen would be a significant step forward in understanding gametophyte development and the control of cell fate in plants. Several molecular-genetic strategies designed to identify such genes have been evaluated. This has led to the identification of three novel classes of gametophytic mutants which affect cell cycle progression and division asymmetry at PMI. The identification of such mutants means that it is feasible to begin to dissect the complex genetic components which are involved in the control of cell polarity and asymmetric cell fate. Furthermore, the gametophytic screening approaches described are likely to be of wider value since they also provide access to genes involved in the control of many key cellular processes such as cell cycle control, cell wall metabolism, cytoskeleton dynamics and intracellular signalling and transport. With increasing resources directed towards map-based cloning and transposon tagging it is realistic to expect that the molecular cloning of key genes involved in generating microspore polarity and differential cell fate is not far away.

Acknowledgements

We gratefully acknowledge The Royal Society for support to DT and BBSRC under the Cell Commitment and Determination Initiative to RH. Work on screening transposon populations in collaboration with Ueli Grossniklaus was supported by a HFSP fellowship to RH. We also thank George Pelletier, Christine Horlow and Sandrine Bonhomme for their willing collaboration on T-DNA screening, Stephen Ohl (MOGEN), Mary Anderson (NASC) and Keith Lindsey for providing seeds of T-DNA populations, and Hugh Dickinson, Bob Pruitt, Melissa Spielman and Rod Scott for preprints and useful discussions on the *stud/tes* mutants.

References

Albertsen, M.C. and Palmer, R.G. (1979). A comparative light- and electron- microscopic study of microsporogenesis in male sterile (*msl*) and male fertile soybeans (*Glycine max* (L.) Merr.). *Amer. J. Bot.* **66**: 253–265.

Benito-Moreno, R.M., Macke, F., Alwen, A. and Heberle-Bors, E. (1988). *In-situ* seed production after pollination with *in-vitro*-matured, isolated pollen. *Planta* **176**: 145–148.

Berger, F., Taylor, A. and Brownlee, C. (1994). Cell fate determination by the cell wall in early *Fucus* development. *Science* **263**: 1421–1423.

Binarova, R., Straatman, K., Hause, B., Hause, G. and van Lammeren, A.A.M. (1993). Nuclear DNA synthesis during the induction of embryogenesis in cultured microspores and pollen of *Brassica napus* L. *Theor. Appl. Genet.* **87**: 9–16.

Blomstedt, C.K., Knox, R.B. and Singh, M.B. (1996). Generative cells of *Lilium longiflorum* possess translatable mRNA and functional protein synthesis machinery. *Pl. Molec. Biol.* **31**: 1083–1086.

Bouget, F-Y., Gerttula, S., Shaw, S. and Quatrano, R.S. (1996). Localization of actin mRNA during the establishment of cell polarity and early cell divisions in *Fucus* embryos. *Pl. Cell* **8**: 189–201.

Brewbaker, J.L. (1967). The distribution and phylogenetic significance of binucleate and trinucleate pollen grains in the Angiosperms. *Amer. J. Bot.* **54**: 1069–1083.

Brown, S.M. and Crouch, M.L. (1990). Characterization of a gene family abundantly expressed in *Oenothera organensis* pollen that shows sequence similarity to polygalacturonase. *Pl. Cell* **2**: 263–274.
Brown, R.C. and Lemmon, B.E. (1991a). Pollen development in orchids. 1. Cytoskeleton and the control of division plane in irregular patterns of cytokinesis. *Protoplasma* **163**: 9–18.
Brown, R.C. and Lemmon, B.E. (1991b). Pollen development in orchids. 2. The cytokinetic apparatus in simultaneous cytokinesis. *Protoplasma* **165**: 155–166.
Brown, R.C. and Lemmon, B.E. (1991c). Pollen development in orchids. 3. A novel generative pole microtubule system predicts unequal pollen mitosis. *J. Cell Sci.* **99**: 273–281.
Brown, R.C. and Lemmon, B.E. (1991d). Pollen development in orchids. 5. A generative cell domain involved spatial control of the hemispherical cell plate. *J. Cell Sci.* **100**: 559–565.
Brown, R.C. and Lemmon, B.E. (1992). Pollen development in orchids. 4. Cytroskeleton and ultrastructure of the unequal pollen mitosis in *Phalaenopsis*. *Protoplasma* **167**: 183–192.
Brumfield, R.T. (1941). Asymmetric spindles in the first microspore division of certain angiosperms. *Amer. J. Bot.* **28**: 713–722.
Carmichael, J.S. and Friedman, W.E. (1995). Double fertilization in *Gnetum gnemon* - the relationship between the cell-cycle and sexual reproduction. *Pl. Cell* **7**: 1975–1988.
Chaudhury, A.M. (1993). Nuclear genes controlling male fertility. *Pl. Cell* **5**: 1277–1283.
Chen, Y-C. and McCormick, S. (1996). Sidecar pollen, an *Arabidopsis thaliana* male gametophytic mutant with aberrant cell division during pollen development. *Development* **122**: 3243–3253.
Dupuis, I. and Pace, G.M. (1993). Factors affecting *in vitro* maturation of isolated maize microspores. *Pl. Cell Rep.* **12**: 564–568.
Eady, C., Lindsey, K. and Twell, D. (1994). Differential activation and conserved vegetative cell-specific activity of a late pollen promoter in species with bi- and tricellular pollen. *Plant J.* **5**: 543–550.
Eady, C., Lindsey, K. and Twell D. (1995). The significance of microspore division and division symmetry for vegetative cell-specific transcription and generative cell differentiation. *Pl. Cell* **7**: 65–74.
Errampalli, D., Patton, D., Castle, L., Mickelson, L., Hansen, K., Schnall, J., Feldman, K. and Meinke, D. (1991). Embryonic lethals and T-DNA insertional mutagenesis in *Arabidopsis*. *Pl. Cell* **3**: 149–157
Evans, T.C., Crittenden, S.L., Kodoyianni, V. and Kimble, J. (1994). Translational control of maternal *glp-1* mRNA establishes an asymmetry in the *C. elegans* embryo. *Cell* **77**: 183–194.
Geitler, L. (1935). Beobachtungen über die Erste Teilung im Pollenkorn der Angiospermen. *Planta* **24**: 361–386.
Gervais, C., Simmonds, D.H. and Newcomb, W. (1994). Actin microfilament organization during pollen development of *Brassica napus* cv. Topas. *Protoplasma* **183**: 67–76.
Glotzer, M. and Hyman, A.A. (1995). The importance of being polar. *Curr. Biol.* **5**: 1102–1105.
Hanson, D.D., Hamilton, D.A., Travis, J.I., Bashe, D.M. and Mashcarenhas, J.P. (1989). Characterization of a pollen-specific cDNA clone from *Zea mays* and its expression. *Pl. Cell* **1**: 173–179.
Haskell, D.W. and Rogers, O.M. (1985). RNA synthesis by vegetative and sperm nuclei of trinucleate pollen. *Cytologia* **50**: 805–809.
Hause, G., Hause, B. and van Lammeren, A.A.M. (1991). Microtubular and actin-filament configuations during microspore and pollen development in *Brassica napus* L. cv. Topas. *Canad. J. Bot.* **70**: 1369–1376.
Havlicky, T., Hause, G., Hause, B., Pechan, P. and van Lammeren, A.A.M. (1996). Subcellular localization of cell-specific gene expression in *Brassica napus* L. and *Arabidopsis thaliana* L. microspores and pollen by non-radioactive *in situ* hybridization. PhD thesis, G. Hause and B. Hause, Wageningen Agricultural University, ISBN 90-5485-614-9.

Heslop-Harrison, J. (1968). Synchronous pollen mitosis and the formation of the generative cell in massulate orchids. *J. Cell Sci.* **3**: 457–475.

Heslop-Harrison, J. (1971). Wall pattern formation in angiosperm microsporogenesis. *Symp. Soc. Exp. Biol.* **25**: 277–300.

Horvitz, R.H., and Herskowitz, I. (1992). Mechanisms of asymmetric cell division: Two Bs or not two Bs, that is the question. *Cell* **68**: 237–255.

Hulskamp, M., Parekh, N.S., Grini, P., Scheitz, K., Zimmermann, I., Lolle, S.J. and Pruitt, R.E. (1997). The STUD gene is required for male-specific cytokinesis after telophase II of meiosis in *Arabidopsis thaliana Developm. Biol.* **187**: 114–124.

Jalouzot, R. (1969). Nuclear and cytoplasmic differentiation in the pollen grain of *Lilium candidum*. *Exp. Cell. Res.* **55**: 1–8. (in French).

Katsuta, J., Hashiguchi, Y. and Shibaoka, H. (1990). The role of the cytoskeleton in positioning of the nucleus in premitotic tobacco BY-2 cells. *J. Cell Sci.* **95**: 413–422.

Kloc, M. and Etkin, L.D. (1995). Two distinct pathways for the localization of RNAs at the vegetal cortex in *Xenopus* oocytes. *Development* **121**: 287–297.

Kraut, R., Chia, W., Jan, L.Y., Jan, Y.N. and Knoblich, J.A. (1996). Role of *inscuteable* in orientating asymmetric cell divisions in *Drosophila. Nature* **383**: 50–55.

Kyo, M. and Harada, H. (1986). Control of the developmental pathway of tobacco pollen *in vitro*. *Planta* **168**: 427–432.

Liu, Y-G., Mitsukawa, N., Oosumi, T. and Whittier, R.F. (1995). Efficient isolation and mapping of *Arabidopsis thaliana* T-DNA insert junctions by thermal asymmetric interlaced PCR. *Plant J.* **8**: 457–463.

Lukowitz, W., Mayer, U. and Jurgens, G. (1996). Cytokinesis in the *Arabidopsis* embryo involves the syntaxin-related KNOLLE gene product. *Cell* **84**: 61–71.

Mascarenhas, J.P. (1966). Pollen tube growth and ribonucleic acid synthesis by vegetative and generative nuclei of *Tradescantia*. *Amer. J. Bot.* **53**: 563–569.

Mascarenhas, J.P. (1975). The biochemistry of angiosperm pollen development. *Bot. Rev.* **41**: 259–314.

Mascarenhas, J.P. (1992). Pollen gene expression. In: S.D. Russell and C. Dumas, (editors). International review of cytology: sexual reproduction in flowering plants, pp. 3–18, Academic Press, San Diego.

Mayer, U., Torres Ruiz, R.A., Berleth, T., Misera, S., and Jurgens, G. (1991). Mutations affecting body organization in the *Arabidopsis* embryo. *Nature* **353**: 402–407.

Mayer, U., Muttner, G. and Jurgens, G. (1993). Apical-basal pattern formation in the *Arabidopsis* embryo: studies on the role of the *gnom* gene. *Development* **117**: 149–162.

McCormick, S. (1993). Male gametophyte development. *Pl. Cell* **5**: 1265–1275.

McCoy, T.J. and Smith, L.Y. (1983). Genetics, cytology, and crossing behaviour of an alfalfa (*Medicago sativa*) mutant resulting in failure of the postmeiotic cytokinesis. *Canad. J. Genet. Cytol.* **25**: 390–397.

Olmedilla, A., Schrauwen, J.A.M. and Wullems, G.J. (1991). Visualization of starch-synthase expression by *in situ* hybridization during pollen development. *Planta* **184**: 182–186

Owen, H.A. and Makaroff, C.A. (1995). Ultrastructure of microsporogenesis and microgametogenesis in *Arabidopsis thaliana* (L.) Heynh. ecotype Wassilewskija (*Brassicaceae*). *Protoplasma* **185**: 7–21.

Pacini, E. (1990). Tapetum and microspore function. In: S. Blackmore and R.B. Knox (editors). Microspores: evolution and ontogeny, pp. 213–237, Academic Press, London.

Palevitz, B.A. (1982). The stomatal complex as a model of cytoskeletal participation in cell differentiation. In: C.W. Lloyd (editor). The cytoskeleton in plant growth and development, pp. 345–376, Academic Press, London.

Pokrywka, N.J. and Stephenson, E.C. (1995). Microtubules are a general component of mRNA localization systems in *Drosophila* oocytes. *Developm.. Biol.* **167**: 363–370.

Preuss, D., Rhee, S.Y. and Davis, R.W. (1994). Tetrad analysis possible in *Arabidopsis* with mutation of the QUARTET (QRT) genes. *Science* **264**: 1458–1460.

Raghavan, V. (1989). mRNAs and a cloned histone gene are differentially expressed during anther and pollen development in rice (*Oryza sativa* L.). *J. Cell Sci.* **92**: 217–229.

Redei, G.P. (1965). Non-Mendelian megagametogenesis in *Arabidopsis*. *Genetics* **51**: 857–872.

Regan, S. and Moffatt, B.A. (1990). Cytochemical analysis of pollen development in wild-type *Arabidopsis* and a male sterile mutant. *Pl. Cell* **2**: 877–889.

Reijnen, W.H., van Herpen, M.M.A., de Groot, P.F.M., Olmedilla, A., Schrauwen, J.A.M., Weterings, K.A.P., and Wullems, G.J. (1991). Cellular localization of a pollen-specific mRNA by *in situ* hybridization and confocal laser scanning microscopy. *Sexual Plant Reprod.* **4**: 254–257.

Reynolds, T.L. and Raghavan, V. (1982). An autoradiographic study of RNA synthesis during maturation and germination of pollen grains of *Hyoscyamus niger*. *Protoplasma* **111**: 177–188.

Rhyu, M.S. and Knoblich, J.A. (1995). Spindle orientation and asymmetric cell fate. *Cell* **82**: 523–526.

Roberts, M. (1996). Transposon-mediated insertional mutagenesis. In: G. Foster and D. Twell (editors). Plant gene isolation: principles and practice, pp. 301–328, John Wiley Chichester.

Sauter, J.J. (1969). Cytochemical investigations on histones in cells differing in RNA and protein synthesis. *Z. Pflanzenphysiol.* **60**: 434–449. (In German).

Schopfer, C.R. and Hepler, P.K. (1991). Distribution of membranes and the cytoskeleton during cell plate formation in pollen mother cells of *Tradescantia J. Cell Sci.* **100**: 717–728.

Schrauwen, J.A.M., de Groot, P.F.M., van Herpen, M.M.A., van der Lee, T., Reynen, W.H., Weterings, K.A.P., and Wullems, G.J. (1990). Stage-related expression of mRNAs during pollen development in lily and tobacco. *Planta* **182**: 298–304.

Shen, J.B. and Hsu, F.C. (1992). *Brassica* anther-specific genes: characterization and *in situ* localization of expression. *Molec. Gen. Genet.* **234**: 379–389.

Shevell, D.E., Leu, W-M., Gilimor, C.S., Xia, G., Feldman, K.A. and Chua, N-H. (1994). *EMB30* is essential for normal cell division, cell expansion, and cell adhesion in *Arabidopsis* and encodes a protein that has similarity to Sec7. *Cell* **77**: 1051–1062.

Spielman, M.L., Preuss, D., Li, F-L., Brown, W.E., Scott, R. and Dickinson, H.G. (1997). *TETRASPORE* is required for male meiotic cytokinesis in *Arabidopsis thaliana*. *Development* **124**: 2645–2657.

Springer, P.S., McCombie, W.R., Sundaresan, V. and Martienssen, R.A. (1995). Gene trap tagging of PROLIFERA, an essential MCM2-3-5-like gene in *Arabidopsis*. *Science* **268**: 877–880.

Stoppin, V., Vantard, M., Schmit, A-C., Lambert, A-M. and (1994). Isolated plant nuclei nucleate microtubule assembly: the nuclear surface in higher plants has centrosome-like activity. *Pl. Cell* **6**: 1099–1106.

Sunderland, N., and Huang, B. (1987). Ultrastructural aspects of pollen dimorphism. In: K.L. Giles and J. Prakash (editors). International review of cytology. pollen: cytology and development, pp. 175–220, Academic Press, London.

Tanaka, I. and Ito, M. (1980). Induction of typical cell division in isolated microspores of *Lilium longiflorum* and *Tulipa gesneriana*. *Pl. Sci. Lett.* **17**: 279–285.

Tanaka, I. and Ito, M. (1981). Control of division patterns in explanted microspores of *Tulipa gesneriana*. *Protoplasma* **108**: 329–340.

Terasaka, O. (1982). Nuclear differentiation of male gametophytes in gymnosperms. *Cytologia* **47**: 27–46.

Terasaka, O. and Niitsu, T. (1987). Unequal cell division and chromatin differentiation in pollen grain cells. I. Centrifugal, cold and caffeine treatments. *Bot. Mag. (Tokyo)* **100**: 205–216.

Terasaka, O. and Niitsu, T. (1990). Unequal cell division and chromatin differentiation in pollen grain cells. II. Microtubule dynamics associated with the unequal cell division. *Bot. Mag. Tokyo* **103**: 133–142.

Terasaka, O. and Niitsu, T. (1995). The mitotic apparatus during unequal microspore division observed by a confocal laser scanning microscope. *Protoplasma* **189**: 187–193.

Terasaka, O., and Tanaka, R. (1974). Cytological studies on the nuclear differentiation in microspore division of some angiosperms. *Bot. Mag. Tokyo* **87**: 209–217.

Theerakulpisut, P., Xu, H., Singh, M.B., Pettitt, J.M. and Knox, R.B. (1991). Isolation and developmental expression of *Bcp*1, an anther-specific cDNA clone in *Brassica campestris*. *Pl. Cell* **3**: 1073–1084.

Thomas, C. (1996). PCR techniques. In: G. Foster and D. Twell (editors). Plant gene isolation: principles and practice, pp. 331–368. John Wiley, Chichester.

Torres Ruiz, R.A. and Jurgens, G. (1994). Mutations in the *FASS* gene uncouple pattern formation and morphogenesis in *Arabidopsis* development. *Development* **120**: 2967–2978.

Tupy, J., Rihova, L. and Zarsky, V. (1991). Production of fertile tobacco pollen from microspores in suspension culture and its storage for *in situ* pollination. *Sexual Pl. Reprod.* **4**: 284–287.

Twell, D. (1992). Use of a nuclear-targeted ß-glucuronidase fusion protein to demonstrate vegetative cell-specific gene expression in developing pollen. *Plant J.* **2**: 887–892.

Twell, D. (1994). The diversity and regulation of gene expression in the pathway of male gametophyte development. In: R.J. Scott and A.D. Stead (editors). Molecular and cellular aspects of plant reproduction, pp. 83–135. Cambridge. (Society for Experimental Biology seminar series, 55).

Twell, D., Yamaguchi, J. and McCormick, S. (1990). Pollen-specific gene expression in transgenic plants: co-ordinate regulation of two different tomato gene promoters during microsporogenesis. *Development* **109**: 705–713.

Twell, D., Patel, S., Sorensen, A., Roberts, M., Scott, R., Draper, J. and Foster, G. (1993). Activation and developmental regulation of an *Arabidopsis* anther-specific promoter in microspores and pollen of *Nicotiana tabacum*. *Sexual Plant Reprod.* **6**: 217–224.

Ueda, K. and Tanaka, I. (1995a). Male gametic nucleus-specific H2B and H3 histones. Designated gH2B and gH3, in *Lilium longiflorum*. *Planta* **197**: 289–295.

Ueda, K. and Tanaka, I. (1995b). The appearance of male gamete-specific histones gH2B and gH3 during pollen development in *Lilium longiflorum*. *Developm. Biol.* **169**: 210–217.

Ursin, V.M., Yamagushi, J. and McCormick, S. (1989). Gametophytic and sporophytic expression of anther-specific genes in developing tomato anthers. *Pl. Cell* **1**: 727–736.

Van Lammeren, A.A.M., Bednara, J. and Willemse, M.T.M. (1989). Organization of the actin cytoskeleton in *Gasteria verrucosa* (Mill.) H. Duval visualized with rhodaminephalloidin. *Planta* **178**: 531–539.

Weterings, K.A.P. (1994). Pollen gene regulation. Ph.D. dissertation, University of Nijmegen. ISBN 90-9006767-1.

Zaki, M.A.M., and Dickinson, H.G. (1991). Microspore-derived embryos in *Brassica*: the significance of division asymmetry in pollen mitosis I to embryogenic development. *Sexual. Pl. Reprod.* **4**: 48–55.

Zhang, D. and Nicklas, R.B. (1993). 'Anaphase' and cytokinesis in the absence of chromosomes. *Nature* **382**: 466–468.

Zhang, G., Gifford, D.J. and Cass, D.D. (1993). RNA and protein synthesis in sperm cells isolated from *Zea mays* L. *Sexual Pl. Reprod.* **6**: 239–243.

Rudall, P.J., Prychid, C.J. and Jones, C. (1998). Intra-ovarian trichomes, mucilage secretion and hollow styles in monocotyledons. In: S.J. Owens and P.J. Rudall (Editors). Reproductive Biology, pp. 219–230. Royal Botanic Gardens, Kew.

INTRA-OVARIAN TRICHOMES, MUCILAGE SECRETION AND HOLLOW STYLES IN MONOCOTYLEDONS

PAULA J. RUDALL, CHRISTINA J. PRYCHID AND CLAIRE JONES

Jodrell Laboratory, Royal Botanic Gardens, Kew, Richmond, Surrey, TW9 3DS, UK

Abstract

Intra-ovarian trichomes are taxonomically widespread in monocotyledons, but especially clustered among the 'basal' groups, such as *Acorus* and Araceae. By means of extensive Golgi activity, they actively secrete a jelly-like mucilage which fills the ovary locules. Since mucilage is known to promote pollen tube growth, intra-ovarian trichomes have a role in the reproductive biology of the plants concerned, as specialised transmitting tissues or obturators. Mucilage-filled ovaries are correlated to some extent with aquatic habitats and orthotropous ovules. This study also indicates a correlation with short styles, which may be either hollow (as in most monocotyledons) or solid (as in most dicotyledons). Secretory intra-ovarian trichomes are rare in dicotyledons, but mucilage-filled ovaries occur in some 'primitive' magnoliid dicot families closely related to monocotyledons.

Introduction

This paper presents new data on intra-ovarian trichomes (IOTs) and mucilage-filled ovaries in monocotyledons, and reviews their occurrence in a systematic context, in order to test their homologies and consider their wider implications in terms of breeding systems. IOTs are hairs that occur within the ovary locules. In monocotyledons they are invariably associated with a jelly-like mucilage which fills the entire ovary cavity (Table 1). It has not hitherto been demonstrated whether the trichomes themselves secrete the mucilage, although this is generally assumed.

There is strong consensus (e.g. Sassen, 1974) that mucilage has a role as a nutrient medium in guiding pollen tubes in both solid and hollow styles, analogous to that of fungal hyphal growth *in vitro*. Ovarian mucilage probably has a similar role to stylar mucilage, presumably with a chemical gradient. French (1987) speculated that in Araceae mucilage bridges the gap for pollen tube growth between the stylar canal and the micropyle. In the few species where IOTs are absent, mucilage is also lacking, and in most of these cases there is a single ovule occupying most of the locule. Similarly, Apparao & Shah (1989) speculated that in *Ottelia alismoides* (Hydrocharitaceae) the intra-ovarian mucilage is a medium for free movement of pollen tubes and their nourishment in the ovarian cavity. Labarca & Loewus (1973) demonstrated that stylar canal exudates are a major source of carbohydrates for pollen tube wall biosynthesis in *Lilium*. Sanders & Lord (1989) introduced latex beads into the stylar transmitting tissue of *Hemerocallis*, *Raphanus* and *Vicia* and demonstrated that they were translocated to the ovules with the same speed as pollen tubes, indicating that pollen tubes are actively guided by the secretory matrix. In plants with solid styles, including most dicotyledons, pollen tubes grow intercellularly between the cells of a central transmitting tissue, or sometimes through the cell walls. In plants with hollow styles (i.e. most monocotyledons), pollen tubes grow along the surface of the secretory stylar canal cells (e.g. in *Gladiolus* and *Lilium*: Clarke *et al.*, 1977), sometimes beneath the cuticle (e.g. in *Hemerocallis*: Sanders & Lord, 1992). In the ovary of the date palm (*Phoenix*), pollen tubes follow a groove of mucilage to the micropyle (Knox, 1984). Mucilage-secreting IOTs therefore have a role in the reproductive biology of the plants concerned, as specialised transmitting tissues or obturators.

Material and Methods

Perhaps the main reason for the paucity of information on IOTs is the notorious difficulty in obtaining good sections of mucilage-filled flowers (e.g. of Araceae: Eyde *et al.* 1967). The carbohydrate component of the mucilage causes the ovaries to harden during dehydration. Aldehydes used in conventional fixation stabilise proteins by forming cross links with them, particularly with the e-NH_2 groups of lysine side chains. Some lipids also appear to be included in the fixation. Osmium tetroxide used as a secondary fixative in electron microscopy work stabilises the majority of lipids and it is hoped that between them some carbohydrates will be fixed. However, if the mucilage is predominantly carbohydrate, as in *Astelia grandis* and *Collospermum hastatum* (this paper), there is nothing to preserve it and it will either flow out of the ovary and be lost or will remain there and act as a barrier to the penetration of the fixative into the ovarian cavity, resulting in degradation of the ovules and IOTs.

Material examined

Material examined was collected from the living collections at the Royal Botanic Gardens, Kew (HK + garden accession number) or from wild-source material.

Acoraceae: *Acorus calamus* L. 'Variegatus' (HK 1969–19633), *A. gramineus* Soland. 'Variegatus' (HK 1969-18000), *A. gramineus* f. *minor* Hort. (HK 1969-17999).

Asteliaceae: *Astelia grandis* Hook. f. ex T. Kirk (HK 1972–3488); *Collospermum hastatum* (Collenso) Skottsb. (HK 1986-3592).

Araceae: *Aglaonema modestum* Schott. ex Engl. (HK 1965–22401), *A. stenophyllum* Merrill (HK 1982-5008); *Alocasia micholitziana* Sander (HK 1980–1232); *Anthurium andraeanum* Linden (HK 1960-31808); *Arisarum vulgare* Targ. Toz. (HK 1972–10427); *Arum dioscoridis* Sibth. & Sm. var. *dioscoridis* (HK 1979-2190), *A. rupicola* Boiss. var. *rupicola* (HK 1985–2128); *Stenospermation ulei* K. Krause (HK 1990–2738).

Hanguanaceae: *Hanguana malayana* Merrill (s.n.).

Transmission Electron Microscopy (TEM)

Gynoecia were cut from the flowers and placed immediately into two different fixatives: Karnovsky's fixative (2% paraformaldehyde and 2.5% glutaraldehyde) (Glauert, 1975) and acrolein (2%)-glutaraldehyde (2.5%) in 0.1M phosphate buffer, pH 7.2. Tissues were dissected whilst under fix, leaving 2mm^2 pieces, and placed under vacuum for 30 minutes. Primary fixation was then continued for 2.5 hours at room temperature and pressure. After washing in buffer the samples were post-fixed in 1% osmium tetroxide for two hours, rinsed, dehydrated in an ethanol series and embedded in LR White resin (medium grade). Silver sections were collected on Formvar-coated slot grids and stained in aqueous uranyl acetate and lead citrate (Reynolds, 1963). Photographs were taken on a JEOL JEM 12-10.

Light Microscopy (LM)

For wax embedding and sectioning, flowers were fixed in Formalin acetic alcohol (FAA) and stored in 70% ethanol. Complete flowers or ovaries were embedded in Paraplast using standard methods of wax embedding, and serially sectioned using a rotary microtome. Sections were stained in safranin and Alcian blue, dehydrated through an alcohol series to 100% ethanol then Histoclear, mounted in Euparal, and examined using normal bright-field optics on a Leitz Diaplan photomicroscope.

For histochemistry the following stains were applied. PAS, periodic acid-Schiff's reaction, was used as a general polysaccharide stain (Hotchkiss, 1948; McManus, 1948). As a control, the periodic acid was omitted. The ninhydrin-Schiff's reaction (Yasuma & Ichikawa, 1953) was used to localise total protein. For controls blocking procedures were applied prior to the ninhydrin step. Starch was localised with the IKI reaction (Johansen, 1940).

Semi-thin sections (c. 0.5μm) taken from material embedded for TEM were collected on slides and stained with the metachromatic stain toluidine blue (0.5% in 0.1M phosphate buffer, pH 7.0) (Feder & O'Brien, 1968).

Intra-ovarian trichomes are taxonomically widespread in monocotyledons (Table 1), occurring among several unrelated groups from each of the major clades (*sensu* Chase *et al.* 1995), indicating that they have evolved more than once. They occur in the commelinoid clade (*Hanguana* and some Zingiberales, such as Musaceae), the asparagoid clade (Asteliaceae) and the dioscoreoid clade (Pandanaceae). However, they are especially clustered among the 'primitive' monocotyledons, such as *Acorus* and Araceae. In contrast to dicotyledons, monocot IOTs are always associated with mucilage secretion. Mucilage-filled ovaries are rare in monocotyledons lacking IOTs (Table 1), apart from some 'basal' taxa (*sensu* Chase *et al.*, 1995), such as *Ottelia* (Hydrocharitaceae), where mucilage is secreted from intracarpellary cells present in clusters in the ovary wall above the micropyle of each ovule (Apparao & Shah, 1989).

Structure and taxonomic distribution of IOTs in monocotyledons

TABLE 1. Records of intra-ovarian trichomes and mucilage-filled ovaries in monocotyledons

Family	Genera	IOTs	Mucilage-filled ovaries	Reference
Acoraceae	*Acorus*	+	+	French 1987; this paper
Araceae	most genera	+	+	Dalmer 1880; Eyde et al. 1967; French 1987
Asteliaceae	*Astelia, Collospermum*	+	+	Skottsberg 1934; this paper
Hanguanaceae	*Hanguana*	+	+	this paper
Hyacinthaceae	*Ornithogalum*	?	+	Tilton & Horner 1980 (& see Dalmer 1880)
Hydrocharitaceae	*Hydrocharis,*	-	+	Troll 1931; Endress 1995
	Ottelia	(+)		Apparao & Shah 1989
Lemnaceae	*Lemna*	?	+	Eyde et al. 1967; Buzgó 1994; Endress 1995
Musaceae	several genera	+	+	e.g. Kirchoff 1992; Alquini 1992
Pandanaceae	genera of Pandananae	+	+	Fagerlind 1940; Endress 1995, Cheah & Stone 1975

In dicotyledons, non-secretory IOTs occur in several families, such as Ericaceae, Rubiaceae, Salicaceae, Styracaceae and Sapotaceae (Dickison, 1993). There is relatively little evidence of secretion from IOTs in dicotyledons, except in rare cases, such as ovarian glandular trichomes of *Acacia* (Kenrick & Knox, 1989), and the placental and integumentary trichomes of some Urticaceae and Euphorbiaceae, which sometimes function as obturators (Maheshwari, 1950). However, mucilage-filled ovaries occur in some Magnoliid dicotyledon families that are considered closely related to monocotyledons, such as Nymphaeaceae, Cabombaceae (Endress, 1990, 1994) and Chloranthaceae (Endress, 1987), where mucilage is secreted directly from the ovary wall rather than from IOTs. IOTs therefore probably evolved from secretory gynoecial epidermal cells (probably several times in monocotyledons), the presence of mucilage-filled ovaries being probably the ancestral (plesiomorphic) state for this character.

According to Endress (1990, 1995) mucilage-filled ovaries occur mainly in aquatic plants or plants of moist habitats, indicating that a constant water supply is necessary for this structure. He also suggested a correlation between orthotropous ovules and mucilage-mediated pollen tube transmission, as there is less constraint on ovules of these species to be curved and to direct the micropyle towards the placenta. He gave examples from the dicot family Nymphaeaceae (*Barclaya*) and the monocot Hydrocharitaceae (*Hydrocharis*). This correlation certainly holds true for many Araceae, and some other monocot taxa, such as *Hanguana*, which is common in swampy areas, and has orthotropous ovules (Fig. 1B). *Astelia* and *Collospermum* are not specifically swamp-dwelling, and indeed include several epiphytes, but may prefer wetter habitats.

We have also observed some congruence between intra-ovarian trichomes and short styles or almost sessile stigmas (lacking styles), especially in Acoraceae (Fig. 1D), Asteliaceae (Fig. 1A), Araceae, Lemnaceae and *Hanguana* (Fig. 1B), although some other taxa with short styles lack IOTs or mucilage-filled ovaries (e.g. *Ruscus* and *Xanthorrhoea*). Perhaps a combination of the two factors (wet habitats and short styles) may encourage the development of mucilage-secreting IOTs, in a group already disposed to the production of mucilage for pollen tube growth.

Structure and distribution of IOTs is very similar in all monocotyledons examined. They are usually unicellular, although French (1987) reported some bicellular or multicellular hairs in a few Araceae. They emerge from the placental region above and below the ovules and sometimes from base of each funicle, and extend around the ovules, overarching the micropyles (Fig. 1). Other epidermal cells of the ovary surrounding the loculi are generally non-secretory, although in some taxa we have noted that they are secretory around the base of the style. In some taxa (e.g. *Astelia*: Fig. 1A) IOTs also extend upwards into the style, effectively forming a 'solid' style, as in most eudicots. This is in contrast to most other monocotyledons, where styles are hollow, although there are occasional exceptions to this; for example solid styles occur in some Poaceae and all Alliaceae *sensu stricto* (Fay and Chase, 1996).

Evidence for secretory activity

Although Engler (1920) considered that in *Anthurium* and *Lagenandra* mucilage formation results from hair breakdown, our results demonstrate that, at least in *Acorus* and Araceae, IOTs actively secrete the mucilage, through extensive Golgi (dictyosome) activity.

In flowers of *Acorus* and Araceae just prior to anthesis (Figs. 2–4), when mucilage is copious, the IOTs have dense cytoplasm, prominent nuclei and small vacuoles often containing a granular or fibrous material (mucilage). Lipid bodies are also present. Large amounts of endoplasmic reticulum, often with a swollen appearance, are present throughout the cytoplasm (Fig. 3C), together with several mitochondria and a number of dense dictyosomes with swollen cisternae on their maturing faces. Along part of their perimeter, the dictyosome cisternae are associated with a population of small vesicles which are coated on their outer surfaces (Figs. 3D, 4A–C). The cisternae bud off from the Golgi stack to form vesicles which then migrate through the cytoplasm and fuse with the plasma membrane to release their secretory contents (Fig. 4D). Mucilage accumulates between the protoplast and the cell wall (Fig. 3A), which is relatively thick, before diffusing into the ovarian cavity (exocytosis). Accumulated mucilage is lacking from other ovary or ovule cells lining the ovary cavity (which are relatively thin-walled), indicating that the IOTs are the primary source of the massive mucilage deposition in the locule.

The presence of large numbers of Golgi bodies associated with small vesicles is consistent with secretory activity, e.g. in legume stigmas (Kenrick & Knox, 1989), stylar canal cells of *Trimezia* (Bystedt & Vennigerholz, 1991) and *Ornithogalum* (Tilton & Horner, 1980) and transmitting tissue of *Primula* (Heslop-Harrison et al, 1981). In mucilage cells of *Opuntia* (Fahn 1979), mucilage is produced in Golgi bodies in the cytoplasm and accumulates between the protoplast and the cell wall.

Rough endoplasmic reticulum (Fig. 3C) is also thought to play a part in the synthesis of mucilage (Bouchet & Deysson, 1974; Horner & Lersten, 1968) and nectar. Swollen cisternae bud off to form vesicles which then fuse with the plasmalemma and eliminate their substances from the cytoplasm. Multivesicular bodies thought to originate from endoplasmic reticulum (Jensen, 1965; Bystedt & Vennigerholz, 1991) and dictyosomes (Sedgley & Clarke, 1986; Miki-Hirosige et al, 1987) are involved in the transportation of secretion in various tissues, such as transmitting tissue of *Lycopersicon* (Cresti et al, 1976) and papillar cells of *Hordeum* and *Secale* (J. Heslop-Harrison & Y. Heslop-Harrison, 1980). It has also been postulated that mitochondria are involved in mucilage production (Werker & Kislev, 1978).

Ovary cells at the base of the trichomes contain numerous starch grains (Fig. 2B) which presumably act as a nutrition reserve and become depleted during mucilage production, as Apparao & Shah (1989) also noted in intra-ovarian mucilage-secreting cells of *Ottelia alismoides*. Starch grains are absent from the trichomes themselves. Loss of starch grains from plastids is thought to result from breakdown into monomeric or dimeric sugars which then leave the cells by eccrine secretion (Kandasamy & Kristen, 1987; Bystedt & Vennigerholz, 1991).

In older IOTs towards the end of their secretory life there are large vacuoles often containing fibrous mucilage. These are surrounded by ribosome-rich cytoplasm containing a single large nucleus, endoplasmic reticulum and dictyosomes, but by this stage there are often relatively few mitochondria. Sometimes hypertophied Golgi bodies occupy a great part of the cytoplasm, togther with occasional spherical bodies (Fig. 2A) that may represent autophagic vesicles, which are commonly found in autolysing secretory cells (e.g. Nair *et al.*, 1983).

Nature of the secretion product

The mucilage itself is often fibrous, indicating that outer wall fibrils are being sloughed off together with the mucilage. This often results in poor definition of the boundary between the exudate and the secretory walls, the wall and the adjacent layer of secretion having a similar structure (see also Bell & Hicks, 1976; Hill & Lord, 1987; Sanders et al, 1990; Sanders & Lord, 1992). At high concentrations mucilage fibrils aggregate laterally to form a web-like network. At lower concentrations globular particles in the secretion aggregate in a linear fashion. Bystedt & Vennigerholz (1991) also reported a separation of the wall fibrils in secretory gynoecial cells of *Trimezia,* as secretion passed through them, and in stylar canal cells of *Lilium* the wall layer becomes a component of the secretion (Hu *et al.*, 1982). The developing testa epidermis of *Linum* seeds produces a fibrillar, predominantly carbohydrate mucilage (Heinz, 1984).

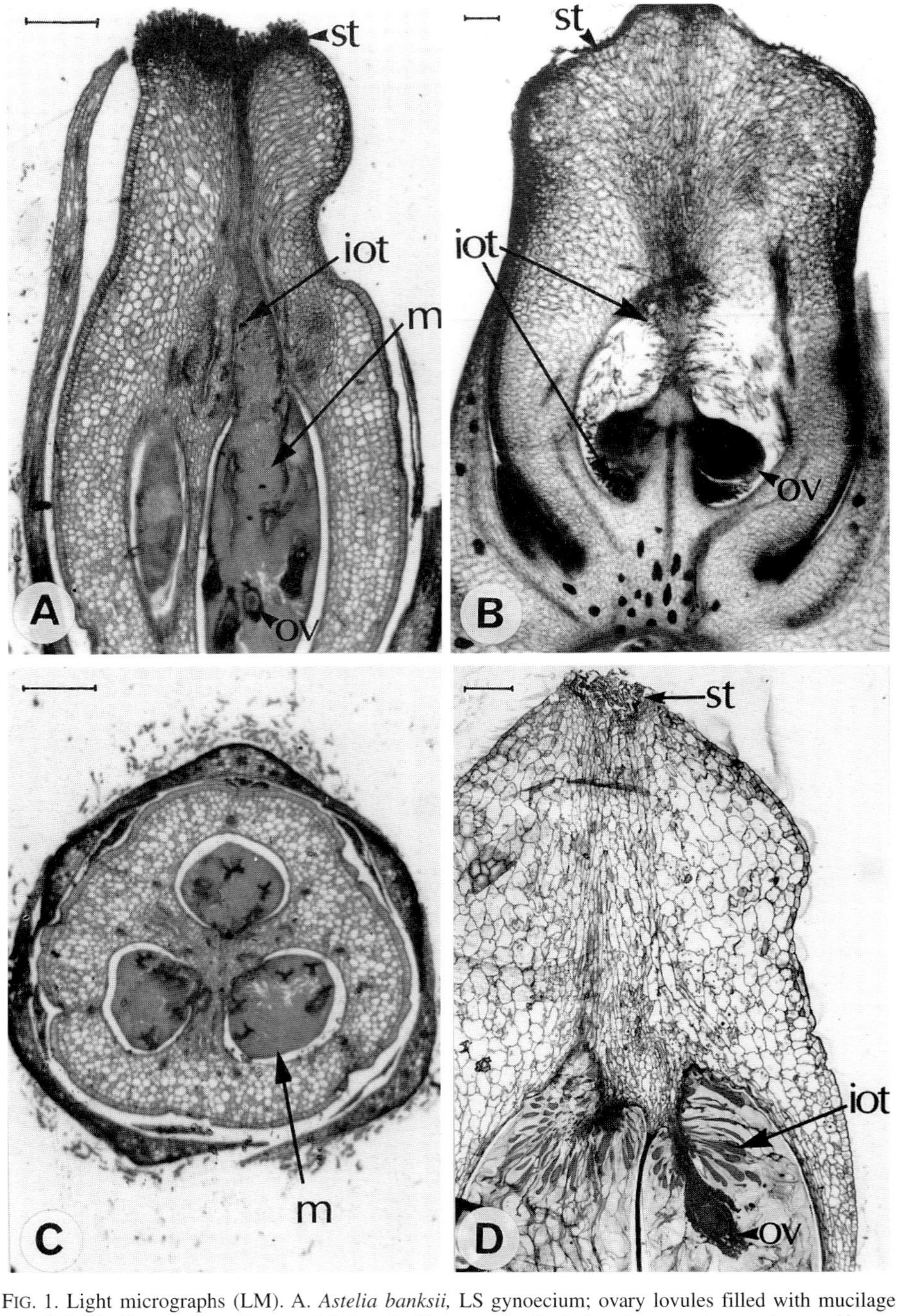

FIG. 1. Light micrographs (LM). A. *Astelia banksii,* LS gynoecium; ovary lovules filled with mucilage (m). B. *Hanguana malayana,* LS female flower with short, broad gynoecium and nectariferous staminodes. C. *Astelia banksii,* TS trilocular ovary with mucilage (m) filling ovary locules. D. *Acorus gramineus* f. *minor,* LS gynoecium. iot = intra-ovarian trichomes. m = mucilage. ov = ovule. st = stigma. Scale bars = 100mm.

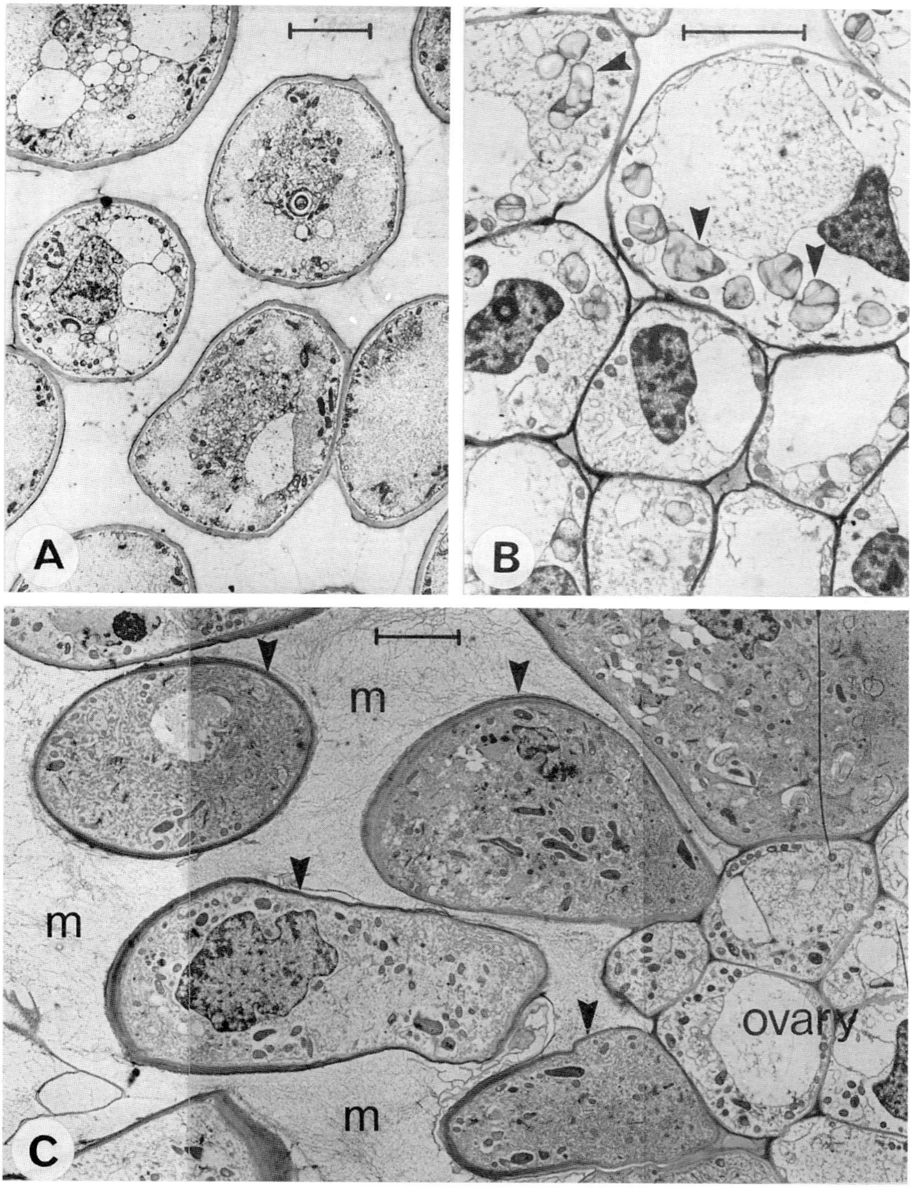

FIG. 2. Transmission electron micrographs (TEM) (composite picture in C). A. *Stenospermation ulei*, TS through IOTs. B., C. *Alocasia micholitziana*. B. Ovarian cells at base of IOTs containing large starch plastids (s). C. IOTs with denser cytoplasm than basal ovary tissue. Fibrous mucilage (m) in ovary locule. Scale bars = 5mm.

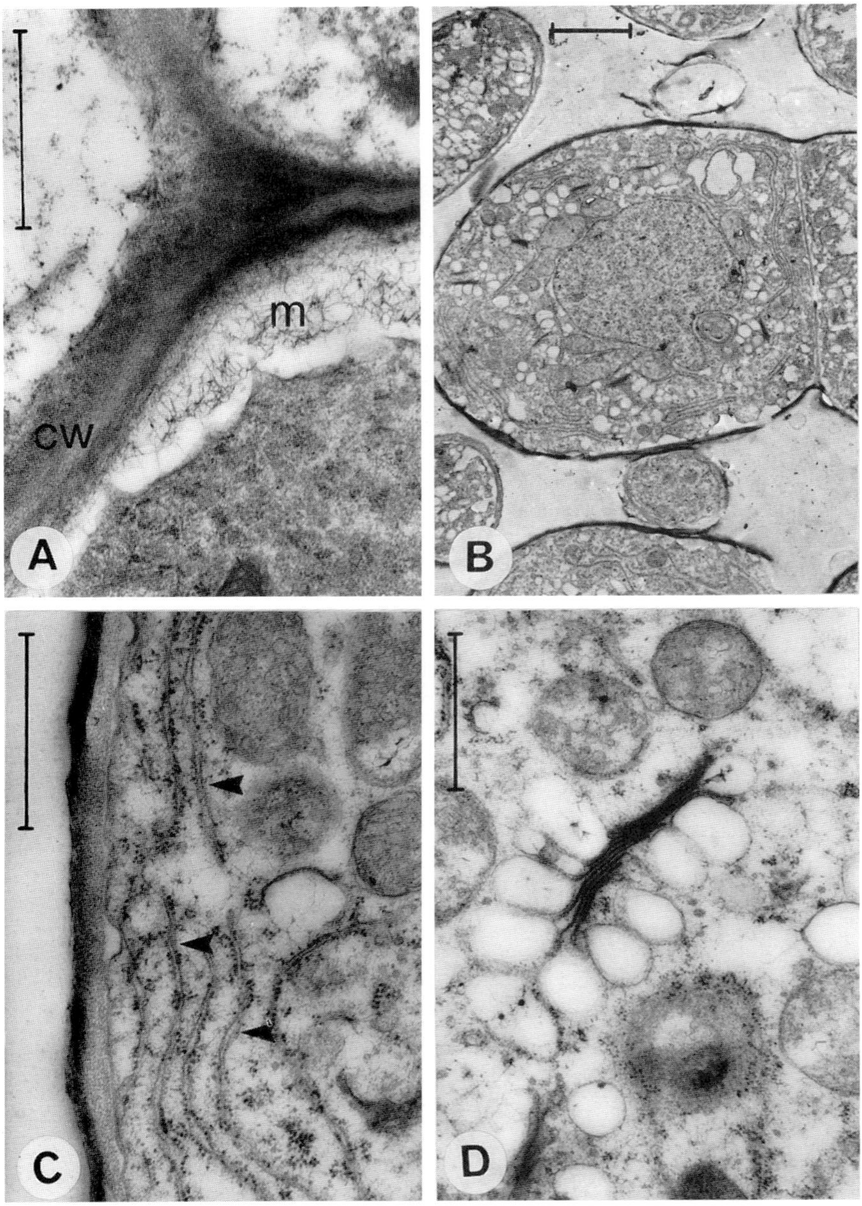

FIG. 3. Transmission electron micrographs (TEM). A. *Stenospermation ulei*, fibrous mucilage (m) present in periplasmic space. B.-D. *Acorus gramineus f. minor.* B. IOT with cytoplasm dominated by secretory vesicles and endoplasmic reticulum. C. Several profiles of endoplasmic reticulum (arrowed) and their associated ribosomes. D. Dictyosome with individual cisternae marginally vesiculated. Some of the swollen cisternae may have already sloughed off to form secretory vesicles. Scale bars: in A., C., D. = 1mm, B. = 5mm.

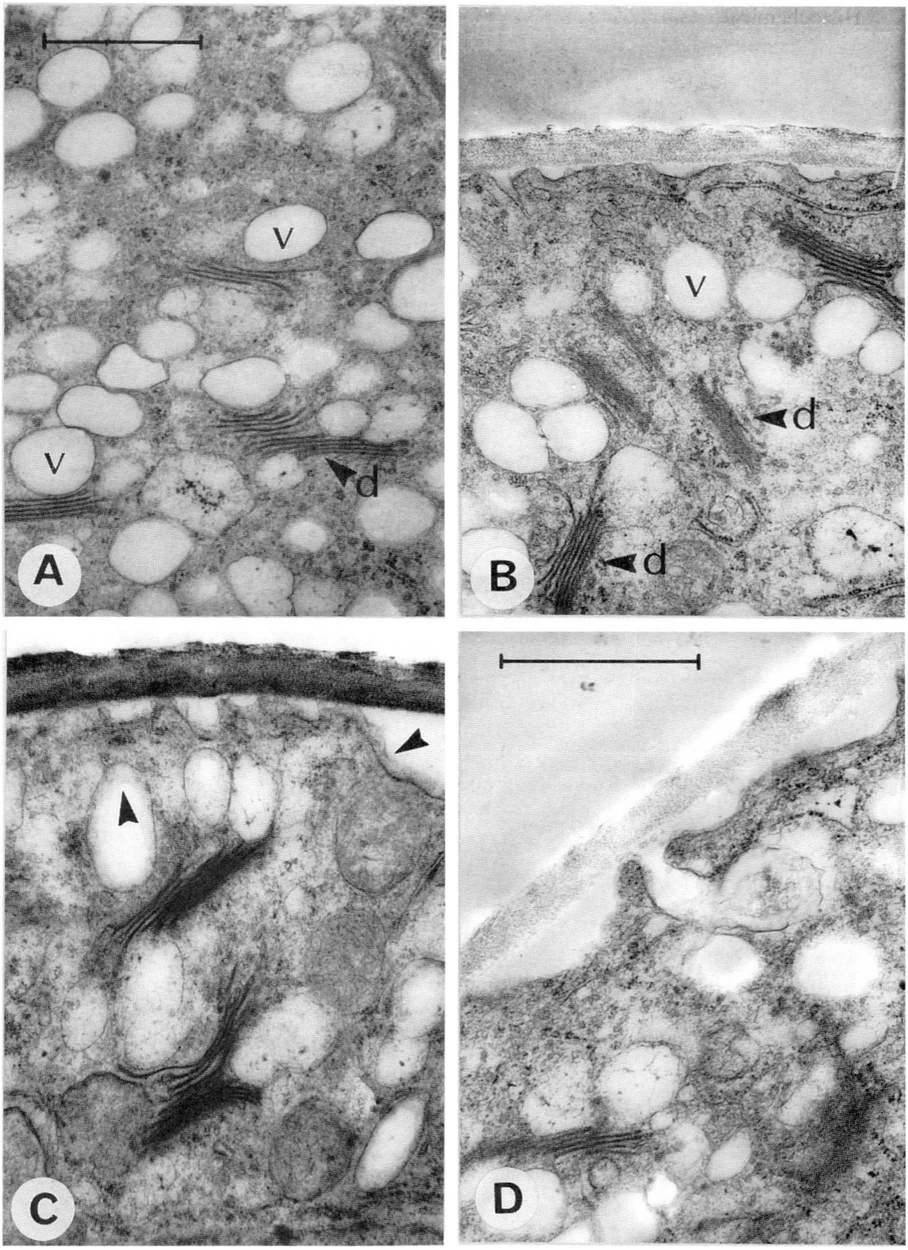

FIG. 4. *Acorus gramineus f. minor* (TEM). A., B. Cytoplasm dominated by dictyosomes (d) and secretory vesicles (v). C. Secretory vesicles and space between plasma membrane and cell wall containing a substance (arrowed), presumably mucilage. D. Secretory vesicle fused with plasma membrane, releasing its contents into the periplasmic space. Scale bars = 1 mm.

Histochemical studies on ovarian mucilage of our material of *Astelia grandis* and *Collospermum hastatum* indicated the presence of insoluble polysaccharides and mucopolysaccharides (carbohydrates) and the absence of proteins and lipids. This coincides with observations by Apparao & Shah (1989) on ovarian mucilage in *Ottelia* (Hydrocharitaceae) and Bystedt & Vennigerholz (1991) on gynoecial secretions of *Trimezia fosteriana* (Iridaceae). Sanders & Lord (1992) and Knox (1984) reported a variety of component compounds of gynoecial secretion products (extracellular matrices: ECMs), including sugars, amino acids, peptides, phenolics, fatty acids, lipids and high molecular weight polysaccharides involving mucilages, proteoglycans and pectic compounds and complex proteins, such as glycoproteins and lipoproteins, the exact composition varying between species. In many plants secretion products are chemically different between stigma and stylar canal cells, indicating a difference in function (Labarca & Loewus, 1973; Kenrick & Knox, 1989). Whereas stylar and ovarian secretions guide the pollen tubes (see below), stigmatic secretions may have a selective role in self-compatible crosses (Sanders & Lord, 1992).

The secretion product from stylar canal cells is often described as mucilage. The transmitting tissue of the solid styles of dicotyledons has an mucilagenous intercellular matrix which Sassen (1974) and Cresti *et al.* (1976) identified as a complex substance containing pectin. This is comparable to the mucilage found in the canal of hollow styles (Labarca *et al.*, 1970).

References

Alquini, Y. (1992). Anatomia dos òrgaos em desenvolvomento de *Musa rosacea* Jacq. (Musaceae). PhD Thesis, unpublished, University of Sao Paulo.

Apparao, B.J. and Shah, C.K. (1989). Intracarpellary secretory cells and their role in *Ottelia alismoides*. *J. Indian Bot. Soc.* **68**: 25–28.

Bell, J. and Hicks, G. (1976). Transmitting tissue in the pistil of tobacco: light and electron microscopic observations. *Planta* **131**: 187–200.

Bouchet, P. and Deysson, G. (1974). Les canaux à mucilage des angiospermes. Etude morphologique et ultrastructurale des cellules constituant les canaux à mucilage du *Sterculia bidwilli* Hook. *Rev. Gén. Bot.* **81**: 369–402.

Bystedt, P.-A. and Vennigerholz, F. (1990). The transmitting tract in *Trimezia fosteriana* (Iridaceae). I. Ultrastructure in the stigma, style and ovary. *Nordic J. Bot.* **9**: 507–518.

Chase, M.W., Stevenson, D.W., Wilkin, P. and Rudall, P. (1995). Monocot systematics: a combined analysis. In: P.J. Rudall, P.J. Cribb, D.F. Cutler and C.J. Humphries (editors). Monocotyledons: systematics and evolution, pp. 685–730, Royal Botanic Gardens, Kew.

Cheah, C.H. and Stone, B.C. (1975). Embryo sac and microsporangium development in *Pandanus* (Pandanaceae). *Phytomorphology* **25**: 228–238.

Clarke, A.E., Considine, J.A., Ward, R. and Knox, R.B. (1977). Mechanism of pollination in *Gladiolus*: roles of the stigma and pollen-tube guide. *Ann. Bot.* **41**: 15–20.

Cresti, M., van Went, J.L., Pacini, E. and Willemse, M.T.M. (1976). Ultrastructure of transmitting tissue of *Lycopersicon peruvianum* style: development and histochemistry. *Planta* **132**: 305–312.

Dalmer, M. (1880). Ueber die Leitung der Pollenschlauche bei den Angiospermen. *Jenaische Z. Naturwiss.* **14**: 530–566.

Dickison, W.C. (1993). Floral anatomy of the Styracaceae, including observations on intra-ovarian trichomes. *Bot. J. Linn. Soc.* **112:** 223–255.

Endress, P.K. (1987). The Chloranthaceae: reproductive structures and phylogenetic position. *Bot. Jahrb. Syst.* **109:** 153–226.

Endress, P.K. (1990). Evolution of reproductive structures and functions in primitive angiosperms (Magnoliidae). *Mem. New York Bot. Gard.* **55:** 5–34.
Endress, P.K. (1994). Floral structure and evolution of primitive angiosperms: recent advances. *Pl. Syst. Evol.* **192:** 79–97.
Endress, P.K. (1995). Major evolutionary traits of monocot flowers. In: P.J. Rudall, P.J. Cribb, D.F. Cutler and C.J. Humphries (editors) Monocotyledons: systematics and evolution, pp. 43–79, Royal Botanic Gardens, Kew.
Eyde, R.H., Nicolson, D.H. and Sherwin, P. (1967). A survey of floral anatomy in Araceae. *Amer. J. Bot.* **54**: 478–497.
Fagerlind, F. (1940). Stempelbau und Embryosackentwicklung bei einigen Pandanazeen. *Ann. Jard. Bot. Buitenzorg* **49**: 55–78.
Fahn, A. (1979). Secretory tissues in plants. Academic Press, London.
Fay, M.F. and Chase, M.W. (1996). Resurrection of Themidaceae for the *Brodiaea* alliance, and recircumscription of Alliaceae, Amaryllidaceae and Agapanthoideae. *Taxon* **45**: 441–451.
Feder, N. and O'Brien, T.P. (1968). Plant microtechnique: some principles and new methods. *Amer. J. Bot.* **55**: 123–142.
French, J.C. (1987). Structure of ovular and placental trichomes of Araceae. *Bot. Gaz.* **148**: 198–208.
Glauert, A.M. (1975). Fixation, dehydration and embedding of biological specimens. North-Holland, Amsterdam. (Practical methods in electron microscopy, vol. 3, pt. 1)
Heinze, U. and Amelunxen, F. (1984). Zur Schleimbildung in *Linum*-Samen. Elektronenmikroskopische und chemische Analysen. *Ber. Deutsch. Bot. Ges.* **97**: 451–464.
Heslop-Harrison, J. and Heslop-Harrison, Y. (1980). The pollen-stigma interaction in the grasses. I. Fine-structure and cytochemistry of the stigmas of *Hordeum* and *Secale*. *Acta Bot. Neerl.* **29**: 261–276.
Heslop-Harrison, Y., Heslop-Harrison, J. and Shivanna, K.R. (1981). Heterostyly in *Primula*. I. Fine-structural and cytochemical features of the stigma and style in *Primula vulgaris* Huds. *Protoplasma* **107**: 171–187.
Hill, J.P. and Lord, E.M. (1987). Dynamics of pollen tube growth in the wild radish *Raphanus raphanistrum* (Brassicaceae). II. Morphology, cytochemistry and ultrastructure of transmitting tissues, and path of pollen tube growth. *Amer. J. Bot.* **74**: 988–997.
Horner, H.T. and Lersten, N.R. (1968). Development, structure and function of secretory trichomes in *Psychotria bacteriophila* (Rubiaceae). *Amer. J. Bot.* **55**: 1089–1099.
Hotchkiss, R.D. (1948). A microchemical reaction resulting in the staining of polysaccharide structures in fixed tissue preparations. *Arch. Biochem.* **16**: 131–141.
Hu, S., Zhu, C. and Xu, L. (1982). The structure of the canal cell in the style of *Lilium regale*. *Acta Bot. Sin.* **24**: 395–406.
Jensen, W.A. (1965). The composition and ultrastructure of the nucellus in cotton. *J. Ultrastruct. Res.* **13**: 112–128.
Johansen, D.A. (1940). Plant microtechnique. McGraw-Hill, New York.
Kandasamy, M.K. and Kristen, U. (1987). Developmental aspects of ultrastructure, histochemistry and receptivity of the stigma of *Nicotiana sylvestris*. *Ann. Bot.* **60**: 427–437.
Kenrick, J. and Knox, R.B. (1989). Pollen-pistil interactions in Leguminosae (Mimosoideae). In: C.H. Stirton and J.L. Zarucchi (editors). Advances in legume biology, pp. 127–156, Missouri Botanical Garden, St. Louis, Mo. (Monographs in systematic botany from the Missouri Botanical Garden, 29)
Kirchoff, B.K. (1992). Ovary structure and anatomy in the Heliconiaceae and Musaceae (Zingiberales). *Canad. J. Bot.* **70**: 2490–2508.
Knox, R.B. (1984). Pollen-pistil interactions. In: H.F. Linskens and J. Heslop-Harrison (editors). Cellular interactions, pp. 508–608, Springer Verlag, Berlin. (Encyclopedia of plant physiology, vol. 17).

Labarca, C. and Loewus, F. (1973). The nutritional role of pistil exudate in pollen tube wall formation in *Lilium longiflorum*. II Production and utilization of exudate from stigma and stylar canal. *Pl. Physiol.* **52**: 87–92.

Labarca C., Kroh M. and Loewus, F. (1970). The composition of stigmatic exudate from *Lilium longiflorum*. Labelling studies with myoinositol, D-glucose, and L-proline. *Pl. Physiol.* **46**: 150–156.

Maheshwari, P. (1950). An introduction to the embryology of angiosperms. McGraw-Hill, New York.

McManus, J.F.A. (1948). Histological and histochemical uses of periodic acid. *Stain Technol.* **23**: 99–108.

Miki-Hirosige, H., Hoek, I.H.S. and Nakamura, S. (1987). Secretions from the pistil of *Lilium longiflorum*. *Amer. J. Bot.* **74**: 1709–1715.

Nair, M.N.B, Shah, J.J. and Subramanyam, S.V. (1983). Ultrastructure and histochemistry of traumatic gum ducts in the wood of *Azadirachta indica* A. Juss. *IAWA Bull.* **4**: 103–112.

Reynolds, E.S. (1963). The use of lead citrate at a high pH as an electron opaque stain in electron microscopy. *J. Cell Biol.* **17**: 202–212.

Sanders, L.C. and Lord, E.M. (1989). Directed movement of latex particles in the gynoecia of three species of flowering plants. *Science* **243**: 1606–1608.

Sanders, L.C. and Lord, E.M. (1992). A dynamic role for the stylar matrix in pollen tube extension. *Int. Rev. Cytol.* **140**: 297–318.

Sanders, L.C., Eckard, K.J. and Lord, E.M. (1990). Divergent pollination systems in the cleistogamous species *Collomia grandiflora* (Polemoniaceae). *Protoplasma* **159**: 26–34.

Sassen, M.M.A. (1974). The stylar transmitting tissue. *Acta Bot. Neerl.* **23**: 99–108.

Sedgley, M. and Clarke, A.E. (1986). Immuno-gold localisation of arabinogalactan protein in the developing style of *Nicotiana alata*. *Nordic J. Bot.* **6**: 591–598.

Skottsberg, C. (1934). Studies in the genus *Astelia* Banks et Solander. *Kongl. Svenska Vetenskapsakad. Handl.*, Ser. 3, **14(2)**: 106pp.

Tilton, V.R. and Horner, H.T. (1980). Stigma, style and obturator of *Ornithogalum caudatum* (Liliaceae) and their function in the reproductive process. *Amer. J. Bot.* **67**: 1113–1131.

Troll, W. (1931). Beiträge zur Morphologie des Gynaeceums. I. Über das Gynaeceum von Hydrocharitaceen. *Planta* **17**: 453–460.

Werker, E. and Kislev, M. (1978). Mucilage on the root surface and root hairs of *Sorghum*: heterogeneity in structure, manner of production and site of accumulation. *Ann. Bot.* **42**: 809–816.

Yasuma, A. and Ichikawa, T. (1953). Ninhydrin-Schiff and alloxan-Schiff staining. A new histochemical staining method for proteins. *J. Lab. Clin. Med.* **41**: 296–299.

Bateman, R.M. and Simpson, N.J. (1998). Comparing phylogenetic signals from reproductive and vegetative organs. In: S.J. Owens and P.J. Rudall (Editors). Reproductive Biology, pp. 231–253. Royal Botanic Gardens, Kew.

COMPARING PHYLOGENETIC SIGNALS FROM REPRODUCTIVE AND VEGETATIVE ORGANS

RICHARD M. BATEMAN[1] AND NICOLA J. SIMPSON[1,2]

[1]**Royal Botanic Garden, 20A Inverleith Row, Edinburgh, EH3 5LR, UK**
[2]**Department of Earth Sciences, Open University, Milton Keynes, MK7 6AA, UK**

Abstract

Linnaeus chose to classify vascular plants primarily according to the number and relative positions of their reproductive organs. The many recently published morphological cladistic matrices offer the opportunity to test the wisdom of his actions, by examining the relative robustness of phylogenetic signals from different plant organs. This can be achieved by two methods: (1) Topological congruence. Treat the consensus tree generated from the data-matrix for all organs as the "true" evolutionary tree, then generate consensus trees for each organ from the same range of coded taxa. Each single-organ tree is compared with the yardstick all-organ tree, thus quantifying any decrease in resolution and any contradictory relationships. (2) Character congruence. Use various statistical indices to compare, among plant organs, patterns of homoplasy (multiple transitions between two states of a single character). Homoplasy reflects evolutionary lability, and thus is an inverse measure of the robustness of specific characters for delimiting "natural" (monophyletic) groups. We exemplified both methods using the rhizomorphic lycopsids as an unusually character-rich case-study, and then applied the computationally simpler index method to a specially compiled database of 129 botanical cladistic matrices. Although differing in detail, both methods yielded similar rankings of organs according to their phylogenetic value. Differences in behaviour among organs can in theory be correlated with variations in physical complexity and scale, and in developmental and functional constraints, but current evidence is circumstantial. In both the case-study and the broader survey, the average performance of reproductive characters is little better than non-reproductive characters, and homoplastic patterns cannot be generalised among clades. Moreover, *a priori* assertions of the best taxonomic characters for each plant group are refuted. Rather, optimal classifications require broadly based, objective descriptions of each organ of the species under comparison, as a necessary precursor to the character congruence test that underpins cladistic analysis.

Introduction: assessing the relative contributions of individual plant organs to a specific phylogeny

In his deservedly popular textbook on systematic botany, Clive Stace (1989: 69–71) wrote: "It is well known that floral characters have been, and still are, those most used in the classification of angiosperms . . . In the diagnostic keys to taxa in Floras, floral characters are those most used, and they figure equally prominently in the descriptions of the taxa which follow the keys. Examples could be given of the use of such features at all taxonomic levels.

The reliance on the flower is remarkable when one considers that for most of the time the majority of angiosperms lack any flowers at all!

From this vast pool of experience certain generalisations can be made with regard to the characters found to be most useful in certain taxa, or the level of conservativeness of certain characters. For example, in the Ranunculaceae the bracts, sepals and petals are particularly important, in the Asteraceae and Poaceae the inflorescence, bracts and general flower-type, in the Fabaceae the stamens and carpels, in the Scrophulariaceae and Lamiaceae the corolla and stamens, and so on. There are, however, very large numbers of exceptions to these generalisations, and too much emphasis on the latter would result in important errors or oversights."

In these paragraphs Stace succinctly summarised the mainstream view of the last two centuries of plant systematics. Reproductive organs are of paramount importance in classifying land-plants in general and angiosperms in particular, and systematists "know" *a priori* which organs – indeed, which particular features of those organs – are of the greatest value for classification within a particular taxonomic group. This position was adopted unashamedly by Linnaeus in his influential *Systema Naturae* (Linné, 1759) (Figs 1–2), though as noted by Stace (1989: 21–26) this was sandwiched in time between the less typological works of John Ray (1703) and Michel Adanson (1763), who opted for broader treatises based on admixtures of reproductive and vegetative characters. In this paper we attempt to step back from this indefensible (but rarely explicitly noted) tautology of *a priori* taxonomic knowledge, seeking instead the most objective means of assessing the relative merits of reproductive and non-reproductive characters in plant systematics.

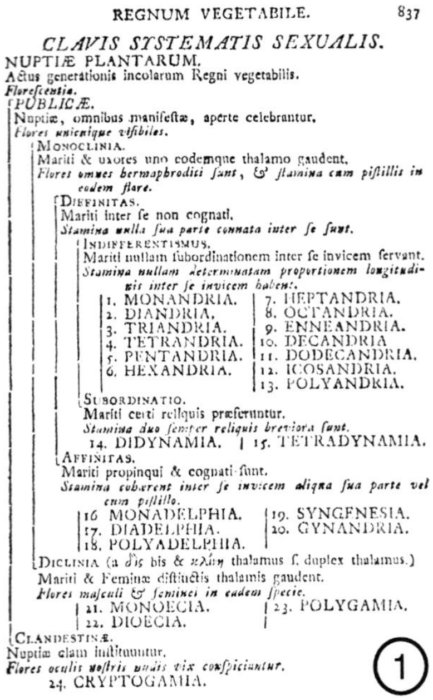

REGNUM VEGETABILE. 837

CLAVIS SYSTEMATIS SEXUALIS.

NUPTIÆ PLANTARUM.
Actus generationis incolarum Regni vegetabilis.
Florescentia.
PUBLICÆ.
Nuptiæ, omnibus manifestæ, aperte celebrantur.
Flores unicuique visibiles.
MONOCLINIA.
Mariti & uxores uno eodemque thalamo gaudent.
Flores omnes hermaphroditi sunt, & stamina cum pistillis in eodem flore.
DIFFINITAS.
Mariti inter se non cognati.
Stamina nulla sua parte connata inter se sunt.
INDIFFERENTISMUS.
Mariti nullam subordinationem inter se invicem servant.
Stamina nullam determinatam proportionem longitudinis inter se invicem habent.
1. MONANDRIA. | 7. HEPTANDRIA.
2. DIANDRIA. | 8. OCTANDRIA.
3. TRIANDRIA. | 9. ENNEANDRIA.
4. TETRANDRIA. | 10. DECANDRIA
5. PENTANDRIA. | 11. DODECANDRIA.
6. HEXANDRIA. | 12. ICOSANDRIA.
| 13. POLYANDRIA.
SUBORDINATIO.
Mariti certi reliquis præferuntur.
Stamina duo semper reliquis breviora sunt.
14. DIDYNAMIA. | 15. TETRADYNAMIA.
AFFINITAS.
Mariti propinqui & cognati sunt.
Stamina cohærent inter se invicem aliqua sua parte vel cum pistillo.
16 MONADELPHIA. | 19. SYNGENESIA.
17. DIADELPHIA. | 20. GYNANDRIA.
18. POLYADELPHIA.
DICLINIA (a δὶς bis & κλίνη thalamus f. duplex thalamus.)
Mariti & Feminæ distinctis thalamis gaudent.
Flores masculi & feminei in eadem specie.
21. MONOECIA. | 23. POLYGAMIA.
22. DIOECIA.
CLANDESTINÆ.
Nuptiæ clam instituuntur.
Flores oculis nostris nudis vix conspiciuntur.
24. CRYPTOGAMIA.

1

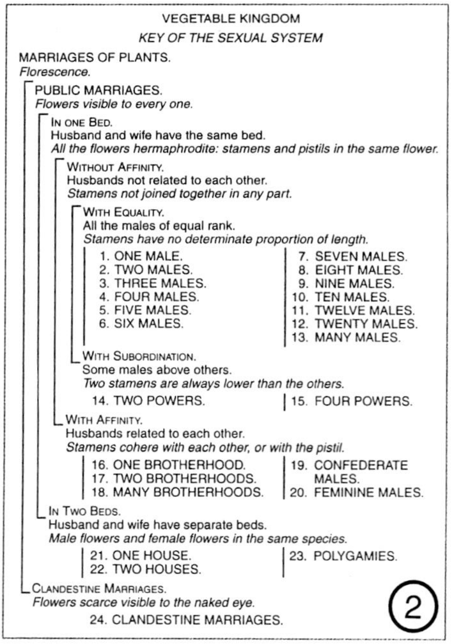

FIGS 1–2. (1) The exclusively reproductive high-level plant classification of Linnaeus (1759: 837), and (2) a relatively free translation into English by Turton (1802) (after Schiebinger, 1996: 101). Although attractively simple, this famous classification is also both highly typological and highly artificial.

We have chosen to address this problem using the now well established cladistic approach to reconstructing phylogenies (evolutionary histories: Hennig, 1966; Farris, 1982; Funk and Brooks, 1990; Wiley, *et al.*, 1991; Forey, *et al.*, 1992). Selected taxa are scored for a range of characters. Molecular characters, notably nucleic acid sequence data, are increasingly used in phylogenetic analysis (e.g. Avise, 1994; Givnish and Sytsma, 1997), but here we discuss only morphological data. This paper is concerned primarily with the phenotype – the basis for both species identification and for the interaction of organisms with other organisms and the environment.

Each morphological cladistic character possesses two states, 0 and 1 (or, increasingly frequently, more than two states; e.g. 0–1–2), that are assumed *a priori* to be homologous (e.g. Patterson, 1982). Their homology is then subjected to the congruence test (Patterson, 1988), using a parsimony programme (e.g. Huelsenbeck and Hillis, 1993) to find the repeatedly dichotomous tree (or set of trees) that requires the fewest transitions between the states of all the characters analysed. One or more coded taxa are selected as an outgroup, and used to simultaneously root the tree and polarise each character into a primitive (plesiomorphic) and a derived (apomorphic) state. The branching pattern (topology) of the resulting tree depicts closeness of evolutionary relationship.

In all but the smallest matrices the tree-building procedure inevitably reveals incongruence among characters; in order to depict some characters as undergoing only one transition others must change more than once, undergoing either parallelism (simplest case 0 > 1, then 0 > 1) or reversal (simplest case 0 > 1, then 1 > 0). Such multiple transitions between the same pair of character states across a single tree are collectively termed **homoplasy** (e.g. Sanderson and Donoghue, 1989; Sanderson and Hufford, 1996; Givnish and Sytsma, 1997).

Here, we use statistical analyses of patterns among lineages of homoplasy as a proxy for relative levels of character conservation among different organs, both within and between lineages (an approach originally advocated, though not attempted, by Sanderson and Donoghue, 1989: 1792). This in turn is viewed as an indicator of the relative taxonomic merits of particular plant organs, both generally across the plant kingdom and within specific clades. The strength and accuracy of the phylogenetic signals from individual organs is also examined by submatrix re-analysis followed by topological comparison. As well as testing the almost universally assumed pre-eminence of reproductive characters in systematics, we also seek evolutionary explanations of the observed patterns of relative frequencies of character change, comparing (1) different organs in the same clade and (2) the same organ among different clades.

Background to the specific case-study

Bateman, *et al.* (1992) presented a detailed cladistic analysis of a largely extinct group of highly derived and ecologically specialised heterosporous pteridophytes, the rhizomorphic lycopsids (Lycopsida: Isoetales s.l.). The coded taxa were 16 conceptual whole-plant species that had been painstakingly reconstructed from fossilised disarticulated organs, anatomically preserved as petrifactions (Chaloner, 1986; Bateman, 1992b; Phillips and DiMichele, 1992; Stewart and Rothwell, 1993), plus a hypothetical ancestor as a *de facto* outgroup. Of the 115 exclusively bistate characters coded, eight were holapomorphic *sensu* Bateman, *et al.* (1992) (delimited the entire ingroup) and 27 autapomorphic (restricted to a single coded taxon). The remaining 80 characters (70%) proved phylogenetically informative at that hierarchical level of analysis.

Subsequent unpublished studies further investigated this cladistic matrix in order to determine (1) the phylogenetic contributions of single-celled spores relative to multicellular tissues (Bateman, 1992a), and (2) the minimum number of organs required to place a previously unknown, disarticulated rhizomorphic lycopsid in the correct phylogenetic position (Bateman, 1992b). Lastly, Bateman (1994, table 4) compared levels of homoplasy in characters reflecting the highly variable overall architectures of the whole plants versus smaller-scale characters reflecting seven individual organs/organ aggregates (including spores), as part of a more elaborate published account of the implications of the preferred phylogeny for radical, developmentally-mediated evolutionary changes in overall plant architecture.

Here, the original rhizomorphic lycopsid matrix of Bateman, *et al.* (1992, table 3) was again divided into non-overlapping subsets of characters, each subset describing either the overall architecture of the plant or one of six definable constituent organs or organ aggregates. Non-reproductive organs were represented by rootstock (rhizomorph plus rootlets), vegetative axes (stems plus lateral and crown branches), and leaves. Reproductive organs were represented by cones *sensu lato* (strictly, the cone axis–sporophyll–sporangium assembly), megaspores, and microspores. Assignment of characters to categories largely followed Bateman, *et al.* (1992), though three characters describing large-scale features of the rootstocks (Ch8–10) were transferred from the rootstock category to architecture/habit in this study. The resulting seven subsets varied greatly in number of characters, from three for the rootstocks to 52 for the vegetative axes; this especially character-rich subset was thus further subdivided into four anatomical cylinders of tissue, listed centrifugally: stele, cortex, periderm, and leaf bases.

Two contrasting methods were developed to investigate the subsets of the original data matrix: the first approach required re-analysis of the subsets via parsimony algorithms, whereas the second approach investigated the homoplastic properties of the subsets using the output of the original whole-matrix parsimony analysis.

Approach 1: Multiple re-analyses of matrix subsets followed by comparison of consensus tree topologies

Method 1 treats the suite of most-parsimonious cladograms generated from the data matrix for all coded organs as a relative truth (or at least an empirical yardstick), on the grounds that it contains the maximum amount of phylogenetic evidence available to the analysts (cf. Bull, *et al.*, 1993; Eernisse and Kluge, 1993; de Queiroz, *et al.*, 1995; Miyamoto and Fitch, 1995; Huelsenbeck, *et al.*, 1996). Subsets of the original data matrix are then re-analysed using the same protocol in the same parsimony program, each subset generating an additional suite of most-parsimonious cladograms whose topologies can be compared with those generated from the original, all-organ matrix (cf. Figs 3–11).

In order to simplify such comparison, each suite of cladograms (i.e. for the all-organ matrix and each submatrix) is summarised as a single consensus tree. Semi-strict consensus (combinable component consensus *sensu* Bremer, 1990) was selected here, as it offers potentially greater topological resolution than the more popular Strict consensus but avoids the greatest weakness of the Adams and Majority Rule algorithms – namely, it cannot yield a consensus topology that contradicts the topologies of at least some of the original constituent cladograms (Bremer, 1990; Wiley, *et al.*, 1991; Forey, *et al.*, 1992).

Here, we focus on two aspects of the submatrix consensus trees: (1) The number of resolved nodes present (spots plus circles on Figs 4–11) indicates the **strength** of phylogenetic information provided by that organ. (2) Whether the clades delimited by those resolved nodes on the submatrix tree also occur in the all-organ tree (spots but not circles in Figs 4–11; cf. Fig. 3) indicates the **accuracy** of phylogenetic information provided by that organ. In this context, an unresolved polytomy (an inevitable factor of a consensus tree, which reflects either the absence of supporting character states in that region of the tree or incongruence between the topologies of the two or more constituent cladograms) is regarded as preferable to an incorrectly resolved node – that is, a node that contradicts the "correct" topology of the all-organ consensus tree. As defined here, a node is deemed incongruent even if only one coded taxon is either missing from, or added to, a large submatrix clade, relative to the corresponding clade in the full-matrix tree.

Although attractively simple, this logical threshold of topological congruence is also undesirably nihilistic. To cite the most extreme example in this case-study, in Fig. 7 the invalid addition of the single *Paralycopodites* species to the valid *Hizemodendron–Lepidodendron–Lepidophloios* clade nullifies the correct phylogenetic placement of the other five species, requiring the shared node to be labelled as wholly incorrect rather than merely 17% imperfect. An additional problem was encountered in the stele category within the vegetative axes subset, which yielded a node that resolved the relationships of the three *Lepidophloios* species shown as unresolved in Fig. 3. This additional node in the stelar tree is neither congruent nor incongruent with the "true tree", and hence was designated neutral.

Approach 2: Single full-matrix analysis followed by comparison of subsets of values for homoplasy indices

Consistency index and retention index

As outlined by Bateman (1994, table 4), the alternative approach to comparing the relative strengths of the phylogenetic contributions of different organs from the same species uses two homoplasy indices that are now output as standard from all mainstream cladistic programs: the consistency index and the retention index. Note that each index is calculated for each character, but that they are also recalculated for combinations of characters – most commonly for the entire matrix – as **ensemble** indices (Wiley, *et al.*, 1991; Forey, *et al.*, 1992; Swofford, *et al.*, 1996).

The **consistency index** (**CI**: Kluge and Farris, 1969) of each character is simply the minimum possible number of steps on the tree (**m**: for a bistate character this always has a value of one) over the actual number of steps on the tree (**s**) (Fig. 12). It is now widely recognised that this index has two main drawbacks:

(1) As noted by Farris (1989), the numerator always has a value of at least one and so CI cannot have a value of zero (e.g. even the poorly performing Ch5 on Fig. 12 has a CI of 0.33). As well as the deleterious inability to span unit variance, this property means that CI values are strongly influenced by the size and structure of the primary matrix – most notably by the number of coded taxa (Archie, 1989; Sanderson and Donoghue, 1989, 1996; Meier, *et al.*, 1991; Archie, 1996).

(2) CI takes no account of the **potential** for homoplasy shown by each character, which depends on the relative frequency of each state of that character in the primary matrix. For example, in Fig. 12, Ch2 delimits a single coded taxon (i.e. it is an autapomorphy) and Ch6 delimits the entire ingroup (it is a holapomorphy *sensu* Bateman, *et al.*, 1992: 519). Both autapomorphies and holapomorphies by definition cannot show homoplasy and thus reliably have a CI of 1.0. They are designated "uninformative" in the phylogenetic package PAUP (Swofford, 1993) and are often deliberately omitted from calculation of ensemble CI's (Brooks, *et al.*, 1986; Carpenter, 1988; Yeates, 1992; Givnish and Sytsma, 1997). More subtly, Ch3 scores "1" for two coded taxa but Ch4 scores "1" for three coded taxa, and thus statistically offers more opportunity for homoplasy (this is greatest for a bistate character such as Ch4 that scores "0" for half the coded taxa and "1" for the other half: see Meacham, 1981; Sanderson and Donoghue, 1989, 1996).

In an attempt to remedy this deficiency, Farris (1989) developed the **retention index** (**RI**). This equation inherited the variables **m** and **s** from the CI, but introduced a new parameter, **g** – the maximum number of possible steps for the character on a fully unresolved tree of the coded taxa (Fig. 12, inset). Thus, on Fig. 12, values of **g** are 1 for Ch2 (the single "1") and Ch6 (the single "0"), 2 for Ch3 (the two "1"s), and 3 for Ch1, Ch4 (the three "1"s) and Ch5 (the three "0"s).

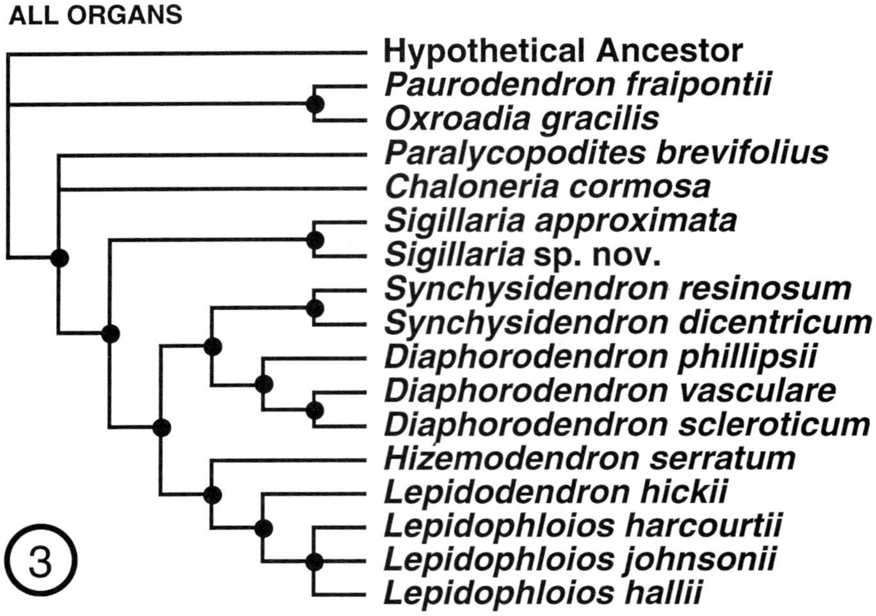

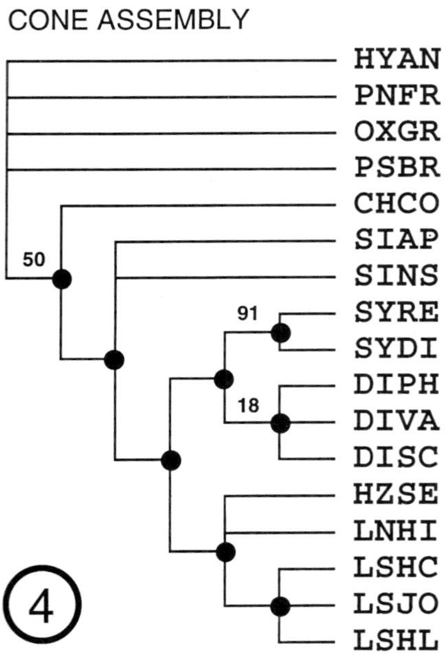

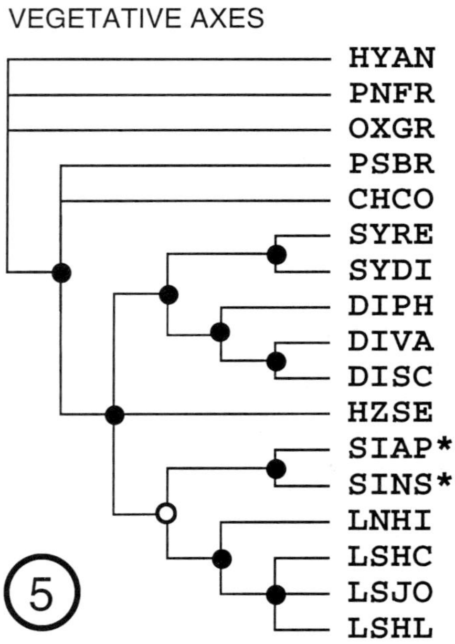

FIGS 3–11. Phylogenies of the rhizomorphic lycopsids (determinate clubmosses: Bateman, *et al.*, 1992, table 3), based on a cladistic matrix of 115 bistate characters describing 16 species of 10 genera. Semi-strict consensus trees summarise all of the most-parsimonious cladograms from the original matrix (3) and from submatrices representing cone assemblies (4), vegetative axes (5), leaf bases (the outermost subset of the vegetative axes set: 6), microspores (7), overall architecture (8), leaves (9), megaspores (10) and rootstock (rhizomorph plus rootlets: 11). Note not only the relative degrees of

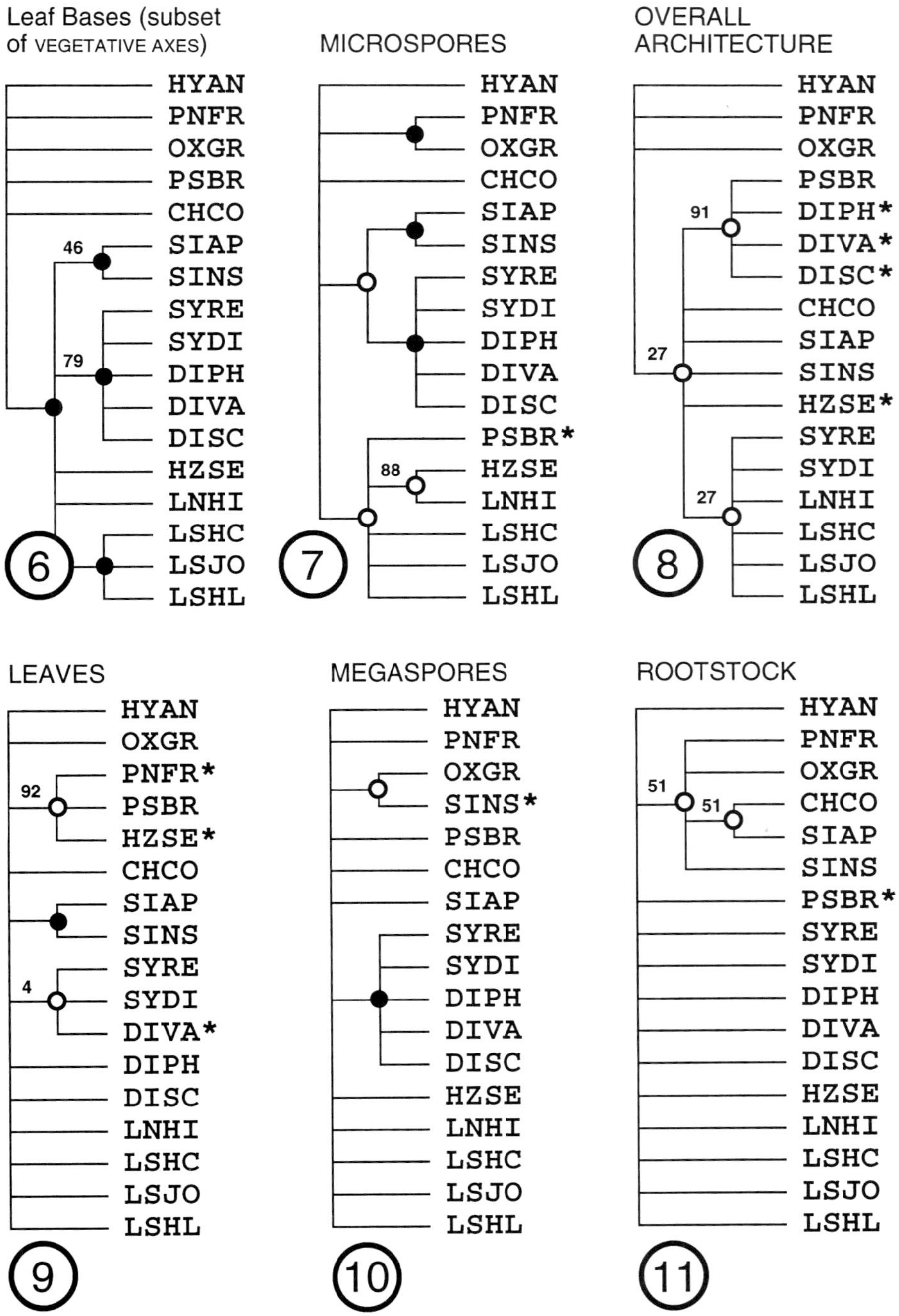

resolution in the submatrix phylogenies (spotted plus circled nodes) but also the relative degrees of topological congruence with the "correct" phylogeny based on the full matrix (spotted nodes only). Figures indicate the percentage of the constituent cladograms containing that clade (resolved nodes lacking percentages occur in all consituent trees). Asterisked species deviate in position from the baseline vertical linear sequence given in Fig. 3.

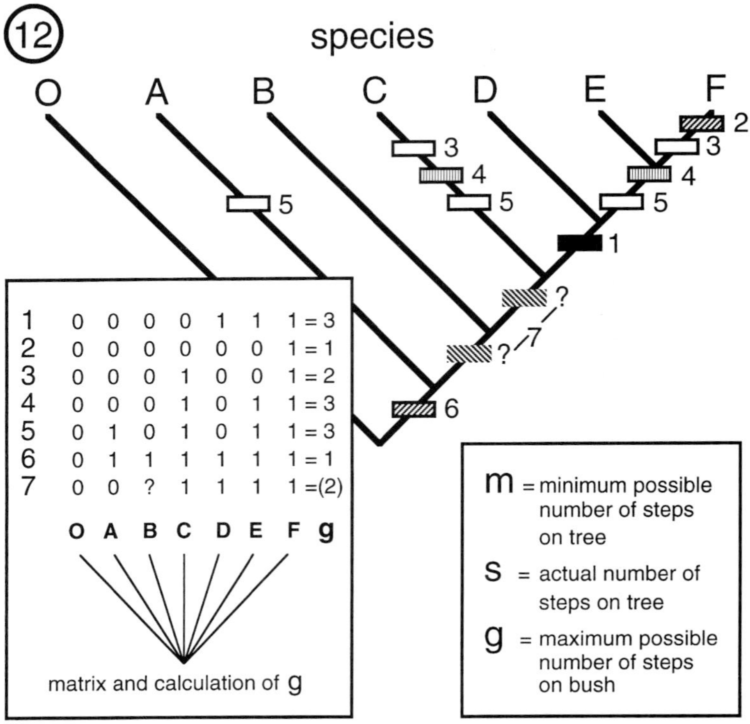

CHARACTER	Consistency Index (CI) = m/s	Retention Index (RI) = (g-s)/(g-m)	Rescaled Consistency (RC) = CI x RI
1	1/1 = 1.0	3-1/3-1 = 1.0	1.0 x 1.0 = 1.0
2	1/1 = 1.0*	1-1/1-1 = 0	1.0 x 0 = 0
3	1/2 = 0.5	2-2/2-1 = 0	0.5 x 0 = 0
4	1/2 = 0.5	3-2/3-1 = 0.5	0.5 x 0.5 = 0.25
5	1/3 = 0.33	3-3/3-1 = 0	0.33 x 0 = 0
6	1/1 = 1.0*	1-1/1-1 = 0	1.0 x 0 = 0
7	1/1 = 1.0	2-1/2-1 = 1.0	1.0 x 1.0 = 1.0

FIG. 12. Hypothetical phylogeny of one outgroup (O) and six ingroup species (A–F) illustrating the behaviour of seven selected bistate characters when measuring the three parameters (**m**, **s** and (inset) **g**) needed to calculate three routine indices of homoplasy: the consistency index (CI), the retention index (RI) and their product, the rescaled consistency (RC). Asterisked characters either delimit a single species (Ch2) or the entire ingroup (Ch6); they can be designated "invariant" in the cladistic program PAUP and thereby excluded from consistency index calculations. In this example, **m** and **g** have maximum values of 1 and 3 respectively for all seven characters. The value of **g** for the ambiguously placed Ch7 is in theory dependent on whether the missing value is optimised as a "0" (when **g** = 3) or a "1" (when **g** = 2).

Missing values, such as that shown for Ch7 in taxon B, complicate the calculation of **g**, as in theory they depend wholly on the choice of optimisation algorithm *sensu* Swofford, *et al.* (1996). For example, Deltran optimisation would replace the "?" in Ch7 with a "0" (giving **g** = 3 based on the three "0"s), whereas Acctran would replace it with a "1" (giving **g** = 2 based on the two "0"s). In practice, the list of **g** values produced by PAUP always minimises **g**, irrespective of the results of optimisation. This leads to potential conflicts with any **g** values calculated by the user from the optimised tree; for example, in the rhizomorphic lycopsid case-study, the leaf-base subset within vegetative axes received an ensemble RI of 0.83 from PAUP but of 0.87 from hand calculations based on the MinF optimisations documented in table 3 of Bateman, *et al.* (1992).

The **g** value is used in both numerator and denominator, beneficially conferring on RI unit variance: unshared invariant characters such as Ch2 and Ch6 (autapomorphies and holapomorphies respectively) reliably score zero and shared characters changing only once on the tree such as Ch1 reliably score unity. Characters showing intermediate levels of homoplasy differ most in behaviour between CI and RI (see also Goloboff, 1991: 224–225; Givnish and Sytsma, 1997). For example, Ch4, which unites taxa E and F but also occurs in taxon C, scores 0.5 for both CI and RI. However, Ch3, occurring as a parallelism in taxon C and taxon F, also scores 0.5 for CI but zero for RI, as it does not unite any coded taxa as a potential synapomorphy; here, the performance of RI seems preferable to that of CI. However, the most revealing character is Ch5, which scores 0.33 for CI and zero for RI; even though it unites taxa E and F, the apparent synapomorphy is regarded as statistically due to chance (Ch5 has a **g** value of 3, the maximum for this matrix, and actually changes three times on the tree). This interpretation of such a character could be regarded as rather stringent – certainly, cladists interpreting cladograms rarely zero weight such partial synapomorphies *a posteriori* on the grounds that they have an RI of zero (for discussions of character weighting in cladistics see Carpenter, 1988; Sharkey, 1989; Wilkinson, 1991; Goloboff, 1993; Chippindale and Wiens, 1994).

Nonetheless, RI has the advantage of being less dependent on primary matrix size than is CI, though RI too is negatively correlated with the number of characters and positively correlated with the number of coded taxa (Meier, *et al.*, 1991). Also, Goloboff (1991: 218) argued that RI is less suitable than CI for comparing homoplasy among (rather than within) primary matrices, because if, for a given matrix, "no tree can require much homoplasy, but the most parsimonious tree requires a considerable fraction of that homoplasy, RI may be low despite low homoplasy" (see also Sanderson and Donoghue, 1996; Givnish and Sytsma, 1997).

Farris (1989) further suggested that RI values should be used to rescale CI values by calculating their product, the mis-named **rescaled consistency** (**RC**). Bateman (1994: 577) argued that this is the best overall measure of homoplasy as it averages out the deficiencies of both its component indices, but Fig. 12 (bottom) demonstrates that this is not necessarily the case. Firstly, the zero values for RI cancel out positive CI values for potentially phylogenetically informative characters such as Ch5. Secondly, characters such as Ch4, with positive but fractional values for both CI and RI, are effectively down-weighted when those values are multiplied to yield the arithmetically reduced RC value. For example, does the performance of Ch4 on the tree in Fig. 12 really justify its down-weighting to one quarter of the phylogenetic value of Ch1?

It now seems to us that CI and RI measure subtly but significantly different aspects of character performance; each has strengths and weaknesses, and they are better compared in the light of these properties rather than subsumed into the RC.

Other indices

Other, more complex indices have been suggested for assessing relative levels of homoplasy (see recent reviews by Archie, 1996; Chang and Kim, 1996; Givnish and Sytsma, 1997). In practice, Archie's (1989) homoplasy excess ratio maximum (HERM) differs little from RI (cf.

Farris, 1989, 1991; Archie, 1990) and remains deleteriously negatively correlated with the number of coded taxa analysed (Meier, *et al.*, 1991: fig. 2). Archie's (1989) homoplasy excess ratio (HER) was based on a randomisation procedure and thus is both computationally expensive and applicable only to matrices rather than individual characters. Inapplicability to individual characters also devalues the alternative measure suggested by Meier, *et al.* (1991), the homoplasy slope ratio (HSR). The homoplasy slope is based on the relationship between the number of coded taxa and ES: the number of extra steps required by the character across the most-parsimonious tree relative to zero homoplasy (ES = 1/[CI–1] for bistate characters). Randomisation tests of the primary matrix can be used to filter out the effect on homoplasy measurement of character number (or, more accurately, the total number of apomorphic states), yielding the HSR, which resembles HER but is less influenced by matrix size. Unfortunately, like the far simpler CI, HSR is insensitive to the internal structure of the data in general and in particular the potential of the matrix to express homoplasy – specifically, the **g** factor that underpins RI (Meier, *et al.*, 1991).

Searching for a measure that reflected the decisiveness (broadly, the ability to select among trees) of a matrix rather than homoplasy *per se*, Goloboff (1991) took the equation for RI (Fig. 12) and substituted **S** for **g**, where **S** is the sum of lengths of all possible trees derived from the matrix over the arithmetically calculated theoretical maximum number of trees which can be generated by that number of coded taxa. The resulting data decisiveness (DD) value measures the relative range of tree lengths, large ranges (and DD values) indicating more decisive matrices. However, DD is fundamentally a tree-comparison rather than character-comparison measure; although an interesting concept, its applicability to comparing phylogenetic signals from different organs of the same coded taxa is unclear. Goloboff (1991) provocatively noted that the decisiveness of a matrix would probably reliably be increased if **both** ensemble CI and ensemble RI increased.

Several authors have attempted to quantify the distribution of homoplasy across the resulting most-parsimonious trees (Sanderson, 1990, 1991; Donoghue and Sanderson, 1992; Sang, 1995). Sang (1995) suggested that ensemble CI values for matrices should be supplemented with mean values for the CI's of the individual characters to yield the average unit character consistency (AUCC). This inevitably exceeds the CI in value. For matrices of equal ensemble CI values, the AUCC value is greatest when the distribution of homoplasy among the characters is most asymmetric (i.e. concentrated in the fewest characters), a logical extension of character compatibility analysis (e.g. Meacham, 1981) into the realm of parsimony. When comparing matrices of different CI values, Sang advocated standardisation of the AUCC by subtracting from it the ensemble CI to generate the homoplasy distribution index (HDI). This in turn can be made independent of absolute levels of homoplasy by dividing HDI by [1-CI], to yield the homoplasy distribution ratio (HDR). However, HDR attempts to summarise so many aspects of homoplasy that it becomes questionably interpretable (e.g. see table 2 of Sang, 1995).

Lastly, Chang and Kim (1996) presented a probabilistic homoplasy measure (*H*), based on a simple Poisson model for bistate characters, that was designed to determine the probability that taxa share character states by chance rather than through shared descent.

Preferred approach

Thus, the student of homoplasy is presented with a bewildering array of indices. HER and HSR require randomisation and thus are properties of the whole matrix rather than of individual characters, as is data decisiveness. They are complex to calculate and at the time of writing are not provided in the routine output from the major cladistics packages. In contrast, CI, RI and their product, RC, are applicable to both whole matrices and individual characters and are routinely produced by cladistic software such as PAUP 3. They are deceptively simple in calculation but their all-important properties relative to matrix size and structure remain surprisingly poorly understood. Moreover, both indices are adversely affected by missing values in the matrix (Meier, *et al.*, 1991; see also Nixon and Davis, 1991; Bateman, *et al.*, 1992: 519–524). Goloboff (1991) argued that CI is best suited to comparing homoplasy among matrices for different sets of coded taxa, whereas RI is most effective for comparing different trees generated from the same matrix. Unfortunately, like all other commentators, he did not address the key question of which algorithm is most appropriate for problems of intermediate hierarchical scale – in this case, comparing relative homoplasy in different submatrices (organs) of the same matrix (whole plants).

For analytical speed and simplicity we have chosen to focus on CI and RI rather than more complex indices; we have not emphasised one at the expense of the other as their properties and constraints differ significantly. However, we have largely ignored their combined product, RC, as it inherits the weaknesses of each of the two contributing indices.

Materials and Methods

The full, all-organ matrix for each case-study was input into MacClade 3.0 (Maddison and Maddison, 1992), retaining the character codings and outgroup selections of the original authors. The entire matrix was then re-analysed using PAUP 3.1.1 (Swofford, 1993), deviating from the default settings only in excluding uninformative characters. Branch-and-bound searches were preferred, but time constraints often obliged us to resort to heuristic searches using the default TBR branch-swapping. In order to conduct Approach 2, lists of **m**, **s** and **g** were printed out for all informative characters, and ensemble CI and RI (plus RC) values were calculated manually for each character subset (i.e. for each organ). Semi-strict consensus trees were also printed as the yardsticks for Approach 1. Each character subset was then defined using the CHARSETS option of PAUP, and subject to a separate parsimony analysis using the same settings as the full-matrix analysis. Each submatrix analysis was also summarised as a semi-strict consensus tree, for comparison with the results of the full matrix analysis.

We also gathered from the literature over the period 1985–1996 a database of 129 morphological cladistic matrices, representing the widest possible range of vascular plant taxa. Each matrix was subjected to Approach 2, using the character coding and outgroups specified by the original authors but imposing our own (often overlapping) organ-based subcategories. Full matrices where the ingroup was not clearly monophyletic were discarded, as were those miraculously devoid of homoplasy. Submatrices that failed to complete on an overnight heuristic search, or contained less than five characters, were also discarded from further analysis, leaving 98 usable studies. Full details of the literary sources of the matrices, analytical methods, and detailed results will be presented by us elsewhere. Here, we offer only a simple comparison of ensemble CI and RI values for the whole matrix relative to mutually exclusive reproductive and non-reproductive submatrices.

Results

Rhizomorphic lycopsid case-study

The results of the rhizomorphic lycopsid analysis are summarised in Fig. 14, together with the total number of characters and the number of phylogenetically informative characters for each subset as Fig. 13.

The semi-strict consensus tree for all organs (Fig. 3) reveals 12 resolved nodes out of a potential maximum of 14 for a fully resolved (i.e. wholly dichotomous) tree; a trivial polytomy

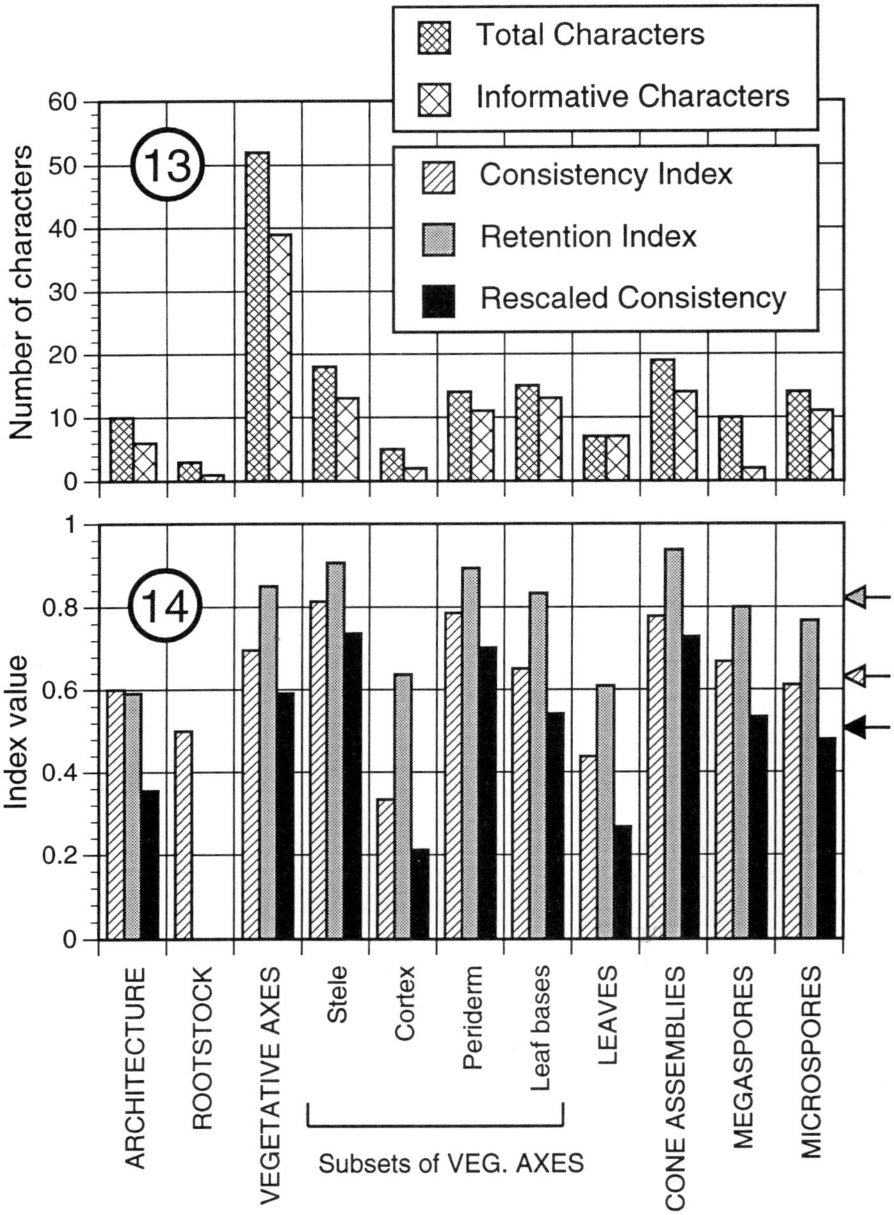

FIGS 13–14. Assessment based on patterns of homoplasy of the relative performance of overall architecture, three vegetative organs (rootstock, vegetative axes [followed by four subsets of this organ], leaves) and three reproductive organs (cones s.l., megaspores, microspores) in the phylogenetic analysis of the rhizomorphic lycopsids. (13) Total number of characters and number of phylogenetically informative characters in each organ category. (14) Performance of each organ category as assessed using the consistency index, retention index and their product, the rescaled consistency. Right-hand arrows indicate ensemble values for these three parameters in the full (whole plant) matrix.

subtends the three *Lepidophloios* species, and a more taxonomically consequential polytomy subtends *Paralycopodites* (the most primitive tree analysed), *Chaloneria* (the only included representative of the extant s.s. isoetalean lineage of shrubs/pseudoherbs), and the bulk of the remaining coded taxa.

The cone assemblies (Fig. 4) yielded eight resolved nodes, all congruent with the relative "true tree" of Fig. 3. They also yielded the highest CI and RI values of the seven main subsets (Fig. 14), despite having a modest total of 14 informative characters (Fig. 13).

The character-rich vegetative axes (39 informative characters: Fig. 13) yielded nine resolved nodes congruent with Fig. 3, plus one node that erroneously linked *Sigillaria* to *Lepidodendron–Lepidophloios*. Vegetative axes gave the second highest CI and RI values (Fig. 14). Analysis of the four additional subsets of the vegetative axes submatrix revealed consistently strong performances from stele, periderm and (to a lesser degree) leaf bases; only the very few informative characters for the cortex showed extensive homoplasy (Fig. 14). The consensus tree generated from the subset for leaf bases (Fig. 6), which traditionally are preferentially used in the taxonomy of the rhizomorphic lycopsids (e.g. Kidston, 1923–1925; Crookall, 1955–1976), revealed only four resolved nodes, though all are congruent with Fig. 3.

Spores yielded similar index values to the whole-organism matrix (Fig. 14, arrows). Microspores provided similar numbers of informative characters to leaf bases (Fig. 13) but marginally lower CI and RI values (Fig. 14). They generated a tree of five nodes, three compatible with Fig. 3 but two incompatible (that linking *Paralycopodites* to *Hizemodendron–Lepidodendron–Lepidophloios* is especially improbable). Megaspores yielded slightly higher CI and RI values than microspores, but these ensemble values reflect only two informative characters: one yields a true synapomorphy of *Synchysidendron* plus *Diaphorodendron*, the other a false synapomorphy of *Oxroadia* plus *Sigillaria* sp. nov. (Fig. 10). In contrast with the other organs, 80% of the megaspore characters were phylogenetically uninformative autapomorphies (Fig. 13).

The six informative characters representing overall plant architecture yielded a similar CI to the spore submatrices, but a significantly lower RI of similar value to the CI (Fig. 14). The resulting consensus tree revealed only three nodes (Fig. 8), none congruent with Fig. 3. The similar number of informative leaf characters yielded a similar RI but a much lower CI. Of three nodes, only one was congruent, and that was merely a trivial link between two closely related species (Fig. 9). Lastly, the rootstock was both under-described and deficient in variation within the group, hence yielding only one informative character out of the three scored, with a CI of 0.5 and an RI of zero. Both nodes evident in Fig. 11 are improbable.

Sanderson and Donoghue (1989: 1792) recommended that homoplasy should routinely be split into reversal and parallelism. Under the two extreme optimisation options (e.g. Swofford, *et al.*, 1996), the rhizomorphic lycopsid matrix yielded 20 parallelisms and 28 reversals with Acctran and 35 parallelisms and 13 reversals under Deltran; these figures average out as 57% parallelism. Ratios of parallelism to reversal are similar among different organs, though reproductive characters are more strongly influenced by the choice of Acctran versus Deltran.

Survey of many morphological cladistic matrices for vascular plants

A total of 65 of the 98 usable matrices in our vascular plant database yielded adequate data to plot reproductive values against whole-matrix values for CI (Fig. 18) and RI (Fig. 19). Points occurring above the diagonal line of equal value represent studies where reproductive characters show less homoplasy than non-reproductive characters. Perhaps the best method of relating the cloud of points to the 1 : 1 line is to calculate a centre of gravity and 95% confidence ellipse, but here we have selected the simpler method of comparing a regression line with the 1 : 1 line.

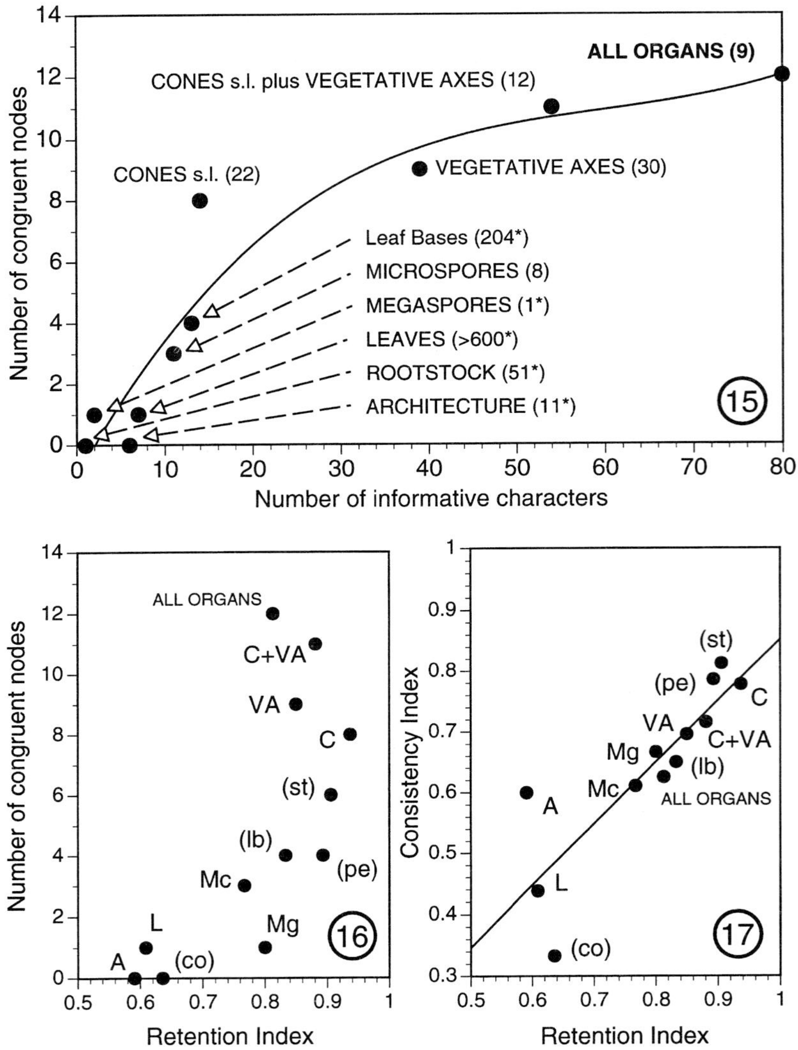

FIGS 15–17. (15) The apparent number of inclusive, wholly congruent clades (nodes) shared between the full matrix semi-strict consensus trees and each submatrix tree, plotted against the number of informative characters in each submatrix to assess the relative phylogenetic performance of each organ. Note than a good approximation of the "true" phylogeny can be obtained with only cones s.l. plus vegetative axes. Fitted curve is a best-fit third degree polynomial ($R3^2 = 0.50$, $R2^2 = 0.63$, $R1^2 = 0.81$, $R0^2 = 0.91$). The number of most-parsimonious trees is given in parentheses; asterisked re-analyses were restricted by time constraints to heuristic rather than branch-and-bound searches. Figs 16 and 17 exclude the single informative rootstock character but include all four zones of the vegetative axes subset: stele (st), cortex (co), periderm (pe) and leaf bases (lb). (16) The number of congruent nodes determined via the comparative topology approach contrasted with the retention index determined via the homoplasy index approach. Note the indifferent ($R^2 = 0.48$) positive correlation. (17) The two main variables resulting from the comparative index approach; note the stronger ($R^2 = 0.74$) positive correlation.

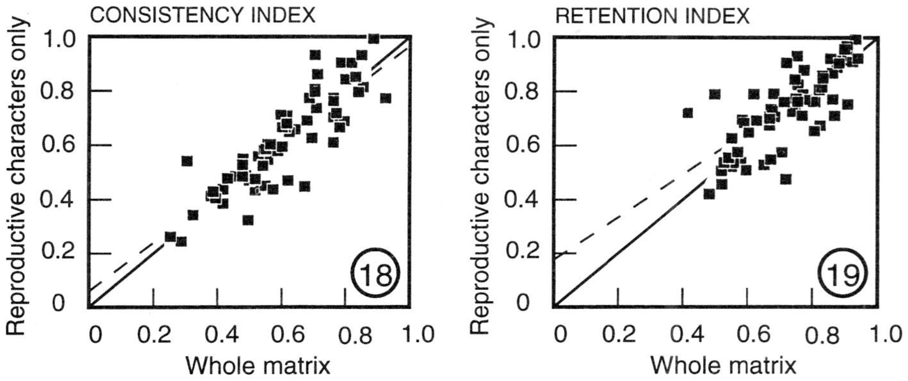

FIGS 18–19. Preliminary results of a homoplasy index survey of 129 published morphological cladistic matrices (65 plotted) of vascular plants. Ensemble index values for the whole matrices (i.e. reproductive plus non-reproductive characters) are plotted against ensemble index values for reproductive characters alone, for consistency index (18) and retention index (19). Note that the clouds of points and regression lines (dashed) occur only just above the 1:1 line (solid) in both cases, suggesting that reproductive characters are on average only slightly less homoplastic, and thus only slightly more phylogenetically conserved, than non-reproductive characters.

CI has smaller minimum and mean values than RI (ca. 0.2 versus ca. 0.4 and ca. 0.55 versus 0.7 respectively) and much less dispersion about the regression line (R^2 = 0.75 versus 0.53), indicating that RI is the more variable (and arguably more sensitive) measure of contrasts in homoplasy **within** matrices (cf. Figs 18, 19; Givnish and Sytsma, 1997, argued that RI is simply more vulnerable to noise in the data-matrix). For both CI and RI, the regression line lies just above, and approximately parallels, the 1 : 1 line (admittedly, the regression line for RI intersects the "y" axis above the origin, due largely to two outlying points with low whole-matrix RI values but average reproductive submatrix RI values). This suggests that, when averaged across all 65 matrices, reproductive characters show slightly less homoplasy, and thus are slightly more evolutionarily conserved, than non-reproductive characters. This relationship was expected, but the trend is surprisingly weak.

Discussion

Methodological observations

The rhizomorphic lycopsid case-study reveals a strong positive correlation between CI and RI among the organ-based subsets (R^2 = 0.74; Fig. 17); also, the mean value per organ for RI is 21±13% greater than for CI (excluding the character-deficient rootstock). This could be taken to indicate the amount of expressed homoplasy (CI) is low relative to the potential for its expression (RI). However, the multi-matrix survey suggests that higher RI than CI values within matrices is a more general trend (cf. Figs 18, 19) and that the behaviour of the rhizomorphic lycopsid matrix is average for both homoplasy indices, given a matrix of that size (see also the surveys of CI in Sanderson and Donoghue, 1989, fig. 1; and of both CI and RI in Meier, *et al.*, 1991, fig. 2; Sanderson and Donoghue, 1996; Givnish and Sytsma, 1997).

Of the three most homoplastic subsets, the microphyllous leaves lie firmly on the regression line (Fig. 17); they are simply poor performers. In contrast, the cortical suite from the vegetative axes subset has a lower than expected CI, but the sample of characters (two) is too small to draw meaningful conclusions. However, the performance of the six informative characters describing overall architecture – the largest scale features of the plants – is of considerable interest. Only overall architecture has similar values for CI and RI (Figs 14, 17); thus, absolute homoplasy (CI) is not especially high relative to other subsets, but the proportion of homoplasy relative to the potential inherent in the characters (RI) lies well to the left of the regression line. We conclude that, among the rhizomorphic lycopsids, architectural characters are especially adept at exploiting opportunities for homoplasy (see also Bateman, 1994, 1997).

The broadly similar patterns evident in Fig. 17 between CI and RI are not surprising, given that both measures were determined during the same suite of parsimony analyses (Approach 2). It is more illuminating to compare the results of Approach 2 with Approach 1. In fact, there is a poorer correlation between the number of congruent nodes identified using Approach 1 and the level of homoplasy (represented by RI in Fig. 16) identified using Approach 2. Both methods highlight overall architecture and leaves as especially phylogenetically unhelpful, but among the remaining organs differences in the homoplasy index values of Approach 2 are relatively small. In contrast, Approach 1 clearly separates cones and vegetative axes as far more phylogenetically valuable than the remaining organs.

Compared with cones and vegetative axes, all other organs/organ aggregates, together with overall architecture, yielded fewer informative characters, lower homoplasy indices, and few if any fully congruent nodes. Moreover, these submatrices were more difficult to analyse in PAUP; those with few informative characters relative to the number of coded taxa and/or with little internal hierarchical structure required long run times and a switch from branch-and-bound searches to heuristic searches with TBR branch-swapping, which are not guaranteed to find all the most-parsimonious trees (e.g. Maddison, 1991; Swofford, 1993). In addition, leaves, leaf bases and rootstocks each yielded many most-parsimonious trees (Fig. 15), reflecting the weak internal structure to the data.

Overall, the two main approaches used here to analyse the submatrices – full matrix re-analysis followed by topological comparison with the product of the full matrix, and calculation and comparison of submatrix ensemble homoplasy indices – yielded broadly similar results in terms of **ranking** the relative contributions of the organs to the whole-plant phylogeny (Fig. 16). However, there are significant **arithmetic** differences between the results of the two methods. Specifically, there is a continuum of variation in ensemble RI values over a fairly narrow range (from 0.59 in overall architecture to 0.94 in cones), whereas there is a much wider range in terms of congruent nodes (from none in architecture and rootstock to nine out of a possible 12 in vegetative axes). Only the nodal approach offers a clear discontinuity between the high-performing cones and vegetative axes on the one hand and far less phylogenetically valuable cluster of additional organs on the other hand (Fig. 16). Note also that the stele and periderm have especially high RI values, but yielded only six and four congruent nodes respectively, performing far less well individually than when combined with the relatively homoplastic leaf bases and cortex to form the vegetative axes submatrix (Fig. 16). The degree of accurate phylogenetic resolution offered by each organ/organ aggregate in Approach 1 appears to be more influenced by the degree of hierarchical structure inherent in the data than by homoplasy *per se*.

We conclude that both methods of analysing the relative strengths of phylogenetic signals from organs yield useful results, though we accept that both would be improved by the advent of more sophisticated comparative algorithms. Approach 1 may benefit from the continued development of software for comparing biogeographic area cladograms (e.g. Page, 1993), and Approach 2 from measures that better combine both potential and realised homoplasy. For any in-depth study of a single well founded morphological cladistic matrix, we recommend that both methods should be applied, as here. However, for a broader survey of phylogenetic patterns among many matrices grazed from the literature, the homoplasy index method, originally pioneered by Sanderson and Donoghue (1989), is more practical; in particular, it is far less time-consuming and more concisely visually summarised.

Causal interpretations and evolutionary scenarios

Despite being largely extinct, the rhizomorphic lycopsids lend themselves to inter-organ comparison because the individual organs are relatively discrete and the ontogeny of the entire organism is effectively determinate, an unusual condition for such large-bodied plants (Bateman, *et al.*, 1992; Phillips and DiMichele, 1992; Bateman, 1994). Although the data matrix could be improved with further study of leaf and rhizomorph anatomy, and of spore ultrastructure (e.g. Hemsley, 1992), it gives a reasonably comprehensive overall morphological and anatomical description of the component organs of each whole-plant species. The study also offers an unusually high ratio of the number of apomorphic states in informative characters over the number of ingroup taxa (80 : 16), so that most organs are represented by statistically acceptable samples of characters for the calculation of subset ensemble homoplasy indices or the production of subset consensus trees.

Among the rhizomorphic lycopsids the lowest homoplasy values – and thus the highest degree of character conservation – occur in the vegetative axes and the cone–sporophyll–sporangium assemblies. Spores of both genders show intermediate levels of homoplasy, whereas leaves and overall architecture are most homoplastic. Several explanations for this pattern are possible (see also Bateman, *et al.*, 1992; Bateman, 1994). Firstly, vegetative axes and cones are the most complex organs of the plants, and hence required the largest number of bistate characters for comprehensive description. Thus, their phylogenetic robustness could simply reflect weight of numbers of characters. This factor undoubtedly contributed to their success in generating resolved topologies from their respective submatrices (Figs 4, 5), but it does not wholly explain the fact that those topologies match closely the full-matrix topology (Fig. 3); this congruence also requires both low homoplasy and strong hierarchical structure at a wide range of taxonomic levels. Alternatively, their high complexity, as measured by the diversity and spatial relationships of constituent tissues, may allow a much greater range of phenotypic expression, reducing the statistical probability of convergence – or at least of making erroneous *a priori* assertions of homology (see also McShea, 1996). In contrast, spores are by definition single cells; their phenotypic expression is confined to aperture morphology (varying little among the rhizomorphic lycopsids), ultrastructure (not considered here) and exine ornamentation. These phenotypic constraints make convergence among distantly related species both highly probable and extremely difficult to detect *a priori* (Bateman, 1992a).

To seek a deeper causal explanation, the relative complexity of these phylogenetically crucial organs could reflect their putatively high functional importance. The vegetative axes perform much of the housekeeping physiology of the plant, particularly given the absence from the clade of megaphyllous leaves and secondary phloem (Phillips and DiMichele, 1992). The cones are both exceptionally complex and functionally specialised by pteridophytic standards, and show a clear evolutionary trend of increasing reproductive sophistication; they control the strongly contrasting life-history strategies of the species (Bateman, *et al.*, 1992; Phillips and DiMichele, 1992). Their vital functions could render experimental morphologies inviable and thus lead to strong character conservation (high burden *sensu* Riedl, 1979), reflected in a strong hierarchical relationship of the characters and low homoplasy. However, one could equally argue that features routinely exposed to the external environment, such as microspores, are at the front line of the plants' battle with the elements and hence **more** likely to undergo repeated and reversible evolutionary modification. Choosing between such contrasting scenarios will require detailed analysis of a much wider range of vascular plant clades.

In this context, it is notable that the more homoplastic discrete organs – spores and leaves – show serial homology; they are small organs that are produced *en masse* by each individual and happily discarded during ontogeny, often during annual cycles. More importantly, they therefore terminate ontogenetic trajectories, and so carry less burden; they can be evolutionarily modified without affecting other organs located downstream in the developmental cascade. Unfortunately, even this hypothesis does not explain the high levels of homoplasy also shown by the **largest** scale features of the plants – the rather heterogeneous category of characters termed overall architecture. Here, homoplasy is genuinely rampant, and it seems that only the characteristic bipolar embryogeny into root and shoot was evolutionarily sacrosanct (Bateman, 1994, 1997).

These observations on the phylogeny of the rhizomorphic lycopsids, and the underlying causal explanations that they have prompted, may apply to a much wider range of taxa. Our survey of other botanical cladistic matrices offers a much greater body of evidence, though meaningful inter-matrix comparison has been hindered by several difficulties, notably profound contrasts in methods of data matrix compilation. More in-depth case-studies of homoplasy, preferably performed by the compilers of the matrices, would be an especially valuable contribution to phylogenetic studies (see also Sanderson and Hufford, 1996).

Implications for evolutionary systematics

There is a significant contrast between characters that are most appropriate for taxonomy, where homoplasy undermines nested hierarchical classifications, and those that are most appropriate for evolutionary studies, where homoplasy can enhance evolutionary inference. Specifically, parallelism is often taken as indicating adaptation to similar functions; in contrast, reversal suggests lack or loss of function, and implies weak evolutionary constraints (e.g. Donoghue, 1989; Lauder, 1990; Sanderson, 1990; Bateman, 1996; Larson and Losos, 1996; Sanderson and Hufford, 1996; Givnish and Sytsma, 1997). Unfortunately, the apparently strong preferences for parallelism or reversal shown by the rhizomorphic lycopsids depend entirely which optimisation algorithm is used. Despite such interpretational ambiguities, we endorse the suggestion of Sanderson and Donoghue (1989: 1792) that relative levels of parallelism and reversal should be assessed in multi-matrix surveys of homoplasy.

Focusing on systematic studies, it is insufficient to simply claim that a particular suite of characters is "best", even within a narrowly delimited clade. Rather, it is necessary to specify at which hierarchical levels in the phylogeny (and taxonomy) the characters are most effective (see also Maddison, 1990; Sanderson, 1991). For example, Bateman, *et al.* (1992: 550) noted that, within the rhizomorphic lycopsids, genera and species are distinguished primarily by features of the largest (architecture) and smallest (spores, leaves) physical scale, whereas the deeper phylogenetic nodes supporting families are distinguished by organs of intermediate scale (cones, vegetative axes). Sanderson and Donoghue (1989) detected no correlation between **levels** of homoplasy and taxonomic rank among different cladistic matrices, but this does not address the generality of **patterns** of homoplasy.

Our study suggests that the strength – or more specifically the taxonomic value – of the phylogenetic signal from a particular organ is greatest if (a) that organ is sufficiently complex to generate a large number of characters, and those characters have (b) a high degree of congruence within the subset and (c) are evenly distributed through the full-matrix phylogeny (i.e. through a wide range of levels in the evolutionary–taxonomic hierarchy).

A consensus tree fully congruent with, and almost as resolved as, the whole-matrix tree for the rhizomorphic lycopsids can be obtained simply by combining submatrices for the two most informative organs: cone assemblies and vegetative axes (Fig. 15). In other words, an additional lycopsid species could be placed phylogenetically with reasonable confidence of accuracy if only its vegetative axes and cones were known. This observation is highly relevant in this case, given the exorbitant period of time needed to fully reconstruct a conceptual fossil species from disarticulated organs (Bateman, 1992b), though note that this conclusion that could be reached only **after** performing the cladistic analysis. Cones show less homoplasy than vegetative axes, though the stele and periderm subsets within vegetative axes yielded CI and RI values similar to those of cones (Fig. 17). However, the greatest difference between cones and vegetative axes lies in the strength of phylogenetic signal as measured per unit character. Specifically, cones yield almost equal resolution and accuracy to vegetative axes (cf. Figs 3–5) on the basis of only 14 informative characters rather than 39. Consequently, cones lie well above the polynomial curve on Fig. 15.

Thus, in the case of the rhizomorphic lycopsids, the best taxonomic characters are a marriage of the most complex reproductive **and** vegetative organs rather than reproductive characters alone. Moreover, the survey of many cladistic matrices (Figs 18, 19) gives equally little support to the *a priori* up-weighting of reproductive characters that has pervaded plant systematics since Linnaeus.

Expert taxonomists generally assume that they "know" which characters are most diagnostic in their specialist group, presumably on the basis of (a) their own instinct, (b) the publications that reflect the instincts of their predecessors, and (c) the apparent success of their classifications when tested empirically. However, we have shown that the leaf base characters used preferentially in the taxonomy of the rhizomorphic lycopsids are poorly evolutionarily conserved, whereas a combination of cones and vegetative axes is sufficient to approximate the correct phylogeny. We suspect that patterns of character conservation in one higher plant taxon cannot be generalised with confidence to even closely related taxa, and that few current taxonomies are in fact based on the optimal combination of the available characters.

In our opinion, the only way to discover the optimal classification for a particular taxonomic group is to describe the broadest possible range of characters of each organism and subject these to a range of numerical analyses (e.g. Hennig, 1966; Sneath and Sokal, 1973). In this context, phylogenetic analyses based on extraordinarily narrow ranges of characters, whether reproductive (e.g. pollen, spermatozoa) or non-reproductive (e.g. apical meristems, leaf hairs), are just as suspect as non-numerical classifications, for they do not make adequate use of the congruence test of homology. This requires that phylogenetic signals from various plant organs should be pitted against each other, not only to identify homoplastic characters but also to reveal potential developmental linkages among characters (Bateman, 1994, 1997). Or, to put it another way, a broad-brush cladistic matrix based **only** on the number and spatial relationships of stamens and carpels would probably generate a highly artificial classification of the angiosperms similar to that compiled by Linnaeus (Figs 1, 2).

Schiebinger (1996: 103) argued that "Linnaeus devised his classification system in such a way that the number of a plant's stamens determined the class to which it was assigned, whereas the number of its pistils determines its order. In the taxonomic tree, class stands above order. In other words, Linnaeus gave male parts priority in determining the status of the organism in the plant kingdom. There is no empirical justification for this outcome; rather, Linnaeus brought traditional tenets of gender hierarchy into science." The accusation of male chauvinism gains little support from probability statistics; after all, in a typological classification, there are only two possible hierarchical options: male above female, or female above male. But the assertion that such blatant typology lacks empirical support is irrefutable. Linnaeus' (1759) error was not that he gave stamens priority over carpels *per se*, but that he generalised this imposed ranking of organs across the entire plant kingdom, rather than allowing the characters to find their own, natural level in a taxonomic hierarchy that was genuinely evolutionary.

Acknowledgements

We thank Dennis Stevenson for critically reading the manuscript. RMB thanks the Linnean Society Palynology Specialist Group for an invitation to speak on the phylogenetic value of spores and pollen in 1992, which ultimately prompted this investigation. NJS gratefully acknowledges the Trustees of the Royal Botanic Garden Edinburgh for the support of a 1995 Summer Studentship to conduct the broader survey of homoplasy indices.

References

Adanson, M. (1763). Familles des plantes. Vincent, Paris.

Archie, J.W. (1989). Homoplasy excess ratios: new indices for measuring levels of homoplasy in phylogenetic systematics and a critique of the consistency index. *Syst. Zool.* **38**: 253–269.

Archie, J.W. (1990). Homoplasy excess ratios and retention indices: a reply to Farris. *Syst. Zool.* **39**: 169–174.

Archie, J.W. (1996). Measures of homoplasy. In: M.J. Sanderson and L. Hufford (editors). Homoplasy: the recurrence of similarity in evolution, pp. 153–188, Academic Press, London.

Avise, J.C. (1994). Molecular markers, natural history, and evolution. Chapman and Hall, London.

Bateman, R.M. (1992a). Can we accurately reconstruct phylogenies using meiospores? Linnean Society Palynology Specialist Group meeting [unpublished abstract].

Bateman, R.M. (1992b). Multiple hierarchies and plurality of species concepts: necessary evils in phylogenetically informative palaeobotanical classifications. In: International Organisation of Palaeobotany Conference (4th: 1992: Paris). Abstracts, pp. 21–22.

Bateman, R.M. (1994). Evolutionary–developmental change in the growth architecture of fossil rhizomorphic lycopsids: scenarios constructed on cladistic foundations. *Biol. Rev. Cambridge Philos. Soc.* **69**: 527–597.

Bateman, R.M. (1996). Nonfloral homoplasy and evolutionary scénarios in living and fossil land plants. In: M.J. Sanderson and L. Hufford (editors). Homoplasy: the recurrence of similarity in evolution, pp. 91–130, Academic Press, London.

Bateman, R. M., DiMichele, W.A. and Willard, D.A. (1992). Experimental cladistic analyses of anatomically preserved arborescent lycopsids from the Carboniferous of Euramerica: an essay in paleobotanical phylogenetics. *Ann. Missouri Bot. Gard.* **79**: 500–559.

Bremer, K. (1990). Combinable component consensus. *Cladistics* **6**: 369–372.

Brooks, D.R., O'Grady, R.T. and Wiley, E.O. (1986). A measure of the information content of phylogenetic trees, and its use as an optimality criterion. *Syst. Zool.* **35**: 571–581.

Bull, J.J., Huelsenbeck, J.P., Cunningham, C.W., Swofford, D.L. and Waddell, P.J. (1993). Partitioning and combining data in phylogenetic analysis. *Syst. Biol.* **42**: 384–397.

Carpenter, J.M. (1988). Choosing among multiple equally parsimonious trees. *Cladistics* **4**: 291–296.

Chaloner, W.G. (1986). Reassembling the whole fossil plant, and naming it. In: R.A. Spicer and B.A. Thomas (editors). Systematic and taxonomic approaches in palaeobotany, pp. 67–78, Oxford University Press, Oxford.

Chang, J.T. and Kim, J. (1996). The measurement of homoplasy: a stochastic view. In: M.J. Sanderson and L. Hufford (editors). Homoplasy: the recurrence of similarity in evolution, pp. 189–203, Academic Press, London.

Chippindale, P.T. and Wiens, J.J. (1994). Weighting, partitioning, and combining characters in phylogenetic analysis. *Syst. Biol.* **43**: 278–287.
Crookall, R. (1955–76). Fossil plants of the Carboniferous rocks of Great Britain (second section). *Mem. Geol. Surv. Gr. Brit. Palaeontol.* **4**, nos. 1–7.
de Queiroz, A., Donoghue, M.J. and Kim, J. (1995). Separate versus combined analysis of phylogenetic evidence. *Annual Rev. Ecol. Syst.* **26**: 657–681.
Donoghue, M.J. (1989). Phylogenies and the analysis of evolutionary sequences, with examples from seed plants. *Evolution* **43**: 1137–1156.
Donoghue, M.J. and Sanderson, M.J. (1992). The suitability of molecular and morphological evidence in reconstructing plant phylogeny. In: P.S. Soltis, D.E. Soltis and J.J. Doyle (editors). Molecular plant systematics, pp. 340–368, Chapman & Hall, London.
Eernisse, D.J. and Kluge, A.G. (1993). Taxonomic congruence versus total evidence, and amniote phylogeny inferred from fossils, molecules, and morphology. *Molec. Biol. Evol.* **10**: 1170–1195.
Farris, J.S. (1982). The logical basis of phylogenetic analysis. In: N. Platnick and V.A. Funk, (editors). Advances in cladistics: proceedings of the second meeting of the Willi Hennig Society, pp. 7–36, Columbia University Press, New York.
Farris, J.S. (1989). The retention index and the rescaled consistency index. *Cladistics* **5**: 417–419.
Farris, J.S. (1991). Excess homoplasy ratios. *Cladistics* **7**: 81–91.
Forey, P.L., Humphries, C.J., Kitching, I.J., Scotland, R.W., Siebert, D.J. and Williams, D.M. (1992). Cladistics: a practical course in systematics. Oxford University Press, Oxford.
Funk, V.A. and Brooks, D.R. (1990). Phylogenetic systematics as the basis of comparative biology. Smithsonian Institution, Washington, DC. (Smithsonian Contributions to Botany, 73.)
Givnish, T.J. and Sytsma, K.J. (1997). Homoplasy in molecular vs. morphological data: the likelihood of correct phylogenetic inference. In: T.J. Givnish and K.J. Sytsma (editors). Molecular evolution and adaptive radiation, pp. 55–101, Cambridge University Press, Cambridge.
Goloboff, P.A. (1991). Homoplasy and the choice among cladograms. *Cladistics* **7**: 215–232.
Goloboff, P.A. (1993). Estimating character weights during tree search. *Cladistics* **9**: 83–91.
Hemsley, A.R. (1992). Evolution of exine ultrastructure in Palaeozoic lycopod megaspores. *Courier Forschungsinst. Senckenberg* **147**: 93–107.
Hennig, W. (1966). Phylogenetic systematics. University of Illinois Press, Urbana.
Huelsenbeck, J.P. and Hillis, D.M. (1993). Success of phylogenetic methods in the four-taxon case. *Syst. Biol.* **42**: 247–264.
Huelsenbeck, J.P., Bull, J.J. and Cunningham, C.W. (1996). Combining data in phylogenetic analysis. *Trends Ecol. Evol.* **11**: 152–157.
Kidston, R. (1923–25). Fossil plants of the Carboniferous rocks of Great Britain. *Mem. Geol. Surv. Gr. Brit. Palaeontol.* **2**.
Kluge, A.G. and Farris, J.S. (1969). Quantitative phyletics and the origin of the anurans. *Syst. Zool.* **18**: 1–32.
Larson, A. and Losos, J.B. (1996). Phylogenetic systematics of adaptation. In: M.R. Rose and G.V. Lauder (editors). Adaptation, pp. 187–220, Academic Press, London.
Lauder, G.V. (1990). Functional morphology and systematics: studying functional patterns in a historical context. *Annual Rev. Ecol. Syst.* **21**: 317–340.
Linné, C. von (1759). Systema naturae. (11th edn). Laur Salvius, Holmiae.
Maddison, D.R. (1991). Discovery and importance of islands of most-parsimionious trees. *Syst. Zool.* **40**: 315–328.
Maddison, W.P. (1990). A method for testing the correlated evolution of two binary characters: are gains or losses concentrated on certain branches of a phylogenetic tree? *Evolution* **44**: 539–557.

Maddison, W.P. and Maddison, D.R. (1992). MacClade, version 3.0: analysis of phylogeny and character evolution. Sinauer, Sunderland, Mass.
McShea, D.W. (1996). Complexity and homoplasy. In: M.J. Sanderson and L. Hufford (editors). Homoplasy: the recurrence of similarity in evolution, pp. 207–225, Academic Press, London.
Meacham, C.A. (1981). A probability measure for character compatibility. *Math. Biosci.* **57**: 1–18.
Meier, R., Kores, P. and Darwin, S. (1991). Homoplasy slope ratio: a better measurement of observed homoplasy in cladistic analyses. *Syst. Zool.* **40**: 74–88.
Miyamoto, M.M. and Fitch, W.M. (1995). Testing species phylogenies and phylogenetic methods with congruence. *Syst. Biol.* **44**: 64–76.
Nixon, K.C., and Davis, J.I. (1991). Polymorphic taxa, missing values and cladistic analysis. *Cladistics* **7**: 233–241.
Page, R.D.M. (1993). Component: tree comparison software for Microsoft Windows, version 2.0. Natural History Museum, London.
Patterson, C. (1982). Morphological characters and homology. In: K.A. Joysey and A.E. Friday (editors). Problems of phylogenetic reconstruction, pp. 21–74, Academic Press, London.
Patterson, C. (1988). Homology in classical and molecular biology. *Molec. Biol. Evol.* **5**: 603–625.
Phillips, T.L. and DiMichele, W.A. (1992). Comparative ecology and life-history biology of arborescent lycopsids in Late Carboniferous swamps of Euramerica. *Ann. Missouri Bot. Gard.* **79**: 560–588.
Ray, J. (1703). Methodus plantarum nova (2nd edn).
Riedl, R. (1979). Order in living organisms. Wiley, New York.
Sanderson, M.J. (1990). Rates of speciation and evolution: a bias due to homoplasy. *Cladistics* **6**: 387–391.
Sanderson, M.J. (1991). In search of homoplastic tendencies: statistical inference of topological patterns in homoplasy. *Evolution* **45**: 351–358.
Sanderson, M.J. and Donoghue, M.J. (1989). Patterns of variation in levels of homoplasy. *Evolution* **43**: 1781–1795.
Sanderson, M.J. and Donoghue, M.J. (1996). The relationship between homoplasy and confidence in a phylogenetic tree. In: M.J. Sanderson and L. Hufford (editors). Homoplasy: the recurrence of similarity in evolution, pp. 67–89, Academic Press, London.
Sanderson, M.J. and Hufford, L. (editors). (1996). Homoplasy: the recurrence of similarity in evolution. Academic Press, London.
Sang, T. (1995). New measurements of distribution of homoplasy and reliability of parsimonious cladograms. *Taxon* **44**: 77–82.
Schiebinger, L. (1996). The loves of the plants. *Sci. Amer.* **274**: 98–103.
Sharkey, M. (1989). A hypothesis-independent method of character weighting. *Cladistics* **5**: 63–86.
Sneath, P.H.A. and Sokal, R.R. (1973). Numerical taxonomy: the principles and practice of numerical classification. Freeman, San Francisco.
Stace, C.A. (1989). Plant taxonomy and biosystematics (2nd edn). Arnold, London.
Stewart, W.N. and Rothwell, G.W. (1993). Paleobotany and the evolution of plants (2nd edn). Cambridge University Press, Cambridge.
Swofford, D.L. (1993). Phylogenetic Analysis Using Parsimony, version 3.1.1. Illinios Natural History Survey, Urbana.
Swofford, D.L., Olsen, G.J., Waddell, P.J. and Hillis, D.M. (1996). Phylogenetic inference. In: D.M. Hillis, C. Moritz and B.K. Mable (editors). Molecular systematics (2nd edn), pp. 407–514, Sinauer, Sunderland, Mass.

Turton, W. (1802). A general system of nature: through the three kingdoms.
Wiley, E.O., Siegel-Causey, D.J., Brooks, D.R. and Funk, V.A. (1991). The compleat cladist: a primer of phylogenetic procedures. Museum of Natural History, University of Kansas, Lawrence, Kansas.
Wilkinson, M. (1991). Homoplasy and parsimony analysis. *Syst. Zool.* **40**: 101–109.
Yeates, D. (1992). Why remove autapomorphies? *Cladistics* **8**: 387–389.

Runions, C.J., and Owens, J.N. (1998). Evidence of pre-zygotic self-incompatibility in a conifer. In: S.J. Owens and P.J. Rudall (Editors). Reproductive Biology, pp. 255–264. Royal Botanic Gardens, Kew.

EVIDENCE OF PRE-ZYGOTIC SELF-INCOMPATIBILITY IN A CONIFER

C. JOHN RUNIONS[1] AND JOHN N. OWENS[2]

[1]**Section of Ecology and Systematics, Corson Hall, Cornell University, Ithaca, New York USA 14853** (Author for correspondence)
[2]**Centre for Forest Biology, University of Victoria, P.O. Box 3020, Victoria, British Columbia, Canada V8W 3N5**

Abstract

Conifers are reported to be self-compatible although relatively low levels of viable seed result from self-fertilisation. Failure of seed maturation has been attributed to accumulation and expression of sub-lethal recessive alleles in the zygote or at later stages of development. A detailed study of controlled cross-, versus self-pollinated ovules of *Picea* was done and development from the time of pollen germination until formation of the early embryo described. Cross-pollinated ovules generally develop normally through fertilisation, and viable embryos result. However, several cases of late or abnormal cell division in developing archegonia resulted in egg cell deterioration even when germinated pollen were abundant. Significantly, two types of pre-zygotic losses occurred only after self-pollination. In one of the clones studied, most pollen tubes aborted after growing most of the way through the nucellus. In a second clone, some pollen tubes penetrated the neck cells but continued to grow into the egg cell without releasing the male gametes. Egg cell abortion occurred shortly afterwards in each case. This is the first report of a partial self-incompatibility mechanism in a conifer. Existence of such a mechanism in a conifer might represent a convergence with, or an early stage of evolution of self-incompatibility mechanisms in angiosperms.

Introduction

Conifers in general produce fewer sound seeds from self-pollination than from cross-pollination (Orr-Ewing, 1957; Sarvas, 1962; Mergen *et al.*, 1965; Koski, 1973). Reduction in seed set is said to be due to abortion during post-zygotic development (Mergen *et al.*, 1965; Fowler and Park, 1983). Pre-zygotic mechanisms which prevent self-fertilisation have not been observed and conifers have therefore been considered self-compatible (Seavey and Bawa, 1986; Zavada and Taylor, 1986; Sage *et al.*, 1994).

We have conducted an ultrastructural study of reproductive development in white spruce, *Picea glauca* (Moench) Voss, which included cross-, and self-pollinations. The rationale for including self-pollinations was that direct observation had rarely been used to evaluate the timing of seed losses in a conifer. Seed loss during post-zygotic stages occurs because of inviability, a manifestation of inbreeding depression. Conifers have high numbers of recessive embryonic lethal alleles (Bishir and Namkoong, 1987) which when expressed in homozygous condition can result in embryo abortion. An assumption that gamete fusion has occurred is required before seed losses after self-pollination can be categorised as post-zygotic. This assumption might not be valid.

Two types of pre-zygotic developmental abnormalities which were unique after self-pollination, and which prevented gamete fusion were observed in this study. Pre-zygotic mechanisms that prevent fertilisation after self-pollination in white spruce give evidence of self-incompatibility acting in a conifer.

Materials and Methods

Tree selection and pollination

Trees of white spruce used in this study had been established as grafts in a clonal seed orchard at the Kalamalka Seed Centre near Vernon, British Columbia in 1982. Individual ramets of three clones were chosen in April 1993. Each clone had an abundant supply of developing pollen-, and seed cone buds. Pollen-cone buds were removed and isolation bags placed over branches bearing seed cones prior to the pollination period. Cross-pollen was collected from 12 clones which were not otherwise included in the study. Cross-, and self-pollen was tested to ensure that it was between 5-10% moisture content prior to artificial pollinations. Pollination was accomplished by injecting 0.5 ml aliquoits of pollen into each bag daily during the five day period when seed cones appeared maximally receptive. Once seed cones had become post-receptive, the paper bags were replaced with mesh bags which excluded insects.

Anatomical study of fertilisation period

Beginning one week after the final pollination, ovules were collected and prepared for the developmental study. Every two days for 18 days, two cross-, and two self-pollinated seed cones were collected from each tree. Median longitudinal sections less than 1 mm in thickness were cut from ten ovules from the centre region of each cone, fixed in glutaraldehyde, post-fixed in OsO_4, dehydrated and embedded in Spurr's resin. A detailed study of development following cross-pollination was completed prior to the study of self-pollinated ovules. Seventy-two cross-pollinated ovules and 15 self-pollinated ovules were examined. In each study, ovules were serially sectioned at 0.8 μm with a Riechert Ultracut E microtome, and stained with toluidine blue (CI 52040) at pH 11.0 for light microscopy (O'Brien and McCully, 1981). For transmission electron microscopy, sections were cut at 0.06 μm, picked up on 300-mesh or 75-mesh copper grids, stained with uranyl acetate and lead citrate and examined with an Hitachi TEM. All of the developmental stages described herein occurred during the eight day period from 7 to 15 days after the final pollination.

Seed production efficiency

Ten seed cones were left within mesh bags on each control pollinated branch until seed maturation in August 1993. Cross-, and self-pollinated cones were collected prior to seed shed. Seeds were extracted and X-rayed to determine whether they were viable or empty. Viable seeds are X-ray opaque. Percent seed efficiency (SEF%) was calculated as the seed to ovule ratio (SEF% = (# seeds ÷ # ovules) × 100).

Results

Seed production after cross-, and self-pollination

Seed production was significantly lower after self-pollination than after cross-pollination in all three clones ($p < 0.001$, Table 1). Large variance in SEF% after self-pollination in clones 1926 and 4752 result from one cone in each case which had unusually high seed production.

TABLE 1. Seed production after cross- and self-pollination in white spruce.
* in all cases, self-pollination results in significantly lower seed production than cross-pollination ($p<0.001$).

Pollen type	SEF% [(# filled seeds ÷ # ovules) × 100]		
	clone 1926	clone 1944	clone 4752
Cross	50.5±20.0	35.0±20.4	39.0±26.7
Self *	13.3±18.8	3.1±2.1	13.7±18.6

Reproductive development after cross-pollination

Development within the ovule was followed between the time of pollen germination and emergence of the early embryo into the megagametophyte. Details of this part of the study will be published separately but, in general, normal development was as described in Owens and Molder (1979). Fig. 1 summarises the normal development of *Picea* during the period of zygote formation.

Pollen germinated without delay and pollen tubes grew vigorously through the nucellus. body cell cytoplasm was dense and contained two male gametes which had formed from mitosis without subsequent cell wall formation (Fig. 2). Male gametes were large (ca. 20 to 30 μm diameter) and surrounded by plastids and mitochondria. Neck cells of each archegonium deteriorated to some extent prior to contact with a pollen tube. The pollen tube grew between them to contact the ventral canal cell, whose cell walls lysed and the pollen tube contents were released into the egg cytoplasm (Fig. 3). Pre-zygotic development continued until one male gamete and the egg nucleus fused. This occurred very quickly (< 1 day) after release of the male gametes from the pollen tube.

Many abnormalities of development were disruptive enough to preclude fertilisation of ovules. Table 2 summarises the types of abnormalities observed in cross-pollinated ovules. In all cases, these abnormalities were independent of the type of pollen applied. Gross deformities of the ovule included: megagametophytes or egg cells which had aborted prior to pollination, megagametophytes incompletely surrounded by nucellar tissue, and atypical mitosis leading to deformed or misplaced neck and ventral canal cells. Pollen germinated in these situations but fertilisation never occurred.

Reproductive development after self-pollination

Many of the abnormalities observed after cross-pollination were also observed after self-pollination but, as described, these were independent of pollen type. Two categories of developmental defect, however, were unique to the self-pollinated situation (Table 2). Both types of defect prevented fertilisation of the egg cell.

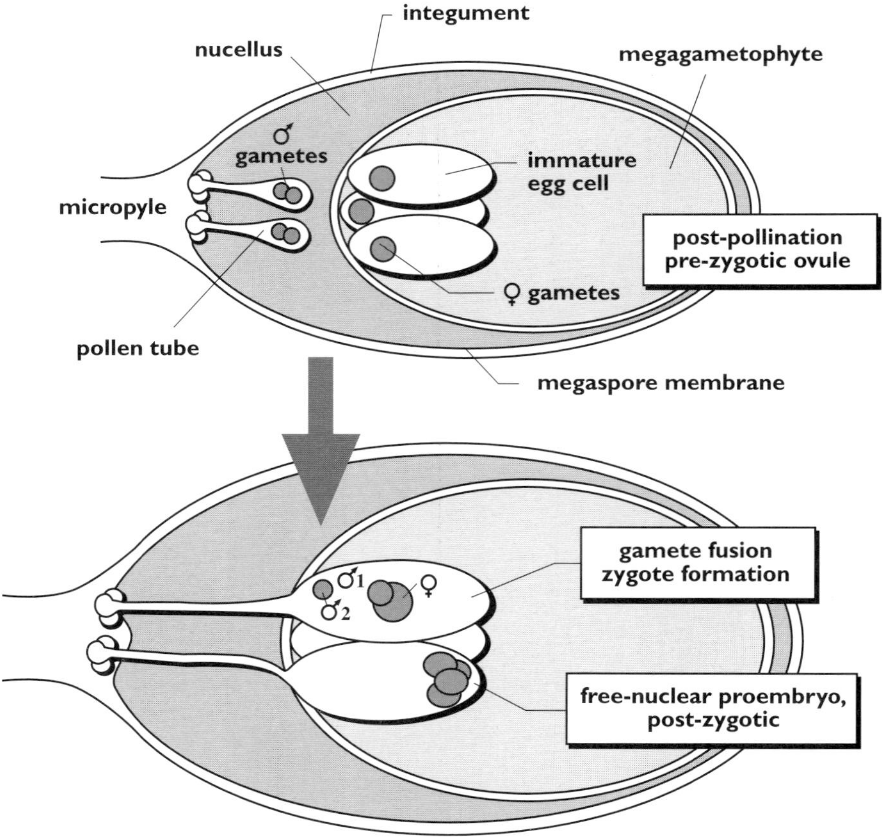

FIG. 1. Normal development within the ovule of *Picea* during the period of zygote formation. Pollen germinates and grows towards the megagametophyte through the nutritive tissue of the nucellus. Pollen tubes release male gametes into an unfertilised egg cell. The pre-zygotic period lasts until one male gamete fuses with the egg nucleus to form the zygote. Two mitoses without cell wall formation result in a proembryo with four free-nuclei. Nuclei migrate to the chalazal end of the egg cell and subsequent mitoses, with cell wall formation produce a four-tiered proembryo.

In three of four ovules of clone 1944, pollen tubes aborted within the nucellus as the tube tip came into proximity with the megagametophyte. In two of these cases, both pollen tubes had aborted, in the third case, one of two pollen tubes had aborted and in the final case, one egg cell had been fertilised. Fig. 4 shows a pollen tube aborted within the nucellus. Male gametes had become small, round and darkly staining and cytoplasm of the body and tube cells appeared deteriorated. No plastids or mitochondria were visible in the body cell cytoplasm.

The second type of developmental anomaly unique to self-pollinated ovules occurred in clone 1926. In two of four ovules studied, pollen tubes grew into the neck of an archegonium but did not then rupture to release male gametes into the egg cell cytoplasm. Pollen tubes continued growing through the ventral canal cell into the egg cell cytoplasm (Fig. 5). Subsequent growth and

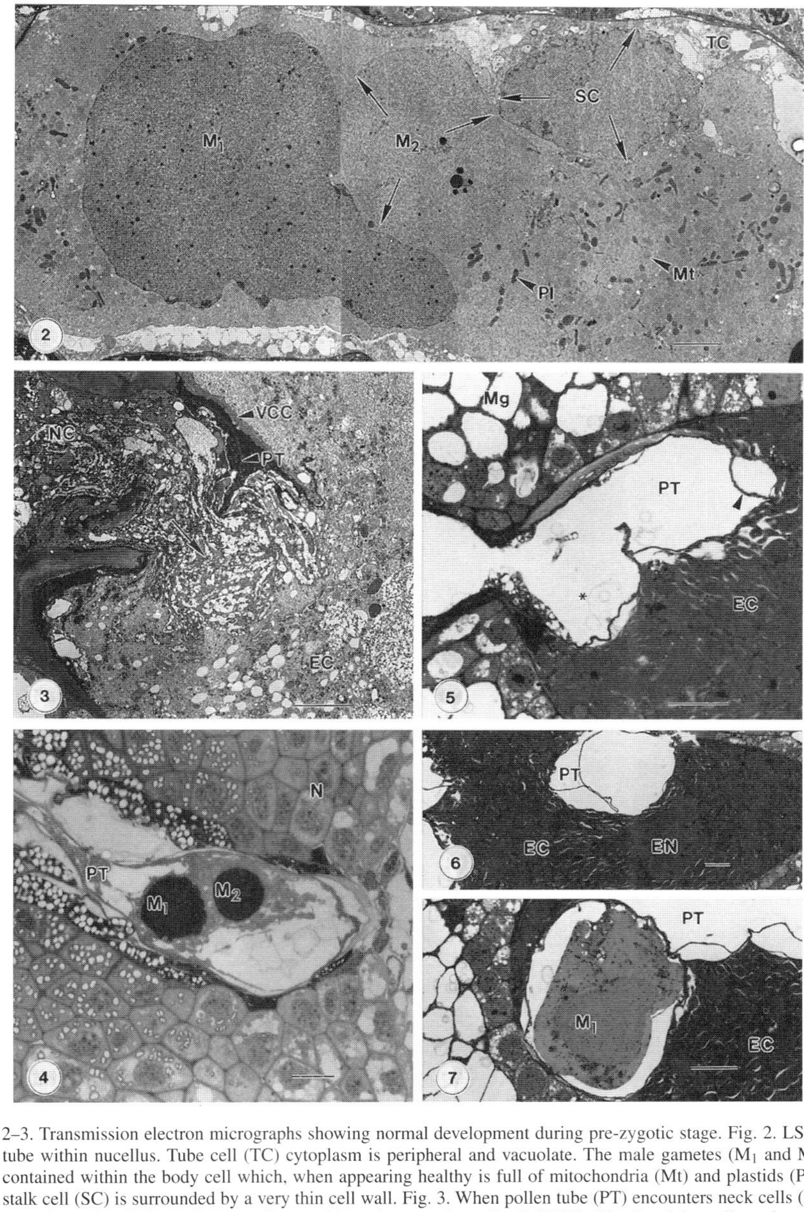

FIGS. 2–3. Transmission electron micrographs showing normal development during pre-zygotic stage. Fig. 2. LS pollen tube within nucellus. Tube cell (TC) cytoplasm is peripheral and vacuolate. The male gametes (M_1 and M_2) are contained within the body cell which, when appearing healthy is full of mitochondria (Mt) and plastids (Pl). The stalk cell (SC) is surrounded by a very thin cell wall. Fig. 3. When pollen tube (PT) encounters neck cells (NC) of an archegonium it forces between them and into the ventral canal cell (VCC). The tip of the pollen tube ruptures and its contents are released into the egg cell (EC) cytoplasm through a small opening (arrow). FIGS. 4–7. Light micrographs which illustrate types of development unique to the self-pollinated case. PT = self-pollen tube; EC = egg cell; Mg = megagametophyte. Fig. 4. Self-pollen tube aborted within nucellus (N). Male gametes (M_1 and M_2) are small, round and dark staining and no organelles can be seen in the body cell cytoplasm. Fig. 5. A self-pollen tube has penetrated into the egg cell cytoplasm without rupturing to release the male gametes. The cell wall of the pollen tube (arrowhead) is visible within the egg cell cytoplasm. *- marks the location of male gametes in adjacent sections (see Fig. 7). Fig. 6. Pollen tube penetration of the egg cell cytoplasm reaches the egg nucleus (EN). The egg cell cytoplasm is deteriorating. Fig. 7. A branch of the pollen tube (* in Fig. 5) within the egg cell-cytoplasm contains the male gametes (M_1) which appear healthy and are surrounded by cellular organelles.

TABLE 2. Types of abnormal development observed within pollinated ovules of white spruce. All types of abnormalities prevented fertilisation.

Abnormal developments independent of type of pollen applied	i)	early abortion of megagametophyte and/or egg cell
	ii)	megagametophyte not completely enclosed by nucellus
	iii)	misplaced ventral canal cell
	iv)	immature egg cell when pollen tube arrives
Abnormal developments unique to self- pollinated case (pre-zygotic SI)	i)	pollen tubes arrested within nucellus and male gametes aborted (clone 1944)
	ii)	pollen tubes penetrate egg cell without releasing male gametes (clone 1926)

branching of the pollen tube throughout the egg cell cytoplasm resulted in deterioration of the egg cell. Fig. 6 shows a long branch of a pollen tube within the egg cell cytoplasm, almost in contact with the egg nucleus. A shorter branch of the same pollen tube (marked by an asterisk in Fig. 5) contains the male gametes (Fig. 7). Male gametes in this case appear healthy (compare Figs. 2 and 7) and are surrounded by plastids and mitochondria.

When abnormalities unique to the self-pollinated situation occurred, egg cells (those adjacent to disrupted ones) appeared mature and competent to be fertilised, i.e. the egg nucleus was large and centrally located within the egg and the neck and ventral canal cells were normally developed.

Discussion

Few studies of conifer seed development have examined the period of gamete fusion in sufficient detail to observe pre-zygotic failure. Orr-Ewing (1957) found no developmental anomalies within the ovule of Douglas fir, *Pseudotsuga menziesii* (Mirb.) Franco. Plym-Forshell (1974) conducted similar studies on Scots pine, *Pinus sylvestris* L., with the same result and Mergen *et al.* (1965) found no barriers to self-fertilisation in white spruce. In all of these cases, early embryo abortion was common after self-fertilisation and our observations bear this out (unpublished). A preliminary survey of paraffin embedded ovules, however, indicated to us that developmental abnormalities such as early deterioration of the megagametophyte without fertilisation could be mistaken for abortion of the early embryo. Deterioration of tissues occurs quickly and remnants of a proembryo or early embryo are difficult, if not impossible, to distinguish. In most cases of apparent embryo abortion we could not say with confidence that gamete fusion had occurred. In this study, tissue preparation for ultrastructural examination preserved structures well enough to allow detailed analysis of pre-zygotic events, with surprising results.

Self-incompatibility

Two types of pre-zygotic departure from normal development were observed. In each case, these abnormalities prevented fertilisation and can be considered the result of self-incompatibility (SI) reactions.

Three categories of SI have been described for angiosperms; gametophytic SI (GSI, Newbigin *et al.*, 1993), sporophytic SI (SSI, Nasrallah and Nasrallah, 1993), and late-acting or ovular SI (Seavey and Bawa, 1986; Sage *et al.*, 1994). The first two types, GSI and SSI are relatively well understood systems (Haring *et al.*, 1990 and references therein) which are under genetic control and mediated by interactions of S-locus gene products. In families with GSI, e.g. Solanaceae, expression of S-locus genes in the haploid genome of the pollen determines the SI response (Newbigin *et al.*, 1993). Pollen germinates on the stigma and pollen tubes begin to grow through the style but abort before reaching the ovule. In families with SSI, e.g. Brassicaceae, S-locus products of the pollen-producing sporophyte determine the SI response (Nasrallah and Nasrallah, 1993). Pollen generally are prevented from germinating on the stigma.

Late-acting or ovular systems of SI are widely reported but not as clearly defined as GSI or SSI systems (Sage *et al.*, 1994). Pollen tubes grow at least to the micropyle prior to arrest, which occurs within the ovary (Seavey and Bawa, 1986). Kenrick *et al.* (1986) found that self-pollen tubes of *Acacia retinodes* arrested within the nucellus. In extreme cases, male gametes of SI pollen may even be released into the embryo sac (Gibbs and Bianchi, 1993). Timing of late-acting SI in the various reports suggests that more than one type of mechanism exists although the way in which these work is not understood.

Timing of the SI reactions observed in this study is illustrated in Fig. 8. Pollen tube arrest within the nucellus appears analogous with the GSI mechanism or with later acting SI mechanisms like that described by Kenrick *et al.* (1986) for *A. retinodes* in which pollen tube arrest is within the nucellus. Pollen tube growth that extends into the egg cell without release of the male gametes can be considered late-acting SI but no description of a similar SI mechanism acting in an angiosperm exists. It is not possible, from these initial observations of SI in white spruce to define a mechanism. The timing of disruption is pre-zygotic in each case but otherwise the two systems are quite different. In the first case, male gametes die in the pollen tube while it is within sporophytic tissue (nucellus). In the second, male gametes appear healthy but they have not been released into the egg cell. We can only speculate about chemical interactions that might result in pollen tube arrest or failure of the pollen tube to lyse in these cases. Pollen tube arrest occurs within the nucellus after some interspecific crosses of *Pinus* (Hagman, 1975). Pollen of the uncrossable species was found to have different polysaccharide composition. Pettitt (1985) was able to demonstrate an array of proteins emitted from growing pollen tubes of a number of conifer species. Interaction of sugars, proteins or other substances secreted from growing pollen tubes with substances that are secreted within the ovule (Takaso and Owens, 1996) are most likely to be the basis of the types of SI response we have observed in white spruce.

Low level seed production after self-pollination means that SI is only a partial barrier to self-fertilisation in white spruce. Self-incompatibility systems need not be absolute in their action, some angiosperm species exhibit partial SI in which low amounts of seed are produced after self-fertilisation (e.g. alfalfa, *Medicago sativa*; Brink and Cooper, 1938; Sayers and Murphy, 1966). Sedgley and Griffin (1989) reported that selection against SI in certain commercial fruit trees can weaken the mechanism so that low levels of seed set are possible. Sage *et al.* (1994) suggested that late-acting SI systems may represent nothing more than 'weak' GSI systems in which pollen tube inhibition is not complete within the style. Weak SI mechanisms may be sufficient to prevent fertilisation in most but not all cases. A few SI systems are mediated by the products of more than a single locus (Hayman and Richter, 1992). If the SI reactions depend on interactions of gene products from more than one locus with differential expression between pollen and ovule or if dominance interactions exist between alleles from a single S-locus (Thompson and Taylor, 1966) then partial SI can result.

Zavada and Taylor (1986) saw the origin of SI mechanisms as one of the significant developments during early angiosperm evolution. Reduced inbreeding depression (Mau *et al.*, 1991) and the possibility of mate selection (Zavada and Taylor, 1986) would have increased the fitness of those species with SI systems and this might explain the success of early angiosperms during the Cretaceous. Primitive SI mechanisms need not have first appeared within basal angiosperms, however. Kenrick *et al.* (1986) downplayed the role of stylar tissue

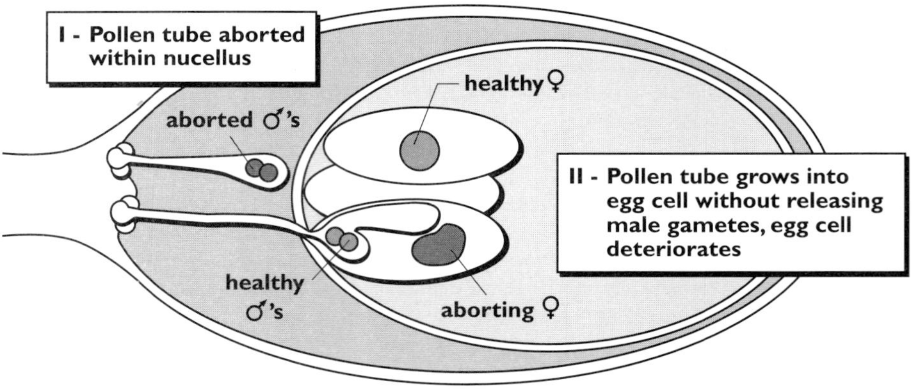

FIG. 8. Timing of self-incompatibility reactions observed in this study. In type I, pollen tube arrest occurs within the nucellus. In type II, pollen tubes penetrate into the egg cell without releasing male gametes.

in the origin of SI systems by suggesting that nucellar tissue or any other structure on which pollen germinated could have provided a site for SI interactions. Our observations of late-acting, partial SI within the nucellus of white spruce give evidence of an SI system which may predate the origin of angiosperms. Whether the various SI systems evolved from a common ancestral system or arose independently is unclear and it is difficult, therefore, to ascribe evolutionary significance to our results.

Post-zygotic seed losses in conifers

Reduced seed production after self-pollination in conifers is well documented (Koski, 1973; Johnsson, 1976; Fowler and Park, 1983; Owens *et al.*, 1990, 1991). Self-pollination in our study resulted in a three- to ten-fold reduction in seed production from the cross-pollinated case. This reduced seed set was quantified by X-ray of seed extracted from cones at the end of the seed maturation period. Timing of seed losses after self-pollination would have been described as post-zygotic had development within the ovule not been carefully observed during the pre-zygotic period.

Self-fertilisation in conifers can result in embryos with high genetic load because of accumulation of sub-lethal recessive alleles in homozygous condition (Johnsson, 1976). When the number of embryonic lethal equivalents is high, empty seeds result (Fowler and Park, 1983). Sub-lethal genetic load is reflected in various levels of inbreeding depression in subsequent development (Williams and Savolainen, 1996). Inbreeding and, in the extreme case, selfing can be used as tools in plant breeding to purge lethal recessive alleles from a breeding population to produce elite, homozygous lines. Williams and Savolainen (1996) suggested that selfing may not be the best strategy in conifer breeding. Because of the unusually high number of embryonic lethal equivalents in most trees, too many desirable gene combinations would be lost in the purge of lethal recessives. If a significant level of seed loss was attributed to pre-zygotic failure, models which predict the number of embryonic lethal equivalents based on post-zygotic seed losses (Savolainen *et al.*, 1992; Namkoong and Bishir, 1987) would need revision. In this scenario, the estimated average number of embryonic lethal equivalents per tree would be reduced. Advocates of selfing in production of elite lines may perceive a benefit from this reduction. Any benefit, however, would likely be outweighed by the risk of losing valuable genes to self-incompatibility.

Conclusion

This report is of preliminary observations that give the first evidence of self-incompatibility acting in a conifer. Two types of barrier to fertilisation that were unique to the self-pollinated case occurred in white spruce. Sample sizes were too small to allow quantification of seed losses due to self-incompatibility in this case. Once self-pollination experiments have been conducted on a larger scale, we will know how common pre-zygotic barriers to fertilisation are in this species. A recent report of egg cell rejection of some spermatozoids in *Encephalartos villosus* (Zamiaceae; Steyn *et al*., 1996) is the first report of a system of pre-zygotic discrimination between intraspecific male gametes in a gymnosperm. Incompatible male gametes were not from self-pollen in that study as *Encephalartos* is dioecious, but the result supports our contention that cell-to-cell signalling and recognition play a role in gymnosperm sexual reproduction.

References

Bishir, J. and Namkoong, G. (1987). Unsound seeds in conifers: estimation of number of lethal alleles and of magnitudes of effects associated with the maternal parent. *Silvae Genet.* **36**: 180–185.

Brink, R.A. and Cooper, D.C. (1938). Partial self-incompatibility in *Medicago sativa. Proc. Nat. Acad. Sci. U.S.A.* **24**: 497–499.

Fowler, D.P. and Park, Y.S. (1983). Population studies of white spruce. I. Effects of self-pollination. *Canad. J. Forest Res.* **13**: 1133–1138.

Gibbs, P.E. and Bianchi, M. (1993). Post-pollination events in species of *Chorisia* (Bombacaceae) and *Tabebuia* (Bignoniaceae) with late-acting self-incompatibility. *Bot. Acta* **106**: 64–71.

Hagman, M. (1975). Incompatibility in forest trees. *Proc. Roy. Soc. London, Ser. B, Biol. Sci.* **188**: 313–326.

Haring, V., Gray, J.E., McClure, B.A., Anderson, M.A. and Clarke, A.E. (1990). Self-Incompatibility: a self-recognition system in plants. *Science* **250**: 937–941.

Hayman, D.L. and Richter, J. (1992). Mutation affecting self-incompatibility in *Phalaris coerulescens* Desf. (Poaceae). *Heredity* **68**: 495–503.

Johnsson, H. (1976). Contributions to the genetics of empty grains in the seed of pine (*Pinus silvestris*). *Silvae Genet.* **25**: 10–15.

Kenrick, J., Kaul, V. and Williams, E.G. (1986). Self-incompatibility in *Acacia retinodes*: site of pollen tube arrest is the nucellus. *Planta* **169**: 245–250.

Koski, V. (1973). On self-pollination, genetic load, and subsequent inbreeding in some conifers. *Commun. Inst. Forest. Fenniae* **78**: 42 pp.

Mau, S.-L., Anderson, M.A., Heisler, M., Haring, V., McClure, B.A. and Clarke, A.E. (1991). Molecular and evolutionary aspects of self-incompatibility in flowering plants. In: G. Jenkins and W. Schuch (editors). Molecular biology of plant development, pp. 245–269, Company of Biologists, Dept. of Zoology, Univ. of Cambridge, Cambridge. (Symposia of the Society for Experimental Biology, no. 45)

Mergen, F., Burley, J. and Furnival, G.M. (1965). Embryo and seedling development in *Picea glauca* (Moench) Voss after self, cross, and wind-pollination. *Silvae Genet.* **14**: 188–194.

Namkoong, G. and Bishir, J. (1987). The frequency of lethal alleles in forest tree populations. *Evolution*. **41**: 1123–1127.

Nasrallah, J.B. and Nasrallah, M.E. (1993). Pollen-stigma signaling in the sporophytic self-incompatibility response. *Pl. Cell.* **5**: 1325–1335.

Newbigin, E., Anderson, M.A. and Clarke, A.E. (1993). Gametophytic self-incompatibility systems. *Pl. Cell.* **5**: 1315–1324.

O'Brien, T.P. and McCully, M.E. (1981). The study of plant structure: principles and selected methods. Termarcarphi Pty. Ltd., Melborne.

Orr-Ewing, A.L. (1957). A cytological study of the effects of self-pollination on *Pseudotsuga menziesii* (Mirb.) Franco. *Silvae Genet.* **6**: 179–185.

Owens, J.N. and Molder, M. (1979). Sexual reproduction of white spruce (*Picea glauca*). *Canad. J. Bot.* **57**: 152–169.

Owens, J.N., Colangeli, A.M. and Morris, S.J. (1990). The effect of self-, cross-, and no pollination on ovule, embryo, seed and cone development in western red cedar (*Thuja plicata*). *Canad. J. Forest. Res.* **20**: 66–75.

Owens, J.N., Colangeli, A.M. and Morris, S.J. (1991). Factors affecting seed set in Douglas-fir (*Pseudotsuga menziesii*). *Canad. J. Bot.* **69**: 229–238.

Pettitt, J.M. (1985). Pollen tube development and characteristics of the protein emission in conifers. *Ann. Bot.* **56**: 379–397.

Plym-Forshell, C. (1974). Seed development after self-pollination and cross-pollination of scots pine, *Pinus sylvestris* L. *Stud. Forest Suec.* **118**: 1–37.

Sage, T.L., Bertin, R.I. and Williams, E.G. (1994). Ovarian and other late-acting self-incompatibility systems. In: E.G. Williams (editor). Genetic control of self-incompatibility and reproductive development in flowering plants, pp. 116–140, Kluwer Academic Publishers, Netherlands.

Sarvas, R. (1962). Investigations on the flowering and seed crop of *Pinus sylvestris*. *Commun. Inst. Forest Fenniae* **53**: 1–198.

Savolainen, O., Kärkkäinen, K. and Kuittinen, H. (1992). Estimating numbers of embryonic lethals in conifers. *Heredity* **69**: 308–314.

Sayers, E.R. and Murphy, R.P. (1966). Seed set in alfalfa as related to pollen tube growth, fertilization frequency, and post-fertilization ovule abortion. *Crop Sci.* (*Madison*) **6**: 365–368.

Seavey, S.R. and Bawa, K.S. (1986). Late-acting self-incompatibility in angiosperms. *Bot. Rev.* **52**: 195–219.

Sedgley, M. and Griffin, A.R. (1989). Sexual reproduction of tree crops. Academic Press, Toronto.

Steyn, E.M.A., Strydom, D.J.F. and Botha, A. (1996). Fertilization and rejection of spermatozoids by egg cells in artificially pollinated ovules of *Encephalartos* (Zamiaceae). *Sexual Pl. Reprod.* **9**: 175–185.

Takaso, T. and Owens, J.N. (1996). Postpollination-prezygotic ovular secretions into the micropylar canal in *Pseudotsuga menziesii* (Pinaceae). *J. Pl. Res.* **109**: 147–160.

Thompson, K.F. and Taylor, J.P. (1966). Non-linear dominance relationships between S-alleles. *Heredity* **21**: 345–362.

Williams, C.G. and Savolainen, O. (1996). Inbreeding depression in conifers: implications for breeding strategy. *Forest Sci.* **42**: 102–117.

Zavada, M.S. and Taylor, T.N. (1986). The role of self-incompatibility and sexual selection in the gymnosperm-angiosperm transition: a hypothesis. *Amer. Naturalist* **128**: 538–550.

Tomlinson, P.B. and Takaso, T. (1998). Hydrodynamics of pollen capture in conifers. In: S.J. Owens and P.J. Rudall (Editors). Reproductive Biology, pp. 265–275. Royal Botanic Gardens, Kew.

HYDRODYNAMICS OF POLLEN CAPTURE IN CONIFERS

P. B. TOMLINSON[1] AND T. TAKASO[2]

[1]Harvard University, Harvard Forest, Petersham, MA 01366 U.S.A.
[2]Dept. Natural Environment Sciences, Kyoto University, Kyoto 606-01, Japan.
(Present address: Dept. Chemistry, Biology & Marine Science,
University of the Ryukyus, 1 Senbaru, Nishihara-cho, Okinawa 903-01, Japan)

Abstract

Pollen in conifers has aerodynamic and hydrodynamic properties. The former relate to wind dispersal, but pollen structure is still highly diversified. This suggests that diversity relates to hydrodynamic properties and this is substantiated by the close functional correlation between pollen structure and features of cone morphology at the time of pollination. The major division between saccate and non-saccate pollen is related to different responses to water, the former shows bubble-like properties and does not sink, the latter sinks and bursts in water, shedding the exine. These types are usually associated with the presence of a pollination drop. A minor category includes taxa that may be derived from one or both of the above major categories but which lack a pollination drop. This small assemblage is significant because several genera show extended siphonogamy of a kind reminiscent of that in angiosperms. There is the possibility that rain may facilitate pollen capture, notably in taxa that have saccate pollen but no pollination drop. The overview demonstrates the complexity of pollen capture mechanisms in conifers, despite their ubiquitous wind pollination.

Introduction

The primary target of pollen in gymno-ovulate plants ('non-flowering seed plants' or Gymnosperms) is most commonly a pollination drop secreted in various ways by the nucellus of the ovule. Drop secretion is little studied, but apparently involves cells of the nucellar apex, which commonly disintegrate (Chesnoy, 1993). The drop is extruded through the micropyle and subsequently retracted so as to ingest some or all of the particles that land on it. The question asked is - how efficient can this simple mechanism become in the absence of any pre-fertilisation pollen selection? We restrict our study to the conifers, since these represent the most diverse group among extant gymnosperms in terms of pollen capture and pollen structure. Although the pollination drop must be accessible to pollen at the time of pollination (hence the gymno-ovulate condition) ovules are often concealed within the scales of the ovulate cone and so are scarcely visible externally. This is particularly the condition in Pinaceae and Taxodiaceae. Otherwise, the pollination drop is visible and accessible for study and manipulation. In Podocarpaceae only the micropyle is visible since the ovule itself is usually enveloped by an epimatium. Of particular interest are those conifers in which a pollination drop does not occur, since these provide a natural experiment whereby changes in ovule function are paralleled by changes in pollen function. In the absence of more complete study the evolutionary implications of these contrasted processes are not clear.

The precise mechanism by means of which pollen eventually becomes incorporated into the ovule and normally lands on the nucellus ('pollen capture') has had only limited study, despite its key to successful seed set. One can study the process easily *in vitro* by placing cut-shoots in water in such a way that cones can be observed directly with a stereomicroscope (e.g. Tomlinson *et al.,* 1997). One considerable limitation is that the period of drop secretion (i.e. pollen reception) for any species in a given location can be quite short (days) and, except in *Cedrus,* always in the spring or early summer (Singh, 1977). Furthermore, the pollination drop seems normally to be secreted at night and is commonly retracted in the early part of the day. Plant reproductive biologists have considered the process of pollen capture in conifers too simple to be worthy of study, compared with the diversity of genetic, physiological and structural processes found in Angiosperms ('angio-ovulate plants'). Conifers are further disregarded by evolutionists because they are a specialised clade within seed plants and considered unlikely to throw light on the evolutionary lineages leading to the flowering plants. This is clearly a mistake because the diverse ways in which conifers can receive pollen include some that approach closely the angio-ovulate condition of Angiosperms. This suggests an orthogenetic trend in seed plant diversification leading to the reproductive mechanisms of flowering plants. We thus have the paradox that although the evolutionary lines leading to modern conifers are well understood from the classical researches of Florin (1954), we have only a limited understanding of functional processes at pollination. In contrast, the origin of angiosperm flowers is not known, and yet we have an enormous amount of information about their pollination mechanisms.

We have had some success in studying pollen capture in conifers by working in both northern and southern hemispheres, effectively allowing two spring pollination periods per year and a large sampling of taxa. Emphasis has been on the Podocarpaceae, appropriate because of their large size (c. 200 species) and wide geographic range (pan-tropical and southern hemisphere). They have been shown to possess distinctive pollen capture mechanisms in relation to complex structural features of both ovulate cones and pollen. In this article we emphasise pollen features of all conifers in relation to pollen capture.

Pollen in air and water

Gymnosperm pollen has both aerodynamic and hydrodynamic features that would appear to be counterproductive, since pollen functions in both air and water. Wind dispersal is an aerodynamic process facilitated by the small size and, in some cases, reduced density of pollen. Water is detrimental to wind dispersal, but usually necessary for pollen capture. When conifer pollen contacts water its hydrodynamic features become evident and important, as can be shown by simple experiment. In saccate pollen hydrodynamic properties may indirectly promote wind dispersal, since sacci reduce grain density. Generally, however, rain is inimical to wind dispersal, but in some instances may promote pollen capture within cones themselves, e.g. in *Picea* (Runions and Owens, 1996). Evolutionary questions may be asked as to how natural selection can work on seemingly antagonistic structures. The story of conifer pollen diversity may explain these paradoxical processes.

Pollen

Pollen structure

Information on pollen structure in conifers has been summarised extensively by Ueno (1960), with a recent up-date and more modern terminology for sporoderm by Kurmann (1991). Table 1 summarises some features of conifer pollen in relation to its reaction to water, as may be observed by adding pollen to small drops of water. All conifer pollen swells

TABLE 1. Structure and hydrodynamics of conifer pollen (families after Page, 1990)

[1]In *Phyllocladus* the exine might be described as alveolate; the sacci are hollow (M.H. Kurmann, pers. comm.)

1. Pollen saccate, floating in water within the surface meniscus. Sporoderm alveolate; intine thin.
    a. Some Pinaceae (*Abies, Cathaya, Cedrus, Keteleeria, Picea, Pinus, Pseudolarix*). Pollen in *Hesperopeuce* with a 'vestigial' or non-functional saccus.
    b. Podocarpaceae (except *Saxegothaea*).
2. Pollen non-saccate, sinking in water.
    a. Sporoderm rupturing and shedding exine within a few minutes of contact with water. Sporoderm granular, intine thick (Cephalotaxaceae, Cupressaceae, Sciadopityaceae, Taxaceae, Taxodiaceae).
    b. Pollen not rupturing on contact with water. Sporoderm various.
        (i) Exine granular, without vestigial sacci (Araucariaceae; *Larix, Pseudolarix, Tsuga* – Pinaceae; *Saxegothaea* - Podocarpaceae).
        (ii) Exine with 'vestigial' sacci, including cavities but without endo-reticulations[1] (Phyllocladaceae; *Hesperopeuce* - Pinaceae)

considerably on first contact with water. Pollen has been described as 'wettable' and 'non-wettable' (Tomlinson, 1994) but perhaps 'sinkable' and 'non-sinkable' is preferred, since this is the basic distinction. Pollen that sinks in water may either subsequently burst within a few minutes, shedding the exine as described later, or remain whole. In the distribution of these hydrodynamic reactions there is incomplete agreement with taxonomic boundaries, but good agreement with the mechanism of pollen capture as indicated later. In addition to the features noted in Table 1 there is a considerable range of grain size, intine thickness and gametophyte elaboration, but emphasis is on sporoderm anatomy since little is known about the functional significance of other features. Grain size ranges over about an order of magnitude, cell number at the time of pollen shedding is from one to over 20 (*Araucaria*) but seemingly quite fixed for each taxon (Ueno, 1960). Intine thickness in dehydrated grains varies two- or three-fold. Initial swelling of the grain is presumably a result of imbibition of water by the intine, demonstrating that, at least locally, the exine is permeable to water.

Saccate pollen occurs in Pinaceae and Podocarpaceae, but not in all genera. Sacci represent extensions of the outer exine, which has an alveolate structure (Kurmann, 1989). Grains are usually bisaccate, but they are trisaccate in *Dacrycarpus* and *Microcachrys* of Podocarpaceae. Sacci have an internal honeycomb of sporopollenin that serves the mechanical function of maintaining the rigidity of this region in both wet and dry conditions i.e., in dry air the body of the grain shrinks, but sacci remain inflated. The exine of the saccus is minutely porous. Sacci when gas-filled allow the pollen to float. A 'vestigial' saccus is reported for *Phyllocladus* and *Hesperopeuce*, but this does not function in flotation. *Dacrydium* (Podocarpaceae) also has reduced sacci, but the hydrodynamics of its pollen is not known.

Non-saccate pollen shows limited surface sculpturing, but this conceals differences in reaction to water. The five families Cephalotaxaceae, Cupressaceae, Sciadopityaceae, Taxaceae and Taxodiaceae (collectively the 'taxad' group) are uniformly consistent in their bursting (exine-shedding) reaction to water. This structural uniformity is particularly evident in the Cupressaceae in which genera cannot be distinguished by pollen features (Kurmann, 1994). This leaves a diverse assemblage of Araucariaceae, Phyllocladaceae and those 'anomalous' genera in Pinaceae and Podocarpaceae that lack sacci, all of which have pollen that sinks in water but without bursting. Despite its many structurally distinctive features, Phyllocladaceae may be close to Podocarpaceae, and some authors (e.g. Quinn, 1987, largely on embryological grounds) do not recognise *Phyllocladus* as a separate family. Similarly *Nothotsuga* and *Hesperopeuce* are often included in *Tsuga*, contrary to the classification by Page (1990). Detailed ultrastructural study of the grains of these taxa may be informative of their evolutionary relationships.

Pollen and pollination drops

Pollen structure cannot be considered independently of cone and ovule morphology, because the two are intimately interrelated in a way that may be referred to as a 'pollen capture syndrome.' Several categories are recognised, and the relationships are summarised in Table 2 but with considerable simplification.

Saccate pollen shows a contrasted hydrodynamic behaviour because grains float within the surface meniscus of a water drop, with bubble-like properties. On any curved water surface they always rise to its highest point, i.e., they will accumulate in the marginal meniscus of an open water-filled container or more dramatically, on the apex of the convex surface of a drop of water. Here they close-pack to form a raft of pollen. Mobility of the pollen raft, or of individual grains, can be demonstrated by propelling them along the surface by gently blowing on it. This mobility within the surface meniscus is essential to pollen capture. Pollen always floats in a water drop with the sacci and distal pole up, or at least towards the outer surface. Sacci are clearly gas-filled, but whether this is just air is not known.

In all members of the 'taxad' group, pollen enters the drop of water directly and sinks, demonstrating its relative density. It swells conspicuously and usually within 10-15 minutes the exine ruptures in a consistent way so that the pollen contents are extruded. The shed exine has a constant configuration because it always ruptures at the region of the obscure germ pore, which is presumably also the site of entry of water. Expansion is the result of hydration of the intine, which swells enormously. The extruded body has about 100 times the volume of the initial grain. Taxodiaceae, at least, are distinguished by the small papilla that represents the pore site. The result of the operation is that a population of hydrated grains comes to be represented by the swollen pollen contents associated with their ruptured exines. Precisely the same process occurs when pollen reaches a pollination drop, as can be observed directly. Clearly with such grains, any contact of dry grains with water before they reach the pollination drop is likely to render them non-functional in pollination.

In a third category are grains of *Phyllocladus*, which sink in water, swelling but not rupturing. This process occurs in pollination drops, as illustrated in Tomlinson *et al.* (1997).

The fourth category includes all taxa without a pollination drop, i.e., Araucariaceae and the anomalous members of Pinaceae (*Larix, Pseudotsuga* and *Tsuga*) on the one hand, and *Saxegothaea* of Podocarpaceae, on the other. Pollen capture does not involve water, since no pollination drop is present, but in water, pollen sinks but does not burst.

TABLE 2. Pollen capture mechanisms in conifers

A. Pollen germinating within micropylar cavity, pollination drop present or absent.

1. Pollination drop present

a. Ovule inverted at time of pollination; pollen scavenged and selected

(i) All Podocarpaceae except *Saxegothaea*

(ii) Some Pinaceae (*Cedrus*, *Picea*, *Pinus*)

b. Ovule without constant orientation at time of pollination; pollen neither scavenged nor selected.

(i) Pollen shedding exine in water - Cephalotaxaceae, Cupressaceae, Sciadopityaceae, Taxaceae, Taxodiaceae

(ii) Pollen not shedding exine in water - Phyllocladaceae

2. Pollination drop absent, pollen engulfed by invagination of integumentary extensions - *Larix*, *Pseudotsuga* (Pinaceae)

B. Pollen germinating outside micropylar cavity (extended siphonogamy), pollination drop absent.

1. Nucellus extruded through micropyle - *Saxegothaea* (Podocarpaceae)

2. Nucellus not extruded, or micropyle absent - Araucariaceae, *Tsuga* (Pinaceae)

How these contrasted hydrodynamic properties relate to pollen capture may be described later in further detail.

Pollen capture - case studies

We have defined pollen capture in gymno-ovulate plants as the process, following the introduction of pollen to the vicinity of the ovule, by means of which pollen is ingested into the ovule via the micropylar cavity and comes to rest on the nucellus. It is a process one can only refer to with precision in taxa that possess a pollination drop. In conifers (and Gnetopsids) the pollen chamber, or cavity immediately above the nucellus, is never as spacious as in cycads and *Ginkgo*, or presumably as in early seed plants, but the same term seems topographically appropriate for all Gymnosperms.

One might also distinguish the two major groups of seed plants (Angiosperms and Gymnosperms) on the basis of the position of pollen germination: inside the pollen chamber in gymnosperms, outside the ovule and on the surface of the stigma in angiosperms, as was first pointed out by Robert Brown. However, there is an intermediate state in Araucariaceae,

Saxegothaea and *Tsuga* in which pollen does not enter a pollen chamber, as in angiosperms, although there is still direct access to ovules at pollination, as in gymnosperms. In these taxa the gymno-ovulate condition is only necessary seemingly because pollen tubes cannot grow through bract tissue. The claim that this was so by Eames (1913) has recently been challenged by Owens *et al.* (1995).

Pollen capture in Podocarpaceae

In this family the size and shape of sacci relative to the body of the pollen grain varies considerably from genus to genus (Tomlinson, 1994). Bisaccate and trisaccate (as in *Dacrycarpus* and *Microcachrys*) grains seem functionally identical. With the exception of *Saxegothaea*, a pollination drop of distinctive shape occurs in all members of the family which have been examined. Despite variation in structural features of the cone, a similar mechanism seems to operate in all taxa. The essential feature is that the ovule is inverted at the time of pollination, the inversion being the result of the activity of the epimatium, on whose adaxial surface the ovule is borne (Tomlinson, 1992). Each ovulate complex consists of a single ovule and its associated, usually enveloping epimatium. The single exception is *Microstrobos*, which lacks an epimatium, but in which inversion of the ovule is achieved by recurvature of the cone axis. Ovulate complexes per cone vary from one to two, as in most *Podocarpus* and *Dacrycarpus* species, up to about twelve in species like *Prumnopitys andinus* and *P. taxifolius*. The overall result is that for each complex the secreted pollination drop contacts an adjacent surface. The complex is thus not only morphologically but also topographically inverted, usually by a strong negative geotropic response of the cone axis. In *Lagarostrobos franklinii* (Tasmania) the ovule is morphologically erect, but becomes inverted by a pendulous cone axis. Therefore, in all taxa the drop itself is essentially inverted and runs down an adjacent surface which is wettable because it is non-waxy. Restriction of the drop to a precise location is determined by abundant wax development on adjacent and therefore non-wettable surfaces.

This construction leads to a process of 'pollen scavenging,' since pollen that falls on the wettable portion of the cone prior to drop secretion can be picked up subsequently when the drop appears i.e., pollen need not land directly on the pollination drop in order to be captured (Tomlinson, 1991; Tomlinson *et al.*, 1991). This extends the pollen capture capability of the ovule in time. Furthermore, since the drop is no longer spherical (of minimal surface area) but of a shape determined by the configuration of the wettable surface on which it falls, pollen capture is extended in space, since the surface area of the drop is now increased. Extension of the initial pollen capture process in both space and time presumably increases its chances of success. The actual configuration of the wettable area that receives the drop varies considerably from genus to genus (Tomlinson *et al.*, 1991), but the overall process seems consistent both with structural features of the ovule and the presence of saccate pollen.

Viewed from a functional perspective, we have an explanation for most of the ovulate cone structures in Podocarpaceae, at the time of pollination, without any initial recourse to a comparison with other conifers. The epimatium can still be interpreted as the homologue of the ovuliferous scale in, for example, Pinaceae, since it actually does produce the ovule, but it can best be seen as an essential feature of the pollen capture mechanism. In its subsequent development the epimatium may or may not continue to enclose the ovule and it may play an important role in rendering the mature seed attractive to animal dispersal agents (usually birds).

Once 'scavenged' by the extended pollination drop, saccate pollen rises up the surface of the meniscus and typically accumulates near the mouth of the inverted micropyle. Doyle (1945) suggested that pollen rises **within** the drop and so floats upwards until it comes to rest with the germination furrow toward the nucellar surface. Direct observation and knowledge of pollen behaviour in water shows that this mechanism is unlikely. Ingestion of saccate pollen into the pollen chamber requires the water surface to withdraw, carrying with it the pollen held in the meniscus. Drop retraction is apparently the result of evaporation in dry air, rather than physiological resorption (Tomlinson *et al*., 1997).

The process of scavenging is rendered selective by the bubble-like properties of saccate pollen, which is the only kind of particle capable of reaching the micropylar orifice. All other pollen, and additional particles that may inadvertently be captured, either sinks in the pollination drop, or remains stranded when the drop evaporates. It is characteristic of drop retraction that the lower portion of the meniscus is rejected by the micropyle as evaporation occurs, a seeming design feature of the mechanism. Selection can be observed directly by adding to a pollination drop mixtures of saccate and non-saccate pollen as well as other particles. The process is not so much a selection for saccate pollen as a rejection of any particle without bubble-like properties. However, the process does not discriminate between different types of saccate pollen. Where species of Podocarpaceae co-occur in some numbers, as in New Zealand, phenological differences minimise but do not prevent pollen contamination. The process may therefore be directed towards the exclusion of pathogen spores or insect eggs.

Pollen capture in Pinaceae

Information about this family is extensive and was initially summarised by Doyle (1945), with an implied typology beginning with the fossil genus *Pseudovoltzia*. Pinaceae uniformly have ovulate cones with numerous spirally arranged fertile bracts. Each seed scale complex consists of a well-developed ovuliferous scale with a pair of inverted ovules, the complex subtended by a bract scale, which is only prominent at the time of pollination. Subsequent research has confirmed and elaborated Doyle's conclusions although details for many taxa are still incomplete. Table 3 summarises information and shows that although saccate pollen occurs in the majority of genera, it is not always associated with a reported pollination drop. However, in the three genera that already lack an involvement of water in pollen capture (*Larix, Pseudotsuga* and *Tsuga*) pollen is non-saccate. In other genera with saccate pollen but without a reported pollination drop (e.g., *Abies,* and possibly *Pseudolarix*) a hydrodynamic mechanism may still operate since there is evidence that pollen, once settled on the vicinity of cone scales, may be redistributed by rain water. This has been reported in *Cedrus* (Takaso and Owens, 1995b) and in *Picea* (Runions and Owens, 1996).

Picea and *Pinus* are analogous to Podocarpaceae in that there is a two-step process of pollen capture. Pollen is initially retained on integumentary extensions (often unfortunately referred to as 'stigmas'), which are rendered sticky by secretion of minute droplets of water. Pollen is subsequently scavenged by secretion of a normal pollination drop and drawn into the micropyle, in this case apparently by active resorption. Whether a particle rejection process could operate because of the distinctive hydrodynamic properties of saccate pollen remains to be established by experiment but Runion and Owens (1996) emphasised the similarity between *Picea* and Podocarpaceae in terms of pollen scavenging. Pollination drops are partly concealed by subtending bracts so that some dissection of cones may be necessary to observe them. Lill and Sweet (1977) manipulated cones in such a way that both saccate and non-saccate pollen was transferred to the nucellus. The secretion of minute droplets by integumentary extensions ocurrs in at least six genera of Pinaceae: *Abies, Cedrus, Hesperopeuce, Picea, Pinus* and possibly *Pseudolarix* (Takaso and Owens, 1995b).

The process of pollen capture in taxa without saccate pollen is described below.

Pollen function in taxads and *Phyllocladus*

The situation in the five families in which pollen bursts, shedding the exine on contact with water, has no immediate functional explanation. It is possible that the mechanism simply guarantees rapid entry of pollen into the pollination drop, since most organic particles are otherwise held on a water surface without penetrating it. However, *Phyllocladus* pollen enters

TABLE 3. Pollen capture mechanisms in Pinaceae

The mechanism in *Cathaya, Keteleeria,* and *Nothotsuga* is not known; *Pseudolarix* possibly lacks a pollination drop and may be similar to *Abies* and *Cedrus*.

1. Pollen with well-developed sacci

   A. Pollination drop present; scavenging pollen previously retained on integumentary arms ('stigmas') - *Cedrus* (?)[a]; *Picea*[b]; *Pinus*[c]

   B. Pollination drop absent; pollen sifting to integumentary extensions by gravity; possibly rain assisted - *Abies*[d]; *Cedrus* (?)[a]

2. Pollen with vestigial or non-functional sacci; pollen retained on bracts or integumentary extensions, germinating and delivering gametes to ovule via a long pollen tube (extended siphonogamy); *Hesperopeuce*.[e]

3. Pollen non-saccate, pollination drop absent

   A. Pollen landing on integumentary extensions and engulfed by their subsequent growth and invagination. - *Larix*[f]; *Pseudotsuga*[g]

   B. Pollen landing on fertile bracts, germinating and delivering gametes to ovule via a long pollen tube (extended siphonogamy); integument not elaborated. - *Tsuga*[h]

Footnotes (representative citations only):
[a]Doyle and O'Leary (1935b); Takaso and Owens (1995a, b).
[b]Owens *et al.* (1987); Runions and Owens (1996).
[c]Brown and Bridgewater (1987); Doyle and O'Leary (1935a); Greenwood (1986); Owens *et al.* (1981b).
[d]Singh and Owens (1982).
[e]Owens and Molder (1975).
[f]Doyle and O'Leary (1935b); Powell and Tosh (1991).
[g]Doyle and O'Leary (1935b); Owens *et al.* (1981a); Takaso and Owens (1995a).
[h]Doyle and O'Leary (1935b).

the pollination drop without bursting. Furthermore drop retraction in all these taxa guarantees that all particles within the drop and on its surface will be ingested and carried into the micropyle. There is clearly much scope for experimental work on this assemblage.

Pollen function in examples without a pollination drop

In *Agathis, Araucaria* (Araucariaceae), *Tsuga* (Pinaceae) and *Saxegothaea* (Podocarpaceae) pollen does not reach a pollen chamber prior to germination. Normally pollen germinates away from the ovule (e.g. on the bract scale surface) and delivery of male gametes to the female gametophyte is effected by extended siphonogamy, although the pollen tubes probably do not penetrate sporophytic tissue before they reach the nucellus. In *Saxegothaea* the nucellus is extended through the micropyle, presumably to facilitate the process. In all these taxa pollen seems totally independent of water in its normal function.

In *Larix* (Powell and Tosh, 1991) and *Pseudotsuga* (e.g. Owens *et al.*, 1981a) there is no pollination drop and pollen lands directly on integumentary extensions, subsequently being engulfed by quite dramatic changes in the integumentary orifice. Pollen is thus brought to the vicinity of the nucellus in the equivalent of a pollen chamber where it shows distinctive behaviour during germination. In both taxa germination may be promoted by secretion of liquid, either by the integument or the nucellus (Takaso and Owens, 1996). Although this fluid may be the homologue of a pollination drop it is produced post-pollination. In *Pseudotsuga* the exine is shed during pollen germination; otherwise, the fact that its pollen sinks in water does not seem relevant to pollen capture.

Drop resorption

A topic that has been little investigated is the response of the drop to added pollen. Clearly in some conifers addition of pollen is a stimulus for drop withdrawal, reported for *Pinus* by Doyle and O'Leary (1935a). Additional examples include *Callitris* (Baird, 1953), *Chamaecyparis* (Owens *et al.*, 1980), *Cryptomeria* (Yokoyama, 1975), *Phyllocladus* (Tomlinson *et al.*, 1997) and *Thuja* (Owens and Molder, 1980). In *Taxus*, on the other hand, added pollen does not stimulate drop withdrawal (Tomlinson and LeRoux, unpublished). This seems also the situation in Podocarpaceae, where drop retraction is the result of its evaporation (Tomlinson *et al.*, 1997). Continued investigation of this phenomenon is required. Taxa in which drop retraction seems to be promoted by the presence of pollen represent an unusual condition in which secretion and resorption may be mediated by the same tissue.

Conclusions

This is a preliminary review of an obviously complex topic, but draws attention to the diversity of mechanisms involved in pollen capture in conifers, with the clear recognition that pollen features are part of the syndrome, especially as conifer pollen shows remarkable responses to water determined by sporoderm structure. The evolutionary implications are considerable, as first suggested by Doyle (1945) but a careful analysis of how pollen capture mechanisms have diversified within conifers should await a more complete understanding of the processes involved. A basic question is whether pollen in modern conifers is ancestrally saccate or non-saccate. At present we have no certain answer. A more useful exercise is to point out directions for future research, as follows:

(i) The mechanism of drop secretion needs to be further investigated, especially in relation to structural and ultrastructural features of the nucellus, if this is indeed the site for drop production.

(ii) The chemical composition of the drop needs to be examined in many more taxa, and especially the influence of these chemicals on pollen development - are the contents of the drop significant other than as the result of cell autolysis?

(iii) The mechanism of drop resorption of taxa should be examined experimentally in a greater diversity of taxa. We have some evidence that pollen-stimulated retraction is a physiological process in some taxa, but not in others. How this relates to efficiency of pollen capture needs to be known.

(iv) The ultrastructure of the pollen sporoderm needs examining in relation to the contrasted mechanisms of pollen capture. Existing correlations need further scrutiny.

(v) A more precise understanding of the physics and mechanics of saccate pollen is needed. This is highly relevant to gymnosperm evolution generally, since saccate pollen is widely distributed in extinct groups.

(vi) The mechanism that directs pollen tube growth toward the nucellus in taxa with extended siphonogamy needs to be better understood, especially as it implies a step to angio-ovuly. In conclusion, it is ironic that we know relatively little of these important processes that relate to efficient seed set in a group of such major economic importance as the conifers. Few taxa lack commercial value and most timber-producing species are grown from seed. Furthermore, the majority of conifers are threatened in some way. Any program of conservation, especially of endangered species must be based on an adequate knowledge of seed production in natural conditions. It is hoped that this kind of investigation can be promoted in the future, and this paper simply draws attention to an important topic.

Acknowledgements

Support for field research on conifers in the South Pacific has been provided to P.B.T. by the National Geographic Society. We thank the Director of the Arnold Arboretum of Harvard University for access to its living conifer collections. Ms. Lulu LeRoux provided unpublished information on pollen capture in a number of gymnosperms.

References

Baird, A.M. (1953). The life history of *Callitris*. *Phytomorphology* **3**: 258–284.

Brown, S.D. and Bridgewater, F.E. (1987). Observations on pollination in loblolly pine. *Canad. J. Forest Res.* **17**: 299–303.

Chesnoy, L. (1993). Les sécrétions dans la pollinisation des gymnospermes. *Acta Bot. Gallica* **140**: 145–156.

Doyle, J. (1945). Developmental lines in pollination mechanisms in the Coniferales. *Sci. Proc. Roy. Dublin Soc.* **24**: 43–62.

Doyle, J. and O'Leary, M. (1935a). Pollination in *Pinus*. *Sci. Proc. Roy. Dublin Soc.* **21**: 181–190.

Doyle, J. and O'Leary, M. (1935b). Pollination in *Tsuga*, *Cedrus*, *Pseudotsuga*, and *Larix*. *Sci. Proc. Roy. Dublin Soc.* **21**: 191–204.

Eames, A.J. (1913). The morphology of *Agathis australis*. *Ann. Bot.* **27**: 1–38.

Florin, R. (1954). The female reproductive organs of conifers and taxads. *Biol. Rev. Cambridge Philos. Soc.* **29**: 367–389.

Greenwood, M.S. (1986). Gene exchange in loblolly pine: the relation between pollination mechanism, female receptivity and pollen availability. *Amer. J. Bot.* **73**: 1443–1451.

Kurmann, M.H. (1989). Pollen wall formation in *Abies concolor* and a discussion on wall layer homologies. *Canad. J. Bot.* **67**: 2489–2504.

Kurmann, M.H. (1991). Exine stratification in extant gymnosperms: a review of published transmission electron micrographs. *Kew Bull.* **47**: 25–39.

Kurmann, M.H. (1994). Pollen morphology and ultrastructure in the Cupressaceae. *Acta Bot. Gallica* **141**: 141–147.

Lill, B.S. and Sweet, G.B. (1977). Pollination in *Pinus radiata*. *New Zealand J. Forest Sci.* **7**: 21–34.

Owens, J.N. and Molder, M. (1975). Sexual reproduction of mountain hemlock (*Tsuga mertensiana*). *Canad. J. Bot.* **53**: 1811–1826.

Owens, J.N. and Molder, M. (1980). Sexual reproduction in western red cedar (*Thuja plicata*). *Canad. J. Bot.* **58**: 1376–1393.

Owens, J.N., Simpson, S.J. and Molder, M. (1980). The pollination mechanism in yellow cypress (*Chamaecyparis nootkatensis*). *Canad. J. Forest Res.* **10**: 564–572.

Owens, J.N., Simpson, S.J. and Molder, M. (1981a). The pollination mechanism and the optimal time of pollination in Douglas-fir (*Pseudotsuga menziesii*). *Canad. J. Forest Res.* **11**: 36–50.

Owens, J.N., Simpson, S.J. and Molder, M. (1981b). Sexual reproduction of *Pinus contorta*. I. Pollen development, the pollination mechanism, and early ovule development. *Canad. J. Bot.* **59**: 1828–1843.

Owens, J.N., Simpson, S.J. and Caron, G.E. (1987). The pollination mechanism of Engelmann spruce (*Picea engelmannii). Canad. J. Bot.* **65**: 1439–1450.

Owens, J.N., Catalano, G.L., Morris, S.J. and Aitken-Christie, J. (1995). The reproductive biology of Kauri (*Agathis australis*). I. Pollination and prefertilization development. *Int. J. Pl. Sci.* **156**: 257–269.

Page, C.N. (1990). Coniferophytina (conifers and ginkgoids). In: K. Kubitzki (editor). The families and genera of vascular plants, 1, pp. 279–361, Springer Verlag, Heidelberg.

Powell, G.R. and Tosh, K.J. (1991). The pollination mechanism and development after bud burst of cones of *Larix laricina. Canad. J. Bot.* **69**: 1179–1187.

Quinn, C.J. (1987). The Phyllocladaceae - a critique. *Taxon* **36**: 559–565.

Runions, C.J. and Owens, J.N. (1996). Pollen scavenging and rain involvement in the pollination mechanism of interior spruce. *Canad. J. Bot.* **74**: 115–124.

Singh, H. (1977). The embryology of gymnosperms. Springer Verlag, Berlin.

Singh, H. and Owens, J.N. (1982). Sexual reproduction in grand fir (*Abies grandis*). *Canad. J. Bot.* **60**: 2197–2241.

Takaso, T. and Owens, J.N. (1995a). Ovulate cone morphology and pollination in *Pseudotsuga* and *Cedrus. Int. J. Pl. Sci.* **156**: 630–639.

Takaso, T. and Owens, J.N. (1995b). Pollination drop and microdrop secretions in *Cedrus. Int. J. Pl. Sci.* **56**: 640–649.

Takaso, T. and Owens, J.N. (1996). Post pollination-prezygotic ovular secretions into the micropylar canal in *Pseudotsuga menziesii* (Pinaceae). *J. Pl. Res.* **109**: 147–160.

Tomlinson, P.B. (1991). Pollen scavenging as a novel reproductive mechanism in Podocarpaceae. *Natl. Geogr. Res. Expl.* **7**: 188–195.

Tomlinson, P.B. (1992). Aspects of cone morphology and development in Podocarpaceae (Coniferales). *Int. J. Pl. Sci.* **153**: 572–588.

Tomlinson, P.B. (1994). Functional morphology of saccate pollen in conifers with special reference to Podocarpaceae. *Int. J. Pl. Sci.* **155**: 689–715.

Tomlinson, P.B., Rattenbury, J.A. and Braggins, J.E. (1991). Pollination drop in relation to cone morphology in Podocarpaceae: a novel reproductive mechanism. *Amer. J. Bot.* **78**: 1289–1303.

Tomlinson, P.B., Braggins, J.E. and Rattenbury, J.A. (1997). Contrasted pollen capture mechanisms in Phyllocladaceae and certain Podocarpaceae (Coniferales). *Amer. J. Bot.* **84**: 214–223.

Ueno, J. (1960). Studies on pollen grains of Gymnospermae. Concluding remarks to the relationship between Coniferae. *J. Inst. Polytechn. Osaka City Univ. Ser. D, Biol.* **11**: 109–136.

Yokoyama, T. (1975). Embryogenesis and cone growth in *Cryptomeria japonica. Bull. Gov. Forest Exp. Sta.* **277**: 1–20. [In Japanese, with English summary].

Stevenson, D.W., Norstog, K.J. and Fawcett, P.K.S., (1998). Pollination biology of cycads. In: S.J. Owens and P.J. Rudall (Editors). Reproductive Biology, pp. 277–294. Royal Botanic Gardens, Kew.

POLLINATION BIOLOGY OF CYCADS

DENNIS WM. STEVENSON[1], KNUT J. NORSTOG[2] AND PRISCILLA K.S. FAWCETT[2]

[1]New York Botanical Garden, Bronx, NY, USA
[2]Fairchild Tropical Garden, Miami, FL, USA

Abstract

Two species of cycad in Florida and one in Mexico are known to be pollinated by beetles. The native Florida *Zamia pumila* is pollinated by a snout weevil, *Rhopalotria slossoni* and a clavicorn beetle, *Pharaxonotha zamiae*. An introduced cycad, *Z. furfuracea*, is in Florida pollinated by the snout weevil, *R. mollis*, apparently introduced from Mexico along with its host. Observations and experiments show that both zamias are dependent upon pollinating beetles for their reproduction and, in turn, the pollinating insects are dependent upon the cycads for such "rewards" as brood places, food, and shelter. *Dioon califanoi*, like *Z. pumila*, is pollinated by a weevil, *Parallocorynus bicolor* and a species of *Pharaxonotha*. Aspects of the pollination syndrome in *D. califanoi* are not completely worked out but are probably much like those in *Zamia*. The importance of these coevolutionary relationships for the ecology and conservation of these cycad species is unquestioned. Neither the insects nor the cycads are separately capable of long-term survival in nature, and unless both are protected and conserved, wild populations of *Zamia* and *Dioon* and no doubt many other cycads are in jeopardy.

Introduction

Although cycads have only recently proven to be insect pollinated, evidence is accumulating that many cycads have symbiotic associations with host-specific insects, mostly beetles. In some way, not completely understood, such insects are attracted to cycad cones, sometimes by fragrances, sometimes for food, shelter, a breeding place, and probably combinations of some or all of these. Oberprieler (1989) remarked that the intimate association of cycads and insects, in this instance weevils, was first noticed and reported by Thunberg in 1773. In addition to pollination interactions, seed cones and seeds of cycads often are preyed upon by insects, including lepidopterans that are capable of damaging or destroying them should the infestation be heavy, as well as by birds, monkeys, etc. which may act as agents of seed dispersal. Involved in all of these plant-animal interactions are the cycad toxins in complex symbiotic relationships only now becoming understood.

C.J. Chamberlain, a widely recognised authority on cycad phenomena (Chamberlain, 1919, 1935) wrote in 1935, "there have been many reports of insect pollination, but in a rather extensive field study in which all the genera have been examined, nothing has been observed which would indicate anything but wind pollination so characteristic of the whole group of gymnosperms." Despite Chamberlain's rather authoritarian opinion to the contrary, considerable evidence for insect pollination of cycads had existed since 1893 (Pearson, 1906; Rattray, 1913) and no experimental data had been obtained for either wind or insect pollination.

At about the time Chamberlain and his students were exploring the morphology and anatomy of the Mexican cycads in the early 1900's, nearly all of the pieces of the cycad-pollination puzzle had been examined and fitted together into a credible explanation for pollination by insects. Pearson (1906), observing beetles in cones of *Encephalartos villosus*, suspected they might in some way be involved in pollination of that species. Pearson wrote "it may be noted that the habitat of *E. villosus* and the position of its cones must, in most cases, render wind-pollination impossible, for the bush is usually so dense that the air a few feet above the ground can rarely move rapidly enough to carry even so light a substance as pollen." He credited Joseph Hooker as first suggesting that cycads must be entomophilous, and goes on to say that male cones of *E. villosus* give off a strong odour at the time pollen is released and that such cones swarm with weevils, which are covered with pollen and are found only on cycads. Pearson did not observe weevils actually transporting pollen to female cones and suggested only that it was probable.

Pearson's observations were subsequently enlarged upon by Rattray (1913) who worked out many of the remarkable interactions between weevils and pollen transport in *E. villosus* and *E. altensteinii*. A weevil, probably the Cossonine *Porthetes zamiae* found in and around Durban by R.A. Crowson (pers. comm. to KJN), was found in great numbers in male cones of *E. altensteinii* at the time pollen is shed in April. At this time the cones emitted a faint but distinct odour and visiting weevils, mating upon the cones and feeding on pollen, become so covered with pollen that much of it adhered to their bodies and remained there for 60 hours or so.

While Rattray failed to present experimental evidence showing that beetles were partially or exclusively responsible for the pollination of the cycad, his comments are worth quoting because they so accurately reflect present thinking on the subject. "The whole life-history [of the weevil] appears so dependent on *Encephalartos* that it would be only reasonable to find that the cycad had demanded a *quid pro quo* and had turned the visits to use."

Rattray also called attention to the major defect in the concept that cycads are wind-pollinated. "Although I have examined over a score of female cones during the time pollen is shed and have had three under daily observation in my own garden I have never noted any separation of the sporophylls. Although I still feel bound to believe that it must occur, yet it must be only to a very limited extent — a condition which, while not altogether excluding anemophily in this species, certainly renders it much less probable; for the exposed parts of the sporophylls stand almost horizontally and do not overlap, so that unless the separation were very marked only wind-borne pollen coming horizontally could effect entrance, and only those ovules on the side of the prevailing winds would be fertilised (see Fig. 1). But in all these cones which I have worked through, even from plants growing on the outside of a group, embryos had been formed in uniform proportions on all sides." One can add here that cycad pollen tends to clump and is not readily airborne.

Later investigators also implicated insects in some aspects of pollination; Marloth (1914) noted insects in cones of *Encephalartos* and Baird (1939) considered that insects might transport pollen within cones of *Macrozamia*, but thought that inter-plant pollen transport required wind. More recently, Chadwick (1993) and Ornduff (1993) presented circumstantial evidence indicating that *Macrozamia communis* and *M. riedlei*, respectively, are weevil pollinated. Connell and Ladd (1993) considered weevil pollination in *M. riedlei* to be a possibility but were unable to produce statistically significant evidence either *pro* or *con*. Other botanists, however, contended that cycads were wind-pollinated, for example Lawson (1926), found nothing to suggest that *Bowenia* required any agent other than wind for its pollination but remarked that pollination in some cases occurred between plants separated by "a great distance." More recently, Wilson (1993) questioned the importance of wind in the pollination of *Bowenia* and emphasised the role of a weevil, *Tranes subopaca* in the pollination of *B. serrulata*.

FIG. 1. Pollen-receptive female cone of *Zamia furfuracea*. Receptivity is indicated by a vertical separation of two adjacent rows of microsporophylls giving the only access to the interior. Bar = 5 mm.

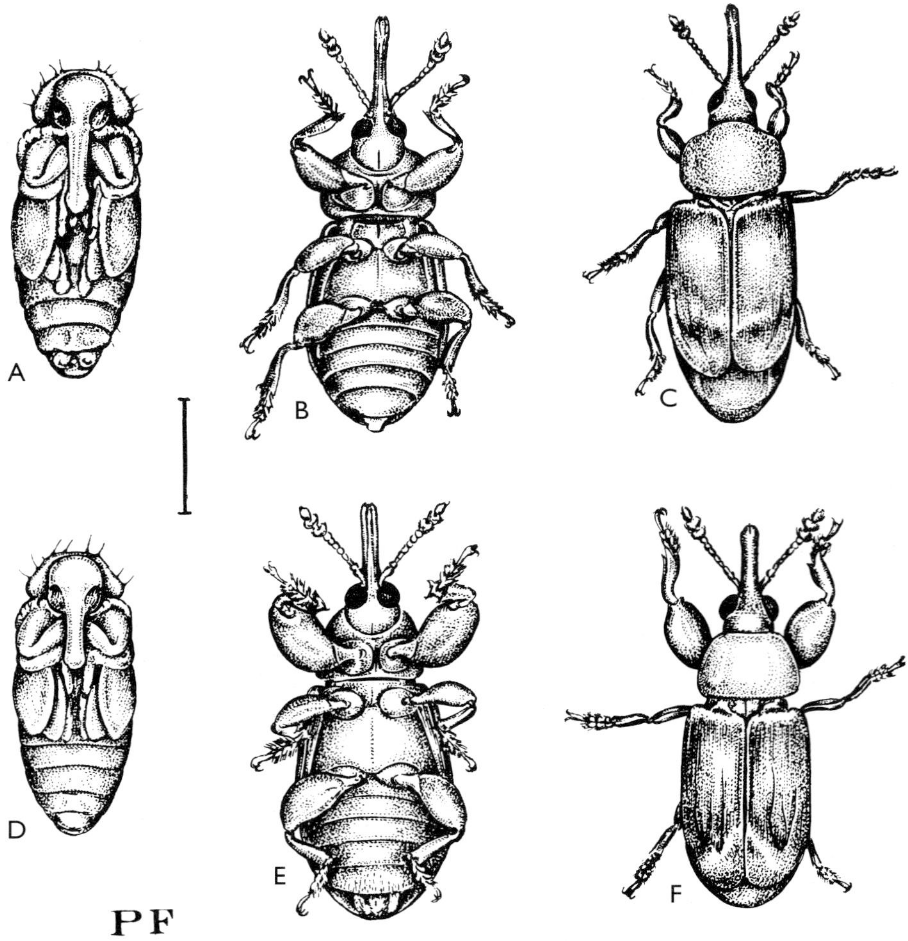

FIG. 2. *Rhopalotria mollis*, a pollinator of *Zamia furfuracea*. A. Female pupa. B. Adult female, ventral view. C. Adult female, dorsal view. D. Male pupa. E. Adult male, ventral view. F. Adult male, dorsal view. Bar = 1.0 mm.

In light of what now is known of the highly specialised relationships between several species of weevils as obligate pollinators of genera and species of cycads, and indirect evidence that others are similarly pollinated, it is difficult to understand why insect pollination was neglected for such a long time after Rattray's remarkable early work. The answer most likely lies in the absent of experimental procedures designed to test the hypotheses of insect or wind pollination and the purely anecdotal evidence that was then available.

Because of the long controversial history of nonconclusive evidence, Norstog and Stevenson (1980) and Norstog *et al.* (1986) designed experiments and established criteria to test both the wind and insect hypotheses of cycad pollination biology. Thus exclusion experiments were designed to test for seed set. The following conditions were established: (1) exclusion of all pollen entry by plastic bagging of prepollination ovulate cones; (2) exclusion of insects by net bagging with a mesh size excluding insects but allowing pollen entry by wind; and (3) exclusion of wind by covering prepollination ovulate cones with inverted cylindrical beakers so that insects could gain access to the ovulate strobilus. Other criteria that were used to determinate if insect pollination was supported included direct observations of potential pollinators and maturing strobili, whether or not the potential pollinators visited other plant genera, the life cycle of pollinators and the pollinated, and the direct observation of pollen being carried from microsporangiate strobili to the micropyles of the ovules of ovulate strobili. Similar exclusion experiments have also been conducted on two species of *Encephalartos* by Donaldson *et al.* (1995) and Donaldson (1997).

Pollination of *Zamia furfuracea*

Zamia furfuracea, a Mexican cycad grown at Fairchild Tropical Garden in scattered plantings is perfectly fertile (i.e., produces viable seed) in cultivation. It was reported by Norstog and Stevenson (1980) to attract swarms of tiny weevils at the time male cones were maturing. The weevils breed and feed upon male cones and, later, adults feeding and sheltering amongst microsporophylls, leave the cones (newly metamorphosed adults chewing exit holes, Figs. 2, 3A-G) and fly to other *Z. furfuracea* cones, including nearby female cones (Fig. 3K). It is during such visitations that pollen is thought to be transferred to ovules. Though circumstantial, the authors considered their observations to be strong evidence that *Z. furfuracea* is pollinated by a specific insect. Also, citing observations that female cones of the native cycad *Z. integrifolia* were fertile despite being buried beneath surface litter, they thought that this species too was insect-pollinated. This set the stage for the implementation of experimental criteria as outlined above.

In the next five years, Norstog *et al.* (1986) succeeded in demonstrating by exclusion experiments, in which either wind or insects were prevented from carrying pollen into female cones, that this weevil, *Rhopalotria mollis*, probably imported inadvertently from Mexico, was the faithful pollinator of *Z. furfuracea*.

The female cone is tightly closed during most of its development, only opening by a narrow crack when it is pollen-receptive (Fig. 1). This crack forms by separation of megasporophylls of adjacent vertical orthostichies and along the phyllotaxic spiral in the apical region of the ovulate strobilus. The crack remains open for several days, then closes completely for the duration of cone maturation. If one opens such closed cones, one often finds live weevils that have been trapped inside. A similar action is described by Chadwick (1993) in female cones of *Macrozamia communis*, in which observations suggest a weevil, *Tranes lyterioides* may be a pollinator. This weevil completes its entire life cycle in male cones of *M. communis* in much the same way as *R. mollis* does in *Z. furfuracea*.

The weevils that start the annual breeding cycle in cones of *Z. furfuracea* emerge from long-dormant larvae that "over-winter" in debris at the base of the cycads they later infest, and they put in an appearance at about the time the male cones approach maturity. The first breeding cycle of the weevil is short, a week or so, and involves just a few individuals, but increasingly large numbers of progeny in succeeding generations soon follow.

Insect attractants

Weevils are attracted to male cones of *Z. furfuracea* at about the time microsporangia are mature and just before they are ready to dehisce (Fig. 3A). At about this time also both male and female cones become aromatic and warm, and a few of the sporophylls begin to loosen and separate. The heat and odours result from developmental and metabolic processes that are of wide occurrence among cycads and function both to cause male and female cones to open and to volatilise fragrances (Tang, 1987b; Tang, 1993; Tang *et al.*, 1987). The male cone scent of *Z. furfuracea* was identified by Pellmyr *et al.* (1991) as consisting of two major components: 1,3-octadiene and linalool. Linalool is known to be an attractant of euglossine bees in orchid flowers. These would appear to have a pheromonal function. Both Pearson (1906) and Rattray (1913) remarked on the odours released by male *Encephalartos* cones at the time of pollination and one suspects these also have a pheromonal function.

Cycad toxicity and pollination by insects

At the time of pollination of both *Z. integrifolia* and *Z. furfuracea,* the microsporophyll parenchyma is rich in starch, whereas megasporophyll tissue is poor in starch (Norstog and Fawcett, 1989). In addition, it can be seen that a type of storage cell (idioblast) appears intact in microsporophylls but modified in content in megasporophylls. Norstog and Fawcett proposed that idioblasts of both species contain sequestered toxins in male-cone tissues, but not in female cone tissues in which toxins are liberated and dispersed at or prior to pollination. The evidence for this scheme is admittedly circumstantial. Idioblasts in cycad tissues are readily stained with ninhydrin and give the same color reaction as does the ninhydrin-positive neurotoxin BMAA, which was first discovered by Vega and Bell (1967) in tissues of *Cycas circinalis*.

At the time of pollination female-cone idioblasts do not stain with ninhydrin, suggesting either degradation or mobilisation of toxin(s) (BMAA?). Other differences in male- and female-cone idioblasts also are apparent at the time of pollination; specifically, male-cone idioblasts are unstained with PAS (periodic-acid Schiff reagent, a general stain for reducing sugars) but those of female cones are PAS positive. This may only indicate an altered structural status for such idioblasts in male vs. female cones. However, another class of toxins, the MAM-glycosides (cycasin, neocycasin, macrozamin) all of which are highly toxic to animal tissues (Whiting, 1963) appear to be PAS positive. This brings up the possibility that female cones of at least some cycads have evolved a differential deployment of these toxins in their idioblasts as a defense against predators. Because male cones must be attractive to certain predatory insects, specifically pollinators, they may have evolved in the direction of lower toxicity levels (or possibly temporarily lowered concentrations of toxins). Female cones, on the other hand, while offering certain cues to attract visitation by pollinators (mimicry of male cone morphology and odours) may employ toxins to prevent seed predation.

Breeding and feeding site selection by weevils

Swarming weevils exhibit cone preferences, selecting only one or two male cones from among as many as 15-20, and remaining upon the chosen cone 1-3 days before abandoning it (Fig. 3A). Male cones of *Z. furfuracea* become mature sequentially thus spreading out the period of pollen production for at least 4-6 weeks, while female cones, which are much fewer per plant, become mature more-or-less synchronously and all are pollen receptive during a 2-3 week interval (Fig. 1). This obviates the difficulty of synchronising cone production that might be encountered were cones of both sexes to be fertile only during similarly short periods of time. A corollary observation is that a dozen or more small pollen cones ripening sequentially, constitute evidence that insect pollination is the mode of reproduction in the species. Even so, male and female cone maturation must be synchronised, if only loosely, for pollination to succeed.

Adult male plants of *Zamia furfuracea* produce up to 20 or so male cones during the reproductive season. These mature in sequence to the point that of several cones visible on a plant, only 1-3 will be ready to shed pollen, a stage indicated by cone elongation, a slight separation of the microsporophylls, and an emanation of a musty fragrance.

Rhopalotria mollis individuals do not collect indiscriminately on male cycad cones, but are attracted only to those cones that are near to dehiscence and pollen release. The weevils swarm on to such cones (Fig. 3A) a day before the sporangia dehisce, and there is constant activity during which weevils probe for food, fight, mate and lay eggs in recently excavated feeding holes. Although the microsporophylls have not become loosely separated, they have begun to separate at the base of the cone. Sometimes the weevils probe into feeding holes which have already been made by other weevils, and sometime they start their own holes. When they start a hole of their own, they do not eat the stiff epidermal trichomes that cover the sporophyll endings, but they pull them out in tufts with their mandibles, which are clawed like hands. Trichomes of this cycad contain idioblasts and, presumably, also contain toxins and this behaviour apparently constitutes toxin avoidance. When a female is going to lay an egg, she feeds very deeply, and her snout goes down into the sporophyll all the way to her eyes (Figs. 2C, 3B).

After the femaies have laid their eggs, both sexes go inside the cone between the separated sporophylls at the base of the cone. Weevil adults and larvae feed on male sporophylls tissue and ingest idioblasts with sequestered BMAA together with other parenchyma but the weevils and larvae do not to digest the idioblasts. Rather, they accumulate and excrete them in their feces. The fecal pellets are used by *Rhopalotria* larvae to construct pupa cases (cocoons) and analysis of these cases reveals that they have a comparatively high content of the cycad neurotoxin, BMAA (Mark Duncan, pers. commun.). The toxins in the pupa cases give added protection to the larvae and pupae within them. While these actions are not direct evidence of either the existence of specific toxins in idioblasts, they do constitute strong circumstantial evidence supporting this assumption.

Larvae of *R. mollis* are legless grubs which hatch within a day or so from eggs laid principally in microsporophyll tips and are quite active (Fig. 3C, D). As many as six eggs can be laid in the tip of a single microsporophyll by various individual females, although 2-4 eggs are more usual, and a corresponding number of larvae can be found in each microsporophyll where they feed voraciously (Norstog and Fawcett, 1989). When, inevitably, feeding larvae meet, one eats the other and eventually only one large and well-fed individual remains. With the exception of this cannibalism, larvae feed exclusively on parenchyma, as indicated by staining reactions of sections of the larval gut. They do not feed at all on pollen, either at this or any other stage of their cycle.

About 3-4 days after hatching, a larva constructs a chamber in a microsporophyll stalk within which it pupates (Fig. 3E, F), its head oriented toward the outer end of the microsporophyll. The pupa case (cocoon) of fecal matter held together with chitin secreted by the larva hardens into a tough, cylindrical capsule. In 2-3 days the adult emerges after metamorphosis, usually boring through the tip of the sporophyll (Fig. 3G). Except toward the end of the season, the entire process from egg to adult takes 7-9 days, and it is possible that during the season a half-dozen, or so, generations may result from a single initial mating. One more larval instar occurs before pupation. Then, after their metamorphosis, adult weevils escape from pupa cases by chewing through the end of the cases and boring distinctive exit holes in the ends of the sporophylls (Fig. 3G).

Rewards to pollinators

Rhopalotria mollis is an excellent flyer and when males and females are attracted to male cones, as related above, where they mate and probe for food and in which the females lay eggs, but whether the emitted volatiles that seem to attract them may be considered as "rewards" is conjectural. A brood place and food thus constitute the chief "reward" the plant gives the weevil for the "service" of pollination.

Pollination by weevils

At the time of pollination some megasporophylls of female cones become separated slightly and give access to the interior via a longitudinal crack (Fig. 1). This is accompanied by

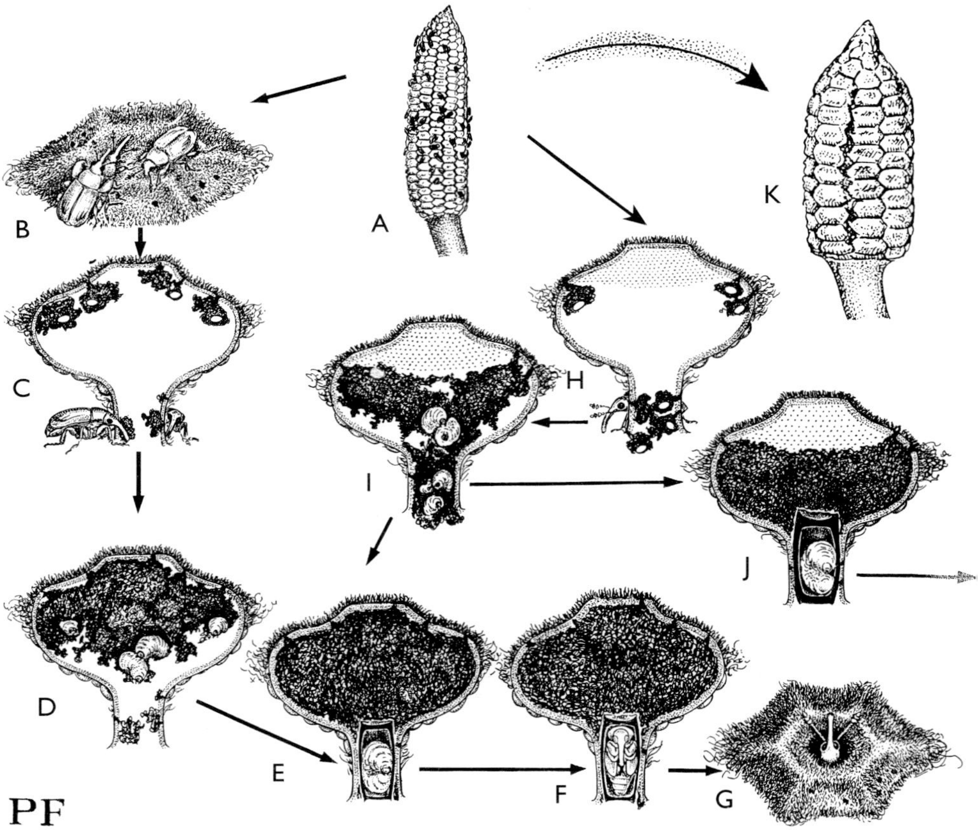

FIG. 3. The reproductive cycle of *Rhopalotria mollis* in *Zamia furfuracea*. A. Male cone with swarming weevils. B. Weevils feed, breed, and lay eggs in microsporophylls. C. Eggs are laid in microsporophyll ends; adults feed at the bases of sporophylls. D. Several larvae feeding in and excavating a microsporophyll. When larvae meet, one eats the other. Commonly only one larva survives in a microsporophyll. E. Larva in pupa case. F. Pupa. G. Emerging adult — from egg to adults usually encompasses about 7-9 days. H.–J. Toward the end of the breeding season, some larvae construct thick-walled pupa cases and enter diapause, remaining dormant until the following year, or longer. K. Pollen-receptive female cone and visitation by weevils.

generation of a considerable amount of heat as well as aroma (Tang *et al.*, 1987) and the cones become attractive to weevils. Throughout the season some weevils, perhaps in error, visit female cones where they attempt to feed, but apparently find the tissues to be unpalatable. The female sporophylls only become separated for a few days when the ovules are receptive as indicated by pollination droplets on the micropyles. We find a few weevils in such cones at this time though most escape.

After pupation, newly emerged adults re-enter male cones where they shelter and feed - a single male cone at this stage may contain a hundred or more weevils. In moving around in the cone they become covered with pollen and it is at this stage that visitation to female cones occurs (Fig. 3K). However, female cones are not fed upon, though interior surfaces sometimes exhibit minute scars, "nibble marks," which are probably made by weevils. The reason for the unpalatability of female cone tissue may be the release of toxin by idioblasts present in both male and female cone tissues. Idioblasts in female cones break down just prior to their pollination and as well as having eroded idioblasts, female cones contain very little starch (Norstog and Fawcett, 1989; Vovides *et al.*, 1993). Weevils that enter female cones do not feed on female cone tissue. Either they are repelled by some noxious compound (toxin) or the fact that no starch is available or both lead to this absence of feeding. In any case, pollen is transported to the cone and at least some of it is deposited upon the micropyles of ovules. As will be noted later, pollination of this cycad by weevils is very efficient and cone fertility may approach 100%.

More recently, in an interesting series of experiments and observations, Donaldson *et al.* (1995) employed both screens and insecticides to exclude the presumed pollinator, an undescribed Langurid beetle, from female cones of *Encephalartos cycadifolius* in habitat in Cape Province, South Africa. The percentage of fertile seed set was drastically reduced from 95% to 30% when the weevils were excluded. One outstanding observation resulted from their use of fluorescent dye to coat pollen-carrying beetles. It was possible to show that beetles, after entering cones, deposited the dye on the micropyles of a large number of ovules. It is the first conclusive evidence for the actual deposition of pollen on cycad ovules by specific insect pollinators.

Dormancy of weevils

Although all the *Rhopalotria* larvae in a male cone construct pupa cases, not all larvae metamorphose directly into pupae and adults. Towards the end of the season, there is a change in the tissue at the ends of the sporophylls which becomes green and tends to be avoided by the weevils. Near the end of the coning season of the cycad, the feeding behaviour of the weevil larvae is therefore a bit different from that occurring earlier in the season. Larvae tend to feed nearer the base of the microsporophylls (Fig. 3H, I). Such late-instar weevil larvae develop thick, opaque cuticles, become fat and waxy, and build very thick-walled pupa cases in which they go into diapause until the next season (Fig. 3C). These constitute the breeding stock for the following year.

Diapause

Diapause, which is a type of dormancy consisting of an arrested state of development occurs in *Rhopalotria* and can last for 2 or more years in this species of weevil (Norstog and Fawcett, unpubl. observ.) This, no doubt, accounts for the long-term efficiency of the relationship between the cycad and its pollinator and its continuity even over several interrupted breeding seasons.

We wondered what precipitated the change in behaviour of the weevils at the end of the season, and whether it was the lower starch content of the axis, or exposure to substances in the green tissue at the ends of the sporophyll, that would account for the initiation of diapause. Because fewer cones are maturing at the end of the season, those remaining continue to attract weevils longer than is the case earlier in the season. It seems to us now that the most likely explanation for the onset of diapause is that due to an increase in weevil numbers in late-season male cones and resulting consumption of all available food, competition among larvae in a microsporophyll becomes fierce. When looking at these late cones, we found larvae which were almost ready to pupate coming up from the base, and down from the end of the sporophyll, as well as younger larvae of all ages, and eggs (Fig. 3I). Often very little edible tissue was left but cannibalism continued apace. We think that the most likely cause of diapause is an increase of protein in the larval diet resulting from increased cannibalism. The greater thickness of the cuticles of diapausic larvae then may be due to increased levels of dietary proteins.

At the end of the season, we collected some late, male cones of *Zamia furfuracea*, and put 20 in one plastic box and 12 in another, then sealed the boxes with paraffin wax and masking tape. A few larvae metamorphosed at the end of December and in January when we were working on the cones of *Zamia pumila*. By March 17th, 11 had metamorphosed in the box with 20 cones, and seven in the box with 12 cones, but a massive metamorphosis took place at the end of March and the beginning of April, starting about a week after a colleague, Andrew Vovides, came to work at Fairchild Tropical Garden bringing with him cones of *Dioon edule*, *D. califanoi*, *Ceratozamia mexicana*, and *Zamia furfuracea* collected in Mexico in heavy plastic bags, pickled in 70% ethanol. The smell of the cycad cones and pickling agents was very strong all over the laboratory and persisted for several weeks and we think emergence from diapause in this instance probably had a pheromonal stimulus. Because live larvae in diapause continued to be present in the stored cones long after this massive emergence, diapause in *R. mollis* can last for more than one year. The end of diapause may either be nutritional and determined by the fat reserves carried by the larvae or the result of stimulants or inhibitors, These could be emitted by the plant which bore the cones containing larvae in diapause.

Pollination of *Zamia integrifolia*

Tang (1987a) conducted observations and exclusion experiments that demonstrated that the native *Zamia integrifolia* also was pollinated by a weevil, in this case *Rhopalotria slossoni*, and also by a second beetle species, *Pharaxonotha zamiae*. Moreover, these observations demonstrated that *R. slossoni*, like *R. mollis*, is a species-specific pollinator as is *P. zamiae*. Neither of these two beetles is active upon *Z. furfuracea*, although they may pollinate Caribbean relatives of *Z. integrifolia* (e.g. *Z. pumila*). This finding has led to the suspicion that the insects that pollinate cycads tend to be host-specific, or at least are faithful to species within a narrow range of taxonomic affinities (Norstog and Fawcett, 1989; Vovides, 1991).

Male cones provide an egg laying site, as well as food for both the adults and larvae of *Rhopalotria slossonae*, a weevil resembling *R. mollis* except for having darker wing covers, and *Pharaxonotha zamiae*, a pollen-eating beetle also present along with *R. slossonae*. In this case, oil of wintergreen (methyl salicylate), also a known insect attractant in orchids, is the major volatile component in male *Z. integrifolia* cones (Pellmyr *et al.*, 1991). Both beetles are able to live only in male cones of the native Florida *Z. integrifolia* or closely related Caribbean forms of this species. In its essentials the life cycle of *R. slossonae* closely resembles that of *R. mollis*, as does its interactions with its host species of *Zamia*, and therefore will not be described further. However, the life cycle of *Pharaxonotha zamiae* is sufficiently different from that of *Rhopalotria* to merit separate treatment.

Life cycle of *Pharaxonotha* in *Zamia*

The place of origin of individuals of *P. zamiae* that first invade the cones of *Z. integrifolia* is uncertain, but it is suspected that they overwinter in soil at the base of the cycad (Fawcett and Norstog, 1993). *Pharaxonotha zamiae* arrives at male cones when pollen is mature (Fig. 4A-C). There they feed and lay eggs between the microsporangia, and sometimes inside dehisced sporangia. Their place of mating is unknown. It is possible that they mate underground, but it is more likely that they mate in the male cones when they go there to feed and lay their eggs.

The eggs are very large in proportion to the size of the female, which carries up to six eggs at a time at various stages of maturity. The adults feed only on pollen. The pollen becomes scattered everywhere, with all of the insects present within the cone, including *R. slossonae*, becoming covered. Unless the sporangia of *Z. integrifolia* are disturbed, pollen tends to remain in them in sticky clumps and is undispersed.

The oval eggs, which are greyish white and translucent, hatch within hours. The larvae eat their way out of the egg and begin feeding on pollen. Larvae in all of the instars look much alike; each body segment has a pair of light-brown patches on the dorsal side, so that the larvae appear striped (Fig. 4D-F).

It is only larvae in the first instars that eat pollen and there is little pollen left in the cone after the first two instars. A larva that is ready to molt eats a hole in the back of one of the sporophylls and crawls inside, later emerging as the next instar (Fig. 4E). Larvae in later instars eat only tissue, entering the cone axis through a microsporophyll, and eating their way down through the cone into the cone peduncle, which becomes completely hollow Fig. 4F, G). When they have finished feeding and are ready to leave the cone they eat a little hole in the peduncle and drop down into the ground where they burrow into the soil (Fig. 4H, I).

After several days in the ground, larvae become pupae and remain in this stage for 4-7 days (Fig. 4J). New adult beetles, and perhaps older dormant ones, leave the soil in search of male cones (Fig. 4A) and, possibly by chance, also visit female cones (Fig. 4K), but late in the season when no new cones are present, the young adults instead of flying away to repeat the breeding cycle, burrow into the soil.

Pharaxonotha zamiae is found in female cones of *Z. integrifolia* in southern Florida, and it is assumed they blunder into them in error as appears to be the case with *Rhopalotria*. In northern Florida another species of *Pharaxonotha* appears to be the only pollinator of *Zamia* and possibly this specificity is related to the subtle genetic differences in allozymes in the northern forms described by Walters and Decker-Walters (1991). The life cycle of *P. zamiae* is not as tightly linked with its host as is that of *R. slossonae* in *Z. integrifolia* or that of *R. mollis* in *Z. furfuracea*.

The case for wind pollination

Informative though the studies described above have been, they leave a partially unanswered question: does wind play any role in cycad pollination?

Most exotic cycads at Fairchild Tropical Garden and elsewhere, do not produce fertile seeds in cultivation unless artificially pollinated. This holds true even when pollen producing male plants are only a meter or so distant from pollen-receptive females. These examples represent the "ultimate pollinator-exclusion experiment," and, at the same time, show that air currents are not pollen vectors for most cycads, otherwise the exotic species should be as fertile or at least have pollen present at the micropyle of the ovules as the native ones. The corollary question of course is that if wind is not the primary agent of pollination in such cycads, does it play any part in a cycad's reproductive cycle?

Cycad pollen grains are large (35-40 μm) and sticky and are not transported beyond a few meters by breezes and light winds. However, to show that even if they were transportable over a distance Norstog and Niklas (1984) used stroboscopic photography to track paths of pollen movements in wind tunnel experiments. Their results revealed that the shapes of the cones were such as to set-up air-current vortices which tended to deposit pollen upon and in those parts of the cone that were open such as at the base of *Dioon* (Fig. 5); a small vertical crack in *Zamia* (Fig. 1), and on the blades of the *Cycas* megasporophylls which at this stage are in erect, cone-like clusters. When these cones were scored for pollen impact points it was found that pollen was not preferentially deposited on the windward side but more-or-less evenly distributed around the circumference, and it piled up on or in the cone openings. Inspection of the interiors of such cones showed that comparatively little pollen actually impacted on the ovules (20-43 ovules per cone in *Z. integrifolia* and *Z. furfuracea*, nil in *Dioon edule* in which only the basal most, sterile sporophylls are open during pollination). In the two species of

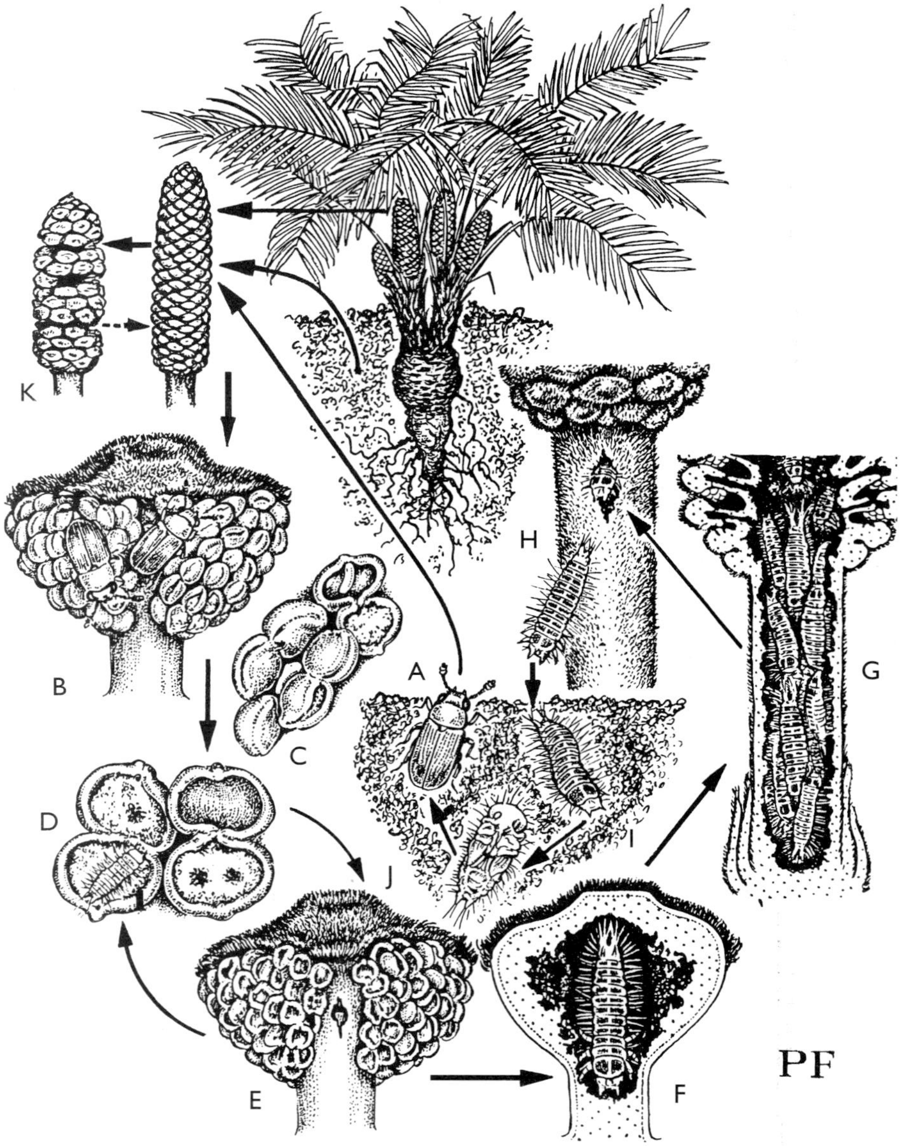

FIG. 4. Reproductive cycle of *Pharaxonotha zamiae* in *Zamia pumila*. A. Young adults leave the soil and visit male cones of *Z. pumila* in which microsporangia are beginning to dehisce. B. Beetles forage in the male cone. C. Beetles feed on pollen in microsporophylls and lay eggs between (arrow) and sometimes in sporangia. D. Eggs hatch and larvae feed on pollen. E. When larvae are ready to moult, they eat a hole in the sporophyll into which they enter to moult. When they emerge in the next instar they return to feeding on pollen. Only the early instars eat pollen and by the third instar much of the pollen is gone. F. Later instars eat only parenchyma tissue, entering the cone axis through a sporophyll. G. Larvae continue feeding until interior tissues of the cone axis and peduncle are empty but do not feed on the stem itself. H. The last larval instar eats a hole in the peduncle and drops to the ground. I. Larvae burrow into soil, pupate (J.) and after 4-7 days emerge as mature beetles. J. Beetles leave the soil in search of a new male cone to feed and breed in. K. Pollination takes place when some individuals enter a female cone instead of a male cone. They neither oviposit in female cones nor feed on them.

Zamia the pollen tended to settle preferentially on ovules near the cracks between megasporophylls. Because these cones contain about 200 ovules per cone, this effect at most would have resulted in pollen striking only about 20% of the ovules assuming the unlikely circumstance that every pollen grain settled upon a micropyle. The area of the micropylar tip is <1 mm^2 while the surface of the whole ovule is about 80 mm^2, so it would have been highly unlikely that more than 1% of the ovules were actually pollinated. Even if such an experiment were continued beyond the time of the experiment so as to mimic pollination in nature it seems unlikely that pollen deposition on micropyles would yield very high seed set. Remarkably, when developing ovules are dissected, pollination by single pollen grains is found to be rare, most ovules contain 3-4 and up to a dozen pollen tubes. In plantings of *Z. furfuracea* that were the sites for the observations reported here, fertility in control cones approached 98%.

The results of wind-tunnel experiments do not rule out wind as a cycad pollen vector, especially in *Cycas*. Pollen blown onto the blades of erect *Cycas* megasporophylls might be washed down to the ovules by dews or rain. However, adjacent male and female specimens of *C. rumphii*, *C. circinalis*, *C. media*, and *C. revoluta* at Fairchild Tropical Garden exhibit very low fertility unless artificially pollinated. In the case of cones of *Dioon* and *Zamia*, Niklas and Norstog (1984) proposed that a combination of wind and insects might function in pollen transport and pollination in such cycads, with air currents depositing masses of pollen upon certain receptive sporophylls and insect visitors distributing it inside the cone. This hypothesis is especially attractive in the case of female cones of *Dioon* species in which the only access is via a basal whorl of sterile megasporophylls (Fig. 5).

Small-scale experiments indicate that wind plays little or no part in the pollination of *Zamia furfuracea* in cultivation (Norstog *et al*., 1992). As noted, in a pollen-receptive cone the single vertical crack, measuring 3-5 cm^2 (Fig. 1), provides the only access to the cone's interior. Such cones were monitored for airborne pollen during the period of pollen receptivity of the cones. It was calculated that about 40 pollen grains entered each female cone. Here as well, in the highly unlikely circumstance that every such pollen grain would have impacted directly on a micropyle of one of the 200 or so ovules in a cone, then 40 seeds might have been produced for a fertility on the order of 20%. Actually, the average fertility of cones in this planting in 1981 was 97%, and in 1988 the fertility of the monitored cones averaged 82% (range 63-93%). Coupled with data from exclusion experiments in this planting, we conclude that the case for a significant contribution to fertility by airborne pollen is very weak in this cycad.

One source of evidence in favour of wind pollination has been fertile seed set, although at a low level ranging from 5-40% in insect exclusion experiments of Norstog *et al.* (1986), Tang (1987a), Donaldson *et al.* (1995) and Donaldson (1997). However, all of these authors have noted that it is virtually impossible to totally exclude weevils whole at the same time as making wind pollination possible. The weevils will chew through mesh bags, burrow around fasteners encircling the peduncle that are coated with lanolin, and even borrow through plastic bags. Our interpretation of this is that total exclusion of insects while allowing wind pollination is not totally possible because of the strong attraction of the weevils to the female cones for weevil sexual reproduction.

Pollination biology of *Dioon califanoi*

Observations in the field on *Dioon califanoi* in its habitat near Teotilan del Camino, Oaxaca, Mexico show that the female-cone life cycle of the cycad lasts about 2 years from initiation of the cone through seed set and cone disintegration. Pollination occurs near the end of the first year. At this stage, the sterile scales at the cone base separate giving rise to interscale apertures which last about 10 days, after which they close up tightly (Fawcett *et al*., 1995). Beetles enter by way of these apertures and exit by way of the apical sporophylls as shown in Fig. 5. Opening of the basal sterile sporophylls coincides with the activity of the pollinating weevil *Parallocorynus bicolor* (= *Rhopalotria bicolor* Oberprieler, 1995) that is emerging from dehiscent male cones and as well as from diapause in old, dried cones of previous reproductive episodes of the cycad. Beetle pupation begins in early August and lasts through October- to mid-November. The female cones mature through the second year and dehisce at the end of this period, or well into the third year.

There actually appear to be two pollinators in *D. califanoi*: *Parallocorynus bicolor* seems to predate the top of the male cone and a species of *Pharaxonotha* does so in its lower part; the former feeding solely on sporophyll tissue and the latter both on sporophyll tissue and pollen, as in the case of *P. zamiae* in *Z. integrifolia* (Fawcett *et al.*, 1995). Dried male cones may persist on the plant well into the second or even third year and contain diapausic weevils. Persistent male cones are not likely to be found when *Pharaxonotha* is active in large numbers because they is very destructive and destroy the integrity of the cone and results in its fragmentation. Individuals of *Parallocorynus bicolor* that emerge from male cones of *D. califanoi* are covered with pollen (Vovides unpubl. observ.) and this pollen is carried by the weevils into the female cone via the spread sterile basal scales. Thus, pollen on the basal cones scales of *Dioon* is not necessarily attributable to wind deposition as proposed by Niklas and Norstog (1984).

Although strict exclusion experiments have not been made with *Dioon* cones, there is sufficient circumstantial evidence to propose that *Parallocorynus bicolor* is the pollinator of *D. califanoi*. This evidence includes the observations that *P. bicolor* feeding and reproductive habits are the same as reported by Norstog and Fawcett (1989) for *R. mollis* on *Z. furfuracea*; that the life cycle of *P. bicolor* is tied to *D. califanoi*; and that the diapause stage is associated with old male cones.

Dioecy in cycads

Cycads are dioecious plants and a function of dioecy may be specialisation of males and females for different functions relating to reproductive fitness. If in fact this is the case in the two species of *Zamia* discussed above, a clear-cut example is seen in the sex-related differential allocation of starch in cones which provides a reward for pollination by cosymbiotic insects. In addition, alternate pathways of idioblast development may have evolved so that toxins sequestered in male cones make them non-poisonous and hence useful as brood places, while mobilised toxins in female cones suppress predation. Thus, the value of dioecy in these species becomes even more apparent.

Species specificity

We have found that *Rhopalotria mollis* will live in *Zamia loddigesii* and *Zamia spartea*, but not *Zamia integrifolia*, the cones of which contain less starch and are more mucilaginous. *Rhopalotria slossoni* has been found in other species similar to *Zamia integrifolia*, but not in *Zamia furfuracea* or similar species. *Zamia furfuracea*, *Zamia loddigesii* and *Zamia spartea* all produce cones in the summer, whereas *Zamia integrifolia* and the similar Caribbean zamias all produce cones in the winter and pollinator specificities may simply be phenological in origin.

Coevolution of cycads and insects

Aside from the value of cycad pollination studies in relation to reproduction and ecology, such studies may shed new light on origins of plant and animal interdependence, not only in pollination but also in protection of seeds during later development and seed dispersal mechanisms as well. Insect pollination is commonly associated with the evolution of flowering plants (Faegri and van der Pijl, 1979), but there seems little doubt that the coevolution of plants and insect pollinators predates the angiosperms (Crepet, 1979; Crowson, 1981, 1989, 1991).

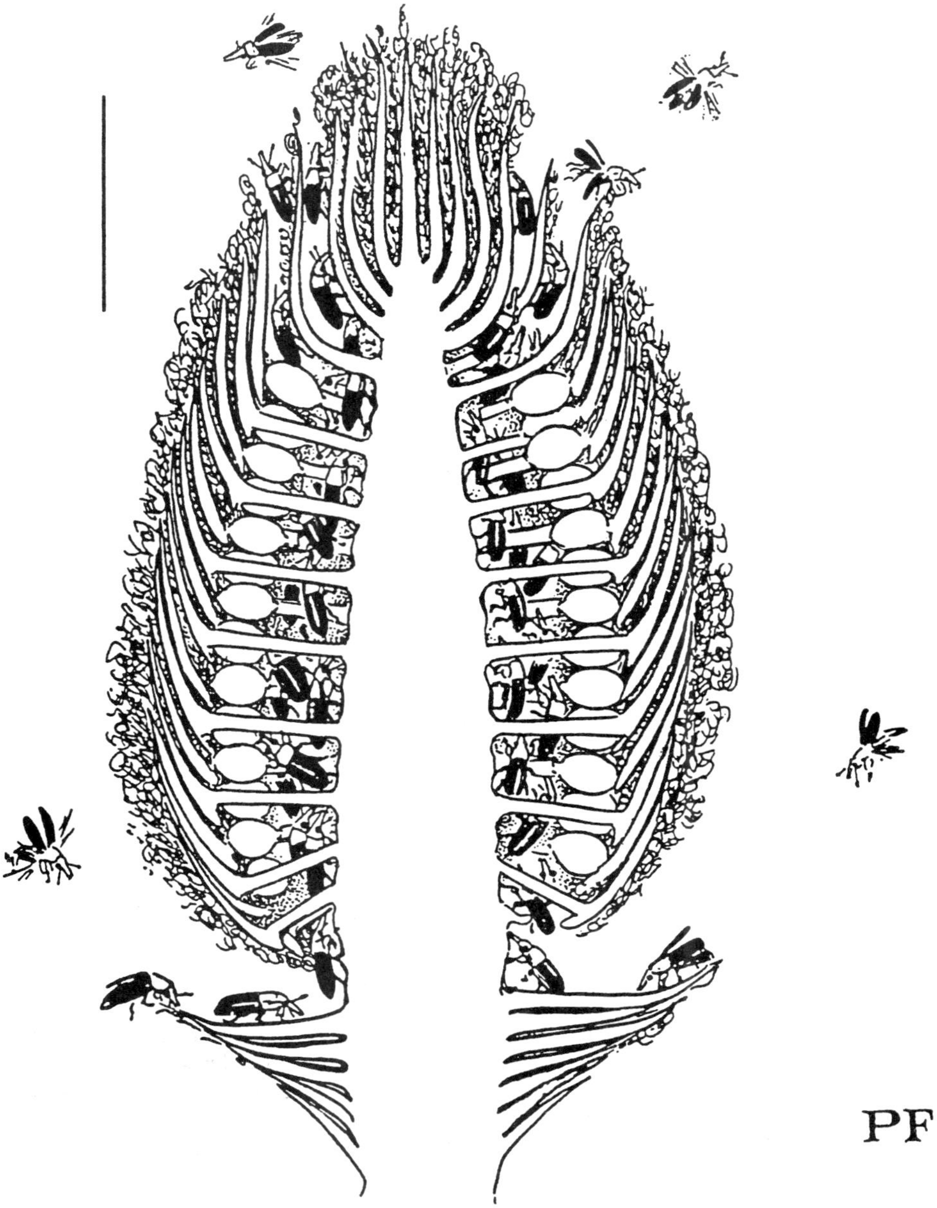

FIG. 5. Pollination of *Dioon califanoi*, a hypothesis. Weevils, probably *Rhopalotria bicolor*, are shown landing on open, sterile sporophylls at the base of a female cone. The basal sporophylls are glabrous and form a kind of "landing platform" where weevils pick up wind-blown pollen, or, more likely, bring it with them from male cones where they breed and feed. The construction of the female cone resembles an old-fashioned fly-trap, in which weevils move upward in the direction of dim light penetrating the felt of hairs that invests all but the basal sporophylls. As they move upward, they rub off pollen on the ends of ovules made sticky by micropylar fluid. Eventually the weevils escape by squeezing past the apical sporophylls.

Coevolution involves an interdependence of two species which have undergone morphological and behavioral modifications of mutual import (Ehrlich and Raven, 1964; Janzen, 1980). The mutualism exhibited by *Rhopalotria mollis* and *Zamia furfuracea* and *R. slossoni* and *Z. integrifolia* well may be an ancient coevolution, although there is no direct evidence that these organisms did indeed evolve together, and curculionid beetles may have parasitised angiosperms before going on to cycads (Crowson, 1991). Even so, these kinds of partnerships might have originated very early in the evolution of both beetles and cycads. Clearly, with each cycad genus having its own genus of curculionid pollinator and with each cycad species within a genus apparently having its own species within a genus of pollinators; coevolution between cycads and weevils has and is occurring.

Conservation

Much remains to be discovered about the relationships of insects and cycads. All attempts at cultivation and maintenance of the plants with a view to their introduction into natural habitats will be completely pointless if the insect pollinators are lost, and it would be of great value if additional workers could be encouraged to study this important area, perhaps rather than concentrating on taxonomic questions. In particular we need to know the life history and environmental requirements of the pollinators for each cycad species. We also need to survey the effects that man-made insecticides used in agriculture and forestry are having on the reproductive capacity of adjacent or incorporated populations of cycads; this may be the primary reason why some species are failing to reproduce in the wild.

References

Baird, A.M. (1939). A contribution to the life history of *Macrozamia riedlei*. *J. Roy. Soc. W. Australia* **25**: 153–169.

Chadwick, C.E. (1993). The role of *Tranes lyterioides* and *T. sparsus* Boh (Col., Curculionidae) in the pollination of *Macrozamia communis* (Zamiaceae). In: D.W. Stevenson and K. J. Norstog (eds). Proceedings of the Second International Conference on Cycad Biology, pp. 77–88, Palm and Cycad Societies of Australia, Milton, Qld.

Chamberlain, C.J. (1919). The living cycads. University of Chicago Press, Chicago.

Chamberlain, C.J. (1935). Gymnosperms, structure and evolution. University of Chicago Press, Chicago.

Connell, S.W. and Ladd, P.G. (1993). Pollination biology of *Macrozamia riedlei* — the role of insects. In: D.W. Stevenson and K.J. Norstog (eds), Proceedings of the Second International Conference on Cycad Biology, pp. 96–102. Palm and Cycad Societies of Australia, Milton, Qld.

Crepet, W. (1979). Insect pollination: a paleontological perspective. *Bioscience* **29**: 102–108.

Crowson, R.A. (1981). The biology of the Coleoptera. Academic Press, London.

Crowson, R.A. (1989). Relationships to cycads. International Congress of Coleopterology, Barcelona. Sept 18-23, 1989 (Abstracts) pp. 13–15, European Association of Coleopterology, Barcelona.

Crowson, R.H. (1991). The relations of Coleoptera to Cycadales. In: M. Zunino, X. Beltes and M. Blas (eds), Advances in coleopterology, pp. 13–28, European Association of Coleopterology, Barcelona.

Donaldson, J.S. (1997). Is there a floral parasite mutualism in cycad pollination? The pollination biology of *Encephalartos villosus* (Zamiaceae). *Amer. J. Bot.* **84**: 1398–1406.

Donaldson, J.S., Nänni, I. and de Wet Bösenberg, J. (1995). Beetle pollination of *Encephalartos cycadifolius*. In: P. Vorster (ed.). Proceedings of the Third International Conference on Cycad Biology, pp. 423–434, The Cycad Society of South Africa, Stellenbosch.

Ehrlich, P.R. and Raven, P.H. (1964). Butterflies and plants: a study in coevolution. *Evolution* **18**: 403–416.

Faegri, K. and Pijl, L. van der (1979). The principles of pollination ecology, ed. 3. Pergamon Press, New York.

Fawcett. P.K.S. and Norstog, K.J. (1993). *Zamia pumila* in South Florida: a preliminary report on its pollinators *R. slossoni*, a snout weevil and *P. zamiae*, a clavicorn beetle. In: D.W. Stevenson and K.J. Norstog (eds). Proceedings of the Second International Conference on Cycad Biology, pp. 109–120, Palm and Cycad Societies of Australia, Milton, Qld.

Fawcett, P.K.S., Norstog, K.J. and Vovides, A.P. (1995). Pollination of two species of *Zamia* in south Florida and its implication for evolution of cycad strobili. In: D.D. Pant, A.N. Bhatnagar, K.R. Surange, M.N. Bose, and P.K. Khare (eds) Proceedings of the International Conference on Global Environment and Diversification of Plants through Geological Time, pp. 111–121, South Asian Publishers, Allahabad, India.

Janzen, D.H. (1980). When is it coevolution? *Evolution* **34**: 611–612.

Lawson, A.A. (1926). A contribution to the life history of *Bowenia*. *Proc. Roy. Soc. Edinburgh* **54**: 357–394.

Marloth, R. (1914). Notes on the entomophilous nature of *Encephalartos*. *Trans. Roy. Soc. South Africa* **4**: 69–71.

Niklas, K.J. and Norstog, K. (1984). Aerodynamics and pollen grain depositional patterns on cycad megastrobili: implications on the reproduction of three cycad genera (*Cycas*, *Dioon*, and *Zamia*). *Bot. Gaz.* **145**: 205–222.

Norstog, K.J. and Fawcett, P.K.S. (1989). Insect-cycad symbiosis and its relation to the pollination of *Zamia furfuracea* (Zamiaceae) by *Rhopalotria mollis* (Curculionidae). *Amer. J. Bot.* **76**: 1380–1394.

Norstog, K.J. and Nicholls, T.J. (1997). *The biology of the cycads*. Cornell University Press, Ithaca, NY.

Norstog, K and Stevenson, D.W. (1980). Wind? Or insects? The pollination of cycads. *Fairchild Trop. Gard. Bull.* **35**(1): 28–30.

Norstog, K.J., Stevenson, D.W. and Niklas, K.J. (1986). The role of beetles in the pollination of *Zamia furfuracea* L. fil. (Zamiaceae). *Biotropica* **18**: 300–306.

Norstog, K.J., Fawcett, P.K.S. and Vovides, A.P. (1992). Beetle pollination of two species of *Zamia*: evolutionary and ecological considerations. *Palaeobotanist* **41**: 149–158.

Oberprieler, R.G. (1989). *Platymerus*, a forgotten weevil. *Péllea* no. **8**: 50–54.

Oberprieler, R.G. (1995). The weevils (Coleoptera: Curculionoidea) associated with cycads 1. Classification, relationships, and biology. In: P. Vorster (ed.). Proceedings of the Third International Conference on Cycad Biology, pp. 295–324, The Cycad Society of South Africa, Stellenbosch.

Ornduff, R. (1993). Features of the reproductive biology of *Macrozamia* species and *Cycas media* in Australia. In: D.W. Stevenson and K.J. Norstog (eds). Proceedings of the Second International Conference on Cycad Biology, pp. 121–124. Palm and Cycad Societies of Australia, Milton, Qld.

Pearson, H.H.W. (1906). Note on the South African cycads. I. *Trans. S. African Philos. Soc.* **16**: 341–354.

Pellmyr., O, Tang, W. Groth, I. Bergstrom, G. and Thien. L.B. (1991). Cycad cone and angiosperm floral volatiles: inferences for the evolution of insect pollination. *Biochem. Syst. & Ecol.* **19**: 623–627.

Rattray, G. (1913). Notes on the pollination of some South African cycads. *Trans. Roy. Soc. South Africa* **3**: 259–270.

Tang, W. (1987a). Insect pollination in the cycad *Zamia pumila* (Zamiaceae). *Amer. J. Bot.* **74**: 90–99.

Tang, W. (1987b). Heat production in cycad cones. *Bot. Gaz.* **148**: 165–174.

Tang, W. (1993). Heat and odour production in cycad cones and their role in insect pollination. In: D.W. Stevenson and K.J. Norstog (eds). Proceedings of the Second International Conference on Cycad Biology, pp. 140–147, Palm and Cycad Societies of Australia, Milton, Qld.

Tang, W., Sternberg, L. and Price, D. (1987). Metabolic aspects of thermogenesis in male cones of five cycad species. *Amer. J. Bot.* **74**: 1555–1559.

Vega, A. and Bell, A.E. (1967). a-amino-b-methylaminopropionic acid, a new amino acid from seeds of *Cycas circinalis*. *Phytochemistry* **6**: 759–762.

Vovides, A.P. (1991). Cone idioblasts of eleven cycad genera: morphology, distribution and significance. *Bot. Gaz.* **152**: 91–99.

Vovides, A.P., Norstog, K.J. Fawcett, P.K.S. Duncan, M.W. Nash, R.J. and Molsen, D.V. (1993). Histological changes during maturation in male and female cones of *Zamia furfuracea* L. fil. (Zamiaceae, Cycadales) and their significance in relation to pollination biology. *J. Linn. Soc., Bot.* **111**: 241–252.

Walters, T.W. and Decker-Walters, D.S. (1992). Patterns of allozyme diversity in the West Indies cycad, *Zamia pumila*. *Amer. J. Bot.* **78**: 436–445.

Whiting, M.G. (1963). Toxicity of cycads. *Econ. Bot.* **23**: 270–302.

Wilson, G.W. (1993). Initial observations of the reproductive behaviour and an insect pollination agent of *Bowenia serrulata* (W. Bull) Chamberlain. *Encephalartos* no. **36** (Dec.): 13–18.

Kite, G.C., Hetterscheid, W.L.A., Lewis, M.J., Boyce, P.C., Ollerton, J., Cocklin, E, Diaz, A and Simmonds, M.S.J. (1998). Inflorescence odours and pollinators of *Arum* and *Amorphophallus* (Araceae). In: S.J. Owens and P.J. Rudall (Editors). Reproductive Biology, pp. 295–315. Royal Botanic Gardens, Kew.

INFLORESCENCE ODOURS AND POLLINATORS OF *ARUM* AND *AMORPHOPHALLUS* (ARACEAE)

GEOFFREY C. KITE[1] *, WILBERT L.A. HETTERSCHEID[2], MERVYN J. LEWIS[3], PETER C. BOYCE[1], JEFF OLLERTON[4], EMMA COCKLIN[4], ANITA DIAZ[5] AND MONIQUE S.J. SIMMONDS[1]

[1]Royal Botanic Gardens, Kew, Richmond, Surrey TW9 3AB, UK
[2]Chrysantenstraat 28, 1214B Hilversum, The Netherlands
[3]IACR - Long Ashton Research Station, Department of Agricultural Sciences, University of Bristol, Long Ashton, Bristol BS18 9AF, UK
[4]School of Environmental Science, Nene College of Higher Education, Park Campus, Boughton Green Road, Northampton NN2 7AL, UK
[5]Farmland Ecology Unit, Scottish Agricultural College, Mill of Craibstone, Craibstone, Bucksburn, Aberdeen AB2 9TS, UK

Abstract

Chemical analyses of the inflorescence odours of 11 species of *Arum*, 20 species of *Amorphophallus* and two species of *Pseudodracontium* are presented and compared to the limited published data that are available on pollinators of these species.

The chemical odour profiles of dung-smelling *Arum* species were complex, with butanoic acid esters, 1-decene, two dimethyloctadienes, *p*-cresol, indole and 2-heptanone being among the most common identified major constituents in species analysed. The fragrant odour of *Arum creticum* consisted mainly of benzyl alcohol, while ethyl acetate dominated the fruity odour of *A. palaestinum*. The odour of *A. rupicola* var. *rupicola* consisted mainly of various sesquiterpenoids.

Experiments on the pollination ecology of *A. maculatum* revealed that its major pollinator, *Psychoda phalaenoides*, was not attracted to the heat generated by the spadix. However, three important components of the odour, *p*-cresol, indole, and 2-heptanone, were attractive to species of *Psychoda* in field tests, with *p*-cresol showing most activity. Data on the phenology of flowering in *A. maculatum* and *Psychoda* flight activity suggested that any synchronisation of flowering and pollinator abundance may be disrupted by physical conditions.

The odours of *Amorphophallus* and *Pseudodracontium* species with 'gaseous' or carrion smells had a simple chemical composition, consisting mainly of dimethyl oligosulphides. The odours of other *Amorphophallus* species having different smells were also generally dominated by one or two compounds: trimethylamine in *A. brachyphyllus*, isocaproic acid in *A. elatus*, 4-methoxyphenethyl alcohol in *A. albispathus*, isoamyl acetate with ethyl acetate in *A. haematospadix*, and ocimene in *A. margaritifer*.

Two odour types associated with carrion-mimicry and dung-mimicry are proposed for the sapromyophilous species of *Arum* and *Amorphophallus*.

* author to whom correspondence should be addressed.

Introduction

Floral odours can be important in attracting pollinators and are therefore subject to the evolutionary selective forces that act to improve a plant's reproductive success (Knudsen, 1993). The diversity of odour cues, arising from both the immense diversity of chemical compounds and the ways in which they can be combined, is potentially greater than can be achieved with other forms of pollinator signalling, such as visual and tactile cues (Knudsen *et al.*, 1993). This diversity allows for the evolution of dedicated plant-pollinator interactions in which a single pollinator type (species or higher taxon) is signalled, thereby excluding vectors that may interfere with the process of cross pollination (Williams and Dodson, 1972). Chemical signals also have the potential to act over great distances, attracting pollinators from a wide catchment area (Steiner, 1952; Williams and Dodson, 1972).

Whilst visual and tactile cues have been much documented in various pollination syndromes, such that potential pollinators can sometimes be predicted from the shape and colour of a flower, attempts to correlate floral odour types with general classes of pollinators have, until recently, relied on subjective descriptions of the scent (Fægri and van der Pijl, 1979). This is mainly due to the technical difficulties associated with the chemical analysis of floral odours and gauging the responses of the pollinators to individual odour components (Dobson, 1994). Recent data on the chemistry of odour types associated with different pollination syndromes have come mainly from extensive analyses of orchid odours (Kaiser, 1993) and surveys of the scents of moth- and bat-pollinated flowers (Knudsen and Tollsten, 1993; 1995). These studies have resulted in some broad generalisations on the chemical nature of odour types but narrower definitions of the scent chemistry have not been possible due to the complex nature of the scents and the lack of information on the sensory preferences of pollinators.

Flowers pollinated by insects that feed or breed on dung or carrion and which attract pollinators by deceit (Dafni, 1984) may provide the clearest examples of well-defined chemical odour syndromes (Dobson, 1994). Unfortunately, the odours of only a few species which potentially show this pollination syndrome have been analysed. The putrid and faecal odours produced by some orchid species in the genera *Cirrhopetalum* and *Masdevallia* appear to be due to various short chain aliphatic acids, such as butyric acid and isovaleric acid and their alcohols and esters (Kaiser, 1993), while the fetid smell of the parasitic plant *Hydnora africana* (Hydnoraceae) contains dimethyl di- and trisulphides and long chain organic acids as the major components (Burger *et al.*, 1988).

Araceae is a family renowned for having unpleasant floral odours (Vogel, 1962), although there are aroids with strongly fragrant or distinctive odours (Willis, 1952). The range of compounds reported so far as the major constituents of aroid odours reflects this diversity: various benzenoids, phenylpropanoids and organic acids in *Spathiphyllum cannifolium* (Lewis *et al.*, 1988); a novel compound, 1,3,6-trimethyl-2,5-dioxabicyclo[3.2.1.]nonane, in *Peltandra virginica* Knuth (= *P. virginica* (L.) Rafin; Patt *et al.*, 1992) and various dimethyl oligosulphides in *Sauromatum venosum* (Ait.) Kunth (reported as S. *guttatum* Schott; Borg-Karlson *et al.*, 1994). Perceived odour variation in aroids can also occur within one genus, such as *Arum* and *Amorphophallus* (Willis, 1952; Boyce, 1993). Thus, as with orchids, the family affords the opportunity to examine the volatile compounds associated with different pollination syndromes among related species, and particularly volatiles associated with pollination by dung and carrion feeding or breeding insects - part of the sapromyophilous syndrome (Fægri and van der Pijl, 1979).

In this paper we present data on the floral odours of 11 species of *Arum,* 20 species of *Amorphophallus* and two species of the related genus *Pseudodracontium.* We review information on potential pollinators in these genera in an attempt to identify some of the odour characteristics associated with sapromyophily.

Odour analysis: techniques and precautions

The technique of headspace trapping followed by thermal desorption - gas chromatography - mass spectrometry (TD-GC-MS) has become the accepted technique for the analysis of floral odours (Kaiser, 1991). This involves passing air from around a flower or inflorescence (the headspace) through an odour trap packed with a porous polymer (e.g. Tenax, Poropak) which retains the volatile organic compounds (Cole, 1980). The odour compounds are then removed from the trap in a stream of hot inert gas and passed either directly, or more usually via some cryofocussing system, onto the column of a gas chromatograph coupled to an on-line mass spectrometer to achieve both separation and identification of the components.

Floral odours contain a wide variety of compounds that may be present at greatly differing concentrations. The composition of the odour may also vary through flowering time (Lewis *et al.*, 1988). Therefore a floral odour cannot be fully described with a single analysis. Ideally, several trapping substances should be employed with different compound retention characteristics; chromatography should be undertaken on different polarity columns to achieve optimal analysis of the various odour components; odour trapping should be performed at various times during scent production; and, if minor but potentially important odour components are to be identified, serial post-chromatography re-trapping and analysis of these is required. Unfortunately, such detailed dissection of a floral scent is often not practicable due to their ephemeral nature. This is particularly true in some aroids where odour production can be brief, sometimes lasting just a couple of hours.

The analyses reported here were obtained using traps packed with Tenax TA and compounds were chromatographed on a non-polar column (for details see Table 1). This method may result in very small molecules going undetected as Tenax TA is reported to have relatively poor affinity for compounds containing fewer than five carbon atoms (Bertsch *et al.*, 1974). In particular, aroid odours are reported to have aminoid components (Smith and Meeuse, 1966) and such components in other floral odours have proved difficult to confirm chemically by headspace trapping followed by TD-GC-MS (Kaiser, 1993). However, it has been shown that C2-C4 monoamines should be revealed by this method if they are major components of aroid odours (Kite, 1995), but ammonia will not be detected. A further technical reservation that needs to be considered when interpreting the results is that polar compounds, such as organic acids, may go undetected at low concentration due to their poor chromatography on the non-polar column used (Adams, 1995).

Inflorescence odours and pollinators in *Arum*

The genus *Arum* comprises 26 currently recognised species of temperate herbs with a distribution through Europe to W. China but centred around the Mediterranean (Boyce, 1993; 1995). It belongs to tribe Areae which is traditionally considered to be one of the most highly evolved in Araceae (Boyce, 1993). In particular, species of *Arum* have evolved complex pollination systems in which pollinating insects are temporarily trapped in a chamber at the base of the spathe containing the fertile part of the spadix (Knoll, 1926; Lack and Diaz, 1991). Their inflorescence odours, which emanate from the appendix of the spadix, are also perceived to be complex and are often described as mimicking various types of dung, although not all are reported to be scented. The species with foul odours are generally those in which the spathe is positioned near the ground (so-called 'cryptic' species): those where the spathe is held above the leaves ('flag species') are said to be scentless, except *A. creticum* which has a pungent 'lemon-freesia' scent (Boyce, 1993).

Various compounds have been implicated in the complex dung odours of *Arum* and it is likely that different analysis techniques will reveal different compounds. Skatole (3-methyl indole), methylamine and ammonia have been detected by paper chromatography of the condensates produced from cut and enclosed warm spadices of *A. italicum* and *A. dioscoridis*; additionally, several other aliphatic amines and putrescine were detected in the condensate of *A. italicum* (Smith and Meeuse, 1966). However, the odour of *A. maculatum* was found to be considerably more complex when it was analysed by non-invasive headspace odour trapping techniques (Kite, 1995). Fifty-six compounds were identified, major components being 2-heptanone, indole (each comprising 8–23% of the volatiles among the inflorescences studied) and bicyclogermacrene[1] (10–14%). *p*-Cresol was also a notable odorous component (0.2–6%) as were several ketones and esters, but simple aliphatic amines were not detected; the majority of the compounds were terpenoids. Vogel (1962) suggested previously that the odour of this species was possibly not due to amines but esters and terpenoids.

Using microheadspace techniques, Kite (1995) also revealed that the bicyclogermacrene component of the total inflorescence odour of *A. maculatum* arose solely from the chamber to give it a more pleasant character than the appendix odour. Microheadspace collections could not be refined to the extent of determining the exact source of the bicyclogermacrene within the chamber but it was assumed to emanate from the fertile region of the spadix, as no odour could be detected from the spathe. Contrasting appendix and chamber odours have also been found in *Sauromatum venosum* in which a lemon odour emanates from the club-shaped sterile organs situated above the pistillate flowers (Meeuse *et al.*, 1984). This sweet odour has been reported to consist mainly of α-pinene and limonene (Borg-Karlson *et al.*, 1994) and it has been suggested that it provides feeding and mating stimuli to the pollinators and serves to keep them inside the chamber until anthesis (Meeuse *et al.*, 1984).

Using non-invasive headspace sampling techniques, we have analysed the odours of ten other species of *Arum* (Table 1). As with *A. maculatum*, most of the odours were very complex and, while the odours of different inflorescences on the same plant had similar chemical profiles, some quantitative and qualitative variation was evident where analysis of different individuals was possible. Nevertheless, with the exception of *A. dioscoridis*, the most important constituents (in terms of absolute levels) that appeared characteristic for each species could be identified (Table 1). The data obtained to date for *A. dioscoridis* suggests that var. *dioscoridis* and var. *philistaeum* have distinct odour profiles.

The appendix odours of species analysed from *Arum* section *Dioscoridea* subsection *Dischroochiton* (Table 1) all had similar chemical profiles that were dominated by three unsaturated hydrocarbons: 1-decene and two dimethyloctadienes (absolute structures could not be determined from the mass spectra). *p*-Cresol was also an important component and methyl butyrate and methyl isobutyrate were present in all except the single analysis performed on *A. apulum*. Indole and skatole could be detected in some analyses of these four species and, apart from 2-heptanone and the ubiquitous ethanol component, the other major constituent was an uncommon sesquiterpenoid that could not be identified from searching available mass spectral libraries. In terms of mass spectrum and relative retention time, this sesquiterpenoid was identical to the major appendix sesquiterpenoid found in the appendix odour of *A. maculatum* (Kite, 1995). A high resolution GC-MS analysis of *A. nigrum* odour on a polar column also revealed numerous esters and organic acids and some alcohols, aldehydes and ketones as trace compounds. The odours of species in subsection *Dischroochiton* are generally described as being reminiscent of horse dung and are reported to attract various dung-breeding flies and beetles (Table 2).

[1] This compound was identified as germacrene B by Kite (1995) according to published GC-MS data (Adams, 1989). The data for germacrene B has since been reassigned to bicyclogermacrene (Adams, 1995).

The appendix odour of *A. italicum* (section *Arum* subsection *Arum*) was very similar to the appendix odours of the species examined from section *Dioscoridea* subsection *Dischroochiton* in containing 1-decene and the two dimethyloctadienes as the major constituents together with methyl butyrate, methyl isobutyrate and the unidentified sesquiterpenoid. Consequently the odour was distinct from that of *A. maculatum*, another species in section *Arum*, in which 1-decene and the two dimethyloctadienes could not be detected. Mono- and sesquiterpenoids were also more important components of the odour of *A. maculatum* than in the odour of *A. italicum*. Psychodid midges are the major pollinators of *A. maculatum* (Lack and Diaz, 1991) and have also been reported to be the main insects attracted to *A. italicum* (Méndez and Obeso, 1992) even though its odour appears distinct from that of *A. maculatum*. Further observations on pollination in *A. italicum* are required.

The appendix odour of *A. creticum*, a flag species, had an entirely different chemical profile to the above cryptic, dung-smelling species, as might be anticipated from its pungent freesia-like odour (Table 1). The dominant component (in excess of 80% composition) was benzyl alcohol but the characteristics of the odour were probably imparted by some of the minor components such as geraniol, geranial and eugenol. Indole, skatole and traces of *p*-cresol were detected in some analyses but the unknown sesquiterpenoid, common to all the above cryptic species, was absent. The species is reported to attract scarab, staphylinid and bruchid beetles, but only bruchids have been observed to carry pollen (Table 2). *Arum idaeum*, a cryptic species thought to be closely related to *A. creticum* (both in section *Dioscoridea* subsection *Creticum*), had a different odour profile. Its appendix odour was composed of *p*-cresol and the *p*-cresol derivatives *p*-methyl anisole and 2-nitro-*p*-cresol. However, only one analysis was made for this species and the quantities of compounds trapped were low. The species is said to be scentless (Boyce, 1993) but Drummond (pers. comm.) reports a sweet odour akin to violets and hyacinth. Another flag species analysed, *A. rupicola* var. *rupicola*, is also claimed to be scentless, but odour analysis here revealed numerous sesquiterpenoids and trace amounts of *p*-cresol, various alcohols, esters and ketones. The species is reported to attract blood-sucking midges belonging to the genus *Cuculoides* that are parasitic on birds (Koach, 1986).

The appendix odour of *A. dioscoridis* var. *dioscoridis* was of the *Dischroochiton* type but an initial analysis of *A. dioscoridis* var. *philistaeum* revealed that the odour contained numerous esters of butanoic acid and lacked 1-decene, dimethyloctadienes, *p*-cresol, and sesquiterpenoids. If confirmed, this divergence of odours amongst these two varieties is surprising given that the varietal status is only based on intergrading staining and patterning of the spathe. An ammonia-like odour could be perceived in the smell of *A. dioscoridis* and the production of free ammonia or amines has been demonstrated by the fumes created with hydrochloric acid placed near to the appendix (Meeuse and Raskin, 1988). As mentioned previously, free ammonia would not be detected by the methods used in the present study. *Arum palaestinum* is placed with *A. dioscoridis* in section *Dioscoridea* subsection *Poeciloporphyrochiton* but it also had a quite distinct appendix odour, consisting almost entirely of ethyl acetate, and lesser amounts of ethanol and acetic acid. The inflorescence had an odour of rotting fruit and the species is reported to attract fruit flies, *Drosophila* spp. (Koach, 1986). *Arum dioscoridis* has been described as having a donkey-dung odour, which again attracts dung-breeding flies and beetles (Table 2).

With the exception of *A. creticum*, the chamber odours of all the species for which collections were performed were found to consist of over 95% bicyclogermacrene (Table 1), as reported for *A. maculatum*. In contrast, the chamber odour of *A. creticum* consisted of a similar percentage of α-farnesene. Given the apparent consistency of the chamber odours among the other species, even between those with very different appendix odours, such as *A. dioscoridis* and *A. palaestinum*, the divergence in the chamber odour of *A. creticum* is likely to be of either taxonomic or biological significance. It would therefore be interesting to determine the chamber odour of *A. idaeum*, the species considered to be most closely related to *A. creticum*.

TABLE 1. Major components* of the appendix and chamber odours of inflorescences of *Arum* species. Odours were collected from inflorescences still attached to plants cultivated at the Royal Botanic Gardens, Kew, except for *A. maculatum* which was growing wild. The appendix odour was collected by sucking air from around a warm appendix through a trap (100 mm long × 3 mm diameter packed with 100 mg of Tenax TA 35/60 mesh) at a rate of 25 ml/min by means of a portable pump. The trap was screwed into one end of a glass tube which covered the appendix. The chamber odour was collected by sucking air at the same rate through a second identical trap attached to a Pasteur pipette that was inserted into the chamber either through the chamber entrance or wall. Sampling times varied depending on the quantity of volatile compounds being produced. The odour collected on the traps was analysed by direct thermal desorption - gas chromatography - mass spectrometry exactly as described by Kite (1995). The results summarise the data obtained from the number of plants, inflorescences and analyses indicated; chamber odours were not collected for *A. apulum*, *A. idaeum* or *A. rupicola*. Compounds that did not constitute more than 1% of the odour in any of the species examined are not listed. Data for *A. maculatum* are from Kite (1995).

Section:		***Arum***		***Dioscoridea***									
Subsection:		*Arum*		*Dischroochiton*				*Poecilophor-phyrochiton*		*Cretica*		*Tenuifilia*
HEADSPACE ANALYSED **Class of compound** Compound name (or type)	identification criteria**	*A. maculatum* L.	*A. italicum* Mill. ssp. *italicum*	*A. apulum* (Carano) P.C. Boyce	*A. cyrenaicum* Hruby	*A. nigrum* Schott	*A. purpureospathum* P.C. Boyce	*A. dioscoridis* Sm. var. *dioscoridis*	*A. palaestinum* Boiss.	*A. creticum* Boiss. & Held.	*A. idaeum* Coust. & Gandoger	*A. rupicola* Boiss. var. *rupicola*
APPENDIX												
Fatty Acid Derivatives												
1-decene	b		+++	+++	+++	+++	+++	+++				
(dimethyloctadiene)	c		+++	++	+++	+++	+++	++				
(dimethyloctadiene)	c		++	+++	+++	+++	+++	++				
(trimethylheptadiene)	c	+								++		
ethyl acetate	a								+++			
methyl butyrate	a	+	+++		++	++	++					+
methyl isobutyrate	a		++		++	++	++					
methyl 2-methyl butyrate	a	+	++		+					++		
2-heptanone	a	+++	+	+	++	++	+	++				
3-methyl butan-2-one	c	+	++									
3-methyl pentan-2-one	c	++										
6-methyl heptan-2-one	c											++
2-nonanone	b	++										
ethanol	a	+	+++	+	++	++	++	++	++	++		
propanol	a						++			++		
acetic acid	a								++			
Benzenoids												
benzyl alcohol	a									+++		
benzaldehyde	a									++		
methyl benzoate	b									++		

TABLE 1 continued

HEADSPACE ANALYSED **Class of compound** Compound name (or type)	identification criteria**	*A. maculatum* L.	*A. italicum* Mill. ssp. *italicum*	*A. apulum* (Carano) P.C. Boyce	*A. cyrenaicum* Hruby	*A. nigrum* Schott	*A. purpureospathum* P.C. Boyce	*A. dioscoridis* Sm. var. *dioscoridis*	*A. palaestinum* Boiss.	*A. creticum* Boiss. & Held.	*A. idaeum* Coust. & Gandoger	*A. rupicola* Boiss. var. *rupicola*
Benzenoids contd.												
methyl salicylate	b	+		+	+					++		
p-cresol	a	++	+	+++	+++	++	++	+++			+++	+++
p-methyl anisole	c										++	
Nitrogen compounds												
indole	a	+++		++	+	++				++		
skatole	b				+	++				+		
2-nitro-*p*-cresol	c										++	
Phenylpropanoids												
eugenol	a									++		
Terpenoids												
α-pinene	a	++										
myrcene	a	++										
limonene	a	++	++									
sabinene	a	++										
β-pinene	a	++										
terpinolene	a	++										
geraniol	b									++		
geranial	b									+		
citronellol	b	+								++		
(unidentified sesquiterpenoid)	c	+++	++	++	+++	++	++	++				
(sesquterpenoids generally)	c	+++	++	++	+	++	++	++		+		+++
CHAMBER												
Terpenoids												
bicyclogermacrene	b	++++	++++	?	++++	++++	++++	++++	++++		?	?
α-farnesene	b			?						++++	?	?
number of analyses performed		3	4	1	7	5	2	9	7	9	1	2
number of inflorescences analysed		3	2	1	3	2	1	6	2	6	1	2
number of plants analysed		3	2	1	2	1	1	1	1	3	1	1

* + signs indicate the average level of compounds in odour (determined by GC peak area):
++++ >90%
+++ 10–90%
++ 1–10%
+ <1% (indicated only if present in another species at higher level)
? chamber headspace not analysed

** compound identification criteria:
a = comparison of mass spectrum and retention time with authentic standard
b = comparison of mass spectrum and relative retention time with published data (Adams, 1995)
c = comparison of mass spectrum with published data (Ausloos *et al.*, 1992)

TABLE 2. Insects reported in inflorescences of *Arum* species. Insect species specified only when they predominated.

Species (Locality/Ecotype)	**Most numerous insects found in inflorescences**	**Other insects found in inflorescences**	**Reference**
Arum concinnatum Schott (Crete)	*Coproica ferruginata* (Stenhammer) (Diptera: Sphaeroceridae)	Diptera: Sphaeroceridae 2 spp. Coleoptera: Staphylinidae 2 spp.	Drummond & Hammond (1993)
A. concinnaum (UK- in cultivation)	*Psychoda phalaenoides* (L.) (Diptera: Psychodidae)	Diptera: Chironomidae 1 sp., Sphaeroceridae 1 sp.	Drummond & Hammond (1993)
A. creticum Boiss. & Held. (Crete)	Coleoptera: nine families including Scarabaeidae, Staphylinidae and Bruchidae	Diptera and Hymenoptera	Drummond & Hammond (1993)
A. dioscoridis Sm. (Lebanon)	Coleoptera: Staphylinidae 2 spp., Scarabaeidae 4 spp., Hydrophilidae 1 sp. Diptera: Scatophagidae 2 spp.		Kullenberg (1953)
A. dioscoridis (Cyprus)	*Coproica ferruginata* (Stenhammer), *C. vagans* (Haliday) (Diptera: Sphaeroceridae) *Anotylus clypeonitens* (Pand.), *A. insutus* (Grav.) (Coleoptera: Staphylinidae)	Diptera: Sphaeroceridae 6 spp., Drosophilidae 3 spp. Coleoptera: Staphylinidae 8 spp.	Drummond & Hammond (1991)
A. dioscoridis (S Turkey / lacked odour)	—	Coleoptera: Staphylinidae 5 spp.	Drummond & Hammond (1991)
A. elongatum Stev. (Israel)	*Cuculoides* spp. (Diptera: Ceratopogonidae)		Braverman & Koach (1982)
A. hygrophilum Boiss. (Israel)	Diptera: Psychodidae		Koach (1986)
A. italicum Mill. (N Spain)	Diptera: Psychodidae	Diptera: Chironomidae, Sciaridae, Sphaeroceridae, Ceratopogonidae, Heleomyzidae	Méndez & Obeso (1992); Méndez, pers. comm.
A. italicum subsp. *neglectum* (F. Towns.) Prime (S England)	Diptera: Psychodidae	Diptera: Sphaeroceridae	Prime (1954)
A. maculatum L.	*Psychoda phalaenoides* (L.) (Diptera: Psychodidae)		Müller (1883)

TABLE 2 continued

A. maculatum (UK)	*Psychoda phalaenoides* (L.)	other *Psychoda* spp.	Sowter (1949)
A. maculatum (UK)	female *Psychoda phalaenoides* (L.)		Prime (1960)
A. maculatum (S England)	female *Psychoda phalaenoides* (L.)		Lack and Diaz (1991)
A. maculatum (N Germany)	Diptera: Psychodidae	Coleoptera: Staphylinidae Other Diptera	Present study
A. maculatum (Hurley, S. England, 1994)	*Psychoda phalaenoides* (L.)	Other *Psychoda* spp.	Kite (1995)
A. maculatum (Hurley, S England, 1996)	*Piophila vulgaris* Fall. (Diptera: Piophilidae)	*Psychoda* spp. including *P. phalaenoides* (L.) Diptera: Cecidomyiidae 1 sp. Diptera: Phoridae 1 sp. *Megaselia* sp. (Diptera: Phoridae) *Fannia canicularis* L. (Diptera: Muscidae)	Present study
A. nigrum Schott (Germany - in cultivation)	Diptera: Sphaeroceridae (many e.g. *Sphaerocera* spp.) Coleoptera: Scarabaeidae (many e.g. *Aphodius* spp.), Staphylinidae (many e.g. *Atheta* spp.)	Other Diptera & Coleoptera	Knoll (1926)
A. sintenisii (Engl.) P.C. Boyce (=*A. orientale* M. Bieb) (Cyprus)	5 *Drosophila* spp. (Diptera: Drosophilidae) 4 *Atheta* spp. (Coleoptera: Staphylinidae)		Drummond & Hammond (1991)
A. palaestinum Boiss. (ecotype 1, fermenting fruit odour) (Israel)	Diptera: Drosophilidae		Koach (1986)
A. palaestinum (ecotype 2, dung and carrion odour) (Israel)	"beetles & flies"		Koach (1986)
A. purpureospathum P.C. Boyce (UK - in cultivation)	*Copromyza equina* Fallén, *C. similis* (Collin) (Diptera: Sphaeroceridae)	Diptera: Psychodidae 1 sp., Chironomidae 1 sp.	Drummond & Hammond (1993)
A. rupicola Boiss. (Germany - in cultivation)	*Cuculoides* spp. (Diptera: Ceratopogonidae),	Diptera: Simulidae	Knoll (1926)

Pollination ecology of *Arum maculatum*: new observations and future work

The pollination of *A. maculatum* by *P. phalaenoides* and the operation of the inflorescence as a pit-fall trap have been studied in detail (Prime, 1960; Lack and Diaz, 1991). Nevertheless observations are still required on the pollination ecology of *A. maculatum* to understand further the factors affecting the success of pollinator trapping.

We have investigated the role of both the heat produced by the spadix appendix (Bermadinger-Stabentheiner and Stabentheiner, 1995) and its odour in attracting *Psychoda* species. In an experiment in which glass tubes heated to 10°C above ambient temperature were used to mimic a warm spadix appendix, *P. phalaenoides* showed avoidance of heat in two out of four experiments and no significant response in the other two (Table 3). Knoll (1926), in a study of *A. nigrum* (see Dormer, 1960), also found that an electrically heated model inflorescence failed to attract that species' pollinators while a cold model charged with stale blood and glycerine did attract various carrion insects. In these two species it appears, therefore, that the heat produced by the spadix does not attract pollinators directly. Its main function is probably to volatilise the odour (Meeuse and Raskin, 1988).

In a field-trapping experiment, *Psychoda* species were attracted to *p*-cresol, indole and 2-heptanone, three of the major chemical components of the odour of *A. maculatum*, either individually or in combination (Table 4). *p*-Cresol was the most effective individual attractant but a mixture of *p*-cresol, indole and 2-heptanone was more active than any single compound or pair of compounds. A key role for *p*-cresol in attracting *P. phalaenoides* was suggested previously from the finding that this compound was an important component of cow-dung odour (Kite, 1995). Cow dung is the normal breeding habitat of *P. phalaenoides* (Withers, 1988) and its odour is presumably mimicked by *A. maculatum*. *p*-Cresol was also present, and generally an important constituent, in the odours of all the dung-smelling species of *Arum* examined so far (Table 1) suggesting that it may be a general attractant for dung-breeding insects.

TABLE 3. Responses of *Psychoda phalaenoides* to heat. Between five and ten midges were introduced into the centre of a rectangular glass tank (35 × 22 × 24 cm) fitted with a cardboard lid through which protruded two boiling tubes 19 cm apart. One tube contained a thermostatic 2.3 kW heating cable which warmed the tube to 10°C above ambient; the surface of the tubes were either left dry or wrapped in wet filter paper. The number of midges in each half of the tank was then recorded every 5 min for 90 min.

Experiment		**Treatment**	**No. midges median (range)**	**Wilcoxon matched pair test**
Dry	1	ambient	4.0 (2 - 6)	z = -1.56
		heated	2.0 (0 - 4)	p = 0.118
	2	ambient	2.0 (1 - 4)	z = -3.00
		heated	1.0 (0 - 2)	p = 0.003
Wet	3	ambient	3.5 (2 - 5)	z = -2.80
		heated	1.0 (0 - 2)	p = 0.005
	4	ambient	1.0 (0 - 5)	z = -1.91
		heated	1.0 (0 - 3)	p = 0.056

TABLE 4. Responses of *Psychoda* species to indole, *p*-cresol and 2-heptanone. Inverted cups, containing cotton wool soaked in the stated concentration of test compound(s), were placed over water traps near to a sewage farm in Reading, Berkshire (UK). The total number of psychodid midges in 10 traps after a 24 h period was recorded.

conc. (%) of compound			
indole	**2-heptanone**	***p*-cresol**	**number of *Psychoda* trapped**
0.1	–	–	49
0.01	–	–	58
–	0.1	–	76
–	0.01	–	53
–		0.1	452
–	–	0.01	353
0.01	–	0.01	38
–	0.01	0.01	496
0.01	0.01	0.01	785

A third factor affecting the success of pollinator trapping in *A. maculatum* is the availability of pollinators. At a site in Harrold & Odell Country Park, Bedfordshire (UK), in 1995, flowering of *A. maculatum* appeared to be synchronised with peak *Psychoda* flight abundance. However, at a site on Nene College campus, Northamptonshire (UK), in 1996 there were few *Psychoda* in flight during peak flowering (Fig. 1). *Arum* flowering at Nene was delayed by about three weeks compared to a typical year (such as 1995), as a result of unusually cold and unsettled weather conditions, which also persisted during the flowering period. *Psychoda* flight activity is reduced by heavy rain or strong winds, thus unfavourable weather conditions may adversely affect pollinator trapping, either by directly reducing *Psychoda* flight activity or uncoupling flowering of *A. maculatum* from any peak in pollinator population density. Similar unsettled weather conditions and a delay in flowering in a population of *A. maculatum* in Hurley, Berkshire (UK), in 1996 may have resulted in inflorescences trapping mainly *Piophila vulgaris* (Table 2), whereas the same population trapped *P. phalaenoides* in 1994 when flowering occurred three weeks earlier (Kite 1995). Further integrated studies on *Psychoda* abundance, *Arum* flowering phenology and climatic conditions are required.

Inflorescence odours and pollinators in *Amorphophallus* and *Pseudodracontium*

Amorphophallus comprises some 170 species distributed mainly in the tropics from West Africa eastward into Polynesia: none are found in the neotropics (Hetterscheid, 1994). They are seasonal herbs in which an underground tuber usually produces one leaf each year and species are found mostly in secondary forests or disturbed spots in primary forests and forest margins. The inflorescence, which may replace the leaf in one season or develop alongside it, is of the typical aroid type. However, *Amorphophallus* is noted for the large size of the inflorescence in some species (such as *A. titanum*) and for the production of strong and obnoxious odours. Odour production is associated with considerable heating of the spadix, notably the male zone and the appendix (Skubatch *et al.*, 1990). In many species the odour consists of a nauseating gaseous stench (like liquid petroleum gas) which in some tends

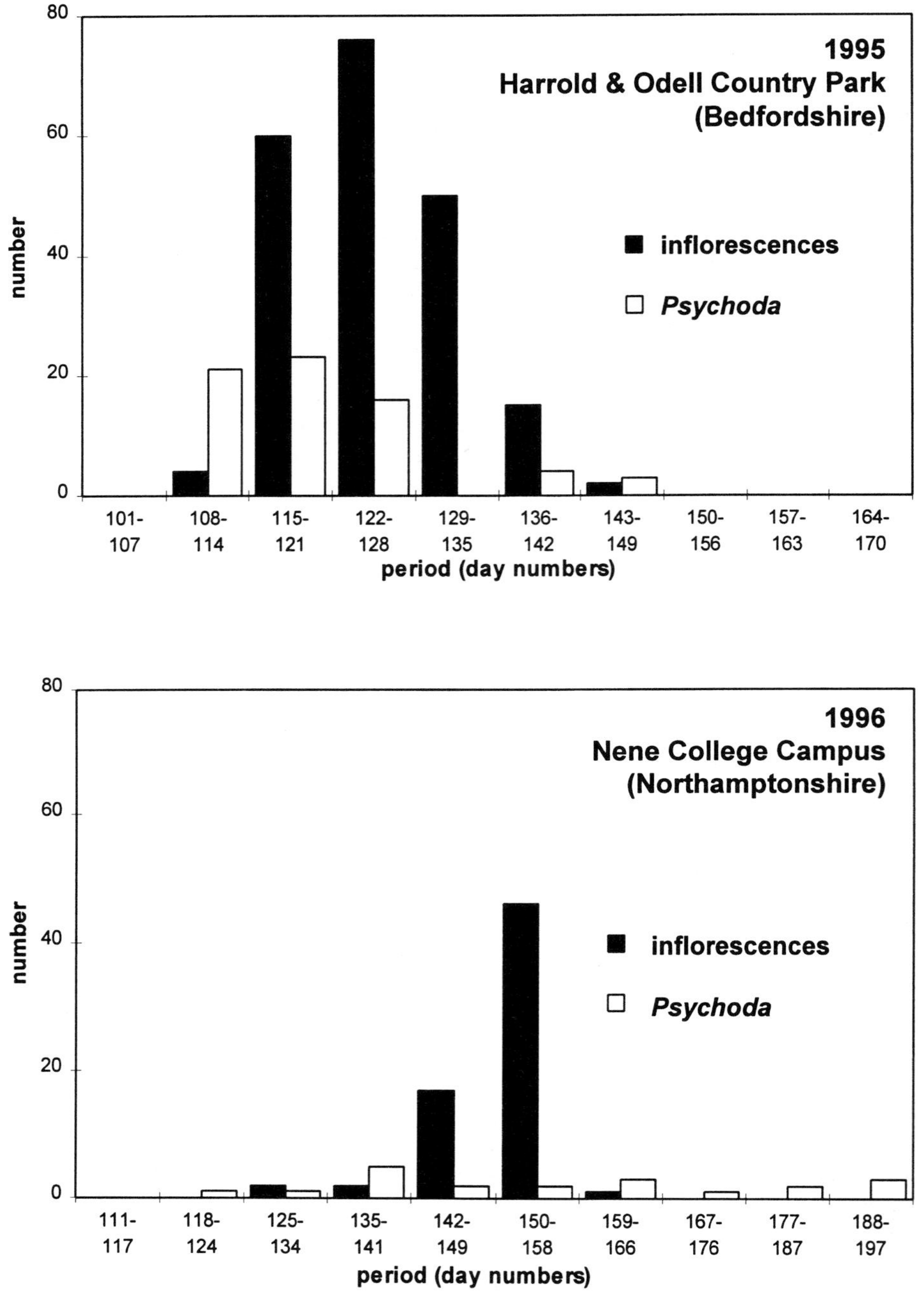

FIG. 1. Phenologies of *Arum maculatum* flowering and *Psychoda* flight activity at two sites in the UK. Numbers refer to number of inflorescences which opened and the number of *Psychoda* caught on yellow sticky traps per 100 m^2 during the stated period of days (day 1 = 1 Jan).

towards the odour of rotting meat: in others a diversity of odours have been noted (Table 5). In carrion-smelling species the olfactory mimicry of decay is often supported visually by the inflorescence being dark brownish and brownish-purple in colour.

The related genus *Pseudodracontium* has been revised recently and contains seven species (Serebryanyi, 1995). They all possess an elongate underground tuber but are otherwise similar to *Amorphophallus*, although much less variable. Typically they flower alongside the leaves and have a rather inconspicuous green spathe. Hetterscheid (1994) argued that *Pseudodracontium* is actually part of *Amorphophallus* and may have to be reduced to it.

We have used headspace techniques to analyse the gaseous odour typical of many *Amorphophallus* species and found it to consist almost entirely of dimethyl oligosulphides (Table 6). Species perceived as having a rotting meat odour, such as *A. konjac* and *A. maximus*, also produced mainly dimethyl oligosulphides. These compounds also predominated in six species that were perceived to have different or additional odours: *A. titanum*, *A. arnautovii*, A. *zenkeri*, *A. eichleri*, *A. cicatricifer* and *A. opertus* (Table 6). The different odours of *A. titanum*, *A. zenkeri* and *A. arnautovii* could not be assigned to any additional compounds detectable by chemical analysis, but the faecal odour in *A. eichleri* was possibly due to the presence of 2-heptanone and indole modifying the smell of the sulphides while acetophenone and 1-phenylethyl acetate may have given the fruity quality to the *A. cicatricifer* odour. In the odour of *A. opertus*, two additional compounds could be detected that could not be identified by searching available mass spectral libraries. The gaseous odours of two species of *Pseudodracontium* were also found to be due to the presence of dimethyl oligosulphides, but both also contained 4-methoxyphenethyl alcohol, which may be responsible for the anise-like character noted in the odour of *P. lacourii* (Table 5).

In other *Amorphophallus* species, in which the inflorescence odour was quite distinct from the typical gaseous or rotting meat smell, the chemical composition was also generally entirely different. The strong fried fish odour of *A. brachyphyllus* consisted of over 90% trimethylamine while isocaproic acid (4-methyl pentanoic acid) was the only compound detectable in the strong cheesy odour of *A. elatus* (isocaproic acid has the odour of pungent cheese). The unusual banana odour of *A. haematospadix* was found to be due to isoamyl acetate ('banana oil') although there was also a major ethyl acetate component. The strong anise-like odour produced by *A. albispathus* was found to be composed of almost pure 4-methoxyphenethyl alcohol, the same compound present in the odour of the two *Pseudodracontium* species examined. Finally, the monoterpenoid ocimene (*cis* or *trans*) dominated the odour of *A. margaritifer*. This species lacks an appendix and therefore the odour presumably emanates from the flowers or staminodes. The odour's terpenoid nature likens it to the chamber odours found in *Arum* species (Table 1) and reported for *Sauromatium venosum* (Borg-Karlson *et al.*, 1994).

Few published observations exist on potential pollinators of *Amorphophallus* (Table 7) and these often pertain to insects seen trapped in the spathe rather than actually pollinating the flowers. Carrion beetles are the insects most often observed in spathes and in *A. johnsonii* and *A. hohenackeri* they have been shown to be effective pollinators (Sivadasan & Sabu, 1991; Beath, 1996). Of the species for which potential pollinator data are available, only the odours of *A. paeoniifolius* and *A. titanum* have been analysed chemically and both consist of dimethyl oligosulphides (Table 6). These sulphur compounds are likely to be effective odour cues for carrion beetles as they are present in the odour of meat (Shankaranarayana *et al.*, 1974). They have also been shown to be attractants for various carrion flies, such as several blowflies (*Calliphora, Lucilia, Sarcophaga* and *Phormia* spp.; Borg-Karlson *et al.*, 1994, Nilssen *et al.*, 1996), as well as *Musca domestica* (Mookherjee *et al.*, 1993). In fact, dimethyl disulphide is a component of the commercial screwworm-fly bait 'Swormlure' (Hall, 1995) and is used for host location by the sheep blowflies *Lucilia sericata* and *L. cuprina* (Ashworth and Wall, 1994). The odour of the stinkhorn fungus *Phallus impudicus*, which consists mainly of dimethyl oligosulphides (Borg-Karlson *et al.*, 1994), also attracts many saprophageous flies (Smith, 1956), including the black blowfly, *Phormia regina* (Stoffolano *et al.*, 1989).

TABLE 5. Perceived odours of species of *Amorphophallus* and *Pseudodracontium*.

Nauseating gaseous odour (like liquid petroleum gas)

A. albus Liu & Wei
A. asterostigmatus Bogner & Hett.
A. bulbifer Blume
A. corrugatus N.E. Br.
A. curvistylis Hett.
A. elegans Ridl. [changing to rotting meat]
A. glossophyllus Hett.
A. krausei Engl.
A. muelleri Blume
A. napalensis (Wall.) Bogner & Mayo
A. opertus Hett. [also burnt rubber]
A. paeoniifolius (Dennst.) Nicolson [also rotting meat].
A. parvulus Gagnep.
A. prainii Hook. f.
A. pygmaeus Hett.
A. sagittarius van Steenis
A. salmoneus Hett.
A. variabilis Blume
P. fallax M.M. Serebryanyi
P. gracilis M.M. Serebryanyi
P. lacourii (Lind. & Andre) N.E. Br. [also anise]
P. latifolium M.M. Serebryanyi

Rotting meat odour

A. angolensis subsp. *maculatus* (Engl.) Ittenb. [becoming fishy]
A. cirrifer Stapf
A. gigas Teijsm. & Binn.
A. kiusianus (Makino) Makino
A. konjac K. Koch
A. maximus (Engl.) N.E. Br.
A. maxwellii Hett.

Anise-like odour

A. albispathus Hett.
A. coudercii (Bogner) Bogner
A. longituberosus Engl. & Gehrm.

Rancid cheese odour

A. annulifer Hett. [also rotting meat]
A. antsingensis Bogner & Hett. (ined.)
A. costatus Hett.
A. elatus Hook f.
A. hayi Hett.
A. hirtus N.E. Br.
A. margaritifer (Roxb.) Kunth [weak odour]

Dung and faecal odours

A. abyssinicus (Rich.) N.E. Br. [cow-dung]
A. eichleri (Engl.) N.E. Br. [faecal changing to shrimp-like]
A. henryi N.E. Br. [cow-dung]

Fishy and urine odours

A. brachyphyllus Hett. (ined.) [fried fish]
A. eburneus Bogner [fishy]
A. lambii Mayo & Widjaja [urine-like, later more fishy]
A. lewallei Malaisse & Bamps [sewage]
A. titanum (Becc.) Becc. ex Arcang. [urine-like or fishy]
A. zenkeri (Engl.) N.E. Br. [shrimp-like]

"Spicy" and chocolate odours

A. ankarana Hett. [spicy and chocolate]
A. arnautovii Hett. [spicy]
A. manta Hett. [chocolate]

Fruit and vegetable odours

A. cicatricifer Hett. [initially like shoe-polish, then changing to fruity]
A. galbra F.M. Bailey [fruity]
A. haematospadix Hook f. [bananas]
A. odoratus Hett. [carrots]
A. yuloensis H. Li [lemon]
A. zenkeri N.E. Br. [mushrooms]

TABLE 6. Major components* of the appendix odours of inflorescences of *Amorphophallus* and *Pseudodracontium* species. Odours were collected from inflorescences still attached to plants cultivated at Leiden Botanic Garden and Aalsmeer. Appendix odours were collected and analysed as described for *Arum* in Table 1 except that the trap was simply positioned near to the appendix which, if possible, was covered by a clear nylon bag. Only one inflorescence of each species has been analysed.

Species	dimethyl disulphide (a)	dimethyl trisulphide (c)	dimethyl tetrasulphide (c)	4-methoxyphenethyl alcohol (a)	trimethylamine (a)	isoamyl acetate (a)	isocaproic acid (a)	ethyl acetate (a)	ocimene (*cis* or *trans*) (b)	2-heptanone (a)	indole (a)	acetic acid (a)	limonene (a)	acetophenone (b)	1-phenylethyl acetate (b)	1,2-dimethoxybenzene (b)	phenylethyl alcohol (a)	Odour
A. konjac K. Koch	+++	+++											++					rotting meat
A. maximus (Engl.) N.E. Br.	+++	+++																rotting meat
A. paeoniifolius (Dennst.) Nicolson	+++	+++	++															gaseous + rotting meat
A. albus Liu & Wei	+++	+++	++															gaseous
A. bulbifer Blume	++	++++	+															gaseous
A. glossophyllus Hett.	+++	+++	++															gaseous
A. pranii Hook. f.	+++	+++	+															gaseous
A. sagittarius van Steenis	+++	+++																gaseous
A. opertus Hett.	+++	+++																gaseous + burnt rubber
A. titanum (Becc.) Becc. ex Arcang.	+++	++																gaseous + urine
A. zenkeri N.E. Br.	+++	++											++					mushrooms
A. arnautovii Hett.	+++	+++											+					spicy
A. eichleri (Engl.) N.E. Br.	+++	+++	++							++	++						++	dung
A. cicatricifer Hett.	+++	+++												+++	++			gaseous + polish, fruity
P. fallax Serebr.	+++	+++	+++	++														gaseous
P. lacourii (Lind. & Andre) N.E. Br.	+++	+++	++	+++														gaseous + anise-like
A. albispathus Hett.				++++														anise-like
A. brachyphyllus Hett. (ined.)					+++													fried fish
A. haematospadix Hook. f						+++		+++										bananas
A. elatus Hook. f.							++++											strong cheese
A. annulifer Hett.	++					++		+++		++		++						strong cheese + rotting meat
A. margaritifer (Roxb.) Kunth***									+++								++	weak cheese

*+ = level of compounds in odour as described in Table 1.
** compound identification criteria as in Table 1.
*** *A. margaritifer* lacks an appendix. The odour emanated from the flowers or the staminodes and also contained various monoterpenoids at less than 1%.

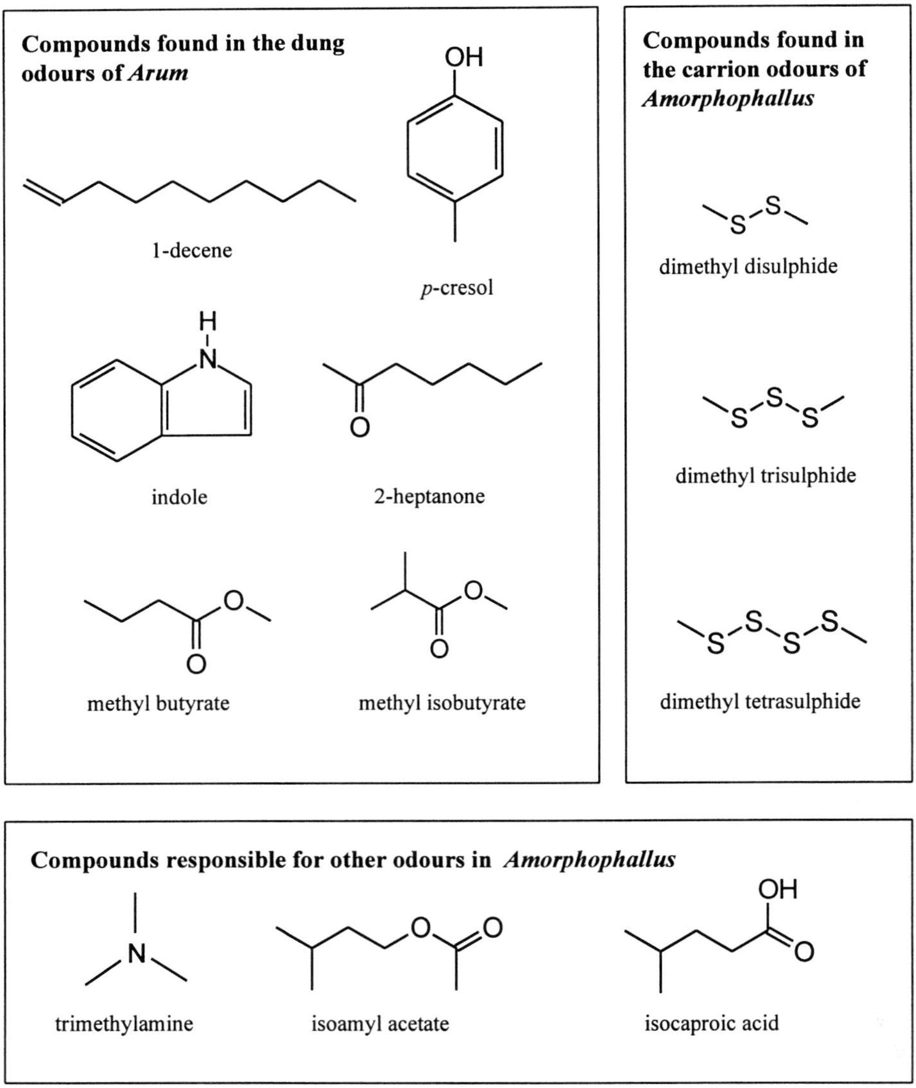

FIG. 2. Structures of odour compounds in *Arum* and *Amorphophallus*.

Perhaps more intriguing than the pollinators of the species having gaseous or rotten meat odours is the identity of the pollinators in species for which odour mimicry is not so obvious, such as the cheesy odour of *A. elatus,* due to isocaproic acid. Indications as to the type of potential pollinator may come from plants in which similar compounds have been detected in the floral odour. For example, isovaleric acid, the lower homologue of isocaproic acid, occurs as the major constituent in the odour of the foul-smelling succulent *Senecio articulatus* (Compositae; Kite and Smith, 1997) and has been reported in the odours of several other foul-smelling flowers such as some *Masdevallia* species (Orchidaceae; Kaiser, 1993), *Deherainia*

TABLE 7. Insects reported in inflorescences of *Amorphophallus* species.

Species	Insects	Reference
A. hohenackeri (Schott) Engl. & Gehrm.	*Haptoncus (=Haptoncurina) motschulshii* Reitter (Coleoptera: Nitidulidae)	Sivadasan & Sabu (1991)
A. johnsonii N.E. Br.	*Phaeochrous amplus* Arrow (Coleoptera: Scarabaeidae)	Beath (1996)
A. maculatus N.E. Br.	*Phaeochrous camerunensis* Arrow (Coleoptera: Scarabaeidae)	Bogner (1976)
A. oncophyllus Prain	Coleoptera: Nitidulidae	van der Pijl (1937)
A. paeoniifolius (Dennst.) Nicolson	*Adoretus* sp. (Coleoptera: Scarabaeidae) *Melipona* sp. (Hymenoptera: Apidae)	Singh & Gadgil (1996)
A. titanum (Becc.) Becc. ex Arcang.	*Diamesus osculans* Vigors (Coleoptera: Silphidae) *Creophilus villipennis* Kraatz (Coleoptera: Staphylinidae)	van der Pijl (1937)
A. variabilis Blume	Coleoptera: Nitidulidae	van der Pijl (1937)

smaragdina, *Theophrasta americana* (Theophrastaceae; Knudsen and Stahl, 1994) and *Leontopodium alpinum* (Erhardt, 1993). Flies have been implicated as pollinators in all these species. Similarly 4-methoxybenzyl alcohol, the lower homologue of 4-methoxyphenethyl alcohol produced by *A. albispathus*, occurs in the inflorescence odour of another aroid, *Spathiphyllum cannifolium* (Lewis *et al.*, 1988). This and other benzenoid compounds have been shown to be attractive to tephritid fruitflies (Chuah *et al.*, 1996). Finally, the trimethylamine produced by *A. brachyphyllus* probably attracts carrion insects; trimethylamine is an indicator of decay in fish (Krzymien and Elias, 1990) and has been shown to be an attractant for houseflies (*Fannia* spp.; Mulla *et al.*, 1984).

The inflorescence odours of *Amorphophallus* may have taxonomic significance. The species producing dimethyl oligosulphides are thought to form two related groups and the anise-like odour of *A. albispathus* is also perceived in two species thought to be closely related to it, *A. longituberosus* and *A. coudercii*. Furthermore, species of *Pseudodracontium* are possibly most closely related to these anise-like scented *Amorphophallus* species, a hypothesis supported by the presence of 4-methoxyphenethyl alcohol in the odours of *P. fallax*, *P. lacourii* and *A. albispathus*.

Conclusions

There appear to be two major odour types among sapromyophilous species of *Arum* and *Amorphophallus* associated with attraction of either dung or carrion insects. Odours of dung-smelling *Arum* species are complex, containing numerous compounds belonging to several classes, while odours of carrion-smelling *Amorphophallus* species have a simple composition, consisting mainly of dimethyl oligosulphides (Fig. 2). Within dung odours of *Arum* a further subdivision is evident, since the odour of *A. maculatum* has a different chemical profile from the other dung-smelling species, whereas all the carrion odours of *Amorphophallus* are consistently dominated by dimethyl oligosulphides. This may reflect the possibility that carrion exhibits less odour variation than dung, or that the semiochemicals used by carrion insects are more consistent than those used by dung insects. Certainly, the dung of different animals have perceptibly different odours and dung odours can also vary according to diet (Aii *et al.*, 1980).

Dung mimicry and carrion mimicry are not mutually exclusive odour types; for example, the odour of *Amorphophallus eichleri* contains both dimethyl oligosulphides and two compounds, indole and 2-heptanone, associated with the dung odours of *Arum*. A similar mixture of sulphides and nitrogen-compounds has been reported in another aroid *Sauromatum venosum* (Borg-Karlson *et al.*, 1994)[2]. Nevertheless, the model of at least two chemical odour types associated with dung or carrion mimicry is a useful one to test in studies of other species pollinated by dung or carrion insects. For example, analysis of the parasitic plant *Hydnora africana* (Hydnoraceae), which is said to smell of rotting hide, has revealed that the odour is dominated by dimethyl oligosulphides and therefore appears to be of the carrion-mimicking type, although the pollinators are not known (Burger *et al.*, 1988). However, not all plants with odours dominated by dimethyl oligosulphides belong to the sapromyophilous syndrome, since the odours of two presumed bat-pollinated flowers, *Crescentia cujete* (Bignoniaceae) and *Pilosocereus tweedyanus* (Cactaceae), have been found to contain high levels of dimethyl trisulphide (Knudsen and Tollsten, 1995).

Another odour type in the syndrome of sapromyophily is possibly the production of alcohols with fungus odours which seems to be associated with pollination by fungus insects; e.g. in *Theophrasta americana* (Knudsen and Stahl, 1994) and orchids of the genus *Dracula* (Kaiser, 1993). Thus, chemical odour types of sapromyophilous plants appear to show greater association with the nature of the substrate being mimicked than any particular taxonomic group of pollinators. Therefore, attempts to consider such plants as either fly or beetle pollinated may be unhelpful. In presumed sapromyophilous species with more unusual odours it is possible that only selected key pollinator-signalling chemicals are being produced and data on the pollinator species and their natural history will be required to determine the identity of the substrate being mimicked.

Acknowledgements

We would like to thank Tony Hall and his staff (RBG Kew) and Art Vogeland and his staff (Leiden Botanic Garden) for expertly cultivating species of *Arum* and *Amorphophallus* respectively. We also thank David Lees (Natural History Museum, London) for assistance with insect taxonomy, Ulrich Meve (University of Bayreuth) for collecting pollinators in Germany and Marcos Méndez for sharing unpublished data.

References

Adams, R. P. (1989). Identification of essential oils by ion trap mass spectroscopy. Academic Press, London.

Adams, R. P. (1995). Identification of essential oil components by gas chromatography / mass spectroscopy. Allured Publishing Corp., Illinois.

Aii, T., Yonaga, M. and Tanaka, H. (1980). Changes in headspace volatiles of feed in the digestive tracts of cattle. *Jap. Soc. Grassland. Sci. J.* **26**: 223–230.

[2] However, a more recent analysis of *S. venosum* suggested that monoterpenes predominated (Skubatz *et al.*, 1996), although dimethyl oligosulphides and indole were detected.

Ashworth, J.R. and Wall, R. (1994). Responses of the sheep blowflies *Lucilia sericata* and *L. cuprina* to odour and the development of semiochemical baits. *Med. and Veterin. Entomol.* **8**: 303–309.

Ausloos, P., Clifton, C., Lias, S. G., Shamin, A. and Stein, S. E. (1992). NIST/EPA/NIH mass spectral database. National Institute of Standards and Technology, US Department of Commerce, Washington, D.C.

Beath, D.D.N. (1996). Pollination of *Amorphophallus johnsonii* (Araceae) by carrion beetles (*Phaeochrous amplus*) in a Ghanaian rain-forest. *J. Trop. Ecol.* **12**: 409–418.

Bermadinger-Stabentheiner, E. and Stabentheiner, A. (1995). Dynamics of thermogenesis and structure of epidermal tissues in inflorescences of *Arum maculatum*. *New Phytol.* **131**: 41–50.

Bertsch, W., Chang, R .C. and Zlatkis, A. (1974). The determination of organic volatiles in air pollution studies: characterization of profiles. *J. Chromatogr. Sci.* **12**: 175–182.

Bogner, J. (1976). *Amorphophallus maculatus* N.E. Br. *Palmengarten* **40**: 83–86.

Borg-Karlson, A-K., Englund, F. O. and Unelius,C. R. (1994). Dimethyl oligosulphides, major volatiles released from *Sauromatum guttatum* and *Phallus impudicus*. *Phytochemistry* **35**: 321–323.

Boyce, P.C. (1993). The genus *Arum*. HMSO, London.

Boyce, P.C. (1995). The genus *Arum* in Greece and Cyprus. *Ann. Mus. Goulandris.* **9**:27–38 (1994 publ. 1995).

Braverman, Y. and Koach, J. (1982). *Cuculoides* species found in the inflorescences of *Arum elongatum* in Israel. *Mosquito News* **42**: 516–517.

Burger, B.V. , Munro, Z.M. and Visser, J.H. (1988). Determination of plant volatiles 1: analysis of the insect-attracting allomone of the parasitic plant *Hydnora africana* using Grob-Habich activated charcoal traps. *HRC&CC, J. High Resolution Chromatogr. & Chromatogr. Commun.* **11**: 496–499.

Chuah, C.H., Ooi, L.E., Yong, H.S. and Goh, S.H. (1996). New male fruitfly attractant from spadix of *Spathiphyllum cannaefolium* (Araceae). *Biochem. Syst. & Ecol.* **24:** 189–193.

Cole, R. (1980). The use of porous polymers for the collection of plant volatiles. *J. Sci. Food Agric.* **31**: 1242–1249.

Dafni, A. (1984). Mimicry and deception in pollination. *Annual Rev. Ecol. Syst.* **15**: 259–278.

Dobson, H.E.M. (1994). Floral volatiles in insect biology. In: E.A. Bernays (editor). Insect-plant interactions, vol. V, pp.47–81, CRC Press, Boca Raton, Fla.

Dormer, K.J. (1960). The truth about pollination in *Arum*. *New Phytol.* **59**: 298–301.

Drummond, D.C. and Hammond, P.M. (1991). Insects visiting *Arum dioscoridis* Sm. and *A. orientale* M. Bieb. *Entomol. Monthly Mag.* **127**: 151–155.

Drummond, D.C. & Hammond, P.M. (1993). Insects visiting *Arum creticum* Boiss. & Heldr., *A. concinnatum* Schott and *A. purpureospathum* Boyce. *Entomol. Monthly Mag.* **129**: 245–252.

Erhardt, A. (1993). Pollination of the edelweiss, *Leontopodium alpinum*. *Bot. J. Linn. Soc.* **111:** 229–240.

Fægri, K. and van der Pijl, L. (1979). The principles of pollination ecology. Pergamon Press, Oxford.

Hall, M.J.R. (1995). Trapping the flies that cause myiasis: their responses to host stimuli. *Ann. Trop. Med. Parasitol.* **89**: 333–357.

Hetterscheid, W.L.A. (1994). Preliminary taxonomy and morphology of *Amorphophallus* Blume ex Decaisne (Araceae). In: M.M. Serebreyanyi (editor), Proceedings Moscow Aroid Conference (1st: 1992: Moscow), pp. 35–48. Moscow.

Kaiser, R. (1991). Trapping, investigation and reconstruction of flower scents. In: P.M. Müller and D. Lamparsky (editors). Perfumes: art, science & technology, pp. 213–250, Elsevier Science Publishers Ltd., London.

Kaiser, R. (1993). The scent of orchids. Elsevier, Amsterdam.

Kite, G.C. (1995). The floral odour of *Arum maculatum*. *Biochem. Syst.& Ecol.* **23**: 343–354.

Kite, G.C. and Smith, S.A.L. (1997). Inflorescence odour of *Senecio articulatus*: temporal variation in isovaleric acid levels. *Phytochemistry* **45**: 1135–1138.

Knoll, F. (1926). Insekten und Blumen IV. Die *Arum*-Blütenstände und ihre Besucher. *Abh K.K. der Zool.-Bot. Ges. Wien* **12**: 383–481.

Knudsen, J.T. (1993). Floral scent and pollination at different systematic levels. Ph.D. thesis, University of Göteborg.

Knudsen, J.T. and Stahl, B. (1994). Floral odours in the Theophrastaceae. *Biochem. Syst. & Ecol.* **22**: 259–268.

Knudsen, J.T. and Tollsten, L. (1993). Trends in floral scent chemistry in pollination syndromes: floral scent composition in moth-pollinated taxa. *Bot. J. Linn. Soc.* **113**: 263–284.

Knudsen, J.T. and Tollsten, L. (1995). Floral scent in bat-pollinated plants: a case of convergent evolution. *Bot. J. Linn. Soc.* **119**: 45–57.

Knudsen, J.T., Tollsten, L. and Bergström, L.G. (1993). Floral scents - a checklist of volatile compounds isolated by head-space techniques. *Phytochemistry* **33**: 253–280.

Koach, J. (1986). Araceae. In: D.N. Feinbrun (editor). Flora Palaestina, 4, pp. 330–340, Israel Academy of Science and Humanities, Jerusalem.

Krzymien, M.E. and Elias, L. (1990) Feasibility study on the determination of fish freshness by trimethylamine headspace analysis. *J. Food Sci.* **55**: 1228–1232.

Kullenberg, B. (1953). Observationer över *Arum*-pollinerare i Libanons kustomåde. *Svensk Bot. Tidskr.* **47**: 24–29.

Lack, A. J. and Diaz, A. (1991). The pollination of *Arum maculatum* L.—a historical review and new observations. *Watsonia* **18**: 333–342.

Lewis, J. A., Moore, C. J., Fletcher, M. T., Drew, R. A. and Kitching, W. (1988). Volatile compounds from the flowers of *Spathiphyllum cannaefolium*. *Phytochemistry* **27**: 2755–2757.

Meeuse, B.J.D., Schneider, E.L., Hess, C.M., Patt, J.M. and Kirkwood, K. (1984). Activation and possible role of the "food-bodies" of *Sauromatum* (Araceae). *Acta Bot. Neerl.* **33**: 483–496.

Meeuse, J.D. and Raskin, I. (1988). Sexual reproduction in the arum lily family, with emphasis on thermogenicity. *Sexual Pl. Reprod.* **1**: 3–15.

Méndez, M. and Obeso, J.R. (1992). Influenca del osmóforo en la producción de infrutesciencias en *Arum italicum* Miller (Araceae). *Anales Jard. Bot. Madrid* **50**: 229–237.

Mookherjee, B.D., Wilson, R.A., Schrankel, K.R., Katz, I. and Butler, J.F. (1993). Semio activity of flavor and fragrance molecules on various insect species. In: R. Teranishi, R.G. Buttery, and H. Sugisawa (editors). Bioactive volatile compounds from plants, pp. 35–48, American Chemical Society, Washington, DC. (ACS symposium series, 525).

Mulla, M.S., Axelrod, H. and Hwang, Y-S. (1984). Field evaluation of chemical attractants against the fly *Fannia femoralis* (Diptera: Muscidae). *J. Chem. Ecol.* **10**: 349–360.

Müller, H. (1883). The fertilization of flowers (translated and edited by D.W.Thompson). MacMillan, London.

Nilssen, A.C., Tømmerås, B.A., Schmid, R. and Evensen, S.B. (1996). Dimethyl trisulphide is a strong attractant for some calliphorids and a muscid but not for the reindeer oestrid *Hypoderma tarandi* and *Cephenemyia trompe*. *Entomol. Exp. et Appl.* **79**: 211–218.

Patt, J. M., Hartman, T. G., Creekmore, R. W., Elliot, J. J., Schal, C., Lech, J. and Rosen, R. T. (1992). The floral odour of *Peltandra virginica* contains novel trimethyl-2,5-dioxabicyclo[3.2.1]nonanes. *Phytochemistry* **31**: 487–491.

Prime, C.T. (1954). *Arum neglectum* (Towns.) Ridley. Biological Flora of the British Isles no. 41. *J. Ecol.* **42**: 241–248.

Prime, C.T. (1960). Lords and Ladies. Collins, London.

Serebryanyi, M.M. (1995). A taxonomic revision of *Pseudodracontium* (Araceae-Aroideae-Thomsonieae). *Blumea* **40**: 217–235.

Shankaranarayana, M.L., Raghavan, B., Abraham, K.O. and Natarajan, C.P. (1974). Volatile sulfur compounds in food flavors. *CRC Crit. Rev. Food Technol.* **29**: 395–435.

Singh, S.N. and Gadgil, M. (1996). Ecology of *Amorphophallus* species in Uttara Kannada District of the Karnataka State, India: implications for conservation. *Aroideana* **18**: 5–20.

Sivadasan, M. and Sabu, T. (1991). Beetle-pollination - Cantharophily - in *Amorphophallus hohenackeri* (Araceae). *Aroideana* **12**: 32–37.

Skubatch, H., Nelson, T.A., Dong, A.M., Meeuwse, B.D.J., and Bendich, A.J. (1990). Infrared thermography of *Arum* lily inflorescences. *Planta* **182**: 432–436.

Skubatz, H., Kunkel, D.D., Howard, W.N., Trankle, R. and Mookherjee, B. (1996). The *Sauromatum guttatum* appendix as an osmophore: excretory pathways, composition of volatiles and attractiveness to insects. *New Phytol.* **134**: 631–640.

Smith, B. N. and Meeuse, B. J. D. (1966). Production of volatile amines and skatole at anthesis in some arum lily species. *Pl. Physiol.* **41**: 343–347.

Smith, K.G.V. (1956). On the Diptera associated with the stinkhorn (*Phallus impudicus* Pers.) with notes on other insects and invertebrates found on this fungus. *Proc. Roy. Entomol. Soc. London, Ser. A, Gen. Entomol.* **31**: 49–55.

Sowter, F.A. (1954). *Arum maculatum* L. Biological Flora of the British Isles no. 27. *J. Ecol.* **37**: 207–219.

Steiner, L.F. (1952). Methyl eugenol as an attractant for oriental fruit fly. *J. Econ. Entomol.* **45**: 241–248.

Stoffolano, J.G., Yin, C.-M. and Zou, B.-X. (1989). Reproductive consequences for female black blowfly (Diptera: Calliphoridae) fed on the stinkhorn fungus, *Mutinus caninus*. *Ann. the Entomol. Soc. Amer.* **82**: 192–195.

van der Pijl, L. (1937). Biological and physiological observations on the inflorescence of *Amorphophallus*. *Recueil Trav. Bot. Neerl.* **34**: 157–167.

Vogel, S. (1962). Duftdrüsen im Dienste der Bestäubung über Bau und Funcktion der Osmophoren. *Akad. Wiss.* (Mainz), *Abh. Math.-Naturwiss. Kl.* (1962): 600–736. Translated in: Vogel, S. (1990). Role of scent glands in pollination. Smithsonian Institution Libraries, Washington.

Williams, N. H. and Dodson, C.H. (1972). Selective attraction of male euglossine bees to orchid fragrances and its importance in long distance pollen flow. *Evolution* **26**: 84–95.

Willis, J.H. (1952). Aroids foul and fragrant. *Victorian Naturalist* **69**: 47–50.

Withers, P. (1988). Revisionary notes on British species of *Psychoda* Latreille (Diptera, Psychodidae) including new synonyms and a species new to science. *Brit. J. Entomol. Nat. Hist.* **1**: 69–76.

Sage, T.L., Sampson, F.B., Bayliss, P., Gordon, M.G. and Heij, E.G. (1998). Self-sterility in the Winteraceae. In: S.J. Owens and P.J. Rudall (Editors). Reproductive Biology, pp. 317–328. Royal Botanic Gardens, Kew.

SELF-STERILITY IN WINTERACEAE

T.L. SAGE[1], F.B. SAMPSON[2], P. BAYLISS[1], M.G. GORDON[2] AND E.G. HEIJ[3]

[1]**Department of Botany, University of Toronto, Toronto ON M5S 3B2, Canada**
[2]**Victoria University of Wellington, School of Biological Sciences, P.O. Box 600, Wellington, New Zealand 6015**
[3]**CSIRO, Division of Horticulture. G.P.O. Box 350, Adelaide 5001, Australia**

Abstract

Species of Winteraceae, *Pseudowintera axillaris* and *Drimys winteri*, exhibit high rates of self sterility. It has been suggested that self sterility in these species is due to self incompatibility (SI) mechanisms acting within the nucellus to prevent self pollen tubes from entering the embryo sac. To date, classification of self sterility in each species as SI within the nucellus is only tentative, however. Interactions between self and cross pollen tubes and carpellary tissues were examined to assess mechanisms of self sterility in each species. Results indicate that self pollen tubes did not undergo discriminatory interactions prior to or after embryo sac entry. Double fertilisation was effected in all cases by three days following self pollination as indicated by the presence of a primary endosperm nucleus and zygote. Rates of double fertilisation following self and cross pollinations were not significantly different. Data do not support the occurrence of SI mechanisms in the nucellus of each species. Temporal aspects of self sterility in these species are currently under investigation to determine whether self sterility is due to SI or inbreeding depression.

Introduction

Self-incompatibility (SI) is defined as the inability of a fully fertile hermaphroditic plant to produce zygotes when self-pollinated (Lundqvist, 1964). It is a genetic mechanism for recognition and rejection of self pollen, thereby functioning to promote outbreeding and heterozygosity. Two types of homomorphic SI, sporophytic (SSI) and gametophytic SI (GSI), have been recognised and studied extensively at the genetic, cytological, and molecular levels (Matton *et al.*, 1994). In comparison, ovarian self-incompatibility (OSI), a third postulated category of SI, has not been characterised in depth (Seavey and Bawa, 1986; Sage *et al.*, 1994). Species exhibiting OSI differ from homomorphic SSI and GSI in the site of failure of self pollen tube function. OSI is characterised by (1) failure of self pollen tube growth within the ovary prior to ovule penetration (Beardsell, 1991); (2) failure of self pollen tube growth within the ovule prior to embryo sac penetration (Kenrick *et al.*, 1986); (3) failure of self-sperm to effect syngamy or double fertilisation (Cope, 1962); or, (4) failure of selfed seed development following double fertilisation (Sage and Williams, 1991; Gibbs and Bianchi, 1993).

It has been postulated that SI evolved on multiple occasions during the history of angiosperms. (Bateman, 1952; Charlesworth, 1985; Gibbs, 1986; Matton *et al.*, 1994). The timing of evolution of SI with respect to the origin of angiosperms is unclear, however. Some authors have reported that some basally rooted angiosperm taxa are self-incompatible and therefore, the first flowering plants were self-incompatible; they note that SI was most likely an important factor contributing to the early success of angiosperms (Whitehouse, 1950; Bernhardt and Thien, 1987). In contrast, other individuals have hypothesised that the original flowering plants were self-compatible (Endress, 1984; Weller *et al.*, 1995). This hypothesis is supported by claims of self-compatibility (SC) in some basally rooted angiosperm taxa and a recent phylogenetic approach to the evolution of self-incompatibility in flowering plants (Weller *et al.*, 1995). Weller *et al.* (1995) also suggested that the evolution of SI was not a major factor in the initial evolutionary radiation of angiosperms. A lack of consensus exists concerning the timing of origin of GSI and SSI, and both GSI and SSI have been implicated as the first SI mechanisms to evolve (Bernhardt and Thien, 1987; Zavada, 1984, 1990; Gibbs, 1986, 1991); GSI may be ancestral to SSI (Dickinson, 1990). An intriguing hypothesis has recently been submitted which suggested primitive SI mechanisms may have been some form of OSI (Kenrick *et al.*, 1986; Barrett, 1988; Lloyd and Wells., 1992). Kenrick *et al.* (1986) hypothesised that SI mechanisms may have evolved and been operative on the adaxial surface of an open carpel in association with any structure which at the time was functionally important in pollen germination and transmission of the male gametophyte to the female gametophyte (i.e., nucellus, integument, and/or placental epithelium). Sage *et al.* (1994) noted that it might be expected that the original site of any genetic recognition/rejection response would be homologous with one or more tissues present in the carpels of primitive extant species, particularly those lacking a distinct style since the style appeared later in the evolution of the angiosperm flower (Friis and Crepet, 1987); taxa showing primitive carpellary structure may still express genetic recognition/rejection responses in tissues homologous with those in which these interactions originally occurred early in the evolution of angiosperms.

Information on breeding systems of basally rooted angiosperm taxa (paleoherbs and woody species) exhibiting primitive carpellary structure is, unfortunately, limited. Claims of SC or SI have not been supported by structural, physiological, genetic and molecular studies. These current gaps in our knowledge on breeding systems of basally rooted angiosperms make it difficult to support hypotheses on the evolution of breeding systems of early angiosperms and the role of any SI mechanisms in the initial evolution of flowering plants. The primary focus of this paper is to provide information on the breeding system of species within the primitive angiosperm family, Winteraceae.

Self-sterility in Winteraceae

Each of the seven genera comprising Winteraceae, a family exhibiting primitive carpellary structure and a relictual Gondwana pattern of distribution, is reported to contain species which are self-sterile. Reports of self-sterility in each genus (*Belliolum*, *Bubbia*, *Drimys*, *Exospermum*, *Pseudowintera*, *Tasmannia* and *Zygogynum*) are based on the failure to produce fruit following self-pollination (Godley and Smith, 1981; Norton, 1980; Thien, 1980; Bernhardt and Thien, 1987; Lloyd and Wells, 1992). Lloyd and Wells (1992) also noted that fruit set was significantly reduced if cross pollen was administered to the stigma 1-3 days following self-pollination. Data on self-sterility in each genus have been interpreted as SI. Qualitative analysis of cross- and self-pollen tube growth in *Pseudowintera colorata*, *P. axillaris* and *Drimys winteri*, indicate the presence of self-pollen tubes in ovules (Norton, 1980; Lloyd and Wells, 1992). Norton (1980) suggested that SI reactions occurred within the ovules of *P. axillaris*, *P. colorata* and *D. winteri*. Lloyd and Wells (1992) concluded that incompatibility reactions occurred prior to double fertilisation, possibly within the nucellus as noted to occur in *Acacia retinoides* (Kenrick *et al.*, 1986).

The assumption that self-sterility in any genus within the Winteraceae represents some form of SI, and perhaps OSI is, however, tentative. Classification of SI in any species requires, in part, an understanding of the detailed structure of transmitting tissues encountered by pollen tubes en route to the embryo sac, analysis of cross- and self-pollen tube growth, and characterisation of ovule/seed development following cross- and self-pollination. Such information is essential to characterise (1) whether or not self-pollen tubes undergo discriminatory interactions within the carpel prior to ovule entry, (2) the site and timing of discriminatory interactions, and (3) the carpellary tissues involved in discriminatory interactions. The purpose of the present paper is to provide such information for the species *D. winteri* and *P. axillaris*.

Structural/functional aspects of the pollen tube pathway

The transmitting tissue of flowering plants is that part of the gynoecium that functions in capture, retention, germination, recognition, and growth of the male gametophyte from the stigma to the embryo sac (Tilton and Horner, 1980). Cells of the transmitting tissue are generally secretory in nature and their exudates are assumed to provide the appropriate physical and nutritional medium for pollen tube growth and extension as well as recognition molecules such as S-gene products (Knox, 1984; Dickinson, 1990; Sanders and Lord, 1992).

The carpels of Winteraceae lack a true style. Winteraceous transmitting tissue consists of a stigma, substigmatic tissue, placental and funicular epithelium, and micropyle lined by inner integument only. The Winteraceous stigma (Figs 1–6) is comprised of stigmatic crests and can be classified as a wet stigma (see Heslop-Harrison and Shivanna, 1977 for stigma types). The protoplast of stigmatic papillae of both *D. winteri* and *P. axillaris* at anthesis did not appear notably secretory in fine structure. Stigmas were composed of papillate cells with large vacuoles and dense peripheral cytoplasm containing numerous mitochondria and sparse profiles of rough endoplasmic reticulum; Golgi bodies were noticeably absent. A prominent exudate histochemically positive for lipids and protein surrounded stigmatic cells. Loosely fibrillar walls of stigmatic cells contained numerous electron dense inclusions presumably destined for secretion. Transfer wall characteristics noted to occur in stigmatic cells of *P. axillaris* were not present in *D. winteri*.

Styles of angiosperms have typically been classified structurally into solid, hollow and semi-solid types. All styles are, however, homologous in that their transmitting tissues are associated with an extracellular matrix which is normally composed of proteins, lipids, and/or insoluble carbohydrates (Knox, 1984). Although the Winteraceous carpel is styleless, substigmatic transmitting tissue resembled a solid style (Figs 7, 8); papillate carpellary cleft cells and surrounding isodiametric cells formed a core of loosely packed cells. Cells of substigmatic transmitting tissue in *D. winteri* were similar in fine structure to stigmatic cells except for the presence of localised regions of dense cytoplasm which contained numerous Golgi bodies, profiles of rough endoplasmic reticulum and lipid containing plastids. Plastids in substigmatic transmitting tract cells of *P. axillaris* contained a well defined internal membrane system. Exudate containing protein and carbohydrate was present in intercellular spaces of substigmatic transmitting tissue. Secretions were amorphous or flocculent in fine-structure.

The secretory nature of placental and funicular epithelial cells has been confirmed at the ultrastructural level in only a small number of derived angiosperms species. The present data represents the first ultrastructural confirmation of the secretory nature of placental and funicular epithelial cells in a primitive angiosperm family. Placental and epithelial cells in both *D. winteri* and *P. axillaris* appeared fine structurally more actively secretory than stigmatic and substigmatic transmitting tract cells (Figs 9–10); Golgi bodies, rough endoplasmic reticulum, mitochondria, amyloplasts, and lipid bodies were abundant. Exudate was apparent between the plasmalemma and cell wall as noted in other species (e.g. *A. exaltata*; Sage and Williams, 1995). A rugose cuticle covered placental and funicular epithelial cells. The cuticle was often uplifted and subtended by exudate partitioned into amorphous and flocculent components. The secretion was again composed of protein and carbohydrates.

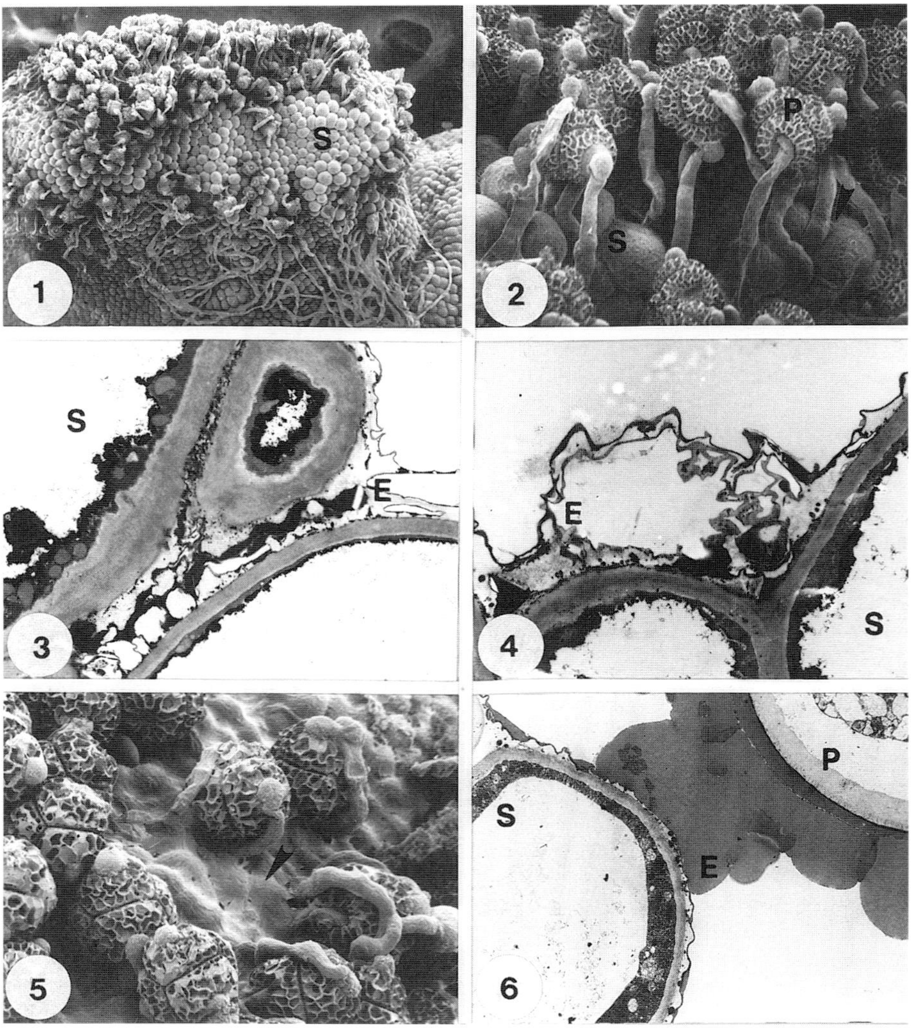

FIGS 1–6. Scanning and transmission electron micrographs of stigmas of *Drimys winteri* and *Pseudowintera axillaris*. Figs. 1, 2. Scanning electron micrograph view of papillate stigma cells and germinating pollen tetrads of *D. winteri*. Fig 1. × 400 . Fig. 2. Arrowhead denotes stigmatic exudate on surface of papillae. × 360. Figs. 3,4. Ultrastructure of stigmatic cells of *D. winteri* illustrating exudate. Fig. 3. × 2586. Fig. 4. × 3848. Figs. 5,6. Stigma of *P. axillaris*. Fig. 5. Arrowhead denotes secretion. × 470. Fig. 6. × 4093. Tissues were prepared as described by Sage and Williams (1995). E = exudate; S = stigma; P = pollen.

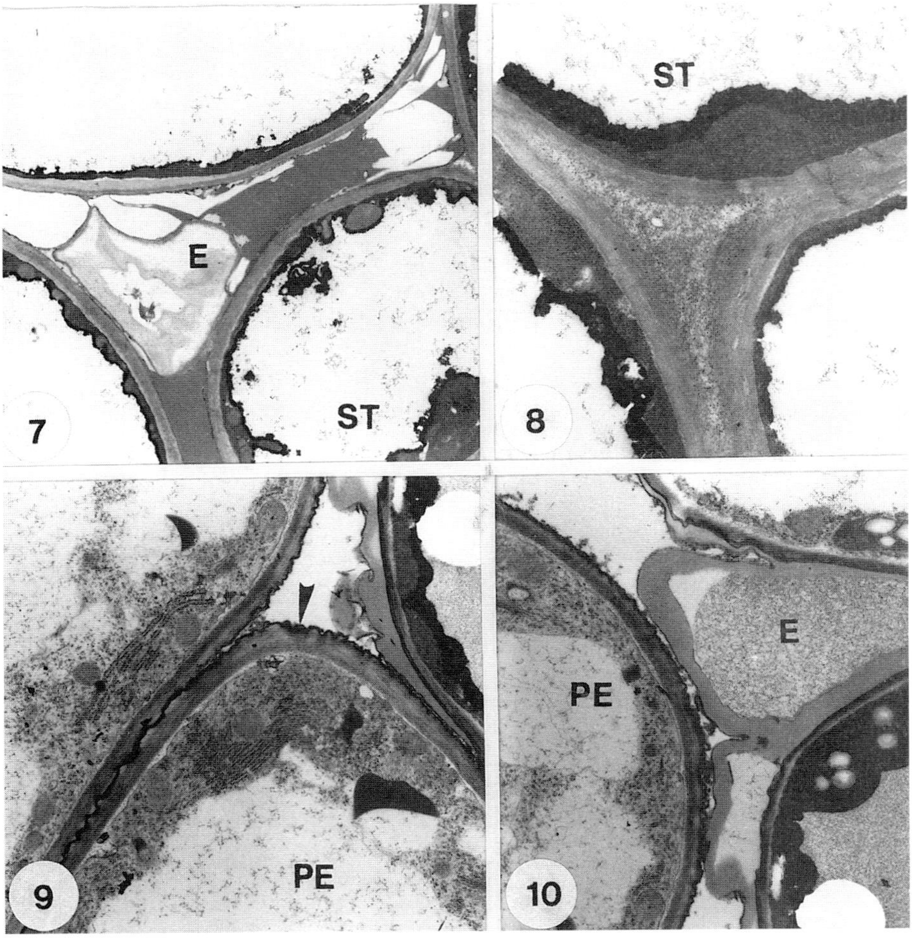

FIGS 7–10. Transmission electron micrographs of substigmatic and placental epithelial cells of *Drimys winteri*. Figs. 7,8. Cross sections through vacuolate substigmatic transmitting cells illustrating exudate. Fig. 7. × 2586. Fig. 8. × 8104. Figs. 9, 10. Longitudinal sections illustrating placental exudate. Endoplasmic reticulum, Golgi, and plastids are more abundant in placental cells than in substigmatic transmitting cells. Fig. 9. Arrowhead denotes cuticle. × 5560. Fig. 10. × 5560. Tissues were prepared as described by Sage and Williams (1995). E = exudate; PE = placental epithelial cell; ST = substigmatic transmitting cell.

Micropylar exudate was apparent at the level of the light microscope in the micropyle of the anatropous ovules of both *D. winteri* and *P. axillaris*. Micropylar exudates have also been reported in *Paspalum* (Chao, 1971), *Ornithogalum* (Tilton and Horner, 1980), *Zea* (Heslop-Harrison *et al.*, 1985), *Rhododendron* (Kual *et al.*, 1986), *Beta* (Brun and Olesen, 1989), *Pennisetum* (Reger *et al.*, 1992), *Gasteria* (Franssen-Verheijen and Willemse, 1993), *Illicium* (Sage *et al.*, 1994), and *Asclepias* (Sage and Williams, 1995). Investigators noting the presence of micropylar exudates in all of the above species indicated that exudates may arise from integumentary and nucellar cells and/or the embryo sac, specifically secreted from synergids.

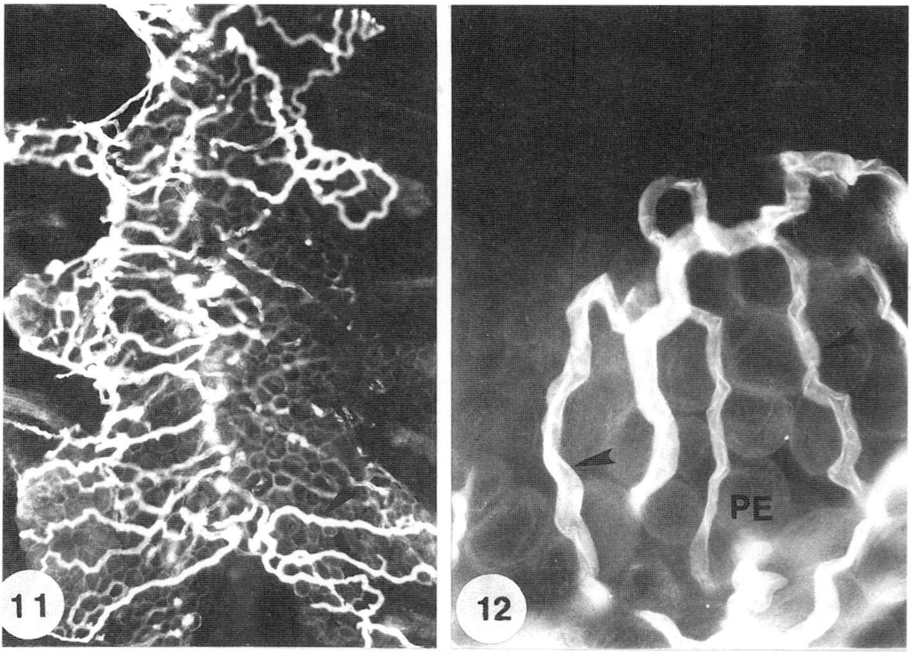

FIGS 11–12. Fluorescence micrographs illustrating growth of pollen tubes (arrowhead) amongst papillate epithelial cells of *Drimys winteri*. Fig. 11. Pollen tubes have emerged from the substigmatic transmitting tissue formed by fused carpellary margins (center) and grown between funiculi of the ovules towards the micropyles. × 300. Fig. 12 × 900. Tissues were prepared as described by Martin (1959). PE = placental epithelial cell.

It has long been hypothesised that micropylar secretions are important in regulating directional growth of pollen tubes into the ovule (Knox, 1984), a function which was likely important early in the evolution of the seed habit. Micropylar secretions may also be important in species with apparent SI reactions in the ovule (Sage *et al.*, 1994). Ultrastructural and physiological studies are not yet available to confirm the origin, nature, and function of micropylar secretions in the ovules of *D. winteri*, *P. axillaris* or any other angiosperm.

Pollen tube growth and seed development following cross- and self-pollination

Qualitative and quantitative analysis of pollen tube growth have traditionally utilised fluorescence microscopy of cleared pistil squashes treated with the fluorochrome decolorised aniline blue which stains the callose (1,3-beta glucan) component of pollen tube walls (aniline blue fluorescence - ABF - technique; Martin, 1959). Qualitatively, self-pollen tube growth in SSI is characterised by inhibition of pollen germination and callose deposition in papillate stigma cells; self-pollen tube growth in stylar GSI is characterised by declined growth rates of pollen tubes within the style accompanied by callose deposition in the pollen tube tip (see Matton *et al.*, 1994 for review). Cross- and self-pollen tubes in most species exhibiting OSI appear to exhibit similar growth characteristics prior to failure of pollen tube function within the ovary (Sage *et al.*, 1994). The presence or absence of callose deposition in ovarian tissues or pollen tubes of species suspected of exhibiting OSI has been documented to occur only in ovules of *Ipomopsis aggregata* (Waser and Price, 1991).

Characterisation of cross- and self-pollen tube growth in *D. winteri* and *P. axillaris* indicated that self-pollen tubes did not undergo discriminatory interactions with the female tissues before entry into ovules. Cross- and self-pollen tetrads readily germinated on stigmas (Figs 1, 2), grew intercellularly amongst exudates of substigmatic transmitting tissue and along papillate placental and funicular epithelial cells (Figs 11, 12) to reach the micropyles which they subsequently entered. Cross- and self-pollen tubes in *P. axillaris* exhibited biphasic growth and initial penetration of ovules occurred within 15 hours post-pollination. Pollen tubes developed a thin callosic wall during the first 10 hours of growth and failed to develop callose plugs prior to 15 hours of growth. Callosic plugs typical of angiosperm pollen tubes developed by 15 hours post-pollination and were initially observed only in the substigmatic transmitting tissue. No differences were observed in callose deposition between cross- and self-pollen tubes in either *D. winteri* or *P. axillaris*. Similarly, callose deposition selective to self-pollinations was not observed in any cells of the transmitting tract.

Double fertilisation in *D. winteri* and *P. axillaris* as determined by viewing serially sectioned carpels was indicated by the presence of (1) one or two degenerated synergids (Figs. 13–15), (2) a cap of penetrated synergid cytoplasm surrounding the chalazal region of a resting zygote (Figs. 13, 14), (3) the absence of unfused sperm within embryo sacs penetrated by a pollen tube, (4) an enlarged nucleus within the zygote relative to the size of the egg nucleus, (5) frequent occurrence of two nucleoli in the resting zygote, and (6) one or more endosperm nuclei (Figs. 13, 14, and 16). Double fertilisation occurred by 3 days following cross- and self-pollination in *D. winteri* and *P. axillaris*. The mean frequency of double fertilisation following cross-pollination in *D. winteri* was not significantly different from the mean frequency of double fertilisation following self-pollination (cross x=53.33 ± 13.3; self x=47.62 ± 11.16: F=0.108; P=0.744). Similar observations were made for *P. axillaris* (cross x=84.56 ± 3.04; self x= 80.00 ± 6.18; F=0.439; P=0.513).

Seed development following cross- and self-pollination proceeded similarly following zygote and endosperm formation except that embryo sac enlargement was less following self-pollination in both species. Significant differences in embryo sac growth were noted by 5 d post-pollination in *D. winteri* (cross median embryo sac length x=97.59 µm ± 4.14 µm; self median embryo sac length x=86.25 µm ± 2.84 µm; F=5.019; P=0.029) and by 15 d post pollination in *P. axillaris* (cross median embryo sac length x=128.3 µm ± 3.08 µm; self median embryo sac length x=116.54 µm ± 1.99 µm; F=11.005; P=0.001). Zygotes remained undivided and 1–2 endosperm nuclei were present at the time differences were noted in embryo sac length following cross- and self-pollination in both species.

Self-sterility in Drimys and Pseudowintera - SI or inbreeding depression?

It is apparent from the present data on cross- and self-pollen tube growth in *D. winteri* and *P. axillaris* that SI mechanisms are not acting at the nucellar boundary to prevent self-pollen tubes from entering embryo sacs and effect double fertilisation as hypothesised by Lloyd and Wells (1992). Rather, self-sterility mechanisms in both species appear to result in failure of embryo sac development after double fertilisation has been effected by self-pollen tubes. Traditionally, pre-zygotic self-sterility has been classified as SI and post-zygotic self-sterility has been categorised as inbreeding depression (de Nettancourt, 1977; Klekowski, 1988). Using these traditional guidelines, self-sterility in *D. winteri* and *P. axillaris* would be interpreted as inbreeding depression. It has been hypothesised, however, that delayed-action OSI systems may exist in which seed development is inhibited after events leading to endosperm and zygote/embryo formation are complete (Charlesworth, 1985; Seavey and Bawa, 1986; Gibbs and Bianchi, 1993; Sage *et al*., 1994). At present, we cannot dismiss the likelihood that self-sterility in *D. winteri* and *P. axillaris* could be due to OSI. We discuss this possibility by addressing some of the criteria which have been posited to indicate OSI versus inbreeding depression.

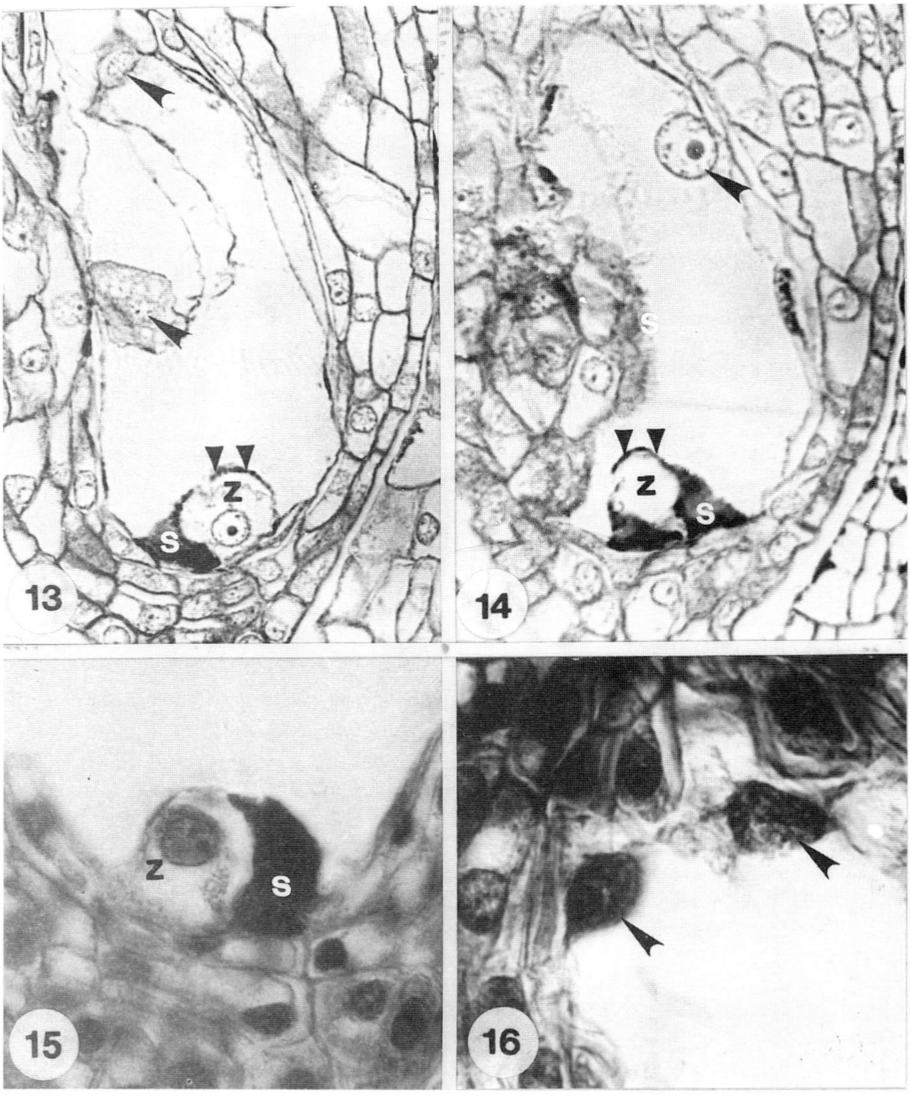

Figs 13–16. Light micrographs illustrating embryo sacs containing zygotes and endosperm following cross- and self-pollination. Fig. 13. *D. winteri* - 7 d following cross-pollination. × 120. Fig. 14. *D. winteri* - 7 d following self-pollination. × 125. Figs. 15, 16. *P. axillaris* - 15 d following self-pollination. Figs. 15, 16 × 175. Double arrowheads denote cytoplasm from penetrated synergid surrounding the zygote. Single arrowheads denote location of endosperm nuclei. S = penetrated synergid. Z = zygote.

It has been also hypothesised that if failure of development following double fertilisation occurs at a single stage of development, then self-sterility may be due to SI. In contrast, a continuum of failure of development throughout the life cycle of the new generation would indicate inbreeding mechanisms are the primary cause of failure (Charlesworth, 1985; Seavey and Bawa, 1986; Wiens *et al.*, 1987; Sage *et al.*, 1994). Temporal aspects of declined embryo sac growth early in seed development, prior to zygotic division in both *D. winteri* and *P. axillaris*, lend support to an hypothesis that self-sterility is due to SI. Failure in seed development following self-pollination, prior to zygotic division, has also been reported in OSI species *Asclepias* (Sparrow and Pearson, 1948; Sage and Williams, 1991), *Rhododendron* (Williams *et al.*, 1984), and *Gasteria* (Sears, 1937).

Waser and Price (1991) presented arguments in support of a post-zygotic SI mechanism of self-sterility versus inbreeding depression based on an analysis of the quantity of lethal equivalents which must be present if inbreeding depression resulted in a seed set approaching zero. The number of lethals (N) for a given level of self-sterility was given by $S=(.75)^N$. Fifteen lethals would generate 99% self-infertility. In addition, the number of loci producing lethals would be 8,000 for a plant to carry 16 lethal equivalents if the mutation rate to lethal alleles was 10^{-6} per locus per generation (assuming the probability of carrying a given number of lethals will follow a Poisson distribution). Although Waser and Price concluded that these values seemed high, there are reports of genetic loads in angiosperms which may be high enough to mimic some of the expected characteristics of a post-zygotic SI (Schmitt and Perry, 1964; Seavey and Carter, 1994; Burbidge and James, 1991). Waser and Price (1991) also noted that a hallmark of inbreeding depression is variation in self-sterility across individuals. Taken as a whole, reports of uniform self-sterility, zero or extremely low levels of self seed set in comparison with cross seed set reported for *D. winteri*, *P. axillaris*, and other species studied to date in the Winteraceae (Godley and Smith, 1981; Norton, 1980; Thien, 1980; Bernhardt and Thien, 1987; Lloyd and Wells, 1992), and the temporal aspects of self-sterility reported in the present study could indicate the operation of OSI in Winteraceae.

The primary function of the pollen tube is delivery of sperm to the female gametophyte, and SSI and GSI act to prevent this primary function. Speculatively, SI in *D. winteri* and *P. axillaris* may be due to both self-recognition and self-rejection occurring after syngamy. Alternatively, if SI recognition reactions in *D. winteri* and *P. axillaris* were to occur at any point within carpellary tissue, but were not sufficiently rapid enough to prevent pollen tubes from completing their primary function, the consequences of recognition (rejection) may not be expressed until after tube entry into the embryo sac. In either instance, self-rejection reactions may affect pollen tube functions other than simply the delivery of sperm to the embryo sac. Signalling events between the male gametophyte and carpellary tissues which appear to be essential for continued development and retention of carpellary tissues (e.g. see O'Neill, 1994; Pimienta and Polito, 1983) may be terminated. Such mechanisms have been proposed for OSI species *Gasteria verrucosa* (Sears, 1937), *Theobroma cacao* (Cope, 1962), *Chorisia* and *Tabebuia* (Gibbs and Bianchi, 1993), *Asclepias exaltata* (Sage and Williams, 1991; Sage *et al.*, 1994), and *Narcissus triandrus* (Sage *et al.*, 1998). In *G. verrucosa* and *A. exaltata*, there is a failure of integument growth of seeds which have been double fertilised by self-pollen tubes. Sears (1937) suggested that failure of integumentary growth after selfing was an indication that SI reactions involve pollen tube-integument interactions; interactions of a compatible pollen tube with the integument might be important for stimulation of normal integumentary growth. Notably, in context of the present observations on *D. winteri* and *P. axillaris*, Pimienta and Polito (1983) and Sage *et al.* (1998) reported embryo sac development in *Prunus dulcis* and *Narcissus triandrus*, respectively, was strongly affected by the type of pollen tubes in the pistil. Cross-pollen tubes had a greater stimulatory effect than self-pollen tubes; irregularities in embryo sac development, to include declined embryo sac elongation, were more frequent following self-pollination.

The nature of any possible stimulatory effects of cross- and self-pollen on embryo sac development in *D. winteri* and *P. axillaris* remains to be characterised at the biochemical and molecular level. It is not clear whether a reduced stimulatory effect on embryo sac development elicited by self- as opposed to cross-pollen tubes in *D. winteri*, *P. axillaris* or any other species, could be classified as an SI mechanism.

Acknowledgments

The authors wish to thank F. Strumas for technical assistance. This research was supported by a Canadian NSERC operating grant awarded to TLS and a grant awarded to FBS by Victoria University of Wellington

References

Barrett, S.C.H. (1988). The evolution, maintenance, and loss of self-incompatibility systems. In: J. Lovett Doust and L. Lovett Doust (editors). Plant reproductive ecology: patterns and strategies, pp. 98–124, Oxford University Press, Oxford.

Bateman, A.J. (1952). Self-incompatibility systems in angiosperms 1. Theory. *Heredity* **6**: 285–310.

Beardsell, D.V. (1991). The reproductive biology of *Thryptomene calycina* (Lindl.) Stapf. PhD. dissertation, University of Melbourne, Australia.

Bernhardt, P. and Thien, L.B. (1987). Self-isolation and insect pollination in the primitive angiosperms: a new evaluation of older hypotheses. *Pl. Syst. Evol.* **156**: 159–176.

Brun, L. and Olesen, P. (1989). A structural investigation of the ovule in sugar beet, *Beta vulgaris*: the micropylar nucellus. *Nordic J. Bot.* **9**: 81–87.

Burbidge, A.H. (1991). Post-zygotic seed abortion in the genetic system of *Stylidium* (Angiospermae: Stylidiaceae). *J. Heredity* **82**: 319–328.

Chao, C.Y. (1971). A periodic acid-Schiff's substance related to the directional growth of pollen tube into embryo sac in Paspalum ovules. *Amer. J. Bot.* **58**: 649–654.

Charlesworth, D. (1985). Distribution of dioecy and self-incompatibility in angiosperms. In: P.J. Greenwood and M. Slatkin (editors). Evolution: essays in honour of John Maynard Smith, pp. 237–268, Cambridge University Press, Cambridge.

Cope, F.W. (1962). The mechanism of pollen incompatibility in *Theobroma cacao* L. *Heredity* **17**: 157–182.

Dickinson, H.G. (1990). Self-incompatibility in flowering plants. *BioEssays*. **12**:155–161.

Endress, P.K. (1984). The flowering process in the *Eupomatiaceae* (Magnoliales). *Bot. Jahrb. Syst.* **104**: 297–319.

Franssen-Verheijen, M.A.W. and Willemse, M.T.M. (1993). Micropylar exudate in *Gasteria* (Aloaceae) and its possible function in pollen tube growth. *Amer. J. Bot.* **80**: 253–262.

Friis, E.M. and Crepet, W.L. (1987). Time of appearance of floral features. In: E.M. Friis, W.G. Chaloner and P.R. Crane (editors). The origins of angiosperms and their biological consequences, pp. 145–170, Cambridge University Press, New York.

Gibbs, P.E. (1986). Do homomorphic and heteromorphic self-incompatibility systems have the same sporophytic mechanisms? *Pl. Syst. Evol.* **154**: 285–323.

Gibbs, P.E. (1991). The 'Zavada' hypothesis: a rebuttal rebutted. *Taxon.* **40**: 583–595.

Gibbs, P.E. and Bianchi, M. (1993). Post-pollination events in species of *Chorisia* (Bombacaceae) and *Tabebuia* (Bignoniaceae) with late-acting self-incompatibility. *Bot. Acta* **106**: 64–71.

Godley, E.J. and Smith, D.H. (1981). Breeding systems in New Zealand plants 5. *Pseudowintera colorata* (Winteraceae). *New Zealand J. Bot.* **19**: 151–156.

Heslop-Harrison, J., and Shivanna, K. (1977). The receptive surface of the angiosperm stigma. *Ann. Bot.* **41**: 1233–1238.

Heslop-Harrison, Y., Heslop-Harrison, J. and Reger, B.J. (1985). The pollen-stigma interaction in the grasses. 7. Pollen-tube guidance and the regulation of tube number in *Zea Mays* L. *Acta Bot. Neerl.* **34**: 193–211.

Kaul V., Rouse, J.L. and Williams, E.G. (1986). Early events in the embryo sac after intraspecific and interspecific pollinations in *Rhododendron kawakamii* and *R. retusum*. *Canad. J. Bot.* **64**: 282–291.

Kenrick, J., Kaul, V. and Williams, E.G. (1986). Self-incompatibility in *Acacia retinodes*: Site of pollen-tube arrest is the nucellus. *Planta* **169**: 245–250.

Klekowski, E.J., Jr. (1988). Mutation, developmental selection and plant evolution. Columbia University Press, New York.

Knox, R.B. (1984). Pollen-pistil interactions. *Annual Rev. Pl. Physiol.* **17**: 508–608.

Lande, R., Schemske, D.W. and Schultz, S.T. (1994). High inbreeding depression, selective interference among loci, and the threshold selfing rate for purging recessive lethal mutations. *Evolution* **48**: 965–978.

Lloyd, D.G. and Wells, M.S. (1992). Reproductive biology of a primitive angiosperm, *Pseudowintera colorata* (Winteraceae) and the evolution of pollination systems in the Anthophyta. *Pl. Syst. Evol.* **181**: 77–95.

Lundqvist, A. (1964). The nature of the two-loci incompatibility system in grasses. IV. Interaction between the loci in relation to pseudo-compatibility in *Festuca pratensis* Huds. *Hereditas* (Lund.) **52**: 221–234.

Martin, F.W. (1959). Staining and observing pollen tubes by means of fluorescence. *Stain Technol.* **44**: 125–128.

Matton, D.P., Nass, N. Clarke, A.E., and Newbigin, E. (1994). Self-incompatibility: how plants avoid illegitimate offspring. *Proc. Natl. Acad. Sci. U.S.A.* **91**:1992–1997.

Nettencourt, D. de (1977). Incompatibility in angiosperms. Springer-Verlag, New York.

Norton, S.A. (1980). Reproductive Ecology of *Pseudowintera* (Winteraceae). M.S. thesis, Victoria University of Wellington, Wellington, N.Z.

Pimienta, E. and Polito, V.S. (1983). Embryo sac development in almond [*Prunus dulcis* (Mill.) D.A. Webb] as affected by cross-, self- and non-pollination. *Ann. Bot.* **51**: 469–479.

Reger, B.J., Chaubal, R. and Pressey, R. (1992). Chemotropic responses by pearl millet pollen tubes. *Sexual Pl. Reprod.* **5**: 47–56.

Sage, T.L. and Williams, E.G. (1991). Self-incompatibility in Asclepias. *Pl. Cell Incompat. Newslett.* **23**: 55–57.

Sage, T.L. and Williams, E.G. (1995). Structure, ultrastructure, and histochemistry of the pollen tube pathway in the milkweed *Asclepias exaltata* L. *Sexual Pl. Reprod.* **8**: 257–265.

Sage, T.L. Bertin, R.I. and Williams, E.G. (1994). Ovarian and other late-acting self-incompatibility systems. In: E.G. Williams, R.B. Knox and A.E. Clarke (editors). Genetic control of self-incompatibility and reproductive development in flowering plants, pp. 116–140, Kluwer, Dordrecht.

Sage, T.L., Strumas, F., Cole, W.W. and Barrett, S.C.H. (1998). Differential ovule development following self- and cross-pollination: The basis of self-sterility in *Narcissus triandrus* (Amaryllidaceae). *Amer. J. Bot.* (in press).

Sanders, L.C. and Lord, E.M. (1992). A dynamic role for the stylar matrix in pollen tube extension. *Int. Rev. Cytol.* **140**: 297–318.

Schmitt, D. and Perry, T.O. (1964). Self-sterility in sweetgum. *Forest Sci.* **10**: 302–305.

Sears, E.R. (1937). Self-sterility in plants. *Genetics* **22**: 130–181.

Seavey, S.R. and Bawa, K.S. (1986). Late-acting self-incompatibility in angiosperms. *Bot. Rev.* **52**: 195–218.
Seavey, S.R. and Carter, S.K. (1994). Self-sterility in *Epilobium obcordatum* (Onagraceae). *Amer. J. Bot.* **81**: 331–338.
Sparrow, I.K. and Pearson, N.L. (1948). Pollen compatibility in *Asclepias syriaca*. *J. Agric. Res.* **77**: 187–199.
Thien, L.B. (1980). Patterns of pollination in the primitive angiosperms. *Biotropica* **12**: 1–13.
Tilton, V.R. and Horner, H.T. (1980). Stigma, style, and obturator of *Ornithogalum caudatum* (Liliaceae) and their function in the reproductive process. *Amer. J. Bot.* **67**: 1113–1131.
Waser, N.M. and Price, M.V. (1991). Reproductive costs of self-pollination in *Ipomopsis aggregata* (Polemoniaceae): are ovules usurped? *Amer. J. Bot.* **78**:1036–1043.
Weller, S., Donoghue, M. and Charlesworth, D. (1995). The evolution of self-incompatibility in flowering plants: a phylogenetic approach. In: P.C. Hoch and A.G. Stephenson (editors). Experimental and molecular approaches to plant biosystematics. pp. 355–382, Missouri Botanical Garden, St. Louis, Mo. (Monographs in Systematic Botany, 53).
Whitehouse, H.L.K. (1950). Multiple-allelomorph incompatibility of pollen and style in the evolution of angiosperms. *Ann. Bot.* **14**: 198–216.
Wiens, D., Calvin, D.L., Wilson, C.A., Davern, D.I., Frank, D. and Seavey, S.R. (1987). Reproductive success, spontaneous embryo abortion, and genetic load in flowering plants. *Oecologia* **71**: 501–509.
Zavada, M.S. (1984). The relation between pollen exine sculpturing and self-incompatibility mechanisms. *Pl. Syst. Evol.* **147**: 63–78.
Zavada, M.S. (1990). Correlations between pollen exine sculpturing and angiosperm self-incompatibility systems - a rebuttal. *Taxon* **39**: 442–447.

Tobe, H., Graham, S.A. and Raven, P.H. (1998). Floral morphology and evolution in Lythraceae *sensu lato*. In: S.J. Owens and P.J. Rudall (Editors). Reproductive Biology, pp. 329–344. Royal Botanic Gardens, Kew.

FLORAL MORPHOLOGY AND EVOLUTION IN LYTHRACEAE *SENSU LATO*

HIROSHI TOBE[1], SHIRLEY A. GRAHAM[2] AND PETER H. RAVEN[3]

[1]Faculty of Integrated Human Studies, Kyoto University, Kyoto 606-8501, Japan
[2]Dept Biological Sciences, Kent State University, Kent, OH 44242, USA
[3]Missouri Botanical Garden, P.O. Box 299, St. Louis, MO 63166, USA

Abstract

Floral morphology in all 31 genera of Lythraceae *sensu lato* is presented. Evolutionary trends for selected floral characters are discussed based on the results of a cladistic analysis employing Onagraceae as the outgroup. Very few of the floral characters are synapomorphic or autapomorphic; most are highly homoplastic. On the basis of where and how often a character state change occurred in the cladograms, floral characters are divided into four categories: (1) characters synapomorphic or autapomorphic in clades that diverged early in the evolution of the family, i.e., ovary position and stigma surface; (2) characters having many states and homoplasy relatively early in the evolution of the family, thus contributing to early radiation of the family, i.e., secondary increase of the androecium, changes in carpel/sepal ratios, changes in the nectary; (3) characters homoplastic late in the evolution of the family, i.e. changes in merism, petal number, number of stamen whorls, anther attachment, ovary stipe and heterostyly; (4) characters providing synapomorphies late in the evolution of the family, i.e. floral symmetry. The change in ovary position provides a basis for recognition of subfamilies: specialised inferior ovary (Punicoideae) and semi-inferior ovary (Duabangoideae, Sonneratioideae). Within the Lythroideae, important floral changes occur in merism and nectary presence, position and shape. Contrary to the current classification which recognises two tribes on the basis of complete vs. interrupted septa in the ovary, all Lythroideae were found to have the septa interrupted above the placenta. The floral characters, when combined with other features, provide distinctive combinations that distinguish the individual genera and small groups of genera but do not alone clarify the major evolutionary lineages within the family. The extensive homoplasy among floral characters indicates dynamic radiation supported by recombination from a diverse but well-defined gene pool.

Introduction

Lythraceae *sensu lato* are a diverse family comprising 31 genera and 600 species in four subfamilies: Punicoideae (*Punica*); Duabangoideae (*Duabanga*); Sonneratioideae (*Sonneratia*); and Lythroideae (28 genera, Lythraceae *sensu stricto*) (Thorne, 1992; see Graham *et al.*, 1993, for list of genera, species numbers and geographical distribution). The family as now delimited lacks any unique apomorphic features. The classification proposed by Koehne (1903) for Lythraceae *sensu stricto* recognised two tribes based on whether ovarian septa were complete above the placenta (Nesaeeae) or not (Lythreae). Evidence from palynology (A. Graham *et al.*, 1985, 1987, 1990) and wood anatomy (Baas and Zweypfenning, 1979), supported by new observations presented in this paper on ovary structure, indicate that the septal character was incorrectly described and natural relationships among the genera do not necessarily correspond to their tribal placements.

Since Koehne's monograph was published, five new genera have been discovered (*Capuronia, Haitia, Hionanthera, Koehneria* and *Lourtella*). *Punica, Duabanga* and *Sonneratia*, which have been treated as satellite families of Lythraceae *sensu stricto* (Cronquist, 1981; Takhtajan, 1997), now tend to be included in the family (Thorne, 1992). A recent molecular study based on the *rbc*L chloroplast gene supports the broader circumscription (Conti *et al.*, 1996). In spite of the diversity in flowers represented by an expanded Lythraceae, no complete investigation of floral morphology has been made. Prior studies have been restricted to a few genera: *Punica* (Sinha and Joshi, 1959; Leins, 1988); *Lagerstroemia* (Tobe and Raven, 1990; Singh, 1992); *Adenaria, Koehneria, Lourtella, Pehria* and *Woodfordia* (Graham *et al.*, 1986, 1987; Graham, 1995). Thus, an overall survey of lythracean floral structure including genera added since the monograph is desirable and timely for clarifying floral evolution and diversification in the family. This paper presents the morphology and structure of the whole flower as well as description of selected floral organs and their relative time of appearance in the evolution of the family. Some systematic implications of the study are also considered.

Materials and Methods

More than 55 species representing all 31 genera were investigated. Fresh buds and mature flowers fixed in FAA (5 parts formalin, 5 parts acetic acid, 90 parts 50% or 70% ethanol) were observed under a stereoscope microscope and at least two flowers per species were sectioned by rotary microtome following standard paraffin methods. Flowers of *Hionanthera* (*H. garciae* Fernandez and Diniz) collected from the herbarium sheet, Biegel 1022, KE-G, were softened in FAA prior to sectioning. Longitudinal and transverse sections cut at 8-12 μm were stained with Heidenhain's hematoxylin, safranin and fastgreen FCF. Data on *Koehneria, Lourtella* and *Lagerstroemia* (*L. excelsa* Lee and Lau and *L. indica* L.) were obtained from sections made for earlier studies (Graham *et al.*, 1986, 1987; Tobe and Raven, 1990).

In order to trace the evolution of individual floral characters within the family and discuss their systematic implications, phylogenetic trees were constructed using parsimony analysis of 29 morphological characters including 12 floral characters. For the cladistic analysis, Onagraceae were employed as an outgroup, a choice supported by results of recent molecular studies (Conti *et al.*, 1996). The characters and coding basically follow Graham *et al.*(1993) but with changes in characters of ovary position and secondary increase of the androecia in place of stamen/sepal ratio and with the new characters of merism and nectary position added (Tables 1 and 2). The cladistic analysis employed PAUP ver. 3.0 (Swofford, 1993) with heuristic search and TBR options in effect. Two most parsimonious trees were obtained having a tree length of 103 steps and a consistency index of 0.447. One of the two most parsimonious trees was used with MacClade ver. 3.04 (Maddison and Maddison, 1993) to follow character evolution of selected floral characters.

Observations

Flowers consist of sepals, petals, stamens and a calyx of deltoid sepals which terminates a perigynous floral tube of varying length. An epicalyx of "appendages" positioned on the floral tube at the sinuses of the sepals is sometimes present. The epicalyx is interpreted by Mayr (1969) as "an emergence of congenitally fused sepals." The petals are borne on the inner rim of the floral tube and alternate with the sepals. Stamen insertion varies from lower to higher levels on the inner wall of the floral tube. In *Duabanga*, *Lagerstroemia* and *Physocalymma*, for example, the stamens are inserted at a rather low level, whereas in some Lythroideae they are inserted higher, and in *Sonneratia* they are at the rim of the floral tube. In *Punica* and *Lagerstroemia* the stamens (or stamen fascicles) are arranged in two whorls, one inserted low, the other higher in the floral tube. In *Cuphea* and *Pleurophora*, with zygomorphic flowers, stamens on the abaxial side may be inserted at a higher level than those on the adaxial side.

TABLE 1. Character coding for Lythraceae cladistic analyses[1]

Floral characters
1. Floral symmetry: 0 = actinomorphic; 1 = zygomorphic.
2. Merism (sepals and petals): 0 = 5-6-mery; 1 = 6+-mery; 2 = 4-mery.
3. Petals: 0 = present; 1 = absent.
4. Stamen whorls: 0 = 2 whorls; 1 = 1 whorl, opposite sepals; 2 = 1 whorl, opposite petals.
5. Secondary increase of androecium: 0 = rare or absent; 1 = common or always present.
6. Heterostyly: 0 = absent; 1 = present.
7. Anther attachment: 0 = versatile; 1 = basifixed.
8. Stigma surface: 0 = wet; 1 = dry.
9. Ovary position: 0 = inferior; 1 = semi-inferior; 2 = superior.
10. Ovary stipe: 0 = absent; 1 = present.
11. Carpel/sepal ratio: 0 = 1/1~1/2; 1 = greater than 1/1; 2 = less than 1/2.
12. Nectary: 0 = at hypanthium-gynoecium junction; 1 = lining on inner surface of lower half of hypanthium; 2 = lining on entire inner surface of hypanthium; 3 = on gynoecium; 4 = discontinuous spots on inner surface of hypanthium; 5 = absent.

Other characters
13. Inflorescence: 0 = blastotelic; 1 = anthotelic.
14. Habit: 0 = woody; 1 = herbaceous.
15. Vascular tracheids: 0 = absent; 1 = present.
16. Spiral thickenings: 0 = absent; 1 = present.
17. Septate wood fibres: 0 = present; 1 = absent.
18. Fibre dimorphism: 0 = absent; 1 = present.
19. Crystalliferous fibres: 0 = absent; 1 = present.
20. Leaf sclereids: 0 = absent; 1 = present.
21. Globose trichomes: 0 = absent; 1 = present.
22. Multicellular, glandular hairs: 0 = absent; 1 = present.
23. Pollen pseudocolpi: 0 = absent; 1 = 3 pseudocolpi; 2 = 6 pseudocolpi.
24. Pollen exine sculpture: 0 = psilate to coarsely scabrate; 1 = verrucate/rugulate; 2 = short fine striate; 3 = long striate; 4 = "onagraceous" (global to rod-like elements and viscin threads).
25. Fruit: 0 = capsule; 1 = berry.
26. Seed: 0 = not compressed; 1 = compressed.
27. Seed wing: 0 = absent; 1 = present, unilateral; 2 = present, encircling seed body.
28. Seed internal epidermal hairs: 0 = absent; 1 = present, straight; 2 = present, spirally twisted.
29. Chromosome base number: 0 = $\underline{x}$=12; 1 = $\underline{x}$=11; 2 = $\underline{x}$=8; 3 = $\underline{x}$=7; 4 = $\underline{x}$=5.

[1]Most characters were used in Graham *et al.* (1993). Floral characters 2, 3, and 12 and character 29 (chromosome base number) were added for this study. Codification for characters 5, 9, and 11 were revised. All characters were treated as "unordered" in cladistic analyses.

TABLE 2. Data matrix for Onagraceae (outgroup) and Lythraceae *sensu lato*. Characters and character codification in individual characters appears in Table 1, but code for 0/1 is presented by "a" and missing data by the symbol "?"

Onagraceae	0 0 0 0 0	0 0 0 0 0	0 0 0 0 0	0 0 0 0 0	0 0 0 4 0	0 0 0 1
Duabanga	0 0 0 0 1	0 0 ? 1 0	0 0 1 0 0	0 1 0 0 1	0 0 0 0 0	0 0 0 0
Sonneratia	0 0 0 0 1	0 0 ? 1 0	1 0 1 0 0	0 0 0 0 1	0 0 1 0 1	1 0 0 0
Punica	0 0 0 0 1	0 0 0 0 0	0 0 1 0 0	0 0 1 1 0	0 0 0 0 1	0 0 0 2
Lagerstroemia	0 0 0 0 1	0 0 0 2 0	0 5 1 0 0	0 0 1 1 0	0 0 2 0 0	0 1 0 ?
Lawsonia	0 2 0 1 1	0 0 ? 2 0	0 5 1 0 0	0 0 1 1 0	0 0 2 0 0	0 0 0 2
Decodon	0 0 0 0 0	1 0 1 2 0	0 3 0 0 0	0 0 0 0 0	0 0 0 0 0	0 0 0 2
Pemphis	0 0 0 0 0	1 0 ? 2 0	2 2 0 0 0	0 1 0 0 0	0 0 2 0 0	0 0 0 2
Capuronia	0 0 0 2 0	0 0 ? 2 0	2 2 0 0 0	0 0 0 1 0	0 0 0 0 0	1 0 0 2
Tetrataxis	0 2 1 2 0	0 0 ? 2 0	0 2 0 0 0	0 0 0 0 0	0 0 0 0 0	1 0 ? ?
Lafoensia	0 1 0 0 0	0 0 ? 2 0	2 2 0 0 0	0 0 0 0 0	0 0 2 0 0	1 2 2 2
Pleurophora	1 0 0 1 0	0 1 1 2 1	2 0 0 0 1	0 0 0 0 0	0 1 0 0 0	1 0 2 3
Cuphea	1 0 0 0 0	0 0 1 2 0	2 0 0 0 0	0 0 0 0 0	0 1 0 3 0	1 0 2 2
Woodfordia	0 0 0 0 0	0 0 1 2 0	2 0 ? 0 1	0 0 0 0 0	1 0 0 0 0	1 0 2 2
Pehria	0 2 0 0 0	0 0 ? 2 1	2 2 0 0 0	0 0 0 1 0	1 0 0 0 0	1 0 2 2
Lourtella	0 2 0 0 0	0 0 ? 2 0	2 4 0 0 0	0 0 0 1 0	1 0 0 0 0	1 0 0 ?
Adenaria	0 2 0 0 0	1 0 ? 2 1	2 1 0 0 0	0 0 0 0 0	1 0 0 0 0	1 0 0 2
Koehneria	0 0 0 0 0	0 0 ? 2 1	0 0 0 0 0	0 0 0 0 0	1 0 1 0 0	1 0 0 ?
Physocalymma	0 1 0 0 1	0 0 ? 2 0	2 5 0 0 0	0 0 1 1 0	0 0 0 1 0	1 2 0 2
Galpinia	0 0 0 2 0	0 0 ? 2 0	2 0 1 0 0	0 0 0 1 0	0 0 0 1 0	1 2 0 2
Crenea	0 2 0 0 0	0 1 ? 2 0	0 5 0 0 0	0 0 0 0 0	0 0 2 1 0	1 0 1 ?
Ginoria	0 0 0 0 1	0 0 1 2 0	0 5 0 0 0	0 0 0 a 0	0 0 2 0 0	1 0 1 ?
Heimia	0 0 0 0 1	0 0 1 2 0	0 5 0 0 1	1 0 0 0 0	0 0 0 0 0	1 0 1 2
Haitia	0 0 0 0 1	0 0 ? 2 0	0 5 0 0 0	0 0 0 0 0	0 0 2 0 0	1 0 1 ?
Diplusodon	0 0 0 0 0	0 0 ? 2 0	2 5 0 0 0	0 1 0 0 0	0 0 0 1 0	1 2 1 4
Peplis	0 0 1 1 0	0 0 1 2 0	2 5 0 1 0	0 1 0 0 0	0 0 1 2 0	1 0 1 4
Lythrum	0 0 0 1 0	a 0 1 2 0	2 0 0 0 0	0 0 0 0 0	0 0 1 2 0	1 0 1 4
Nesaea	0 0 0 0 0	a 0 1 2 0	0 0 0 0 0	1 1 0 0 0	0 0 2 2 0	1 0 1 4
Ammannia	0 2 0 1 0	0 0 1 2 0	0 0 0 0 0	0 1 0 0 0	0 0 2 2 0	1 0 1 2
Rotala	0 2 0 1 0	a 0 1 2 0	0 0 0 1 0	0 1 0 0 0	0 0 2 0 0	1 0 1 2
Hionanthera	0 2 0 1 0	0 0 1 2 0	0 0 0 1 0	0 1 0 0 0	0 0 2 2 0	1 0 1 ?
Didiplis	0 2 1 1 0	0 0 1 2 0	2 0 0 1 0	0 1 0 0 0	0 0 0 2 0	1 0 1 2

Floral symmetry and merism

Flowers are actinomorphic and more or less campanulate in most genera; *Cuphea* and *Pleurophora* are zygomorphic with elongated floral tubes. Flowers of *Woodfordia* and some species of *Lythrum* may also be weakly zygomorphic.

The number of floral parts varies widely in the family. Koehne (1903) found that in Lythroideae about 70% of the species had 6-merous flowers and about 18% 4-merous flowers. However, merosity varies among different organs in a single flower, particularly in the frequent reduction of carpels and the amplification of stamen number, and is not always consistent within a genus or even within a species. Petal and sepal number always correspond but their number may not coincide with the number of stamens or carpels. In the cladistic analysis, floral merism is coded on the basis of the most frequent number of petals and sepals, rather than on the more variable stamen or carpel number. In the case of *Lagerstroemia*, for example, 100 flowers of *L. indica* were examined to determine the most frequent number of sepals and petals. Of the total, 87 were 6-merous, five were 5-merous and one was 8-merous; therefore, the genus was coded as having 6-merous flowers. Floral merism of 10 genera is 4-merous, one is 5-merous, and 18 are 6-merous (Table 2; Figs. 1-12). *Lafoensia* (8- to 16-merous) and *Physocalymma* (8-9-merous) are coded as 6+-merous.

Petals

Petals are white, various shades of red to purple, or yellow. Shape ranges from oblong to elliptical, lanceolate to oval, or oblanceolate to obovate. They are usually sessile but may be clawed, as in *Lagerstroemia*. Petals are present in most species, are reduced in number or absent in some species of *Ammannia*, *Cuphea*, *Nesaea* and *Rotala*, and are always absent in the genera *Didiplis* (Fig. 11), *Peplis* and *Tetrataxis* (Fig. 10).

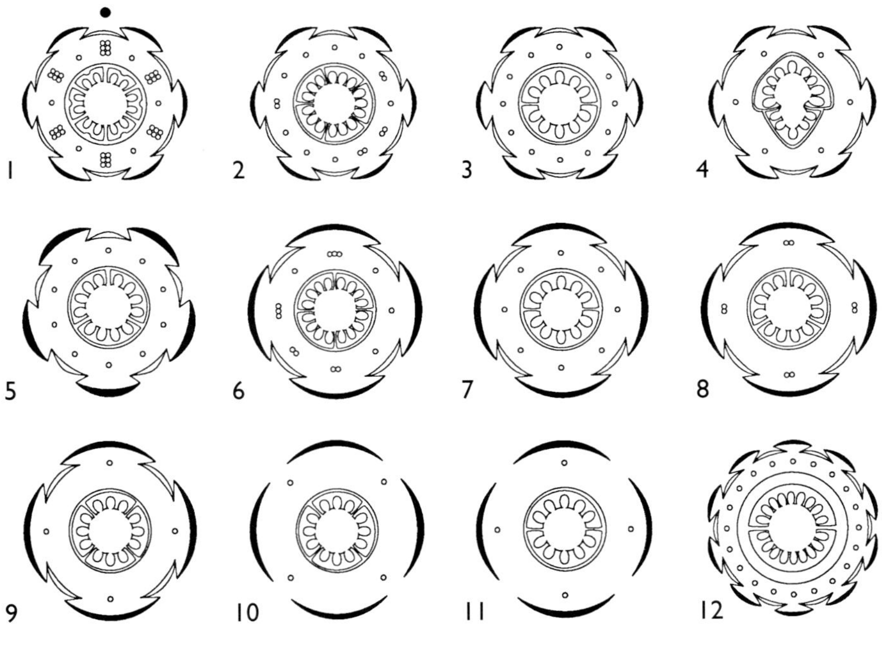

FIGS. 1–12. Floral diagrams of some Lythroideae showing diversity in merism, presence or absence of petals, number and arrangement of stamen whorls, presence or absence of the secondary increase of the androecium, and carpel/sepal ratio. 1: *Lagerstroemia indica* L. 2: *Ginoria glabra* Griesb., *Haitia buchii* Urb., *Heimia salicifolia* Link & Otto, *Koehneria madagascariensis* (Baker) S.A. Graham, Tobe & Baas and *Nesaea ongipes* A. Gray. 3: *Diplusodon virgatus* Gardn. ex Koehne, *Pemphis acidula* Forst. and *Woodfordia fruticosa* Kurz. 4: *Pleurophora saccocarpa* Koehne. 5: *Decodon verticillatum* Ell. 6: *Crenea patentinervis* (Koehne) Standl. 7: *Lourtella resinosa* S.A. Graham., Baas & Tobe. 8: *Lawsonia inermis* L. 9: *Ammannia coccinea* Rottb. and *Hionanthera garciae* Fernandez & Diniz. 10: *Tetrataxis salicifolia* Baker. 11: *Didiplis diandra* Wood. 12: *Lafoensia punicaefolia* DC.

Androecium

The androecium consists of stamens with small anthers and long distinct filaments. The anther is typically versatile except in *Crenea* and *Pleurophora* where it is basifixed. Flowers are basically diplostemonous with the stamens opposite the sepals forming the outer whorl and those opposite the petals, the inner whorl. Typical diplostemony is seen in *Lagerstroemia* (e.g. *L. indica*, Fig. 1). In *Lagerstroemia*, the episepalous stamens are inserted at lower levels in the floral tube than the epipetalous ones. In *Punica* (*P. granatum*) as well, flowers may be diplostemonous, apparently having two whorls of stamen fascicles which are inserted at different levels of the floral tube (Sinha and Joshi, 1959). Leins (1988) however shows that developmentally *Punica* appears to be obhaplostemonous with the first stamen in front of a petal. In the genera *Cuphea*, *Decodon*, *Diplusodon*, *Ginoria*, *Haitia*, *Heimia*, *Lythrum*, *Nesaea* and *Pemphis* (Figs. 2, 3, 5), flowers are considered diplostemonous; the stamens appear to be arranged in a single whorl and all inserted at nearly the same level in the floral tube, but are always positioned opposite both petals and sepals. This is also the case for *Duabanga*, *Sonneratia*, *Lafoensia* and *Physocalymma*, although the stamens are more numerous. In contrast, haplostemonous flowers having only episepalous stamens are found in *Didiplis*, *Hionanthera*, *Lawsonia* and some species of *Ammannia*, *Lythrum*, *Nesaea*, *Peplis*, *Pleurophora*, *Rotala* (Figs. 4, 8, 9, 11), whereas obhaplostemonous flowers having only epipetalous stamens occur in *Capuronia*, *Galpinia*, *Tetrataxis* (Fig. 10) and in a some species of *Nesaea*.

In *Punica*, *Duabanga* and *Sonneratia* stamen increase results from the secondary multiplication of androecia. A number of stamens arise at various levels of the floral tube. Duke and Jackes (1987) described flowers of *Sonneratia* with approximately 300 stamens. In *Lagerstroemia indica*, the secondary increase occurs only in the epipetalous androecia, with each initial divided into six or seven stamens (Fig. 1; Singh, 1992). In *Lawsonia*, the secondary increase is seen in all four initials, with each forming two stamens and all arranged in a single whorl (Fig. 8). In a few other genera such as *Crenea*, *Nesaea*, *Heimia* and *Koehneria*, some but not all initials are forked, forming two stamens (Figs. 2, 6).

Gynoecium

The gynoecium is composed of a stigma, distinct style and a compound ovary containing many anatropous, bitegmic ovules. The stigma is not divided and is small-capitate to punctiform in most genera, although it is enlarged in *Pemphis*. *Lagerstroemia*, *Punica* and *Sonneratia* have a wet stigma whereas most (all?) Lythroideae have a dry stigma (Graham *et al.*, 1993). Heterostyly is known in *Adenaria*, *Decodon*, *Lythrum*, *Nesaea*, *Pemphis* and *Rotala*.

The ovary is inferior in *Punica* (Fig. 13), semi-inferior in *Duabanga* (Fig. 14) and *Sonneratia* (Fig. 15), and superior in all the genera of Lythroideae. Sinha and Joshi (1959; p. 43) interpreted the inferior ovary of *Punica granatum* as composed of 4-15 carpels in one to three whorls. Although placentation is axile with a solid axis and erect carpels at the basal end of the ovary (Fig. 16), it is parietal at the distal end. This unusual situation is described by them as the result of the shifting of some outer carpels "inside-out and upside-down." Thus, the inferior ovary is not homologous to the usual type formed by congenital fusion of the outer floral organs to the ovary wall.

In both *Duabanga* (Fig. 17) and *Sonneratia* (Fig. 18), placentation is axile except in the uppermost level where it becomes parietal. The central axis is hollow, in marked contrast to the solid axis of *Punica* and *Lagerstroemia*. The semi-inferior ovary of *Duabanga* and *Sonneratia* appears to be enclosed by congenitally fused outer organs. In fruiting stages the upper, exposed portion of the ovary develops rapidly, producing a fruit which is superior in position. All Lythroideae have a superior ovary with axile placentation and a solid central axis (Figs. 19, 21). The placenta is often enlarged and intrudes into the locules. In *Lagerstroemia* and *Physocalymma*, the base of the ovary appears to be slighly embedded in the receptacle, but it is uncertain whether or not this should be considered a synapomorphy.

The ovary of the Lythroideae is ellipsoid, ovoid, or obovoid and has a more or less distinct stipe in *Adenaria*, *Koehneria*, *Pehria* and *Pleurophora*. In *Pleurophora* the stipe is enclosed by a tubular nectary (Fig. 28). The ovary is divided by longitudinal septa into two or more locules equivalent to the number of constituting carpels (except in the unilocular *Ammannia microcarpa* DC. fide Koehne, 1903: p. 56). The septa are distinct over nearly the entire length of the ovary. However, they are always interrupted above the level of the placenta, i.e., at the upper end of the locule, in all genera. This observation negates Koehne's tribal divisions which separated genera with complete septa (Nesaeeae: *Adenaria*, *Crenea*, *Decodon* [Figs. 21, 22], *Ginoria*, *Heimia*, *Lagerstroemia*, *Lawsonia*, *Nesaea*, *Pehria* and *Tetrataxis*) from those with incomplete septa (Lythreae: *Ammannia*, *Cuphea*, *Diplusodon*, *Galpinia*, *Lafoensia*, [Fig. 20] *Lythrum*, *Pemphis*, *Pleurophora*, *Physocalymma*, *Rotala* and *Woodfordia*).

The number of carpels is variable in the family. It corresponds to the number of sepals and petals in *Ammannia* (*pro parte*), *Crenea*, *Duabanga*, *Lagerstroemia* (*pro parte*) and *Tetrataxis*. In most genera, however, there is a tendency to fewer carpels than sepals or petals (Fig. 3, 4, 7, 11, 12). The number of carpels is always two, irrespective of the number of sepals and

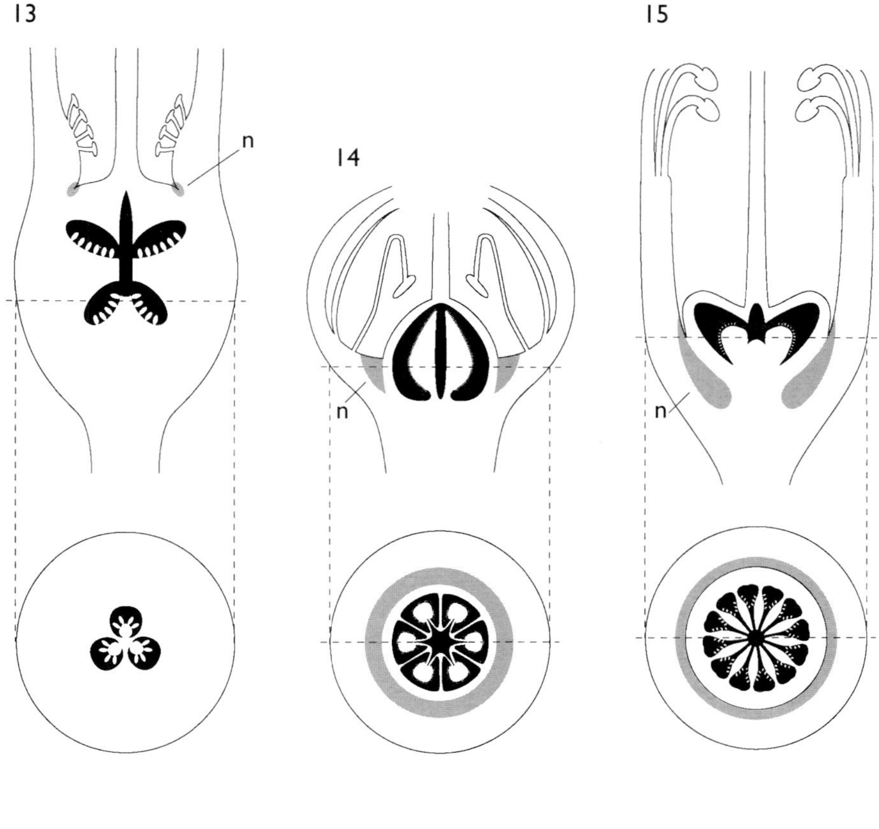

FIGS. 13–15. Diagrams illustrating longitudinal and transverse sections of flowers. n = nectary (areas shaded grey). 13: *Punica granatum* L. (Punicoideae). 14: *Duabanga grandiflora* Walp. (Duabangoideae). 15: *Sonneratia alba* J. Smith (Sonneratioideae).

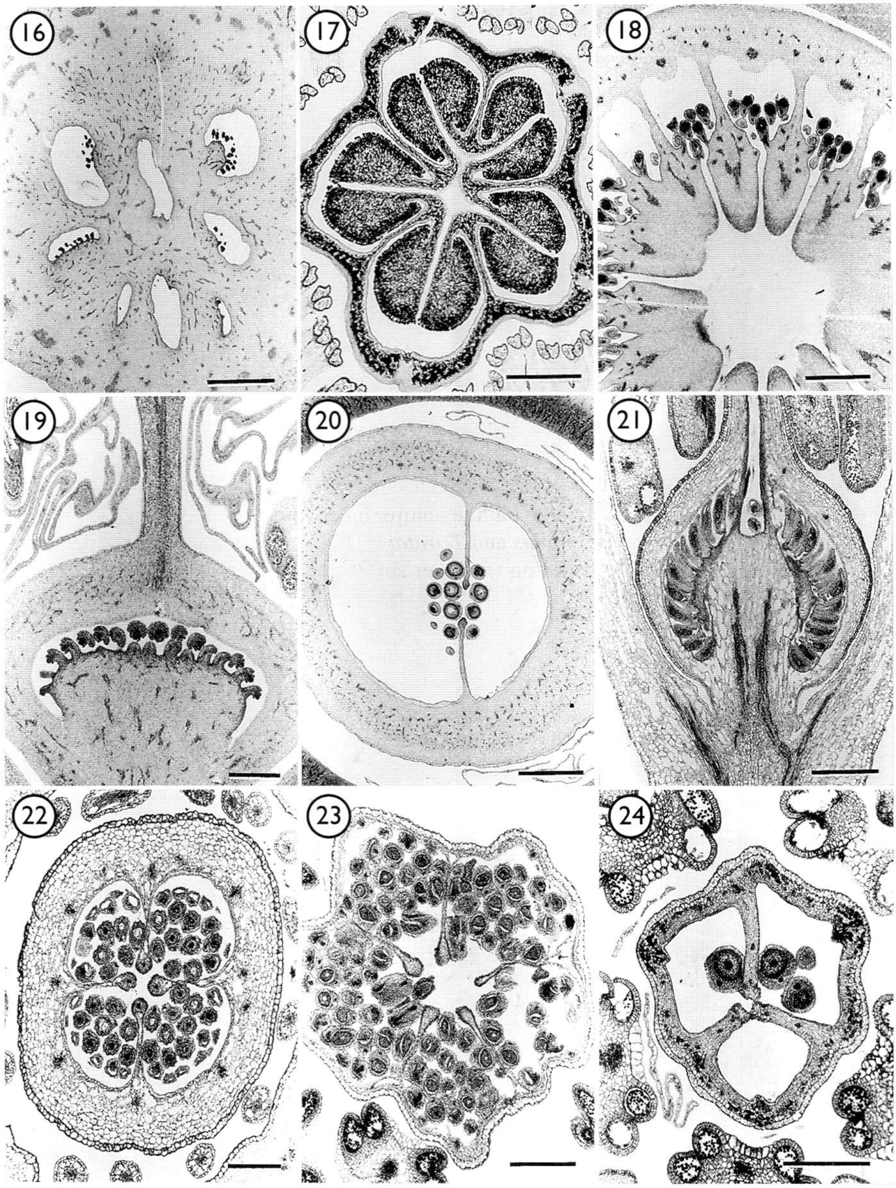

FIGS. 16–24. Floral anatomy of Lythraceae *sensu lato*. 16: Transverse section (TS) of flower of *Punica granatum* (voucher: Takaso s. n., MAK 196920), showing placenta at lower level of inferior ovary. 17: TS flower of *Duabanga grandiflora* (voucher: Stone *et al.* 15064, KLU, MO), showing that the central axis of placenta is hollowed. 18: TS flower of *Sonneratia alba* (voucher: Ashton 4635, MO), a parietal placenta at higher level of ovary. 19: Longitudinal section (LS) of flower of *Lafoensia punicaefolia*, (voucher: Steyermark 125143, VEN) showing ovary and placenta. 20: TS flower of *Lafoensia punicaefolia*, showing that the septa of the ovary are interrupted above the placenta. 21: LS flower of *Crenea patintinervis* (voucher: Jentry & Juncosa 38542, MO), showing ovary and placenta. 22: TS flower of *Crenea patintinervis*, showing that the septa of ovary are interrupted above the placenta. 23: TS flower of *Nesaea longipes* (voucher: Powell & Powell 4055, MO), showing that the septa of ovary are interrupted above the placenta. 24: TS *Decodon verticillatum* (voucher: Graham 917, KE-G), showing that the septa of ovary are interrupted above the placenta.

petals, in *Adenaria*, *Capuronia*, *Cuphea*, *Didiplis*, *Diplusodon*, *Galpinia*, *Lafoensia*, *Lythrum*, *Pemphis*, *Pleurophora* and *Woodfordia*.

Nectary

Among the 31 genera, nine lack floral nectaries (*Crenea*, *Diplusodon*, *Ginoria*, *Haitia*, *Heimia*, *Lagerstroemia*, *Lawsonia*, *Peplis* and *Physocalymma*) (Fig. 30). The remaining 22 genera display diverse nectary forms and positions. Two of the most conspicuous are the nectaries of *Cuphea* (Fig. 29) and *Pleurophora* (Fig. 28). In *Cuphea*, the nectary enlarges to a globular structure present only on the adaxial side of the junction of the floral tube and the ovary; in *Pleurophora*, it develops as a tube from the junction of the floral tube and ovary to enclose the elongated ovarian stipe. At least in *Cuphea*, stomata-like pores are present on the surface of the nectary and probably exude nectar. Nectariferous tissue that is invisible externally but easily detected as a densely stained tissue in microtome sections of flowers occurs in eleven genera at the floral tube-gynoecium junction: *Ammannia*, *Didiplis*, *Duabanga*, *Galpinia*, *Koehneria*, *Lythrum*, *Nesaea*, *Punica*, *Rotala*, *Sonneratia* and *Woodfordia* (Fig. 25). Nectariferous tissue occurs along the inner surface of the lower half of the floral tube in *Adenaria* and on the entire inner surface below the stamens in *Capuronia*, *Lafoensia*, *Pehria*, *Pemphis* and *Tetrataxis* (Fig. 26). In *Lourtella*, the nectary is distributed as discontinuous patches on the inner surface of the floral tube (Graham *et al.*, 1987). *Decodon* is the only genus in which the nectary occurs on the lower part of the ovary (Fig. 27). The nectaries of *Cuphea* and *Decodon* are highly vascularised. The morphology and position of the nectary is generally consistent within a genus. Exceptions are *Lythrum* and *Rotala* (paper in prep.).

Discussion

Evolution of floral characters

Results of the cladistic analysis yielded two most parsimonious trees which differ only in the position of *Didiplis*. *Didiplis* occurs either in a clade with *Ammannia*, *Rotala* and *Hionanthera* or with *Lythrum* and *Peplis*. The topology of the two trees is similar to that of a strict consensus tree found by Graham *et al.* (1993, Fig. 4) based on 26 characters and after successive weighting (Fig. 31). The trees all agree in having one clade comprising (*Duabanga* + *Sonneratia*) + (*Punica* (*Lagerstroemia* + *Lawsonia*)) and a second large clade that includes all other members of the Lythroideae. The *Duabanga-Lawsonia* clade is supported by an anthotelic inflorescence, which also occurs once in the second clade, in *Galpinia*. The evolution of the floral characters hypothesised by this study is shown in Fig. 31. Few characters are synapomorphic or autapomorphic. Most are highly homoplastic. On the basis of where and how often a character state change occurred, the 12 floral characters were classified into four categories: (1) synapomorphic or autapomorphic characters in clades that diverged early in the evolution of the family; (2) multistate characters that were homoplastic relatively early in the evolution of the family; (3) homoplastic characters arising late in the lythracean evolution; (4) synapomorphic characters arising late in lythracean evolution.

In the first category are stigma surface and ovary position (characters 8 and 9). Ovary position, in particular, deserves comment, because the superior ovary of the Lythraceae *sensu lato* combined with numerous other presumed plesiomorphic features suggests that Lythraceae "take a central position in the order" (Dahlgren and Thorne, 1984). With Onagraceae as the outgroup, the superior ovary of Lythraceae is treated as a synapomorphy and both the inferior and semi-inferior ovaries of *Punica* and *Duabanga*/*Sonneratia* are then derived. The ancestral

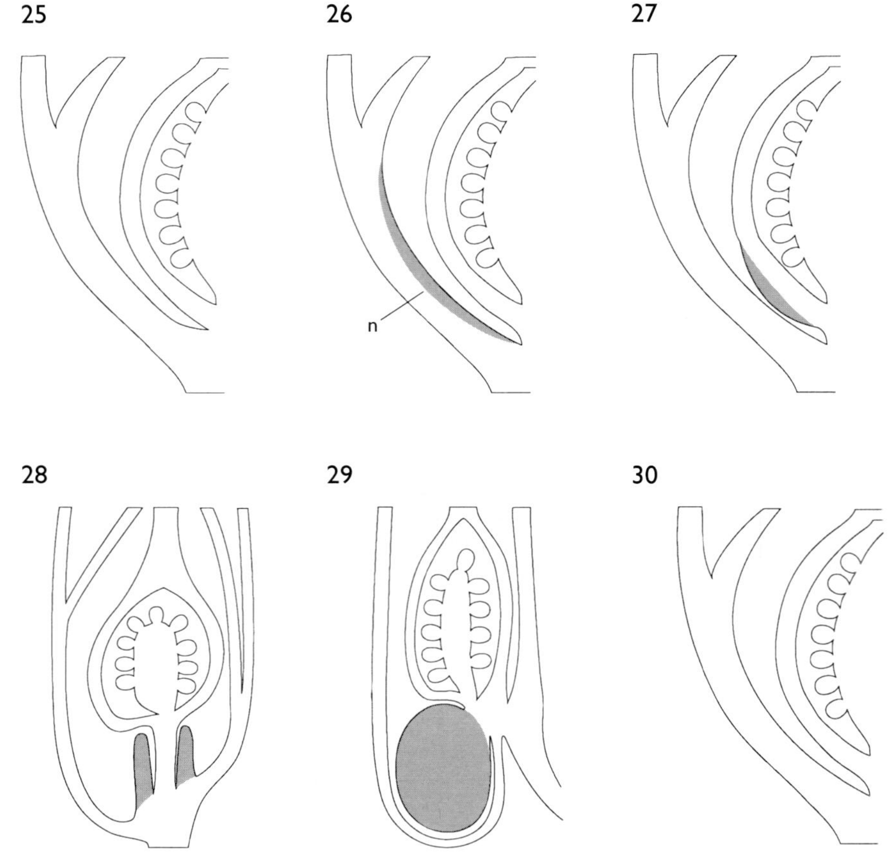

FIGS. 25–30. Diagrams illustrating the position and morphology of the floral nectary in some Lythroideae. n = nectary (areas shaded grey). 25: *Ammannia*, *Didiplis*, *Galpinia*, *Hionanthera*, *Lythrum*, *Nesaea*, *Rotala* (*pro parte*) and *Woodfordia*. 26: *Capuronia*, *Lafoensia*, *Pehria*, *Pemphis* and *Tetrataxis*. 27: *Decodon*. 28: *Pleurophora*. 29: *Cuphea*. 30: *Crenea*, *Ginoria*, *Heimia*, *Lagerstroemia* and *Lawsonia*.

state of the ovary in Lythraceae, however, is still unclear and will remain so until ordinal levels of angiosperm phylogeny are better known, and the ancestor of the onagracean-lythracean lineage can be hypothesised with greater certainty.

The inferior ovary of *Punica granatum* greatly differs structurally from that of Onagraceae. In Onagraceae, four (to six) carpels constructing the ovary are arranged in an erect whorl (Eyde, 1977, Fig. 10), and their inferior condition is interpreted as the probable result of congenital adnation of the outer floral organs. In the little-known *Punica protopunica* Balf. f., the carpels are said to be arranged in one to three whorls and semi-inferior; the species was not available for study. The fully inferior ovary of *P. granatum* is considered derived by pronounced growth of a receptacle that shifted the outer carpels "inside-out and upside-down" (see Sinha and Joshi, 1959, Figs. 18-24). Such structural differences suggest that the inferior ovaries of *Punica*

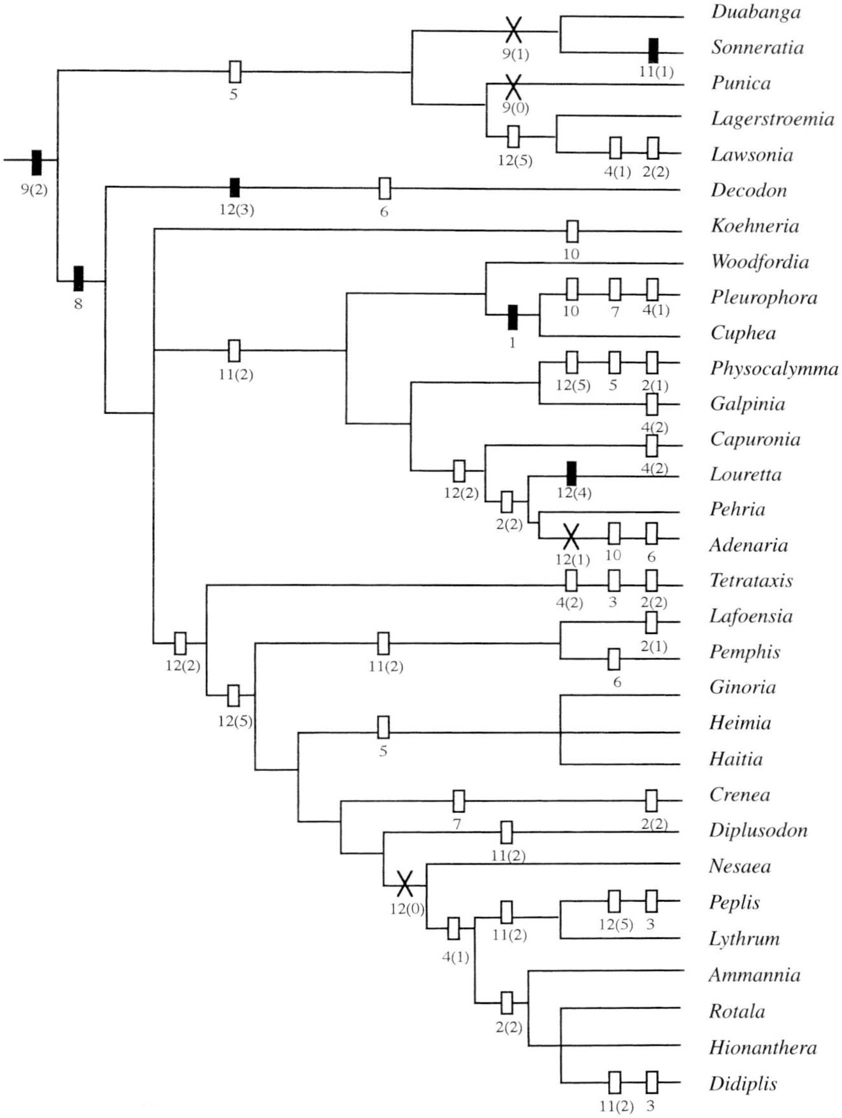

FIG. 31. One of the two most parsimonious trees generated on the basis of 29 morphological characters. Tree length is 103 steps and consistency index 0.447. Characters and character-codification are presented in Table 1 and 2. Character states in multistate characters are indicated in brackets. Solid rectangles represent apomorphies, open rectangles homoplasies and crosses reversals.

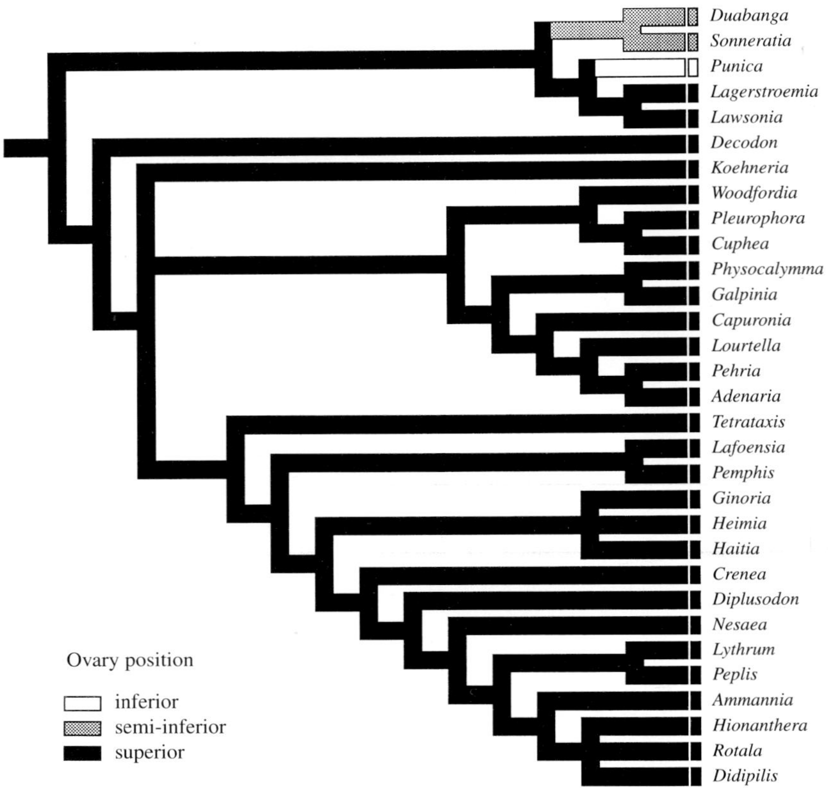

FIG. 32. Tree showing evolution of ovary position in Lythraceae *sensu lato*. Topology is the same as that in Fig. 31.

granatum and Onagraceae are not homologous. The inferior ovary of *Punica granatum*, although represented on the cladogram (Fig. 32) as derived by reversal from a superior ovary to an onagraceous-type inferior ovary, is more likely a form specialised from the superior ovary.

In *Duabanga* and *Sonneratia* all carpels appear erect as in Onagraceae and Lythroideae but the placentation is most similar to that of Onagraceae. In *Duabanga/Sonneratia* and Onagraceae, the placentation is parietal rather than axile at the uppermost level of the locule (for Onagraceae, see Eyde, 1977, Fig. 10, 14, 15). The semi-inferior ovary of *Duabanga* and *Sonneratia* may represent an intermediate form derived by reversal from the superior to the inferior onagraceous-type ovary.

The second category includes presence or absence of the secondary increase of the androecium, changes in carpel/sepal ratio and changes in the nectary (characters 5, 11 and 12). These multistate characters are homoplastic relatively early in the evolution of the family, thus contributing to its early radiation (Fig. 31). The nectary, for example, is plesiomorphically positioned at the floral tube-gynoecium junction in genera such as *Punica*, *Duabanga*, *Sonneratia* (Fig. 33). Its position shifted once to the gynoecial base, in *Decodon*, and twice to the entire inner surface of the floral tube, in the *Capuronia-Adenaria* clade and in *Tetrataxis-Lafoensia-Pemphis* (Fig. 33). Subsequent evolution involves independent loss of the nectary four times, in *Lagerstroemia-Lawsonia*, *Physocalymma*, *Ginoria-Diplusodon* and *Peplis*. A reversal from no nectary to gain of a nectary at the floral tube-gynoecium junction occurs in the seven-member clade *Nesaea-Didiplis*. Further changes involve reduction of the nectary lining the floral tube to discontinuous nectary patches (*Lourtella*) or restriction to the lower half of the floral tube (*Adenaria*). Other autapomorphies are the tubular nectary of *Pleurophora* and the large, adaxial, globular gland of *Cuphea*. The diversification of the nectary is thus closely tied to radiation of Lythraceae *s.l.*

Characters of the third category, those which are homoplastic late in the evolution of the family, include merism, presence or absence of petals, number of stamen whorls and stamen position, appearance of heterostyly, anther attachment and development of an ovarian stipe

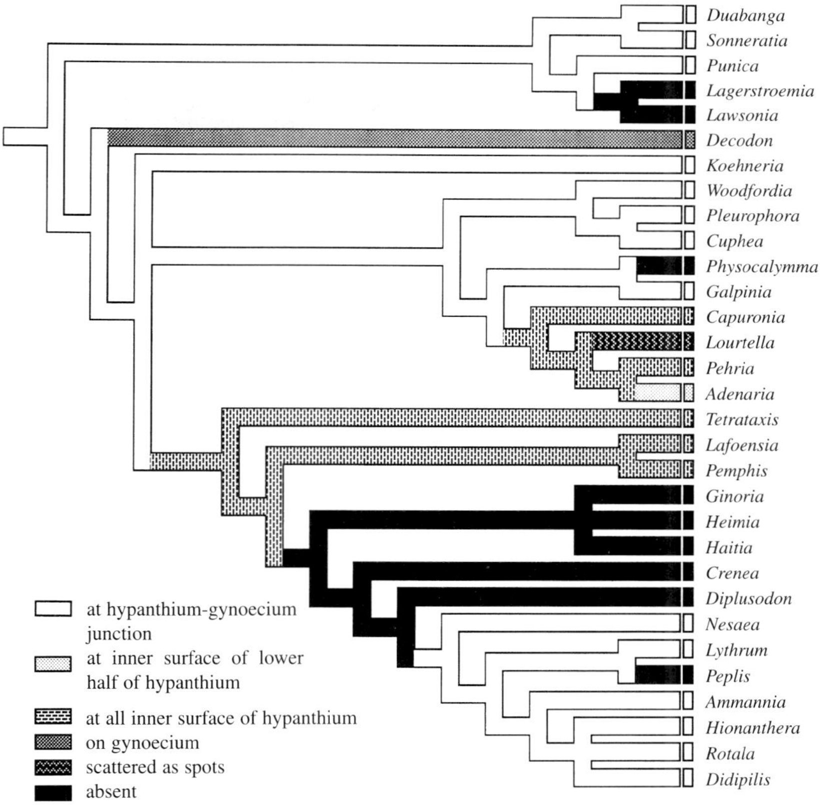

FIG. 33. Tree showing evolution of floral nectary in Lythraceae *sensu lato*. Topology is the same as that in Fig. 31.

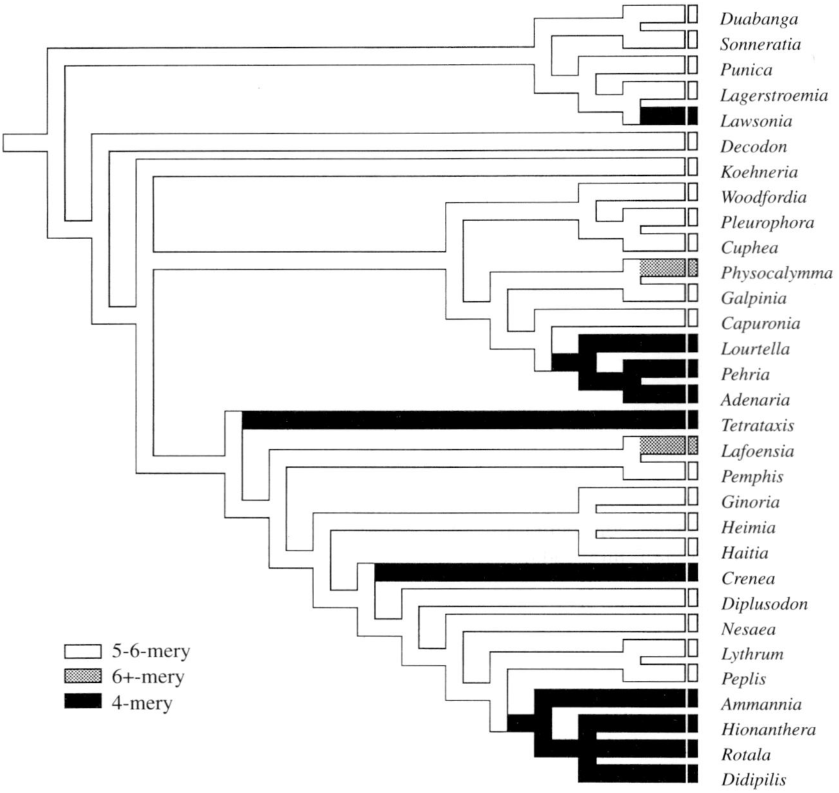

FIG. 34. Tree showing evolution of floral merism in Lythraceae *sensu lato*. Topology is the same as that in Fig. 31.

(Table 1; characters 2-4, 6, 7 and 10). Based on the cladistic analysis, the 5-6-merous floral condition is interpreted as plesiomorphic in the family and 4-merosity is separately derived five times (Fig. 34; see Eyde, 1977 for discussion of merism changes in Onagraceae). The 6+-merosity is also homoplastic, derived twice, in 8(-9)-merous flowers of *Physocalymma* and 8-16-merous flowers of *Lafoensia*.

Change in symmetry of flowers (character 1) is the only character in the fourth category, synapomorphies arising late in the evolution of the family. The change is from actinomorphic to zygomorphic flowers for the clade of South American herbaceous perennial genera *Pleurophora-Cuphea*.

Systematic implications of floral characters

As demonstrated in the cladistic analysis, most floral characters in Lythraceae show multiple homoplasies. Only a few, such as ovary position and stigma surface, are synapomorphic or autapomorphic. In the latest classifications (e.g. Cronquist, 1981; Dahlgren and Thorne, 1984; Takhtajan, 1997; Thorne, 1992), *Duabanga* and *Sonneratia* are distinguished from the rest of the family by their semi-inferior ovary and *Punica granatum* by its specialised form of inferior ovary, which differs from the type occurring in the outgroup Onagraceae. Most recently, *Duabanga* and *Sonneratia* are treated as distinct families (Takhtajan, 1997) or as subfamilies of Lythraceae (Thorne, 1992). Floral morphologically, *Sonneratia* is distinguished from *Duabanga* by having the stamens inserted at a higher level in the floral tube and by a multicarpellate ovary. However, *Duabanga* and *Sonneratia* are very similar in ovary position, placentation and nectary position, as discussed above. They also share x = 12 with similar chromosome morphology (Graham *et al.*, 1993), instead of x = 8 common to Punicoideae and Lythroideae. Floral morphology and chromosome number suggest that *Duabanga* and *Sonneratia* be treated within the same subfamily; they occur consistently as sister genera in analyses, although overall they are dissimilar due to accumulated autapomorphies.

The present study, as well as the cladistic analysis of Graham *et al.* (1993), place *Punica* as sister to *Lagerstroemia-Lawsonia*, these genera as sister to *Duabanga-Sonneratia* and this larger clade as sister to the rest of the Lythraceae. The five-member monophyly *Duabanga - Lawsonia* is supported by the anthotelic inflorescence, the secondary increase of the androecium, unreduced carpel number and possibly by wet stigma surfaces (unknown in *Duabanga*, probably wet in at least one species of *Sonneratia*, and dry in *Lawsonia*). However, *Lagerstroemia* and *Lawsonia* are clearly different from *Punica*, *Duabanga* and *Sonneratia* by absence of a nectary and by their superior ovary. If subfamilies are recognised, *Lagerstroemia* and *Lawsonia* should be considered for separate subfamily status. Therefore, particular attention must be paid in future studies to the relationships of these two genera to each other and to the rest of the family.

With respect to the Lythroideae, the present study found that all genera have interrupted or incomplete septa, invalidating Koehne's (1903) recognition of two tribes based on septa complete above the ovary (Nesaeeae) vs. septa incomplete (Lythreae). Further, no other floral characters are known to divide the subfamily in a similar way. Floral morphology suggests relationships between several pairs or clusters of genera, but in cladistic analysis, no major lineages are well-supported by non-homoplastic floral characters. Among the generic clusters, *Ginoria-Heimia-Haitia* are monophyletic based on the synapomorphies of the secondary increase of the androecium and absence of a nectary; *Peplis-Lythrum-Ammannia-Rotala-Hionanthera-Didiplis* are related by haplostemonous flowers; *Pehria-Lourtella-Adenaria* are related by 4-merous flowers, and the relationship of *Pehria* and *Adenaria* is further supported by the shared ovarian stipe. 4-merous flowers also characterise the clade *Ammannia-Didiplis*. *Cuphea* and *Pleurophora* are related to each other by the synapomorphy of zygomorphic flowers.

In conclusion, floral morphology is greatly diversified throughout Lythraceae. The ovary position and stigma surface underwent major evolutionary changes and the ovary changes, in particular, provide a basis of recognition of subfamilies Duabangoideae/ Sonneratioideae, Punicoideae and Lythroideae. The Lythroideae are paraphyletic and together *Lagerstroemia* and *Lawsonia*, on the basis of floral morphology, could be recognised at the subfamily level, making them equivalent to *Duabanga/Sonneratia* and *Punica*. Within Lythroideae, important floral changes are changes in merism and in the nectary presence, position and shape. The characters of heterostyly, the loss of petals and appearance of basifixed anthers each evolved independently in unrelated genera and, although diagnostic for the genera in which they appear, are not indicative of natural generic groups.

Acknowledgements

We are grateful to many people for collecting materials used for the present study. This study is also supported by funds from the Missouri Botanical Garden.

References

Baas, P. and Zweypfenning, R.C.V.J. (1979). Wood anatomy of the Lythraceae. *Acta Bot. Neerl.* **28:** 117–155.

Conti, E., Litt, A. and Sytsma, K. J. (1996). Circumscription of Myrtales and their relationships to other Rosids: evidence from rbcL sequence data. *Amer. J. Bot.* **83:** 221–233.

Cronquist, A. (1981). An integrated system of classification of flowering plants. Columbia University Press, New York.

Dahlgren, R. and Thorne, R. F. (1984). The order Myrtales: circumscription, variation and relationships. *Ann. Missouri Bot. Gard.* **71:** 633–699.

Duke, N. C. and Jackes, B. R. (1987). A systematic revision of the mangrove genus *Sonneratia* (Sonneratiaceae) in Australasia. *Blumea* **32:** 277–302.

Eyde, R. (1977). Reproductive structures and evolution in *Ludwigia* (Onagraceae). I. Androecium, placentation, merism. *Ann. Missouri Bot. Gard.* **64:** 644–655.

Graham, A., Nowicke, J., Skvarla, J., Graham, S., Patel, V. and Lee, S. (1985). Palynology and systematics of the Lythraceae. I. Introduction and genera *Adenaria* through *Ginoria*. *Amer. J. Bot.* **72:** 1012–1031.

Graham, A., Nowicke, J., Skvarla, J., Graham, S., Patel, V. and Lee, S. (1987). Palynology and systematics of the Lythraceae. II. Genera *Haitia* through *Peplis*. *Amer. J. Bot.* **74:** 829–850.

Graham, A., Graham, S., Nowicke, J., Patel, V. and Lee, S. (1990). Palynology and systematics of the Lythraceae. III. Genera *Physocalymma* through *Woodfordia*, addenda and conclusions. *Amer. J. Bot.* **77:** 159–177.

Graham, S.A. (1995). Systematics of *Woodfordia* (Lythraceae). *Syst. Bot.* **20:** 482–502.

Graham, S.A., Tobe, H. and Baas, P. (1986). *Koehneria*, a new genus of Lythraceae from Madagascar. *Ann. Missouri Bot. Gard.* **73:** 788–809.

Graham, S.A., Baas, P. and Tobe, H. (1987). *Lourtella*, a new genus of Lythraceae from Peru. *Syst. Bot.* **12:** 519–533.

Graham, S.A., Crisci, J. and Hoch, P. (1993). Cladistic analysis of the Lythraceae *sensu lato* based on morphological characters. *Bot. J. Linn. Soc.* **113:** 1–33.

Koehne, E. (1903). Lythraceae. In: A. Engler (editor) Das Pflanzenreich, Bd. IV, 216. (Heft 17), pp. 1–326, Engelmann, Berlin.

Leins, P. (1988). Das zentripetale Androecium von *Punica*. *Bot. Jahrb. Syst.* **109:** 555–561.

Maddison, W. P. and Maddison, D. R. (1993). MacClade, version 3.04: analysis of phylogeny and character evolution. Sinauer Press, Massachusetts.

Mayr, B. (1969). Ontogenetische Studien an Myrtales-Blüten. *Bot. Jahrb. Syst.* **89:** 210–271.

Sinha, S. G. and Joshi, B. C. (1959). Vascular anatomy of the flower of *Punica granatum* L. *J. Indian Bot. Soc.* **38:** 35–45.

Singh, V. (1992). Floral organogenesis of *Lagerstroemia indica* L. (Lythraceae). *Phytomorphology* **42:** 7–14.

Swofford, D. L. (1993). PAUP: phylogenetic analysis using parsimony, version 3.1. Illinois Natural History Survey, Champaign.

Takhtajan, A. L. (1997). Diversity and classification of flowering plants. Columbia University Press, New York.

Thorne, R. F. (1992). An updated phylogenetic classification of the flowering plants. *Aliso* **13:** 365–389.

Tobe, H. and Raven, P. H. (1990). Comparative reproductive morphology of "Orias" and *Lagerstroemia* (Lythraceae). *Flora, Morphol. Geobot. Oekophysiol.* **184:** 177–185.

Prychid, C.J., Owens, S.J. and Rudall, P.J. (1998). Fruit and seed set in *Strongylodon macrobotrys*. In: S.J. Owens and P.J. Rudall (Editors). Reproductive Biology, pp. 345–352. Royal Botanic Gardens, Kew.

FRUIT AND SEED SET IN *STRONGYLODON MACROBOTRYS*

CHRISTINA J. PRYCHID, SIMON J. OWENS AND PAULA J. RUDALL

Royal Botanic Gardens, Kew, Richmond, Surrey, TW9 3AB, UK

Abstract

The jade vine, *Strongylodon macrobotrys* A. Gray, is a rare, decorative papilionoid legume that grows at low to medium altitudes in damp, mountainous forests in the Philippine islands. The plant growing in the Palm House at Kew since 1963 has flowered regularly each spring, but until 1995 had never produced seed. This was thought to be due to the absence of its natural pollinator and/or a self-incompatibility mechanism. A revised study of flower structure at the macro- and microscopic (including ultrastructural) levels, has now resulted in successful pollination and viable seed set, and has indicated late acting self incompatibility (LSI).

Introduction

The Philippine jade vine, *Strongylodon macrobotrys* A. Gray, is one of twenty species in the genus *Strongylodon* (Fabaceae - Papilionoideae - Erythrininae). Its distribution includes Papua New Guinea in Australasia; El Salvador, Central America; Fiji and Hawaii in the Pacific Ocean; some Caribbean Islands and Singapore, Hong Kong and the Philippine Islands of Luzon, Mindoro and Catanduanas in Asia (Polhill, 1972). It is a rare climbing shrub that grows at low to medium altitudes in damp, mountainous forests. The large, attractive pea-like flowers on pendulous pseudoracemose inflorescences (Figs. 1, 2) have a luminous blue-green colour which is thought to attract pollinators such as birds and bats, as in the related genus *Mucuna* (van der Pijl, 1941; Faegri & van der Pijl, 1971). It is grown as an ornamental in tropical gardens and warm greenhouses.

The plant in the Palm House at Kew was grown from seed sent from Hawaii in 1963. Although it has flowered regularly each spring, it has never produced seed, presumably due to the absence of natural pollinators needed to trigger the flowers for both self- and cross-pollinations and/or to the presence of a self incompatibility system (SI). SI has been found in other species in the Papilionoideae (Shivanna and Owens, 1989). A series of hand pollinations, together with microscopical analysis of the reproductive structures, was undertaken to see if a cause for the lack of seed set could be determined and from this a possible remedy to the problem found. This paper includes an anatomical and morphological description of the stigma of *Strongylodon macrobotrys*, both pre- and post-pollination, with some discussion of breeding systems in other legume taxa.

Material and Methods

A mature plant of *Strongylodon macrobotrys* (Kew accession number 1963-72801) growing in the Palm House at Kew, was used for this investigation.

Light microscopy

For light microscopy (LM), semi-thin sections (c. 0.5μm) of pistils, taken from material embedded for TEM, were collected on slides and stained with the metachromatic stain toluidine blue (0.5% in 0.1M phosphate buffer, pH 7.0; Feder & O'Brien, 1968). Pollen germination and pollen tube growth in the stigma, style and ovary were checked by mounting fresh hand sectioned pistils in aniline blue (Linskens & Esser, 1957), two, four, eight, twenty-four and forty-eight hours after pollination. Observations were made using an Ernst Leitz Wetzlar GUBH UV fluorescence microscope with a Leica filter block A. Ovule fertilisation was checked by clearing fixed, dissected, partially flattened ovaries in a modified version of Herr's clearing fluid (lactic acid : chloral hydrate : phenol : clove oil : Histoclear, in proportions 2 : 2 : 2 : 2 : 1 by weight; Rudall and Clark, 1992). The slides were examined using differential interference contrast (DIC) optics on a Leitz Dialux 20 photomicroscope.

Electron microscopy

For scanning electron microscopy (SEM), pistils were fixed in FAA for three days (Johansen, 1940) and then dehydrated through an alcohol series. Styles and ovaries were torn open longitudinally with surgical syringe needles, which give less tissue damage than scalpel blades. The specimens were then critically point dried and platinum coated. Observations were carried out using a Cambridge Stereoscan 240 SEM. For transmission electron microscopy (TEM), pistils were cut into small transverse slices (2-3mm) using a double edged razor blade and transferred as quickly as possible into fresh Karnovsky's fixative (2% paraformaldehyde and 2.5% glutaraldehyde; Glauert, 1975) in 0.05M phosphate buffer for three hours. After washing in buffer, the samples were post-fixed in 1% osmium tetroxide for two hours, rinsed, dehydrated in an ethanol series and embedded in LR White resin (medium grade). Gold sections were collected on copper grids and stained in aqueous uranyl acetate and lead citrate (Reynolds, 1963). Photographs were taken on a JEOL JEM 12-10 TEM.

Pollinations

Techniques

Flowers show a development of the keel into a strongly beaked structure. Anthesis occurs when the standard petal folds back on itself. Self pollinations were carried out between 8.30am and 9.30am, the majority of stigmas being selfed on the first day of anthesis, fewer on the day following anthesis. Uncontrolled pollinations, if any, had not set seed in the lifetime of the plant. Flowers are protandrous with anthers generally dehiscing two days prior to anthesis. The ten anthers are nearly uniform with alternate filaments inserted a little higher. The vexillary stamen is free. Within the keel the anthers lie below the stigma. The large amount of pollen liberated by the anthers becomes compacted around and above the stigma in the keel apex. Pollen is bicellular, as in other Papilionoideae (Brewbaker, 1967; Owens & Stirton, 1986).

Stigma structure is relevant to successful pollination. Stigmas are wet, with a papillate surface covered by a sculptured layered cuticle of varying thickness (1-4μm) and enclosed by a fringe of dense elongate thin hairs (Fig. 3). These stigmatic hairs seem to restrict the stigma from being automatically self pollinated, in the same way as the thick cuticle prevents self pollen germination in *Trifolium pratense* (Heslop-Harrison & Heslop-Harrison, 1983). At the basal end of the papillae there is a subcuticular space filled with stigmatic exudate (Fig. 4) which is released following cuticle rupture upon pollination. In TEM the secretion consists largely of lipidic droplets although in some cases it has a foam-like appearance that is less

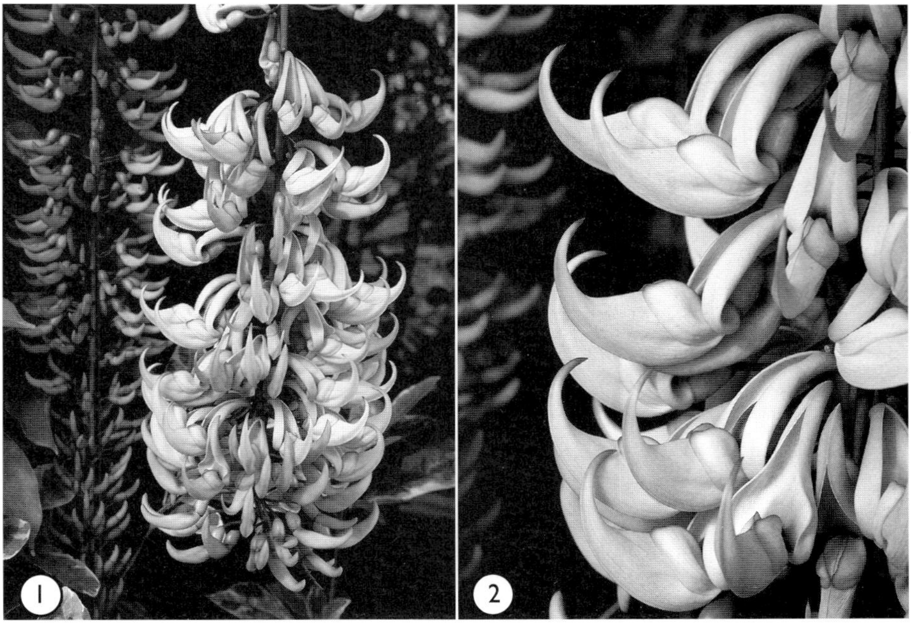

FIGS. 1, 2. *Strongylodon macrobotrys*, pendulous inflorescences with pea-like flowers.

electron opaque. As in most other Papilionoideae (Owens & Stirton, 1989), there is a small area of solid style tissue near the stigma, adjacent to an equally small lysigenous area above the stylar canal. Lysigenous cells are thin walled, as opposed to the thick walled, largely vacuolised outer cortical cells, possibly due to degradation of the cell contents and a breakdown of the plasma membrane. The uppermost layer of stigmatic cells appears to degenerate prior to pollination, consistent with findings for Papilionoideae in general (Owens & Stirton, 1989).

For pollination, stigmatic hairs need to be brushed to one side of the receptive region and pollen forced through the cuticle into the exudate beneath. This was originally attempted by dissecting off a portion of the keel surrounding the anthers and stigma and then using a paint brush to move aside the hairs and to dab pollen on the stigmatic surface. However, the humidity in the air caused the exposed anthers and style to become flaccid and rot, so seed set was not accomplished. Thus, a pollination technique was needed that mimicked the natural piston-type pollination mechanism, in which a plug of pollen is pushed out through the keel apex by the stigma when pressure is applied to the keel and wing petals by the pollinator. The stigmatic hairs are pulled back as the stigma emerges from the keel exposing the receptive region to pollen on the pollinator's body. When the pollinator leaves the flower, the stigma is drawn back down into the keel. Individual flowers were therefore hand-held so that as the base of the keel and wings were depressed the stigma and pollen were forced through the tip. This process alone was enough to move the stigmatic hairs away from the receptive region, leaving it exposed (Fig. 5), i.e. the hairs were not brushed aside by the operator. Dehisced pollen was collected and dabbed onto the stigma surface. Upon release of the lower petals the reproductive structures withdrew back into the keel.

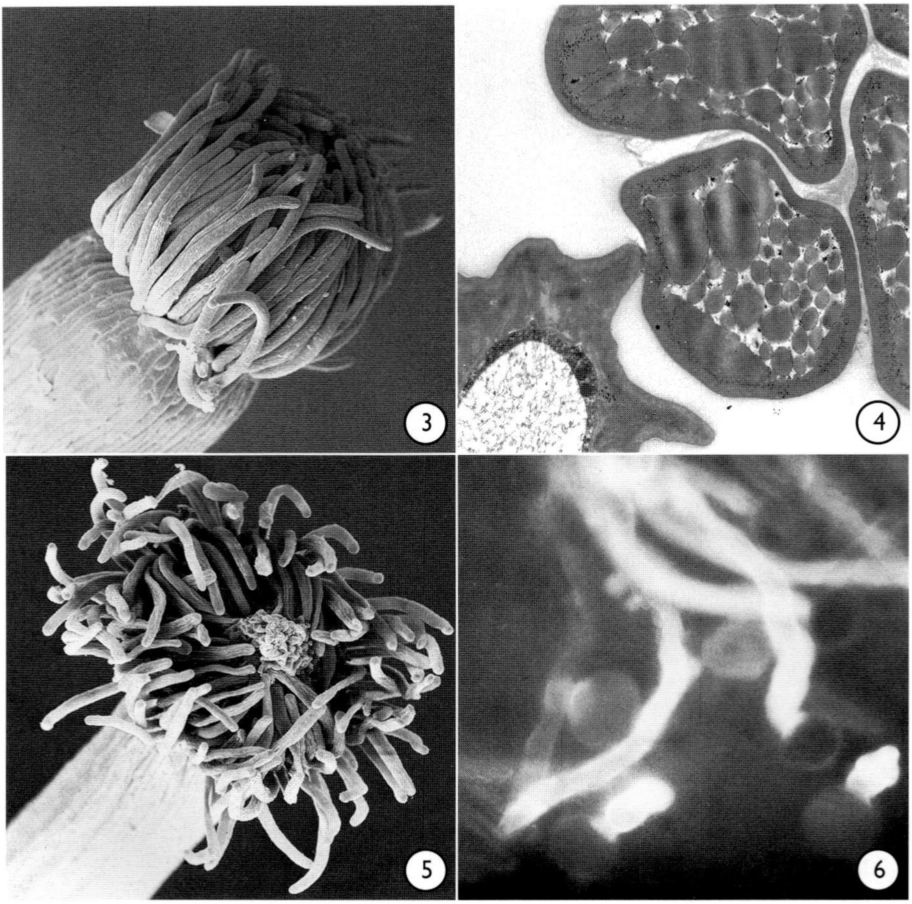

FIGS. 3–6. Stigmatic images of *Strongylodon macrobotrys*. Fig 3. Before pollination numerous hairs cover the receptive region (SEM); Fig 4. Stigmatic epidermis and exudate. A hair is on the bottom left of the picture (TEM); Fig 5. At pollination the hairs have moved uncovering the stigma surface (SEM); Fig 6. Pollen germination and tube growth, (aniline blue stain with UV illumination).

Results of pollinations

Following hand pollination, numerous pollen grains were brought into close contact with the stigmatic surface and its associated secretions. Pollen hydration and germination followed pollen attachment (Fig. 6), with pollen tubes entering the neck of the style four hours after pollination. Pollen tubes reached the ovarian cavities in most selfed pistils, and had penetrated most ovular micropyles two days after pollination, resulting in some initial seed and fruit set. No pollen tubes were found in unpollinated two day-old pistils. The number of pollen tubes arriving in the ovary was approximately equal to ovule number. Thus there was little competition between pollen tubes to enter micropyles; male gamete selection and maximization of seed set did not occur through an excess of male gametophytes.

Seed and fruit set

Although a total of 316 flowers were self pollinated from twelve inflorescences, only four pods were set, each on separate inflorescences. A variable number of seeds (6-8 seeds out of 10-14 ovules) were set per pod, even though most ovules had been penetrated by pollen tubes, indicating that not all of the ovules were mature or receptive. Seeds were much more likely to mature at or near the distal or stylar end of the pod than the proximal or pedicel end. The majority of flowers (93.9%) that were not pollinated dropped between days 2-5, with most abscissions occurring on days 2 and 3. Presumably ovules would be receptive before the flower was abscissed, i.e. before day 2, at the time the ovules were being penetrated. Mulcahy (1979) suggested that an excess of ovules within an ovary allows the maternal parent to selectively abort ovules that have been fertilised at different times with pollen grains from different sources. This would optimise gamete competition. There was generally no difference in the numbers of developing seeds in aborted fruits and matured fruits. These results suggest that fertilisations occur and initial embryo development begins. Thus abortion cannot be due to an insufficient number of pollen tubes reaching the ovules, i.e. due to an insufficient number of fertilisations. Indeed, the results show that after fertilisation seeds and fruits are continually being aborted, so there is a post-fertilisation mechanism.

Mature pods are large (200mm long, c. 0.5kg) and fleshy (Fig. 7) and dehisce four months after pollination. Large fruit-eating bats are thought to be the agents of dispersal, as they favour fruits of this type (van der Pijl, 1957). There are c. 7 seeds per pod, each c. 50mm long, with a long hilum and thin testa. The funiculus of each developing seed swells and assumes a puffy, undulating appearance. Ovary walls also swell, acting as cushioning for the ovules. Seeds survive only if suitable conditions for germination are available immediately. In contrast, other species of *Strongylodon* have papery pods with one or two large hard subglobose or discoid seeds which can float in sea water and are viable even after a year (Polhill, 1972). In *S. macrobotrys* shoots appeared about a week after potting.

FIG. 7. Dehisced pod of *Strongylodon macrobotrys*.

Mature pods were distributed at random among infructescences. However, within an infructescence, immature pods in the upper third were more likely to be aborted than those in the lower and middle thirds. According to Bawa & Webb (1984), a significantly greater investment in pedicel and peduncle thickening is required to support upper rather than lower pods. It is more likely, however, that in this case the plant rejects selfed fruits up to a certain stage of flowering, after which if no crossed fruits have formed, it becomes more tolerant of selfing.

Flower and fruit abortion

The plant may produce an excess of flowers to cope with the uncertainties of the natural pollination technique. Some of the excess flowers may serve primarily as a source of male gametes (Charnov, 1979).

Pollinated flowers showed a bimodal pattern of abscission, with a large number of flowers dropping around day 2 (the first peak) and a second peak between days 6-9. Apart from fruits which survived to maturity and one fruit that survived to day 20, all immature fruits abscissed by day 12. The time at which the first peak occurs may be the maximum time allowed for fertilisation, i.e. if fertilisation has not occurred by day 2 then the flower is abscissed. This would appear to be the case here, and matches UV and DIC data. The first peak is therefore due to abscission of unsuccessfully fertilised flowers. As only 37.4% of pollinated flowers had aborted before day 5, the pollination technique had been successful on the whole.

If flowers were retained after day 3, their ovaries and ovules began to enlarge. In legumes, only fertilised ovules are known to enlarge (Cooper & Brink, 1940). The second peak could be due to post-fertilisation incompatibility with endosperm formation up to a certain point. Thus flower abortions occurred in the first three days and fruit abortions thereafter. The second peak could also be due to competition for maternal resources, a major factor in fruit abortion (Lloyd, 1980; Stephenson, 1981; Lee & Bazzaz, 1982a). However, as mentioned previously, fruits set by flowers which opened first in an inflorescence, thus initially having access to more maternal resources than later opening flowers, had a lower chance of being retained than those set by flowers that open subsequently.

Genetically inferior offspring may be selectively aborted by the maternal parent (Janzen, 1977b; Stephenson, 1981; Lee & Bazzaz, 1982a). However, there is no evidence here that aborted fruits were genetically inferior.

Large-scale flower and immature fruit abortion is influenced by both internal factors (incompatibility mechanisms, senescence processes, assimilate supplies and partitioning of minerals) and external factors (temperature, mineral nutrition and water availability: Krabel, 1991). As with many other legumes the percentage of pollinated flowers that develop into mature fruits is small (Bawa & Webb, 1984; Wiens, 1984). In our material of *S. macrobotrys* four seed pods, one per inflorescence, developed out of 316 pollinated flowers, denoting a fruit : flower ratio of 0.0127. Only a certain proportion of ovules developed into seeds (seed : ovule ratio = 0.6364). Thus only 1.27% of flowers set fruit, even though 63.64% of ovules initially started developing into seeds. This occurs within the realms of self incompatibility. However, the plant is largely self-sterile, not because of a conventional SI mechanism which blocks self pollen tubes in the style but because most selfed pistils are rejected post-fertilisation.

Naturally self-pollinated species tend to have higher fruit : flower and seed : ovule ratios (Bawa, 1974; Sutherland & Delph, 1984). As the fruit : flower ratio for *Strongylodon macrobotrys* is low, cross-pollination in the wild is more likely. However, there are exceptions to this. In Mimosoideae, the proportion of aborted fruits and flowers is high, whereas seed : ovule abortion in mature fruits is low, regardless of breeding system (Bawa & Buckley, 1989). Other species have a negatively skewed distribution with low fruit : flower and high seed : ovule ratios (Bawa *et al.*, 1989).

Acknowledgements

We thank Dr. Peter Gibbs for critical reading of the manuscript.

References

Bawa, K.S. (1974). Breeding systems of tree species of a lowland tropical community. *Evolution* **28**: 85–92.

Bawa, K.S. and Buckley, D.P. (1989). Seed:ovule ratios, selective seed abortion, and mating systems in Leguminosae. In: C.H. Stirton and J.L. Zarucchi (editors). Advances in legume biology, pp. 243–262, Missouri Botanical Garden, St. Louis, Mo. (Monographs in Systematic Botany, Missouri Botanical Garden, 29)

Bawa, K.S. and Webb, C.J. (1984). Flower, fruit and seed abortion in tropical forest trees: implications for the evolution of paternal and maternal reproductive patterns. *Amer. J. Bot.* **71**: 736–751.

Bawa, K.S. Hedge, S.G., Ganeshaiah, K.N. and Uma Shaanker, R. (1989). Embryo and seed abortion in plants. *Nature* **342**: 625–626.

Charnov, E.L. (1979). Simultaneous hermaphroditism and sexual selection. *Proc. Natl. Acad. Sci. USA* **76**: 2480–2484.

Cooper, D.C. and Brink, R.A. (1940). Partial self-incompatibility and the collapse of fertile ovules as factors affecting seed formation in alfalfa. *J. Agric. Res.* **60**: 453–472.

Faegri, K. and Pijl, L. van der (1971). The principles of pollination ecology. 2nd ed. Pergamon, Oxford.

Feder, N. and O'Brien, T.P. (1968). Plant microtechnique: some principles and new methods. *Amer. J. Bot.* **55**: 123–142.

Glauert, A.M. (1975). Fixation, dehydration and embedding of biological specimens. North Holland, Amsterdam. (Practical methods in electron microscopy, 3).

Heslop-Harrison, J. and Heslop-Harrison, Y. (1983). Pollen-stigma interaction in the Leguminosae: the organization of the stigma in *Trifolium pratense* L. *Ann. Bot.* **51**: 571–583.

Janzen, D.H. (1977a). A note on optimal mate selection by plants. *Amer. Naturalist* **111**: 365–371.

Janzen, D.H. (1977b). Developmental demography of *Bauhinia pauletia* Pers. seeds and ovules. *Brenesia* **12/13**: 105–111.

Johansen, D.A. (1940). Plant microtechnique. McGraw Hill Book Co., New York.

Krabel, D. (1991). Fruit abortion in *Hibiscus rosa-sinensis* L. *Angew. Bot.* **65**: 133–138.

Lee, T.D. and Bazzaz, F.A. (1982a). Regulation of fruit and seed production in an annual legume, *Cassia fasciculata*. *Ecology* **63**: 1364–1373.

Lee, T.D. and Bazzaz, F.A. (1982b). Regulation of fruit maturation pattern in an annual legume, *Cassia fasciculata*. *Ecology* **63**: 1374–1388.

Linskens, H.F. and Esser, K. (1957). Über eine spezifische Anfarbung der Pollenschlauche im Griffel und die Zahl der Kallosepropfen nach Selbstung und Frembung. *Naturwissenschaften* **44**: 1–2.

Lloyd, D.G. (1979). Paternal strategies of angiosperms. *New Zealand J. Bot.* **17**: 595–606.

Lloyd, D.G. (1980). Sexual strategies in plants. I. An hypothesis of serial adjustment of maternal investment during one reproductive session. *New Phytol.* **86**: 69–79.

Mulcahy, D.L. (1979). The rise of the angiosperms: a genecological factor. *Science* **206**: 20–23.

Owens, S.J. and Stirton, C.H. (1989). Pollen, stigma, and style interactions in the Leguminosae. In: C.H. Stirton and J.L. Zarucchi (editors). Advances in legume biology, pp. 105–112, Missouri Botanical Garden, St. Louis, Mo. (Monographs in Systematic Botany, Missouri Botanical Garden, 29).

Pijl, L. van der (1941). Flagelliflory and cauliflory as adaptations to bats in *Mucuna* and other plants. *Ann. Bot. Gard. Buitenzorg* **51**: 83–93.

Pijl, L. van der (1957). The dispersal of plants by bats. (Chiropterochory). *Acta Bot. Neerl.* **6**: 291–315.

Polhill, R.M. (1972). *Strongylodon macrobotrys*. Leguminosae. *Curtis's Bot. Mag.* **179**: 627.

Reynolds, E.S. (1963). The use of lead citrate at a high pH as an electron opaque stain in electron microscopy. *J. Cell Biol.* **17**: 208–212.

Rudall, P. and Clark, L. (1992). The megagametophyte in Labiatae. In: R.M. Harley and T.R. Reynolds (editors). Advances in Labiate Science, pp. 65–84, Royal Botanic Gardens, Kew.

Shivanna, K.R. and Owens, S.J. (1989). Pollen-pistil interactions (Papilionoideae). In: C.H. Stirton and J.L. Zarucchi (editors). Advances in legume biology. pp. 157–182, Missouri Botanical Garden, St. Louis, Mo. (Monographs in Systematic Botany, Missouri Botanical Garden, **29**).

Stephenson, A.G. (1981). Flower and fruit abortion: Proximate causes and ultimate functions. *Annual Rev. Ecol. Syst.* **12**: 253–279.

Sutherland, S. and Delph, L.F. (1984). On the importance of male fitness in plants: patterns of fruit-set. *Ecology* **65**: 1093–1104.

Wiens, D. (1984). Ovule survivorship, brood size, life history, breeding systems, and reproductive success in plants. *Oecologia* **64**: 47–53.

Sornsathapornkul, P. and Owens, J.N. (1998).Fertilisation and early embryo development in a tropical *Acacia* hybrid (*A. mangium* Willd. and *A. auriculiformis* A. Cunn. ex Benth.). In: S.J. Owens and P.J. Rudall (Editors). Reproductive Biology, pp. 353–362. Royal Botanic Gardens, Kew.

FERTILISATION AND EARLY EMBRYO DEVELOPMENT IN A TROPICAL ACACIA HYBRID (*A. MANGIUM* WILLD. AND *A. AURICULIFORMIS* A. CUNN. EX BENTH.)

PRASERT SORNSATHAPORNKUL AND JOHN N. OWENS

Centre for Forest Biology, University of Victoria, Victoria, B.C., Canada

Abstract

Development of the *Acacia* hybrid ovule (*A. mangium* Willd. and *A. auriculiformis* A. Cunn. ex Benth.) at fertilisation and early zygotic embryo development were investigated using light and electron microscopy. At anthesis, the hybrid ovule has features similar to those of other angiosperms but the embryo sac is only partly enclosed by the inner and outer integuments. Ultrastructurally, the egg apparatus, central cell and antipodal cells contain numerous cell organelles, large vacuoles and stored reserves. Because the hybrid ovule lacks a micropyle, the pollen tube directly penetrates the micropylar nucellus and this occurs by 24 hours after pollination. Synergid degeneration is independent of pollination and possibly triggered by the arrival of the pollen tube. The pollen tube penetrates and discharges its contents into the degenerating synergid via the filiform apparatus. No sperm cytoplasmic body was observed in the degenerate synergid. It presumably remains in the pollen tube. This indicates that the hybrid possibly possesses maternal cytoplasmic inheritance. The polar nuclei do not fuse until arrival of the sperm nucleus. Polar nuclear fusion is associated with endoplasmic reticulum, whereas the fusion of the sperm nucleus with the polar nuclei and with the egg may occur by alignment and contact of both nuclei. The primary endosperm undergoes first division shortly after karyogamy but the first zygotic division did not occur until after nearly complete endosperm cellularisation. The hybrid embryo lacks a distinct suspensor. Endosperm formation is of the nuclear type. During early development, the endosperm is nourished by stored products from the central cell, digested nucellus and other maternal tissues through different pathways, including wall ingrowths, haustorial-like endosperm cells, and the endothelium. The endosperm then becomes the main nutrient source for the growing embryo during late embryo development.

Introduction

Acacia hybrids are fast growing trees. Both parental species are native to Queensland, Australia and Papua New Guinea (Doran and Skelton, 1982; Turnbull, *et al.*, 1986) but have been widely introduced to most of Southeast Asia for industrial plantations and reforestation programs. In Thailand, the hybrids show great potential in their adaptability and growth performance, thus interest has increased in using the hybrid for reforestation. Preliminary studies show that low seed production in the hybrids partially results from insufficient pollination. However, limitations to seed production can occur through other reproductive processes. Some studies have been done on various aspects of reproductive biology of the hybrids, mostly emphasising phenology and pollination biology of the parental species

(Ibrahim and Awang, 1992; Josue, 1992; Sedgley *et al.*, 1992). Detailed anatomical and ultrastructural studies have been conducted on various aspects of pre- and post- fertilisation of other leguminous plants (Folsom and Cass, 1989; Chamberlin *et al,*. 1994; Johansson and Walles, 1994) but no studies have concentrated on pre- and post- fertilisation development in the hybrid. This study describes the anatomical changes of the embryo sac from fertilisation through the early embryo and emphasises the mechanism of double fertilisation and structural adaptations of the ovule involved in the embryo nutrition during early embryo development using light and electron microscopy.

Materials and Methods

The study was carried out using an experimental plot of 4-year-old *Acacia* hybrids at the ASEAN- Forest Tree Seed Centre, Muak-lek, Saraburi, Thailand, located at latitude N 14∞ 40' and longitude E 101∞ 17' and about 200 m elevation. The hybrid trees were planted in 1988 and first flowered in late 1990. The hybrid seeds were obtained from a plantation of *A. mangium* adjacent to a stand of *A. auriculiformis*. Parental species overlap in their flowering period in October-November and appear to have the same pollinators, therefore the hybrids resulted from natural hybridisation. About 10 hybrid flowers and 20 seeds at each developmental stage were collected from two trees. Flowers were collected 1 day before anthesis and every 6 to 8 hours for 1 week after pollination, then during fruit set, spikes were collected every 7 to 10 days for 3 months. Measurements were taken on the number of pods per spike, and pod size and colour.

Light and electron microscopy

To ensure penetration of fixatives, ovules and small portions of pods during early seed development (up to 65 days after pollination, DAP) were used, whereas during later development, developing seeds were dissected from pods. Specimens were fixed in FAA (50% ethanol) for light (LM) and in 2.5% glutaraldehyde in 0.075 M phosphate buffer (pH 7.2 to 7.4) for LM and transmission electron microscopy (TEM). FAA-fixed-specimens for LM were dehydrated through a series of alcohol solutions (Johansen, 1940), embedded in TissuePrep and sectioned at 6 to 8 mm. Sections were stained with 1% safranin and 1% aniline blue. Specimens fixed in 2.5% glutaraldehyde were rinsed in the same buffer and postfixed for 1 h in 1% osmium tetroxide in 0.075 M phosphate buffer (pH 7.2 to 7.4). Specimens were then dehydrated in a series of acetone solutions, embedded in Spurr's resin (Spurr, 1969) and sectioned at 0.5 to 1 mm for LM and 60 nm for TEM using an ultramicrotome. Semithin sections were stained with 0.5 % toluidine blue. Ultrathin sections were collected on single hole grids coated with 0.3 % Formvar in chloroform, then stained with 2% aqueous uranyl acetate for 15 min and 0.2% Sato's lead citrate (Sato, 1968) for 15 min and viewed using a Hitachi 7000 transmission electron microscope (TEM) at 75 kV.

Results and Discussion

Before pollen tube penetration of nucellus

Prior to fertilisation the egg apparatus consists of the egg and two synergids as in most angiosperms (Fig. 1; Johri *et al.*, 1992). All three cells are polarised, rich in cytoplasmic organelles and separated from each other by thin cell walls and no plasmodesmata were observed. Compared to the egg apparatus of *Lilium longiflorum* in which all three cells are similar in ultrastructure (Janson and Willemse, 1995), the hybrid egg apparatus appears similar to that of *Nicotiana tabacum* (Mogensen and Suthar, 1979). The hybrid egg cell is distinguished from both synergids. The egg cell usually contains a large micropylar vacuole, whereas each synergid has a large chalazal vacuole and prominent filiform apparatus (Fig. 1). Both synergids appear identical and their micropylar walls are attached to the micropylar embryo sac and egg-cell wall. In contrast to the egg cell, the synergid contains a large chalazal vacuole and a few small plastids, some with small starch grains (Fig. 1).The latter feature presumably functions for secretion of chemotropic substances to direct pollen-tube growth (Huang and Russell, 1992). The presence of dictyosomes and endoplasmic reticulum (ER), is thought to be associated with such a secretion. Compared to the egg apparatus, the central cell appears to be highly metabolically active as evidenced by abundant cytoplasmic organelles such as mitochondria, dictyosomes and ER (Fig. 2). Starch does not aggregate to form multigrain packets as reported in other leguminous species such as soybean (Folsom and Peterson, 1984). The central cell has wall ingrowths along the embryo sac wall. Prior to fertilisation the polar nuclei have a double nuclear membrane, single nucleolus and are separate (Fig. 2). The antipodal cells are rich in cytoplasmic organelles and each contains one large and several small vacuoles (Fig. 3).

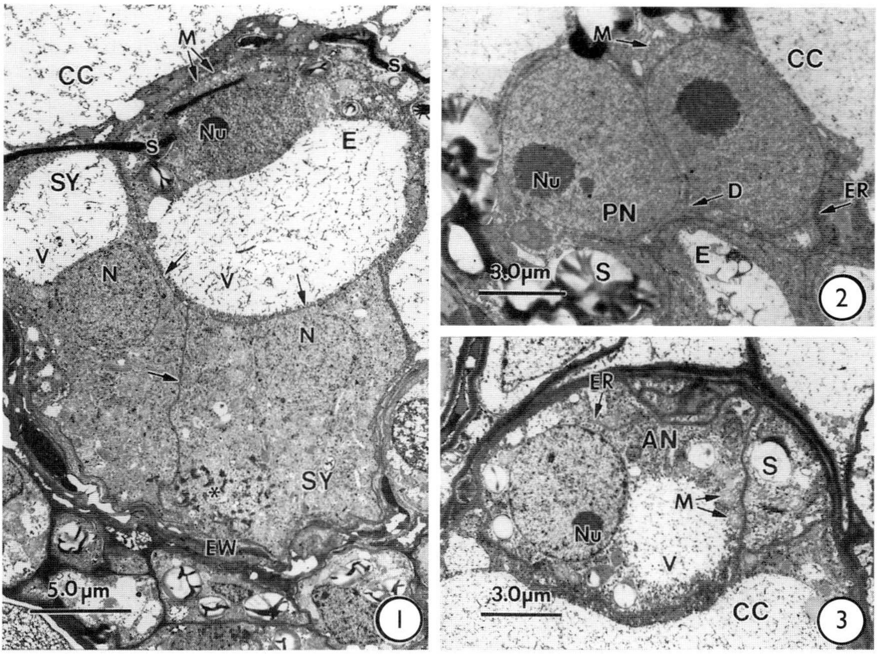

FIGS. 1–3. The embryo sac of the *Acacia* hybrid at anthesis. 1. Egg apparatus showing egg and two synergids. Three cells are separated from each other by thin cell walls (arrows); small part of filiform apparatus (asterisk). 2. Central cell showing polar nuclei and dense cytoplasm with abundant cytoplasmic organelles. 3. Antipodal cells showing dense cytoplasm with abundant cytoplasmic organelles. Abbreviations: AC, apical cell; AN, antipodal cells; BC, basal cell; CC, central cell; CE, cellular endosperm; D, dictyosome; DS, degenerate synergid; E, egg cell; EM, embryo; ENT, endothelium; ER, endoplasmic reticulum; EW, embryo sac wall; FA, filiform apparatus; II, inner integument; M, mitochondria; N, nucleus; NC, nucellus; NE, nuclear endosperm; Nu, nucleolus; OI, outer integument; P, polysome; PE, proembryo; PN, polar nuclei; PT, pollen tube; PS, persistent synergid; S, starch grain; SN, sperm nucleus; SY, synergid; V, vacuole; VB, vascular bundle; Z, zygote.

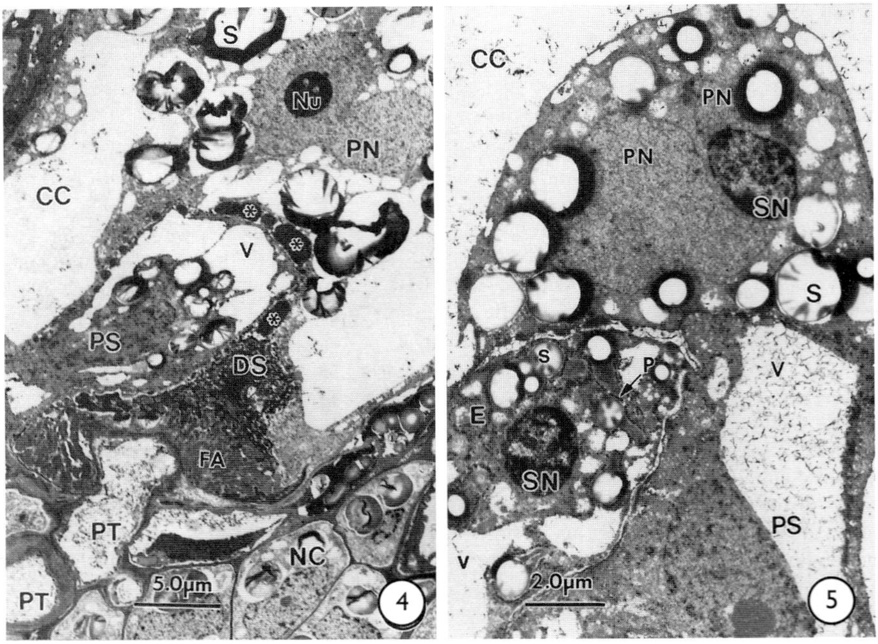

FIGS. 4–5. The embryo sac of the *Acacia* hybrid at anthesis. 4. Hybrid embryo sac at about 2 DAP; the pollen tube has penetrated the micropylar nucellus and entered the degenerate synergid at the filiform apparatus. Electron dense materials (asterisks) accumulate at the micopylar region of the persistent synergid. 5. Hybrid embryo sac after pollen-tube penetration (about 3 DAP). A sperm nucleus is deposited in the egg cell and in the central cell. Note the accumulation of polysomes in the egg cell. Abbreviations as in Figs. 1–3.

During and after pollen tube penetration of nucellus

Several studies have shown variation in timing of synergid degeneration in relation to pollination (Johri *et al.,* 1992). Synergid degeneration in the hybrid is independent of pollination. In all ovaries containing pollen tubes, unfertilised ovules show intact synergids. Synergid degeneration does not occur until penetration of the pollen tube into the nucellus, suggesting that synergid degeneration may be triggered by the arrival of the pollen tube.

Unlike most angiosperms, the mature hybrid ovule lacks a distinct micropyle due to the partial development of both integuments. In all fertilised ovules observed, pollen tubes usually penetrate the micropylar nucellus by 24 hours after pollination. Only one pollen tube penetrates and discharges its contents into the degenerating synergid via the filiform apparatus (Fig. 4) as in soybean (Tilton *et al.*, 1983; Dute *et al.*, 1989). In some ovules, the pollen-tube tip branches and penetrates both the degenerating and persistent synergid. It is unlikely that a certain synergid is preprogrammed to degenerate during the pollen-tube penetration since either one of the synergids may show degeneration in different ovules. In the serial sections of fertilised embryo sacs, no cytoplasmic bodies from sperm cells were observed in the degenerate synergid. We presume that they remain in the pollen tube. The remnants of two nuclei in the degenerate synergid are likely the large synergid nucleus and the small pollen tube vegetative nucleus as in *L. longiflorum* (Janson and Willemse, 1995). The remaining cytoplasm in the pollen tube contains many mitochondria near the tube tip and plastids lining the tube wall.

By 3 days after pollination (DAP), the two sperm nuclei are observed, one in the egg cell and one in the central cell (Fig. 5). Both sperm nuclei are almost identical, being disc-shaped and electron dense because of condensed chromatin as described in soybean (Folsom and Cass 1992). In *Plumbago* which possesses biparental cytoplasmic inheritance (Hagemann and Schroder 1989), preferential fertilisation was reported in which the sperm cell containing more plastids, fused with the egg cell (Russell, 1993). Sperm dimorphism in the hybrid, contrary to *Plumbago*, is difficult to determine during fertilisation since the sperm cytoplasm is already excluded.

Prior to karyogamy, the fertilised egg shows ultrastructural changes, including an increased number of starch grains and polysomes (Fig. 5) similar to that in *L. longiflorum* (Janson and Willemse, 1995) and cotton (Jensen, 1968). In the same embryo sac, the fusion of polar nuclei with a sperm nucleus occurs shortly before that of the egg and the sperm nucleus. In soybean (Folsom and Cass, 1992), the polar nuclei of the central cell fuse before the arrival of the sperm nucleus. In the hybrid, the polar nuclei remain unfused until migration and alignment of the sperm nucleus between the polar nuclei and triple fusion occurs shortly before the fusion of the egg and the other sperm nucleus. Polar nuclear fusion starts with the linkage of the nuclear envelopes along the margins of the bowl-shaped nuclei. The polar nuclei start to fuse, then appear to engulf the sperm nucleus. The ER observed in the fusion site may be involved in the fusion of the polar nuclei (Jensen, 1964; van Went, 1970). Jensen (1964) showed that polar nuclei are linked by fused ER and shortening of ER enables them to come into contact and merge. Because the sperm nucleus is engulfed by two polar nuclei, further polar nuclear fusion enables the sperm nucleus to contact both polar nuclei simultaneously then fusion among polar nuclei and sperm nucleus occurs. Janson and Willemse (1995) suggested that the fusion of the egg nucleus with the sperm nucleus is less associated with the ER as the egg cell contains little ER. The fusion process between the egg and sperm nucleus occurs by alignment and contact of both nuclei. This may be true in the fusion process between the egg and the sperm nucleus as no ER has been observed in the fusion site. No cytoplasmic fusion occurs among the sperm and the egg cell and the sperm and the polar nuclei since the sperm cytoplasm is already excluded. This indicates that the hybrid possibly possess maternal cytoplasmic inheritance as reported in most angiosperms (Russell, 1992; Mogensen, 1996).

The zygote, endosperm and early embryo development

The newly formed zygote is observed 5 days after pollination (Fig. 6). Shortly after fusion of the sperm nucleus with the polar nuclei the primary endosperm undergoes its first division forming two free nuclei. Compared to the primary endosperm stage, there is a reduction in number of starch grains at the two nuclear endosperm stage. This lysis of starch may provide the nutrients required for division. In soybean, the nutritional reserves which break down before fertilisation are needed for future development of the egg, zygote, embryo and endosperm (Folsom and Cass, 1992). In contrast to soybean, the great amounts of metabolites stored in the central cell of the hybrid seem to be used only for development of endosperm since the zygote remains dormant for over two months until the endosperm nearly completes development (Fig. 13). Further development of the free nuclear endosperm does not occur until the outer integument completely encloses the embryo sac and one of the nucellar layers abutting the embryo sac degenerates. We assume that this anatomical change supports the expanding embryo sac and provides nutrition necessary for development of the free nuclear endosperm (Fig. 13).

Hybrid endosperm development is of the nuclear type. Cellularisation of the free nuclear endosperm begins during the late zygote in the periphery of the chalazal region and progresses toward the micropylar end of the embryo sac (Fig. 7). At this stage, the large cellular endosperm is formed in the chalazal region of the embryo sac (Fig. 7) and may possess some features similar to the endosperm haustoria observed in some leguminous plants (Dute and Peterson, 1992, also reviewed by Johri *et al.*, 1992). This large cellular endosperm, which is chalazally connected with the integument, probably has a function similar to the endosperm haustoria in which it facilitates nutrient transport from the integument into the embryo sac (Jakobsen *et al.*, 1994). Ultrastructurally, the cellular endosperm, at the two-cell proembryo stage, has a large vacuole, dense cytoplasm with many organelles and storage products (Fig. 9).

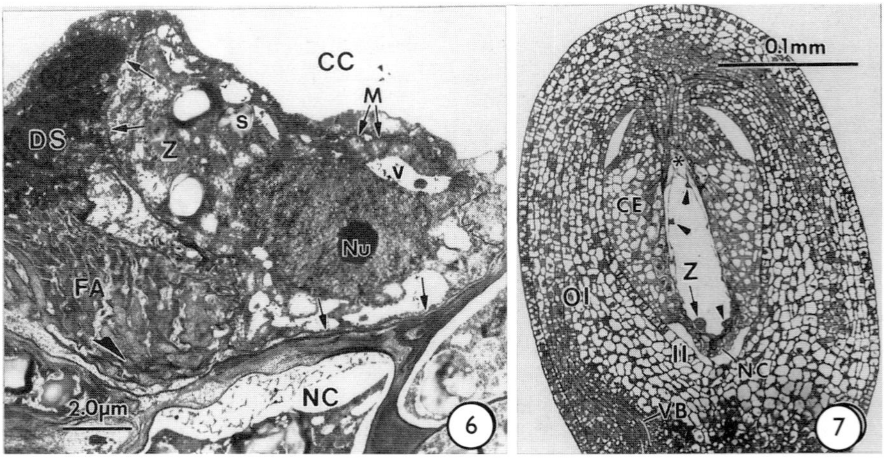

FIGS. 6–7. The zygote stage in the hybrid. 6. Newly- formed zygote and degenerate synergid. Note that the zygote is separated from the embryo sac wall and the degenerate synergid by thin cell walls (arrows). 7. Developing seed at the late zygote stage (about 65 DAP) showing cellularization of the endosperm and development of the integument; nuclear endosperm (arrowheads); endosperm haustoria (asterisk). Abbreviations as in Figs. 1–3.

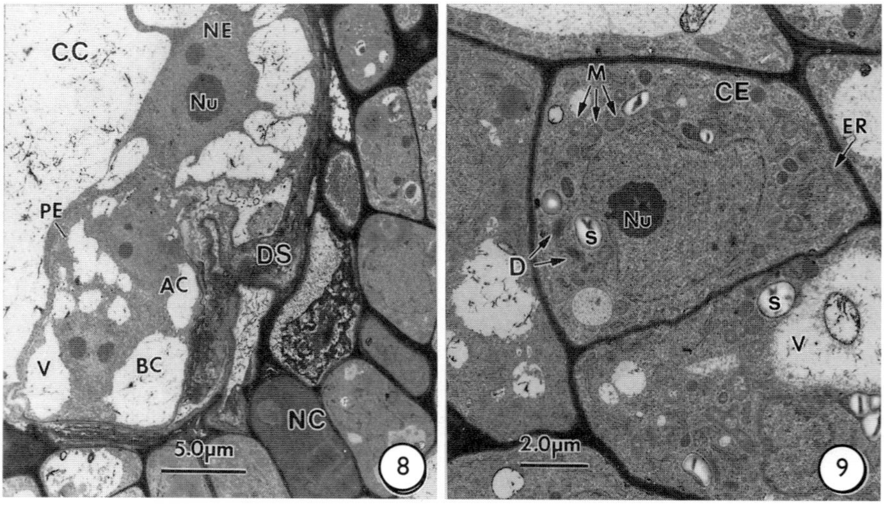

FIGS. 8–9. The two-celled proembryo stage in the hybrid. 8. Section showing two-celled proembryo, nuclear endosperm and remnants of degenerate synergid. 9. Cross-section of cellular endosperm showing dense cytoplasm with abundant cytoplasmic organelles. Abbreviations as in Figs. 1–3.

When compared to the egg cell, the zygote is reduced in size and lacks a large vacuole. Many plastids and mitochondria are present but ER and dictyosomes are much less abundant. Degeneration of the persistent synergid is evident. The zygote remains unchanged for about 65 DAP and this coincides with a significant period of pod abortion. It is postulated that the delay in zygote development into the proembryo relates to final fruit set which is partially determined by nutrient resource allocation by the mother plant (Stephenson, 1981).

The first zygotic division occurs when the endosperm cellularisation is almost completed. Transverse division of the zygote results in two equal-size cells which remain attached to the embryo sac wall (Fig. 8). There are no ultrastructural differences between the apical and basal cells. This differs from most angiosperm proembryos in which the apical cell usually contains more cytoplasm and differentiates into the embryo proper (Folsom and Cass, 1992; Johri *et al.*, 1992; Taylor and Vasil, 1995). The hybrid apical and basal cells are vacuolate, densely cytoplasmic and contain abundant ER, mitochondria and ribosomes which aggregate to form polysomes. The great reduction in number of starch grains in the two-cell proembryo may be due to their breakdown during the zygote stage to provide nutrients for development from the zygote to the two-cell proembryo (Fig. 13).

Endosperm cellularisation is completed when the embryo reaches the early globular stage (Fig. 10). No distinct suspensor is observed in the hybrid proembryo. This has also been reported in several members of Mimosaceae, such as *Acacia baileyana*, *Albizzia lebbek* and *Mimosa hamata*. (Johri *et al.*, 1992). The suspensor plays an important role in supporting embryo growth (Yeung and Meinke, 1993). As the hybrid embryo lacks a suspensor, the cup-like nucellus, which remains from nucellus degeneration during the zygote stage, may function to support embryo growth. This feature is also reported in *Acacia apiculata* (Souza, 1993).

At the late globular stage, an embryo protoderm forms (about 76 DAP; Fig. 12) and the endothelium, which forms from the inner integument during the early globular stage (Fig. 11), degenerates and disappears. Up to the late globular stage, development of the hybrid embryo sac has features similar to soybean and broad bean (Johansson and Walles, 1993a, 1993b, 1994; Chamberlin *et al.*, 1994) in which digestion of the stored products abundant in the nucellus may provide nourishment to the growing embryo and endosperm (Fig. 13). Also, the endothelium may facilitate nutrient transport from maternal tissues into the embryo sac (Fig. 13) as indicated by histochemical (Sehgal and Gifford, 1979) and autoradiographic studies (Chamberlin 1993). Obviously, the endosperm is a sink during the early embryo development and presumably during later development provides nutrition to the growing embryo.

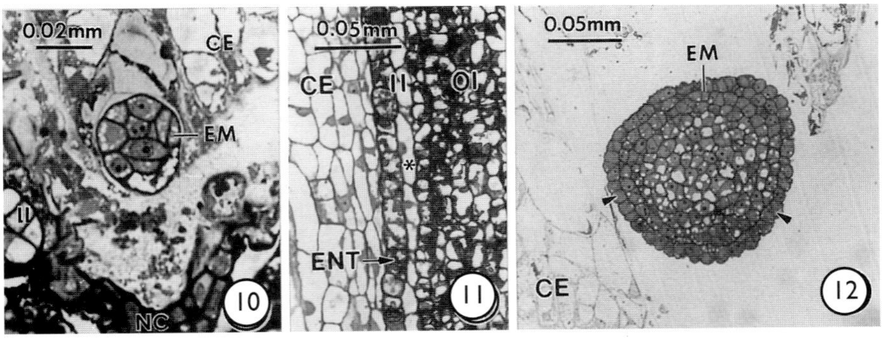

FIGS. 10–12. The globular embryo stage in the hybrid. 10. The early globular embryo which lacks a suspensor. 11. Developing integument showing the endothelium and inner integumentary cells (asterisk) during the early globular stage. 12. The late globular embryo showing formation of the protoderm (arrowheads). Abbreviations as in Figs. 1–3.

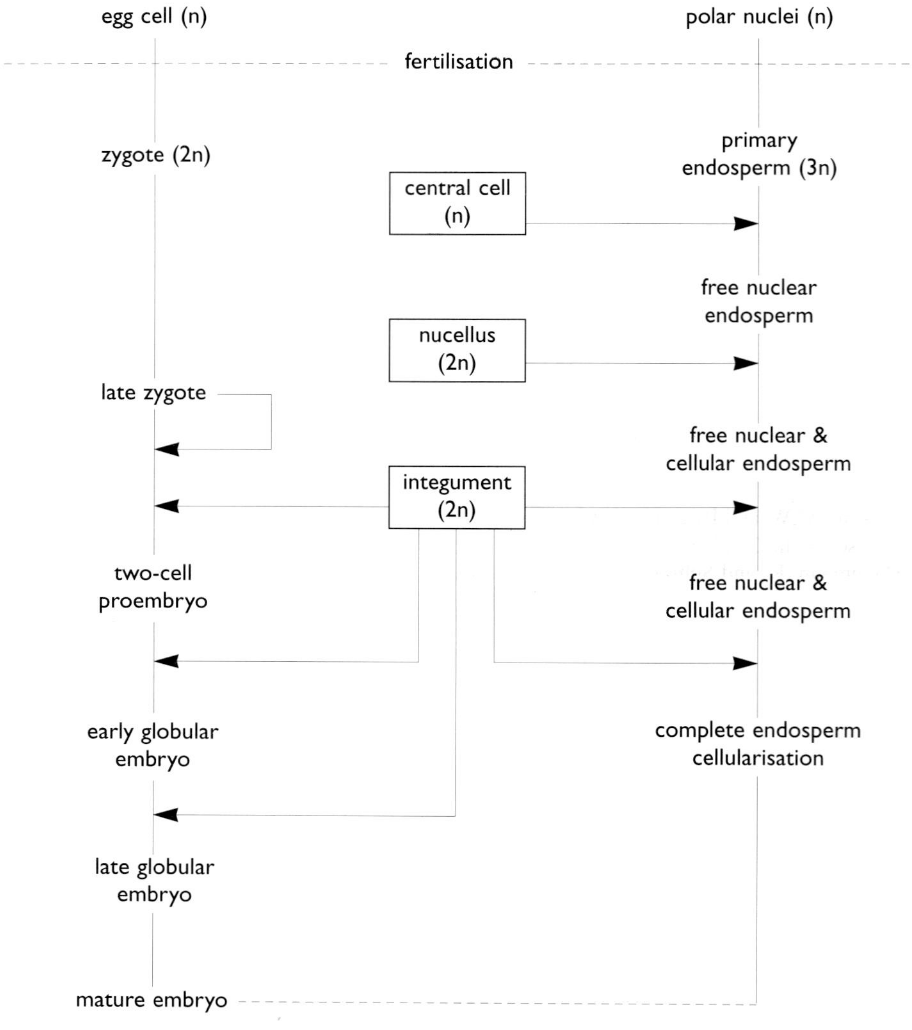

Fig. 13. Schematic diagram showing proposed pathway of nutrient transport (arrows)

Acknowledgements

The research was financially supported by the Canadian International Development Agency (CIDA) with the Canadian Forest Service (Petawawa National Forestry Institute) as Canadian Executing Agency. We thank ASEAN Forest Tree Seed Centre for providing facilities on field collections.

References

Chamberlin, M.A., Horner, H.T. and Palmer, R.G. (1993). Nutrition of ovule, embryo sac and young embryo in soy bean: an anatomical and autoradiographic study. *Canad. J. Bot.* **71**: 1153-1168.

Chamberlin, M.A., Horner, H.T. and Palmer, R.G. (1994). Early endosperm, embryo, and ovule development in *Glycine max* (L.) Merr. *Int. J. Pl. Sci.* **155**: 421–436.

Doran, J.C. and Skelton, D.J. (1982). *Acacia mangium* seed collections for international provenance trials. pp. 47–53, FAO, Rome. (Forest Genetic Resource Information, no. 11).

Dute, R.R. and Peterson, C.M. (1992). Early endosperm development in ovules of soybean, *Glycine max* (L.) Merr. (Fabaceae). *Ann. Bot.* **69**: 263–271.

Dute, R.R., Peterson, C.M. and Rushing, A.E. (1989). Ultrastructural changes of the egg apparatus associated with fertilization and proembryo development of soybean, *Glycine max* (Fabaceae). *Ann. Bot.* **64**: 123–135.

Folsom, M.W. and Cass, D.D. (1989). Embryo sac development in soybean: cellularization and egg apparatus expansion. *Canad. J. Bot.* **68**: 2135–2147.

Folsom, M.W. and Cass, D.D. (1992). Embryo sac development in soybean: the central cell and aspects of fertilization. *Amer. J. Bot.* **79**: 1407–1417.

Folsom, M.W. and Peterson, C.M. (1984). Ultrastructural aspects of the mature embryo sac of soybean, *Glycine max* (L.) Merr. *Bot. Gaz.* **145**: 1–10.

Hagemann, R. and Schroder, M.B. (1989). The cytological basis of the plastid inheritance in angiosperms. *Protoplasma* **152**: 57–64.

Huang, B.Q. and Russell, S.D. (1992). Synergid degeneration in *Nicotiana*: a quantitative, fluorochromatic chlorotetracycline study. *Sexual Pl. Reprod.* **5**: 151–155.

Ibrahim, Z. and Awang, K. (1992). Flowering and fruiting phenology of *Acacia mangium* and *A. auriculiformis* in Peninsular Malaysia. In: J.W. Turnbull (editor). Breeding technologies for tropical *Acacia* research, pp. 45–48, ACIAR, Canberra.

Jakobsen, H.B., Martens, H. and Lyshede, O.B. (1994). Accumulation of metabolites during seed development in *Trifolium repens* L. *Ann. Bot.* **74**: 409–415.

Janson, J. and Willemse, T.M. (1995). Pollen tube penetration and fertilization in *Lilium longiflorum* (Liliaceae). *Amer. J. Bot.* **82**: 186–196.

Jensen, W.A. (1964). Observations on the fusion of nuclei in plants. *J. Cell Biol.* **23**: 669–672.

Jensen, W.A. (1968). Cotton embryogenesis: the zygote. *Planta* **79**: 346–366.

Johansen, D.A. (1940). Plant microtechnique. McGraw-Hill, New York.

Johansson, M. and Walles, B. (1993a). Functional anatomy of the ovule in broad bean (*Vicia faba* L.). I. Histogenesis prior to and after pollination. *Int. J. Pl. Sci.* **154**: 80–89.

Johansson, M. and Walles, B. (1993b). Functional anatomy of the ovule in broad bean (*Vicia faba* L.). II. Ultrastructural development up to early embryogenesis. *Int. J. Pl. Sci.* **154**: 535–549.

Johansson, M. and Walles, B. (1994). Functional anatomy of the ovule in broad bean (*Vicia faba* L.): ultrastructural seed development and nutrient pathways. *Ann. Bot.* **74**: 233–244.

Johri, B.M., Ambegaokar, K.B. and Srivastava, P.S. (1992). Comparative embryology of angiosperms. Vol.1 & 2. Springer-Verlag, London.

Josue, J. (1992) Preliminary observations on the flowering phenology and seed production in a seedling hybridization orchard of *Acacia mangium* and *Acacia auriculiformis*. In: J.W. Turnbull (editor). Breeding technologies for tropical *Acacia* research, pp. 49–50, ACIAR, Canberra.

Mogensen, H.L. (1996). The hows and ways of cytoplasmic inheritance in seed plants *Amer. J. Bot.* **83**: 383–404.

Mogensen, H.L. and Suthar, H.K. (1979). Ultrastructure of the egg apparatus of *Nicotiana tabacum* (Solanaceae) before and after fertilization. *Bot. Gaz.* **140**: 168–179.

Russell, S.D. (1992). Double fertilization. *Int. Rev. Cytol.* **140**: 357–385.
Russell, S.D. (1993). The egg cell: development and role in fertilization and early embryogenesis. *Pl. Cell* **5**: 1349–1359.
Sato, T. (1968). A modified method for lead staining of thin sections. *J. Electron Microscopy* **17**: 158–159 (in Japanese).
Sedgley, M., Harbard, J., Smith, R.M., Wickneswari, R. and Griffin, A.R. (1992). Reproductive biology and interspecific hybridisation of *Acacia mangium* Wild and *Acacia auriculiformis* A. Cunn. ex Benth. (Leguminosae : Mimosoideae). *Austral. J. Bot.* **40**: 37–48.
Sehgal C.B. and Gifford, E.M. (1979). Developmental and histochemical studies of the ovules of *Nicotiana rustica* L. *Bot. Gaz.* **140**: 180–188.
Souza, L.A. (1993). Morfo-anatomia do desenvolvimento do fruto de *Acacia paniculata* Willd. (Leguminosae). *Arq. Biol. Tecnol.* **36**: 851–871.
Spurr, A.R. (1969). A low-viscosity epoxy resin embedding medium for electron microscopy. *J. Ultrastruct. Res.* **26**: 31–43.
Stephenson, A.G. (1981) Flower and fruit abortion: proximate causes and ultimate functions. *Annual Rev. Ecol. Syst.* **12**: 253–79.
Taylor, M.G. and Vasil, I.K. (1995). The ultrastructure of zygotic embryo development in pearl millet (*Pennisetum glaucum*; Poaceae). *Amer. J. Bot.* **82**: 205–219.
Tilton, V.R., Palmer, R.G. and Wilcox, L.E. (1983). The female reproductive system in soybeans, *Glycine max* (L.) Merr. (Leguminosae). In: O. Erdelska (editor). Fertilization and embryogenesis in ovulated plants. Proceeding of the 7th International Cytoembryological Symposium, June 1982, pp. 35–37, Veda, Bratislava, Czechoslovakia.
Turnbull, J.W., Martensz, P.N. and Hall, N. (1986). Notes on lesser-known Australian trees and shrubs with potential for fuelwood and agroforestry. In: J.W. Turnbull (editor). Multipurpose Australian trees and shrubs: lesser-known species for fuelwood and agroforestry, pp. 81–313, ACIAR, Canberra.
van Went, J.L. (1970). The ultrastructure of the egg and central cell of *Petonia*. *Acta Bot. Neerl.* **19**: 313–322.
Yeung, E.C. and Meinke D.W. (1993). Embryogenesis in angiosperms: development of suspensor. *Pl. Cell* **5**: 1371–1381.

Nic Lughadha, E. (1998). Preferential outcrossing in *Gomidesia* (Myrtaceae) is maintained by a post-zygotic mechanism. In: S.J. Owens and P.J. Rudall (Editors). Reproductive Biology, pp. 363–379. Royal Botanic Gardens, Kew.

PREFERENTIAL OUTCROSSING IN *GOMIDESIA* (MYRTACEAE) IS MAINTAINED BY A POST-ZYGOTIC MECHANISM

EIMEAR NIC LUGHADHA

Herbarium, Royal Botanic Gardens, Kew, Richmond, Surrey, TW9 3AE, UK

Abstract

Field and laboratory studies on the breeding system of four species of *Gomidesia* were carried out in parallel with a taxonomic revision of the genus. Experimental pollinations indicate that most species show some degree of self-compatibility but that seed-set after selfing is reduced relative to that after outcrossing. Hand-pollinated flowers fixed up to 28 days after pollination were examined using fluorescence microscopy and differential interference contrast optics. Pollen tube growth, ovule penetration and early stages of embryo development are similar in self- and cross-pollinated flowers. Thus the mechanism maintaining preferential outcrossing in these species must be post-zygotic in operation. Late-acting self-incompatibility and inbreeding depression are discussed as alternative explanations for the phenomena observed in *Gomidesia*. The results of breeding system studies in other myrtaceous genera are surveyed and compared.

Introduction

Gomidesia is a genus of some forty species of shrubs and trees with a predominantly Brazilian distribution. It is one of a number of genera of the fleshy fruited neotropical Myrtaceae which have never been monographed since they were first described more than 150 years ago. In terms of species numbers Myrtaceae are often the most important woody element in habitats as diverse as the Atlantic forests of the Eastern Brazilian seaboard, the inland savannas or *cerrados*, two of Brazil's most threatened habitats, as well as in *campo rupestre* montane vegetation and in the high altitude relict forests in the *caatinga* domain of NE Brazil. In many of these areas the Myrtaceae are the single most speciose family, exceeding the Leguminosae and the Orchidaceae in their contribution to the biodiversity, yet the taxonomy and biology of this important group have been largely neglected. Among ecologists and taxonomists the neotropical Myrtaceae have a reputation as a 'difficult' group. The complex variation patterns which give rise to this notoriety have been attributed at different times to apomixis or to hybridisation but in fact the reproductive biology of the group is even more poorly understood than the systematics. A recent review of literature on the reproductive biology of the Myrtoideae yielded only fifteen neotropical species for which sufficient data were available to estimate the Index of Self-Incompatibility (Nic Lughadha and Proença, 1996). Eight of these records derive from a single study by Proença and Gibbs (1994) on Central Brazilian species. These authors followed self-pollen tubes in the style and ovary for up to three days after pollination, reported that selfed pistils were shed one week to one month after pollination, and postulated the existence of late-acting self-incompatibility in three species. Their work represents the most detailed study of 'self-incompatibility' in the Myrtoideae published to date but, as no material was fixed beyond 72 hours after pollination, we can only speculate as to the nature of the rejection mechanism in operation. The present paper outlines some of the results of a subsequent study of the reproductive biology of four taxa of *Gomidesia*. The project was designed to complement a taxonomic revision of the genus (Nic Lughadha, in prep.) and included surveys of flowering phenology and pollination biology which will be published elsewhere. This paper focusses on the evidence for the post-zygotic maintenance of preferential outcrossing in *Gomidesia* and elsewhere in the Myrtaceae and assesses the relative merits of inbreeding depression and late-acting self-incompatibility as explanations for the phenomena reported.

Materials and Methods

Populations of four taxa of *Gomidesia* were chosen to represent a broad spectrum of habitats and taxonomic groupings within the genus (Table 1). All species presented small (ca 1 cm diam.) flowers with reflexed sepals and petals and numerous (50+) stamens surrounding a single style. The ovary is inferior and bilocular except in *G. fenzliana* where it is trilocular. In all species the locules are biovulate. Natural fruiting success was estimated from flowers which were inconspicuously tagged with lengths of white cotton immediately after anthesis and then left for five to eight weeks to fall or set fruit. The breeding system was assessed by means of self and cross hand-pollinations, and emasculations to test for apomixis. Inflorescences in bud were tagged and bagged. Greaseproof paper bags were used when weather conditions permitted but in particularly wet and windy weather they were substituted by fine-mesh nylon bags which proved more resistant to tearing. Bags were fastened one or two internodes below the inflorescence using lengths of plastic-coated wire. Bagged inflorescences were monitored on a daily basis. When anthesis occurred open flowers were individually tagged with lengths of coloured thread. Unopened buds in the same inflorescence were removed only where this was necessary to permit unambiguous tagging of the open flowers. Open flowers were then hand-pollinated or emasculated before the bags were replaced.

TABLE 1. Location and habitat of *Gomidesia* populations included in this study.
BA = Bahia, CVRD = Companhia Vale do Rio Doce, DF = Distrito Federal, ES = Espírito Santo, IBGE = Instituto Brasileiro de Geografia e Estatística, JBB = Jardim Botânico de Brasília.

Species	Study Site	near	State	Location	Habitat	Habit	Study Period
G. cerqueiria Nied.	Reserva CVRD	Linhares	ES	19°12'S40°02'W	Atlantic Forest	Treelet	Jan - Mar 1993
G. fenzliana O.Berg	Catolés	Abaira	BA	13°15'S41°55'W	High altitude gallery forest	Tree	Jan - Mar 1992
G. lindeniana O.Berg	Catolés	Abaira	BA	13°14'S41°54'W	Campo Rupestre	Shrub	Feb - Apr 1992
G. lindeniana O.Berg	Reserva JBB	Brasília	DF	15°52'S47°50'W	Cerrado	Tree	Jan - Mar 1994
G. pubescens (DC.) Kausel	Reserva IBGE	Brasília	DF	15°57'S 47°52'W	Planalto gallery forest	Tree	Jan - Mar 1994

Hand pollinations were carried out by brushing dehisced anthers against the stigma. In the case of *G. cerqueiria* the apically dehiscent anthers were removed from the donor flower and opened on a watch glass using a scalpel blade which was then used to apply pollen to the stigma of the recipient flower. The time of each pollination and the pollen source(s) were recorded. As all species presented dry stigmas, receptivity could not easily be assessed under field conditions. Pollinations were carried out at least 30 minutes after anthesis and before midday i.e. several hours before the first flowers began to show signs of wilting. For self-pollinations the anthers were taken from the same flower. For cross-pollinations the anthers were taken from recently collected flowers from plants at least 50m (and sometimes up to 1km) away from the recipient plant. Emasculations were carried out using fine scissors immediately after anthesis and before anther dehiscence. Hand pollinated flowers were not emasculated.

Treated flowers were monitored on a daily basis and samples of hand-pollinated flowers were fixed in cold 70% FAA at intervals of 2-28 days after pollination in the case of selfs and 2-25 days after pollination in the case of crosses.

The younger fixed flowers were used to study pollen tube growth in the style and ovary using leuco-aniline blue stain and fluorescence microscopy (Martin, 1959). Some of the older material was sectioned by rotary microtome, either at 14µm embedded in paraffin wax, or at 5µm in historesin. Sections were stained with Delafield's haemotoxylin or alcian blue/safranin (Johansen, 1940). In addition, whole ovules/developing seeds were observed using clear-squash techniques (Herr, 1971; Stelly *et al.*, 1984) and differential interference contrast (DIC) optics (Herr, 1973).

The remaining hand-pollinated flowers and all the emasculated flowers were left *in situ* to determine fruit set. Daily observations on all trees continued for three to seven weeks after the onset of flowering in the population. Some five to eight weeks later the tagged inflorescences were collected and numbers of buds, persisting ovaries and developing fruits were recorded. Developing fruits were soaked in dilute detergent solution and dissected, and the dimensions of the ovules (and/or young seeds) within were recorded.

Descriptive indices

To aid in the interpretation of data and, in particular, to facilitate between-species comparisons, indices of pre-emergent reproductive success and self-incompatibility were calculated as follows:

Pre-emergent reproductive success (hereafter PERS) was estimated by measuring the fruit/flower (Fr/Fl) ratio and the seed/ovule (S/O) ratio in each species. PERS is the product of (Fr/Fl) and (S/O) and is the percentage of all ovules maturing into seeds. Wiens *et al.* (1987) demonstrated that among outcrossing species PERS averaged about 22%, whereas in inbreeding species the average PERS was approximately 90%. They interpreted PERS as a measure of selective forces acting on the embryo phenotype during development. In outcrossing, genetically heterozygous organisms a greater proportion of potential progeny is lost during the pre-emergent phase of the life-cycle than the post-emergent phase.

The index of self-incompatibility (calculated as % fruit set in selfed flowers / % fruit set in crossed flowers) reflects the relative fruiting success of selfed and crossed flowers. Fruit set results rather than seed counts were used as the basis of estimates of ISI following Bawa (1974) and Bullock (1985). Bawa (1974) and Ruiz and Arroyo (1978) considered species with an ISI of 0.2 or less to be self-incompatible while Bullock (1985) considered 0.25 the threshold value above which species were considered self-compatible.

Results

Natural fruit set

Ovary retention at the time of collection of flowers left to set fruit naturally ranged from 9% to 35%. Fruit dissections revealed that ovules within these ovaries varied considerably in size and shape and that those from a single ovary could usually be divided into two classes. Ovules which had mostly convex surfaces and a pale colour were generally termed developing seeds while darker structures with some concave surfaces were termed non-developing. Non-developing ovules often equalled developing seeds in length but were considerably smaller in total volume. All developing fruits of *G. cerqueiria* were found to contain only a single developing seed. More variation was observed in the other species: the modal number of developing seeds per retained ovary was 1 in *G. fenzliana* and *G. pubescens* and 2 in *G. lindeniana* in cerrado while in the campo rupestre population of *G. lindeniana* fruits with one, two and three developing seeds were equally common. Developing seeds tended to be evenly distributed across the locules. The mean number of developing seeds per ovary ranged from one to 2.3 (see Table 2). Resulting PERS values ranged from 3.5% to 19%.

TABLE 2. Results for natural fruit set and hand pollinations.
1. Pre-emergent reproductive success - see Methods.
2. Data collected @ 8 days after hand-pollination at which point daily monitoring of this population ceased.
3. Retention rates of ovaries after hand-pollination expressed as a percentage of the number pollinated. Differences between pairs of results presented in bold are statistically significant. Chi-square tests were performed on the raw data before conversion to percentages. Values for *G. fenzliana* @ collection include a second cohort not monitored daily but left to set fruit.
4. Index of self-incompatibility - see Methods.
5. SC = Self-compatible, SI = Self-incompatible - see Methods.
* = missing data.

		G. cerqueiria		*G. fenzliana*		*G. lindeniana*				*G. pubescens*	
						Catolés		Brasília			
% Natural fruit set		29		9		31		35		10	
Mean developing ovules/ seeds per fruit		1		2.3		2.2		2.2		1.5	
PERS[1]		7.2		3.5		17		19		3.9	
Crosses : Selfs		Cross	Self	Cross	Self	Cross	Self	Cross	Self	Cross	Self
persisting at	@ 10 days	24	29	**60**	**52**	**48**[2]	**78**[2]	**53**	**28**	**82**	**72**
intervals after	@ 20 days	22	29	39	35	*	*	**46**	**3**	**72**	**56**
hand-pollination[3]	@ collection	13	3	**46**	**20**	**39**	**5**	**30**	**0**	**39**	**5.5**
ISI[4]		0.23		0.43		0.13		~ 0		0.14	
Breeding system[5]		SC/SI		SC		SI		strict SI		SI	

Breeding system

Fruit and seed-set. In all species studied substantial loss of hand-pollinated flowers (both self- and cross-pollinated) occurred (2-)3-6(-10) days after pollination (Fig. 1). Typically this first intense period of fall-off was followed by a plateau period of several days during which fall off was low or negligible. After the plateau period, fall-off usually resumed and tended to be more acute in self-pollinated flowers than in cross-pollinated flowers (*G. pubescens*,

Fig. 1). The divergence in fall off rates of self- and cross-pollinated flowers was particularly marked in *G. lindeniana* (Fig. 1) where loss of cross-pollinated flowers remained negligible after the plateau period when fall-off of self-pollinated flowers had resumed. By four weeks after pollination the situation had generally stabilised again with cross-pollinated flowers being present in greater numbers than self-pollinated flowers. Rates of loss of hand-pollinated flowers varied widely between species (Table 2) but, within species, differences between losses of self- and cross-pollinated flowers were statistically significant in all cases except that of *G. cerqueiria* where sample sizes were small and differences were nearly statistically significant. The resulting ISI values ranged from 0.43 to close to zero (Table 2). Thus, although fruit set after cross-pollination was always higher than fruit set after self-pollination, one or two of the populations studied would rate as self-compatible depending on the criteria applied. No fruit set was observed from emasculated flowers.

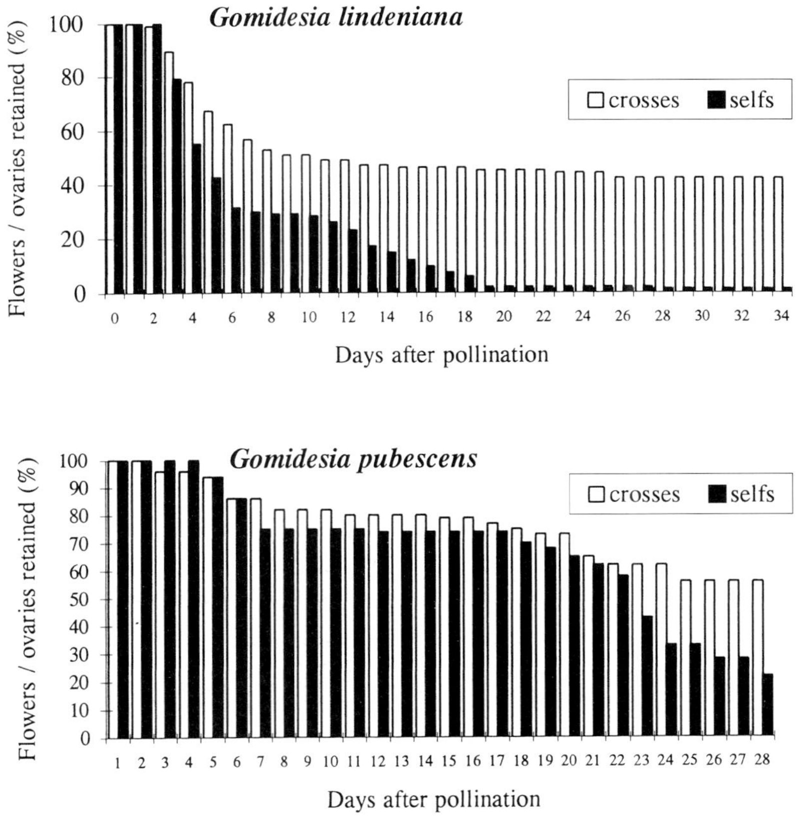

FIG. 1. Histograms charting loss of hand-pollinated flowers with time in *G. lindeniana* (above) and *G. pubescens* (below).

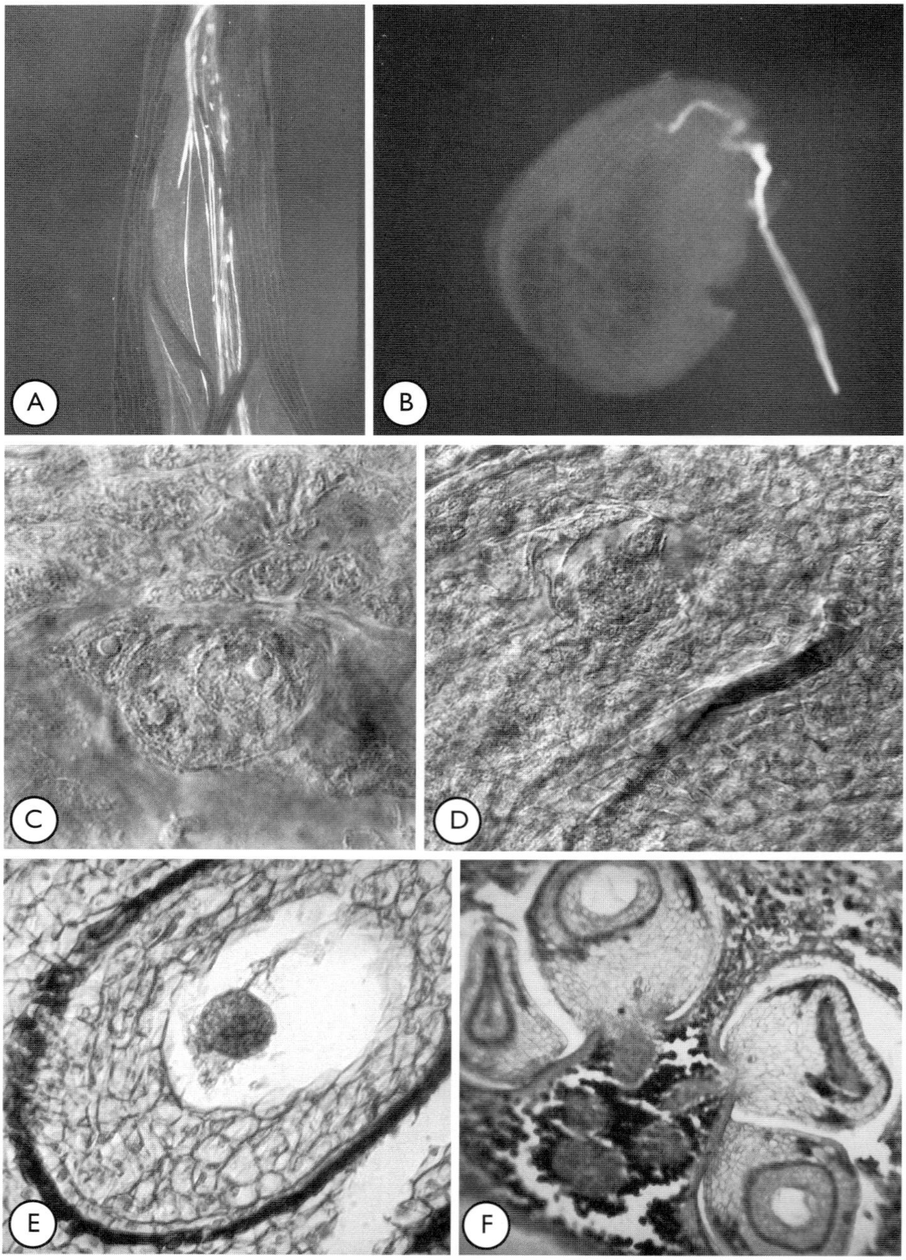

FIG. 2. Pollen tube growth, ovule penetration and embryo development in *Gomidesia*. A, B. *G. lindeniana*; C-F. *G. fenzliana*. A. Self pollen tubes in style fixed 48 hours after anthesis. B. Self pollen tube penetrating ovule fixed 72 hours after anthesis. C. Proembryo at 13 days after cross-pollination. D. Proembryo at 18 days after cross-pollination. E, F. T.S. ovary at 26 days after self-pollination showing developing embryo and endosperm (in E) and occluded and normally developing ovules/seeds (in F).

Pollen tube growth and ovule penetration. Fluorescence microscopy revealed multiple pollen tubes in the styles of hand-pollinated flowers at 2-3(-4) days after cross- or self-pollination (Fig. 2A). Ovule penetrations could be observed from one to two days after pollination (Fig. 2B). The results were not always suitable for statistical analysis since the thickness of the integuments made it difficult to follow the course of the pollen tubes with certainty, but no consistent differences were noted between self- and cross-pollinated material with respect to rate of pollen tube growth or numbers of ovule penetrations.

Proembryo and endosperm development. In *G. fenzliana* DIC study of 10- and 13- day fixations of selfed and crossed ovaries revealed a globular pro-embryo in most ovules/developing seeds for which a clear image could be obtained (Fig 2C). Due to the limitations of the technique, failure to discover an embryo in a particular ovule/seed could not be interpreted as proof of its absence as it could equally be ascribed to the faint nature of the images being studied. It can, however, be stated with certainty that in at least some selfed and crossed ovaries of this age class all six seeds contained developing embryos. DIC images of older ovules/seeds were poorer still. In 15-day fixations proembryos were detected in two (or occasionally three) out of six ovules/seeds and in 18-day samples most ovaries had just two developing embryos. The embryos observed at this point were still globular, generally more than eight-celled and had a clearly defined suspensor (Fig. 2D). In some ovules in which proembryos were not discovered, highly refractive areas were encountered within the embryo sac. Their shape, size and position relative to the synergids suggested that they might be interpreted as abortive embryos. In *G. cerqueiria* DIC microscopy revealed the presence of a proembryo in a single ovule/seed of most crossed ovaries at 15 days (6/8). Proembryos were detected in only 2/10 selfed ovaries. Possible degenerating embryos were detected in the ovules/seeds of a further two selfed ovaries. Similar results were obtained for 20 day fixations; only a single embryo was encountered in the six selfed ovaries screened while four of the six crossed ovaries screened contained at least one embryo and one had two seeds containing developing embryos. Histological sections of self- and cross-pollinated flowers of *G. lindeniana* fixed at nine days after pollination revealed the presence of pro-embryos and free nuclear endosperm. Sections of 25-27 day old selfed ovaries of *G. fenzliana* revealed the deposition of tannins in the integuments and in the synergids. Developing embryos and endosperm (Fig. 2E) were clearly visible in two (or occasionally three) seeds per ovary. In the other ovules the embryo sac appeared to have collapsed and there was sometimes heavy deposition of tannin (Fig. 2F).

Discussion

Natural fruiting success

The seed/ovule ratios (PERS values) obtained for *Gomidesia* are all below the average for outcrossing species (22%, Wiens *et al.*, 1987) and for woody plants in general (33 %, Wiens *et al.*, 1989). The range of values reported here for four species (3.5-19 %) is comparable to the range of estimates for eight Central Brazilian species of Myrtaceae (0.3 -12.7 %) by Proença and Gibbs (1994). Low seed/ovule fecundity levels appear to be characteristic of the Myrtaceae in general, independent of breeding system. The highly self-incompatible *Thryptomene calycina* produces 4-7 % viable seed per ovule, while similar fecundity levels are typical of the partially self-compatible *Callistemon citrinus* (Beardsell *et al.*, 1993a). In self-incompatible *Melaleuca* species only 1-2% of ovules form seeds (Barlow & Forrester, 1984). Still lower fecundity levels are characteristic of *Syzygium aromaticum* where one-seeded fruits are produced from ovaries containing c. 100 ovules in about 25% of the flowers borne by the plant (Pool and Bermawie, 1987) giving a PERS of 0.25. Beardsell *et al.* (1993a) considered that the similarity in fecundity of various members of the Myrtaceae with differing breeding systems suggests that similar limitations to seed set occur in diverse members of this family, despite differences in reproductive behaviour. Nic Lughadha and Proença (1996) discussed the apparent superfluity of ovules in fruits of the Myrtoideae. Possible explanations include historical factors, insurance against the wastage of ovules inherent in a post-zygotic SI system or the provision of scope for female choice. Clearly an improved understanding of the basis of this phenomenon will require more detailed studies of the non-developing ovules in a range of species to establish whether they contain aborted embryos (at similar or different stages of development) or are unfertilised. In the meantime extreme caution should be exercised in inferring predominant outbreeding on the basis of low PERS values alone.

Breeding system

Although most of the trees studied showed some degree of self-compatibility, the range of ISI values obtained (0-0.43) indicates the operation of some mechanism to maintain preferential outcrossing in these populations. That this mechanism is post-zygotic has been established without doubt. However, the nature of the mechanism in operation remains open to discussion. Phenomena of this kind, loosely referred to as late-acting self-incompatibility (hereafter LSI), have been reviewed in detail by Seavey and Bawa (1986) and Sage *et al.* (1994). Seavey and Bawa (1986) classified LSI systems according to the site of rejection. The *Gomidesia* species studied here would be assigned to their third category 'post-zygotic rejection' of which only eleven instances (in seven families) were known to the authors at the time of their review. In the intervening decade numerous additional examples have been published of species which exhibit phenomena of this type, e.g. *Vellozia squamata*, (Oliveira *et al.*, 1991); *Chorisia chodatii, C. speciosa, Tabebuia caraiba* and *T. ochracea* (Gibbs and Bianchi, 1993) and other examples discussed below.

Distinguishing between true LSI and inbreeding depression

Seavey and Bawa (1986) discussed the importance and the difficulties of distinguishing between active self-rejection (i.e. the operation of an LSI system) and the effects of inbreeding depression. Four criteria were listed which might help to determine whether zygote or embryo abortion is the result of delayed action SI or of inbreeding depression determined by the genotype of the zygote.

(1) The timing of abortion: uniform embryo failures at a particular developmental stage would suggest LSI whereas embryos blocked at a variety of stages could more easily be interpreted as the result of inbreeding depression, with different recessive lethals taking effect at different times during embryo or seed development.

(2) Between-plant variation in selfing success: where genetic load is responsible for low seed set after selfing one would expect plants with fewer detrimental recessives to approach normal seed-production. Uniformly low seed-set after selfing may be evidence of active self-rejection.

(3) Dependence of abortion on the paternal rather than the progeny genotype: where SI is controlled by one or a few specific loci, cross-compatible and cross-incompatible groups should be detectable within the population.

(4) Embryo rescue: embryos homozygous for lethal genes would not survive in tissue culture while those aborting because of their interaction with their female parent might be rescued by such means.

Klekowski (1988) challenged the first and third of these criteria, arguing that the mechanisms of self-sterility are insufficiently understood to permit predictions as to the timing

of self-rejection and that inbreeding depression is consistent with the presence or absence of clear segregation ratios in progeny after selfing (because the lethal equivalent concept is based on the homozygosities of many genes with small detrimental effects). Klekowski added a fifth criterion to the list:
(5) Induced mutations in ovarian self-incompatibility systems (or indeed in any SI system) should favour a shift from self-sterility to self-fertility whereas the opposite should be the case where inbreeding depression is involved.

Case studies of post-zygotic rejection
Sage *et al.* (1994) reviewed recent progress in the study and interpretation of post-zygotic rejection. Several researchers have attempted to apply some of the criteria listed above to distinguish between post-zygotic SI and inbreeding depression in species with ovarian self-infertility. The occurrence of self-rejection at a continuum of embryonic stages was cited as evidence to support inbreeding depression as an explanation for decreased fertility after self-pollination in *Epilobium angustifolium* (Wiens *et al.* 1987), *Campsis radicans* (Bertin *et al.* 1989), and *Crinum erubescens* (Manasse and Pinney 1991). In contrast, post-zygotic failure in *Liquidambar styraciflua* occurred in a fairly restricted developmental phase when embryos were 4-6 celled (Schmitt and Perry 1964). Although this could be construed as evidence in support of the operation of LSI in this species, Sage *et al.* (1994) commented that the synchronized embryo failure could equally be due to an inbreeding depression system that mimics ovarian LSI by virtue of the involvement of a large enough number of early-acting embryonic lethals. Gribel (1995) favoured genetic load as an explanation for self-sterility in *Ceiba pentandra* and *Pseudobombax munguba* because of the observed diversity of developmental stages shown by the ovules in self-pollinated flowers just before pistil abscission and the variation in levels of self-sterility between trees in the study populations. Weller and Ornduff (1991) exploited the distylous breeding system of *Amsinckia grandiflora* to demonstrate unambiguously that self-sterility in this species is attributable to inbreeding depression.

Can inbreeding depression mimic LSI?
With so many case studies tending to support inbreeding depression rather than SI as the explanation for post-zygotic self-rejection one might wonder why the problem is not generally considered to have been resolved. The objection most often offered against such an explanation for species showing high levels of self-sterility is the 'unrealistically high' number of genetic lethals required to produce near-uniform abortion of young embryos (Seavey and Bawa, 1986; Waser and Price, 1991; Sage *et al.* 1994). The level of self-fertility (S) corresponding to a given number of embyronic lethals (N) is given by $S = (0.75)^N$. Thus at least ten embryonic lethals (equivalent to 94% self-infertility) would be necessary to produce a situation where inbreeding depression caused embryo death with sufficient frequency to mimic SI. It has been argued that these levels of embryonic lethals are unrealistic, both from an empirical point of view and also on theoretical grounds.

Empirical evidence. The empirical argument is based simply on comparison with published estimates of mean numbers of lethal equivalents in angiosperms. These are rather few and range in value from less than 1 to 12 (Levin, 1984; Klekowski, 1988). This would seem to suggest that only in species bearing genetic loads at the higher limits of the known range could preferential outcrossing be maintained by inbreeding depression. However, Sage *et al.* (1994) outlined several additional factors that should be taken into account when assessing genetic loads and their potential role in the maintenance of preferential outcrossing. Firstly, numbers of genetic lethals have been reported for a rather small and unrepresentative sample of the angiosperms but results for the few woody species studied show some of the highest estimates of genetic load to date (e.g. 9.6 in *Vaccinium corymbosum* (Krebs and Hancock 1991), 13.8 in *Pseudobombax munguba* (Gribel, 1995), 11-12 in *Liquidambar styraciflua* (Schmitt and Perry 1964). Large perennial species can be expected *a priori* to have much higher mutation rates per generation than small annual species because somatic mutations can accumulate during many mitotic cell divisions between meiotic events. On the basis of preliminary data on chlorophyll deficiencies Klekowski and Godfrey (1989) suggested that genomic lethal mutation rates may be more than an order of magnitude higher in trees than in annual plants. Secondly, it seems likely that numbers of lethal equivalents are underestimated in species that never mature more than a certain (small) fraction of their ovules. In such species the seed to ovule ratio may act as a soft-selection sieve resulting in artificially low estimates of numbers of genetic lethals especially where the load levels are genuinely rather low (Klekowski, 1988). Finally, as Sage *et al.* (1994) emphasized, an embryo which has the capability to survive to maturity may fail to do so simply because the ovary which contains it is aborted, perhaps because of the low number of other viable embryos present. Thus the ratio of seeds resulting from selfing to the total number of ovules in all selfed flowers underestimates the number of selfed-embryos free from developmental lethals. In summary, the numbers of lethal recessives carried by individuals of woody species may be higher than previously estimated AND the genetic load necessary for a population to simulate SI may be lower than has been generally supposed. Together these conclusions may be interpreted as support for inbreeding depression as a plausible explanation for high levels of outcrossing in woody species with low PERS values. However, it should be noted that, since estimates of genetic load are based on relative survivorship of selfs and crosses, the argument becomes somewhat circular as in many cases the possibility that low self-survivorship is due to active self-rejection rather than to homozygosity for recessive lethals cannot be discounted.

Theoretical objections. The theoretical grounds on which the numbers of required lethals are deemed 'unrealistically high' have been mentioned often but are rarely discussed thoroughly. Seavey and Bawa (1986) comment that theoretical models of genetic load do not appear to anticipate genetic loads of a magnitude great enough to mimic SI, citing Dobzhansky (1970), Wallace (1970) and Wright (1977). Waser and Price (1991) calculated that about 8000 loci producing lethals would be required to explain a situation in which only 1% of selfed offspring survive and considered this number unrealistic, citing Lewontin (1974). Lande and Schemske (1985) concluded that even a moderate degree of selfing in a population would greatly reduce the level of inbreeding depression by purging recessive or nearly recessive lethal mutations. However, recent studies on *Arabidopsis thaliana,* which has the smallest known plant genome, have generated estimates of about 5000 loci producing recessive lethal mutations (Jurgens *et al.*, 1991). In this context, the Waser and Price estimate of 8000 does not seem extraordinary. Furthermore, these latter authors' calculations were based on an assumed mutation rate of 10^{-6} per locus per generation to lethal alleles whereas recent studies on loci producing recessive chlorophyll deficiencies suggest that mutation rates are several orders of magnitude greater (Klekowski and Godfrey, 1989; Willis, 1992). Extrapolating from the chlorophyll deficiency data to lethals in general, Lande *et al.* (1994) suggested that the long-lived red mangrove may have a genomic mutation rate to recessive lethals as high as 0.2 per generation. These authors modeled the evolutionary dynamics of recessive or slightly dominant lethal mutations in partially self-fertilizing plants and concluded that with genomic mutation rates to (nearly) recessive lethals of the order of 0.1 to 1.0 per generation, completely outcrossing populations can maintain large average numbers of lethals and very high early inbreeding depression. In such situations, where the mean number of lethals per mature plant exceeds 10, selective interference among loci creates a sharp discontinuity in the mean number of lethals maintained as a function of selfing rate. Purging of lethals is negligible unless the selfing rate approaches or exceeds a threshold level at which there is a sharp drop in the number of lethals maintained. Thus, Lande *et al.* (1994) have established that with high mutation rates to recessive lethals plant species may display very high inbreeding depression (sufficient to mimic LSI) despite intermediate selfing rates.

Since neither the empirical nor the theoretical objections to inbreeding depression as an explanation for post-zygotic rejection of selfs can be sustained in general terms, inbreeding depression must be considered along with active LSI as a potential explanation for observed post-zygotic preferential outcrossing phenomena. The relative explanatory power of these alternatives should be assessed in relation to the data obtained for individual species.

Inbreeding depression or LSI in Gomidesia?

Returning to the particular case of *Gomidesia* we can attempt to apply some of the general considerations discussed by Seavey and Bawa (1986), Klekowski (1988) and Sage *et al.* (1994). Of the criteria outlined only two are relevant to the data available for *Gomidesia*; nonetheless the exercise is not completely uninformative.

Is abortion of selfed embryos synchronous? The data on ovary retention suggest that abortion is not synchronous but that it is particularly common at certain intervals after pollination and absent or negligible at others. Typically, a period when fall-off is high immediately after pollination, is followed by a plateau period where fall-off is negligible, before the onset of another phase of gradual loss of selfed ovaries. Such a pattern could be considered consistent with genetic load hypotheses, with embryo abortion occurring at various developmental stages. However, the histological data show the error involved in interpreting fall-off patterns as a direct reflection of embryo abortion patterns. In *G. cerqueiria*, for instance, DIC microscopy of flowers sampled at the middle (day 15) and towards the end (day 20) of the plateau period (during which ovary loss was negligible), revealed preferential abortion of embryos in selfed flowers during this period. Considerable embryo abortion also took place in *G. fenzliana* in developing seeds of a similar age, during a period when ovary fall-off was very gradual. Surviving embryos at 18 days after pollination were at the globular stage of development, generally more than eight-celled with a clearly defined suspensor. Preferential shedding of selfed ovaries occurred at some time after this date and whether it was due to the abortion observed cannot now be established.

Does self-sterility vary from tree to tree within a population? Unfortunately the experimental pollinations were not designed to test this possibility. There is, however, some circumstantial evidence to suggest that self-sterility varies between trees within a population (particularly in *G. pubescens*).

The data discussed so far indicate that self-sterility is incomplete and variable within and between populations of *Gomidesia* species and also that the timing of embryo abortion and rejection of selfed ovaries is variable. These observations provide some support for the genetic load model of post-zygotic discrimination rather than for active self-rejection. The situation is further complicated by the fact that embryo abortion must also occur in ovaries which ultimately mature into fruit, since in all cases studied the average number of ovules fertilised per ovary was greater than the average number of seeds per mature fruit. The basis for embryo abortion in successful fruit has not been investigated in the present study but it is of interest to explore whether such abortion can be accounted for by invoking the same factors used to explain the post-zygotic maintenance of preferential outcrossing.

A role for competition. Competition among developing seeds seems, *a priori*, to represent the most parsimonious explanation for embryo abortion in outcrossed ovaries. Where developing embryos in the same ovary are competing for resources (as is presumed to be the case in *Gomidesia*) a positive feedback situation may develop with more vigorous embryos representing a more effective sink for resources and thus widening the gap between them and the less vigorous embryos which would ultimately abort. Broadening this resource allocation model to consider competition between ovaries it is not difficult to envisage a situation whereby the abortion of just one or two embryos due to homozygosity for recessive lethals could render a selfed ovary a less effective sink for resources and result in the ultimate abortion of that ovary even though it may have contained potentially viable embryos. The even distribution of developing embryos among the locules of an ovary is consistent with competition between developing seeds, assuming that competition is particularly intense between those occupying the same locule. However, this even distribution is also consistent with explanations which depend on maternal choice, e.g. the plant may 'allow' a maximum of one embryo per locule and cause the abortion of all but the most vigorous of the developing seeds in each locule. The fact that the non-developing ovules in an individual ovary are usually similar in size suggests that they fail more or less simultaneously rather than being gradually outcompeted one by one

Thus 'female choice' or maternal control represents a non-exclusive alternative to competition as a determinant of the fates of ovules/developing seeds. Just as LSI is difficult to distinguish from inbreeding depression so the operation of female choice within outcrossed ovaries may be difficult to distinguish from the unmodified effects of competition. However, one of the species studied here, *G. cerqueiria*, seems to represent a strong candidate for the operation of some degree of maternal control. In this species all mature fruits examined contained only a single seed, although in many ovaries two or three ovules were penetrated by pollen tubes and, in at least some cases, embryo development was initiated in more than one ovule. The regularity with which the number of maturing seeds is reduced by abortion to one per fruit is suggestive of maternal control. Although simple competition between developing seeds for limited resources could also account for reductions in the number of seeds maturing per ovary, if this were the case one would expect occasional two-seeded fruits where two embryos of comparable vigour had acted as equally effective sinks for resources, each failing to gain a competitive edge over the other.

It seems that the results presented here for *Gomidesia* are best explained in terms of inbreeding depression operating in tandem with or followed by inter- and intra-ovarian competition and, perhaps, maternal selection. The data are consistent with a situation in which some embryos fail due to genetically based lethality and others lose out in the competition between the remaining potentially viable embryos. Once some embryos have gained the competitive edge the maternal plant may induce early abortion in the other developing seeds, thus producing cohorts of similarly sized abortive seeds. As the number and vigour of developing seeds is likely to influence fruit survivorship, a fairly low level of homozygosity for recessive lethals in selfed ovaries may be amplified into a marked preferential outcrossing effect. There is nothing in the data to suggest the operation of a late-acting self-incompatibility system such as was suggested by Proença and Gibbs (1994) for three of the eight Central Brazilian Myrtoideae species studied by them.

Preferential outcrossing elsewhere in the Myrtaceae

Given the limited nature of the data available on SI in the Myrtoideae it seems appropriate to broaden the discussion to consider and compare reports of SI elsewhere in the Myrtaceae. This can be justified on the grounds that repeated sampling within a family has usually revealed identical mechanisms controlling SI (Charlesworth 1985). Beardsell *et al.* (1993b) considered it likely that the same mechanism of SI should operate throughout the Myrtaceae. However, among the several myrtaceous breeding system studies cited by these authors the variation in the timing of operation of reproductive barriers after self-pollination is striking. Self pollen tubes were arrested at the base of the style in *Melaleuca hypericifolia* and *M. thymifolia* and in the ovarian cavity in *M. capitata* (Barlow and Forrester, 1984). In *Thryptomene calycina* self-pollen tubes were arrested on the placenta or at the micropyle (Beardsell *et al.*, 1993a). Even within a genus the timing of discrimination against self pollen may vary considerably. In *Eucalyptus woodwardii* reduced seed crops following self-pollination were attributed to reduced ovule penetration by self pollen tubes (Sedgley and Smith, 1989). In contrast post-zygotic SI has been reported in *E. regnans* (Griffin *et al.*, 1984, 1987), *E. spathulata, E. cladocalyx* and *E. leptophylla* (Ellis and Sedgley, 1992). Potts and Savva (1988) also concluded that some late-acting mechanism of SI was likely to be in operation in *E. morrisbyi*.

The ovarian SI reported in *Melaleuca* and *Thryptomene* has been tentatively interpreted as a gametophytic SI system (Barlow and Forrester, 1984; Beardsell, 1991) although no genetic analyses were undertaken. Another apparent candidate species for SI is *Melaleuca alternifolia* for which an extremely high outcrossing rate of 93% has been reported (Butcher *et al.*, 1992). The genetic bases for post-zygotic discrimination are not so easily fathomed however. The species most thoroughly studied to date is *Eucalyptus regnans* F. Muell., a large mass-flowering forest tree with hermaphrodite flowers (Griffin and Cotterill, 1988; Griffin and Hand, 1979; Griffin *et al.*, 1987; Sedgley *et al.*, 1989). Isozyme analysis of mixed-pollinated material showed this species to be a preferential outcrosser (Griffin *et al.*, 1987) yet studies employing bright field, fluorescence, phase contrast, Nomarski (DIC) and scanning electron microscopy found no differences in rates of ovule penetration or in development of embryos between self- and cross-pollinated material up to 16 weeks after pollination (Sedgley *et al.*, 1989). At this time embryos with two to sixteen cells were present in some developing seeds but some undivided zygotes were also observed. The timing of abortion of selfed seeds could not be established in this study but the authors considered it unlikely to occur before the eight-celled stage. There was no evidence for a self-incompatibility reaction, rather the evidence suggests that the control mechanisms for the preferential outcrossing observed in this species are dependent upon both embryo genotype and maternal resource allocation (Griffin *et al.*, 1987). Individual trees showed a range of self-fertility after self-pollination and there was a reduction in mean seed yield relative to cross pollination consistent with the expression of recessive deleterious alleles in the developing inbred embryos. However, even in outcrossed capsules, the seed:ovule ratio was only about 20% (Griffin *et al.*, 1987), and as this was much lower than the numbers of penetrated and fertilised ovules at 16 weeks it was inferred that considerable selection occurs during the later stages of seed development (Sedgley *et al.*, 1989). Thus a combination of inbreeding depression and maternal choice could be sufficient to explain the results obtained for *E. regnans*. Inbreeding depression had previously been demonstrated for growth and survival of seedlings of this species (Eldridge and Griffin, 1983).

Three species of *Eucalyptus* Section *Bisectaria, E. cladocalyx, E. leptophylla* and *E. spathulata* have also been studied in some detail and found to be highly self-incompatible (Ellis and Sedgley, 1992). The preferential outcrossing observed in these species was found to be maintained by a post-zygotic mechanism operating in *E. cladocalyx* and *E. leptophylla* while the existence of mixed pre-zygotic and post-zygotic mechanisms was inferred in *E. spathulata*. Microscopic studies of embryo development were not undertaken for these species so the timing of operation of the post-zygotic mechanism cannot be ascertained from the data provided. Although the authors refer to the post-zygotic mechanism in operation as SI, the data available are also consistent with explanations based on inbreeding depression.

Other mating system studies of *Eucalyptus* have concentrated on estimating outcrossing levels in populations without investigating the mechanisms by which these levels are maintained. In general, such studies have indicated outcrossing with a significant amount of selfing. Outcrossing estimates for mass-flowering species of *Eucalyptus* tend to fall within the range 0.70-0.86 (Moran and Bell, 1983; James and Kennington, 1993) and thus are generally lower than might be expected in species with an active SI system but consistent with levels of inbreeding depression which could be produced by several lethal alleles. Barrett and Eckert (1990) commented on the uniformity of outcrossing rates within and between species in *Eucalyptus* and suggested that this might reflect a lack of genetic variation within *Eucalyptus* for floral mechanisms that would increase outcrossing rates. They considered that this phylogenetic constraint could have arisen during the evolution of this group's specialized floral morphology. James and Kennington (1993) interpreted their data for *Eucalyptus camaldulensis* in terms of selection against recessive homozygotes and maternal control of resource allocation. In their model, post-zygotic seed selection depends on each plant carrying a large number of seed-aborting recessive lethal factors in heterozygous condition. The intensity of selection operating against any particular genotype will be determined by the genotype of its neighbours and by the maternally controlled availability of resources. They demonstrated the use of this model in the successful interpretation of much of the published mating system data for *Eucalyptus*, including data for some species and situations previously considered anomalous.

Curiously James and Kennington (1993) did not discuss the results obtained by Potts and Savva (1988). These latter authors reported zero or near zero self seed set in most individuals of *Eucalyptus morrisbyi*. Though such high levels of self-sterility might be interpreted as evidence to support the existence of an active self-rejection mechanism in this species, Ellis and Sedgley (1992) provide an equally convincing alternative explanation: *E. morrisbyi* is a rare Tasmanian endemic, very restricted in range and population size, and the high degree of self-incompatibility may be due to an accumulation of deleterious alleles in the population.

Conclusions

Although ovarian SI has been postulated in some Myrtaceae, there is no evidence to indicate the existence of a post-zygotic self-incompatibility system in any species of Myrtaceae studied to date. A number of species, including the members of the genus *Gomidesia* treated here and many (most?) species of *Eucalyptus* exhibit preferential outcrossing which is maintained by some post-zygotic mechanism. On the basis of the data available to date, inbreeding depression seems to offer the most satisfactory explanation for this phenomenon. Estimates of genetic load for some of these species could represent a first step towards testing this hypothesis.

Acknowledgements

For the support and facilities which made this study possible I am indebted to my Brazilian counterparts during three field seasons Ana Maria Giulietti, Inês Machline Silva and Carolyn Proença; to the staff and students of the Departamento de Botânica, Instituto de Biociências, Universidade de São Paulo; to Wilson Ganev of Catolés; to Renato Jesus Moraes, Alvaro Garcia and Gilson Farias of the Companhia Vale do Rio Doce at Linhares; to Iracema Gonzales, Maria Aparecida da Silva and the technical staff at the Reserva Ecológica of the Instituto Brasileira de Geografia e Estatística, and to Alba Resende, Glaucia Jordão Zerbini and the herbarium staff of the Jardim Botânico in Brasília. My warmest thanks also to Graziela Barroso, Peter Gibbs, Ray Harley, Thomas Læssøe, Renato de Mello-Silva, Ariane Luna Peixoto, Helia Piedade, Marli Pires Morim de Lima, Rogério Ribeiro de Oliveira, Paula Rudall, Paulo Sano, Genise Somner and Brian Stannard for their company, advice and encouragement. This study represents part of a doctoral project funded by the Royal Botanic Gardens, Kew.

References

Barlow, B.A. and Forrester, J. (1984). Pollen tube interactions in *Melaleuca*. In: E.G. Williams and R.B. Knox (editors). Pollination '84, pp 154–160, School of Botany, University of Melbourne, Parkville, Victoria.

Barrett, S.C.H. and Eckert, C.G. (1990). Variation and evolution of mating systems in seed plants. In: S. Kawano: (editor) Biological approaches and evolutionary trends in plants, pp. 229–254, Academic Press, London

Bawa, K.S. (1974). Breeding systems of tree species of a lowland tropical community. *Evolution* **28**: 85–92.

Beardsell, D.V. (1991) The reproductive biology of *Thryptomene calycina* (Lindl.) Stapf. PhD. dissertation, University of Melbourne, Australia.

Beardsell, D.V., Knox, R.B. and Williams, E.G. (1993a). Breeding system and reproductive success of *Thryptomene calycina* (Myrtaceae). *Austral. J. of Bot.* **41**: 333–353

Beardsell, D.V., O'Brien, S.P., Williams, E.G., Knox, R.B. and Calder, D.M. (1993b). Reproductive biology of Australian Myrtaceae. *Austral. J. Bot.* **41**: 511–26.

Bertin, R.I., Barnes, C. and Guttman, S.I. (1989). Self-sterility and cryptic self-fertility in *Campsis radicans* (Bignoniaceae). *Bot. Gaz.* **150**: 397–403.

Bullock, S.H. (1985). Breeding systems in the flora of a tropical deciduous forest. *Biotropica* **17**: 287–301.

Butcher, P.A., Bell, J.C. and Moran, G.F. (1992). Patterns of genetic diversity and nature of the breeding system in *Melaleuca alternifolia* (Myrtaceae). *Austral. J. of Bot.* **40**: 365–375.

Charlesworth (1985). Distribution of dioecy and self-incompatibility in angiosperms. In: J.J. Greenwood and M.Slatkin (editors) Evolution - essays in honour of John Maynard Smith, pp 237–268 Cambridge University Press, Cambridge, U.K.

Dobzhansky, T. (1970). Genetics of the evolutionary process. Columbia University Press, New York.

Ellis, M.F. and Sedgley, M. (1992). Floral morphology and breeding system of three species of *Eucalyptus*, section *Bisectaria* (Myrtaceae). *Austral. J. of Bot.* **40**: 249–262.

Eldridge, K.G. and Griffin, A.R. (1983). Selfing effects in *Eucalyptus regnans*. *Silvae Genet.* **32**: 216–221.

Gibbs, P.E. and Bianchi, M. (1993). Post-pollination events in species of *Chorisia* (Bombacaceae) and *Tabebuia* (Bignoniaceae) with late-acting self-incompatibility. *Bot. Acta* **106**: 64–71.

Gribel, R. (1995). Reproductive biology of two Bombacaceous trees in the Brazilian Central Amazon. Unpublished PhD. thesis, University of St. Andrews, Scotland.

Griffin, A.R. and Cotterill, P.P. (1988). Genetic variation in growth of outcrossed, selfed and open-pollinated progenies of *Eucalyptus regnans* and some implications for breeding strategy. *Silvae Genet.* **37**: 124–31.

Griffin, A.R., and Hand, F.C. (1979). Post-anthesis development of the flowers of *Eucalyptus regnans* F. Muell. and the timing of artificial pollination. *Austral. Forest. Res.* **9**: 9–15.

Griffin, A.R., Moran, G.F. and Fripp, Y.J. (1984). Preferential outcrossing in *Eucalyptus* - a post-zygotic mechanism. In: E.G.Williams and R.B. Knox (editors) Pollination '84, pp. 145, School of Botany, University of Melbourne, Parkville, Victoria.

Griffin, A.R., Moran, G.F. and Fripp, Y.J. (1987). Preferential outcrossing in *Eucalyptus regnans* F.Muell. *Austral. J. Bot.* **35**: 465–475.

Herr, J.M. (1971). A new clearing-squash technique for the study of ovule development in angiosperms. *Amer. J. Bot.* **58**: 785–790.

Herr, J.M. (1973). The use of Nomarski interference microscopy for the study of the structural features in cleared ovules. *Acta Bot. Indica* **1**: 335–340.

James, S.H. and Kennington, W.J. (1993). Selection against homozygotes and resource allocation in the mating system of *Eucalyptus camaldulensis* Dehnh. *Austral. J. Bot.* **41**: 381–391.

Johansen, D.A. (1940). Plant microtechnique, McGraw Hill, New York.

Jurgens, G., Mayer, U., Ruiz, R.A.T., Berleth, T. and Misera, S., (1991). Genetic analysis of pattern formation in the *Arabidopsis* embryo. *Development*, **suppl. 1.** (1991): 27–38.

Klekowski, E. J., Jr. (1988). Mutation developmental selection and plant evolution. Columbia University Press, New York.

Klekowski, E.J., Jr. and Godfrey, P.J. (1989). Ageing and mutation in plants. *Nature* **340**: 431–433.

Krebs, S.L. and Hancock, J.F. (1991). Embryonic genetic load in the highbush blueberry, *Vaccinium corymbosum* (Ericaceae). *Amer. J. Bot.* **78**: 1427–1437.

Lande, R. and Schemske, D.W. (1985). The evolution of self-fertilisation and inbreeding depression in plants. I. Genetic models. *Evolution* **39**: 24–40.

Lande, R., Schemske, D.W. and Schultz, S.T. (1994). High inbreeding depression, selective interference among loci and the threshold selfing rate for purging recessive lethal mutations. *Evolution* **48**: 965–978.

Levin, D.A. (1984). Inbreeding depression and proximity dependent crossing success in *Phlox drummondii*. *Evolution* **38**: 116–127.

Lewontin, R.C. (1974). The evolutionary basis of genetic change. Columbia University Press, New York.

Manasse, R.S. and Pinney, K. (1991). Limits to reproductive success in a partially self-compatible herb: fecundity depression at serial life-cycle stages. *Evolution* **45**: 712–720.

Martin, F.W. (1959). Staining and observing pollen tubes in the style by means of fluorescence. *Stain Technology* **34**: 125–128.

Moran, G.F. and Bell, J.C. (1983). *Eucalyptus*. In: S.D. Tanksley and T.J. Orton (editors). Isozymes in plant genetics and breeding, Part B, pp 423–441, Elsevier, Amsterdam.

Nic Lughadha, E. and Proença, C. (1996). A survey of the reproductive biology of the Myrtoideae (Myrtaceae). *Ann. Missouri Bot. Gard.* **83**: 480–503.

Oliveira, P.E., Gibbs, P.E. and Bianchi, M. (1991). Pollination and breeding system of *Vellozia squamata* (Liliales: Velloziaceae): a species of the Brazilian cerrados. *Bot. Acta* **104**: 392–398.

Pool, P.A. and Bermawie, N. (1987). Procedure for artificial cross pollination in clove (*Syzygium aromaticum*). *Euphytica* **36**: 479–482.

Potts, B.M. and Savva, M. (1988). Self-incompatibility in *Eucalyptus*. In: R.B. Knox, M.B. Singh and L.F. Troiani (editors). Pollination '88, pp. 165–170. School of Botany, University of Melbourne, Parkville, Victoria.

Proença, C.E.B. and Gibbs, P.E. (1994). Reproductive biology of eight sympatric Myrtaceae from Central Brazil. *New Phytol.* **126**: 343–354.

Ruiz, T.Z. and Arroyo, M.T.K. (1978). Reproductive ecology of a secondary deciduous forest. *Biotropica* **10**: 221–230.

Sage, T.L., Bertin, R.I. and Williams, E.G. (1994). Ovarian and other late-acting self-incompatibility systems. In: E.G. Williams, A.E. Clarke and R.B. Knox (editors). Genetic control of self-incompatibility and reproductive development in flowering plants, pp 116–140, Kluwer Dordrecht.

Schmitt, D. and Perry, T.O. (1964). Self-sterility in sweetgum. *For. Sci.* **10**: 302–305.

Seavey, S.R. and Bawa, K.S. (1986). Late-acting self-incompatibility in angiosperms. *Bot. Rev.* **52**: 195–218.

Sedgley, M. and Smith, R.M. (1989). Pistil receptivity and pollen tube growth in relation to the breeding system of *Eucalyptus woodwardii* (Symphyomyrtus: Myrtaceae). *Ann. Bot.* **64**: 21–31.

Sedgley, M., Hand, F.C., Smith, R.M. and Griffin, A.R. (1989). Pollen tube growth and early seed development in *Eucalyptus regnans* F. Muell. (Myrtaceae) in relation to ovule structure and preferential outcrossing. *Austral. J. of Bot.* **37**: 397–411.

Stelly, D.M., Peloquin, S.J., Palmer, R.G. and Crane, C.F. (1984). Mayer's haemalum-methyl salicylate: a stain-clearing technique for observation within whole ovules. *Stain Technol.* **59**: 155–161.

Wallace, B. (1970). Genetic load, its biological and conceptual aspects. Prentice-Hall, Inc., New Jersey.

Waser, N. and Price, M.V. (1991). Reproductive costs of self-pollination in *Ipomopsis aggregata* (Polemoniaceae): are ovules usurped. *Amer. J. Bot.* **78**: 1036–1043.

Weller, S.G. and Ornduff, R. (1991). Pollen tube growth and inbreeding depression in *Amsinckia grandiflora* (Boraginaceae). *Amer. J. Bot.* **78**: 801–804.

Wiens, D., Calvin, C.L., Wilson, C.A., Davern, C.I., Frank D. and Seavey, S.R. (1987). Reproductive success, spontaneous embryo abortion and genetic load in flowering plants. *Oecologia* **71**: 501–509.

Wiens, D., Nickrent, D.L., Davern, C.I., Calvin, C.L. and Vivrette, N.J. (1989). Developmental failure and loss of reproductive capacity in the rare palaeoendemic shrub *Dedeckera eurekens*. *Nature* **338**: 65–67

Willis, J.H. (1992). Genetic analysis of inbreeding depression caused by chlorophyll-deficient lethals in *Mimulus guttatus*. *Heredity* **69**: 562–572.

Wright, S. (1977). Evolution and the genetics of populations. Vol. 3. Experimental results and evolutionary deductions. University of Chicago Press, Chicago.

Oliveira, P.E. (1998). Reproductive biology, evolution and taxonomy of Vochysiaceae in central Brazil. In: S.J. Owens and P.J. Rudall (Editors). Reproductive Biology, pp. 381–393. Royal Botanic Gardens, Kew.

REPRODUCTIVE BIOLOGY, EVOLUTION AND TAXONOMY OF VOCHYSIACEAE IN CENTRAL BRAZIL

PAULO EUGÊNIO OLIVEIRA

Departamento de Biociências, Universidade Federal de Uberlândia, Brazil

Abstract

The Vochysiaceae (Rosidae) are a mainly neotropical family of woody species with spurred, single-stamened flowers. The family consists of approximately 200 species in two markedly different tribes. The largest tribe, Vochysieae, is well represented in forest and savanna (cerrado) areas in Central Brazil. A review of the reproductive biology of the taxa in this tribe is presented. The main trend in the group is towards fewer petals and zygomorphic flowers associated with melittophily. The archetypal floral morphology consists of pentamerous flowers, as in *Salvertia convallariodora*, a hawkmoth pollinated cerrado species. Both *Vochysia* and *Qualea*, the main genera in this tribe, are largely bee pollinated. *Vochysia* species show secondary pollen presentation, as in *Salvertia*, but with a distinctly larger spurred calyx lobe. Hummingbirds and diurnal sphingid moths are secondary pollinators, but truly ornithophilous species may occur. *Qualea* have *Salvertia*-like calyx lobes but are mostly bee-pollinated. Hawkmoth pollination in *Q. grandiflora* seems to be a secondary shift. The one *Callisthene* species studied is pollinated by small bees. Breeding systems observed range from self-compatibility, self-incompatibility with stylar arrest of pollen tubes, and ovarian, late-acting type rejection.

Introduction

Vochysiaceae are a mainly neotropical family usually divided in two groups: the tribe Erismeae, with two genera, one of them occurring in Africa (ca. 25 spp; see Kawasaki 1997 for a recent review of this group); and Vochysieae, with five neotropical genera (ca. 175 spp.). The tribe Vochysieae has a centre of diversity in tropical Amazonian forest but species of each genus occur in the *cerrados*, neotropical savannas of Central Brazil (Stafleu, 1948, 1952, 1953). Although of negligible economic importance the Vochysiaceae are important elements in neotropical plant formations. In the *cerrados*, for example, *Qualea grandiflora* is probably the most widespread tree (Ratter and Dargie, 1992). Vochysieae species are also recognised as aluminium accumulating plants of the mineral poor, low pH soils of Central Brazil (Araújo and Haridasan, 1988; Haridasan and Araújo, 1988), although species of *Callisthene* occur on better soils of calcareous outcrops (Araújo and Haridasan, 1988; Furley and Ratter, 1988).

The revisions of Stafleu (loc. cit.) although taxonomically complete and thorough, do not discuss evolutionary or ecological aspects of the family. Even the specialisation trends in floral morphology and the easily recognisable links between the genera were not discussed clearly in this revision.

The reproductive biology of Vochysiaceae and the floral specialisation which characterise the family have been studied for some species and groups in the tribe Vochysieae from cerrados of Central Brazil (Silberbauer-Gottsberger and Gottsberger, 1975; Barbosa, 1983; Oliveira, 1991; Oliveira and Gibbs, 1994; Oliveira, 1996; Santos *et al.*, 1997). I intend here to put together and discuss these recent results, comparing pollination and breeding systems with the different levels of floral specialisation in the tribe Vochysieae. Furthermore, I suggest how these pollination and breeding features may contribute to an understanding of the taxonomy of the group.

Taxonomy of the family

Vochysiaceae are tree or shrub species characterised by spurred flowers with a single stamen, opposite or verticilate leaves. They have been loosely included in the order Polygalales (Cronquist, 1981, 1988) though it has been increasingly difficult to link the family with the other groups in the order (Litt *et al.*, 1995, Conti *et al.*, 1996). Even the spur ontogenesis which could be a link with Trigoniaceae seems to be originated from the axis of the flower and not from calyx or corolla (Kopka and Weberling, 1983). Stafleu (1948, 1952, 1953) in the latest revision of the family confirmed its separation in the tribe Erismae, with almost epigynous flowers, unilocular ovary and indehiscent fruits (Stafleu, 1952) and Vochysieae, with hypogynous flowers, dehiscent, capsular fruits and winged seeds. The tribe Erismae is basically restricted to tropical rain forest while the Vochysieae is well represented in open savannah areas though most species also occur in forest formations. Vochysieae include well defined genera which have been recognised since last century. The monospecific *Salvertia* has a pentamerous corolla. *Vochysia* species have zygomorphic flowers with a large calyx-lobe and usually three petals (sometimes one or no petals). *Qualea* species typically have a single petal and the stamen is outside the plane of symmetry, whilst *Callisthene* species shows flower morphology similar to *Qualea* but with a long *Vochysia*-type calyx-lobe and different fruit structure. A possible fifth genus, *Ruizterania*, was created by Marcano-Berti (1969) from *Qualea* section *Trichantera* which has a longer spurred calyx-lobe and barbate anther. A summary of these characteristics is presented in Fig. 1.

Although well defined, the taxonomic delimitation of Vochysiaceae, especially the tribe Vochysieae, was based on morphological characters, mainly, or sometimes only, from herbarium material. Characters such as pubescence and flower size, have proved difficult to use in the field to separate sympatric species. Other important characters including enantiostily, petal morphology, different calyx types and stamen structure were only mentioned or, no comment on their ecological or evolutionary meaning was attempted. Decisions on groups such as the new genus *Ruizterania* were made without any clue about how important were the distinctive characters justify the new taxa.

Studies on breeding biology

Studies on pollination and breeding systems for the genera in the tribe Vochysieae are presented below. Table 1 summarises the data for the species studied. Floral diagrams for each genus and distinctive characters are depicted in Fig. 1. Since there are no data for *Ruizterania*, its distinctive characters are discussed together with *Qualea*.

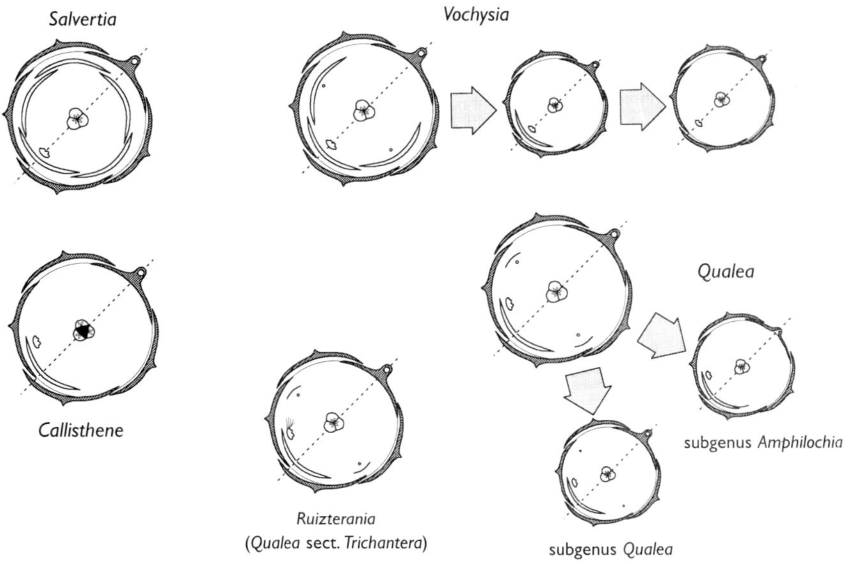

FIG. 1. Floral diagrams of the tribe Vochysieae flowers (Vochysiaceae). Based on Stafleu (1948, 1952, 1953) and Kopka & Weberling (1983).

Salvertia

The only species of this monotypic genus, *Salvertia convallariodora* is endemic to Brazilian cerrado vegetation. Its floral biology was studied and compared with other species of Vochysiaceae in the region (Oliveira, 1996). This species is a tree up to 10 m, which is widely distributed in cerrado regions although it is rarely common. Flowering is mainly in May and June. Trees produce terminal racemose inflorescences with 2-3 flowered cicinni arranged in whorls. Flowers are pentamerous, with actinomorphic symmetry and a spurred calyx lobe. The nocturnal anthesis is slow and can take ca. two hours after sunset to be complete. Two or three flowers open each night per inflorescence. Pollen is shed before anthesis and deposited on the style surface in a secondary pollen presentation of the pseudostamen-type (Yeo, 1993). At the beginning of anthesis the wide stigmatic surface is turned to the petals. Later, the style becomes erect in the centre of the flower and only then may flower visitors touch the stigma. The stigmatic surface may be turned to the right or to the left (enantiostyly or enantiomorphy). Nectar is produced and held in the spur in small amounts (5-20 μl) and with 9 to 35% of sucrose equivalents (*sensu* Inouye *et al.*, 1980). The flower is visited by medium sized to large nocturnal hawk moths from the start of anthesis. *Erynnis ello* (L.) was the most common species observed visiting the flowers. The species is self-compatible with self-pollinated fruits which developed normally and produced well-formed seeds.

Vochysia

Few species of this large genus have been studied for their pollination biology. Six *Vochysia* species occurring in gallery forest or cerrado vegetation were studied in central Brazil, including large forest trees up to 25 m, open cerrado small trees and subshrub species (Oliveira

TABLE 1. Some floral features, pollination and breeding biology of the species of the tribe Vochysieae. Number of petals, *Salvertia* or *Vochysia* calyx type, type of pollen presentation, presence of enantiostyly, main pollinators, time of anthesis, self-compatible (SC), or self-incompatible (SI) breeding system and corresponding index of self-incompatibility (ISI) are given together with the source of information.

Species	Petals	Calyx	Pollen presentation	Enantiostyly	Pollination	Anthesis	Breeding System	ISI	Source
Callisthene fasciculata (Spr.) Mart	1	Voch.	normal	yes	–	–	–	–	1
Callisthene major Mart.	1	Voch.	normal	yes	small bees	day	SI	0.18	2
Qualea cordata Spreng.	1	Salv.	normal	yes	large bees	day	SI	0.0	3, 4
Qualea dichotoma (Mart.) Warm.	1	Salv.	normal	yes	large bees	day	–	–	5
Qualea grandiflora Mart.	1	Salv.	normal	no	hawkmoths	night	SI	–	5, 6, 7
Qualea multiflora Mart.	1	Salv.	normal	yes	large bees	day	SI	0.0	7
Qualea parviflora Mart.	1	Salv.	normal	yes	large bees	day	SI	0.06	7
Salvertia convallariodora St. Hil.	5	Salv.	secondary	yes	hawkmoths	night	SC	0.50	8
Vochysia cinnamomea Pohl	3	Voch.	secondary	no	large bees	day	SC/SI	0.26	9
Vochysia elliptica Mart.	3	Voch.	secondary	no	large bees	day	SC/SI	0.26	10
Vochysia guatemalensis Donn. Smith	3	Voch.	secondary	no	large bees	day	SC	–	11
Vochysia ferruginea Mart.	3	Voch.	secondary	no	large bees	day	SC	0.95	10, 12
Vochysia pumila Pohl	3	Voch.	secondary	no	large bees	day	SI	0.16	10
Vochysia pyramidalis (Spr.) Mart.	3	Voch.	secondary	no	large bees	day	SI	0.18	10
Vochysia rufa Mart.	3	Voch.	secondary	no	large bees	day	SI	0.05	10
Vochysia thyrsoidea Pohl	3	Voch.	secondary	no	large bees	day	SI	0.0	10
Vochysia tucanorum (Spr.) Mart.	3	Voch.	secondary	no	large bees	day	SI	0.11	10

Sources: 1. Under study. 2. Afonso and Oliveira, manuscript in preparation. 3. Ficher and Gordo 1993. 4. S. Godoy-Aveiro, unpublished studies. 5. Personal observations. 6. Silberbauer-Gottsberger and Gottsberger, 1975. 7. Barbosa, 1983. 8. Oliveira, 1996. 9. Santos *et al.*, submitted. 10. Oliveira and Gibbs, 1994. 11. Flores, 1993. 12. Bawa *et al.*, 1985.

and Gibbs, 1994). Another cerrado treelet was recently studied (Santos *et al.*, 1997). There is also sparse information on two Central American tree species (Bawa *et al.*, 1995; Flores, 1993). The cerrado species show rather different flowering phenologies, but present a relatively uniform floral morphology and attract a mostly similar spectrum of visitors even though they range from large gallery forest trees to small hemixyle sub-shrubs in open cerrado. Anthesis is diurnal but sometimes occurs in mid-morning. The spurred calyx-lobe (*sensu* Stafleu, 1948) curves backwards during anthesis and breaks the filament, ejecting the anther before the flower is totally open. Pollen is shed before anthesis and remains loosely deposited on the style surface, in a secondary pollen presentation similar to that in *Salvertia*. However, the stigmatic surface is terminal and no enantiostyly was observed. Nectar is produced in the spur mainly during the first day and flowers wilt the second or third day. Visitors and pollinators comprise mainly large Anthophoridae bees and hummingbirds, but at least four species are visited by sphingid moths. Species in Costa Rica are also visited and pollinated by large bees and hummingbirds but information was not detailed. Some observations indicate that the slight differences in flower size or habitat between species may influence the relative importance of bees *versus* hummingbirds or sphingids as effective pollinators. The relatively larger flowers of some species of *campo rupestre* (upland areas in Central Brazil) may be more dependent on hummingbirds for their pollination. Most *Vochysia* species are self-incompatible with self-pollen tube rejection, usually accompanied by callose deposition occurring in the mid-style. *Vochysia cinnamomea* and *V. elliptica* seem to be borderline cases with self-incompatibility index (ISI *sensu* Bullock, 1985) at the limit usually used to separate self-compatible from self-incompatible species. Costa Rican species are truly self-compatible (Flores, 1993).

Qualea

Species of *Qualea* show a more heterogeneous pollination biology and floral morphology. *Qualea multiflora* individuals are usually small trees in cerrado though huge specimens have been observed in cerradões and gallery forest margins (Ratter, 1985). It flowers in November and December. Flowers have a single white petal with a purple nectar guide in the centre and a single stamen. Anthesis is late in the morning and the flowers seems to be functional for one day only. Nectar in the spur is the only reward. During the second day the petal becomes yellow and the stamen usually wilts though the stigma is still receptive. Pollinators are primarily large bees although butterflies and hummingbirds were also observed visiting the flowers (Barbosa, 1983, Oliveira, 1991). *Qualea parviflora* is another cerrado species studied by Barbosa (1983). It is a tree up to 15 m high, which flowers early in the rainy season (October-November). It is neither massive nor prolonged but the violet flowers are very conspicuous. The structure of the flower is similar to that of *Q. multiflora* but smaller. Visitors were mainly Euglossinae and other long-tongued medium to large bees. Hummingbirds and butterflies were occasionally observed. Flowers last one day and anthesis is diurnal. Both species are self-incompatible although pollen tubes from self- as well as cross-pollinations reach the ovary in 24 hours or so. Tubes seems to entangle in a cluster on the top of the ovary (Fig. 2) and ovule penetrations follow only after a week (Oliveira, 1991). Fertilisation was not directly observed but ovule penetration after self- and cross-pollination suggests a late-acting self-incompatibility mechanism (Seavey and Bawa 1986).

Qualea grandiflora is a cerrado tree up to 12 m high, which was studied by Silberbauer-Gottsberger and Gottsberger (1975), Barbosa (1983) and Oliveira (1991). Flowering is later in comparison with the two species described before (mainly December). Flowers have anthesis at dawn and a rather different structure when compared with the other large-bee pollinated congeneric species. The single petal is yellow and curved to form a kind or tube which encloses the single anther and the style. The relatively large anther is positioned in the upper part of the tube and the style is positioned under it, with the small stigma in-between the thecae of the anther (Fig. 3). Flowers last one or two nights and seem to be protandrous, with the stigma receptive later on the first or second night. The main visitors and pollinators are sphingis moths, sometimes large ones such as *Cocytius* sp. which land quickly on the petal (Silberbauer-Gottsberger and Gottsberger, 1975 and personal obs.). The species seems to be basically self-incompatible but very few fruits were obtained from hand-pollination treatments to give a clear-cut result (Barbosa, 1983 and personal obs.). Pollen tube growth shows a pattern similar to the melittophilous species discussed above.

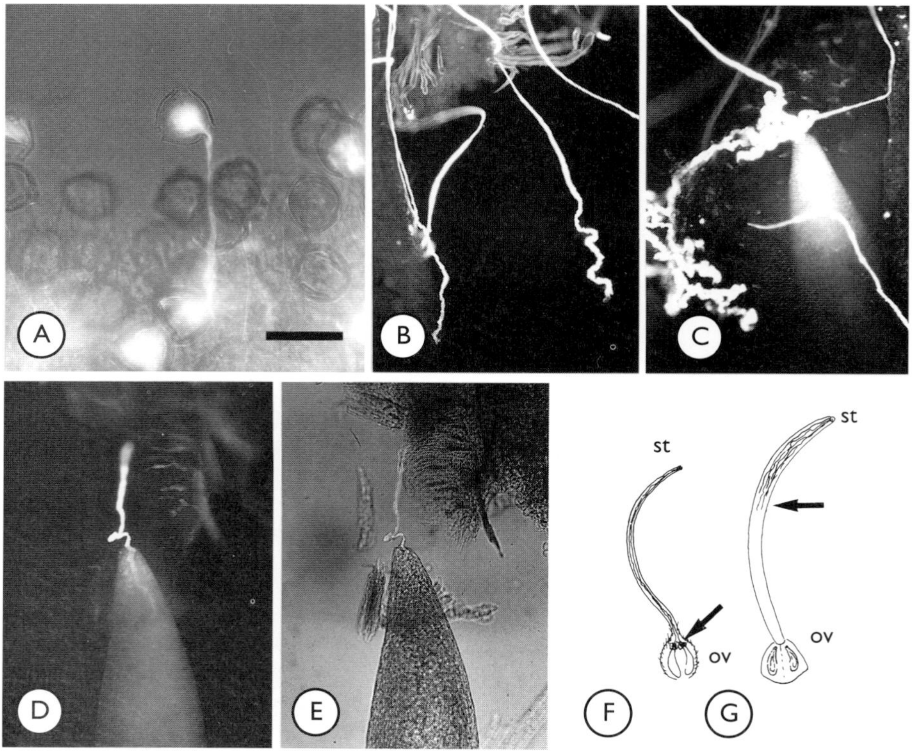

FIG. 2. Pollen tube growth and fertilisation in *Qualea* spp. Fluorescence microscopy of pistils stained with aniline blue (A-E) A. Germination and growth was rapid for both cross and self-pollination (self-pollinated pistil of *Qualea grandiflora* - scale bar = 40 µm). B. Tubes grow normally down to the ovary where they become characteristically zig-zag (SP pistil of *Q. multiflora* fixed 48hs after pollination). C. In the ovary, pollen tubes seem to entangle on the tip of the ovules forming clusters some days before any ovule penetration happens (In this case *Q. multiflora* SP + 96hs). D, E. Ovule penetration was observed both in self- and cross-pollinated pistils after one week only (In this case cross-pollinated pistil of *Q. multiflora*. Note fluorescent pollen tube into the micropyle in D. Tungsten backlighting in E. shows the contour of the ovule) - For B, C, D and E, scale bar = 200 µm. F. Schematic view of pistils of *Qualea* showing pollen tube growth straight to the ovary after self-pollination (arrow). G. Pistil of *Vochysia* sp. showing pollen tube arrest at the style (arrow). st - stigma, ov - ovary. Line drawings not in scale.

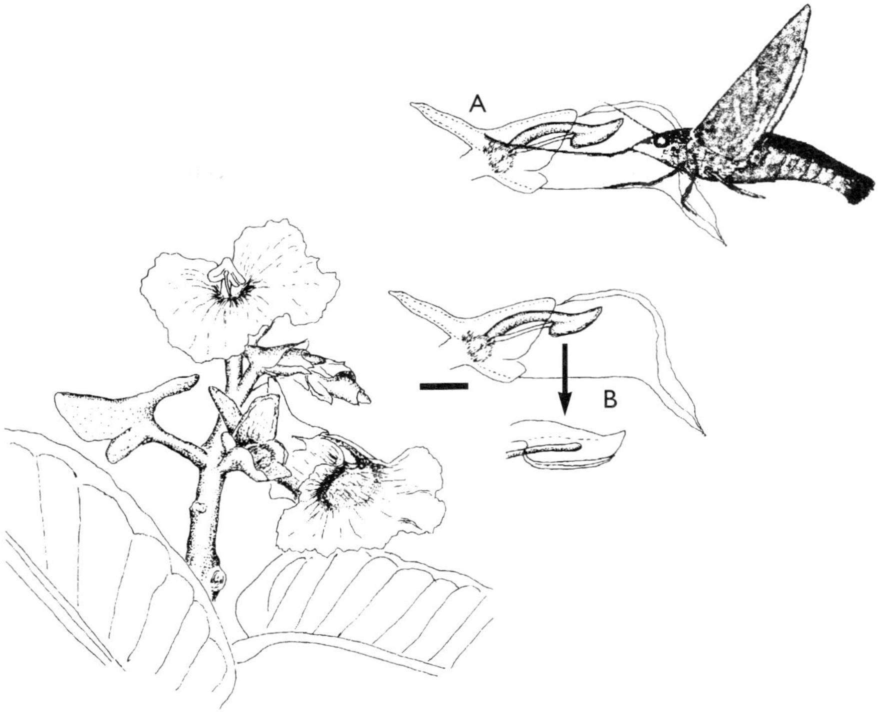

FIG. 3. *Qualea grandiflora* flowers. Note the curled petal forming a tube-like structure. A. A section of flower in detail and a scheme of hawkmoth visit. B. Position of the style and stigma between the thecae of the anther.

Qualea cordata is a small tree (6 m or less) in cerrado which also grows in rocky *campo rupestre* of Central Brazil and flowers in October. Although the species has been studied both for pollination (Fischer and Gordo, 1993) and breeding biology (S. Godoy-Aveiro, unpublished studies), little attention was given to floral morphology, which is probably the most interesting aspect of this species and of the subgenus *Amphilochia*. The species has a gibbous nectary at the calyx base, similar to that found in *Qualea dichotoma*, another species of this group which I observed in cerrado. This gibbous nectary seems to be similar in structure and function to the spur in the other species of *Qualea* (R. Oliveira pers. comm.) and although the flowers allow visits by bees with a shorter proboscis, the main pollinators are not significantly different from the spurred species. Enantiostyly is also present and as in other species, left- and right-hand flowers can be found in similar numbers in a tree (Fischer and Gordo, 1993). *Qualea cordata* is self-incompatible with late-acting self-incompatibility similar to the other species studied, and initial embryo development seems to follow both self- and cross-pollination treatments (Godoy-Aveiro, unpublished data). *Qualea dichotoma* also belongs in subgenus *Amphilochia* and has flowers very similar to *Q. cordata*, to a point that the two species can be confused. *Q. dichotoma* occurs in cerrado woodlands and mesophilous forest in Central Brazil and may grow to a large, 25 m tree with a very characteristic bark. Field observations also indicate large bee-pollination but detailed studies with this species are still incomplete.

Callisthene

As far as I know there are floral biology data for only one species of *Callisthene* (Afonso and Oliveira, unpublished work). *Callisthene major* is a cerradão (cerrado woodland) or mesophilous forest tree up to 20 m or more. It flowers in August and September. Trees present a swift blooming period with a relatively inconspicuous floral display. The small fragile flowers occur in two-flowered cicinni at the leaf base. They present a single white petal with a yellow nectar guide. The *Vochysia*-type spurred calyx-lobe is longer than half the size of the petal and covers the stamen and pistil. The spur is short (ca. 3 mm). Enantiostyly similar to *Qualea* species is also found in *Callisthene*. Flowers of *Callisthene* resemble those of *Ruizterania* Marcano-Berti (section *Trichanthera* of *Qualea*) which also share the same kind of innate anthers, though they are glabrous in *Callisthene*. The main visitors in the areas studied were social bees, *Trigona* (Meliponinae) and *Apis* (Apidae), but small Anthoporidae bees of the genus *Exomalopsis* were also observed. Bees land on the petal and enter under the calyx-lobe to reach the nectar in the spur. Nectar is present in small amounts. Butterflies were also observed but seem to be poor pollinators of the species. Hand-pollination treatments showed self-incompatibility, and pollen tube growth had a pattern similar to that observed in *Qualea* species, with a cluster of pollen tubes near the micropyle of the ovules and ovule penetration in both self- and cross-pollination.

Fig. 4 summarises the pollination relationships for the four genera of Vochysieae discussed here. No data are available on *Ruizterania* pollination systems.

Evolutionary trends in Vochysiaceae: ideas from pollination biology of cerrado species

Some evolutionary trends are suggested by the pollination data summarised here, and they can indicate new lines of research for the pollination biology and taxonomy of the group.

From moth to bees and then moth again

It seems likely that trends of floral specialisation in Vochisieae, with reduction in number of petals and change in floral symmetry, are linked to melittophily. If *Salvertia* flowers represent an archetypal condition, with a pentamerous, actinomorphic corolla and nectary in a long spur, the other flower types can be viewed as stages of adaptation to large bee and sometimes social bee pollination. The zygomorphic flowers, diurnal anthesis, shorter spur, nectar guides and one-day flowers observed for the *Vochysia* and *Qualea* species studied in cerrado areas may be associated with a large bee pollination syndrome (Wyatt, 1983, Frankie *et al.*, 1983). The two largest genera in the family seem to have inherited different characteristics from the basic *Salvertia*-like flowers. *Vochysia* has similar secondary pollen presentation but no stigma torsion and enantiostyly, while *Qualea* has no secondary pollen presentation and a rather different enantiostyly. Stafleu (loc. cit) and earlier authors suggested that *Qualea's* asymmetric stamen originated from a different whorl, but there may be a more parsimonious explanation. The stamen position may be the result of torsion during bud development and thus similar to the enantiostily in *Salvertia* (A. Litt, pers. comm.).

If so, *Qualea grandiflora* is a particularly important species since its adaptation to sphingophily seems to be secondary. The solitary petal and other floral features are similar to the melittophilous species and even though the large anther has a completely different position in the flower, style and filament position reflect an earlier stage of bilateral position and enantiostyly. Since diurnal sphingids are secondary visitors of the melittophilous species of *Vochysia* and *Qualea*, a secondary shift to nocturnal sphingids may always have been possible. Yet, it is difficult to know exactly how, since there is no described intermediate floral morph.

The problem is that the distinctive features of flower structure found in *Qualea grandiflora* are difficult to analyse from dried material. Other species of *Qualea* with large flowers and flat anthers like those of *Qualea grandiflora* appear in different sections in Stafleu's revision (e.g. *Q. macropetala* and *Q. magna* in section *Qualea*) and they may have the intermediate floral morphology which would help to understand the evolution of this secondary sphingophily in the genus.

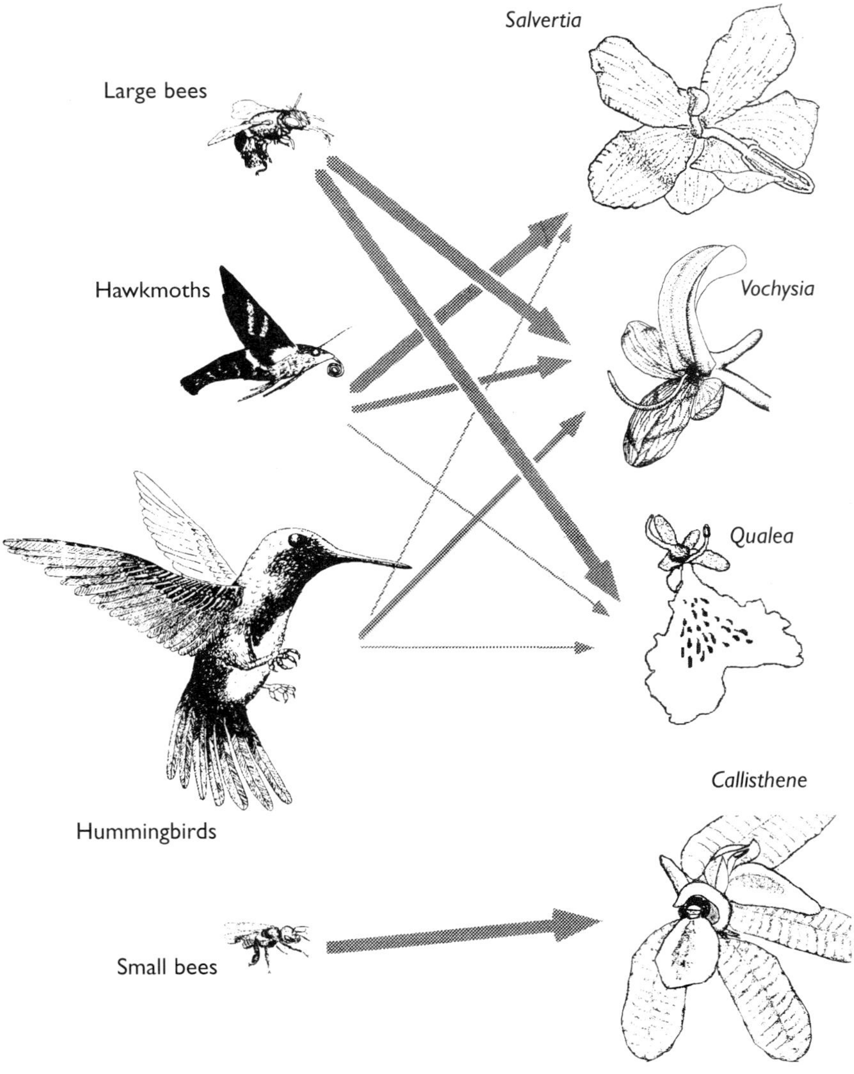

FIG. 4. Summary of the relationships between visitors and the flowers of the genera of the tribe Vochysieae (Vochysiaceae). Thick arrows indicate main pollinators, medium arrows secondary pollinators and thin arrows opportunistic visitors.

Bees to hummingbirds in upland areas

Another interesting trend is suggested by *Vochysia* flower studies. Differences in flower size observed in species of *Vochysia* may be important for their effective pollination and perhaps linked with their distribution throughout the cerrado regions and perhaps in the neotropics as a whole. Forest trees such as *V. tucanorum* and *V. pyramidalis*, with their small fragile flowers, seem to be well adapted to bee pollination, since even relatively small Euglossinae bees can occasionally touch the stigma during their visits. On the other hand, *V. elliptica* and *V. thyrsoidea*, which occur in open areas and have relatively larger flowers, seem to receive poorer pollination service from bees which are otherwise efficient pollinators for both forest species. Contact with the stigma in *V. elliptica* and *V. thyrsoidea* seems to be common only for larger *Epicharis* bee species. Occasional sphingid moths and even small hummingbirds could be increasingly important pollinators. This trend may be important especially in the case of *V. thyrsoidea* which is a species of relatively high altitude cerrado or campo rupestre areas where bee pollination and particularly large bee occurrence can be less important and predictable, and ornithophily and diurnal moth pollination are usually more common (Cruden, 1972; Sazima and Sazima, 1990). This switch from bee pollination may have occurred in other taxa since *V. aurantiaca*, described by Stafleu (1948) for high altitude cloud forest in Colombia, has relatively large, 'orange-red flowers', which may indicate a further adaptation to bird pollination. This and other cloud forest species of *Vochysia* would be very interesting subjects for pollination and breeding studies.

Spurs and spurless taxa

A spurred calyx-lobe is a feature of the family Vochysiaceae and it is surprising that a subgenus of *Qualea* has no or only a weakly developed spur. Flowers of *Amphilochia* have gibbous calyx-lobe with nectariferous structure similar to the nectaries in the other spurred species (Renata Oliveira, pers. comm.). This structure can be viewed as a shallow spur which has a similar function in the flowers of *Qualea cordata* (Fischer and Gordo, 1993) and *Q. dichotoma* (pers. obs.). These species are pollinated by the same spectrum of large bees and other mellitophilous spurred flowered *Qualea* species. This shallow spur may make the flower less selective to visitors and allow access to bees with short proboscises. Shallow spurs also permit access to the nectar by social bees of Meliponinae and *Apis mellifera*. A small spur or gibbous calyx-lobe is also present in other sections of the genus: *Qualea calantha* and *Q. gestasiana* are forest species from the section *Qualea* cited by Stafleu (1953) with a short spur (2 mm) or with a bag-shaped base to the fourth calyx-lobe and no spur. Size of the spur may be a variable character selected by the interaction with different bee assemblages, and spurless or gibbous flowers, as found in *Amphilochia*, may represent more generalist melittophilous floral morph than the long spurred ones.

Flowering phenology in relation to social bee pollination

There is another complementary trend from large bee-pollinated *Qualea* species to the more fragile flowers of the genus *Callisthene* which are adapted to pollination by small bees, some of them social and with recruitment foraging strategies. Social bees such as *Trigona* spp. and the introduced *Apis mellifera* are able to mark foraging routes and recruit other workers from the colony to use a good food source (Roubik, 1989). In *Callisthene major*, the swift blooming period may limit the possibilities of pollination service to bees which can react quickly to changes in the amount of resources available by recruiting workers. Although *Callisthene major* trees may flower for several days, they are not exactly synchronised, and this can create a kind of pulsing resource in the environment similar to the phenological pattern described for *Vellozia squamata* (Oliveira *et al*., 1991).

Breeding systems

Vochysiaceae have a variety of breeding systems ranging from self-compatibility in *Salvertia* and some *Vochysia* species to the "classic" self-incompatibility mechanism in *Vochysia* species and "late-acting self-incompatibility" in *Qualea* and *Callisthene*. The presence of a somewhat "classic" pollen tube arresting mechanism in the style, which has been related to gametophytic self-incompatibility, and late-acting mechanisms in the same family, suggests that the two types of mechanism may have similar genetic control. The late-acting self-incompatibility in *Qualea* and *Callisthene* is intriguing. Since the style is fragile and remains attached to the ovary for a short time, pollen tube growth has to be very quick in these species. This could have forced pollen tube arrest, as in *Vochysia*, to be delayed to the ovary in *Qualea* and *Callisthene*. After reaching the ovary pollen tubes seem to wait entwining the tip of ovules before fertilisation, perhaps until ovule maturation. Embryo sac differentiation studies are necessary to confirm this idea. After some days, ovule penetration and fertilisation follow, irrespective of pollen tube origin suggesting a true late-acting self-incompatibility mechanism. However, embryo development in *Qualea cordata* self-fertilised ovules does not rule out the possibility of ovary abortion due to inbreeding depression (Wiens *et al.*, 1987).

General trends, genera delimitation and pollination biology

Various lines of research on taxonomy and morphology may be suggested based on the breeding data and evolutionary trends outlined here.

The occurrence of enantiostyly in *Salvertia* caused by torsion of the stigmatic surface suggests that enantiostyly in *Qualea* can have a similar origin. Confirmation of this idea by observing bud initiation and flower ontogeny, a line of work A. Litt and S. Mori are carrying out in New York Botanical Garden would unify further the tribe Vochysieae in a evolutionary trend of flower specialisation from possibly sphingophilous, actinomorphic *Salvertia*-like flowers to mellitophilous zygomorphic flowers in *Vochysia* and *Qualea*.

The separation of *Ruizterania* as a distinct genus is somewhat arbitrary since pubescent thecae also occur in some *Vochysia* species and the flower is basically the same as in other non-pubescent large bee-pollinated species of *Qualea*. However, bud shape and the *Vochysia*-like spurred calyx-lobe in this taxon do provide a link between *Vochysia*, *Qualea* and *Callisthene* flowers. Pollination biology data on this group would be important. On the other hand, large flowered species of the *Q. grandiflora* type, with secondary adaptation to sphingophily, may be worth reconsidering in a taxonomic context.

Besides secondary sphingophily, there are two other pollination systems derived from mellitophilous flowers: the possibly ornithophilous upland and cloud-forest *Vochysia* species and the pollination of *Callisthene* species with phenology and flower morphology adapted to small social bees.

Floral structure links *Qualea* and *Callisthene* species which also share a common outcrossing breeding system of the "late-acting self-incompatibility" type, at least for the species studied so far. This last feature is distinct from the "classic" pollen tube arresting mechanism observed for the *Vochysia* species. Again, as intermediate morphs, flowers of *Ruizterania* will be important to understand the evolution of these distinct mechanisms.

Acknowledgement

I am grateful to Peter E. Gibbs for the revision of the manuscript and supervision during most of the field work which provided data for this paper. I thank Silvia Godoy-Aveiro and Renata C. Oliveira who allowed me to cite some of their unpublished data, and Ana Angélica Barbosa, Amy Litt and Kikyo Yamamoto for discussion of some ideas on taxonomy and floral morphology of Vochysiaceae. A PhD and later a research grant from CNPq permitted most of my own work cited here. I thank FAPEMIG - Minas Gerais Research Fund for the support to attend the Reproductive Biology 96 Conference and present this paper.

References

Araújo, G.M. and Haridasan, M. (1988). A comparison of the status of two forest communities on mesotrophic and distrophic soils in central Brazil. *Commun. Soil Sci. Pl. Analysis* **19**: 1075–1089.

Barbosa, A.A. (1983). Aspectos da ecologia reprodutiva de três espécies de *Qualea* (Vochysiaceae) de um cerrado de Brasília. - D.F. Tese de Mestrado, Universidade de Brasilia.

Bawa, K.S. Perry, D.R. and Beach, J.H. (1985). Reproductive biology of tropical lowland rain forest trees. I. Sexual systems and incompatibility mechanisms. *Amer. J. Bot.* **72**: 331–345.

Bullock, S.H. (1985). Breeding systems in the flora of a tropical deciduous forest. *Biotropica* **17**: 287–301.

Conti, E., Litt. A. and Sytsma, K.J. (1996). Circumscription of Myrtales and their relationships to other Rosids: evidence from RbcL sequence data. *Amer. J. Bot.* **83**: 221–233.

Cronquist, A. (1981). An integrated system of classification of flowering plants. Columbia University Press. New York.

Cronquist, A. (1988). Evolution and classification of the Flowering Plants. New York Botanical Garden, New York.

Cruden, R.W. (1972). Pollinators in high-elevation ecosystems: relative effectiveness of birds and bees. *Science* **176**: 1439–1440.

Fischer, E.A. and Gordo, M. (1993). *Qualea cordata*, pollination by the territorial bee *Centris tarsata* in the "Campos Rupestres", Brazil. *Ci. & Cult.* **45**: 144–147.

Flores, E.M. (1993). *Vochysia guatemalensis* y *Vochysia ferruginea*. *Arboles Semillas Neotrop.* **2**: 1–52.

Frankie, G.W., Haber, W.A., Opler, P.A. and Bawa, K.S. (1983). Characteristics and organization of the large bee pollination system in the Costa Rican dry forest. In: C.E. Jones and R.J. Little (editors). Handbook of experimental pollination biology, pp. 411–447. Van Nostrand Reinhold Co., New York. p. 411–447.

Furley, P.A. and Ratter, J.A. (1988). Soil resources and pant communities of the central Brazilian cerrado and their development. *J. Biogeogr.* **15**: 97–108.

Haridasan, M. and Araújo, G.M. (1988). Aluminium-accumulating species in two forest communities in the cerrado region of Central Brazil. *Forest Ecol. Managem.* **24**: 15–26.

Inouye, D.W., Favre, N.D., Lanun, J.A., Levine, D.M., Meyers, J.B., Roberts, M.S., Tsao, F.C. and Wang, Y.Y. (1980). The effect of non-sugar nectar constituents on estimates of nectar energy content. *Ecology* **61**: 992–995.

Kawasaki, M.L. (1998). Systematics of *Erismae* (Vochysiaceae). *Mem. New York Bot. Gard.* **81:** 1–40. 1998.

Kopka, S. and Weberling, F. (1983). Zur Morphologie und Morphogenese de Blüte von *Vochysia acuminata* Bong. subsp. *laurifolia* (Warm.) Stafleu (Vochysiaceae). *Beitr. Biol. Pflanzen* **59**: 273–302.

Litt, A., Conti, E., Chase, M.W. and Sytsma, K.J. (1995). The systematic position of *Euphronia* (Polygalales) based on molecular and morphological evidence. *Amer. J. Bot.* **82** (suppl.): 146–147.

Marcano-Barti, L. (1969). *Ruizterania*. *Pittieria* **2**: 6–27.
Oliveira, P.E. (1991). The pollination and reproductive biology of a cerrado woody community in Brazil. PhD, University of St. Andrews (Scotland).
Oliveira, P.E. (1996). Polinização e sistema de reprodução de *Salvertia convallariodora* (Vochysiaceae): uma espécie de cerrado polinizada por mariposas (Lepidoptera-Sphingidae). *Revista Brasil. Bot.* **19**: 49–56.
Oliveira, P.E. and Gibbs, P.E. (1994). Pollination and breeding systems of six *Vochysia* species (Vochysiaceae). *J. Trop. Ecol.* **10**: 509–522.
Oliveira, P.E., Gibbs, P.E. and Bianchi, M. (1991). Pollination and breeding biology of *Vellozia squamata* (Liliales-Velloziaceae): a species of the Brazilian cerrados. *Bot. Acta* **104**: 392–398.
Ratter, J.A. (1985). Notas sobre a vegetação de Fazenda Agua Limpa. Royal Botanic Garden, Edinburgh.
Ratter, J.A. and Dargie, T.C.D. (1992). An analysis of the floristic composition of 26 cerrado areas in Brazil. *Edinburgh J. Bot.* **49**: 235–250.
Roubik, D.W. (1989). Ecology and natural history of tropical bees. Cambridge University Press, Cambridge.
Santos, M.L., Afonso, A.P. and Oliveira, P.E. (1997). Biologia floral de *Vochysia cinnamomea* (Vochysiaceae) em cerrados do Triângulo Mineiro. *Revista Brasil. Bot.* **20:** 127–132.
Sazima, M. and Sazima, I. (1990). Hummingbird pollination in two species of *Vellozia* (Lilliflorae: Velloziaceae) in Southeastern Brazil. *Bot. Acta* **103**: 83–86.
Silberbauer-Gottsberger, I. and Gottsberger, G. (1975). Uber sphingophile Angiospermen Brasiliens. *Pl. Syst. Evol.* **123**: 157–184.
Stafleu, F.A. (1948). A monography of the Vochysiaceae. I. *Salvertia* and *Vochysia*. *Meded. Bot. Mus. Rijks Univ. Utrecht* **95**: 397–540.
Stafleu, F.A. (1952). A monography of the Vochysiaceae. II. *Callisthene*. *Acta Bot. Neerl.* **1:** 222–242.
Stafleu, F.A. (1953). A monography of the Vochysiaceae. III. *Qualea*. *Acta Bot. Neerl.* **2:** 144–217.
Wiens, D., Calvin, C.L., Wilson, C.A. Davern, C.I., Frank, D. and Seavey, S.R. (1987). Reproductive success, spontaneous embryo abortion and genetic load in flowering plants. *Oecologia* **71**: 501–509.
Wyatt, R. (1983). Pollinator-plant interactions and the evolution of breeding systems. In: L. Real (editor) Pollination biology, pp. 51–95, Academic Press, Orlando.
Yeo, P.F. (1993). Secondary pollen presentation: form, function and evolution. Springer Verlag, Vienna.

Vieira, A.O.S. and Shepherd, G.J. (1998). Breeding systems in *Ludwigia* (Onagraceae) in southeast Brazil. In: S.J. Owens and P.J. Rudall (Editors). Reproductive Biology, pp. 395–406. Royal Botanic Gardens, Kew.

BREEDING SYSTEMS IN *LUDWIGIA* (ONAGRACEAE) IN SOUTHEAST BRAZIL

ANA ODETE SANTOS VIEIRA[1,2] AND GEORGE JOHN SHEPHERD[1]

[1]Departamento de Botânica, Universidade Estuadual de Campinas, 13083-970, Campinas, São Paulo, Brazil
[2]Permanent Address: Departamento de Biologia Animal e Vegetal, Universidade Estadual de Londrina, 86051-970, Londrina, Paraná, Brazil

Abstract

A group of ten *Ludwigia* species, occurring in a marshy area in Campinas, Southeast Brazil, were studied during 1995 and 1996, for their phenology, breeding systems and germination. The species showed rather uniform flowering and fruiting phenologies, with flowering maximum during autumn, and fruiting mainly at the end of this season. Some of the open capsules retain the seeds inside until the next rainy season. The seeds are dormant or germination is delayed until the beginning of the next hot rainy season. Germination is higher among the large seeds, comparing populations or species. Agamospermy was not detected. Three species have gametophytic self-incompatibility systems, with self-tube rejection occurring in the style. Two diploid species are autogamous. *L. octovalvis* has autogamous and outcrossed populations. Autogamy is associated with herbaceous habitat, smaller seeds and flowers, and anthers shedding pollen on the stigma. The self-incompatible species are shrubby and this is also the case for species with vegetative reproduction. Floral morphology and the behaviour of oligoletic bees may increase outcrossing.

Introduction

The family Onagraceae is considered to be a natural family and particularly interesting for studies of breeding systems because of its moderate size, clear demarcation of taxa and relative abundance of morphological, systematic, molecular and phytochemical information (Raven, 1979, 1988). Onagraceae consists of 7 tribes, 16 genera and approximately 652 species and has a worldwide distribution (Hoch *et al.*, 1993). Members of this family share a peculiar set of characters: viscin threads or ektaxinous strands on the proximal pollen wall (Patel *et al.*, 1984), a 4-nucleate embryo sac (Tobe and Raven, 1986a), presence of abundant calcium oxalate raphides in the vegetative cells (Carlquist, 1975) and the presence of septa dividing the sporogenous tissue (Eyde, 1977; Tobe and Raven, 1986b).

The genus *Ludwigia* occupies a rather isolated position in the family Onagraceae (Hoch *et al.*, 1993), being the only member of the tribe Jussiaeeae. It contains more than 80 species with a predominantly south American distribution (Raven, 1963; Ramamoorthy and Zardini, 1987), generally occurring in moist habitats (Munz, 1947).

Approximately 42% (of 674 species) of Onagraceae are outcrossing, with another 6% including some partly outcrossing populations (Raven, 1979). *Ludwigia* has 26 outcrossing species (about 33%), with 8 of these self-incompatible - 3 diploid and 5 polyploid (Raven, 1979). *Ludwigia* is also interesting because of its pollen, released in tetrads or polyads united by viscin threads, ample stigmatic surfaces (Cruden and Jensen, 1979), massive ovuliferous placentas (Eyde, 1977), absence of a constriction in the floral tube beyond the ovary, gynoecial nectaries (Eyde, 1981, 1982; Raven, 1988) and pollination by oligoletic bees (Gimenes, 1988, 1991, 1993; Martins and Antonini, 1994; Schlindwein, 1995).

The present study was undertaken to investigate the breeding systems of a series of sympatric species occurring in a limited area in Southeast Brazil, and to examine correlations between germination, weight of seeds and habit.

Material and Methods

The species and populations studied are shown in Table 1. Vouchers (Vieira numbers 451-478) are deposited in the herbaria FUEL and UEC.

L. octovalvis has a spectrum of morphological variation which is not represented by current taxonomic treatments that recognise only one species (Raven and Tai, 1979). We found three forms (Fig. 1): the first, named SESSI, resembles the now synonymised *L. octovalvis* subsp. *sessiliflora* (Raven 1963): perennial shrubs, leaves often subovate, with larger flowers; the second, denominated OCTO, similar to *L. octovalvis* subsp. *octovalvis* (Raven 1963), with smaller plants and flowers, and lanceolate leaves; the third form is a morphological intermediate but had too few individuals flowering at the same time to be tested. *L. tomentosa* has two different forms: one with orbicular, tomentose leaves, produced during the end of autumn and winter (TOM) and one with lanceolate almost glabrous leaves which are puberulous on the veins, developing new leaves during the winter like the first form (GLA). *L. hyssopifolia* is a pantropical weed and its origin is difficult to ascertain (Raven, 1963). The remaining species are all native and morphologically more homogeneous.

Field studies were made at the Centro Experimental de Campinas - Instituto Agronômico de Campinas (22° 54' S, 47° 05' W, 674 m above sea level) in the municipality of Campinas, São Paulo state, Brazil, from March 1995 to August 1996. Floral biology and breeding systems required daily visits during the flowering period and phenology was recorded with a monthly record for each species. The region has a Köppen type Cwa climate with a humid spring/summer and a dry winter and annual average temperature of 20.3° C with summer/autumn the hottest seasons (Mello *et al.*, 1994). The species grew in two marshy areas around lakes, 600m from each other. In the first site (750m^2), we studied *L. elegans* and *L. laruotteana*, and in the second (36,000m^2) the remaining species. There was almost no spatial or ecological separation between the species in the second area.

The experiments used different individuals (Table 2) except in *L. elegans*, whose decumbent branching stems made it difficult to separate individuals. In this case, three separate sites were treated as individuals.

Morphological characteristics were observed from fresh and fixed material. Breeding systems were determined by means of standard bagging and pollination techniques (Radford *et al.* 1974; Dafni 1992). We used fluorescence microscopy for pollen tube observation (Martin, 1959), acetic carmine staining for pollen viability (Radford *et al.*, 1974) and presence of esterase for stigma receptivity (Dafni, 1992). Artificial crosses were made by emasculation, bagging and manual pollination, using pollen from at least two different individuals in each cross. Fruits were monitored and the results used to calculate indices of Self Incompatibility (ISI) and Automatic Self-pollination (IAS). ISI was calculated from the ratio between selfed and crossed fruit-set (Bullock, 1985) or seed production (Zapata and Arroyo 1978) and IAS as the ratio between spontaneous versus cross-pollinated fruits (Zapata and Arroyo, 1978). Seeds

TABLE 1. Species studied and their populations, habitat, chromosome number, seed weight and germination

Section	Species	Chromosome number (2n=)	Population	Habit	Weight of seed (mg)	Germination (%) expt. 1 (4 months)	Germination (%) expt. 2 (8 months)
Fissendocarpa (Haines) Raven	*Ludwigia hyssopifolia* (G. Don) Exell	16[1]		annual	(S) 0.001	1	30
					(L) 0.012[4]	5	41
Macrocarpon (Mich.) Hara	*Ludwigia octovalvis* (Jacq.) Raven	32,48[1]	OCTO	annual	0.029	58	12
			SESSI	perennial	0.067	86	78
Myrtocarpus (Munz) Hara	*Ludwigia elegans* (Camb.) Hara	64[2]		perennial	0.037	7	43
	Ludwigia sericea (Camb.) Hara	32[2]		perennial	0.021	37	42
	Ludwigia nervosa (Poir.) Hara	16[2]		perennial	0.014	21	14
	Ludwigia tomentosa (Camb.) Hara	64[2]	GLA	perennial	0.045	65	42
			TOM	perennial	0.067	31	50
	Ludwigia laruotteana (Camb.) Hara	64[2]		perennial	0.047	63	57
Pterocaulon T.P. Ramamoorthy	*Ludwigia decurrens* Walt.	16[2]		annual	0.002	1	20
	Ludwigia erecta (L.) Hara	16[2]		annual	0.002	39	12
Seminuda Raven	*Ludwigia leptocarpa* (Nutt.) Hara	32[1,3]		perennial	0.136[4]	40	60

[1] Raven and Tai, 1979
[2] Ramamoorthy and Zardini, 1987
[3] Ormond *et al.*, 1978 b
[4] Including endocarps

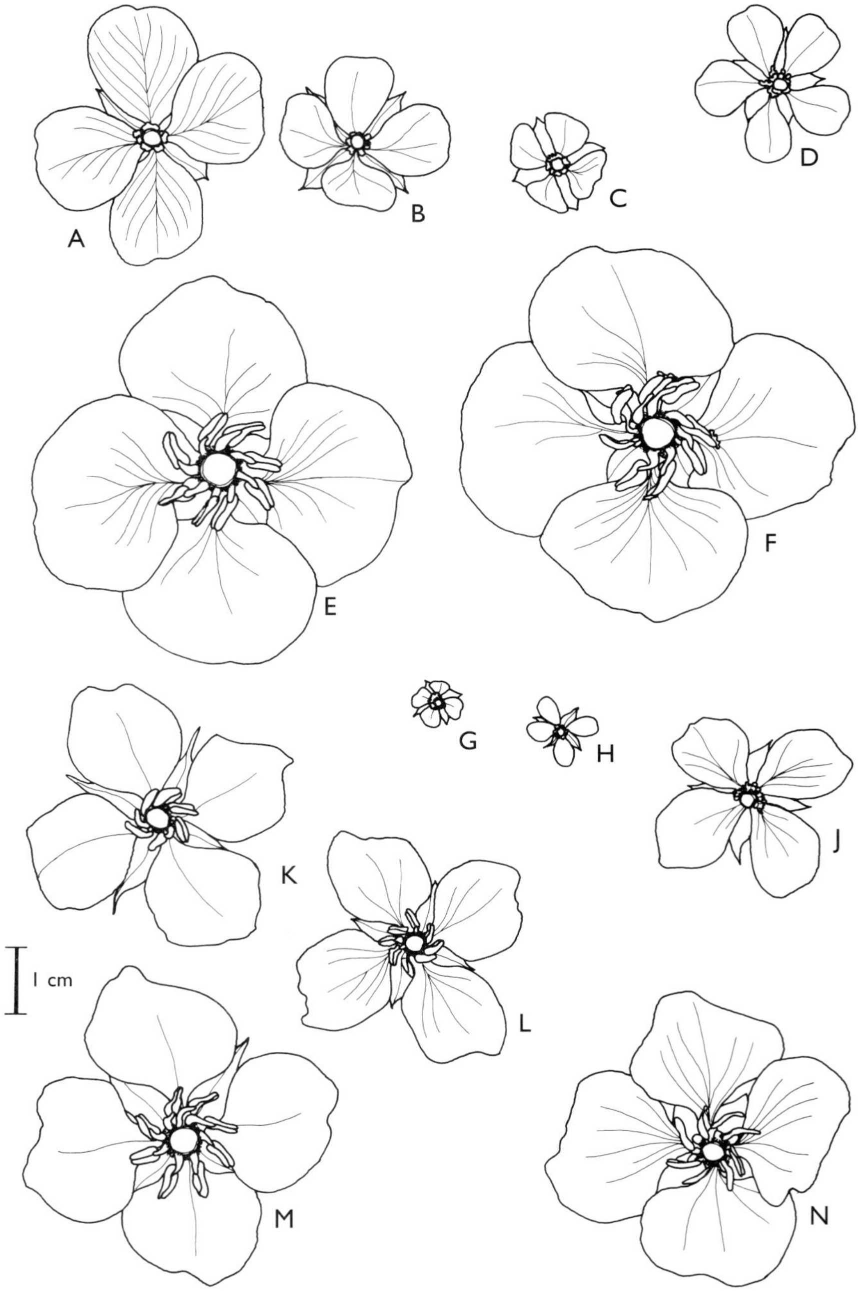

FIG. 1. Flowers of different populations of *Ludwigia*: A. *L. octovalvis* - SESSI. B. *L. octovalvis* - intermediate. C. *L. octovalvis* - OCTO. D. *L. leptocarpa*. E. *L. tomentosa* - GLA. F. *L. tomentosa* - TOM. G. *L. erecta*. H. *L. hyssopifolia*. J. *L. decurrens*. K. *L. elegans*. L. *L. sericea*. M. *L. laruotteana*. N. *L. nervosa*.

TABLE 2. Results of controlled pollination and breeding systems

Species	No of plants	Control	Agamospermy	Spontaneous	Selfed	Crossed	IAS	ISI
L. sericea	45	94 (34/32)	0 (34/0)	0 (55/0)	0 (33/0)	88 (34/30)	0	0
L. elegans	3*	61 (31/19)	0 (39/0)	0 (27/0)	0 (27/0)	86 (29/25)	0	0
L. nervosa	40	75 (37/28)	0 (43/0)	12 (50/6)	15 (32/5)	84 (33/28)	0.8	0.18
L. laruotteana	40	90 (21/19)	0 (30/0)	49 (21/13)	88 (26/23)	97 (33/32)	0.56	0.91
L. tomentosa - TOM	9	86 (21/18)	0 (17/0)	62 (16/10)	86 (14/12)	100 (17/17)	0.72	0.86
L. tomentosa - GLA	15	95 (20/19)	0 (27/0)	62 (29/18)	94 (17/16)	100 (20/20)	0.66	0.94
L. erecta	31	84 (31/26)	0 (40/0)	82 (34/28)	76 (33/25)	79 (33/26)	1.08	0.96
L. decurrens	58	80 (30/24)	0 (44/0)	72 (51/37)	90 (30/27)	86 (29/25)	0.8	1.04
L. octovalvis - SESSI	43	93 (30/28)	0 (36/0)	40 (37/15)	73 (30/22)	81 (32/26)	0.55	0.9
L. octovalvis - OCTO	47	63 (30/19)	0 (40/0)	75 (32/28)	80 (30/24)	80 (25/20)	0.94	1
L. leptocarpa	42	75 (36/27)	0 (42/0)	8 (59/5)	57 (28/16)	71 (28/20)	0.14	0.8
L. hypossifolia	31	84 (31/26)	0 (40/0)	82 (34/28)	75 (33/25)	79 (33/26)	1.09	0.95

Fruit set as percentage of total flowers pollinated (absolute numbers in parenthesis)
IAS - Index of automatic self-pollination
ISI - Index of self-compatibility
* - 3 sites treated as individuals

were collected from one nearly dehiscent fruit from five different individuals from May to July 1995 and used for two germination experiments. The first began after 4 months of storage at room temperature, and the second after 8 months. A sample of seeds was weighed and we observed germination in water and moist vermiculite placed in controlled environment chambers at 25° C and 12-12 hour light-dark photoperiod. The resulting seedlings were cultivated in a greenhouse.

Phenology, floral biology and breeding systems

Some species flower during the whole year, but flowering was concentrated mainly in the autumn, like *L. peruviana* in Australia (Jacobs *et al.*, 1994). The majority of individuals, flowering at the end of the wet season and the beginning of the dry season (autumn), diminish blossom production during winter. Frutification occurs simultaneously and open fruits were found almost all year, but predominantly at the end of autumn, except in *L. decurrens*, *L. elegans*, and *L. octovalvis* population OCTO which did not produce any flowers during the dry season, and only one or two capsules with seeds were recorded for a few plants. *L. laruotteana* only flowers from the beginning of autumn till the end of winter and open fruits retained some seeds till the first rains.

Flowers open during the morning, near noon during the winter, and the petals fall during the afternoon. One of the factors that determine the flowering period may be floral herbivory (Bawa, 1983). In these species, one or few flowers are produced by each branch every day, possibly to avoid coleoptera that feed on the ovaries and fruit. Sankaran and Krishna (1966) and Ott (1993) also documented coleoptera living in *Ludwigia* flowers.

Ludwigia flowers have an inferior ovary with many ovules, four (5-6 in *L. leptocarpa*) yellow, flat (excluding *L. nervosa*), free petals, and the 8 stamens are free with extrorse anthers. The solitary flowers vary in size and morphology (Fig. 1). The smallest are those of *L. erecta* and *L. hyssopifolia* (0.5 cm diameter) and the largest is *L. tomentosa* (6 cm). The stigma is capitate, covered with papillae and becomes humid and shiny when receptive. This type of stigma - receptive cells papillate with moderate to light stigma secretion - has been noted for other species of *Ludwigia* (Heslop-Harrison, 1990). The esterase reaction indicates that for all species the whole of the stigmatic area is receptive, differing from *L. peploides* with the upper two-thirds receptive (Raven, 1963). Receptivity is synchronised with the liberation of pollen. Pollen viability was high for all species, the lowest value being found in *L. sericea* (82%). The pollen was shed singly in *L. hyssopifolia*, and in tetrads in the remaining species.

The results of the breeding system experiments are summarised in Table 2. They show that agamospermy did not occur (the possibility of pseudogamy was not tested, however).

The self-incompatible (SI) species, *L. elegans* and *L. sericea*, produced no fruit on selfing, suggesting the presence of a very strong incompatibility barrier. In *L. elegans* and *L. sericea* the filaments lay flat on the petals, differing from the arrangement described by Ramamoorthy and Zardini (1987), and have a rotational movement which results in very little pollen being deposited on the stigma. Pollen tube growth halted in the upper style, forming small transparent vesicles in self-pollinated flowers, exhibiting behaviour typical of a gametophytic system (Nettancourt, 1977). As in other self-incompatible species of *Ludwigia*, the incompatible pollen tube inhibition site is different from other genera in the family (Heslop-Harrison, 1990). Although the self-incompatibility reaction observed appears to be that of a gametophytic system, it showed unusual features, with different ploidy levels being SI (Table 1). Another population of *L. sericea* in the same region was also shown to be self-incompatible by Sazima and Santos (1982). *L. nervosa* produces some fruits (Table 2) and seeds from self-pollination, although its flowers have reflexed petals and filaments arched around the style but projecting the anthers away from the stigma. However the ISI calculated for fruits (Table 2) and seeds (average of 249 seeds on selfing versus 850 in crossed fruit) include it among the self-incompatible species (0.18 / 0.20). *L. nervosa* is reported to be autogamous in Venezuela (Ramirez and Brito, 1990) maturing both selfed (40% fruit, 926.25 seeds average) and crossed fruit (66.67% fruit, 1,157.19 seeds). This number of seeds was higher than the number found in the Campinas population, and more populations need to be tested to explain these results. The three SI species studied here belong to section *Myrtocarpus* which contains other self-incompatible species (Ramamoorthy and Zardini, 1987; Heslop-Harrison, 1990).

Among the self-compatible species, the ISIs were greater than or equal to 0.8 (Table 2). The rate of spontaneous self-pollination was largely determined by the degree of spatial separation of anthers and stigma and the liberation of the pollen away from the stigmatic surface, since there is little temporal separation between stigma receptivity and anther dehiscence. *Ludwigia* flowers in selfing species, as in other self-fertilised taxa, are usually smaller than those produced by outcrossers or show a reduction in the degree of stigma exsertion (Stebbins, 1970; Wyatt, 1988). The anthers shed pollen over the stigma in the three smaller flowers (*L. erecta*, *L. hyssopifolia* and *L. octovalvis* - OCTO), which can be considered autogamous, because there is no evident difference between selfed and spontaneous fruit production. Fruit formation was higher with spontaneous selfing than with any other treatment and lower rates in experimental treatments may have resulted from damage to the flower or stigma caused by emasculation or manipulation of these small flowers. The similar fruit production observed in control and spontaneous treatments may be the result of the low number of pollinator visits observed in these species. Population SESSI of *L. octovalvis* showed completely different behaviour from population OCTO, with larger flowers, and the filaments lower than the style, and appears to be outcrossed. Its larger flowers may attract more insects, as observed in the larger flowered subspecies of *L. peploides* which has a more elevated average of visits per hour (Flores, 1989). *L. octovalvis* is also reported self-compatible in other populations (Ormond 1973; Ormond *et al.*, 1979).

L. leptocarpa is also self-compatible (Ormond *et al.*, 1978a and b), and it has the stamens disposed in two whorls and the filaments do not reach stigma level. The presence of pollinators is very important in this species (IAS = 0.14) as described in *L. peploides* (Estes and Thorp, 1974).

In the other species the stamens are at the same level as the stigma although in *L. laruotteana* and *L. tomentosa*, a space between stigma and anthers is maintained by thicker, more rigid filaments. If pollination did not occur, pollen was observed to fall from the anthers in strands linked by viscin threads, but did not cover the whole stigma. *L. decurrens* exhibits an eccentric position of stamens whose anthers are projected to one side of the flower. In all of these species, manual pollen transfer to the stigma enhanced fruit production and if the transference is from other plants, fruit-set is higher (Table 2), suggesting greater success for cross pollen. In at least one *Clarkia* species, self pollen apparently suffered a strong disadvantage when competing with outcross pollen during fertilisation (Jones, 1994), and it is possible that a similar effect occurs here.

A correlation between breeding systems and habit is reported for many families as well as in Onagraceae (Raven, 1979). The perennial habit can be correlated with strategies which increase cross pollination. *L. octovalvis* is an example where the perennial population has lower selfed fruit production because of differences in floral morphology. Among the species of *Ludwigia* investigated here, the self-incompatible species are all perennial shrubs (Tables 1 and 2). Some of the shrubs, such as *L. elegans*, *L. laruotteana*, *L. tomentosa* both populations, and *L. peruviana* (Jacobs *et al.*, 1994), may grow from shoots budding from the rootstock and not from perennial above-ground stems. The annual species are diploid (Table 1), excluding the *L. octovalvis* complex. Autogamy is associated with herbaceous forms and potential colonisation of open areas, and although all the *Ludwigia* species studied here tend to follow this strategy, the annual herbs regularly occupy the most open habitats and receive less pollinators. Self-pollination occurs in many conditions where pollinators are limited as in colonisation of isolated islands (Lloyd, 1992).

Ludwigia is melitophilous and was originally described with polylectic or generalist bees playing a predominant role in pollination (Estes and Thorp, 1974; Raven 1988). More recently, a series of specialised and oligoletic bees was observed visiting south-american species. Species of solitary bees - Anthophoridae: *Diadasina*, *Ptilothrix*; Colletidae: *Tetraglossula*; Halictidae: *Pseudagapostemon* - visit different species of *Ludwigia* (Gimenes, 1988, 1991, 1993; Martins and Antonini, 1994; Schlindwein, 1995). Some of them have morphological adaptations to collect *Ludwigia* pollen (Gimenes, 1991), and have simultaneous temporal patterns (Gimenes *et al.*, 1993). The adults emerge from the ground at the end of rainy season while *Ludwigia* is flowering (Martins and Antonini, 1994). The outcrossed *L. octovalvis* population SESSI has its flowering peak after OCTO, during the drier months. Species of *Pseudagapostemon*, *Tetraglossula* and *Heterosarellus* observed on *Ludwigia* can exhibit aggressive behaviour directed against the same or other species, while they were feeding in flowers (Gimenes, 1988). This behaviour can result in shorter but more numerous visits to plants that produce few flowers per day. The flowers and nectary areas of *L. nervosa* are larger than the other incompatible species and attract large bees like *Xylocopa* (Ramires and Brito, 1992) and *Bombus* that can disturb other visitors (pers. obs.) and are known to fly for many kilometers (Jansen, 1974; Frankie *et al.*, 1976). All of these strategies, allied to few open flowers per plant every day, may be very important to avoid contamination of stigma by pollen of the same plant. This is essential for self-incompatible species and to increase visits and bring more pollen from different individuals providing an increase in efficiency of pollination in the other species.

Fruits and seeds

The fruits are capsules that mature and open after approximately one month, a fast maturation rate possibly associated with low mass of fruit and seeds (Ramirez and Brito, 1987). *L. leptocarpa* capsules dry and remain closed during their reproductive season, but they open slowly when the plants dry. *L. hyssopifolia* has two kinds of seeds (Raven, 1963; Eyde, 1978) an apical small free seed (called S) and a bigger basal seed included in an endocarp (L). When the capsule dries, the apical seeds are liberated immediately, but the basal seeds remain in the fruit for some weeks. The seeds of *Ludwigia* can be spread adhering to feathers or in mud on birds' feet and by water (Jacobs *et al.*, 1994). The fruit and seeds of *Ludwigia* were studied by Eyde (1978) who showed that the raphe that is common in sections *Myrtocarpus* and *Macrocarpon* can aid floatation. All the species considered here have seeds that float on water for several weeks and germinate both in water and on moist vermiculite. Not all the seeds are dispersed at the same time, and some remain inside the open fruit. The seeds may pass the dry season inside the capsule and can float for several weeks, in both cases permitting them to survive the winter or drier periods. The dispersal of closed fruits of *L. leptocarpa* till the beginning of the dry and cold season may be a similar strategy. In this species (Christy and Charitz, 1980) as in *L. peruviana* (L.) Hara (Jacobs *et al.*, 1994) the initiation of germination may be delayed in colder temperatures. In this manner the seeds would be protected and continue dispersal till the beginning of the following rainy season when the temperature increases.

All of the species required light to germinate as found in *L. peruviana* (Jacobs *et al.*, 1994), behaviour typical of colonising species. The mean seed masses (Table 1) varied from 0.001 mg (*L. hyssopifolia* apical seed) to 0.136 mg (*L. leptocarpa*). Germination rates decreased in the smallest seeds and the annual species (*L. decurrens*, *L. erecta* and *L. hyssopifolia*) had smaller seeds than the perennials (Table 1). Among the two kinds of seeds of *L. hyssopifolia*, the smaller seeds had lower rates of germination. The non-autogamous *L. octovalvis* SESSI had higher germination percentages than the autogamous OCTO population. The same relationship was observed between the subspecies of *L. peploides*, with the outcrossing taxon germinating more seeds than the selfer (Flores, 1989). Similar differences may occur between populations since small-sized seeds of *L. leptocarpa* have a low germination rate (Dolan, 1984). The average weight of *L. leptocarpa* seed found here is lower (Table 1) than in other populations which range from 0.138 to 0.285 mg (Christy and Sharitz, 1980; Dolan, 1984). The germination rate is inferior in the annual species *L. decurrens* and *L. hyssopifolia*. These species and *L. elegans*, which appear to express some dormancy (Table 1), released the seeds before winter. The other annual species, *L. erecta*, lost seed viability (Table 1), and this also occurred in *L. octovalvis* (SESSI and OCTO), *L. nervosa*, *L. tomentosa* (population GLA) and *L. laruotteana*. Seed viability also decreased in *L. peruviana* (Jacobs *et al.*, 1994).

Conclusions

In general, the phenology of flowering and fruiting is synchronous, flowering in the autumn at the end of the rainy season when some oligoletic bees appear. *Ludwigia* sets fruit during the autumn and the capsules open while the seeds are dormant or germination is delayed until the beginning of next hot rainy season.

Many species occupy this restricted site, showing a variety of habits and breeding systems, from autogamous herbaceous species to self-incompatible shrubs. Outcrossed species tend to have vegetative reproduction. A series of self-compatible species avoid self-pollination by modifications of flower morphology associated with pollinator behaviour. In this set of *Ludwigia* species, autogamy can be associated with reduction of floral size, but polyploidy, increased flower dimensions, loss of self-incompatibility and strategies to increase cross-pollination can all be found within one evolutionary line in section *Myrtocarpus*.

Acknowledgements

A.O.S. Vieira thanks Fundação Margaret Mee / Margaret Mee Amazon Trust for travel grants to participate in the conference in Kew and CAPES for funding during this project. We are grateful to Centro Experimental de Campinas - Instituto Agronômico de Campinas for permitting access to the field site, Maria Augusta Santos Vieira who drew the figure and E. Zardini who confirmed some identifications of *Ludwigia* species.

References

Bawa, K. (1983). Patterns of flowering in tropical plants. In: C.E. Jones and R.J. Little (editors). Handbook of experimental pollination biology. Van Nostrand Reinhold, Scientific and Academic Editions, New York.

Bullock, S.H. (1985). Breeding systems in the flora of a tropical deciduous forest. *Biotropica* **17:** 287–301.

Carlquist, S. (1975). Wood anatomy of Onagraceae, with notes on alternative modes of photosynthate movement in dicotyledon woods. *Ann. Missouri Bot. Gard.* **62:** 386–424.

Christy, E.J. and Sharitz, R.R. (1980). Characteristics of three populations of a swamp annual under different temperature regimes. *Ecology* **61:** 454–460.

Cruden, R.W. and Jensen, K.G. (1979). Viscin threads, pollination efficiency and low pollen-ovule ratios. *Amer. J. Bot.* **66:** 875–879.

Dafni, A. (1992). Pollination ecology. A practical approach. pp. 25–58. Oxford University Press, New York.

Dolan, R.W. (1984). The effect of seed size and maternal source on individual size in a population of *Ludwigia leptocarpa* (Onagraceae). *Amer. J. Bot.* **71:** 1302–1307.

Estes, J.R. and Thorp, R.W. (1974). Pollination in *Ludwigia peploides* ssp. *glabrescens* (Onagraceae). *Bull. Torrey Bot. Club* **101:** 272–276.

Eyde, R.H. (1977). Reproductive structures and evolution in *Ludwigia* (Onagraceae) I. Androecium, placentation, merism. *Ann. Missouri. Bot. Gard.* **64:** 644–655.

Eyde, R.H. (1978). Reproductive structures and evolution in *Ludwigia* (Onagraceae) II. Fruit and seed. *Ann. Missouri. Bot. Gard.* **65:** 656–675.

Eyde, R.H. (1981). Reproductive structure and evolution in *Ludwigia* (Onagraceae) III. Vasculature, nectaries, conclusions. *Ann. Missouri, Bot. Gard.* **68:** 379–412.

Eyde, R.H. (1982). Evolution and systematics of the Onagraceae: floral anatomy. *Ann. Missouri, Bot. Gard.* **69:** 735–747.

Flores, S.I. (1989). Reproductive biology of *Ludwigia peploides peploides* and *Ludwigia peploides glabrescens*. Ph.D. Thesis, University of Chicago, Chicago.

Frankie, G.W., Opler, P.A. and Bawa, K.S. (1976). Foraging behaviour of solitary bees: implications for outcrossing of a neotropical forest tree species. *J. Ecol.* **64:** 1049–1057.

Gimenes, M. (1988). Interações entre *Tetraglossula bigamica* (Strand; Hymenoptera, Colletidae), *Pseudagapostemon brasiliensis* Cure (Hymenoptera, Halictidae), outras espécies e flores de *Ludwigia elegans* (Camb.) Hara (Onagraceae). MSc. Thesis. Universidade de São Paulo, Ribeirão Preto.

Gimenes, M. (1991). Some morphological adaptations in bees (Hymenoptera, Apoidea) for collecting pollen from *Ludwigia elegans* (Onagraceae). *Revista Brasil. Entomol.* **35:** 413–422.

Gimenes, M. (1993). Estudo dos padrões temporais de *Ludwigia elegans* (Camb.) Hara (Onagraceae) e as abelhas (Apoidea) visitantes em duas localidades do Estado de São Paulo (Brasil). Doctoral Thesis, Universidade de São Paulo, São Paulo.

Gimenes, M., Benedito-Silva, A.A. and Marques, M.D. (1993). Chronobiologic aspects of a coadaptive process: the interaction of *Ludwigia elegans* flowers and their more frequent bee visitors. *Chronobiol. Int.* **10:** 20–30.

Heslop-Harrison, Y. (1990). Stigma form and surface in relation to self-incompatibility in the Onagraceae. *Nordic J. Bot.* **10:** 1–19.

Hoch, P.C., Crisci, J.V., Tobe, H., Berry, P.E. (1993). A cladistic analysis of the plant family Onagraceae. *Syst. Bot.* **18:** 31–47.

Jacobs, S.W.L., Perret, F., Sainty, G.R., Bowmer, K.H. and Jacobs, B.J. (1994). *Ludwigia peruviana* (Onagraceae) in the Botany Wetlands near Sydney (Australia). *Austral. J. Mar. Freshwater Res.* **45:** 1481–1490.

Janzen, D.H. (1974). Euglossine bees as long - distance pollinators of tropical plants. *Science* **171:** 203–205.

Jones, K.N. (1994). Nonrandom mating in *Clarkia gracilis* (Onagraceae): a case of cryptic self-incompatibility. *Amer. J. Bot.* **81:** 195–198.

Lloyd, D. (1992). Self- and cross-fertilization in plants. II. The selection of self-fertilization. *Int. J. Pl. Sci.* **153:** 370–380.

Martin, F.N. (1959). Staining and observing pollen tubes by means of fluorescence. *Stain Technol.* **34:** 125.

Martins, R.P. and Antonini, Y. (1994). The biology of *Diadasina distincta* (Holmberg, 1903) (Hymanoptera: Anthophoridae). *Proc. Entomol. Soc. Wash.* **96:** 553–560.

Mello, M.H. de A., Pedro Junior, M.J., Ortolani, A.A. and Alfonsi, R.R. (1994). Chuva e temperatura: cem anos de observações em Campinas. Instituto Agronômico de Campinas, Campinas. (Boletim Técnico 154).

Munz, P.A. (1947). Onagraceae. In: F.C. Hoehne, Flora Brasilica, 41 (1), Instituto de Botânica, São Paulo.

Nettancourt, D. (1977). Incompatibility in angiosperms. Springer-Verlag, Berlin.

Ormond, W.T. (1973). Contribuição ao estudo biossistemático e ecológicos de *Ludwigia octovalvis* (Jacq.) Raven (Onagraceae). *Revista Brasil. Biol.* **33:** 87–107.

Ormond, W.T., Castells, A.R.C. and Correia, M.C.R. (1978a). Contribuição ao estudo biossistemático e ecológico de *Ludwigia leptocarpa* (Nutt.) Hara. *Rodriguésia* **30:** 345–357.

Ormond, W.T., Castells, A.R.C. and Pinheiro, M.C.B. (1978b). Ecologia da reprodução de *Ludwigia leptocarpa* (Nutt.) Hara (Onagraceae). *Bol. Mus. Nac. Rio de Janeiro, Bot.* **48:** 1–11.

Ormond, W.T., Pinheiro, M.C.B., Castells, A.R.C. de and Correia, M.C.R. (1979). Hibridação natural em populações de *Ludwigia octovalvis* (Jacq.) Raven (Onagraceae). *Bol. Mus. Nac. Rio de Janeiro, Bot.* **54:** 1–7.

Ott. J.R. (1993). Host plant-mediated variation in overwinter site quality: implications for the size and composition of populations of *Acanthoscelis alboscutellatus* (Coleoptera, Bruchidae). *Oecologia* **96:** 493–499.

Patel, V.C., Skarvala, J.J. and Raven, P.H. (1984). Pollen characters in relation to the delimitation of Myrtales. *Ann. Missouri Bot. Gard.* **71:** 858–969.

Radford, A.E., Dickinson, W.C., Massey, J.R. and Bell. C.R. (1974). Vascular plant systematics. pp. 211–222. Harper and Row Pub., New York.

Ramamoorthy, T.P. and Zardini, E.M. (1977). The systematics and evolution of *Ludwigia* sect. *Myrtocarpus* sensu latu (Onagraceae). Missouri Botanical Garden, St. Louis. (Monographs in Systematic Botany vol. 19).

Ramirez, N. and Brito, Y. (1987). Patrones de floracion y frutificacion en una comunidad pantanosa tipo morichal (Calabozo - Guarico, Venezuela). *Acta Ci. Venez.* **38:** 376–381.

Ramirez, N. and Brito, Y. (1990). Reproductive biology of a tropical palm swamp community in the Venezuelan llanos. *Amer. J. Bot.* **77:** 1260–1271.

Ramirez, N. and Brito, Y. (1992). Pollination biology in a palm swamp community in the Venezuelan central plains. *Bot. J. Linn. Soc.* **110:** 277–302.

Raven, P.H. (1963). The old world species of *Ludwigia* (including *Jussiaeae*), with a synopsis of the genus (Onagraceae). *Reinwardtia* **6:** 327–427.

Raven, P.H. (1979). A survey of reproductive biology in Onagraceae. *New Zealand. J. Bot.* **17:** 575–593.

Raven, P.H. (1988). Onagraceae as a model of plant evolution. In: L.D. Gottlieb and S.K. Jain (editors). Plant evolutionary biology, pp. 85–107, Chapman and Hall, New York.

Raven, P.H. and Tai, W. (1979). Observations of chromosomes in *Ludwigia* (Onagraceae). *Ann. Missouri Bot. Gard.* **66:** 682–879.

Sankaran, T. and Krisha, K. (1966). The biology of *Nanophyes* sp. (Col., Curculionidae) infesting *Jussiaea repens* in India. *Bull. Entomol. Res.* **47:** 337–341.

Sazima, M. and Santos, J.U.M. (1982). Biologia floral e insetos visitantes de *Ludwigia sericea* (Onagraceae). *Bol. Mus. Paraense Emilio Goeldi, N.S., Bot.* **54:** 1–10.

Schlindwein, C. (1995). Wildbienen und Trachtpflanzen in einer sudbrasilianiaschen Buschlandschaft: Fallstudie Guaritas, Bestäubung bei Kakteen und Loasaceen. Verlag Urich E. Graver, Stuttgart.

Stebbins, G.L. (1970). Adaptative radiation in angiosperms. I. Pollination mechanism. *Annual Rev. Ecol. Syst.* **1:** 307–326.

Tobe, H. and Raven, P.H. (1986a). A comparative study of the embryology of *Ludwigia* (Onagraceae): characteristics, variation, and relationships. *Ann. Missouri Bot. Gard.* **73:** 768–787.

Tobe, H. and Raven, P.H. (1986b). Evolution of polysporangiate anthers in Onagraceae. *Amer. J. Bot.* **73:** 475–488.

Wyatt, R. (1988). Phylogenetic aspects of the evolution of self-pollination. In: L.D. Gottlieb and S.K. Jain (editors). Plant evolutionary biology, pp. 109–131, Chapman and Hall, New York.

Zapata, T.R. and Arroyo, M.T.K. (1978). Plant reproductive ecology of a secondary deciduous tropical forest in Venezuela. *Biotropica* **10:** 221–230.

Riveros, M., Palma, B. and Barlow, P. (1998). Reproductive biology of three *Nothofagus* species in the south of Chile. In: S.J. Owens and P.J. Rudall (Editors). Reproductive Biology, pp. 407–416. Royal Botanic Gardens, Kew.

REPRODUCTIVE BIOLOGY OF THREE *NOTHOFAGUS* SPECIES IN THE SOUTH OF CHILE

MAGALY RIVEROS[1], BEATRIZ PALMA[2] AND PETER BARLOW[3]

[1]Instituto de Botánica, Universidad Austral, Casilla 567, Valdivia, Chile
[2]Instituto de Biología, Universidad Católica de Valparaíso, Casilla 4059, Valparaíso, Chile
[3]Long Ashton Research Station, University of Bristol, Bristol BS18 9AF, UK

Abstract

Aspects of the reproductive biology of three Chilean *Nothofagus* tree species, *N. dombeyi, N. nitida* and *N. obliqua*, were studied. In addition to the fact that *Nothofagus* species are generally monoecious, self-incompatible and anemophilous, a number of other biotic factors limit their fertility. Firstly, the three investigated species show an intrinsically low fertility, as judged by the number of seeds produced following pollination procedures that should maximise the probability of fertilisation. Secondly, under natural conditions, the number of viable and compatible pollen grains which land on receptive stigmas may also be limiting. The humid weather conditions that often prevail at flowering time are one of the abiotic factors affecting pollination since such conditions reduce the amount and viability of the pollen that is released from the anther for dispersal in the wind. Under optimal conditions, on the other hand, pollen is usually available to the first cohort of receptive stigmas because male flowering generally precedes female flowering. The purity and density of the stands of trees also favour maximal pollination with compatible pollen. However, there may be no great necessity for high fertility in a stable *Nothofagus* forest environment because the longevity of the trees does not require their frequent replacement.

Introduction

Species of the genus *Nothofagus* play an important role in the structure of the temperate rain forests in the Southern Hemisphere, notably in Chile, Argentina and New Zealand (Wardle, 1984; Poole, 1987; San Martín and Ramírez, 1987). Owing to the excellent quality of their wood, certain species of this genus are now being exploited extensively in the south of Chile with consequent ecological dangers resulting from the deforestation of the *Nothofagus* environment (Armesto and Smith-Ramírez, 1994). Although attempts are being made to replant areas denuded of trees, one problem frustrating these efforts is the low rate of viable seed production in the Chilean *Nothofagus* species (Riveros *et al.*, 1995a). This low productivity is exacerbated by the fact that trees do not flower regularly and in some years widespread non-flowering occurs, as was found also for New Zealand *Nothofagus* species (Wardle, 1984).

In the light of these biological and practical problems, our principal interest has been to establish which factors limit, or regulate, the formation of viable seed (Palma *et al.*, 1996). While important observations on the genecology of both Chilean and New Zealand *Nothofagus* species have been collected over a number of years (Wardle, 1984; Donoso, 1987), more information is required concerning the reproductive biology of *Nothofagus*, especially for the Chilean species. In particular, it proved necessary (1) to investigate their reproductive system, focussing on three of the economically most important Chilean species, and (2) to characterise the flow of pollen between anthers and stigmas in these species. (1) deals with the biological processes mediating the interaction of male and female gametes within the flower and the subsequent embryogenesis which follows fertilisation, whereas (2) deals with both biological and physical processes relating to pollen release and its subsequent movement culminating in pollen capture on receptive stigmatic surfaces. This paper addresses these two areas of the reproductive biology of *Nothofagus*.

Methods

The species studied were *Nothofagus dombeyi* (Mirb.) Oerst., *N. nitida* (Phil.) Krasser and *N. obliqua* (Mirb.) Oerst. growing in forests in the vicinity of Valdivia (39°S 73°W), southern Chile. Some gymnosperm species of trees were also present in these forest stands. The first two species are evergreen (section *Calusparassus*), whereas *N. obliqua* is deciduous (section *Nothofagus*). All three species are monoecious and exclusively anemophilous. Three or five trees of these species were sampled in attempts to estimate their fertility. Results of the various observations from the individual trees of a given species were usually pooled.

Assessment of fertility

Three types of pollination regimes, or tests, were employed to assess the levels of both potential and natural fertility in the three *Nothofagus* species mentioned. Flowers on accessible branches up to 20m from the ground were selected to be donors of pollen (♂ flowers) and to receive pollen and for the subsequent sampling of infructescences (♀ flowers). Fertility was assessed by reference to the proportion, N, of fruits (nuts) containing embryos (seeds) produced per flower:

1. Manual self-pollination — giving N_{MS}
2. Manual cross-pollination — giving N_{MC}
3. Open pollination —— giving N_F.

Manual pollination (tests 1 and 2) should maximise the probability of fertilisation. Each of the resulting values N_{MS} and N_{MC} can therefore be taken as an estimate of the maximum intrinsic fertility where the gametes are, respectively, from either the same or different trees. An index, $I_S = 1 - (N_{MS}/N_{MC})$, derived from these values, represents a measure of self-infertility. An index value of 1.0 could indicate complete self-infertility, whereas a value of 0 would indicate complete self- and cross-fertility.

When pollination involves only pollen of different trees of the same species (as it may in exceedingly pure and isolated stands), an N value, N_{FC}, can be used to estimate natural fertility (or infertility). The index, $I_C = 1 - (N_{FC}/N_{MC})$, might then be used to represent the degree of departure from maximum fertility due to the anemophilous mode of pollination. In practice, however, such open cross-pollination always admits the possibility of some self-pollination since there can be no control over whether the pollen landing on a stigma is from the same or from a different tree(s) of that species. Nevertheless, there is another useful value — from test 3 — in which fruit set is estimated from tagged flowers left for open pollination in the natural state. Hence, the index $I_N = 1 - (N_F/N_{MC})$ represents the degree of infertility under natural conditions. N_F and N_{MC} also estimate the natural relative fertility, I_F, where $I_F = N_F/N_{MC}$.

Any of these indices can be assessed for either an individual of a given species, or for a population of that species providing that sufficient numbers of representative trees can be sampled.

Assessment of pollen flow

The rate of pollen flow amongst flowering branches of *Nothofagus* trees was estimated by counting the number of pollen grains captured on strips of sticky tape (18mm x 18mm) placed close to the exserted stigmas of flowers within the canopies of selected trees (Riveros et al., 1995a). The number of grains was estimated at 10 min intervals throughout the hours of daylight and for varying numbers of days during the flowering period. The influx of pollen to the tapes was also noted in relation to the changes in weather conditions throughout any given day (i.e. temperature, humidity, wind-speed) which not only affect pollen efflux from the anther but also its subsequent transmission and capture by the stigma. The pollen captured on the tapes was also inspected by light microscopy to assess the degree of ingress of pollen from other (foreign) species as well as pollen from the particular *Nothofagus* species under study: in the latter case, the pollen could be either from the same tree near whose flowers the tapes were placed, or the pollen could be from different trees of that species. Viability of the pollen was assessed as soon as possible after its capture by recording the ability of the grains to reduce colourless triphenyltetrazolium to the red formazan salt. Although it is realised that there may be differences between the way in which pollen comes to adhere to the tape as compared to a stigma (due to the local aerodynamics), sampling of pollen can be repeated at a particular site with the tape method, whereas repeated sampling of a stigma (or observation of its surface) is impossible.

The approximate mean surface area of a receptive stigma was evaluated using an ocular micrometer to measure length and breadth of the stigmatic surfaces. Stigmas were curved, however, so their total length was estimated using the lengths of 2-3 successive stigma segments. Ten stigmas per tree (3-4 trees being used) were measured in this way and a mean area was calculated. Dimensions of pollen grains were also estimated by micrometry and the mean projected area of a pollen grain, as it might lie on the surface of a stigma, was calculated. Together, the two estimates permit a calculation of the approximate maximum number of pollen grains which could ‘fit’ onto a stigmatic surface. This value can then be compared with the number of pollen grains that landed on an area of sticky tape similar to that of a stigmatic surface within a given period of time. This measure may have some biological importance because not all pollen grains which land on a stigma can effect fertilisation.

Results

The results of the pollination tests for the three species of *Nothofagus* are shown in Tables 1 and 2. First, it is important to note that all flowers, when pollinated according to tests 1 and 2, developed a fruit. Fruits are the result of *pollination* and the subsequent growth of the pericarp wall which this event stimulates; the outer portion of the fruit therefore develops regardless of the formation, as a result of *fertilisation*, of an embryo (or seed) contained within it. Thus, some fruit may contain an embryo, others may not. Second, following test 3 there is evidently a diminution in the number of embryos, relative to the results from test 2. Fewer ovules are fertilised under natural conditions than is possible by manual cross-pollination. However, the proportion of seeds formed from test 3 is not as low as that following test 1. From test 3 it is interesting to note that the number of fruits was always equal to the number of flowers, indicating that pollen was not limiting for fruit formation (evidence only of pollination), but was only limiting for seed formation (evidence of fertilisation). These results from open pollination indicate that there may naturally be a scarcity of pollen which can fertilise the ovule, or that there is some embryo abortion. Thus the already low intrinsic fertility, as judged

TABLE 1. Results of three pollination tests [in terms of the number of fruits and fruits with embryos (seed) produced with respect to the number of flowers initially tagged] performed on three species of *Nothofagus*. The resulting indices of self- and natural-infertility (I_S and I_N, respectively) are given. *n* = number of trees sampled.

Species	Test	No. flowers	No. fruits	No. seeds	I_S	I_N	n
N. dombeyi	1	119	119	6	0.911 (tests 1–2)		1
	2	1393	1393	785		0.775 (tests 2–3)	5
	3	567	567	72			5
N. nitida	1	954	954	21	0.966 (tests 1–2)		4
	2	165	165	107		0.921 (tests 2–3)	4
	3	879	879	45			4
N. obliqua	1	353	353	6	0.952 (tests 1–2)		2
	2	757	757	266		0.533 (tests 2–3)	2
	3	469	469	77			2

by the low values of N_{MC} for the three species, is diminished still further. The degree of agamospermy (i.e. the proportion of embryos forming in the absence of pollen) was also assessed since if this event occurred frequently it could influence the interpretation of the values relating to fertility. The numbers of possibly agamospermic seeds in each species were few (< 1.3%) and could have been the result of accidental contamination, with pollen, of the bagged flowers used for the assessment of this variable (Riveros *et al.*, 1995b). In the light of these findings, no correction of the infertility indices was deemed necessary.

All three species of *Nothofagus* clearly show a high level of infertility after self-pollination, with indices of I_S between 0.91 and 0.97 (Table 1). Successful sexual reproduction therefore almost exclusively depends upon an interaction with pollen from neighbouring trees of the same species. However, even when the opportunity for cross-pollination was greatest (test 2), there was still a relatively high degree of infertility, as reflected in the values N_{MC}, which were appreciably <1.0 in the three species (Table 2). These results, when considered in conjunction with the low values of N_F for each of the species (Table 2), ensure low indices of relative fertility (I_F): that is, the levels of natural embryo production are much less than is potentially possible with controlled manual fertilisation.

The data in Table 1 for the numbers of fruits and embryos (seed) following test 3 indicate that only between 5% and 16% of fruit contain an embryo. A further sampling of fruits from another set of openly pollinated trees also indicated that many fruits lacked an embryo (Table 3). Collections of fruiting cupules from these trees indicated that, in this case, embryogenesis failed in 66-94% of fruit. What proportion of embryos aborted after initially commencing development was not established.

TABLE 2. Proportions of fruits with embryos produced in three species of *Nothofagus* following open or manual cross-pollination (respectively tests 3 and 2 from which N_F and N_{MC} are estimated). The quotient of the two values (N_{MC} being the divisor) is the index (I_F) of relative fertility.

Species	N_F	N_{MC}	I_F
N. dombeyi	0.127	0.564	0.225
N. nitida	0.051	0.648	0.079
N. obliqua	0.164	0.351	0.467

Relative fertility might be explained in terms of pollen flow. If there is not enough pollen from a compatible (i.e. different) tree, then fertility, in terms of seed/embryo production, might inevitably fall, even though there is enough pollen to induce fruit formation. Further information in this direction can be obtained from estimates of pollen influx. Estimates of the rate of pollen capture at sites near to receptive stigmas are shown in Fig. 1. The pollen was categorised as being either of foreign origin or originating from the species in question. Although no assessment was possible of the source of the same-species pollen (i.e. from the same tree as the reference stigmas, or from another tree), pollen could be categorised as being either viable or non-viable. The reason for differential viability of the captured pollen is the subject of further investigations, but probably the timing of pollen release from the anther in relation to meteorological factors (e.g. mist) could affect viability. In practice, it would be difficult to establish a relation between flux and relative fertility, I_F. What the present flux studies do show, however, is that pollen availability is usually greatest in the later part of the day (Fig. 1). The reason for this is that mornings in the *Nothofagus* forest environment are usually misty and pollen is not released from the anther until the humidity diminishes and the temperature rises. Only then is pollen liberated and the stigmas become receptive, as judged by their sticky, glistening appearance. In general, flowering of *Nothofagus* in the Valdivian region coincides with a part of the year when rainfall is diminishing and air temperature is increasing (Riveros et al., 1995b).

TABLE 3. Results of a further survey of fruits from open pollinated trees of three *Nothofagus* species collected and examined for the presence or absence of embryos (recorded as a percentage). n = number of fruits examined.

Species	With embryo	Without embryo	n
N. dombeyi	31.8	68.2	600
N. nitida	6.3	93.8	400
N. obliqua	33.8	66.2	1400

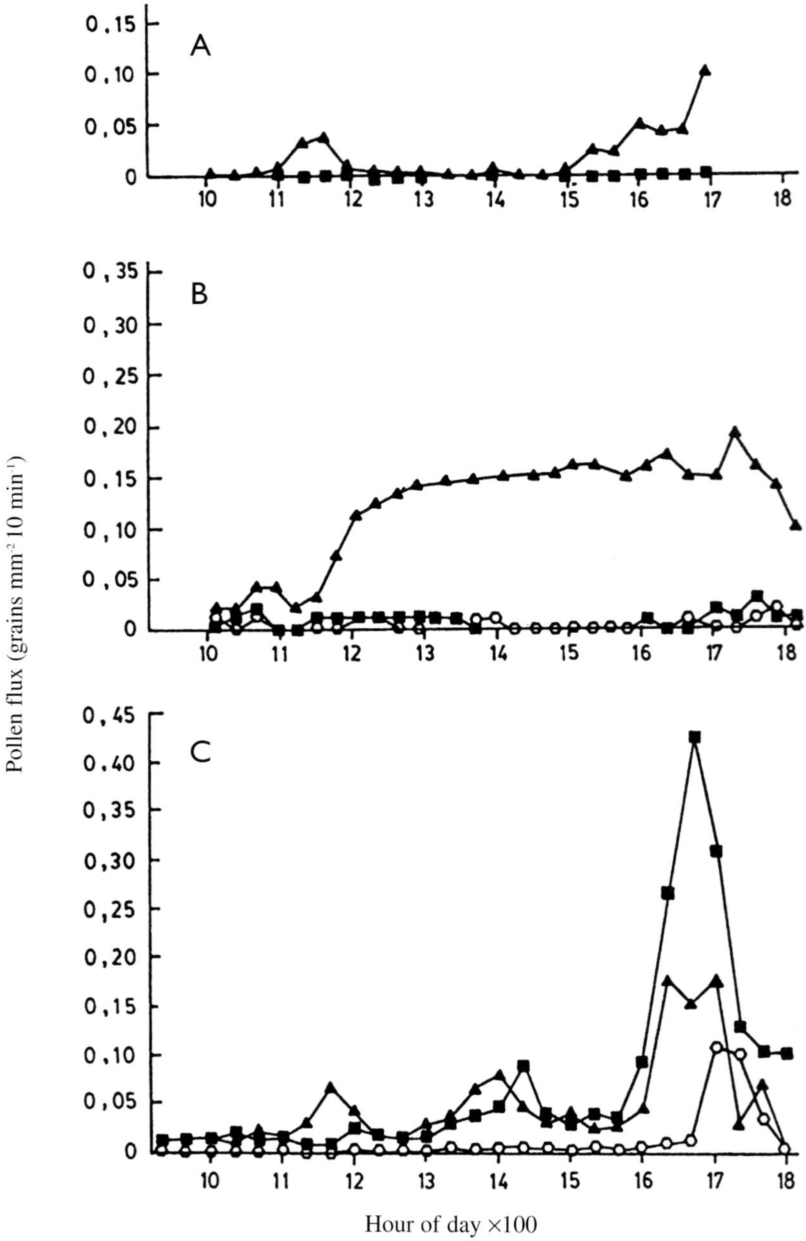

FIG. 1. Pollen capture within the canopy of three Chilean forest tree species of *Nothofagus*: (A) *N. dombeyi*, (B) *N. nitida* and (C) *N. obliqua*, as measured by the capture of grains placed near stigmas of female flowers (number of grains per mm^2 adhering to sticky tape in successive 10-min intervals at various times throughout the day). Captured grains were also examined for their viability (▲ = viable, ○ = non-viable) by means of a tetrazolium reaction. Influx of foreign pollen (■) was assessed by microscopic examination of the captured sample.

Our estimates of pollen grain and stigma surface areas suggest that stigmas could theoretically accept approximately not more than 100 pollen grains. About a hundred times this number of grains is released from each male flower (Riveros *et al.*, 1995a). However, the opportunity for self-pollination is diminished not only by the distinctly different locations of male and female flowers along the branches, but also by disparities in their times of opening (Palma et al., 1996). Counts of pollen grains and ovules per flower are shown in Table 4. What is not yet known is the relative numbers of male and female flowers on a branch, or within a canopy, to give a complete estimate of the relative numbers of male and female gametes per tree. An excess of male flowers would seem to be advantageous in off-setting the relatively small number of pollen grains that might land upon the stigmas of a female flower owing to the enormous spatial dispersion of pollen which follows its release.

Timing of flowering was observed in natural *Nothofagus* populations in order to complement the studies of pollen flow. In all three species, the male flowers open and begin to shed pollen before the female flowers of the species begin to open (Fig. 2). Female flowers remain open for a short time after male flowering has finished and are thus able to capture the last of the pollen to be liberated and found floating within the canopy. In each species, the life-spans of the individual flowers (male or female) were similar, as were the average periods of male and female flowering in any given tree (Fig. 2; Riveros *et al.*, 1995a). The fruiting period commenced mid-way through the female flowering period which may or may not coincide with the termination of the male flowering period (Fig. 2). The disparity in the timing of flowering between the three *Nothofagus* species also reduces opportunities for interspecific hybridisation, although only rarely do individuals of one of the other species coexist in what are virtually monospecific stands of trees.

TABLE 4. Mean numbers of male and female structures per flower in three species of *Nothofagus*. Five flowers from three or five individuals of each species were examined.

Species	Mean number ± SD		
	Pollen grains/ male flower ($x10^3$)	Ovules/ female flower	Pollen grains/ ovule ($x10^3$)
N dombeyi	58.8 ± 3.8	4.7 ± 0.6	12.4 ± 1.7
N. nitida	207.1 ± 24.7	4.6 ± 1.0	44.5 ± 6.3
N. obliqua	147.4 ± 19.5	4.5 ± 0.7	33.4 ± 1.3

Discussion

Anemophily is a hazardous pollination strategy for any species to pursue. Hence, it is probable that evolution has provided mechanisms which compensate for the apparent inability to target pollen to stigmas following its release from the anthers into the air current. Some such compensatory mechanism clearly might be desirable in the *Nothofagus* species of our study in order to alleviate their intrinsically low fertility. The ready availability of pollen at times when the stigmas first become receptive diminishes the chances of non-fertilisation and this, together

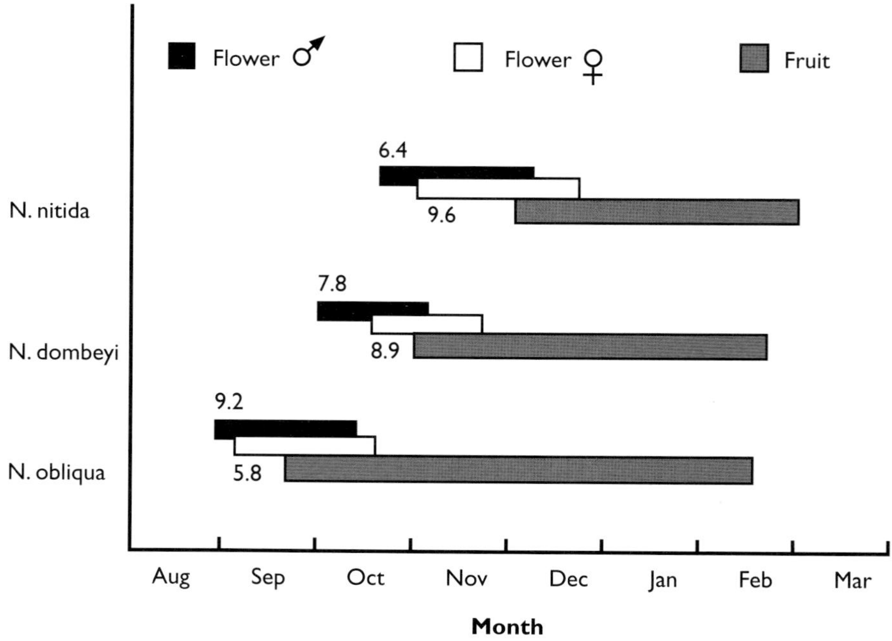

FIG. 2. Flowering and fruiting durations, denoted by horizontal bars spanning months of the year (August 1991 to March 1992), observed for three Chilean species of *Nothofagus*. The mean lifespans (in days) of individual male and female flowers are indicated above or below the respective horizontal bars. (Redrawn from Riveros et al., 1995b)

with the observed disparity of flowering times between individual trees of a population (Riveros *et al.*, 1995a; Palma, 1996), may be reasons, in evolutionary terms, for male flowering to be in advance of female flowering. Moreover, not only is male flowering in advance in the population as a whole, but the branches of any given tree show this same flowering pattern too. Such features might also tend to diminish the competition for space on the surface of a stigma between self-incompatible and compatible pollen. This is one advantage of monoecy when pollination is anemophilous.

A further factor in an anemophilous strategy is the distance that pollen has to travel to effect pollination and/or fertilisation. The temperate *Nothofagus* forests are notable for consisting largely of single species of tree (San Martín and Ramírez, 1987). Accordingly, foreign pollen is often scarce (see Fig. 1A and 1B, though contrast with Fig. 1C) and hence stigmatic surfaces are available mostly for the compatible and incompatible pollen of the *Nothofagus* species in question. Moreover, the density of trees in the forest is high and there can be extensive overlap of branches of neighbouring (compatible) trees. Thus, both the purity of the stand and the density of the trees help to maximise reproductive success. These two last-mentioned factors are of importance in management strategies for the *Nothofagus* forests, especially in cases where there is exploitation by felling and a compensatory necessity for regrowth and regeneration.

The relatively low fertility of *Nothofagus* represents a problem in reforestation programmes because only a relatively small seed stock can be collected from which a genetically-diverse population of saplings can subsequently be raised (Palma *et al.*, 1996). Probably, one of the major factors affecting fertility is the small amount of compatible, viable pollen which actually reaches receptive stigmatic surfaces. This may account for the low values of both N_F and the relative fertility indices, I_F. Another factor appears to be some inherent obstacle to fertilisation which is reflected in the relatively low proportion of embryos that form, even in response to favourable conditions for fertilisation (see the values of N_{MC} in Table 2). Incompatible pollen (from the same individual) as well as compatible but 'weak' pollen from different individuals may provoke fruit development but only occasionally lead to embryo formation (Riveros et al., 1995b). In contrast, there may be insufficient 'strong', compatible pollen from different individuals to effect fertilisation and give viable seed. The distinction between putative weak and strong pollen (M. Riveros, unpublished) requires further clarification, as does the relative amounts of incompatible and compatible pollen arriving on the stigmas. Studies of the last-mentioned feature would necessitate use of markers specific for pollen from different individuals within the interbreeding population. Analysis of DNA sequences, using PCR techniques, of the DNA from single pollen grains would be useful here. Infertility is further compounded by the degeneration of some of the fertilised ovules and their embryos (see also Poole, 1952). However, details of this aspect of reproductive wastage also remain to be established in the species of *Nothofagus* studied here. Clearly, answers in these problematic areas are urgently required, especially if a genetically diverse *Nothofagus* population is to be preserved in the face of exploitation.

The 'low fertility' of *Nothofagus* is of course a relative matter. Fertility is low in comparison to the level which theoretically should be possible (when $N_{MC} = 1.0$). It might also be perceived as 'low' relative to the contemporary needs of a human economic system which now requires a rapid, artificial regeneration or re-establishment of the *Nothofagus* forests. However, in the *Nothofagus* forest itself, the level of fertility and the variation in the rate of seedling germination may be perfectly adjusted to meet the need for a 'natural' replacement of dead or dying members of the population. Given that flowering and fertilisation are inevitable consequences of tree development, a certain level of reproductive wastage may have evolved in order to ensure that there is an adequate level of tree replacement (bearing in mind that *Nothofagus* trees can live hundreds of years) and that the energy which is expended in flowering and fruit formation in the population as a whole is kept to a minimum. Moreover, the shedding of fruits, whether they be viable or non-viable, might also be a way of regularly returning nutrients to the soil of the forest floor.

Replacement of an individual occurs only when conditions in the forest floor and forest canopy become suitable (following the death and decay of a tree, for example) for the germination of seeds stored within the seed bank on the forest floor and for subsequent seedling establishment. The dynamics of the reproductive process, pollen flow, germination, etc., are all integral components of a much larger natural dynamic of the population and its ecosystem as a whole — the maintenance of a certain density of trees and a certain age structure of the population being two processes found at the very end of this more extensive sexual reproduction/plant replacement chain.

The *Nothofagus* forests in the south of Chile, besides being a valuable part of the national patrimony, can also be regarded as a natural laboratory where fundamental questions can be answered concerning reproductive strategies in relation to the life cycles of the individual trees and the structure of the ecosystem of which they are a part.

Acknowledgements

This work is supported by the Biodiversity Support Program of the World Wildlife Foundation (MS-34), the Dirección de Investigación y Desarrollo, Universidad Austral de Chile (S 91-37) and the Dirección de Investigación de la Universidad Católica de Valparaíso, Chile. IACR receives grant-aided support from the Biotechnology and Biological Research Council of the United Kingdom.

References

Armesto, J.J. and Smith-Ramírez, C. (1994). Propuesta de la Sociedad de Ecología de Chile respecto al proyecto de ley "Recuperación del Bosque Nativo y de Fomento Forestal". Una proposición para integrar los valores ecológicos del bosque. *Not. Biol. Soc. Biol. Chile* **2**: 2–8.

Donoso, C. (1987). Variación natural en especies de *Nothofagus* en Chile. *Bosque* **8**: 85–97.

Palma, B., Riveros, M. and Barlow, P. (1996). Pollen flow in three species of *Nothofagus* in the south of Chile. *Acta Hort.* **416**: 73–79.

Poole, A.L. (1952). The development of *Nothofagus* seed (including a preliminary account of the embryogeny, etc.). *Trans. Roy. Soc. New Zealand* **80**: 207–212.

Poole, A.L. (1987). Southern beeches. Science Information Publ. Centre, DSIR, Wellington, N.Z. (New Zealand DSIR Information Series, 162).

Riveros, M., Palma, B., Erazo, S. and O'Reilly, S. (1995a). Fenología y flujo de polen en especies del género *Nothofagus*. *Phyton* (Buenos Aires) **57**: 45–54.

Riveros, M., Parades, M.A., Rosas, M.T., Cardenas, E., Armesto, J., Arroyo, M.T.K. and Palma, B. (1995b). Reproductive biology in species of the genus *Nothofagus*. *Environm. Exp. Bot.* **35**: 519–524.

San Martín, J. and Ramírez, C. (1987). Fitosociología de los *Nothofagus* de la zona mesomórfica chilena. *Bosque* **8**: 121–125.

Wardle, J. (1984). The New Zealand beeches: ecology, utilization and management. New Zealand Forest Service, Wellington, N.Z.

Baudoin, J.P., Degreef, J., Hardy, O., Janart, F. & Zoro Bi, I. (1998). Development of an *in situ* conservation strategy for wild lima bean (*Phaseolus lunatus* L.) populations in the central valley of Costa Rica. In: S.J. Owens and P.J. Rudall (Editors). Reproductive Biology, pp. 417–426. Royal Botanic Gardens, Kew.

DEVELOPMENT OF AN *IN SITU* CONSERVATION STRATEGY FOR WILD LIMA BEAN (*PHASEOLUS LUNATUS* L.) POPULATIONS IN THE CENTRAL VALLEY OF COSTA RICA

J.P. Baudoin, J Degreef., O. Hardy, F. Janart & I. Zoro Bi

Phytotechnie des Régions Intertropicales, Faculté Universitaire des Sciences Agronomiques, 2, Passage des Déportés, B-5030 Gembloux, Belgium

Abstract

A study was conducted on the dynamics of wild Lima bean (*Phaseolus lunatus* L.) populations in the Central Valley of Costa Rica, characterised by a mixed mating system. Genetic structure, gene flow and demography were investigated in representative populations of the Valley with a view to develop an *in situ* conservation strategy. First results show a predominantly self-pollinating system, a high interpopulation variability, a small effective neighbourhood size and the occurrence of spatial genetic structure at intrapopulation level. The establishment of a preliminary life cycle graph allows identification of key factors in the survival and growth of wild populations, which will usefully guide management operations.

Introduction

In the conservation of plant genetic resources, particular attention has been given to wild ancestral populations of cultigens with potential value as large stores of genetic variation and sources of novel variants for plant breeding. However, wild materials are not easy to maintain *ex situ* (in gene banks) due to limited available information on their ecogeographic distribution and consequently the difficult task of assembling sufficiently representative genetic stocks of the wild populations. Additional constraints also include seed dispersal and variation in both seed ripening and dormancy among individual plants and populations (Marshall, 1990; Debouck *et al.*, 1993). In order to circumvent the limitations of *ex situ* gene banks, *in situ* conservation is regarded as a dynamic conservation system which enables the continuing evolution of the crop gene pool and maintains the genetic integrity and the potential of adaptation of each population (Ingram & Williams, 1984; Brush, 1991; Astley, 1992). This type of conservation therefore complements *ex situ* preservation very well.

In an *in situ* conservation programme, the major goal is to preserve as many variants as possible for the target species, which means conserving locally common and rare alleles over a maximum number of sites. Indeed, each variant of wild populations may assume significance under unforeseen conditions and respond to unpredictable needs of plant breeding (Marshall & Brown, 1975; Chambers, 1983; Singh & Williams, 1984).

Planning *in situ* conservation programmes involves several operations concerning not only the target species or group of plants but also the management of nature reserves where the wild populations are scattered. A key objective is to monitor population dynamics, which requires a full understanding of the following interrelated factors: breeding system, gene flow (depending upon vegetative growth, pollen and seed dispersal) and demography (describing and explaining changes in number of individuals within a population) (Lande, 1988; Brown, 1990; Ennos, 1990; Kresovich & McFerson, 1992; Maquet *et al.*, 1996).

As a model to develop a strategy for *in situ* conservation, a study was conducted on the dynamics of wild Lima bean (*P. lunatus*) populations in the Central Valley of Costa Rica, a centre of diversity for this neotropical food legume. This material is made of short-lived perennial self-compatible plants with a mixed mating system, e.g. predominantly self-pollinating. Around 400 wild *P. lunatus* populations have been recorded in the target area, covering 2,100 km^2, in variants of premontane and lower montane humid forests and altitudes ranging from 500 to 1,800 m a.s.l. The region is characterised by a mean annual rainfall varying between 1,200 and 3,000 mm, with dry and humid seasons. Wild populations are usually found in open and disturbed areas with grasses and scattered trees or bushy thickets. Lima bean also colonises coffee plantations from long-lived fences (usually *Erythrina* and euphorbs) bordering the plots. Unfortunately this valuable material is now seriously threatened by several land uses: growing urbanisation, severe grazing, seasonal fires in pasture lands and sugar cane plantations and the replacement of traditional small-scale coffee plantations intermixed with legume trees by modern high input demanding plantations.

In order to study population dynamics of wild Lima bean populations and to develop an *in situ* conservation strategy, three major components were considered: genetic structure among and within populations, gene flow and demography in selected materials of the Central Valley.

Materials and methods

Genetic structure

Genetic structure was investigated within and among 20 wild Lima bean populations uniformly distributed throughout the Central Valley. In each selected population, isozyme electrophoresis was carried out with a seed sample coming either from randomly harvested individuals or from plants plotted at each node of a grid (in case of bidimensional populations) or of a transect (in case of linear populations). The detailed procedure adopted for electrophoretic analysis and seed sampling is described in Maquet *et al.* (1994, 1996) and Zoro Bi *et al.* (1997). Allozyme variation was assessed at the species level and within and among population levels on the basis of several population genetics parameters well defined in the studies of Nei (1973, 1987), Brown (1990) and Hoelzel (1992). Some results were compared with data obtained by Hamrick & Godt (1990) in a reference work covering 473 species of different life forms and breeding systems.

Gene flow

The pattern of genetic variation of wild Lima beans is mainly determined by gene flow within and among populations. According to Wright (1943, 1946), Crawford (1984) and Gliddon *et al.* (1987), estimation of gene flow is based upon the number of individuals in a local random breeding unit, i.e. a « neighbourhood », defined more precisely as the genetic neighbourhood area (NA) and the effective neighbourhood size (Nb). These two parameters are determined from two equations:

1. $NA = 4.\Pi(\frac{1}{2}.t.\sigma_p^2 + \sigma_s^2 + \sigma_v^2)$
2. $Nb = NA.d\ (1+t)/2$

σ_p^2, σ_s^2 and σ_v^2 = dispersal variances respectively for pollen, seed and flower (or vegetative growth) ; t = outcrossing rate; d = adult plant density.

Gene flow was measured in three selected Lima bean populations of the Valley. The methods used to estimate pollen and flower dispersals were described by Hardy *et al.* (1997). The third component, seed dispersal, was measured by choosing in each population four individuals at the pod ripening stage. On each plant we selected 4 or 5 racemes and for each of them stained seeds inside the drying pods with a distinct fluorescent dye. After pod dehiscence (which in wild Lima bean violently projects the seeds) the pigmented seeds were easily detected on the ground during the night using a UV lamp. To this end, we cleaned around each sampled individual a sufficiently large area (about 1.5 times the height of the plant) and measured the horizontal distances between the mother foot and seeds dropped on the soil on one hand, and the vertical distances between the pods on the marked racemes and the ground on the other.

Demography

The demographic study conducted at the population level is based on the determination of the vital rates including fecundity, survival, growth and mortality of each individual. From these parameters, matrix population models are elaborated, following the works of Caswell (1986, 1989), and van Groenendael *et al.* (1988). These models provide a complete interpretation of demographic data and even the expectation of growth rate and stable structure of wild populations. They also help to determine the most constraining factors in the population maintenance, which is useful for *in situ* conservation proposals.

To undertake our demographic study, we selected six populations in undisturbed sites and different climatic regions of the Valley. In each site, several 1 m^2 quadrats were demarcated for the sampling method. Within each population, vital rates of individuals, seed reserves present in the soil (the soil seed bank) and proportion of seeds contributing each year to the next generation (the seed bank dynamics) were determined on the basis of monthly observations, regular soil sampling and seed germination tests, as described by Degreef *et al.* (1997). All the data accumulated during two years were used to establish demographic models and a life cycle graph, following the method of Hubbel & Werner (1970).

Results

Parameters of population genetics

The isozyme analysis provided results for 10 loci, which are developed in Maquet *et al.* (1996). Some alleles were distributed uniformly in the Valley while others appeared to have a narrow distribution in some climatically different regions. This suggests a selection pressure of the environment, at least for some alleles.

At the species level, the percentage of polymorphic loci (P = 50.0) and mean genetic diversity (H=0.146) were similar to values reported for annual species with an outcrossing-animal breeding system. The mean number of alleles per locus (A = 1.50) was less than the value reported by Hamrick & Godt (1990) for the numerous reviewed species. This A-value was closer to that of a short-lived perennial species and a selfing or an animal-pollinated mixed-mating species. On the other hand, at the within-population level, P (= 13.0), A (= 1.14) and H (= 0.029) values were lower than those reported for other species whatever the life forms and breeding systems (Hamrick & Godt, 1990). At the among-population level, total genetic diversity (Ht = 0.296) was similar to that of short-lived perennial species and animal-pollinated mixed-mating systems. Intrapopulation diversity was very low (Hs = 0.058), markedly inferior to values reported for both annuals and short-lived perennial forms and for selfing species. Indeed 80% of total diversity was observed between wild Lima bean populations, as expressed by Dst (= 0.238). Consequently the coefficient of gene differentiation (Gst = 0.803) obtained from our material was largely superior to that of an annual and an autogamous species, on the basis of Hamrick & Godt's data (1990). Such values for Dst and Gst might be due to limited pollen and seed dispersal. Mean fixation index (F = 0.790) was very close to the value reported by Brown (1979) for the predominant inbreeders. The allogamy rate ranged from 0 to 21% according to the populations. Clearly, we noticed for the analysed populations a lack of heterozygotes.

In order to study spatial genetic structure patterns at the intrapopulation level, collected maternal genotypes were plotted in five populations of different shapes. Figs 1 and 2 show how the alleles were distributed in two populations: E25 and E84, respectively for the locus Mdh-2 and for the loci Adh-2 and Mdh-2. The distribution indicates patches of alternative alleles in the two sites and a spatial genotypic pattern distinct from randomness. Data (not shown) from other populations and loci confirm this structure at the intrapopulation level.

Gene flow in selected populations

In the Central Valley, Lima bean blooms during the dry season, from about mid-November to mid-February. Flowers open gradually between 7 a.m. and 11 a.m, and wilt during the night. Mean pollen : ovule ratio is about 863 (Hardy *et al.* 1997). According to Cruden (1977), this ratio is typical of species with a mating system qualified as facultative allogamy. Mechanisms favouring partial outcrossing in *P. lunatus* are well explained by Webster *et al.* (1979). In the target area, the flower visitors belong to *Thysanoptera, Coleoptera, Lepidoptera* and *Hymenoptera*; the major pollinator being *Apis mellifera*. In all investigated populations, pollen dispersal was limited as shown by Hardy *et al.* (1997). Most pollen transfers occurred across distances of less than one metre. This result confirms that common bees disperse pollen mostly over short distances.The frequency of pollen transfers drops quickly around one metre but, afterwards, the decrease slows down. The longest pollen transfer observed reached 5.5 metres. The corresponding dispersal variance for pollen is $\sigma_p^2 = 1.7$ m^2. This value is probably somewhat underestimated because it does not take into account pollen transfers over distances higher than 6.5 metres.

Measures of flower dispersal show great variability in the vegetative growth of wild Lima bean individuals. Distances separating each inflorescence from its respective plant base ranged between 0.37 and 6.5 metres. This variation could be explained by the surrounding vegetation that constitutes a support allowing the plant to climb or preventing any vertical propagation. In the tested populations, the mean flower dispersal variance is $\sigma_v^2 = 2.7$ m^2 with extremes varying between 0.35 and 10.96 m^2. Seeds of wild Lima beans are too heavy to be carried by air, and since they are perfectly smooth and highly toxic (presence of cyanogenetic components) they are not expected to be dispersed by animals like rodents or birds. The most important contribution to seed dispersal then occurs when dehiscent pods open and project their seeds on the ground. The mean dispersal variance is $\sigma_s^2 = 1.68$ m^2, with values ranging from 1.34 to 2.46 m^2. On the basis of our observations, the maximum distance of seed dispersal was 5.5 metres from the pod to the site on the ground.

In the Valley, Maquet *et al.* (1996) and Hardy *et al.* (1997) calculated a mean outcrossing rate and a mean adult plant density of respectively t = 0.1 and d = 0.235 plants/m^2. From the two equations (1) and (2), we obtain therefore a neighbourhood area NA of 56 m^2 and an effective neighbourhood size Nb of 7.23 individuals.

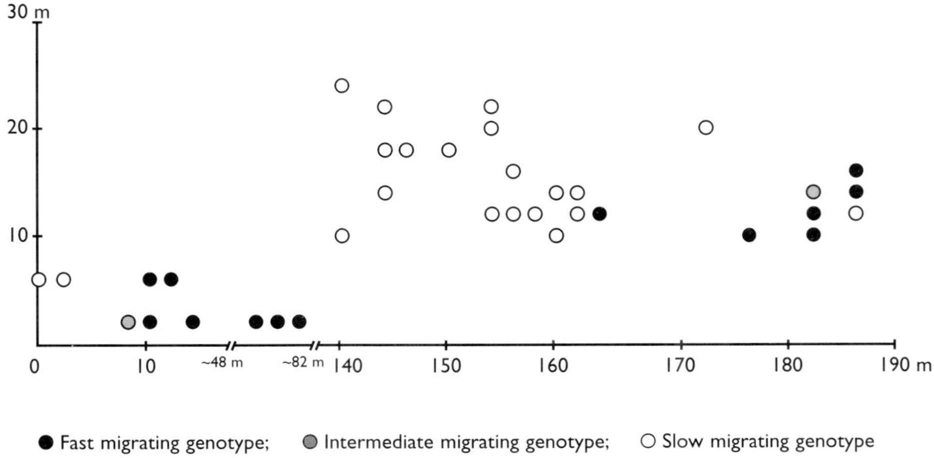

FIG. 1. Spatial distribution of genotypes in population E25 for the locus (Mdh-2.).

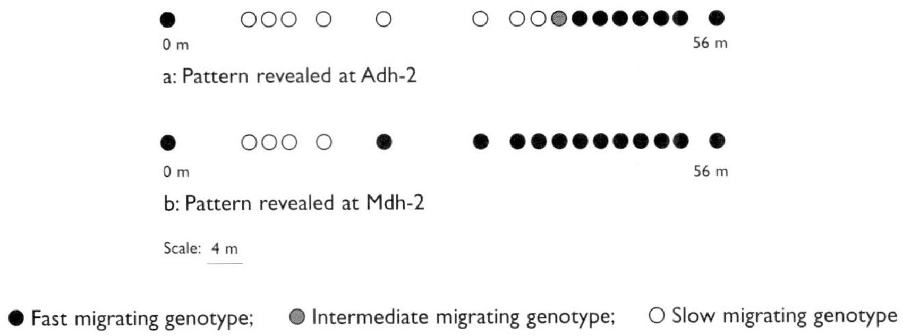

FIG. 2. Spatial distribution of genotypes in population E84 for two loci (Adh-2 and Mdh-2).

Demographic behaviour

Results obtained from the soil seed bank study in the six sampled populations showed clearly the occurrence of a persistent seed bank. As indicated by Degreef *et al.* (1997), seeds are found in the soil at least one year after their dissemination. With respect to the germination tests with seeds taken directly from the plants or from the soil seed bank, we did not observe any primary dormancy. Germination rates reached 88 % after 30 days from seeds directly picked from the adult plants and 81 % after 6 months from seeds collected in the soil bank at different depths (Degreef *et al.*, 1997). A germination delay was however noticed with seeds from the 0–1 cm soil depth; this probably results from the direct exposure of these seeds to the very high temperature and moisture deficit conditions occurring particularly at this soil level during the dry season. Such severe conditions could induce micropyle obstruction, considered as the main water entrance during seed imbibition.

The bulk of vital rates data obtained during two years in the six populations allowed the establishment of a first demographic model, describing the behaviour of 3,000 wild Lima bean individuals. Fig. 3 shows the life cycle graph of an hypothetical mean wild population. To construct it, Degreef *et al* (1997) divided the life cycle of any *P. lunatus* individual into six stages: G_0 for the seeds produced the year 0; G_1 for the seeds remaining in the soil after one year, J for the juvenile individuals showing stem diameter inferior to 1.4 mm; L_1, L_2 and L_3 for the lignified individuals, showing stem diameter respectively between 1.4 and 3.6 mm, between 3.6 and 9.7 mm and equal or superior to 9.7 mm. In Fig. 3, each node is associated with a developmental stage and the arrows linking two stages express the transition from one stage to the other in one year.

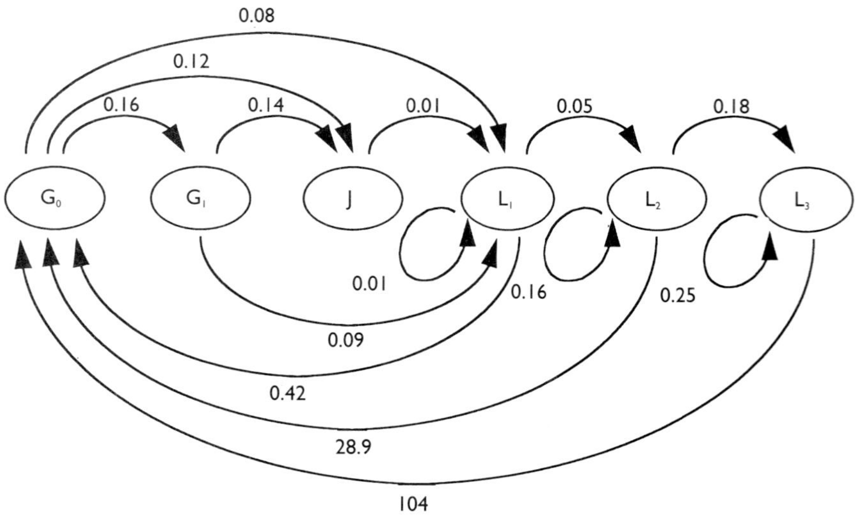

FIG. 3. Life cycle graph of the hypothetical << mean wild Lima bean population >> (Degreef *et al.*, 1997).

All yearly possible transitions are reported on the graph and the corresponding probabilities allow construction of a projection matrix (Silvertown *et al.*, 1993; Bullock *et al.*, 1994; Shea *et al.*, 1994) whose dominant eigenvalue is equal to the population growth rate (= 0.7593). This value, corresponding to a mean wild population, confirms the endangered status of many Lima bean populations. A sensitivity analysis revealed two key factors for the *in situ* maintenance of the wild materials (Degreef *et al*, 1997): early germination of fresh seeds after the maturing period and plant lignification. As the mortality rate is particularly important during the dry season, the direct transition from a seed G_0 to a ligneous individual L_1 contributes significantly to preserve wild populations in the Central Valley.

Discussion

Parameters of population genetics indicate that wild Lima beans behave like a predominant selfing species, even though several values are intermediate between a selfing and a mixed-animal mating system. As explained by Maquet *et al.* (1996), departure from the Hardy-Weinberg equilibrium could be the consequence of several factors: overlap of distinct generations during the same period and on the same site, presence of subpopulations in a single natural population with low natural crosses between them, and/or founder effect generated by anthropic destruction.

At the intrapopulation level, the occurrence of a spatial genetic structure pattern, as shown in the two populations E25 and E84, could result from two neighbourhoods that would have diverged under limited gene flow. Another explanation lies in a demographic event. The populations studied might have been founded by the recent junction of two subpopulations presenting different allelic frequencies. New heterozygotes have appeared from natural hybridisations between the two subpopulations. This situation, typical of other wild populations in the Valley, is probably the result of colonisation and extinction events occurring frequently in such human disturbed areas. For example, in the target area, many wild Lima bean individuals are found in coffee plantations, fallow lands or along hedges, so that weeding practices contribute to the extinction of several plants. Recolonisation of the cleaned sites could be due to any nearby plants, to new individuals emerging from the soil seed bank or to human activities (such as seed transportation over long distances by means of shoes or tools). In a single population or in a group of nearby populations, as gene dispersal is probably restricted, a genetic subdivision will appear with time and will be maintained for a long period. This will allow for seed sampling to select individuals in several locations throughout the population with a small number of seeds per location.

According to Wright (1943, 1946), the low value obtained for the effective neighbourhood size (Nb inferior to 20 individuals) corresponds to a high probability of random local genetic differentiation within the population. As our populations were spread over areas ranging from about 100 m^2 to more than 1,000 m^2, they could contain several to many neighbourhood areas. Therefore allelic distribution in a single population is expected to be highly structured. This shows again the need to sample systematically at many sites of the population for collecting the whole genetic diversity.

With regard to demography, the presence of an important seed reservoir in the soil explains population survival in years of unfavourable environmental conditions. This represents a mechanism of adaptation at the population level.

The life cycle graph of our hypothetical mean wild population and the sensitivity analysis point out the importance of seed production, early germination and rapid plant lignification to maintain *in situ* populations in the Valley. The seed production of the Lima beans will be favoured by delaying the weeding of the coffee plantations till the beginning of the wet season.

That practice will avoid the severe impact of a dry season management on the Lima bean population dynamics as it generally occurs in the traditional coffee plantations. An early germination will be induced by cleaning the soil at the vicinity of adult productive plants. The survival of the ligneous individuals, constituting the renewal potential of the population, will be ensured by cutting around the juvenile plants, reducing the competition for water and nutrients.

Conclusion

Conservation of plant genetic resources aims to maintain a broad genetic diversity. Therefore, populations and individuals are sampled to best preserve the existing allelic richness. *In situ* conservation strategies will enable the evolution of protected populations in their natural sites. Nevertheless, management will be necessary to maintain these endangered wild populations as is illustrated by our study in the Central Valley of Costa Rica.

Isozyme electrophoresis indicates a genetic variability occurring mainly at the interpopulation level and, at least for some alleles, a narrow distribution likely to be influenced by ecological factors. In the sampling strategy it is therefore essential to identify the most representative populations per ecological zone of the Valley. As subdivisions may occur with some extended populations, collectors should take care to sample in different sites within a population.

The gene flow study shows that neighbourhood areas are limited in wild Lima bean populations of the Central Valley. Genetic drift is therefore important and genetic uniformity might directly threaten some alleles. The reintroduction of the threatened genotypes at regular time intervals will allow the maintenance of genetic integrity in some endangered populations.

Finally, the benefits of an appropriate management of the populations are indicated by analysis of the demographic model. Seed production of single individuals, early seed germination and rapid lignification process are all factors to be favoured by the management of wild materials in protected areas.

References

Astley, D. (1992). Preservation of genetic diversity and accession integrity. *Field Crops Res.* **29**: 205–224.

Brown, A.H.D. (1979). Enzyme polymorphism in plant population. *Theor. Populat. Biol.* **15** : 1–42.

Brown, A.H.D. (1990). Genetic characterization of plant mating systems. In: A.H.D. Brown, M.T. Clegg, A.L. Kahler and B.S. Wein (editors). Plant population genetics, breeding and genetic resources, pp. 145–162, Sinauer Associates Inc. Publishers, Sunderland, Massachusetts.

Brush, S.B. (1991). A farmer-based approach to conserving crop germplasm. *Econ. Bot.* **45**: 153–165.

Bullock, J.M., Clear Hill, B. and Silvertown, J. (1994). Demography of *Cirsium vulgare* in a grazing experiment. *J. Ecol.* **82** : 101–111.

Caswell, H. (1986). Life cycle models for plants. *Lectures Math. Life Sci.* **18**: 171–233.

Caswell, H. (1989). Matrix population models. Construction, analysis and interpretation. Sinauer Associates Inc. Publishers, Sunderland, Massachusetts.

Chambers, S.M. (1983). Genetic principles for managers. In: C.M. Schonewald-Cox., S.M. Chambers, B. MacBryde and W.L. Thomas (editors). Genetics and conservation, pp. 15–46. Benjamin/Cummings, Menlo Park, California.

Crawford, T.J. (1984). The estimation of neighbourhood parameters for plant populations. *Heredity* **52**: 273–283.

Cruden, R.W. (1977). Pollen-ovule ratios: a conservative indication of breeding systems in flowering plants. *Evolution* **31**: 32–46.
Debouck, D.G., Toro, O., Paredes, O.M., Johnson, W.C. and Gepts, P. (1993). Genetic diversity and ecological distribution of *Phaseolus vulgaris* (Fabaceae) in northwestern South America. *Econ. Bot.* **47**: 408–423.
Degreef, J., Baudoin, J.P. and Rocha, O.J. (1997). Case studies on breeding systems and its consequences for germplasm conservation. 2. Demography of wild Lima bean populations in the Central Valley of Costa Rica. *Genet. Resources Crop Evol.* **44**: 1–10.
Ennos, R.A. (1990). Detection and measurement of selection: genetic and ecological approaches. In: A.H.D. Brown, M.T. Clegg, A.L. Kahler and B.S. Wein, (editors). Plant population genetics, breeding and genetic resources, pp. 200–214, Sinauer Associates Inc. Publishers, Sunderland, Massachusetts.
Gliddon, C., Belhassen, E. and Gouyon, P.H. (1987). Genetic neighbourhoods in plants with diverse systems of mating and different patterns of growth. *Heredity* **59**: 29–32.
Hamrick, J.L. and Godt, M.J.W. (1990). Allozyme diversity in plant species. In: A.H.D. Brown, M.T. Clegg, A.L. Kahler and B.S. Wein (editors). Plant population genetics, breeding and genetic resources, pp. 43–63, Sinauer Associates Inc. Publishers, Sunderland, Massachusetts.
Hardy, O., Dubois, S., Zoro, Bi I. and Baudoin, J.P. (1997). Gene dispersal and its consequences on the genetic structure of wild populations of Lima bean (*Phaseolus lunatus*) in Costa Rica. *Pl. Genet. Resources Newslett.* **109**: 1–6.
Hoelzel, A.R. (1992). Molecular genetic analysis of populations - a practical approach. IRL Press, Oxford.
Hubbel, S.P. and Werner, P.A. (1979). On measuring the intrinsic rate of increase of populations with heterogeneous life histories. *Amer. Naturalist.* **113**: 277–293.
Ingram, C.B. and Williams, J.T. (1984). *In situ* conservation of wild relatives of crops. In: J.H.W. Holden and J.T. Williams, (editors). Crop genetic resources : conservation and evaluation, pp.163–179, George Allen and Unwin, London.
Kresovich, S. & McFerson J.R. (1992). Assessment and management of plant genetic diversity: considertations of intra- and interspecific variation. *Field Crops Res.* **29**: 185–204.
Lande, R. (1988). Genetics and demography in biological conservation. *Science* **241**: 1455–1460.
Maquet, A., Wathelet, B. and Baudoin, J.P. (1994). Etude du réservoir génétique de la légumineuse alimentaire *Phaseolus lunatus* L. par l'analyse électrophorétique d'isozymes. *Bull. Rech. Agron. Gembloux* **29**: 369–381.
Maquet A., Zoro Bi I., Rocha O.J. and Baudoin J.P. (1996). Case studies on breeding systems and its consequences for germplasm conservation. 1. Isoenzyme diversity in wild Lima bean populations in central Costa Rica. *Genet. Resources Crop Evol.* **43**: 309–318.
Marshall, D.R. (1990). Crop Genetic resources: current and emerging issues. In: A.D.H. Brown, M.T. Clegg and B.S. Wein (editors). Plant population genetics, breeding and genetic resources, pp. 367–388, Sinauer Associates Inc. Publishers, Sunderland, Massachusetts.
Marshall, D.R. and Brown, A.D.H. (1975). Optimum sampling strategies in genetic conservation. In: O.H. Frankel and J.G. Hawkes (editors). Crop genetic resources for today and tomorrow, pp. 53–80, Cambridge University Press, Cambridge.
Nei, M. (1973). Analysis of gene diversity in subdivided populations. *Proc. Natl. Acad. Sci., USA* **70**: 3321–3323.
Nei, M. (1987). Molecular evolutionary genetics. Columbia University Press, New York.
Shea, K., Rees, M. and Wood, S.N. (1994). Trade-offs, elasticities and the comparative method. *J. Ecol.* **82** : 951–957.
Silvertown, J., Franco, M., Pisanty I. and Mendosa A. (1993). Comparative plant demography - relative importance of life-cycle components to the finite rate of increase in woody and herbaceous perennials. *J. Ecol.* **81** : 465–476.

Singh, R.B. and Williams, J.T. (1984). Maintenance and multiplication of plant genetic resources. In: J.H.W. Holdes, and J.P. Williams (editors). Crop genetic resources: conservation and evaluation, pp. 120–130. George Allen and Urwin, London.
Van Groenendael, J., de Kroon, H. and Caswell, H. (1988). Projection matrices in population biology. *Trends Ecol. Evol.* **3** : 264–269.
Webster, B.D., Lynch, S.P. and Tucker, P.L. (1979). A morphological study of the development of reproductive structures of *Phaseolus lunatus* L. *J. Amer. Soc. Hort. Sci.* **104**(2): 240–243.
Wright, S. (1943). Isolation by distance. *Genetics* **28**: 114–128.
Wright, S. (1946). Isolation by distance under diverse mating systems. *Genetics* **31**: 39–59.
Zoro, Bi I., Maquet, A. and Baudoin, J.P. (1997). Spatial patterns of allozyme variants within three wild populations of *Phaseolus lunatus* L. from the Central Valley of Costa Rica. *Belg. J. Bot.* **129**: 149–155.

Prance, G.T. (1998). The pollination of three Amazonian plant species of economic importance and its relevance to conservation. In: S.J. Owens and P.J. Rudall (Editors). Reproductive Biology, pp. 427–438. Royal Botanic Gardens, Kew.

THE POLLINATION OF THREE AMAZONIAN PLANT SPECIES OF ECONOMIC IMPORTANCE AND ITS RELEVANCE TO CONSERVATION

GHILLEAN T. PRANCE

Royal Botanic Gardens, Kew, Richmond, Surrey, TW9 3AB, UK

Abstract

The floral biology of three Amazonian plants of economic importance is discussed: the Brazil nut (*Bertholletia excelsa*), cupuaçu (*Theobroma grandiflorum*) and the peach tomato (*Solanum sessiliflorum*). An understanding of their floral biology yields useful information for both the management and conservation of these plants. Each of these species is dependent upon a range of insect pollinators and the future of the crop, depends upon the maintenance of adequate habitat to sustain these insects. The peach tomato is the most domesticated of these crops and the domestication is probably related to the simple genetics of spinyness and the maternal inheritance of fruit characteristics. Plants with less complicated breeding systems and which are easily propagated vegetatively are more likely to become domesticated.

Introduction

The study of the floral biology of wild plants of economic use is important but has often been neglected. These are plants that are likely eventually to enter into cultivation or other systems of management and so knowledge about their reproductive biology is vital for their management. It is also much easier to convince those responsible for conservation policies by providing data from plants of known economic use. They serve to demonstrate the web of interactions that hold the ecosystem together and upon which they depend for their survival.

Several authors have recently pointed out that the world is facing an impending pollination crisis in which both wild and managed pollinators are disappearing at alarming rates due to habitat loss, pesticide poisoning, disease and pests (Ingram *et al.*, 1996). It is important to both study pollination of useful plants and to strive to conserve their pollinators.

I have, therefore, selected three Amazonian species on which I and my students and research associates have worked. The Brazil nut, *Bertholletia excelsa* Humb. & Bonpl. is still mainly harvested from wild trees and is only just entering plantations. However, it has been managed, transported and probably even improved by indigenous populations for many centuries. The cupuaçu, *Theobroma grandiflorum* (Willd. ex Spreng.) Schum., a wild relative of the cacau, is exploited from both wild sources and small plantations. It is also grown in dooryard gardens throughout Amazonian Brazil and its selection was probably also influenced by indigenous peoples. The third example is the cubiu, peach tomato or cocona, *Solanum sessiliflorum* Dunal which is a semi-domesticated crop of Brazil, Peru and Colombia. In all three cases it is the fruit or seeds that are exploited and it is in this type of crop where the understanding of the reproductive biology is likely to be of most importance.

The Brazil nut (*Bertholletia excelsa* Humb. & Bonpl.)

Pollination

The Brazil nut is a member of the Neotropical subfamily Lecythidoideae of the pantropical family Lecythidaceae. Within the Lecythidaceae it belongs to the more advanced group of genera with zygomorphic flowers and nectar as a pollinator reward (see Prance and Mori, 1979; Mori and Prance, 1990a; Tsou, 1994). In this group the androecium consists of a basal ring bearing the fertile stamens in a circle around the style and a lateral extension consisting of a ligule and a hood that completely covers and is pressed to the staminal ring (Fig. 1A, B). The inside of the hood is covered by sterile appendages and nectar is secreted at their base. In less advanced members of the subfamily the hood may also bear stamens (in some species of *Lecythis* and in *Corythophora* and *Couroupita*) where in most cases it is not adpressed to the staminal ring (Fig. 2E). In *Couroupita guianensis* Aubl. and *Lecythis pisonis* Camb. it has been shown that the hood pollen is a sterile pollinator reward that does not germinate whereas the ring pollen is fertile and germinates normally. In *Couroupita guianensis* the staminal ring and hood pollen are also morphologically distinct and that of the ring is presented in monads and has a relatively smooth exine, while that of the hood is in tetrads and has a rougher exine (Jacques, 1965; Mori *et al.*, 1980; Ormond *et al.*, 1981; Yacsirk *et al.*, 1986). In most of the species with pollen producing hood stamens, except for *Corythophora*, there is a gap between the hood and the staminal ring, but in *Bertholletia* and the other nectar producing species in the genera *Couratari* and *Eschweilera* the hood is tightly appressed to the staminal ring and the style. In this case only those bees that are large and strong enough to pry open the hood are able to pollinate these zygomorphic nectar-producing species. The bees land on the hood and enter in a sternotribic position so that their backs rub against the ring stamens while they forage for nectar in the hood (Fig. 3).

Bertholletia excelsa is unique among the Lecythidaceae in having styles of two different lengths and various studies have shown it to be an obligate outcrosser (Moritz, 1984, Müller *et al.*, 1980). Moritz (1984) showed that, although self pollination led to the initial formation of seeds, no self pollinations produced mature fruit. Samples from six clones were crossed and there were statistically significant differences in fertility among the different crosses made. This has considerable importance for plantation use of the Brazil nut where care will have to be taken to mix clones that cross well.

Styles of both lengths occurred in all clones studied by Moritz (1984) but it has not yet been shown whether there is an incompatibility system between them. This merits further investigation.

The pollinators are large bees mainly *Xylocopa* carpenter bees, *Bombus*, *Centris* and Euglossine orchid bees (Müller *et al.*, 1980; Nelson *et al.*, 1985), especially in the genus of larger bees *Eulaema*. These male euglossine bees are well known for their relationship with orchid flowers where they gather the odours and place them in their rear legs as part of the process to attract female bees for mating (Dressler, 1982; Dodson, 1966, 1967). This means that various epiphytic orchids play an important role in the maintenance of the Brazil nut harvest.

Fruit development and dispersal

Following successful pollination by pollen from another tree the Brazil nut fruits gradually begin to develop. In Central Amazonia flowering takes place in October and November. The fruits then take 14 months to mature and are ready in January and February of the second year after pollination. When the fruit ripens it is a large secondarily indehiscent woody capsule or pyxidium of 10-16 cm diameter containing between 12 and 20 hard shelled nuts, arranged

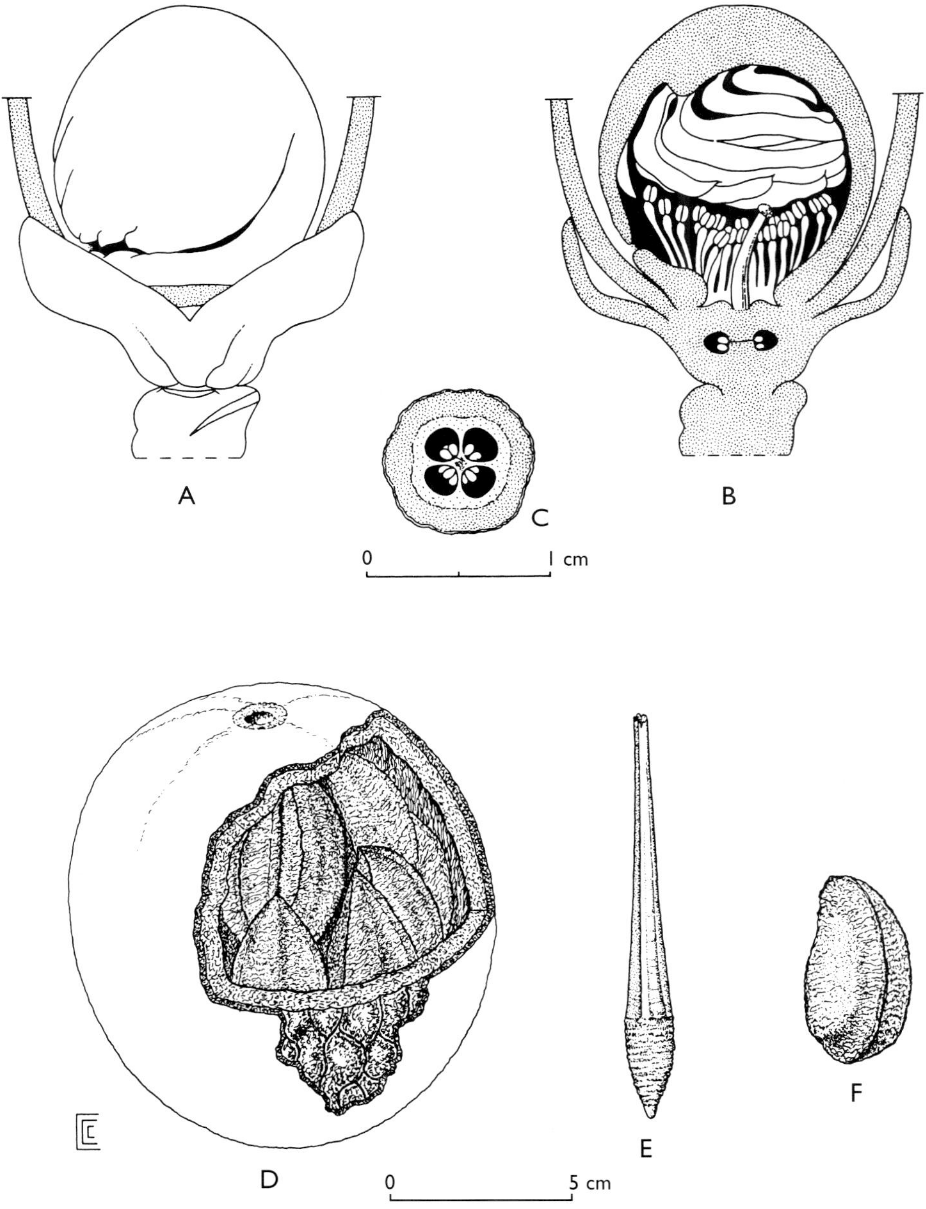

FIG. 1. The flower and fruit of the Brazil nut *Bertholletia excelsa* Humb. & Bonpl. A. Flower with petals removed to show the androecium; B. Section of androecium showing ring stamens and sterile appendages in hood, note this is of a long-styled flower; C. Ovary section; D. Fruit with part of pericarp removed; E. Operculum which falls inward into fruit; F. Seed known as Brazil nut. Drawn by Charles C. Clare.

rather like segments of an orange, hence their angled shape (Fig. 1D, F). The whole fruits are dropped from the trees in January and February and are either collected by Brazil nut gatherers who open the outer shell with a machete and pour out the nuts into their baskets, or they lie on the forest floor until found by agoutis (*Dasyprocta* sp.) or squirrels, the only rodents with teeth sharp enough to gnaw open the outer case of the fruit. Agoutis are scatter-hoarding rodents that remove the nuts and bury them in caches some distance from the trees (Huber, 1909). The nuts that are not relocated by the rodents are thereby safely dispersed some distance from the parent tree. For further details see Mori and Prance (1990b) and Mori (1992) who reported the activity of squirrels for the first time.

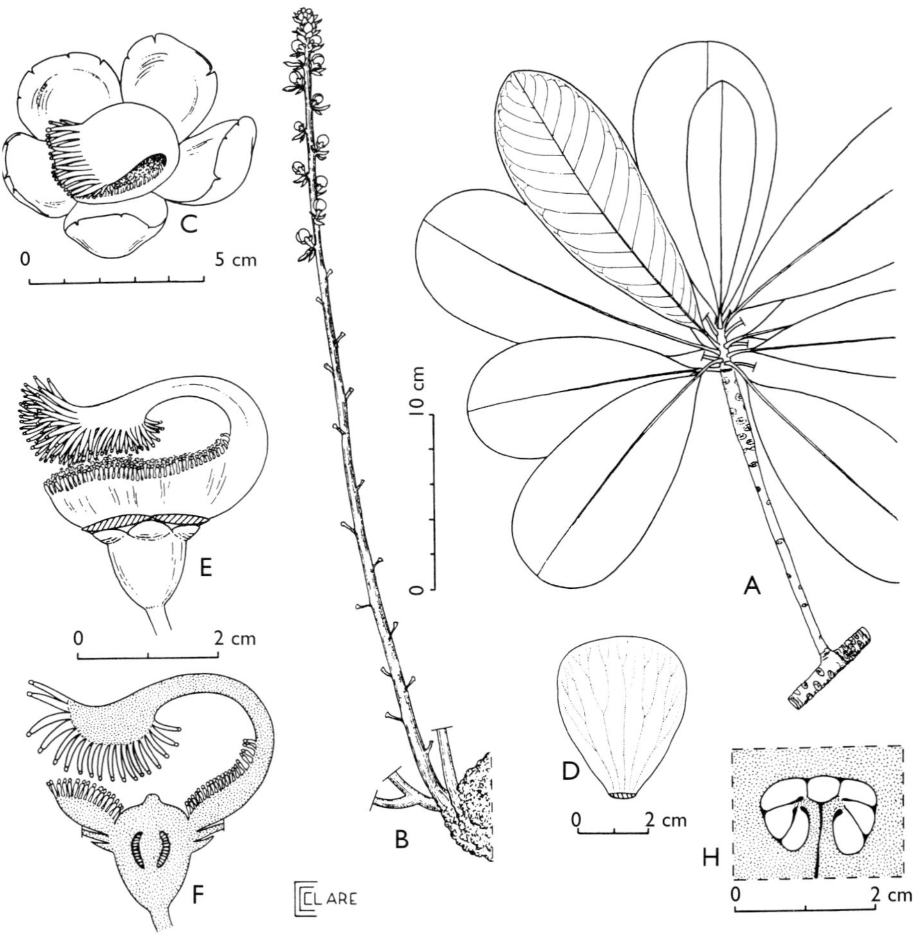

FIG. 2. *Couroupita guianensis* Aubl. A. Habit; B. Cauliflorous inflorescence; C. Flower from above; D. Petal; E. Androecium showing the gap between fertile ring stamens and sterile hood stamens; F. Flower section; H. Ovary. Drawn by Charles C. Clare.

Crop Management

Brazil nuts are currently harvested almost entirely from wild trees and so are one of the most important extraction crops from the Amazonian forest of Brazil, Bolivia and Peru (Sánchez, 1973). There have been a few unsuccessful attempts to establish Brazil nut plantations and there is one fairly large young plantation near to the town of Itacoatiara in Amazonas, Brazil. Since it is illegal to fell Brazil nut trees they are often left in fields and pastures following deforestation. The trees in these areas are notably unproductive in terms of their fruit and this is in large part due to lack of pollination. These isolated trees also often die quickly in such an exposed situation and this could be connected with lack of nutrients. In one small experimental Brazil nut plantation near to Manaus, Brazil the health of the trees and their productivity increased markedly when cacau was planted between the rows and this was due to the fertiliser used to establish the cacau trees.

The management of Brazil nuts outside their native forest requires a situation where pollinators are still available. Since the pollinators are largely forest bees it is being found that successful plantations are small ones surrounded by or bordering forest or strips in the forest. It is interesting to note the current patchy distribution of Brazil nuts within the Amazon forest. In various forest inventories they are often found in distinct clumps, rather than a more random distribution. This is probably because previously Brazil nuts have been planted in small groups by the indigenous inhabitants of the forest. Their management technique is the one that will work best today because it maintains the access to an adequate number of pollinators. The maintenance of pollinators depends also upon the availability of epiphytic orchids and so the Brazil nut story is a helpful argument for the conservation of Amazonian biodiversity.

FIG. 3. A bee entering the flower *Bertholletia excelsa* Humb. & Bonpl. Photo A. Henderson

FIG. 4. Fruit of cupuaçu (*Theobroma grandiflorum* (Willd. ex Spreng.) Schum.)

Cupuaçu (*Theobroma grandiflorum* (Willd. ex Spreng.) Schum.)

The fruit of the cupuaçu, a relative of the cacau, produces the most delicious of all Amazonian flavours for juices and deserts. The brown fruits which are produced on the woody branches of the tree resemble small rugby footballs. The flavour is produced from the pulp which surrounds the seeds, and the seeds can also be used to produce a high quality chocolate (Venturieri and Aguiar, 1988). Cupuaçu is a native of the rainforests of eastern Amazonia (Fig. 4) but it has been spread throughout and beyond Amazonia as a cultivated plant. Unlike the Brazil nut, there are now many small plantations of cupuaçu and it is grown in the gardens and yards of almost all homes in lowland Amazonia. Fruits are harvested from both cultivated and wild trees. It is unfortunate that cupuaçu is a native of the forest of Pará and Maranhão in an area which has been much deforested and so it is becoming rarer in the wild. Another unfortunate practice by local peoples has been to cut down the larger trees in order to obtain the fruit.

Floral biology

The flowers of cupuaçu have a very similar structure to those of cacau (*Theobroma cacau*) but are much larger. The trees flower abundantly at the height of the dry season in June, July and August in Central Amazonia (Falcão and Lleras, 1983) and they fruit during the rainy season from December to March (Calzavara *et al.*, 1984; Retto Junior, 1986; Venturieri, 1993).

The flowers of cupuaçu open during the day, mostly around noon, but they tend to remain only partially open until late afternoon when they open fully. Unfertilised flowers remain on the trees for one or two days following opening but have generally abscissed by the second afternoon. The stigmas are receptive from anthesis until about 10 a.m. of the following day. Venturieri (1993) found that in the case of hand pollination experiments better results were obtained if fertilisation was performed before 8 a.m. Therefore successful natural pollination can take place between late afternoon of the day of anthesis and the morning of the following day. Venturieri (1992, 1993) has shown that cupuaçu is a predominantly allogamous species with a genetic incompatibility similar to that of *Theobroma cacau* governed by three loci called A, B and S by Knight and Rogers (1995) and Cope (1962). Addison and Tavares (1951) found that different individual plants had different capacities to fertilise other plants. One clone used (GR-11) was either autogamous or allogamous. Interestingly this clone was the one with the greatest capacity for hybridisation with other species of *Theobroma* including *T. cacau*, suggesting that the inactivation of the incompatibility mechanism between the species tested had a biochemical basis. Wild plants of cupuaçu are generally self incompatible but many cultivated ones are self compatible. This shows that the domestication process of cupuaçu has progressed far enough to break down incompatibility mechanisms as has happened with many other crop plants.

Pollination of cupuaçu is by small stingless bees (Silva, 1976; Falcão and Lleras, 1983; Venturieri, 1993 and personal observations). Pollinators which have been observed include *Plebeia minima*, *Trigonisca* sp. and *Ptilotrigona lurida*. Falcão and Lleras observed the African honey bee *Apis mellifera adansonii* visiting the flowers, but this introduced species is not the legitimate pollinator and is probably too large to be effective. Other species observed by Falcão and Lleras included *Frieseomelitta silvestris faceta* and *Paratrigona impunctata*. Both Silva (1976) and Falcão and Lleras (1983) found that the small bee *Tetragona clavipes* perforated the lower part of the petals that conceal the anthers and was obviously a pollen robber rather than a pollinator.

One of the problems for the cutlivators of cupuaçu trees is the low productivity per tree. For example, Falcão and Lleras (1983) found that the average production of flowers per tree was 3,500 and the average fruit set was 17. This is not surprising considering the size of the individual fruits. Various farmers have experimented with artificial pollination in order to increase productivity. Interestingly the same problem as in cacau occurs where the average production per tree does not increase. Many more young fruits form, but they abort. There seems to be some sort of genetic control of total production probably limited by nutrient resources available in the poor soils of the Amazon region. The way to improve the crop yield may be through adding nutrients to the soil rather than pollination. Although in all cases pollinators have been observed in low numbers at cupuaçu trees, there appears to be sufficient insects to obtain the maximum possible harvest, however Young (1994) showed that fruit set can be increased in cacao (*Theobroma cacau*) plantations by increasing pollinators. Young increased the breeding habitats of the midge pollinators and thereby increased fruit set.

The peach tomato (*Solanum sessiliforum* Dunal)

This species, which is known as cubiu in Brazil and cocona in Peru, is a small short-lived shrub up to 1.5 m tall that belongs to the spiny section *Lasiocarpa* of *Solanum*. Its fruit is a large berry with a fleshy pulp which is eaten raw, in salads or as a tropical fruit juice. It is also cooked in many local dishes throughout Amazonia. It is mainly grown in dooryard gardens, but is cultivated in small numbers commercially in both Peru and Brazil. It is closely related to the Andean species the naranjillo or *S. quitoense* of Colombia. There has been considerable selection for larger fruited spineless varieties. Wild varieties are common in secondary disturbed habitats and along river banks and so it is easy to compare the differences between wild and cultivated plants.

FIG. 5. A plant of *Solanum sessiliflorum* Dunal, the peach tomato.

Floral Biology

The peach tomato has typical flowers of *Solanum* with poricidal anthers. They are open from 07.00 to 16.00 and last for 2 days. It is interesting because it is an andromonoecious species. The male flowers have a reduced style and a rudimentary ovary but are otherwise similar to the hermaphrodite flowers. Each plant has up to nine inflorescences each bearing from 5-9 flowers, but with only one or two flowers opening on any one day. Flowering in Central Amazonia is mostly between September and November towards the end of the dry season. The hermaphrodite flowers have a wet stigma. The flowers have a faint scent and the anthers and petals have a pattern in ultra violet light (Storti, 1988).

The legitimate pollinators are large bees which vibrate the poricidal anthers to extract the pollen, as is common in many species of *Solanum*. The principle pollinators appear to be female euglossine bees in search of pollen. Storti (1985) observed the euglossine bees *Eulaema (Apeulaema) nigrita* Lepetelier and *Euglossa* sp. pollinating the flowers as well as *Paratrigona impunctata* Ducke. She termed these large euglossine species where body size is 22 mm and 12 mm respectively as primary pollinators and indicated the *Paratrigona* and *Exomalopsis* sp. as occasional pollinators. *Eulaema nigrita* visits the flowers from 7.00 to 16.00 when they close, but is most active between 9.00 and 11.00 and *Euglossa* sp. visits between 7.00 and 15.00 but is most active between 8.00 and 9.00. Both species vibrate the pollen from the anthers. *Paratrigona impunctata* is found visiting the flowers between 7.30 and 15.00 but has an entirely different behaviour. It scoops the pollen out of the anthers with its middle pair of legs with the help of its proboscis and then transfers it to the ventral part of the thorax and later to the corbiculae. Pollen on the lower part of the thorax contacts the style and so this smaller bee of 5 mm length was considered to be a secondary pollinator by Storti (1988). She made two observations of the bee *Exomalopsis* sp. visiting flowers and vibrating the anthers. She also observed *Trigona* cf *fulviventris* and *T. dorsalis* Smith collecting pollen which had fallen onto the petals as a result of the activity of vibrating bees, but neither of these species came into contact with the stigma. Salick (1992) collected the large bee species *Eulaema cingulata* (Fabricius) and *Bombus transversalis* (Olivier) and considered them to be the pollinators. She also observed the small bees *Trigona* sp and *Pseudogocholropsis graminea* (Fabricius) but doubted that these species were pollinators considering their small size.

Storti (1988) conducted pollination tests and obtained no fruit set from self fertilised flowers. Fruits were formed by both geitonogamous and xenogamous crosses indicating that the species is self-compatible as was suggested by Whalen and Anderson (1981), who showed that self-incompatibility in *Solanum* is largely restricted to species of subgenus *Potatoe*.

The principal pollinators of the peach tomato are forest bees which indicates that like the Brazil nut it is necessary to have an area of forest relatively nearby to maintain the pollinators.

Domestication

The domestication of the peach tomato in Peru was the topic of an interesting study by Salick (1992). In lowland Peru on the eastern foothills of the Andes the peach tomato is both cultivated and wild. The fields of selected varieties tend to have a number of wild forms which are quickly eliminated by the cultivators. One easy characteristic to distinguish the cultivated from the wild types is the absence or presence of spines. Salick found that when a non spiny mother was fertilised with pollen from a spiny plant the F_1 progeny were all spiny. When these F_1 were self pollinated the ratio of spiny plants to non spiny were approximately 3:1 as would be expected for a single gene dominance. The loss of spines during domestication appears to have involved a mutation at a single locus. As a consequence Salick was able to conduct an experiment on gene flow by placing a single spiny plant in a field of non spiny ones and studying their progeny. The results showed that gene flow decreases linearly with the square root of distance and that this did not differ with different densities of planting. In cultivation there is a strong human selection for the recessive allele against spiny plants whereas, in contrast, in nature there is a strong selection for spiny plants which afford considerable protection against predators.

Another interesting conclusion of Salick's work came from the results of a complete diallele cross between seven varieties of individuals with different fruit characteristics where all varieties were crossed with all other varieties including self. Progeny showed that there was an overwhelming predominance of maternal inheritance for the fruit characters measured. For example, a large fruited mother produces large fruited offspring regardless of the fruit type of the pollen provider. The result of this maternal inheritance, whatever the mechanism, is that artificial selection is simple since the preferred phenotype is transferred to the next generation without modification which is similar to vegetative reproduction. There is no need for the cultivator to eliminate the undesirable recombinants. For a review of maternal inheritance see Roach and Wulff (1987). It is probably the simplicity of inheritance in the peach tomato that has led to easy domestication through the selection against one dominant gene for spinyness and the use that is made of maternal inheritance of many fruit characters. Plants that have less complicated breeding systems and inheritance patterns are more likely to become domesticated than those with complicated systems which might make selection of characteristics much harder. This also accounts for the domestication of so many species that are easy to propagate vegetatively.

Conclusion

All three species considered here are of considerable economic importance in Amazonia. The production of these three fruit crops depends upon a wide range of bee pollinators. A more convincing argument for conservation can be made when the examples used are familiar plants which yield products of economic importance. In all three cases pollination is effected by bees that require natural areas of vegetation for their continued survival. Many different studies are demonstrating declines in pollinators. For example Buchmann and Nabhan (1996) showed that in Costa Rica wild bee diversity in degraded forest land dropped from 70 to 37 species in just 14 years. They hypothesise that habitat fragments do not provide the diversity of resources needed by pollinators. This could well happen to the pollinators of the three species discussed here. Information about pollination of such important species as the Brazil nut, the peach tomato and the cupuaçu is needed to preserve the crop. This information provides strong reasons for ecosystem conservation. I urge that we all use our data about reproductive biology to promote the conservation of these processes upon which the world depends to maintain our food crops, beautify our gardens and maintain wild ecosystems.

Acknowledgements

I would like to thank the Instituto Nacional de Pesquisas da Amazonia where most of this work was based. I thank Scott A. Mori and Bruce W. Nelson for collaboration with work on the Brazil nut, Marta Falcão, Jan Salick and Eduardo Lleras for work on the peach tomato, Marlene F. da Silva and Giorgini Venturieri for their studies of cupuaçu. I also thank Scott Mori for reviewing an earlier draft of this paper.

References

Addison, G.O. and Tavares, R.M. (1951). Observações sobre as espécies do gênero *Theobroma* que ocorrem no Amazônia. *Bol. Técn. Inst. Agron. N.* **25**: 1–20.

Buchmann, S.L. and Nabhan, G.P. (1996). The forgotton pollinators. Island Press, Washington, D.C.

Calzavara, B.B.G., Müller, C.H. and Kahwage, O.N.C. (1984). Fruticultura tropical: o cupuaçuzeiro. Cultivo, beneficiamento e utilização do fruto. EMBRAPA/CPATU, Belém.

Cope, F. (1962). The mechanism of pollen incompatibility in *Theobroma* L. *Heredity* **17**: 157–182.

Dodson, C.H. (1966). Ethology of some bees of the tribe Euglossini (Hymenoptera: Apidae). *J. Kansas Entomol. Soc.* **39**: 607–629.

Dodson, C.H. (1967). Relationship between pollinators and orchid flowers. In: H. Lent (editor). Atas da Simposio sobre a Biota Amazônica, Belem, Para, Brasil, Junho 6–11, 1996. Vol. 5. Zoologia, pp. 1–72, Conselho Nacional de Pesquisas, Rio de Janeiro.

Dressler, R.L. (1982). Biology of the orchid bees (Euglossine) *Annual Rev. Ecol. Syst.* **13**: 373–394.

Falcão, M.A. and Lleras, E. (1983). Aspectos fenológicos, ecológicos e de produtividade do cupuaçu (*Theobroma grandiflorum* (Willd. ex Spreng.) Schum.). *Acta Amazôn.* **13**: 725–735.

Huber, J. (1909). Mattas e madeiras Amazônicas. *Bol. Mus. Goeldi Paraense Hist. Nat. Ethnogr.* **6**: 91–225 (1909 publ. 1910).

Ingram, M., Nabhan, G.P. and Buchman, S. (1996). Impending pollination crisis threatens biodiversity and agriculture. *Tropinet* **7(2)**:1–2.

Jacques, F. (1965). Morphologie du pollen et des ovules de *Couroupita guianensis* Aubl. (Lécythidacées). *Pollen et Spores* **7**: 175–180.

Knight, R. and Rogers, H.H. (1955). Incompatibility in *Theobroma cacau*. *Heredity* **9**: 69–72.

Mori, S.A. (1992). The Brazil nut industry - past, present and future. In: M. Plotkin and L. Famolare (editors). Sustainable harvest and marketing of rain forest products, pp. 241–251, Island Press, Washington, D.C.

Mori, S.A and Prance, G.T. (1990a). Lecythidaceae: part II. The zygomorphic-flowered New World genera (*Couroupita, Bertholletia, Couratari, Eschweilera* and *Lecythis*). *Fl. Neotrop. Monogr.* **21(2)**: 1–376.

Mori, S.A. and Prance, G.T. (1990b). Taxonomy, ecology and economic botany of the Brazil nut (*Bertholletia excelsa* Humb. & Bonpl. : Lecythidaceae). *Advances Econ. Bot.* **8**: 130–150.

Mori, S.A., Orchard, J.E. and Prance G.T. (1980). Intrafloral pollen differentiation in the New World Lecythidaceae subfamily Lecythidoideae. *Science* **209**: 400–403.

Moritz, A. (1984). Estudos biológicos da floração e da fruitificação da castanha-do-brasil (*Bertholletia excelsa* H.B.K.). EMBRAPA-CPATU, Belém. (Documentos, 29).

Müller, C.H., Rodrigues, I.A., Müller, A.A., and Müller, N.R.M. (1980). Castanha-do-brasil; resultados de pesquisa. EMBRAPA-CPATU, Belém. (Miscelânea, 2).

Nelson, B.W., Absy, M.L., Barbosa, E.M. and Prance, G.T. (1985). Observations on flower visitors to *Bertholletia excelsa* H. & B. and *Couratari tenuicarpa* A.C. Sm. (Lecythidaceae). *Acta Amazon. Supl.* **15 (1–2)**: 225–234.

Ormond, W.T., Pinheiro, M.C.B. and Cortella de Castells, A.R. (1982). A contribution to the floral biology and reproductive system of *Couroupita guianensis* Aubl. (Lecythidaceae). *Ann. Missouri Bot. Gard.* **68**: 514–523 (1981 publ. 1982).

Prance, G.T. and Mori, S.A. (1979). Lecythidaceae: Part I. The actinomorphic flowered New World Lecythidaceae. *Fl. Neotrop. Monogr.* **21(1)**: 1–270.

Retto Junior, A da S. (1986). Estudos preliminares sobre a biologia floral do cupuaçuzeiro (*Theobroma grandiflorum* (Willd. ex Spreng.) Schum.). Thesis, Fundação Universidade do Amazonas, Manaus.

Salick, J. (1992). Crop domestication and the evolutionary ecology of cocona (*Solanum sessiliflorum* Dunal). *Evol. Biol.* **26**: 247–285.

Sánchez, J.S. (1973). Explotación y comercialización de la Castaña en Madre de Dios. Ministerio de Agricultura, Dirección General de Forestal y Caza, Peru. (Informe 30.)

Silva, M.F. (1976). Insetos que visitam o "cupuaçu" (*Theobroma grandiflorum* (Willd. ex Spreng.) Schum.) (Sterculiaceae), e índice de ataque nas folhas. *Acta Amazôn.* **6**: 49–54.

Storti, E.F. (1988). Biologia floral de *Solanum sessiliflorum* Dun. var. *sessiliflorum*, na região de Manaus, AM. *Acta Amazôn.* **18(3–4)**: 55–65.

Tsou, C.H. (1994). The embryology, reproductive morphology and systematics of Lecythidaceae. *Mem. New York Bot. Gard.* **71**: 1–110.

Venturieri, G.A. (1992). Floral biology of cupuassu (*Theobroma grandiflorum* (Willdenow ex Sprengel) Schumann. Report to Ph.D. programme at School of Plant Sciences, Reading University.

Venturieri, G.A. (1993). Cupuaçu: a espécie, sua cultura, usos, e processamento. Clube do cupu, Belém.

Venturieri, G.A. and Aguiar, J.P.L. (1988). Composição do chocolate de amêndoas de cupuaçu (*Theobroma grandiflorum*). *Acta Amazôn.* **18(1–2)**: 3–8.

Whalen, M.P. and Anderson, G.J. (1981). Distribution of gametophyic self-incompatibility and infrageneric classification in *Solanum*. *Taxon* **30**: 761–767.

Yacsirk, S., Xena, N., Ramírez, N. and Agostini, G. (1986). Notes on the floral biology of *Couroupita guianensis*. Aubl. (Lecythidaceae). *Ann. Missouri Bot. Gard.* **73**: 99–101.

Young, A.M. (1994). The chocolate tree: a natural history of cacau. Smithsonian Institution Press, Washington, D.C.

Chamberlain, J.R. (1998). *Calliandra calothyrsus:* Reproductive biology in relation to its role as an important multipurpose tree. In: S.J. Owens and P.J. Rudall (Editors). Reproductive Biology, pp. 439–448. Royal Botanic Gardens, Kew.

CALLIANDRA CALOTHYRSUS: REPRODUCTIVE BIOLOGY IN RELATION TO ITS ROLE AS AN IMPORTANT MULTIPURPOSE TREE

JOANNE R. CHAMBERLAIN

Oxford Forestry Institute, Department of Plant Sciences, University of Oxford, South Parks Road, Oxford OX1 3RB

Abstract

Calliandra calothyrsus Meisn. is a small leguminous tree native to Central America and Mexico, but cultivated widely throughout the humid tropics for the provision of fuelwood, livestock fodder and shade in small farming communities. Growing interest in the species and the results of a provenance trial network co-ordinated by the Oxford Forestry Institute (OFI) have meant that the demand for seed of the species is great. In exotic environments, however, problems with the production of seed have been experienced that may be the result of several factors. *C. calothyrsus* possesses a gametophytic self-incompatibility system that has a low rate of failure. The species is andromonoecious and the development of fruit has a significant effect on whole plant sex expression. *C. calothyrsus* flowers at night and is pollinated by small nectivorous bats and less commonly, hawkmoths. The reproductive characteristics of the species are explored and their role in limiting seed production discussed.

Introduction

Calliandra calothyrsus Meisn. (Leguminosae - Mimosoideae) is a small tree that is widely distributed throughout Central America and southern Mexico. The species occurs across a considerable variety of ecological conditions from very wet (over 5000 mm rainfall per year) to seasonally dry (as low as 900 mm rainfall per year), and from sea-level to 1500 m altitude in southern Guatemala (Macqueen, 1992; 1993). *C. calothyrsus* is a riverine coloniser in natural undisturbed forest, but is now a very common element of secondary vegetation in areas such as roadsides and fallow agricultural land. It often appears in thickets during early colonisation, but with time tends to be outcompeted and has a scattered distribution in older stands. The species is bat-pollinated and assumed to be outcrossing (Macqueen, 1992).

Calliandra calothyrsus has been cultivated and used widely for the provision of fuelwood, animal fodder, green manure, shade and soil conservation (National Research Council, 1983; Macqueen, 1992; 1993). The species was introduced to Indonesia from Guatemala in the 1930's (Verhoef, 1939) and is now naturalised in many parts of Java. Seed from these naturalised populations has been, until recently, the major source of material for subsequent planting in south-east Asia, Australia and Africa. In 1990, a research programme was initiated at the Oxford Forestry Institute (OFI) to map the natural distribution of *C. calothyrsus* and collect seed for an international network of provenance trials that aims to determine the best seed sources for biomass production (Macqueen, 1993). It is anticipated that demand for seed of the best provenances will be great once the final results are obtained by trial collaborators. The distribution and use of selected material by researchers, NGO's and farmers will be dependent on the ability to produce large quantities of seed. Seed production in the native range is, however, limited by both the small size of *C. calothyrsus* populations and low seed set per tree. The maximum number of individuals in any of the OFI seed collections was 65 (Macqueen, 1993), while an average collection from around 30 individuals will yield less than 2 kg of seed (17000 seeds per kg). In some cases, repeated visits to the same group of trees across the fruiting season are necessary to collect an adequate quantity of seed.

In order to improve seed production in *C. calothyrsus*, and anticipate demand for seed of the best provenances, seed orchards have been established in both the native range, and where the species is cultivated (Chamberlain and Rajaselvam, 1996a). Low seed production has, however, often been obtained in orchards established in exotic environments (Boland and Owour, 1996; B.M. Kamondo, pers. comm.[1]; B. Duguma, pers. comm.[2]). *C. calothyrsus* flowers precociously, but sets relatively little seed in comparison to the abundance of flowers it produces, and hence, there has been the perception that an individual tree should produce more seed (Boland and Owour, 1996). Exceptionally low seed production has, however, been obtained in some cases (B.M. Kamondo, pers. comm.; B. Duguma, pers. comm.), and there may be one or more of a range of factors responsible for this including the species' breeding and/or sexual system, low visitation rates by pollinators, physical/spatial restrictions on the inflorescence, or adverse climatic conditions. This paper summarises recent research on the reproductive biology and pollination system of *C. calothyrsus*, and discusses possible limitations to fruit set in the species.

Flowering and sex expression

Calliandra calothyrsus has a flowering period that can extend to up to twelve months per year if sufficient moisture is available. Across the native range, flowering reaches a peak during October and November, and will cease where a dry season is experienced (January to April). The inflorescence of *C. calothyrsus* is a paniculate raceme in which the flowers open acropetally over a period of between 60–90 days. The flowers are held within subumbels and the number of flowers open per subumbel ranges from 4–14 (Fig.1; Macqueen, 1992). The flowers are characterised by their long red or pink staminal filaments that form the brush-type floral display typical of many mimosoid legumes, and which serves to attract pollinators. Anthesis occurs between 15:30 and 17:00 h and is followed by anther dehiscence once the flowers are fully open. The stigma is receptive from 19:00 h, but by 06:00 h the following morning receptivity has been lost and the flowers begin to wilt. From pollination to seed reaching maturity, a period of between 90–120 days is required.

Many outcrossing plants produce more flowers than fruits, and *C. calothyrsus* is no exception with fruit-flower ratios in the region of 0.15 – 0.20 (Chamberlain, in prep.). This phenomenon may serve to maximally attract pollinators (Stephenson, 1981) and increase the amount of pollen relative to the number of ovules in a given plant population (Cruden, 1977). Low fruit-flower ratios in *C. calothyrsus* can, however, be partially explained by the fact that the species is andromonoecious (Macqueen, 1992), i.e. an individual plant produces both hermaphrodite and staminate flowers. In the latter, the gynoecium is completely reduced and the flower is functionally male. There are a number of hypotheses for the

[1]Kenya Forest Research Institute, P.O. Box 20412, Nairobi, Kenya
[2]IRA/ICRAF Project, P.B. 2067, Yaounde, Cameroon

FIG. 1. *Calliandra calothyrsus*, inflorescence.

evolution of andromonoecy in plants, but the most plausible is that it is a means of optimally allocating limited reproductive resources to both the male and female functions (Maynard Smith, 1978; Bawa and Beach, 1981; Bertin, 1982; Charnov, 1982; Willson, 1983; May and Spears, 1988). Alternatively, the possession of staminate flowers could increase plant fitness by promoting outcrossing and regulating genetic recombination (Mather, 1973), or occur in response to the dynamics of the pollination system (Bawa and Beach, 1981). Decreased nectar rewards from staminate flowers are suggested to promote repeated visits by pollinators and therefore outcrossing (Cruden, 1976), as are increased quantities of pollen per flower (Traveset, 1995).

Sex expression was studied in a trial plot of *C. calothyrsus* at San Juan, the experimental station of ESNACIFOR (Escuela Nacional de Ciencias Forestales), Siguatepeque, Honduras. The plants were randomly split into two treatments. In the first treatment, all open flowers on each plant were hand-pollinated every day (weather permitting) for eight weeks. Flower position, sex expression and fruit production were recorded for each inflorescence on the plant. In the second treatment, the position and sex expression of all open flowers were recorded for each inflorescence per plant, also on a daily basis. In this treatment, however, all hermaphrodite flowers were removed after recording position and sex to monitor the plasticity of sex expression when maternal investment is high (with fruit - first treatment), and when maternal investment is low (no fruit - second treatment). In addition, buds were dissected from inflorescences both with and without fruits, and the variation in nectar and pollen production (nectar volume, nectar sugar concentration, number of anthers per flower, polyad number per anther) with sex was also studied. A full description of the methods and results is given elsewhere (Chamberlain, submitted).

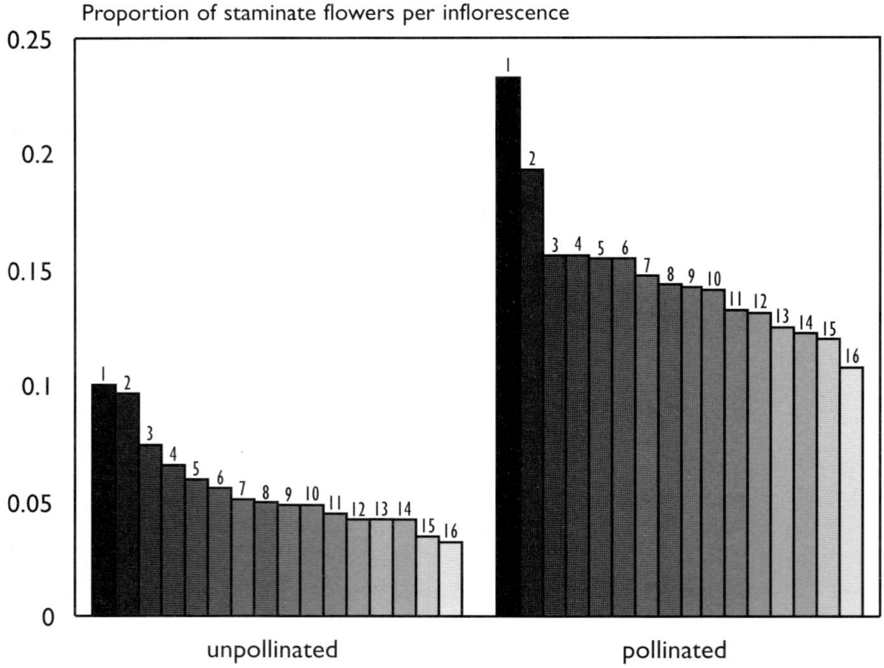

FIG. 2. Mean proportion of staminate flowers per inflorescence produced in pollinated (+ fruit) and unpollinated (no fruit) treatments for 16 *C. calothyrsus* trees. Numbers refer to individual trees.

The presence of developing fruit had a significant effect on whole plant sex expression (Fig. 2), with the mean proportion of staminate flowers per inflorescence on pollinated plants being higher than that on unpollinated plants. A striking pattern of sexual allocation along the inflorescence was also observed in both experimental and wild plants. Once an inflorescence starts developing fruit in the lower part of the inflorescence, there was a gradual increase in the proportion of staminate flowers towards the apex of the inflorescence. Often a complete switch from hermaphrodite to staminate flower production could be observed. Buds dissected from inflorescences that had just started to initiate flowering were almost without exception hermaphrodite, but inflorescences with developing fruits were associated with a much larger proportion of staminate buds. T-tests comparing the sex of flowers with nectar and pollen production gave consistently high P-values revealing that there were no significant differences between staminate and hermaphrodite flowers for nectar or pollen production in the species.

Measures of resource availability, e.g. soil nutrients (Solomon, 1985; Diggle, 1993), photosynthetic leaf area (Emms, 1996) and plant size (Dudash, 1991; Emms 1996) have been experimentally manipulated in a number of andromonoecious plant species. Although such investigations were not possible for the study of sex expression in *C. calothyrsus*, the pattern of increased staminate flower production along the inflorescence and with fruit development was comparable to that obtained in other andromonoecious species where resource availability has been manipulated, e.g. *Solanum carolinense* (Solomon, 1985), *Passiflora incarnata* (May and Spears, 1988) and *Solanum hirtum* (Diggle, 1993). This suggests that a resource allocation mechanism may also be operating in *C. calothyrsus* which limits the species' capacity to produce fruit. The production of staminate flowers may, however, also benefit the integrity of the species' pollination system, promote outcrossing and increase the overall efficiency of resource allocation.

Breeding system

The breeding system of *C. calothyrsus* was studied using controlled pollinations on trees from each of two provenances in a trial plot at Cuyuta, the experimental station of ICTA (Instituto de Ciencia y Tecnologia Agricolas), Guatemala. The inflorescences chosen were bagged once the flowers began to open, and any previously fertilised flowers removed. The flowers were then emasculated prior to anther dehiscence and re-bagged. Pollen was collected from the same tree (selfing), from flowers on several other trees of the same provenance (within-provenance crosses) and from flowers on several other trees of the alternative provenance (between-provenance crosses). The pollen was applied with a fine brush to receptive stigmas from 19:30 h onwards. All flowers on an inflorescence received the same type of pollen. A number of inflorescences on the same trees were tagged and left unbagged to monitor the amount of successful fertilisation (fruit set) under natural conditions. In addition, emasculated and pollinated flowers were fixed after 12 to 36 hours, and the fixed material was examined with fluorescent microscopy to determine compatibility. A full description of the methods and results is given by Chamberlain (in prep.).

The results of both controlled crosses and microscope observations showed the species to be weakly self-incompatible. Self pollinations resulted in a small amount of fruit set (mean 4.1%). Fruit set from within-provenance crosses was higher (mean 24.2%) than from crosses made between-provenances (mean 14.8%), although these differences were not significant. There were no significant differences in the seed-ovule ratio per fruit between the different pollination treatments. Microscope observations show pollen grains germinated readily following both cross- and self-pollination. After 18–21 h the pollen tubes had reached the base of the style and could be observed in the ovary, suggesting that the self-incompatibility system in *C. calothyrsus* is late-acting (see Seavey and Bawa, 1986). Further work is continuing to quantify the number of penetrated ovules that result from selfs and determine at what point rejection of selfs occurs.

Outcrossing rate and mating system

Although knowledge of a species breeding system is important, it does not indicate actual levels of outcrossing within populations. Isozymes may be used as genetic markers to provide estimates of the mating system and gene flow, although the interpretation of such data is made more meaningful when combined with field observations on the spatial distribution of trees, their flowering phenology and reproductive biology. A natural population of *C. calothyrsus* found within a river system (Rio Bonito, Colón, Honduras) was used to study the mating system of the species using starch gel electrophoresis. Open-pollinated seed was collected from individual trees distributed across four sites within the population, which were separated by a minimum distance of 570 m, and a maximum distance of 3600 m. General descriptions of the enzyme systems observed in *C. calothyrsus* over a range of populations are detailed elsewhere (Chamberlain, 1998).

Population outcrossing rates (t_m - multilocus and t_s - single locus) have been estimated for trees in two of the four sites using the MLTR computer program (Ritland, 1994), which fits a mixed mating model (Ritland and Jain, 1981). The outcrossing rate between trees in these two sites was estimated using ten isozyme loci as $t_m = 0.741 \pm 0.085$ and $t_s = 0.716 \pm 0.082$. The lack of a significant difference between t_m and t_s suggests there is no biparental inbreeding and that the inbreeding observed is due to selfing. Outcrossing rate estimates for each individual tree sampled range from 0.00 to 2.00. Four trees were estimated to be completely selfed, and were either among the first to flower, or in one case very late to flower. This latter tree had a peak flowering period some 18 days later than the peak flowering period of the majority of the trees. The selfed trees set a similar quantity of fruit as those trees estimated to be either partially or wholly outcrossing. The results of this initial investigation appear to correlate well with the possession of a weak self-incompatibility system, and suggest that the breeding system of *C. calothyrsus* is unlikely to limit seed production in the species.

Pollination system

The lack of suitable pollinators may be a factor causing low fruit set in plants. Several researchers have implicated the lack of successful pollination as a major cause of flower abortion (Willson and Schemske, 1980; Bierzychudek, 1981; Rathcke, 1983; Koptur, 1984). This may either result from a shortage of pollinators, or the inadequate transfer of compatible pollen by pollinators. A study to determine the most common pollinator of *C. calothyrsus* and its foraging behaviour was carried out within the Rio Bonito (Colón, Honduras), a river valley within a low-lying hill area of degraded tropical wet forest (Chamberlain, 1996; Chamberlain and Rajaselvam, 1996b).

Bats were caught with mist nets (6 m long by 3 m high) set to extend from ground level beside and across streams where *C. calothyrsus* occurred, or on riverbanks in front of swathes of *C. calothyrsus*. Bats were trapped at four separate sites within the Rio Bonito, separated from one another by a minimum distance of 570 m, and a maximum distance of 3600 m. Hawkmoths were also trapped, occasionally in the mist nets, but generally by the use of hand-held insect kite nets. Bats and hawkmoths were caught over 23 nights, and once caught were tested for the presence of pollen, their morphological characteristics noted, the bats marked by removing a small patch of fur from the underside, head or back and then released.

The most common pollinator of *C. calothyrsus* were *Glossophaga* spp., bats commonly known as nectar-feeding or long-tongued bats (Glossophaginae). These small bats are characterised by the possession of long, narrow, extensible tongues covered in hair-like papillae that are specialised for nectar-feeding (Emmons, 1990). *Glossophaga* spp. have been found to possess restricted foraging ranges (<600 m in this study; 360 m - Fleming *et al.*, 1972; 380 m - Heithaus *et al.*, 1975) relative to other species, e.g. *Phyllostomus discolor* forages over distances of up to 1600m (Heithaus *et al.*, 1975). In this study, however, no data on the foraging distances of bat species other than *Glossophaga* could be obtained due to low capture rates of *Phyllostomus*, *Carollia*, *Sturnia*, *Artibeus* and *Uroderma*. These larger, primarily fruit-eating bats, were significantly less common visitors to *C. calothyrsus* than *Glossophaga* spp. (Fig. 3), and were mainly captured in sites with a higher proportion of fruit-producing trees. Hawkmoths were not observed to be major pollinators of *C. calothyrsus* at this site.

Where *C. calothyrsus* is cultivated outside its native range, small bat species adapted for nectar-feeding are likely to be the most efficient pollinators. Amongst the Old World fruit bats (Pteropodidae), the members of subfamily Macroglossinae are characterised by the possession of a long, extensible tongue covered in bristle-like papillae, much as in the Glossophaginae (Mickleburgh *et al.*, 1992). These bats are native to west and central Africa, south-east Asia, the Pacific Islands and northern Australia. Macroglossinae bats are likely to be the most efficient pollinator of *C. calothyrsus* when it is planted in exotic environments, and were found to pollinate the species in Indonesia (Chamberlain, unpublished data). In countries where these bats are not native, but where *C. calothyrsus* is cultivated, e.g. east Africa, India and Sri Lanka,

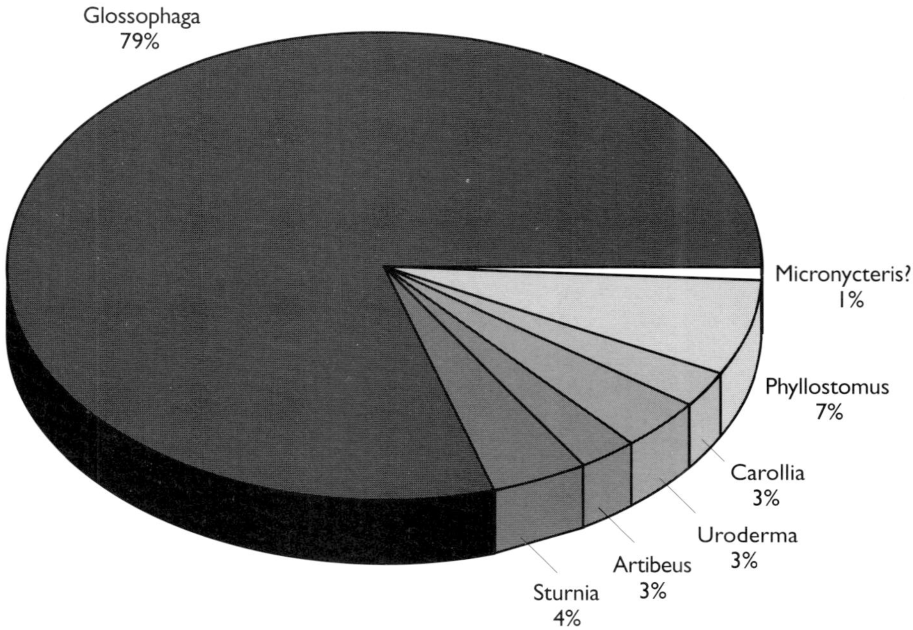

FIG. 3. Proportion of neotropical bat species caught visiting *C. calothysus* in Colón, Honduras.

other species of bats visit *C. calothyrsus*. In Sri Lanka, flowers of *C. calothyrsus* in a provenance trial at Dodangolla, experimental station of the University of Peredeniya, Kandy, were visited by two genera of fruit bat, *Cynopterus* and *Rousettus* (Pteropodinae) (Chamberlain and Rajaselvam, 1996b). These bats appear to visit *C. calothyrsus* flowers much less frequently than bats in the native range and alter their feeding habits according to seasonal food availability, but are effective pollinators when they visit.

The red or pink staminal filaments that characterise the flowers of *C. calothyrsus* and form the floral display do not immediately indicate that the species is pollinated by bats. Bat-pollinated plants tend to possess flowers that have a strong, musky-scented nectar and are white or cream in colour. The bat species that pollinate *C. calothyrsus* in the native range, however, possess smallish eyes, and use echolocation and olfaction as their primary means of locating food resources (Nowak, 1991). The faint musky scent of the *C. calothyrsus* nectar and the size of the floral display may therefore be more important than the colour of the flowers. Where the species is planted in exotic locations, however, colour maybe of greater importance. Generally, the Old World bats do not have the ability to echolocate, but have large eyes and can distinguish colour (Nowak, 1991), hence coloured flowers and a large floral display may be more important to a bat-pollinated species in theses areas.

Physical/spatial restrictions

If pollinators are not specific to *C. calothyrsus* but visit other plants as well, clogging of the stigma with foreign pollen could occur (Waser, 1978) and may serve to prevent the germination of *C. calothyrsus* pollen. Microscope observations of open-pollinated pistils often show the presence of both foreign pollen and *C. calothyrsus* pollen. Pollen tube growth of the former is arrested in the style, and appears not to inhibit the subsequent or simultaneous pollen tube growth of *C. calothyrsus* pollen, though the relative size of the pollen grains may be important in this situation (Chamberlain, in prep.). *C. calothyrsus* possesses pollen grains aggregated into large, 8-grained polyads. In many Mimosoids with polyads, one polyad perfectly fits in the stigmatic chamber. This is not the case in *C. calothyrsus*, although generally one or two polyads can be observed in the stigmatic cup. Foreign pollen grains, or polyads, of a similar size could cause clogging of the stigma, whereas pollen of a much smaller size is likely to have no effect.

Once fertilisation has occurred, the plant may be unable to mature every fruit because of resource or physical restrictions. Seed-ovule ratios average 0.67 in *C. calothyrsus*, hence, either the ovary does not receive sufficient male gametes to fertilise every ovule, or abortion occurs. Microscope observations suggest that the former is the more likely cause of such low seed-ovule ratios. Physical restrictions arise at the stigma, where the number of polyads that can fit in the stigmatic cup is limited to four, but more commonly, only one or two are observed. The number of pollen grains that actually germinate is often limited to between 6–8, as only those pollen grains in contact with the stigmatic surface will germinate. As the ovary contains between 8–12 ovules, seed-ovule ratios are therefore likely to be low, and very rarely are 10 or 12-seeded fruits obtained. Physical/spatial restrictions will also occur at the inflorescence level. The average number of fruits per sub-umbel that have the physical space to develop may be limited, i.e. to one or two.

Environmental restrictions

Adverse climatic conditions could serve to limit fruit set. In its native range, *C. calothyrsus* is found commonly in regions with mean annual rainfalls of up to 3000 mm and which have no true dry season (Macqueen, 1993). The timing and pattern of rainfall across the flowering and fruiting season could, however, affect seed production. For example, heavy rain during the early part of the evening may cause both the anthers and pollen to be washed from the flowers and severely restrict subsequent pollen flow. Precipitation in the form of mist or fog may also limit anther dehiscence and the effective transport of pollen by pollinators, especially if it occurs during the early evening.

Conclusions

Investigations into the breeding system and outcrossing rate in *C. calothyrsus* suggest that the species possesses a mixed mating system, a factor unlikely to limit reproductive success. Andromonoecy and the mechanism controlling it will, however, limit the proportion of flowers that can set fruit. The effects of this mechanism may be compounded by physical or spatial restrictions at the inflorescence or flower level. Fruit set in the species may reach abnormally low levels, however, if there are a lack of suitable pollinators, or the dispersal of pollen is limited by unfavourable climatic conditions. Such investigations have allowed a greater understanding of the factors that control reproductive success in *C. calothyrsus* and form a basis on which practical decisions concerned with managing the species for seed production can be made.

Acknowledgements

This publication is an output from research projects funded by the Department for International Development of the United Kingdom. However, the Department for International Development can accept no responsibility for any information provided, or view expressed. R5728 and R6535 Forestry Research Programme. Valuable field assistance has been provided by Modesto Castillo, Andrew Matthews, Gus Hellier, John Roberts, Douglas Gibbs and Jason Hubert. CONSEFORH in Siguatepeque, Honduras are acknowledged for their valuable support, as are ESNACIFOR and ICTA, for the use of their *Calliandra* trial material. Kermit Ritland provided an updated version (September 1994) of his multilocus mixed mating system program. David Boshier provided valuable comments on the manuscript.

References

Bawa, K.S. and Beach, J.H. (1981). Evolution of sexual systems in flowering plants. *Ann. Missouri Bot. Gard.* **68**: 254–274.

Bertin, R.I. (1982). The ecology of sex expression in red buckeye. *Ecology* **63**: 455–456.

Bierzychudek, P. (1981). Pollinator limitation of plant reproductive effort. *Amer. Naturalist* **117**: 837–840.

Boland, D.J. and Owour, B. (1996). Some aspects of floral biology and seed production in exotic *Calliandra calothyrsus* at Maseno, Kenya. In: D.O. Evans (editor). Proceedings of the International Workshop on the Genus *Calliandra*, pp. 49–61, Winrock International, Morrilton, Arkansas. (Forest, Farm and Community Tree Research Reports, special issue).

Chamberlain, J.R. (1996). *Calliandra calothyrsus* - pollinator behaviour and seed production. Oxford Forestry Institute, Oxford. (Oxford Forestry Institute Research Highlights, no. 3).

Chamberlain, J.R. (1998). Isozyme variation in *Calliandra calothyrsus* (Leguminosae): its implications for species delimitation and conservation. *Amer. J. Bot.* **85**: 37–47.

Chamberlain, J.R. (in prep.). Floral and reproductive biology of *Calliandra calothyrsus* Meisn.

Chamberlain, J.R. (submitted). Sex expression in the andromonoecious mimosoid legume, *Calliandra calothyrsus* (Leguminosae). Submitted to *Canad. J. Bot.*

Chamberlain, J.R. and Rajaselvam R.J. (1996a). *Calliandra* seed production - a problem or not? In: D.O. Evans (editor) Proceedings of the International Workshop on the Genus *Calliandra*, pp. 29–33, Winrock International, Morrilton, Arkansas. (Forest, Farm and Community Tree Research Reports, special issue).

Chamberlain, J.R. and Rajaselvam R.J. (1996b). *Calliandra calothyrsus* - pollinator behaviour and seed production. In: D.O. Evans (editor) Proceedings of the International Workshop on the Genus *Calliandra*, pp. 34–40, Winrock International, Morrilton, Arkansas. (Forest, Farm and Community Tree Research Reports, special issue).

Charnov, E.R. (1982). The theory of sex allocation. Princeton University Press, Princeton, N.J.

Cruden, R.W. (1976). Fecundity as a function of nectar production and pollen-ovule ratios. In: J. Burley and B.T. Styles (editors). Tropical trees; variation, breeding and conservation, pp. 171–178, Academic Press, London.

Cruden, R.W. (1977). Pollen-ovule ratios: a conservative indicator of breeding systems in flowering plants. *Evolution* **31**: 32–46.

Diggle, P.K. (1993). Developmental plasticity, genetic variation, and the evolution of andromonoecy in *Solanum hirtum* (Solanaceae). *Amer. J. Bot.* **80**: 967–973.

Dudash, M.R. (1991). Plant size effects on female and male function in hermaphroditic *Sabatia angularis* (Gentianaceae). *Ecology* **72**: 1004–1012.

Emms, S.K. (1996). Temporal patterns of seed set and decelerating fitness returns on female allocation in *Zigadenus paniculatus* (Liliaceae), an andromonoecious lily. *Amer. J. Bot.* **83**: 304–315.

Emmons, L.H. (1990). Neotropical rainforest mammals: a field guide. University of Chicago Press, Chicago.

Fleming, T.H., Hooper, E.T. and Wilson, D.E. (1972). Three Central American bat communitites: structure, reproductive cycles and movement patterns. *Ecology* **53**: 555–568.

Heithaus, E.R., Fleming, T.H. and Opler, P.A. (1975). Foraging patterns and resource utilization in seven species of bats in a tropical forest. *Ecology* **56**: 841–854.

Koptur, S. (1984). Outcrossing and pollinator limitation of fruit set: breeding systems of neotropical *Inga* trees (Fabaceae: Mimosoideae). *Evolution* **38**: 1130–1143.

Macqueen, D.J. (1992). *Calliandra calothyrsus*: implications of plant taxonomy, ecology and biology for seed collection. *Commonw. Forest Rev.* **71**: 20–34.

Macqueen, D.J. (1993). *Calliandra* series Racemosae: taxonomic information, OFI seed collections and trial design. Oxford Forestry Institute, Oxford.

Mather, K. (1973). Genetical structure of populations. Chapman and Hall, London.

May, P.G. and Spears, E.E. (1988). Andromonoecy and variation in phenotypic gender of *Passiflora incarnata* (Passifloraceae). *Amer. J. Bot.* **75**: 1830–1841.

Maynard Smith, J. (1978). The evolution of sex. Cambridge University Press, Cambridge.

Mickleburgh, S.P., Hutson, A.M. and Racey, P.A. (1992). Old World fruit bats: an action plan for their conservation. IUCN, Gland, Switzerland.

National Research Council. (1983). *Calliandra*: a versatile small tree for the humid tropics. National Academy Press, Washington D.C.

Nowak, R.M. (1991). Walker's Mammals of the world. The John Hopkins University Press, Baltimore, Maryland.

Rathcke, B.J. (1983). Competition and facilitation among plants for pollination. In: L. Real (editor) Pollination biology, pp. 305–329, Academic Press, New York.

Ritland, K. (1994). Multilocus mating system program (MLTR), Version 0.9. Unpublished user manual, Department of Botany, University of Toronto, Toronto.

Ritland, K. and Jain S.K. (1981). A model for the estimation of outcrossing rate and gene frequencies using n independent loci. *Heredity* **47**: 35–52.

Seavey, S.R. and Bawa, K.S. (1986). Late-acting self-incompatibility in angiosperms. *Bot. Rev.* **52**: 195–219.

Solomon, B.P. (1985). Environmentally influenced changes in sex expression in an andromonoecious plant. *Ecology* **66**: 1321–1332.

Stephenson, A.G. (1981). Flower and fruit abortion: proximate causes and ultimate functions. *Annual Rev. Ecol. Syst.* **12**: 253–279.

Traveset, A. (1995). Reproductive ecology of *Cneorum tricoccon* L. (Cneoraceae) in the Balearic Islands. *Bot. J. Linn. Soc.* **117**: 221–232.

Verhoef, L. (1939). *Calliandra calothyrsus* Meisn. Boschbouwproefstation, Bogor, Indonesia. (Korte Mededeelingen, 79).

Waser, N.M. (1978). Competition for hummingbird pollination and sequential flowering in two Colorado wild flowers. *Ecology* **59**: 934–944.

Willson, M.F. (1983). Plant reproductive ecology. John Wiley, New York.

Willson, M.F. and Schemske, D.W. (1980). Pollinator limitation, fruit production and floral display in pawpaw (*Asimina triloba*). *Bull. Torrey Bot. Club* **107**: 401–408.

Knox, B., Taylor, P., Ladiges, P., Nelson, G. and Suphioglu, C. (1998). Pollen allergens: molecular and immunological analyses and implications for systematics. In: S.J. Owens and P.J. Rudall (Editors). Reproductive Biology, pp. 449–463. Royal Botanic Gardens, Kew.

POLLEN ALLERGENS: MOLECULAR AND IMMUNOLOGICAL ANALYSES AND IMPLICATIONS FOR SYSTEMATICS

BRUCE KNOX[1], PHILIP TAYLOR[1], PAULINE LADIGES[1], GARY NELSON[1] AND CENK SUPHIOGLU[2]*

[1]**School of Botany, The University of Melbourne, Parkville, Victoria 3052, Australia**
[2]**Department of Allergy and Clinical Immunology, Monash Medical School, Alfred Hospital, Prahran 3181, Victoria, Australia**

Abstract

Pollen grains house the male gametes (or their progenitor cells) and in many plants are released into the air for fertilisation. Consequently pollen is a natural part of the airflora and is most abundant in spring and early summer in temperate climates, when grass pollen is the principal source in the atmosphere. Certain pollen proteins are allergenic, interacting with the immune system in susceptible humans and binding to specific IgE antibodies, which provide probes for molecular cloning of the allergen genes. Allergens occur in a number of cellular sites: several occur in the cytosol, one in ER cisternae, two in starch granules of amyloplasts, several in the pollen wall layers, at the surface or in the orbicules.

The natural function of allergens remains largely unknown, but some have been associated with cell wall proteins, for example, ragweed Amb a 1 shows some amino acid sequence similarity to pectate lyase, Amb a 5 to an avirulence gene product of fungi, rye-grass Lol p 1 to expansins. Lol p 1 has a N-terminal cysteine-rich region responsible for formation of disulfide bridges and distinctive loops ideal for performing roles in recognition and specificity, while the C-terminal region contains the allergenic epitope. Group 1 allergen genes have been cloned and sequenced from the pollen of 8 grass genera and show considerable amino acid sequence similarity. Parsimony analysis of their amino acid sequences shows the close relationship of Pooid genera, distinct from Panicoid, Bambusoid, and Chloridoid genera. These environmental proteins appear to be allergenic because of the presence of specific B- and T-cell epitopes, which have recently been identified in a few allergens, for example, Lol p 5.

Introduction

In conifers and many flowering plants, pollen grains, which house the male gametes (or their progenitor cells), are released into the air for fertilisation. Consequently pollen is a natural part of the airflora. There, it is well known to provoke the miseries of hayfever and allergic asthma, especially in spring when the grasses are flowering or in autumn during the ragweed season. This symposium is to honour Professor Jack Heslop-Harrison FRS, who has pioneered the study of the structure and function of pollen grains and their interactions with the pistil. In particular, his research initiated in the 1960s re-defined the structures of the pollen wall, the

* Corresponding author

exine and intine, in functional terms for the first time (Heslop-Harrison, 1968). In collaborative research in the early 1970s, we showed that proteins and glycoproteins diffused out of the intine and exine when pollen grains were placed in water. First, a range of hydrolytic enzymes were shown by light and electron microscope cytochemistry to be associated with the pollen wall layers, the intine and exine (Knox and Heslop-Harrison, 1969). Second, these diffusible wall proteins acted as antigens in eliciting the allergic response in susceptible humans. We detected allergens in the pollen wall of ragweed by immunofluorescence methods specially adapted for the first time for plant cells (Knox, Heslop-Harrison and Reed, 1970). One of us, Bruce Knox, was privileged to collaborate in this research and recalls Heslop-Harrison's extraordinary gift of perceiving what is novel and exciting in routine research observations. The initial work on pollen allergens stemmed from a collaboration that Heslop-Harrison built up with his then colleague and director of the Allergy Clinic at the University of Wisconsin, Charles Reed. That research laid the foundations for research into pollen biology and allergens that has been carried on over the subsequent 25 years. During this period, there has been a revolution in progress in understanding the biology of pollen allergens through application of the techniques of molecular biology. It is the purpose of this article to review our current understanding of the biology of pollen allergens in the light of Heslop-Harrison's pioneering contributions.

Pollen allergens

An allergen is an environmental antigen, for example, those proteins diffusing from pollen grains, that can provoke the allergic reaction by interacting with the immune system in susceptible humans (Brostoff and Hall,1996). Type 1 hypersensitivity occurs almost immediately following contact with the allergen. This is a two part process. First, on exposure to an allergen, the immune system produces specific Immunoglobulin E (IgE) antibodies, and T cells play a pivotal role in regulating this process, recognising the allergenic peptides presented in association with the major histocompatibility complex (MHC) class II molecules. IgE antibodies migrate and bind to the surfaces of mast cells and basophils in epithelial tissues. The allergic response can later be initiated when on subsequent exposure to the same allergens, these molecules bind and crosslink specific IgE antibodies present on the surface of mast cells and basophils, which then release the mediators of inflammatory responses, histamine, etc. Allergen molecules bind to adjacent IgE molecules by means of specific regions, epitopes, at the surface of the protein.

Specific recognition of pollen allergens by IgE antibodies means that they can be defined as IgE-binding proteins and human IgE from subjects sensitised to pollen can be used as a specific probe for the allergen. Allergic patients typically have elevated levels of specific IgE in their serum, which provides a specific probe for the identification of allergens, their cross-reactivity with allergens of related species of plants, the isolation of genes encoding the allergens and identification of the allergenic epitopes.

Pollen grains, like other plant cells, contain different types of proteins, which are located in the cytoplasm and the outer wall and only a small number of the proteins are allergenic. Pollen allergens have a molecular mass corresponding to that of common plant enzymes, 5 to 70 kDa. A difficulty in working with allergens is that the identity of the particular protein is usually not known and it is known only as an IgE-binding protein. Accordingly, allergens are named by World Health Organisation rules after the genus and species of the plant of origin (King *et al.*, 1995). For example, the major ragweed pollen allergen is Amb a 1 (from *Ambrosia artemisiifolia*), denoted by the first three letters of the genus, space, the first letter of the species, space, followed by a number, reserving 1 for the major allergen. In particular families, such as Poaceae, where many genera possess very similar allergens, these are referred to as Group 1 allergens, etc.

We have recently reviewed the environmental and molecular biology of pollen allergens (Knox and Suphioglu, 1996). In this chapter, the focus is on the nature and function of pollen allergens in reproduction, their potential for evolutionary studies of plant families, and what characteristics make an environmental protein allergenic.

Cellular sites of pollen allergens

Efforts to locate the subcellular sites of pollen allergens were among the first reported applications of immunofluorescence techniques in plants (Knox *et al.*, 1970).

Amb a 1 is located in pollen walls

Initial studies used rabbit polyclonal antibodies directed against the soluble extract of pollen proteins, but were refined in a series of subsequent studies with polyclonal antibodies specifically raised against highly purified Amb a 1 (= Antigen E). Developmental studies employed unfixed cryostat sections of anthers, prepared within 30 s of contact with an aqueous gelatine embedding medium, to minimise diffusion artifacts (Knox and Heslop-Harrison, 1971a). Results indicated that during pollen development, Amb a 1 at first accumulates during the late vacuolate period, when it is incorporated in the intine polysaccharide wall layer as it is being laid down. In mature pollen, the fluorescent probe bound very strongly to the intine, with some labelling on the surface exine, but little or no labelling in the cytoplasm (Table 1; Knox and Heslop-Harrison, 1971b).

The subcellular site of Amb a 1 in the intine was later confirmed using two techniques in which contact with aqueous media was prevented: after formaldehyde vapour fixation and after freeze-substitution followed by paraffin embedding (Howlett *et al.*, 1973). The location in the exine cavities was confirmed by immunoprinting of dry pollen on agar films (Howlett *et al.*, 1973). Unfortunately, there have been no subsequent immunogold localisation studies of Amb a 1 at the resolution of the transmission electron microscope, nor has any work been done on other ragweed allergens. We conclude that the major ragweed allergen Amb a 1 is synthesised in the cytoplasm during the vacuolate period and incorporated in the intine wall when it is laid down.

Ole e 1 is located in ER

Developmental studies showed that the major allergen of olive pollen, Ole e 1 is located by immunogold labelling within cisternae of the endoplasmic reticulum from the mid-bicellular stage to maturity (Table 1; Rodriguez-Garcia *et al.*, 1995). However, the grains had been aldehyde-fixed, and diffusion of cytoplasmic or wall-located allergens may have occurred; the location in the ER may be more indicative of site of synthesis rather than the cellular site.

Cry j 1 is located in orbicules

In Japanese red cedar, *Cryptomeria japonica*, Cry j 1 has been located by immunogold labelling in developing anthers after both aldehyde fixation (together with microwave irradiation treatment) and freeze-substitution in osmium. Gold labelling occurred in cytoplasmic sites in golgi bodies of pollen grains, and in extracellular sites on the exine-like orbicules (Table 1; Miki-Hirosige *et al.*, 1994).

Group 1 allergens of grasses are located in the cytosol and secreted to the surface

Immunofluorescence studies of grass pollen showed that allergens could not be detected after aldehyde fixation, probably because the allergens diffused from their cellular sites into the fixative (Knox, 1973). Rapid solubility is a characteristic of pollen allergens (Knox *et al.*, 1973; Vrtala *et al.*, 1993). In order to prevent diffusion from sites of allergen synthesis and accumulation, several different modes of fixation were tested by various workers. Developmental immunofluorescence studies of freeze-dried anthers of rye-grass pollen showed that Lol p 1 is synthesised in the cytoplasm of the grains at early maturation period, immediately after division of the microspore nucleus and before the intine had been laid down (Howlett *et al.*, 1981). With polyclonal IgG, fluorescent label is concentrated at the periphery of the cytoplasm. Late in the maturation period, about 24 h before dehiscence, fluorescent label is intense throughout the cytosol and is especially evident around the starch granules, which are not labelled, and in the intine. (Fig. 1)

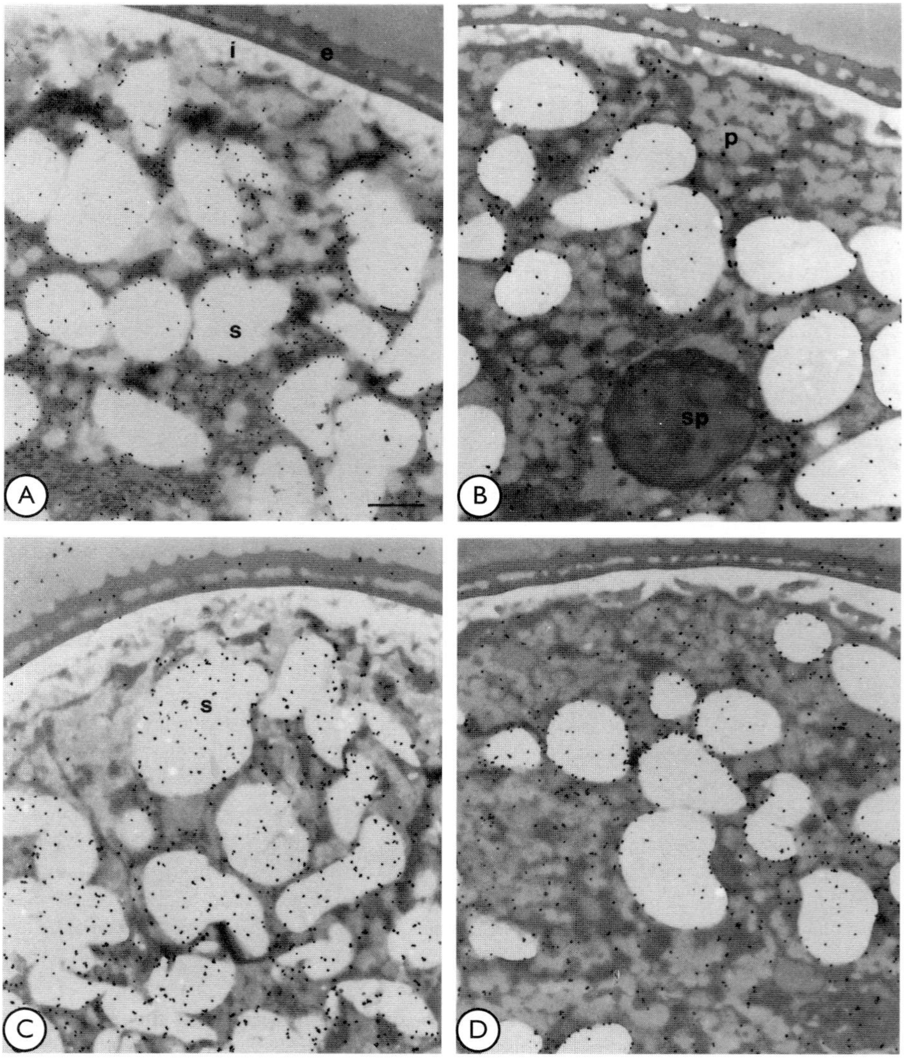

FIG. 1 Transmission electron micrographs of rye-grass pollen after aqueous or anhydrous fixation, processing, sectioning and immunocytochemistry with gold probes and silver enhancement (from Taylor et al. 1993). e exine, i intine, p plastid, s starch granules, sp sperm cell. A. B. Localisation of Lol p 1 using anhydrous fixation by combining aldehydes with DMP (A) and aqueous aldehyde fixation (B). C. D. Localisation of Lol p 5 using anhydrous fixation (C) and aqueous aldehyde fixation (D). Bar = 1μm.

Taking advantage of the higher resolution of transmission electron microscopy, immunoferritin labelling confirmed that Lol p 1 occurred in wall sites: in vesicles and plasma membrane evaginations in the intine and in cavities of the exine as well as cytoplasmic sites. Since then, there have been two major developments: use of monoclonal antibodies (mAbs) for improved molecular specificity, and development of an anhydrous fixation technique to prevent relocation of allergens from their native sites. Anhydrous processing of rye-grass pollen occurs in a mixture of a precipitating agent, DMP (2,2-dimethoxypropane) which withdraws water from the tissue, and aldehydes to improve fixation for immunogold labelling. Using mAb FMC-A1, specific for Lol p 1 with mature pollen, the allergen is located in the cytosol (Fig. 1; Table 1). Abadie *et al.* (1988) used methanol as a fixative for allergen localisation in orchard grass, *Dactylis glomerata*, using allergen-specific IgE as probe. Allergens were concentrated in the pollen wall at maturity, while cytoplasmic labelling was considered an artifact. Grote *et al.*, (1994) used acrolein vapour fixation followed by DMP and immunogold labelling to localise allergens in timothy grass, *Phleum pratense*. Phl p 1 is located in the cytosol (Table 1).

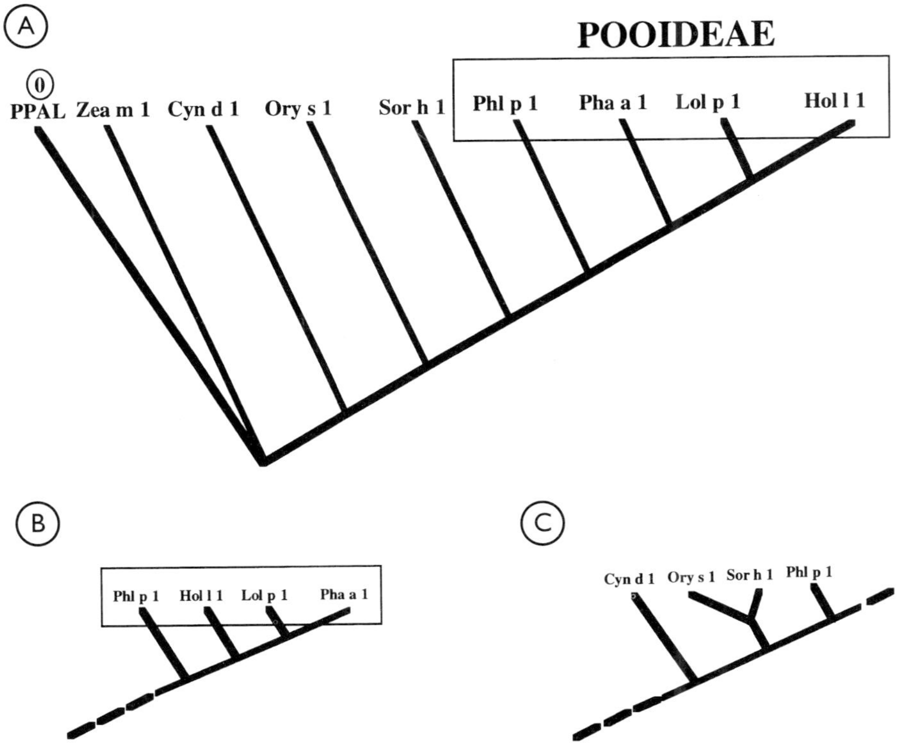

FIG. 2. Phylogenetic analysis of eight taxa of Poaceae obtained using the deduced amino acid sequences of cDNA clones to Group 1 pollen allergens. A and B show the two equally most parsimonious trees obtained using PAUP, Hennig86 or NONA, which differ in the relationships of members of Pooideae (inset). Length is 315 for 126 unordered characters. The tree is rooted by including sequence of an outgroup, PPAL, a gene from *Nicotiana* styles encoding a protein with about 50% sequence similarity to Lol p 1 (courtesy Prof C. Mariani). Consistency index is 0.86 (0 - 1). Classification of genera within Poaceae according to Clayton & Renvoize (1986) is shown in Table 2. C shows tree generated when coded data were used for parsimony analysis by the method of 3-item analysis (Nelson & Platnick, 1991); here, Ory s 1 and Sor h 1 diverge from a single branch rather than two separate branches of the tree.

TABLE 1. Cellular sites in anthers, pollen or pollen tubes of various allergenic species that have been identified by immunocytochemical studies to date. Sites of allergen expression are listed for pollen grains unless otherwise stated. Abbreviations: Aq., aqueous; MAb, monoclonal antibody; PAb, polyclonal antibody; DMP, 2,2-dimethoxypropane.

Species	Allergen	Method	Probe	Cellular site	Reference
Ambrosia artemisiifolia short ragweed	Amb a 1	unfixed, frozen section	PAb, fluorescent	intine, exine	Knox *et al.*, 1971a
		aldehyde vapour or freeze-substitution	PAb, fluorescent	intine, exine	Howlett *et al.*, 1973
Betula verrucosa birch	Bet v 1	cracked, air dried, aldehyde vapour,	MAb, immunogold	cytosol	Grote, 1991
		aldehyde vapour, DMP	MAb, immunogold	cytosol	Vrtala *et al.*, 1993; Grote, 1993
Corylus avelana hazel	Cor a 1	aldehyde + cuprolinic blue	IgE, immunogold	exine surface	Völker *et al.*, 1986
Cryptomeria japonica Japanese red cedar	Cry j 1	Aq. aldehyde + microwave or freeze-substitution in osmium	MAb, immunogold	golgi, exine, orbicules	Miki-Hirosige *et al.*, 1994
Dactylis glomerata orchard grass	Dac g 1	methanol or aldehyde + cetylpiridium chloride	IgE, immunogold	exine	Abadie *et al.*, 1988
Lolium perenne rye-grass	Lol p 1	aldehyde vapour, DMP	Mabs, IgE, immunogold	cytosol, exine	Staff *et al.*, 1990
		aldehyde + DMP	MAbs, immunogold	cytosol	Taylor *et al.*, 1994
	Lol p 5	aldehyde + DMP	MAbs, immunogold	starch granues	Singh *et al.*, 1991; Taylor *et al.*, 1994

Nicotiana tabaccum tobacco	profilin	aldehyde vapour, DMP	PAb, immunogold	cytosol	Grote, 1993
		pollen tube, in vitro, aldehyde vapour, DMP	PAb, immunogold	cytosol	Grote *et al.*, 1995
Oleaceae species (eg. *Olea europaea,* olive)	Ole e 1	Aq. aldehyde	MAb, immunogold	endoplasmic reticulum	Rodriguez-Garcia *et al.*, 1995a,b
Parietaria judaica wall pellitory	Par j 1	Hydrated, aldehydes	IgE, PAb, immunogold	intine, oncus	Casas *et al.*, 1996
Phleum pratense timothy grass	Phl p 1	Acrolein vapour, DMP	PAb, immunogold	cytosol	Grote *et al.*, 1994
	Phl p 4	Acrolein vapour, DMP	MAb, immunogold	starch granules, cytosol, exine	Fischer *et al.*, 1996
	Phl p 5	Acrolein vapour, DMP	MAb, immunogold	starch granules	Grote *et al.*, 1994

IgE was also used to locate allergens in mature pollen of birch and showed labelling concentrated in the exine of the pollen wall (Fromme *et al.*, 1985, Grote and Fromme, 1986). Volker *et al.* (1986) reported the localisation of allergens only on the exine surface of mature hazel pollen after fixation with cuprolinic blue. Later studies on birch using vapour fixation with acrolein over 7 days followed by DMP treatment of mature pollen and immunogold labelling with monoclonal antibodies located Phl p 1 in the cytosol (Grote *et al.*, 1994 - in Phl p 4). Grote (1993) and Grote *et al.* (1995) used similar methods to locate allergens in the cytosol of both pollen and pollen tubes of transgenic tobacco (Table 1).

Grass pollen walls are unusual in having micropores in both the tectum and foot layer of the exine, and the intine is notably traversed by plasma membrane evaginations (Knox, 1984). This means that there is a physical route available for allergen molecules to be secreted to the surface in living pollen. Since most grass pollen grains have only a short life of a few hours as living cells, most release allergens when dead. Dead pollen grains have porous and ineffective plasma and organelle membranes. When wetted, as during rainfall in the atmosphere, the pollen grains resemble sponges and allergen molecules can flow out readily through the pollen wall into the environmental aerosol, where they can become dispersed in microdroplets of water.

Group 4 and 5 allergens of grasses are located in starch granules

A major grass pollen allergen, Lol p 5 (previously designated Lol p 9 [Ong *et al.*, 1993] or Lol p 1b[Singh *et al.*, 1991]), has been located using anhydrous fixation and specific monoclonal antibodies predominantly within starch granules within the pollen grains (Singh *et al.*, 1991; Taylor *et al.*, 1993). Subsequently, Grote *et al.* (1994) showed that a second Group 5 allergen, Phl p 5, is also located in starch granules. The significance of such a cellular location for an allergen was obscure until we showed that starch granules can be released into the aerosol when pollen grains encounter rainfall (Suphioglu *et al.*, 1992; Knox, 1993). Recently, Fischer *et al.* (1996) have shown that Phl p 4 also occurs within the starch granules of timothy pollen, as well as in other cytoplasmic and wall sites.

We conclude that pollen allergens may originate from different sites within the anther and that the transfer and presentation of allergens on the pollen surface may involve several mechanisms.

Structure and function of pollen allergens

The extracellular sites of pollen allergens, especially in the exine cavities and in the intine, implicated allergens in recognition reactions at the stigma surface during pollen attachment, hydration and germination (Howlett *et al.*, 1973). In two plants with allergenic pollen, canary grass and ragweed, intine-held antigens diffused from the grains, binding to the adjacent stigma surface (Knox and Heslop-Harrison, 1971; Knox, 1973). In canary grass and in *Cosmos*, a genus related to *Ambrosia*, the same response occurred following self-incompatible and cross-compatible pollinations, suggesting that the outflow of pollen antigens is part of a general ‘dialogue’ between pollen and stigma. During the last two decades, there has been a revolution in our ability to probe for particular molecules, particularly with monoclonal antibodies with their precise specificities and molecular cloning. This has enabled us to both isolate genes and ‘silence’ a particular gene during development and to dissect its function. Such techniques now need to be applied to gain an understanding of the natural role of pollen allergens in initiating fertilisation.

Molecular cloning of allergen genes has led to identification of the putative natural function for a few allergens, but for the majority, the natural function remains unknown, although analysis of their IgE interactions has shown why these molecules are allergenic (Knox and Suphioglu, 1996). Molecular cloning has led to the scrutiny of the amino acid sequences of allergens to determine if they share characteristic motifs that may provide a clue to the natural function of these IgE binding proteins.

Sequence similarity of allergens to anti-microbial molecules

Similarity of amino acid sequences of cloned allergens to other known proteins has suggested possible functions. For example, the major allergen from birch, *Betula verrucosa,* is Bet v 1 (M_r 17 kDa), which shows sequence similarity to a family of pathogenesis-related proteins from garden pea (Swoboda *et al.*, 1994). These plant pathogenesis-related genes form a secondary part of the disease resistance response by encoding enzymes, like chitinase, that show anti-microbial activity and prevent disease (Briggs, 1995). The arrival of a pathogenic microbe is perceived by the plant signal transduction system,which induces the expression of these proteins. Homologues occur in other tree pollen of the same family (Betulaceae), including hornbeam and hazel. These allergens show IgE cross-reactivity with extracts of fruits, such as apple and pear as well as vegetable proteins, such as celery, carrot and potato (Ebner *et al.*, 1995). Bet v 2 (M_r 14 kDa) also shows IgE cross-reactivities with fruit and vegetable proteins and sequence similarity with profilins, actin binding proteins of the plant cytoskeleton (Valenta *et al.*, 1995).

The ragweed allergen, Amb a 5, a small (M_r 15kDa) cysteine-rich protein, resembles the avirulence gene products of fungi (Templeton *et al.*, 1994), as well as certain toxins, such as snake venom.

Group 1 grass pollen allergens show sequence similarity to cell wall proteins

Group 1 allergens have a high level of sequence similarity in pollen of many grass genera. Lol p 1 (M_r 35kDa), is a glycoprotein with a cysteine-rich N-terminal region, whose disulfide bonds link pairs of C-residues resemble the lectin wheat germ agglutinin, creating distinctive loops ideal for performing roles in recognition and specificity. The hydrophobic signal peptide region shows similarities to that of α-amylase, suggesting Lol p 1 is also a secreted protein (Griffith *et al.*, 1991). Curiously, the 99 amino acid residues at the C-terminus share sequence similarity with Lol p 2 and Lol p 3 (Griffith *et al.*, 1991) and 28 of these residues constitute the only identified allergenic epitope (Esch and Klapper, 1989; van Ree *et al.*, 1993). Also, the presence in this region of four tryptophan residues is characteristic of the cellulose binding domains of cellulase (Shcherban *et al.* 1995); while the pair of phenylalanine residues and five tyrosine residues are also characteristic of sugar-binding regions in other proteins. Based on sequence similarity data, Lol p 1 appears to have a C-terminal carbohydrate-binding domain.

Lol p 1 shows only low levels of sequence similarity with other known proteins in the databases (except for other Group 1 allergens). Lol p 1 showed about 14% sequence similarity with a family of cell wall proteins, known as expansins, from cucumber hypocotyls (Shcherban *et al.*, 1995). Expansins mediate acid growth of plant cell walls, apparently by promoting "sustained slippage" between wall polymers in the absence of hydrolytic activity (Shcherban *et al.*, 1995). The cucumber sequences showed sequence similarity with anonymous clones (ESTs) from *Arabidopsis*, rice and garden pea. Lol p 1 and other group 1 allergens have seven conserved cysteine residues at the N-terminus, whereas expansins have eight in similar positions, but the 4 C-terminal tryptophan residues are conserved in both proteins (Shcherban *et al.*, 1995). Whether this sequence similarity indicates anything other than commonality of structure with these cell wall proteins remains to be determined. Interestingly, the immunofluorescence studies reviewed earlier in this article showed that both Amb a 1 and Lol p 1 are stored in the intine, the pecto-cellulosic wall of the pollen grain and thus have a cell wall association. The fact that Lol p 1 accounts for a significant amount of the total pollen proteins (Marsh, 1975), and diffuses out over the surface of the stigma (Knox and Heslop-Harrison, 1971), suggests a role in pollen tube penetration of the stigma and intercellular growth through the walls of the transmitting tissue.

Phylogenetic analysis of Group 1 allergens

Since Group 1 allergens are widespread in the Poaceae, attempts have been made to use them for evolutionary studies. In an early study of eight genera, the Group 1 allergens were partially purified and subjected to partial proteolysis and IgE binding patterns of the peptides correlated with their taxonomic subfamily (Suphioglu *et al.*,1993). Results showed that Group 1 allergens from *Zea mays* and *Cynodon dactylon* showed the greatest divergence from *Lolium perenne* . This fits in with the classification of grasses (Table 2), which shows that these genera belong to different subfamilies.

In recent years, cDNA clones encoding Group 1 allergens from eight genera of grasses have been isolated and sequenced (Table 2). Sequence analyses of Pooid genera show a limited number of amino acid substitutions, which become more extensive in Chloridoid and Andropogonoid genera (Knox and Suphioglu, 1996b). These sequences offer the potential for use in phylogenetic analysis as they represent genera from four of the five subfamilies of Poaceae recognised by Clayton and Renvoize (1986). Parsimony analysis using PAUP, Hennig86 (Farris, 1988) and NONA (Goloboff, 1993) of these deduced amino acid sequences of seven full length clones and one partial clone of Group 1 allergens shows the Pooid genera clustering together as a group, distinct from Bambusoid, Chloridoid and Panicoid genera (Fig. 2). Molecular data have already been used to provide additional characters for phylogenetic analysis, for example, chloroplast DNA sequences (Davis and Soreng,1993) and ndhF DNA sequences (Clark *et al.*, 1995) each of which has given results in general agreement with those from morphological characters.

When cDNA clones to the Group 1 allergens from further taxa become available, the sequence variations may provide further clues to evolutionary relationships among Poaceae. Major gaps urgently needing to be filled include allergens from the subfamily Arundinoideae, and from other tribes within the subfamily Pooideae.

TABLE 2. List of cloned Group 1 allergens from grass pollen, family Poaceae, based on classification of Clayton and Renvoize (1986)

Subfamily Tribe *Species*	Group 1 Allergen	Reference
Bambusoideae		
Oryzeae		
Oryza sativa L.	Ory s 1	Xu *et al.*,1995
Chloridoideae		
Chlorideae		
Cynodon dactylon	Cyn d 1	Smith *et al.*,1996
Panicoideae		
Andropogoneae		
Sorghum halepense	Sor h 1	Avjioglu *et al.*,1993
Zea mays	Zea m 1 (partial sequence)	Broadwater *et al.*,1993
Pooideae		
Aveneae		
Holcus lanatus	Hol l 1	Unpublished
Phalaris aquatica	Pha a 1	Suphioglu & Singh, 1995
Phleum pratense	Phl p 1	Peterson *et al.*, 1995
Poeae		
Lolium perenne	Lol p 1	Griffith *et al.*,1991; Perez *et al.*,1990

Enzymic activity associated with major pollen allergens

Both the major grass pollen allergen Phl p 5 and birch allergen Bet v 1 have been shown on the basis of experiments with SDS-PAGE to possess ribonuclease activity (Bufe *et al.*, 1995, 1996) despite the fact that neither of these allergens possesses sequence similarity to any known RNases. Three allergens show similarity to enzyme inhibitors: ragweed cystatin, Lol p 11 and Ole e 1 (Kunitz trypsin inhibitor). Lol p 10 (cytochrome c), Amb a 3, Amb a 6 and Amb a 7 have been shown to be transport proteins (see review by Stewart and Thompson, 1996). Cry j 1 has been shown to possess pectate lyase activity (Taniguchi *et al.*,1995). Minor birch tree and timothy grass allergens have been shown to have sequence similarity to profilins and calmodulins.

What makes an environmental protein allergenic?

Allergens are proteins that contain epitopes capable of eliciting the formation of specific IgE antibodies and binding to them. Allergens contain two types of epitopes, T- and B-cell epitopes. T-cell epitopes, as peptides, bind to the surface of antigen-presenting cells through T-cell recognition (McCluskey *et al.*, 1991) and regulate B-cell response to the allergen. B-cell epitopes are surface regions of an allergen that bind to IgE on B-cells, causing their proliferation and IgE production. Both T- and B-cell epitopes have recently been identified for the major rye-grass pollen allergen Lol p 5 (Fig. 3). Identification of these regions has been made possible by synthesis of dodecapeptides covering the entire molecule. Blaher *et al.*, (1996) showed that two T-cell epitopes occur at amino acid positions (105-116) and (193-204), while Suphioglu *et al.*, (1998) showed that B-cell epitopes occur at seven sites mainly in the C-terminal region of the molecule (Fig. 3). Previously, multiple T-cell epitopes of birch pollen allergen Bet v 1 were identified (Ebner *et al.*, 1993).

```
MAVQKYTVALFLAVALVAGPAASYA          -25
ADAGYTPAAAATPATPAATPAAAGGKATTD      30
EQKLLEDVNAGFKAAVAAAANAPPADKFKI      60
FEAAFSESSKGLLATSAAKAPGLIPKLDTAY     90
DVAYKAAEGATPEAKYDAFVTALTEALRVIA    120
GALEVHAVKPATEVPAAKIPTGELQIVDKIDA   150
AFKIAATAANAAPTNDKFTVFESAFNKALNE    180
CTGGAYETYKFIPSLEAAVKQAYAATVAAAP    210
EVKYAVFEAALTKAITAMTQAQKAGKPAAA     240
AATGAATVATGAATAAAGAATAAAGGYKA      269
```

FIG. 3. T- and B-cell epitopes of Lol p 5. T-cell epitopes function as peptides on the surface of antigen presenting cells where they are involved in T-cell recognition. B-cell epitopes bind to the surface of B-cells and are involved in IgE production. The amino acid sequence of Lol p 5 is shown with the two dodecapeptides that are involved in T-cell recognition indicated in broken line boxes, while the seven peptides involved in IgE-binding are indicated in solid line boxes. Based on data of Blaher *et al.* (1996) and Suphioglu *et al.* (1998).

Conclusions and future perspectives

Pollen allergens are soluble proteins (often glycoproteins) that diffuse usually rapidly from their cellular sites in the grains. Allergens are generally synthesised in the cytosol, and then targeted for secretion to the surface or incorporated in the intine wall of the grain when it is laid down. In the case of two grass pollen allergens, targeting is to the starch granules, which can be subsequently released into the atmosphere if grains encounter rainfall. In red cedar, the major allergen has been associated with the extracellular orbicules.

Genes encoding pollen allergens that are clinically important have mostly been isolated by molecular cloning during the past decade. Through sequence similarity or biochemical studies, their functions include cell wall enzymes, both hydrolytic and non-hydrolytic; enzyme inhibitors, transport and regulatory proteins. Pollen allergens from conifers are likely to follow Japanese red cedar in showing pectin-degrading enzymic activity. Tree pollen allergens may prove to be like Ole e 1, the olive tree allergen, in showing similarity to Kunitz trypsin inhibitor.

Since some major grass pollen allergen genes have been cloned from several genera of the large family Poaceae, these cloned allergens provide amino acid sequences that can be useful in phylogenetic analysis, providing a new molecular source of information to complement morphological characters.

References

Abadie, M., Hideux, M. and Bury, E. (1988). Détection immunochimique et ultrastructurale d'antigènes chez le pollen de *Dactylis glomerata* L. *Ann. Sci. Nat. Bot. Biol. Vég.* (13 sér) **9**: 209–223.

Avjioglu, A., Creaney, J., Smith P., Taylor, P.E., Singh, M.B. and Knox, R.B. (1993). Cloning and characterization of the major allergen of *Sorghum halepense*, a subtropical grass. In: D. Kraft, and A. Sehon (editors) Molecular biology and immunology of allergens, pp. 161–164, CRC Press, Boca Raton.

Blaher, B., Suphioglu, C., Knox, R.B., Singh, M.B., McCluskey, J. and Rolland, R. (1996). Identification of T-cell epitopes of Lol p 9, a major allergen of ryegrass (*Lolium perenne*) pollen. *J. Allergy Clin. Immunol.* **98**: 124–132.

Briggs, SP (1995). Grand unification theory in sight. *Curr. Biol.* **5**: 128–131.

Broadwater, A.H., Bedinger, P.A., Rubinstein, A.L., Chay, C.H. and Klapper, D.G. (1993) *Zea m* I, the maize homolog of the allergen-encoding *Lol p* I gene rye-grass. *Gene* **131**: 227–230.

Brostoff, J and Hall, T (1996). Hypersensitivity - Type I. In: I. Roitt, J. Brostoff and D. Male (editors) Immunology. 4th ed, pp. 221–217, Mosby, London.

Bufe, A., Schramm, G., Keown, M.B., Schlaak, M. and Becker, W.-M. (1995). Major allergen *Phl p* Vb in timothy grass is a novel pollen RNase. *FEBS Lett.* **363**: 6–12.

Bufe, A., Spangfort, M.D., Kahlert, H., Schlaak, M. and Becker, W.-M. (1996). The major birch pollen allergen, Bet v 1, shows ribonuclease activity. *Planta* **199**: 413–415.

Casas, C., Márques, J., Suárez-Cervera, M. and Seoane-Camba, J.A. (1996). Immunocytochemical localization of allergenic proteins in *Parietaria judaica* L. (Urticaceae) pollen grains. *Eur. J. Cell Biol.* **70**: 179–188.

Clark, L.G., Zhang, W. and Wendel, J.F. (1995). A phylogeny of the grass family (Poaceae) based on ndhF sequence data. *Syst. Bot.* **20**: 436–460.

Clayton, W.D. and Renvoize S.A. (1986). Genera Graminum. HMSO, London.

Davis, J.I. and Soreng, R.J. (1993). Phylogenetic structure in the grass family (Poaceae) as inferred from chloroplast DNA restriction site variation. *Amer. J. Bot.* **80**: 1444–1454.

Ebner, C., Szepfalusi, Z., Ferreira, F. *and others*, (1993). Identification of multiple T-cell epitopes on Bet v 1, the major birch pollen allergen, using specific T-cell clones and overlapping peptides. *J. Immunol.* **150**: 1047–1454.

Ebner, C., Hirschwehr, R., Bauer, L., Breiteneder, H., Valenta, R., Ebner, H., Kraft, D. and Scheimer, O. (1995). Identification of allergens in fruits and vegetables: IgE cross-reactivities with the important birch pollen allergens Bet v 1 and Bet v 2 (birch profilin). *J. Allergy Clin. Immunol.* **95**: 962–969.

Esch, R.E. and Klapper, D.G. (1989). Isolation and characterization of a major cross-reactive grass group 1 allergenic determinant. *Molec. Immunol.* **26**: 557–561.

Farris, J.S. (1988) Hennig86, version 1.5. Published by the author.

Fischer, S., Grote, M., Fahlbusch, B., Muller, W.D., Kraft, D. and Valenta, R. (1996). Characterization of Phl p 4, a major timothy grass (*Phleum pratense*) pollen allergen. *J. Allergy Clin. Immunol.* **98**: 189–198.

Fromme, H.G., Grote, M, Sinclair, N.J. and Kalveram, K. (1985). Immunoautoradiographic and protein-A/gold labelling experiments for localization of pollen allergens using antisera from atopic human individuals. *Histochemistry* **82**: 391–396.

Goloboff, P.A. (1993). NONA version 1.1. Published by the author.

Griffith, I.J., Smith, P., Pollock, J., Theerakulpisut. P., Avjioglu, A., Davies, S., Hough, T., Singh, M.B., Simpson, R.J., Ward, L.D. and Knox, R.B. (1991). Cloning and sequencing of *Lol p* I, the major allergenic protein of ryegrass pollen. *FEBS Lett.* **279**: 210–215.

Grote, M. (1991). Immunogold electron microscopy of soluble proteins: localization of Bet v I major allergen in ultra-thin sections of birch pollen after anhydrous fixation techniques. *J. Histochem. Cytochem.* **39**:1395–1401.

Grote, M. (1993). Ultrastructural detection of the allergens Bet v I in birch pollen and profilin in tobacco pollen by special techniques of immuno gold electron microscopy. In: D. Kraft and A. Sehon (editors). Molecular biology and immunology of allergens, pp. 201–203, CRC Press, Boca Raton.

Grote, M. and Fromme, H.G. (1986). Visualization of birch pollen allergens using IgE-containing sera from human atopic individuals in immunogold labelling experiments. *Histochem. J.* **18**: 24–28.

Grote, M., Dolecek, C., van Ree, R. and Valenta, R. (1994). Immunogold electron microscopic localization of timothy grass (*Phleum pratense*) pollen major allergens Phl p I and Phl p V after anhydrous fixation in acrolein vapor. *J. Histochem. Cytochem.* **42**: 427–431.

Grote, M., Swoboda, I., Meagher, R.B. and Valenta, R. (1995). Localization of profilin- and actin-like immunoreactivity in vitro-germinated tobacco pollen tubes by electron microscopy after special water-free fixation techniques. *Sexual Pl. Reprod.* **8**: 180–186

Heslop-Harrison, J. (1968). The pollen grain wall. *Science* **161**: 230–237.

Heslop-Harrison, J. (1975). Physiology of the pollen grain surface. *Proc. Roy. Soc., London Ser. B, Biol. Sci.* **190**: 275–300.

Howlett, B.J., Knox, R.B. and Heslop-Harrison, J. (1973). Pollen-wall proteins: release of the allergen Antigen E from intine and exine sites in pollen grains of ragweed and *Cosmos. J. Cell Sci.* **13**: 603–619.

Howlett, B.J., Vithanage, H.I.M.V. and Knox, R.B. (1981). Immunofluorescence localization of two water-soluble glycoproteins, including the major allergen from pollen of ryegrass, *Lolium perenne. Histochem. J.* **13**: 461–480.

King, T.P., Hoffman, D., Lowenstein, H., Marsh, D.G., Platts-Mills, T.A.E. and Thomas, W. (1995). Allergen nomenclature. *J. Allergy Clin. Immunol.* **96**: 5–14

Knox, R.B. (1973). Pollen-wall proteins: pollen-stigma interactions in ragweed and *Cosmos* (Compositae). *J. Cell Sci.* **12**: 421–443.

Knox, R.B. (1984). The pollen grain. In: M.B. Johri (editor). Embryology of plants, pp. 197–271, Springer-Verlag, Berlin.

Knox, R.B. (1993). Grass pollen, thunderstorms and asthma. *Clin. Exp. Allergy* **23**: 354–359.

Knox, R.B. and Heslop-Harrison, J. (1969). Cytochemical localization of enzymes in the wall of the pollen grain. *Nature* **223**: 92–94.

Knox, R.B. and Heslop-Harrison, J. (1970). Pollen-wall proteins: cytochemical localization and enzymatic activity. *J. Cell Sci.* **6**: 1–27.

Knox, R.B. and Heslop-Harrison, J. (1971a). Pollen-wall proteins: localization of antigenic and allergenic proteins in the pollen grain walls of *Ambrosia* spp. (ragweeds). *Cytobios* **4**: 49–54.

Knox, R.B. and Heslop-Harrison, J. (1971b). Pollen-wall proteins: the fate of intine-held antigens on the stigma in compatible and incompatible pollinations of *Phalaris tuberosa* L. *J. Cell Sci.* **9**: 239–251.

Knox, R.B. and Suphioglu, C. (1996a). Environmental and molecular biology of pollen allergens. *Trends Pl. Sci.* **1**: 156–163.

Knox, R.B. and Suphioglu C. (1996b). Pollen allergens: development and function. *Sexual Pl. Reprod.* **9**: 318.

Knox, R.B., Heslop-Harrison, J. and Reed, C.E. (1970) Localization of antigens associated with the pollen grain wall by immunofluorescence. *Nature* **225**: 1066–1068.

Marsh, D.G. (1975) Allergens and the genetics of allergy. In: M. Sela (editor) The antigens. Vol 3, pp. 271–359, Academic Press, London.

McCluskey, J. (editor) (1991) Antigen processing and recognition. CRC Press, Boca Raton.

Miki-Hirosige, H., Nakamura, S., Yasueda, H., Shida, T. and Takahashi, Y. (1994). Immunocytochemical localization of the allergenic proteins in the pollen of *Cryptomeria japonica*. *Sexual Pl. Reprod.* **7**: 95–100.

Ong, E.K., Griffith, I.J., Knox, R.B. and Singh, M.B. (1993). Cloning of a cDNA encoding a group V (group IX) allergen isoform from rye-grass pollen that demonstrates specific antigenic reactivity. *Gene* **134**: 235–240.

Perez, M., Ishioka, G.Y., Walker, L.E. and Chesnut, R.W. (1990). cDNA cloning and immunological characterization of rye grass pollen allergen Lol p I. *J. Biol. Chem.* **265**: 16210–16215.

Peterson, A., Schramm, G., Bufe, A., Schlaak, M. and Becker, W.M. (1995). Structural investigations of the major allergen Phl p I on the complementary DNA and protein level. *J. Allergy Clin. Immunol.* **95**: 987–94.

Ree, R. van, Leeuwen, W.A. van and Aalberse, R.C. (1993) Demonstration of human IgE and IgG antibodies against the Lol p I C-terminus that cross-react with the C-terminus of Lol p II and Lol p III. *J. Allergy Clin. Immunol.* **91**: 187

Rodriguez-Garcia, M.I., Fernández, M.C. and Alché, J.D. (1995a) Immunocytochemical localization of allergenic protein (*Ole e I*) in the endoplasmic reticulum of the developing pollen grain of olive (*Olea europea* L.). *Planta* **196**: 558–563.

Rodriguez-Garcia, M.I., Fernández, M.C., Alché, J.D., Olmedilla, A. (1995b) Endoplasmic reticulum as a storage site for allergenic proteins in pollen grains of several Oleaceae. *Protoplasma* **187**: 111–116.

Shcherban, T.Y., Shi, J., Durachko, D.M., Guiltinaan, M.J., McQueen-Mason, S.J., Shieh, M. and Cosgrove, D.J. (1995) Molecular cloning and sequence analysis of expansins - a highly conserved, multigene family of proteins that mediate cell wall extension in plants. *Proc. Nat. Acad. Sci.* USA **92**: 9245–9249.

Singh, M.B., Hough, T., Theerakulpisut, P., Avjioglu, A., Davies, S., Smith, P.M., Taylor, P., Simpson, R.J., Ward, L.D., McClusky, J., Puy, R. and Knox, R.B. (1991) Isolation of cDNA encoding a newly identified major allergenic protein of rye-grass pollen: intracellular targeting to the amyloplast. *Proc. Natl. Acad. Sci. USA* **88**: 1384–1388.

Smith, P.M., Suphioglu, C., Griffith, I.J., Theriault, K., Knox, R.B. and Singh, M.B. (1996). Cloning and expression in yeast *Pichia pastoris* of a biologically active form of Cyn d 1, the major allergen of Bermuda grass. *J. Allergy Clin. Immunol.* **98**: 331–343.

Staff, I.A., Taylor, P.E., Smith, P., Singh, M.B. and Knox, R.B. (1990). Cellular localization of water soluble, allergenic proteins in rye-grass (*Lolium perenne*) pollen using monoclonal and specific IgE antibodies with immunogold probes. *Histochem. J.* **22**: 276–290.

Stewart, G.A. and Thompson, P.J. (1996). The biochemistry of common aeroallergens. *Clin. Exp. Allergy* **26**: 1020–1044.

Suphioglu, C. and Singh, M.B. (1995) Cloning, sequencing and expression in *Escherichia coli* of Pha a 1 and four isoforms of Pha a 5, the major allergens of canary grass pollen. *Clin. Exp. Allergy* **25**: 853–865.

Suphioglu, C., Singh, M.B., Taylor, P.E., Bellomo, R., Holmes, P., Puy, R. and Knox, R.B. (1992). Mechanism of grass-pollen-induced asthma. *Lancet* **339**: 569–572

Suphioglu, C., Singh, M.B. and Knox, R.B. (1993) Peptide mapping analysis of group I allergens of grass pollens. *Int. Arch. Allergy Immunol.* **102**:144–151.

Suphioglu, C., Blaher, B., Rolland, J.M., McCluskey, J., Kenrick, J., Singh, M.B. and Knox, R.B. (1998). Molecular basis of IgE-recognition of natural, recombinant and synthetic peptides of Lol p 5, a major allergen of rye-grass pollen. *Mol. Immunol.* (In press).

Swoboda, I., Scheiner, O., Kraft, D., Breitenbach, M., Heberle-Bors, E. and Vicente, O. (1994). A birch gene family encoding pollen allergens and pathogenesis-related proteins. *Biochim Bophys Acta* **1219**: 457–464.

Taniguchi, Y., Ona, O., Sawatani, M., Nanba, M., Kohno, K., Usui, M., Kurimoto, M. and Matuhasi, T. (1995). Cry j 1, a major allergen of Japanese red cedar pollen has pectate lyase activity. *Allergy* **50**: 90–93.

Taylor, P.E., Singh, M. and Knox, R.B. (1993). Strategies for the immunocytochemical localization of rapidly diffusible proteins in pollen. *J. Computer-Assisted Microscop.* **5**: 53–56.

Taylor, P.E., Staff, I.A., Singh, M.B. and Knox, R.B. (1994). Localization of the two major allergens in rye-grass pollen using specific monoclonal antibodies and quantitative analysis of immunogold labelling. *Histochem J.* **26**: 392–401.

Templeton, M.D., Rikkerink, E.H.A. and Beever, R.E. (1994). Small, cysteine-rich proteins and recognition in fungal-plant interactions. *Molec. Pl.-Microbe Interact.* **7**: 320–325.

Valenta, R. *and others*, (1995). Profilin: a novel pan allergen and actin-binding protein in plants. In: S.S. Mohapatra and R.B. Knox (editors). Pollen biotechnology: gene expression and allergen characterization, pp. 269–278, Chapman and Hall, Baltimore.

Volker, W., Sinclair, N.J., Kalverman, K.J. and Robenek, H. (1986) Is the surface layer from hazelnut pollen, which is precipitated by cuprolinic blue, an effective antigen in hay-fever patients? *Histochemistry* **84**: 57–60.

Vrtala, S., Grote, M., Duchêne, M., van Ree, R., Kraft, D., Scheiner, O. and Valenta, R. (1993). Properties of tree and grass pollen allergens: reinvestigation of the linkage between solubility and allergenicity. *Int. Arch. Allergy Immunol.* **102**:160–169.

Xu, H., Theerakupisut, P., Goulding, N., Suphioglu, C., Singh, M.B. and Bhalla, P. (1995). Cloning, expression and immunological characterisation of Ory s 1, the major allergen of rice pollen. *Gene* **164**: 255–259.

Murphy D.J. (1998). Why is rapeseed pollen such a sticky subject? In: S.J. Owens and P.J. Rudall (Editors). Reproductive Biology, pp. 465–474. Royal Botanic Gardens, Kew.

WHY IS RAPESEED POLLEN SUCH A STICKY SUBJECT?

DENIS J MURPHY

Department of Brassica & Oilseeds Research, John Innes Centre, Norwich, NR4 7UH, United Kingdom

Introduction

The purpose of this paper is to examine some of the latest work on various aspects of the biology of rapeseed pollen. Rapeseed, *Brassica napus* is the major oilseed crop in large areas of the temperate climatic regions, ranging across from Canada and Northern Europe, to the Indian sub-continent, Eastern Asia and Australia. The pollen of this crop plant can be considered as a sticky subject, both literally and metaphorically. Rapeseed pollen is literally sticky, since it is of the entomophilous variety, i.e. it has a sticky coating in order to allow it to adhere to insect and other vectors which transfer it from one plant to another.

Rapeseed pollen can also be considered as metaphorically "sticky" for two other reasons. Firstly, rapeseed is one of the first major crops to undergo genetic engineering. Large-scale commercial releases of genetically modified rapeseed plants have occurred during the last few years (Murphy, 1996). There is much concern about the possibility for transfer of the transgenes from the genetically modified rapeseed varieties, either to non-transgenic varieties, or indeed to the many weedy relatives of rapeseed which occur naturally. This genetic transfer would have to be mediated via the rapeseed pollen, which means that the potential of rapeseed pollen to travel and to fertilise other plants is the subject of considerable interest both in the research community and with the general public.

The second area of public concern about rapeseed pollen involves its possible role in precipitating or potentiating allergic responses, leading to conditions such as hay fever and asthma. The reported incidents of allergies in both urban and rural populations has increased enormously over the past few decades. Rapeseed, particularly during the flowering season, is a prominent and relatively novel feature of our landscape. It impacts on our senses, both due to its bright colour and its distinctive odour. This has led to some concern about the possible allergenic nature of rapeseed pollen. In this paper I will examine some recent results from our laboratory on the composition and formation of the substances responsible for the stickiness of rapeseed pollen. I will also review recent research findings on the potential of rapeseed pollen to disperse transgenes in the environment and finally, discuss the possible role of rapeseed pollen as an allergen.

Pollen structure

Pollen grains are the male gametophytes of higher plants. In the family *Brassicaceae*, the pollen grain normally contains three cells, two of which, the sperm cells, are mostly made up of nuclear material, while the third much larger cell, the vegetative cell, contains the cytoplasmic contents of the pollen grain. The pollen grains are surrounded by two boundary layers, the innermost layer, the intine, has a cellulose-based composition, similar to most other plant cell walls. Immediately outside the intine is a further layer termed the exine. The exine is the highly sculptured structure so characteristic of pollen grains. It is uniquely resistant to most methods of chemical degradation and intact exines have been found in fossilised pollen grains that are many hundreds of millions of years old. It is therefore the most durable known biological material. Unfortunately, these fascinating properties have also rendered it extremely difficult to study the composition and properties of the exine. It is known that the exine is made up of a complex biopolymer termed sporopollenin, which may contain fatty acid, carotenoid and amino acid derivatives (Guilford et al, 1988; Wiermann & Gubatz, 1992, Wilmesmier et al, 1993).

In the *Brassicaceae* and other plants bearing entomophilous pollen, the exine is covered with a waxy material, variously termed pollen coat, pollenkitt, or tryphine. It is the tryphine that is responsible for the stickiness of rapeseed pollen. It is important for rapeseed pollen to have such an adhesive coating for several reasons. Firstly, it enables the pollen grains to stick both to each other and to insect vectors, hence facilitating the transfer of relatively large quantities of pollen between plants. This is particularly important in most Brassicas, because their flowers are self-incompatible, i.e. the stigma of a given plant will reject pollen originating from the same plant. Rapeseed however, is a relatively new species formed by the hybridisation of two diploid species, *Brassica rapa* and *Brassica oleracea*, to create the amphidiploid species, *Brassica napus*. In the course of this hybridisation event, the self-incompatibility mechanism of the parental species has become suppressed and most modern crop varieties of rapeseed are self-compatible. Nevertheless, it is evolutionarily advantageous for rapeseed plants to be able to outcross in order to maximise their genetic variability. Hence, insect pollination is still important even in this self-compatible species.

The second role of the tryphine is to enable pollen grains to adhere to the dry surface of the rapeseed stigma. Once a pollen grain has landed on the stigmatic surface, the tryphine flows out within less than 1 min to form a "foot", which then provides a medium for pollen stigma communication (Dickinson & Elleman, 1985).

Thirdly, the tryphine appears to be important in enabling pollen hydration to occur. The pollen is dehydrated when it lands on the stigma, and can normally only hydrate by drawing in water from the adjacent stigmatic cells. This occurs once the tryphine has undergone a process known as "coat conversion", whereby the tryphine appears to change its structure, presumably allowing the formation of aqueous channels through which pollen rehydration may proceed (Elleman & Dickinson, 1986).

Finally, the tryphine is required for the adhesion of extracellular proteins to the pollen grain. In *Arabidopsis* mutants which have a lesion in tryphine biosynthesis, the pollen extracellular proteins are initially targeted to the exine, but are unable to remain there in the absence of an adhesive tryphine layer (Preuss et al, 1993). The tryphine proteins include structural components and signalling molecules involved in pollen stigma interactions.

Although it is a pollen component, the tryphine is synthesised in the tapetum, which is a sporophytic glandular tissue responsible for nourishing the developing pollen grains. Lipids are an important product of the tapetum in members of the tribe, *Brassiceae* (which includes the genera *Brassica, Raphanus* and *Sinapis*). These lipids accumulate particularly during the later stages of tapetal development. Tapetal cells form several types of lipid bodies, some of which may be secreted to immature pollen grains early in development and some of which are only released to form the tryphine following tapetal cell death (Dickinson & Lewis, 1973; Murgia et al, 1991).

Tryphine composition

Despite its importance in pollination, relatively little is known about the composition of tryphine in higher plants. It has been reported that rapeseed tryphine has a very different fatty acid composition to that of the intracellular pollen lipids (Evans et al, 1987). In the present study, mature rapeseed pollen grains were briefly flushed with cyclohexane in order to remove the tryphine layer without affecting the integrity of the intracellular contents. Ultrastructural analysis confirmed that >95% of tryphine was removed by a single cyclohexane wash (Piffanelli et al, 1997). The tryphine lipids were separated into classes by thin-layer chromatography (TLC) and quantitative fatty acid compositions were determined by gas-liquid chromatography. The results are summarised in Table 1. The major lipid classes of the intracellular lipids are the polar lipids (mostly phospholipids) that form the various membrane systems, and the triacylglycerols that make up the storage lipid bodies found in pollen grains. In contrast, the tryphine mostly consists of neutral esters and free fatty acids. The neutral esters are a very hydrophobic fraction that separate between the wax/hydrocarbon fraction and the triacylglcyerol fraction on TLC plates. The neutral esters are probably either terpene esters or sterol esters - triterpene esters are major components of maize pollen grains (Bianchi et al, 1990).

The fatty acid composition of the tryphine is dominated by C_{14}, C_{16} and C_{18} saturated species, whereas the intracellular pollen lipids are mostly C_{18} polyunsaturates. More recent preliminary analysis of the volatile fraction of freshly collected intact rapeseed pollen shows that the major components are C_6-C_{12} free fatty acids (Murphy, D.J. and Mithen, R.F. unpublished results). This is distinct from the volatiles released from rapeseed flowers which included organic sulphides, C_4 substituted compounds, aldehydes and terpenes (Robertson et al, 1993).

The polypeptide components of rapeseed tryphine are dominated by an abundant and highly variable family termed the oleosin-like proteins (OLPs; Ross & Murphy, 1996). The genes encoding these proteins are expressed specifically in the tapetum and the structures of some representative OLPs are shown in Fig. 1. All anther-specific OLPs contain a highly-conserved

TABLE 1. Acyl lipid composition of rapeseed pollen

I. Lipid Classes

	Neutral Esters	Fatty Acids	Polar Lipids	Triacylglycerols
		% w/w		
Tryphine	42	16	10	1
Intracellular lipids	5	5	40	39

II. Fatty Acids

	14:0	16:0	18:0	18:1	18:2	18:3
			% w/w			
Neutral esters	35	23	23	3	3	8
Fatty acids	17	44	28	tr	tr	3
Polar lipids	0	38	tr	2	7	51
Triacylglycerols	0	27	tr	2	5	63

FIG. 1. Structure of the predicted mRNA translation products of tapetal-specific oleosin-like genes. The proteins all contain a conserved 70-residue central hydrophobic domain that is highly homologous to the corresponding domain of seed oleosins, which encapsulate storage lipid bodies in oil-bearing seeds. The C-terminal domains of the tapetal oleosin-like proteins are highly variable, but are characterised by extensive repeats rich in Lys/Gly, Pro/Ala or Gly. Such motifs are common in structural proteins.

70 residue hydrophobic domain that is very similar to the seed-specific oleosins that surround storage lipid bodies found in oilseeds (Murphy, 1993). However, the anther-specific OLPs also contain characteristic C-terminal domains of varying length that are largely made up of various repetitive motifs. Such repeats, including K/G, P/A or G, are typical of extracellular structural proteins, such as cell wall components.

Direct amino sequencing of the major tryphine proteins has revealed that they correspond to the C-terminal domains of the tapetal OLPs. We have now sequenced nine of the most abundant tryphine polypeptides and, in each case, they appear to result from a cleavage of the original OLPs, at or near the junction of the oleosin-like hydrophobic domain and the C-terminal structural-like domain (Fig. 2). These results have been confirmed by raising antibodies against one of the major OLPs which is initially synthesised as two isoforms of 42 and 38 kDa. These

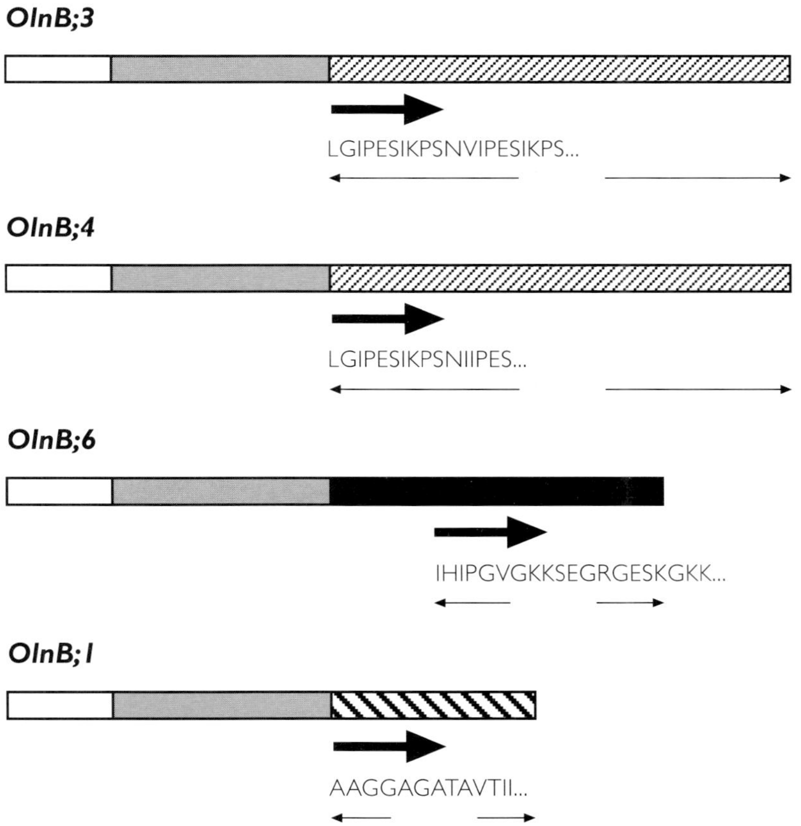

FIG. 2. Sequences of the major tryphine polypeptides. Note that each of these sequences corresponds to the C-terminal region of a full-length oleosin-like protein. The mature tryphine polypetides are therefore cleavage products of OLPs which have lost their oleosin-like domains, but retained their structural-like C-terminal regions. Shading is the same as Fig. 1.

antibodies recognise proteins of 42 and 38 kDa in whole anther or tapetum-enriched fractions from immature flower buds. The same antibodies recognise lower molecular weight proteins of about 37 and 32 kDa in tryphine extracts. The same 37 and 32 kDa proteins, when sequenced, corresponded to the C-terminal domains of the OLPs. (Murphy and Ross, 1997)

We conclude that the major class of tryphine proteins are initially synthesised in the tapetum as an oleosin domain fused to a structural-like protein. The purpose of the oleosin domain is unclear, but it may be involved in the accumulation and/or transfer of the large quantities of lipid material that eventually makes up the bulk of the tryphine. A specific cleavage event removes the oleosin domain, leaving the structural domain as the major protein in the tryphine. The subsequent fate of the oleosin domain and the function of the mature OLPs in the tryphine have yet to be determined. They may have a structural role in ensuring that the tryphine lipids are drawn into the interstices of the exine. They may also be involved in facilitating pollen rehydration on the stigma. Initial results from transgenic rapeseed plants exhibiting down-regulation of OLP genes indicate that there are resulting abnormalities in both tapetal and pollen function (Ross, 1996).

The combination of highly saturated neutral esters, free fatty acids and waxes with the characteristic structural proteins (OLP) and other minor components gives rise to the material known as tryphine, that is responsible for the stickiness of rapeseed pollen. Despite recent advances, however, much remains to be discovered about the structure and function of this fascinating material.

Pollen dispersal and its consequences

The extent of pollen dispersal from rapeseed crops is of interest for several reasons. Firstly, the increase in cultivation of genetically distinct varieties of rapeseed makes it important to establish minimum isolation distances so that pollen from one cultivar does not fertilise different varieties in neighbouring fields. This is particularly important in assessing the risk of the spread of pollen from genetically modified rapeseed varieties in view of the recent granting of permission for the large-scale commercial cultivation of several transgenic rapeseed varieties in the USA and parts of Europe (Murphy, 1996). Secondly, widely-broadcast rapeseed pollen is at least potentially able to fertilise some of the many related crop and weed species within the *Brassiceae* (Scheffler & Dale, 1994). This may result in the formation of hybrid plants containing genes from the original rapeseed pollen. Thirdly, there is a public perception that rapeseed, and possibly rapeseed pollen, is implicated in certain allergic responses (Tickell, 1996).

In view of these concerns, there is a small but growing body of research into the extent and impact of pollen dispersal from field-grown rapeseed either to other rapeseed varieties or to related members of the *Brassiceae* and to humans in the neighbouring environment. As part of the PROSAMO (Planned Release of Selected and Modified Organisms) programme, the extent of pollen dispersal from transgenic rapeseed was examined in several sites in Europe. In this study, pollen dispersal *per se* was not measured, but rather its consequences, i.e. the successful fertilisation by transgenic pollen of neighbouring non-transgenic rapeseed plants. This is an important point since it is possible that non-viable pollen grains could be transmitted over larger distances, but this would obviously have no consequences for gene dispersal. The conclusion from these studies was that the frequency of pollen dispersal from transgenic rapeseed plants is extremely low (Scheffler et al, 1993). The frequency was estimated to be 1.5% at a distance of one metre, dropping sharply to 0.02% at 12 metres and only 0.0003% at 47 metres. Honey bees were free to forage for pollen throughout the experimental area, but were found to have a relatively restricted range. More recently, Timmons et al (1996), have estimated somewhat larger dispersal distances for rapeseed pollen, whereby the characteristic dispersal distance is estimated at between 128 and 172 metres as opposed to between 2.3 and 8.8 metres in the study of Scheffler et al (1993). Clearly further research is needed to clarify the extent of pollen dispersal from commercial rapeseed varieties in a genuine agricultural environment.

Rapeseed is a member of the tribe *Brassiceae*, which also includes numerous other crop plants and several highly invasive and undesirable weed species, such as charlock (*Sinapis arvensis*). It is known that rapeseed is able to hybridise with several other members of the *Brassiceae*, including *B. rapa*, *B. juncea*, *B. adpressa* and *Raphanus raphanistrum* by open pollination. The potential for gene transfer from rapeseed to related species has recently been reviewed by Scheffler and Dale (1994). There have also been several recent studies of hybridisation between oilseed rape and weedy relatives including *B. rapa* (Mikkelson et al, 1996) and *Hirschfeldia incana* (Lefol et al, 1996). In most of these cases, rapeseed pollen was able to fertilise these related species albeit at very low frequencies. Moreover, it was found that the fitness (measured as seed germination rate) of hybrids between rapeseed and *H.incana* was only 10^{-6} compared to wild-type *H.incana*. These studies show that rapeseed pollen has the potential to fertilise other related species, and that there is therefore a statistical possibility that transgenes could be transferred by such a mechanism from rapeseed into other plants. Nevertheless, the frequency of such transfers and the viability of the resulting hybrid plants appears very low. More work is needed to assign more rigorous statistical probabilities to these events in order to inform decisions about the risks of releasing transgenic rapeseed into the environment.

Allergic responses

Over the past 20 years, the incidence of asthma reported by general practitioners in the UK has increased more than 5-fold. Over the same period, the reported incidences of hay fever and other allergic conditions have also increased greatly. The complexity of these phenomena and their likely multifactorial origins is underlined by the fact that similar symptoms are found in both urban and rural populations from locations as diverse as Melbourne, Australia, and Norfolk, UK. Although rapeseed is by no means a new crop, it has only been cultivated on a large-scale in Europe and North America during the last 25 years. With its bright yellow colour and characteristic odour, the crop is highly distinctive, particularly during its flowering season in April to May in Europe. The coincidence of the appearance of such a high profile feature of the landscape with the increase in allergic conditions has led to a general perception that rapeseed and particularly rapeseed pollen, may be implicated in the triggering of such allergies.

However, recent rigorous epidemiological studies of populations living in the vicinity of rapeseed crops have failed to show specific allergic responses to rapeseed pollen in anything but a tiny fraction (0.2%) of the population (Fell et al, 1992). This is perhaps not surprising in view of the relatively low dispersal rate of rapeseed pollen (see above). Rapeseed pollen is both heavy and sticky and relies on insect vectors for its transmission. Unlike anemophilous pollen, e.g. grass or birch pollen, rapeseed pollen cannot be borne very far by the wind. It is therefore unlikely that human populations would be exposed to rapeseed pollen, unless they are in direct physical contact with rapeseed plants during the brief Spring flowering season. Nevertheless, rapeseed flowers do produce many volatile compounds which are responsible for the characteristic smell of the crop. It is possible that some of these floral volatiles, including aldehydes, terpenes and organic sulphides, which could act as respiratory irritants. Recent preliminary analyses in our laboratories have indicated that the tryphine of rapeseed pollen (var. Topaz) itself also contains a characteristic cocktail of volatile chemicals. These pollen volatiles are completely different from the flower volatiles and appear mainly to consist of short-chain (C_6-C_{12}) fatty acids, although more analysis is necessary in order to establish whether these differences apply in other rapeseed cultivars. Similar differences between pollen and flower volatiles have also been reported recently for several other plant species (Dobson et al, 1996). It is therefore unlikely that rapeseed pollen *per se* is implicated in the reported increases in allergic conditions in the general population. It is possible that non-pollen rapeseed volatiles could affect people who are exposed to the flowering crop. Although these volatiles are normally only irritants at much higher concentrations than found in the environment, it is possible that they could sensitise people to other allergens, ranging from wind-borne pollen and diesel fuel emissions to the excreta of house dust mites (Robertson et al 1993; Knox et al, 1996; Knox & Suphioglu, 1996).

Although rapeseed pollen is probably not responsible for widespread allergies in the general population, there is increasing evidence that long-term direct exposure to flowering rapeseed plants is implicated in the development of allergies. This kind of chronic exposure is unlikely to occur in farm workers growing rapeseed crops, since the latter only flower for several weeks of the year. However, many research workers work year-round at close-quarters with flowering rapeseed plants cultivated in glasshouses. In a study at the Institute of Arable Crops Research at Rothamsted, it was found that 29 out of 37 researchers (78%) who regularly handled rapeseed exhibited allergic responses, although only 5 individuals (17%) showed allergic responses to rapeseed pollen itself and all but one of these people were additionally allergic to other types of pollen (Fell et al, 1996). It appears that type I hypersensitivity to rapeseed pollen is relatively uncommon and only occurs in atopics who are already allergic to other airborne pollens. We have observed similar incidences of allergic responses amongst research staff working with rapeseed at our Institute in Norwich. The fact that only one person was specifically allergic to rapeseed pollen in the Rothamsted study, indicates that there are probably other causative agents involved in these allergic responses. Possible culprits range from volatiles from the rapeseed flowers themselves to other plants growing nearby, or indeed other non-biological agents including pest control chemicals routinely used in glasshouse management. It is likely that a combination of these factors contributes to glasshouse allergies. Further epidemiological studies are now underway to investigate the causes of these allergies amongst research staff and to suggest possible remedial methods that can be adopted.

Conclusions

Rapeseed pollen is coated by a sticky lipid-rich layer, the tryphine, that is responsible for its adhesion to insect vectors and to the stigmatic surface. Tryphine lipids mostly comprise C_{14}-C_{18} saturated neutral esters and free acids, plus numerous volatile components dominated by C_6-C_{12} fatty acids. The most abundant tryphine proteins are a novel class of structural-like polypeptides derived from tapetally-synthesised oleosin-like proteins. The interactions of these and other tryphine components are responsible for its adhesive qualities and its additional roles in facilitating pollen-stigma contact and pollen rehydration.

Despite its public perception as a potential allergen, most available evidence suggests that, due to its sticky and heavy nature, rapeseed pollen is unlikely to be widely dispersed in the environment. Rapeseed pollen is also unlikely to be an important allergen in rural populations.

Acknowledgements

I am grateful to all members of the Brassica & Oilseeds Research Department who contributed to this work and in particular Dr R F Mithen for the initital analysis of pollen volatiles. I also thank Dr M Naldrett for protein sequence analysis and Dr M Blight for advice on pollen allergies. I also acknowledge the support of the Biotechnology and Biological Sciences Research Council competitive strategic grant to the John Innes Centre.

References

Bianchi, G., Murelli, C. and Ottaviano, E. (1990). Maize pollen lipids. *Phytochemistry* **29**: 739–744.

Dickinson, H.G. and Elleman, C.J. (1985). Structural changes in the pollen grain of *Brassica oleracea* during dehydration in the anther and development on the stigma as revealed by anhydrous fixation techniques. *Micron Microscop. Acta* **16**: 255–270.

Dickinson, H.G. and Lewis, D. (1973). Cytochemical and ultrastructural differences between intraspecific compatible and incompatible pollinations in *Raphanus. Proc. Roy. Soc. London, Ser. B, Biol. Sci.* **183**: 21–38.

Dobson, H.E., Groth, I. and Bergstrom, G. (1996). Pollen advertisement: chemical contrasts between whole-flower and pollen odors. *Amer. J. Bot.* **83**: 877–885.

Elleman, C.J. and Dickinson, H.G. (1986). Pollen stigma interactions in *Brassica*. 4. Structural reorganisation in the pollen grains during hydration. *J. Cell Sci.* **80**: 141–157.

Evans, D.E., Rothnie, N.E., Palmer, M.V., *and others*. (1987). Comparative analysis of fatty acids in pollen and seed of rapeseed. *Phytochemistry* **26**: 1895–1897.

Fell, P.J., Soulsby, S., Blight, M.M. and Brostoff, J. (1992). Oilseed rape - a new allergen? *Clin. Exp. Allergy* **22**: 501–505.

Guilford, W.J., Schneider, D.M., Labovitz, J. and Opella, S.J. (1988). High resolution solid state 13C NMR spectroscopy of sporopollenins from different plant taxa. *Pl. Physiol.* **86**: 134–136.

Knox, B. and Suphioglu, C. (1996). Environmental and molecular biology of pollen allergens. *Trends Pl. Sci* **1:** 156–164.

Knox, R.B., Suphioglu, C., Taylor, P., Peng, J.L. and Bursill, L.A. (1996). Asthma and air pollution: two major grass pollen allergens bind to diesel exhaust particles (DEPs). *J. Allergy Clin. Immunol.* **97**: 378.

Lefol, E., Fleury, A. and Darmency, H. (1996). Gene dispersal from transgenic crops. II. Hybridization between oilseed rape and the wild hoary mustard. *Sexual Pl. Reprod.* **9**: 189–196.

Mikkelsen, T.R., Andersen, B. and Jorgensen, R.B. (1996) The risk of crop transgene spread. *Nature* **380**: 31.

Murgia, M., Charzynska, M., Rougier, M. and Cresti, M. (1991). Secretory tapetum of *Brassica oleracea* L.: polarity and ultrastructural features. *Sexual Pl. Reprod.* **4**: 28–35.

Murphy, D.J. (1993). Structure, function and biogenesis of storage lipid bodies and oleosins in plants. *Progr. Lipid Res.* **32**: 247–280.

Murphy, D.J. (1996). Biotechnological improvement of oil crops. *TIBTECH* **14**: 206–213.

Murphy, D.J. and Ross, J.H.E. (1997) Biosynthesis and processing of the major pollen coat proteins of *Brassica napus*. *Plant J.* (in press).

Piffanelli, P., Ross, J.H.E. and Murphy, D.J. (1996). Intra and extracellular lipid composition and associated gene expression patterns during pollen development in *Brassica napus*. *Plant J.* **11**: 549–562.

Preuss, D., Lemieux, B., Yen, G. and Davis, R.W. (1993). A conditional sterile mutation eliminates surface components from *Arabidopsis* pollen and disrupts cell signaling during fertilization. *Genes & Developm.* **7**: 974–985.

Robertson, G.W., Griffiths, D.W., MacFarlane Smith, W. and Butcher, R.D. (1993). The application of thermal desorption-gas chromatography-mass spectrometry to the analyses of flower volatiles from five varieties of oilseed rape (*Brassica napus* spp *oleifera*). *Phytochem. Analysis* **4**: 152–157.

Ross, J.H.E. (1996). Oleosin-like genes and proteins in the tapetum and pollen coat of *Brassica napus*. PhD Thesis.

Ross, J.H.E. and Murphy, D.J. (1996). Characterisation of anther-expressed genes encoding a major class of extracellular oleosin-like proteins in the pollen coat of *Brassicaceae. Pl. J. Cell. Molec. Biol.* **9**: 625–637.

Scheffler, J.A. and Dale, P.J. (1994). Opportunities for gene transfer from transgenic oilseed rape (*Brassica napus*) to related species. *Transgenic Res.* **3**: 263–278.

Scheffler, J.A., Parkinson, R. and Dale, P.J. (1993). Frequency and distribution of pollen dispersal from transgenic oilseed rape (*Brassica napus*). *Transgenic Res.* **2**: 356–364.

Tickell, O. (1996). Winnowing out the yellow peril. *Independent on Sunday* 12th May: 46.

Timmons, A.M., Charters, Y.M., Crawford, J.W., *and others.* (1996). Risks from transgenic crops. *Nature* **380**: 487.

Wiermann, R. and Gubatz, S. (1992). Pollen wall and sporopollenin. *Int. Rev. Cytol.* **140**: 35–72.

Wilmesmeier, S., Steuernagel, S. and Wiermann, R. (1993). Comparative FTIR and 13C CP/MAS NMR spectroscopic investigations on sporopollenin of different systematic origins. *Z. Naturf., C.* **48**: 697.

SUBJECT INDEX

INDEX OF PLANT NAMES